Methods in Enzymology

Volume 138
COMPLEX CARBOHYDRATES
Part E

METHODS IN ENZYMOLOGY

EDITORS-IN-CHIEF

Sidney P. Colowick   Nathan O. Kaplan

Methods in Enzymology

Volume 138

Complex Carbohydrates

Part E

EDITED BY

Victor Ginsburg

NATIONAL INSTITUTE OF DIABETES
AND DIGESTIVE AND KIDNEY DISEASES
NATIONAL INSTITUTES OF HEALTH
BETHESDA, MARYLAND

1987

ACADEMIC PRESS, INC.
Harcourt Brace Jovanovich, Publishers
Orlando   San Diego   New York   Austin
Boston   London   Sydney   Tokyo   Toronto

COPYRIGHT © 1987 BY ACADEMIC PRESS, INC.
ALL RIGHTS RESERVED.
NO PART OF THIS PUBLICATION MAY BE REPRODUCED OR
TRANSMITTED IN ANY FORM OR BY ANY MEANS, ELECTRONIC
OR MECHANICAL, INCLUDING PHOTOCOPY, RECORDING, OR
ANY INFORMATION STORAGE AND RETRIEVAL SYSTEM, WITHOUT
PERMISSION IN WRITING FROM THE PUBLISHER.

ACADEMIC PRESS, INC.
Orlando, Florida 32887

United Kingdom Edition published by
ACADEMIC PRESS INC. (LONDON) LTD.
24–28 Oval Road, London NW1 7DX

LIBRARY OF CONGRESS CATALOG CARD NUMBER: 54-9110

ISBN 0–12–182038–6 (alk. paper)

PRINTED IN THE UNITED STATES OF AMERICA

87 88 89 90   9 8 7 6 5 4 3 2 1

Table of Contents

Section I. Analytical Methods

Section II. Preparations

Section III. Carbohydrate-Binding Proteins

Section IV. Biosynthesis

Section V. Degradation

Contributors to Volume 138

Article numbers are in parentheses following the names of contributors. Affiliations listed are current.

JUNKO AMANO (64), *Department of Biochemistry, The Institute of Medical Science, The University of Tokyo, 4-6-1 Shirokanedai, Minato-ku, Tokyo 108, Japan*

KALYAN R. ANUMULA (67), *Department of Biological Sciences, State University of New York at Buffalo, Buffalo, New York 14260*

JANICE AU-YOUNG (56), *Center for Cancer Research, Massachusetts Institute of Technology, Cambridge, Massachusetts 02139*

OM P. BAHL (27, 28, 67), *Department of Biological Sciences, State University of New York at Buffalo, Buffalo, New York 14260*

KAREN J. BAME (52), *Department of Biochemistry, The University of Alabama at Birmingham, Birmingham, Alabama 35294*

SAMUEL H. BARONDES (42, 43), *Department of Psychiatry, University of California at San Francisco, San Francisco, California 94143*

MANJU BASU (51), *Department of Chemistry, Biochemistry, Biophysics, and Molecular Biology Program, University of Notre Dame, Notre Dame, Indiana 46556*

SUBHASH BASU (51), *Department of Chemistry, University of Notre Dame, Notre Dame, Indiana 46556*

EDWARD A. BAYER (35), *Department of Biophysics, The Weizmann Institute of Science, Rehovot 76100, Israel*

PAUL D. BISHOP (60), *Program in Biochemistry and Biophysics, Washington State University, Pullman, Washington 99164*

CHERYL A. BORGMAN (9), *Department of Pathology, University of Virginia Medical Center, Charlottesville, Virginia 22908*

ENRICO CABIB (55, 56), *National Institute of Diabetes and Digestive and Kidney Diseases, National Institutes of Health, Bethesda, Maryland 20892*

VICTOR J. CHEN (33), *Lilly Research Laboratories, Indianapolis, Indiana 46285*

DAVID T. CHEUNG (31), *Bone and Connective Tissue Research Laboratory, Orthopedic Hospital of Los Angeles, Los Angeles, California 90007-2697*

HUNG-CHE CHON (51), *Department of Pathology, St. Louis University Medical Center, St. Louis, Missouri 63104*

ERNST CONZELMANN (66), *Institut für Organische Chemie und Biochemie, Universität Bonn, D-5300 Bonn 1, Federal Republic of Germany*

RICHARD D. CUMMINGS (18), *Department of Biochemistry, School of Chemical Sciences, University of Georgia, Athens, Georgia 30602*

J. DAKOUR (24), *Department of Clinical Chemistry, University Hospital, S-221 85 Lund, Sweden*

PETER F. DANIEL (7), *Department of Biochemistry, E. K. Shriver Center, Waltham, Massachusetts 02254*

KAMAL K. DAS (51), *Department of Chemistry, University of Notre Dame, Notre Dame, Indiana 46556*

TRIPTI DE (51), *Indian Institute of Chemical Biology, Calcutta 700032, India*

JACK J. DISTLER (41), *Department of Internal Medicine, The University of Michigan Medical School, Ann Arbor, Michigan 48109*

ALAN D. ELBEIN (58), *Department of Biochemistry, The University of Texas Health Science Center, San Antonio, Texas 78284*

JUKKA FINNE (22), *Department of Biochemistry, University of Kuopio, SF-70211 Kuopio, Finland*

ALFRED P. FISHMAN (19), *Cardiovascular-Pulmonary Division, Department of Medicine, Hospital of the University of Pennsylvania, Philadelphia, Pennsylvania 19104*

HAROLD M. FLOWERS (29), *Department of Biophysics, The Weizmann Institute of Science, Rehovot 76100, Israel*

BAIBA GILLARD (23), *Department of Medicine, Baylor College of Medicine, Houston, Texas 77030*

VICTOR GINSBURG (14), *Laboratory of Structural Biology, National Institute of Diabetes and Digestive and Kidney Diseases, National Institutes of Health, Bethesda, Maryland 20892*

R. C. GOLDMAN (20), *Anti-infective Research Division, Abbott Laboratories, Abbott Park, Illinois 60064*

I. J. GOLDSTEIN (48, 49), *Department of Biological Chemistry, University of Michigan, Ann Arbor, Michigan 48109*

GARY R. GRAY (3), *Department of Chemistry, University of Minnesota, Minneapolis, Minnesota 55455*

SEN-ITIROH HAKOMORI (1, 2), *Program of Biochemical Oncology, Fred Hutchinson Cancer Research Center, and Department of Pathobiology, Microbiology, and Immunology, University of Washington, Seattle, Washington 98104*

PATRICK C. HALLENBECK (12), *Department of Food Science and Technology, University of California at Davis, Davis, California 95616*

G. HANSSON (24), *Department of Medical Chemistry, University of Göteborg, S-400 33 Göteborg, Sweden*

VINCENT C. HASCALL (21), *Bone Research Branch, National Institute of Dental Research, National Institutes of Health, Bethesda, Maryland 20892*

RALPH HEIMER (19), *Department of Biochemistry, Thomas Jefferson University, Philadelphia, Pennsylvania 19107*

CARLOS B. HIRSCHBERG (59), *E. A. Doisy Department of Biochemistry, St. Louis University School of Medicine, St. Louis, Missouri 63104*

C. A. HOPPE (30), *Department of Biology, The Johns Hopkins University, Baltimore, Maryland 21218*

THOMAS C.-Y. HSIEH (13), *Department of Food Science and Department of Biochemistry, Louisiana State University and A & M College and The Louisiana State University Agricultural Center, Baton Rouge, Louisiana 70803*

GEORGE W. JOURDIAN (41), *Departments of Biological Chemistry and Internal Medicine, The University of Michigan Medical School, Ann Arbor, Michigan 48109*

MOHINDER S. KANG (55, 56), *National Cancer Institute, National Institutes of Health, Bethesda, Maryland 20892*

REIJI KANNAGI (1), *Department of Laboratory Medicine, Kyoto University School of Medicine, Kyoto, Japan*

KARL-ANDERS KARLSSON (16, 17), *Department of Medical Biochemistry, University of Göteborg, S-400 33 Göteborg, Sweden*

K. KAWAGUCHI (30), *Department of Biology, The Johns Hopkins University, Baltimore, Maryland 21218*

AKIRA KOBATA (6, 64), *Department of Biochemistry, The Institute of Medical Science, The University of Tokyo, 4-6-1 Shirokanedai, Minato-ku, Tokyo 108, Japan*

THEODORE A. W. KOERNER (4), *Department of Pathology, College of Medicine, University of Iowa, Iowa City, Iowa 52242*

ROSALIND KORNFELD (46), *Departments of Internal Medicine and Biological Chemistry, Division of Hematology-Oncology, Washington University School of Medicine, St. Louis, Missouri 63110*

BARTŁOMIEJ KWIATKOWSKI (65), *Department of Physics and Biophysics, Institute of Medical Biology, Medical Academy, PL-80-211 Gdánsk, Poland*

JOHN W. KYLE (51), *Department of Biochemistry, St. Louis University Medical Center, St. Louis, Missouri 63104*

STEPHAN LADISCH (23), *Department of Hematology/Oncology, Department of Pediatrics, School of Medicine, University of California at Los Angeles, Los Angeles, California 90024*

ROGER A. LAINE (13), *Department of Biochemistry and Department of Chemistry, Louisiana State University and A & M College and The Louisiana State University Agricultural Center, Baton Rouge, Louisiana 70803*

REIKO T. LEE (34), *Department of Biology and McCollum-Pratt Institute, The Johns Hopkins University, Baltimore, Maryland 21218*

Y. C. LEE (30, 34), *Department of Biology and McCollum-Pratt Institute, The Johns Hopkins University, Baltimore, Maryland 21218*

HAKON LEFFLER (42), *Department of Psychiatry, University of California at San Francisco, San Francisco, California 94143*

L. LEIVE[1] (20), *Laboratory of Structural Biology, National Institute of Diabetes and Digestive and Kidney Diseases, National Institutes of Health, Bethesda, Maryland 20892*

STEVEN B. LEVERY (2), *Program of Biochemical Oncology, Fred Hutchinson Cancer Research Center, University of Washington, Seattle, Washington 98104*

P. LIN (30), *Department of Biology, The Johns Hopkins University, Baltimore, Maryland 21218*

HALINA LIS (47), *Department of Biophysics, The Weizmann Institute of Science, Rehovot 76100, Israel*

A. LUNDBLAD (24), *Department of Clinical Chemistry, University Hospital, S-221 85 Lund, Sweden*

JOHN L. MAGNANI (14, 15, 39), *Laboratory of Structural Biology, National Institute of Diabetes and Digestive and Kidney Diseases, National Institutes of Health, Bethesda, Maryland 20892*

FRANK MALEY (62), *Wadsworth Center for Laboratories and Research, New York State Department of Health, Albany, New York 12201*

M. MALIARIK (49), *Department of Biological Chemistry, University of Michigan, Ann Arbor, Michigan 48109*

ROBERT M. MAYER (57), *Department of Chemistry, The Ohio State University, Columbus, Ohio 43210*

R. H. MCCLUER (8), *Department of Biochemistry, E. K. Shriver Center, Waltham, Massachusetts 02254*

RONALD D. MCCOY (12, 54), *Department of Biological Chemistry, School of Medicine, University of California at Davis, Davis, California 95616*

ARLENE J. MENCKE (31), *Central Research Laboratories, 3M Company, St. Paul, Minnesota 55133*

ROBERT K. MERKLE (18), *Department of Biochemistry, School of Chemical Sciences, University of Georgia, Athens, Georgia 30602*

RONALD J. MIDURA (21), *Bone Research Branch, National Institute of Dental Research, National Institutes of Health, Bethesda, Maryland 20892*

RONALD L. MILLER (45), *Department of Biochemistry, Medical University of South Carolina, Charleston, South Carolina 29425*

JAAKKO PARKKINEN (22), *Department of Medical Chemistry, University of Helsinki, SF-00170 Helsinki, Finland*

JAMES C. PAULSON (11, 44), *Department of Biological Chemistry, School of Medicine, University of California at Los Angeles, Los Angeles, California 90024*

[1] Deceased.

MARY PEREZ (59), *Department of Immunology, Scripps Clinic and Research Foundation, La Jolla, California 92037*

JERZY PETRYNIAK (48), *Department of Biological Chemistry, University of Michigan, Ann Arbor, Michigan 48109*

THOMAS H. PLUMMER, JR. (63), *Wadsworth Center for Laboratories and Research, New York State Department of Health, Albany, New York 12201*

JAMES H. PRESTEGARD (4), *Department of Chemistry, Yale University, New Haven, Connecticut 06511*

M. H. RAVINDRANATH (44), *Department of Surgery/Oncology, University of California at Los Angeles, Los Angeles, California 90024*

VERNON N. REINHOLD (5), *Department of Nutrition, Harvard School of Public Health, Boston, Massachusetts 02115*

APRIL R. ROBBINS (37), *Genetics and Biochemistry Branch, National Institute of Diabetes and Digestive and Kidney Diseases, National Institutes of Health, Bethesda, Maryland 20892*

MARIE M. ROBERSON (43), *Department of Psychiatry, University of California at San Diego, La Jolla, California 92093*

D. ROBERTS (38, 49), *Laboratory of Structural Biology, National Institute of Diabetes and Digestive and Kidney Diseases, National Institutes of Health, Bethesda, Maryland 20892*

CALVIN F. ROFF (37), *Genetics and Biochemistry Branch, National Institute of Diabetes and Digestive and Kidney Diseases, National Institutes of Health, Bethesda, Maryland 20892*

GARY N. ROGERS (11), *Department of Pharmacology, College of Medicine, University of Iowa, Iowa City, Iowa 52242*

LEONARD H. ROME (52), *Department of Biological Chemistry, School of Medicine, University of California at Los Angeles, Los Angeles, California 90024*

CLARENCE A. RYAN (60), *Institute of Biological Chemistry, Washington State University, Pullman, Washington 99164-6340*

PHYLLIS M. SAMPSON (19), *Cardiovascular-Pulmonary Division, Department of Medicine, Hospital of the University of Pennsylvania, Philadelphia, Pennsylvania 19104*

BO E. SAMUELSSON (50), *Department of Medical Biochemistry, University of Göteborg, S-400 33 Göteborg, Sweden*

KONRAD SANDHOFF (26, 66), *Institute für Organische Chemie und Biochemie, Universität Bonn, D-5300 Bonn 1, Federal Republic of Germany*

ROBERT J. SCHAEPER (51), *Department of Chemistry, University of Notre Dame, Notre Dame, Indiana 46556*

ROLAND SCHAUER (10, 53), *Biochemisches Institüt, Christian-Albrechts-Universität zu Kiel, D-2300 Kiel, Federal Republic of Germany*

K. SCHROER (24), *Laboratory of Pathology, National Cancer Institute, National Institutes of Health, Bethesda, Maryland 20892*

GÜNTER SCHWARZMANN (26), *Institut für Organische Chemie und Biochemie, Universität Bonn, D-5300 Bonn 1, Federal Republic of Germany*

NATHAN SHARON (47), *Department of Biophysics, The Weizmann Institute of Science, Rehovot 76100, Israel*

HAKIMUDDIN T. SOJAR (27), *Department of Biological Sciences, State University of New York at Buffalo, Buffalo, New York 14260*

SARAH SPIEGEL (25), *Membrane Biochemistry Section, Developmental and Metabolic Neurology Branch, National Institute of Neurological and Communicative Disorders and Stroke, National Institutes of Health, Bethesda, Maryland 20892*

STEVEN L. SPITALNIK (14, 40), *Department of Pathology and Laboratory Medicine, University of Pennsylvania, Philadelphia, Pennsylvania 19104*

PAMELA STANLEY (36), *Department of Cell Biology, Albert Einstein College of Medicine, Bronx, New York 10461*

STEPHAN STIRM (65), *Biochemisches Institut am Klinikum der Universität, D-6300 Giessen, Federal Republic of Germany*

NICKLAS STRÖMBERG (17), *Department of Medical Biochemistry, University of Göteborg, S-400 33 Göteborg, Sweden*

KUNIHIKO SUZUKI (61), *Departments of Neurology and Psychiatry, Biological Sciences Research Center, School of Medicine, University of North Carolina, Chapel Hill, North Carolina 27514*

SEIICHI TAKASAKI (6), *Department of Biochemistry, The Institute of Medical Science, The University of Tokyo, 4-6-1 Shirokanedai, Minato-ku, Tokyo 108, Japan*

ANTHONY L. TARENTINO (63), *Wadsworth Center for Laboratories and Research, New York State Department of Health, Albany, New York 12201*

NAGESWARA R. THOTAKURA (28), *Department of Biological Sciences, State University of New York at Buffalo, Buffalo, New York 14260*

SHERIDA TOLLEFSEN (46), *Department of Pediatrics, Washington University School of Medicine, St. Louis, Missouri 63110*

ROBERT B. TRIMBLE (62), *Wadsworth Center for Laboratories and Research, New York State Department of Health, Albany, New York 12201*

FREDERIC A. TROY (12, 54), *Department of Biological Chemistry, School of Medicine, University of California at Davis, Davis, California 95616*

ROBERT J. TRUMBLY (62), *Department of Biochemistry, Medical College of Ohio, Toledo, Ohio 43699*

M. D. ULLMAN (8), *Research Service and Geriatric Research, Education and Clinical Center, ENRM Veterans Hospital, Bedford, Massachusetts 01730*

ERIC R. VIMR (12), *Department of Veterinary Pathobiology, College of Veterinary Medicine, University of Illinois, Urbana, Illinois 61801*

PREMANAND V. WAGH (67), *Department of Biological Sciences, State University of New York at Buffalo, Buffalo, New York 14260*

KIYOHIRO WATANABE (1), *Department of Biochemistry, Shigei Medical Research Institute, Okayama, Japan*

MEIR WILCHEK (35), *Department of Biophysics, The Weizmann Institute of Science, Rehovot 76100, Israel*

FINN WOLD (31), *Department of Biochemistry and Molecular Biology, University of Texas Medical School, Houston, Texas 77227*

KATSUKO YAMASHITA (6), *Department of Biochemistry, Kobe University School of Medicine, 7 Kusunoki-cho, Chuo-ku, Kobe 650, Japan*

SAU-CHI BETTY YAN (32), *Biochemical Research Division, Lilly Research Laboratories, Eli Lilly & Company, Indianapolis, Indiana 46285*

MASAKI YANAGISHITA (21), *Bone Research Branch, National Institute of Dental Research, National Institutes of Health, Bethesda, Maryland 20892*

WILLIAM W. YOUNG, JR. (9), *Department of Pathology, University of Virginia Medical Center, Charlottesville, Virginia 22908*

ROBERT K. YU (4), *Departments of Neurology and Molecular Biophysics and Biochemistry, School of Medicine, Yale University, New Haven, Connecticut 06510*

D. ZOPF (24), *Laboratory of Pathology, National Cancer Institute, National Institutes of Health, Bethesda, Maryland 20892*

Preface

New methods for studying the composition, synthesis, and degradation of complex carbohydrates are rapidly being developed. This volume contains some of the advances that have appeared in the literature since the publication in 1982 of Volume 83. As were the previous volumes in this series that deal with complex carbohydrates, this volume is divided into five sections: Analytical Methods, Preparations, Carbohydrate-Binding Proteins, Biosynthesis, and Degradation. I would welcome suggestions for articles on important areas of research for future supplements in this series.

VICTOR GINSBURG

METHODS IN ENZYMOLOGY

EDITED BY

Sidney P. Colowick and Nathan O. Kaplan

VANDERBILT UNIVERSITY
SCHOOL OF MEDICINE
NASHVILLE, TENNESSEE

DEPARTMENT OF CHEMISTRY
UNIVERSITY OF CALIFORNIA
AT SAN DIEGO
LA JOLLA, CALIFORNIA

I. Preparation and Assay of Enzymes
II. Preparation and Assay of Enzymes
III. Preparation and Assay of Substrates
IV. Special Techniques for the Enzymologist
V. Preparation and Assay of Enzymes
VI. Preparation and Assay of Enzymes (*Continued*)
Preparation and Assay of Substrates
Special Techniques
VII. Cumulative Subject Index

METHODS IN ENZYMOLOGY

EDITORS-IN-CHIEF

Sidney P. Colowick and Nathan O. Kaplan

Section I

Analytical Methods

[1] Isolation and Purification of Glycosphingolipids by High-Performance Liquid Chromatography

By Reiji Kannagi, Kiyohiro Watanabe, and Sen-itiroh Hakomori

High-performance liquid chromatography (HPLC) has been applied in the separation of underivatized glycosphingolipids for preparative purposes, as originally described by Watanabe and Arao,[1] employing a column of porous silica gel Iatrobeads 6RS-8010 or 6RS-8060. The method has been extensively utilized in separation of a number of glycosphingolipids.[2-11] The quantity of the glycolipid mixture to be separated, the size of the column, the conditions of elution, particularly the composition and slope of gradient elution, and the speed of elution can be varied from one case to another, and these conditions have been published in various papers in which the separation of specific types of glycolipids have been described. This article describes the general rules of the method applied for separation of different types of glycolipids by modifications of this method. However, glycolipids with close chromatographic mobilities are coeluted and cannot be separated by a simple application of this method. Often, those components can be separated as acetylated derivatives in a similar, but less polar, solvent mixture. These conditions will also be described in this article.

Materials

Porous silica gel. Iatrobeads 6RS-8060 (diameter 60 μm) and 6RS-8010 (diameter 10 μm) purchased from Iatron Chemical Co., Hi-

[1] K. Watanabe and Y. Arao, *J. Lipid Res.* **22**, 1020 (1981).

[2] R. Kannagi, M. N. Fukuda, and S. Hakomori, *J. Biol. Chem.* **257**, 4438 (1982).

[3] R. Kannagi, E. Nudelman, S. B. Levery, and S. Hakomori, *J. Biol. Chem.* **257**, 14865 (1982).

[4] R. Kannagi, S. B. Levery, and S. Hakomori, *J. Biol. Chem.* **259**, 8444 (1984).

[5] R. Kannagi, S. B. Levery, F. Ishigami, S. Hakomori, L. H. Shevinsky, B. B. Knowles, and D. Solter, *J. Biol. Chem.* **258**, 8934 (1983).

[6] R. Kannagi, N. A. Cochran, F. Ishigami, S. Hakomori, P. W. Andrews, B. B. Knowles, and D. Solter, *EMBO J.* **2**, 2355 (1983).

[7] R. Kannagi, S. B. Levery, and S. Hakomori, *FEBS Lett.* **175**, 397 (1984).

[8] E. G. Bremer, S. B. Levery, S. Sonnino, R. Ghidoni, S. Canevari, R. Kannagi, and S. Hakomori, *J. Biol. Chem.* **259**, 14773 (1984).

[9] Y. Okada, R. Kannagi, S. B. Levery, and S. Hakomori, *J. Immunol.* **133**, 835 (1984).

[10] K. Murayama, S. B. Levery, V. Schirrmacher, and S. Hakomori, *Cancer Res.* **46**, 1395 (1986).

[11] S. K. Kundu and D. D. Scott, *J. Chromatogr.* **232**, 19 (1982).

gashi-Kanda, Chiyoda-ku, Tokyo, Japan. Several brands of porous silica gel preparations were compared by Watanabe and Arao.[1] Only this brand showed satisfactory reproducibility and durability for many repeated operations.

HPLC equipment and columns. HPLC equipment capable of composing a gradient solvent mixture from three solvents. We use a Varian HPLC model 5020, but other vendors' equipment should be equally satisfactory. The columns should be packed in special packing equipment with high pressure (6000 psi) for homogenous packing. Packing of the columns can be performed by various vendors, such as Analytical Sciences, Inc., Santa Clara, CA 95050.

Hydrophobic porous silica gel columns. C_{18} alkylated porous silica gel columns of various sizes can be purchased from various vendors (e.g., Whatman Company, Kent, England). A small prepacked column in a disposable plastic syringe is also available from various vendors (e.g., Bond-Elut, Analytichem International, Harbor City, CA).

Solvents and reagents. Isopropyl alcohol (isopropanol), hexane, toluene, and absolute ethanol are reagent grade; water is deionized by electroendosmosis. Acetic anhydride is dehydrated with calcium hydride and distilled, and pyridin is dehydrated by reflexing over barium oxide and distilled. Primulin and orcinol are purchased from Aldrich Chemical Co. (Milwaukee, WI). HPTLC plates are made by Merck Chemical Co. (Darmstadt, West Germany) or J. T. Baker Chemical Co. (Phillipsburg, NJ).

Procedure 1: Separation of Underivatized Glycolipids by
  High-Performance Liquid Chromatography

Intact glycosphingolipids without derivatization can be readily separated by HPLC with elution by gradient solvents composed of isopropanol–hexane–water in various ratios. Typical examples are shown below for separation of (i) neutral glycolipids from whole human blood cell membranes, and (ii) gangliosides of whole human blood cell membranes.

Preliminary Purification of Glycosphingolipid Fraction. Nonlipid components, such as salts, mono- or oligosaccharides, nucleotides, amino acids and polypeptides, present in the glycolipid fraction hamper the efficiency of chromatographic separation of glycolipids. Therefore, it is desirable to eliminate these contaminants before the fraction is subjected to chromatography. Nonlipid components were eliminated by treating the glycolipid aqueous solution with alkylated porous silica gel. The glycolipid fraction dissolved in water–methanol–chloroform (1 : 1 : 0.1, v/v/v) was passed through a C_{18} alkylated silica gel column that had been equilibrated in the same solvent, washed with 5 column volumes of the same solvent, and the lipid adsorbed on the column was eluted with 3 column volumes of chloroform–methanol (2 : 1, v/v). In a small-scale operation (<100 μg), a Bond-Elut column is convenient.

Separation of Neutral Glycosphingolipids from Whole Human Blood Cell Membranes. Approximately 100–300 mg of the upper neutral glycolipid fraction obtained from 500 g of packed membrane fraction of whole human blood was dissolved in 2 ml of chloroform–methanol (1 : 1), and the mixture was warmed and sonicated. The solution was applied to a column of Iatrobeads 6RS-8060 with dimensions of 1 × 100 cm, which was preequilibrated with isopropanol–hexane–water (55 : 40 : 5, v/v/v). The column was eluted with a gradient of isopropanol–hexane–water from 55 : 40 : 5 to 55 : 30 : 15 (v/v/v). Fractions were collected with a fraction collector at 2-min intervals and the total volume of eluate was 400 ml. The flow rate was set at 2 ml/min by applying approximately 210–280 psi (15–20 atm). A small aliquot (10–12 μl) of each fraction was analyzed on high-performance thin-layer chromatography (HPTLC) developed in chloroform–methanol–water (60 : 35 : 8, v/v/v). Glycolipid spots were revealed by spraying with 0.2% orcinol dissolved in 2 M sulfuric acid and heating in a 120° oven for 5–7 min. Glycolipids as well as phospholipids and polypeptides that often occur in the glycolipid fraction were visualized by spraying with 0.01% Primulin solution in acetone–water (4 : 1, v/v) and observed under ultraviolet light. The elution pattern of glycolipids is shown in Fig. 1. The fractions eluted between H1 and H2 glycolipids, those eluted between H2 and H3 glycolipids, and those eluted after H3 glycolipid were tentatively called "x fraction," "y fraction," and "z fraction," respectively. Namely, in Fig. 1, tubes 70–89 were the x fraction, tubes 103–107 were the y fraction, and tubes later than 118 were the z fraction. Further fractionation of both x and y fractions was performed using HPLC as described below; i.e., x and y fractions were separated into components x_1, x_2, x_3, x_4, and y_1, y_2, y_3, and y_4, respectively, by chromatography on Iatrobeads 6RS-8010 (diameter 10 μm) with dimensions of 1 × 50 cm or 1 × 30 cm, with a gradient elution from isopropanol–hexane–water 55 : 40 : 5 to 55 : 38 : 7 (v/v/v), followed by elution with a shallower gradient from 55 : 38 : 7 to 55 : 36 : 9 (v/v/v), applying about 140–280 psi (10–20 atm) to obtain a flow rate of 0.5 ml/min. The separation of x and y components is shown in Fig. 2. Some of these components were homogeneous, but others were found to be mixtures of glycolipids having different carbohydrate structure. Separation of these fractions into components having a single carbohydrate can be performed after acetylation (acetylated derivatives or acetates) and will be described subsequently.

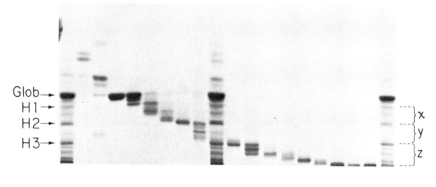

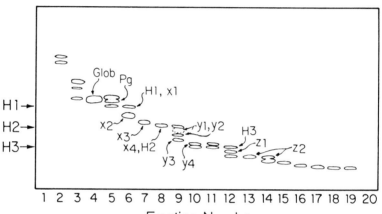

Fraction Number

FIG. 1. Separation pattern of neutral glycolipids of human blood group O cell membranes through low-pressure HPLC on Iatrobeads 6RS-8060 column. The conditions of elution are as described in the text (Procedure 1). The upper panel shows thin-layer chromatography of each fraction separated in chloroform–methanol–water (60 : 35 : 8, v/v/v) and visualized by orcinol reagent. The lower panel represents a schematic drawing and identification of each spot separated. Lane numbers are identified as follows: lane 2 is a pool of fractions 22–30; 3, 33–40; 4, 41–57; 5, 58–69; 6, 70–80; 7, 81–89; 8, 90–102; 9, 103–109; 11, 110–117; 12, 118–130; 13, 131–140; 14, 141–152; 15, 153–164; 16, 165–175; 17, 176–186; 18, 187–200; 19, insoluble material. Lanes 1, 10, and 20 are total upper neutral glycolipids. Glob, globoside. Lanes 6 and 7 (pooled fraction tubes 70–89) represent the x fraction, and lane 9 (tubes 103–107) represents the y fraction. Components identified: x_1, Le^a glycolipid (III^4FucLc_3) and other unidentified components; x_2, $IV^3GalNAcnLc_4$; x_3, a mixture of globo-H ($IV^3Fuc\alpha1 \rightarrow 2GalGb_4$) ($x_{3a}$) and para-Forssman ($IV^3\beta$-GalNAcGb$_4$) ($x_{3b}$); x_4, lactonorhexaosylceramide (nLc$_4$). y_1, $IV^2FucIII^3FucnLc_4$ (Le^y glycolipid) and other unidentified components; y_2, $V^3FucnLc_6$ (Le^x-active heptaosylceramide); y_3, $VI^2FucV^3FucnLc_6$ (Le^y-active hexaosylceramide) and other unidentified components; y_4, a mixture of lactoisooctaosylceramide ($IV^6Gal\beta1 \rightarrow 4GlcNAcnLc_6$) ($y_{4a}$; I-active glycolipid), Le^a-active heptaosylceramide ($IV^3Gal\beta1 \rightarrow 3[Fuc\alpha1 \rightarrow 4]GlcNAcnLc_4$) ($y_{4b}$), and monofucosyllactoisooctaosylceramide (y_4c; I-active glycolipid). z_1, Le^a-active nonaosylceramide ($VII^3FucnLc_8$); z_2, Le^x-active decaosylceramide ($V^3FucVII^3FucnLc_8$); z_3, H4 glycolipid ($VIII^2FucIV^6Fuc\alpha1 \rightarrow 2Gal\beta1 \rightarrow 4GlcNAcnLc_8$).

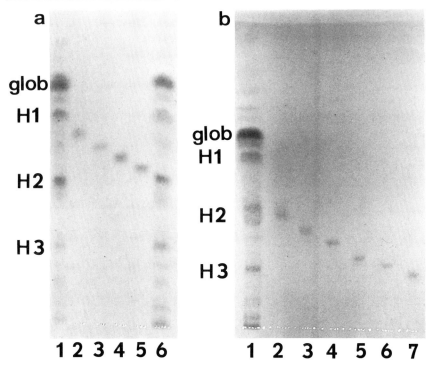

FIG. 2. HPTLC patterns of fractionated x series glycolipids (a) and y series glycolipids (b). Separation was performed on HPTLC with a solvent mixture of chloroform–methanol–water (60 : 35 : 8, v/v/v) and visualized by orcinol reagent. (a) Lanes 1 and 6, unfractionated total upper neutral glycolipids of whole blood cell membranes; 2, x_1; 3, x_2; 4, x_3; 5, x_4. (b) Lane 1, unfractionated total upper neutral glycolipids; 2, H2; 3, y_1; 4, y_2; 5, y_3; 6, y_4; 7, H3; glob, globoside.

Separation of Monosialogangliosides of Human Whole Blood Cell Membranes into Components. The monosialoganglioside fraction weighing 150 mg was prepared from 500 g of packed membrane fraction of whole human blood cells. This fraction was dissolved in 2 ml of chloroform–methanol (1 : 1, v/v) and injected onto a column of Iatrobeads 6RS-8010 with dimensions of 1 × 100 cm, which was preequilibrated with isopropanol–hexane–water (55 : 40 : 5, v/v/v), and eluted with a gradient from the same solvent to isopropanol–hexane–water (55 : 25 : 20, v/v/v) with a flow rate of 2 ml/min by applying 700–840 psi pressure. Fractions were collected at 2-min intervals with a fraction collector, and the total volume of the eluate was 400 ml. From each fraction, a small aliquot (10–20 μl) was analyzed on HPTLC. The separation pattern of gangliosides

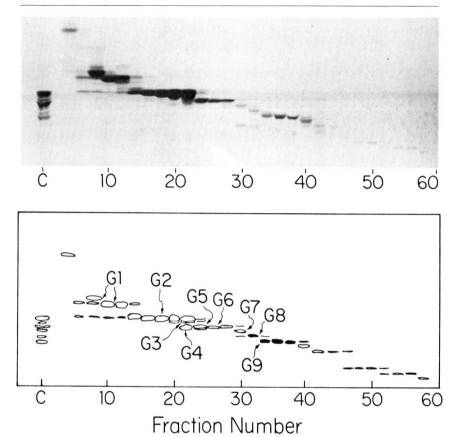

Fraction Number

Fig. 3. Separation pattern of gangliosides of human blood group O cell membranes after HPLC on Iatrobeads 6RS-8010 column. The conditions of elution are described in the text. The upper panel shows the HPTLC pattern of each separate fraction developed with chloroform–methanol–water (60 : 35 : 8, v/v/v) and visualized by orcinol reagent. The lower panel represents a schematic drawing of each spot and identification of each ganglioside. Components identified: G1, G_{M3}, G2, IV^{3}NeuAcnLc$_4$. G3, IV^{3}NeuAcα2 → 3GalNAcnLc$_4$. G4, IV^{6}NeuAcnLc$_4$. G5, IV^{3}NeuAcα2 → 3GalGb$_4$. G6, VI^{3}NeuAcnLc$_6$. G7, VI^{6}NeuAcnLc$_6$. G8, I-active ganglioside (monofucosyllactoisooctaosylceramide; IV^6 Galβ1 → 4GlcNAc-VI^{3}NeuAcnLc$_6$). G9, H-active ganglioside (fucosylsialosyllactoisooctaosylceramide; IV^6-Fucα1 → 2Galβ1 → 4GlcNAcVI^{3}NeuAcnLc$_6$).

derived from monosialogangliosides of adult blood group O erythrocytes is shown in Fig. 3. Each of the gangliosides G1 to G9 were characterized as previously described[12,13] and are listed in the legend for Fig. 3.

[12] K. Watanabe, M. E. Powell, and S. Hakomori, *J. Biol. Chem.* **254**, 8223 (1979).
[13] K. Watanabe, M. E. Powell, and S. Hakomori, *J. Biol. Chem.* **253**, 8962 (1978).

Procedure 2: Separation of Glycosphingolipids as Acetylated Derivatives

Glycolipids having close chromatographic mobilities are only separable as acetates. An apparently homogeneous glycolipid band revealed by thin-layer chromatography (TLC) can be further separated into multiple components after acetylation followed by HPLC or HPTLC. A few examples are described below.

Acetylation of Glycolipids. A crude glycolipid fraction, or a semipurified fraction, preliminarily separated by HPLC as above, could be acetylated by the following procedure. The fraction, in most cases 100 μg or less, was placed in chloroform–methanol solution in a small Teflon-lined, screw-capped test tube (1 × 10 cm) and evaporated under nitrogen to dryness. Glycolipid fractions in larger quantity were evaporated repeatedly in a rotary evaporator in a round-bottom flask with absolute ethanol. The residue was placed in a desiccator over phosphorus pentoxide *in vacuo.* The dried residue was dissolved in 3–5 parts (w/v) of dried pyridine and sonicated well; 1.5–2.5 parts of acetic anhydride was added and the mixture was allowed to stand overnight at room temperature. To this was added 10 parts of toluene, and the mixture was evaporated under nitrogen stream or in a rotary evaporator at 30–40°. For effective acetylation, the glycolipid fraction should be free from salts, which can be eliminated by hydrophobic column chromatography as described above.

Separation of Total Neutral Glycolipids Present in Folch's Lower Phase Extracted from Human Blood O Erythrocytes. The majority of glycolipids with ceramide mono- to pentasaccharides are found in Folch's lower phase. These glycolipids were freed from phospholipids and cholesterol by chromatography on a Florisil column eluted with 1,2-dichloroethane–acetone (1 : 1, v/v).[14] A mixture of such glycolipid acetates was directly separated on HPLC. The glycolipid acetate mixture was dissolved in chloroform and injected onto a Iatrobeads 6RS-8010 column (dimensions 0.5 × 30 cm), which was preequilibrated with 100% hexane. The gradient elution was performed from 100% hexane to isopropanol–hexane–water (55 : 45 : 0, v/v/v) during 50 min, followed by another gradient elution to isopropanol–hexane–water (55 : 45 : 5, v/v/v). The flow rate was adjusted to 1 ml/min with an approximate pressure of 400–450 psi applied. The acetate of the ceramide monohexoside was eluted at 5–10 min, that of the ceramide dihexoside at about 18–20 min, and that of the ceramide trihexoside at about 30 min. The acetate of the major glycolipid, globoside, was eluted between 40 and 55 min. The acetates of lactoneotetraosylceramide, H1 glycolipid, and x₂ glycolipid were eluted between 50 and 60 min, as shown in Fig. 4A.

[14] T. Saito and S. Hakomori, *J. Lipid Res.* **12,** 257 (1971).

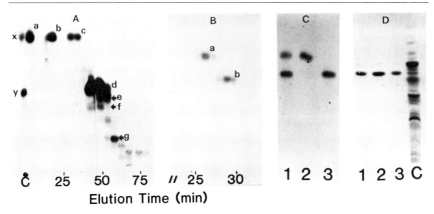

Elution Time (min)

FIG. 4. Separation pattern of glycolipid acetates by HPLC on Iatrobeads 6RS-8010 column and by HPTLC. (A) Total neutral glycolipids of O erythrocyte membranes present in Folch's lower phase. Total glycolipids were acetylated and separated by HPLC in an isopropanol–hexane–water system as described in the text. Fractions separated on a fraction collector were analyzed on HPTLC developed in a solvent of 1,2-dichloroethane–acetone–water (60:40:1, v/v/v), and detected by orcinol–sulfuric acid reaction. Spots a–g were identified as acetates of ceramide monohexoside (a), ceramide dihexoside (b), ceramide trihexoside (c), globoside (d), H1 glycolipid (e), paragloboside nLc$_4$ (f), x$_2$ glycolipid (IV^{3}GalNAcnLc$_4$) (g), respectively. (B) Separation pattern of acetylated H1 glycolipid (IV^{2}FucnLc$_4$) (spot a) and acetylated IV^{3}β-GalnLc$_4$ (spot b) on HPLC. Numbers on the abscissa indicate the elution time in minutes in both (A) and (B). (C) HPTLC pattern of acetylated glycolipids separated under the conditions described for (B). Lane 1, a mixture of IV^{2}FucnLc$_4$ (upper spot) and IV^{3}β-GalnLc$_4$ acetates; lane 2, a purified acetate of IV^{2}FucnLc$_4$; lane 3, a purified acetate of IV^{3}β-GalnLc$_4$. (D) Glycolipids after deacetylation. Lane 1, a mixture of IV^{2}FucnLc$_4$ and IV^{3}β-GalnLc$_4$; lane 2, purified deacetylated IV^{2}FucnLc$_4$; lane 3, purified deacetylated IV^{3}β-GalnLc$_4$; C, total upper neutral glycolipids of O erythrocytes.

Separation of Acetylated Glycolipid Mixture by HPLC: Separation of IV^{2}FucnLc$_4$ and IV^{3}β-GalnLc$_4$. Two glycolipids, IV^{2}FucnLc$_4$ (H1 glyco-lipid) and IV^{3}β-GalnLc$_4$ (β-galactosylparagloboside), showed an identical TLC mobility and were not separated on HPLC under the conditions described above. These glycolipids were, however, able to be separated as acetylated derivatives on HPLC with much less polar solvents. The glycolipid acetates were dissolved in chloroform, and the solution was injected onto an HPLC column packed with Iatrobeads 6RS-8010 with dimensions of 0.5 × 30 cm, which was preliminarily washed and equili-brated with 100% hexane. The gradient elution started immediately from 100% hexane to isopropanol–hexane–water (20:79:1, v/v/v) during 10 min, followed by another gradient elution to isopropanol–hexane–water (43:56:1, v/v/v) during 90 min. The flow rate was adjusted to 1 ml/min by applying 400–450 psi pressure.

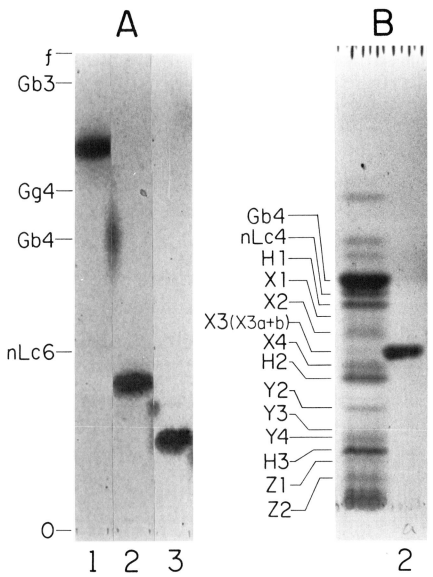

FIG. 5. Separation of x_3 glycolipid into components by HPTLC. (A) Purified glycolipid acetates. Lane 1, x_3a (globo-H); lane 2, x_2; lane 3, x_3b (para-Forssman antigen). The position of the acetylated derivative of each glycolipid is indicated in the left margin. Developed in a solvent of 1,2-dichloroethane–acetone–water (60 : 40 : 0.1, v/v/v). Spots were visualized by orcinol–sulfuric acid reaction. (B) Lane 1, upper neutral glycolipids of blood group O erythrocytes; lane 2, purified deacetylated x_3a glycolipid (globo-H antigen). Left margin indicates the position of characterized glycolipids. Note that the x_3 component is very minor. HPTLC in B and C developed in chloroform–methanol–water (60 : 35 : 8, v/v/v).

As shown in Fig. 4B, $IV^2FucnLc_4$ (H1 glycolipid) was eluted between 25 and 27 min, whereas $IV^3\beta$-GalnLc$_4$ (β-galactosylparagloboside) was eluted between 28 and 30 min. Each fraction collected on the fraction collector was pooled, and the separation pattern on HPLC is shown in Fig. 4C. Separation of these two components was complete. After deacetylation, these two components migrated to the same positions on regular HPTLC (Fig. 4D).

Separation of Acetylated x_3 Components ($IV^3\beta$-GalNAcGb$_4$ and $IV^3Fuc\alpha1 \rightarrow 2GalGb_4$) by HPTLC. The x_3 components, which were separated by HPLC from the upper neutral glycolipids prepared from O erythrocytes, were further separated into x_3a and x_3b after acetylation. The x_3 fraction, which also contained a small quantity of x_2 glycolipid, was acetylated as described above, applied on an HPTLC plate (Merck, Darmstadt, West Germany), and developed in a solvent mixture of 1,2-dichloroethane–acetone–water (60 : 40 : 0.1, v/v/v). Acetylated derivatives of Gb$_3$, globoside, Gg$_4$, and nLc$_6$ were run as controls. One component, x_3a, migrated between the acetylated Gb$_3$ and Gg$_4$, and another component, x_3b, migrated very slowly close to the origin. In contrast, acetylated components of x_2 glycolipid (GalNAc$\beta1 \rightarrow 3$Gal$\beta1 \rightarrow 4$GlcNAc$\beta1 \rightarrow 3$Gal$\beta1 \rightarrow 4$Glc$\beta1 \rightarrow 1$Cer) migrated faster than the acetylated derivative of x_3b, as shown in Fig. 5. These components were separated on HPTLC plate, visualized by Primulin spray, and observed under ultraviolet light; the silica gel in the bands was then scraped off the plate and extracted in a solvent mixture of chloroform–methanol (2 : 1, v/v) by sonication for 5 min, followed by centrifugation. The silica gel pellet was extracted in the same way three times. The extracts were combined, pooled, and evaporated under nitrogen, and the residue was deacetylated in chloroform–methanol (2 : 1, v/v) containing 0.2% sodium methoxide.[14] Thus, these components, x_3a and x_3b, were isolated in a pure state (see Fig. 5). The x_3a component was identified as globo-H, i.e., an H-active glycolipid reactive with monoclonal antibody MBr1 having the structure Fuc$\alpha1 \rightarrow 2$Gal$\beta1 \rightarrow 3$GalNAc$\beta1 \rightarrow 3$Gal$\alpha1 \rightarrow 4$Glc$\beta1 \rightarrow 1$Cer.[6,7] The x_3b component was identified as GalNAc$\beta1 \rightarrow 3$GalNAc$\beta1 \rightarrow 3$Gal$\alpha1 \rightarrow 4$Gal$\beta1 \rightarrow 4$Glc$\beta1 \rightarrow 1$Cer, i.e., β-N-acetylgalactosaminylgloboside (termed "para-Forssman").[15] These components were not separated at all without acetylation.

[15] S. Ando, K. Kon, Y. Nagai, and T. Yamakawa, *in* "New Vistas in Glycolipid Research" (A. Makita, S. Handa, T. Taketomi, and Y. Nagai, eds.), p. 71. Plenum, New York, 1982.

[2] Microscale Methylation Analysis of Glycolipids Using Capillary Gas Chromatography–Chemical Ionization Mass Fragmentography with Selected Ion Monitoring

By STEVEN B. LEVERY and SEN-ITIROH HAKOMORI

Unambiguous information regarding the position of glycosidic linkages and the position of branching in complex carbohydrates and polysaccharides can only be obtained by methylation analysis. The current methodology is based on (1) an effective methylation method using the Corey–Chaykovsky base,[1,2] and (2) gas chromatography–mass spectrometric (GC-MS) identification of partially O-methylated hexitols or hexosaminitols.[3–6] Applications of these methods have been reviewed previously.[7–9] More recently, however, the methodology has undergone numerous improvements in resolution and sensitivity, particularly through the use of capillary columns and in the identification of each component by chemical ionization mass spectrometry.[10–14] A modified procedure has been used in the linkage analysis of glycolipids, and many new structures have been determined through its application.[15–19] De-

[1] S. Hakomori, *J. Biochem. (Tokyo)* **55**, 205 (1964).

[2] P. A. Sandford and H. E. Conrad, *Biochemistry* **5**, 1508 (1966).

[3] H. Björndal, B. Lindberg, and S. Svensson, *Carbohydr. Res.* **5**, 443 (1967).

[4] K. Stellner, H. Saito, and S. Hakomori, *Arch. Biochem. Biophys.* **155**, 464 (1973).

[5] T. Tai, K. Yamashita, and A. Kobata, *J. Biochem. (Tokyo)* **78**, 679 (1975).

[6] W. Stoffel and P. Hanfland, *Hoppe-Seyler's Z. Physiol. Chem.* **354**, 21 (1973).

[7] H. Björndal, C. G. Hellerqvist, B. Lindberg, and S. Svensson, *Angew. Chem., Int. Ed. Engl.* **9**, 610 (1970).

[8] B. Lindberg, this series, Vol. 28, p. 178.

[9] P.-E. Jansson, L. Kenne, H. Liedgren, B. Lindberg, and J. Lönngren, *Chem. Commun., Univ. Stockholm* **8**, 1 (1976).

[10] M. McNeil and P. Albersheim, *Carbohydr. Res.* **56**, 239 (1977).

[11] R. Laine, *Anal. Biochem.* **116**, 383 (1981).

[12] R. Geyer, H. Geyer, S. Kühnhardt, W. Mink, and S. Stirm, *Anal. Biochem.* **133**, 197 (1983).

[13] M. E. Lowe and B. Nillson, *Anal. Biochem.* **136**, 187 (1984).

[14] J. Lomax, A. H. Gordon, and A. Chesson, *Carbohydr. Res.* **138**, 177 (1985).

[15] S. Hakomori, E. Nudelman, S. B. Levery, and C. M. Patterson, *Biochem. Biophys. Res. Commun.* **113**, 791 (1983).

[16] R. Kannagi, S. B. Levery, F. Ishigami, S. Hakomori, L. H. Shevinsky, B. B. Knowles, and D. Solter, *J. Biol. Chem.* **258**, 8934 (1983).

[17] S. Hakomori, E. Nudelman, S. B. Levery, and R. Kannagi, *J. Biol. Chem.* **259**, 4672 (1984).

tailed here is the practical procedure that has been used in these studies, which currently allows us to determine routinely all intersugar linkages of a 10–20 nmol sample of glycolipid (allowing for multiple sample injection and reduced yield of polysubstituted N-acetylhexosamines) using bonded-phase fused silica capillary columns and methane chemical ionization mass spectrometry (CI-MS) with selected ion monitoring (SIM). Because of the high resolution and reproducibility of the columns used, identification of partially methylated alditol acetates and acetylated neuraminic acid methyl ester methylglycosides can, with the help of standards, be reliably based on a combination of retention times and characteristic ions produced. The methodology described is specifically applied to analysis of glycosphingolipids, but all procedures and standard data are applicable for the analysis of glycoprotein oligosaccharides.[20] The numerous alternative methods of analysis are outside the scope of this article.

General

Reactions are conveniently performed in a 13-mm Teflon-lined, screw-capped Pyrex test tube, the bottom of which has been formed into a conical shape by heating about 60 mm from the top with a two-sided torch until the lower half drops off (Fig. 1). Triangular magnetic stirring bars (1.0 ml size, Pierce Chemical Co., Rockford, IL) conveniently fit into these conical tubes. Reagents should be of high quality, i.e., acetic anhydride (Baker, reagent grade, distilled over sodium acetate before use), "ultrapure" sulfuric acid and acetic acid (Alfa Products, Danvers, MA), chloroform (Baker, nanograde), dimethyl sulfoxide (Pierce, silylation grade), hexane and methanol (Baker, HPLC grade), methyl iodide (Baker, reagent grade, distilled before use), sodium tetradeuteridoborate, $NaBD_4$ (Alfa Products), sodium hydride (Alfa Products, 80% dispersion in oil, washed with hexane before use), and sodium hydroxide (Baker "Dilut-it," free of carbonate).

Preparation of Methylsulfinyl Carbanion Reagent

Methylsulfinyl carbanion (MSC) solution is prepared essentially as described by Corey and Chaykovsky.[21] Several adaptations of this proce-

---

[18] R. Kannagi, S. B. Levery, and S. Hakomori, *J. Biol. Chem.* **259**, 8444 (1984).

[19] R. Kannagi, S. B. Levery, and S. Hakomori, *J. Biol. Chem.* **260**, 6410 (1985).

[20] H. Rauvala, J. Finne, T. Krusius, J. Kärkkäinen, and J. Järnefelt, *Adv. Carbohydr. Chem. Biochem.* **38**, 389 (1981).

[21] E. J. Corey and M. Chaykovsky, *J. Am. Chem. Soc.* **87**, 1345 (1965).

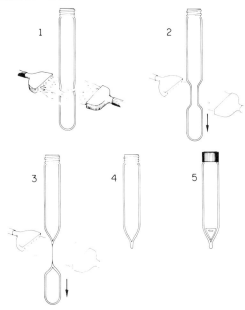

FIG. 1. Steps in the formation of a conical bottom using a "disposable" Pyrex screw-capped test tube: (1) 13 × 100 mm test tube is held vertically in a double burner torch until the glass begins to pull; (2) the test tube is rotated ~90° to ensure even heating around the circumference of the tube; best results are obtained by letting gravity pull the bottom piece rather than grasping with forceps; (3) the bottom is allowed to fall away, the connecting glass thread is severed, and the bottom is heated further to allow glass to contract back and form a thicker wall with a less fragile point at the bottom, as depicted in (4); the conical bottom is compatible with either a triangular stirring bar as shown in (5), or with a micro-stir "flea."

dure, for use with larger amounts of glycoconjugates, have been described.[2,6] Sodium hydride (8 g, 80% suspension in mineral oil) is introduced into a dry 500-ml three-necked round-bottom flask with a magnetic stirring bar. All further operations are conducted under a dry nitrogen atmosphere maintained in the flask. The NaH is washed 4 times by suspending in 200 ml dry hexane, allowing it to settle, and decanting the supernatant. Residual hexane is then removed under vacuum, dimethyl sulfoxide (100 ml) is added, and the flask is fitted with a thermometer and $CaCl_2$ drying tube for exit of nitrogen and hydrogen. The magnetically stirred reaction mixture is heated slowly to 50° with an oil bath and maintained at that temperature until hydrogen evolution has ceased and the solution has become transparent green in color (~90 min). Best results are obtained if the temperature is not allowed to exceed 50° during this period. The solution is then aliquoted in 1- to 2-ml portions under nitrogen into 13 × 100 mm Teflon-lined screw-capped test tubes and frozen at an angle to prevent cracking due to expansion of the reagent. MSC reagent can be stored thus for over a year at $-20°$ and thawed just prior to use, with no loss in reactivity. Basicity can be ascertained at any stage using triphenylmethane (Aldrich, Milwaukee, WI) as an indicator, as described by Rauvala.[22] In the past few years, potassium *tert*-butoxide in DMSO[23] and potassium methylsulfinyl carbanion[24] have been used for methylation. Recently, an adaptation using lithium methylsulfinyl carbanion has been introduced and proposed to significantly reduce the amounts of impurities encountered in the analysis.[25]

Permethylation Procedure[1]

Glycolipid samples (15–20 μg for less than pentaglycosylceramide; 25–50 μg for hexaglycosylceramide and larger) are dried in the bottom of conical-bottom test tubes under N_2 stream, then overnight *in vacuo* over P_2O_5. DMSO (100 μl) is added by dry syringe, followed by MSC reagent (100 μl). After flushing with dry N_2, the tube is sealed and the reaction mixture stirred at room temperature. Sonication is avoided, as this appears to promote formation of degradation products. After 1 hr, the reaction mixture is cooled in an ice bath, and methyl iodide (100 μl) is added. After flushing with dry N_2 and resealing, stirring is continued at room temperature for 2 hr, after which time the tube is flushed with N_2 in a warm ($\sim 35°$) water bath for a half-hour to remove excess methyl iodide. The remaining DMSO can then be removed almost completely by drying *in vacuo* over P_2O_5 overnight. The residue is partitioned between $CHCl_3$ and H_2O (0.5 ml each). Following removal of the H_2O layer, the $CHCl_3$ is washed twice more with H_2O, the combined H_2O washes are backwashed with $CHCl_3$, and the combined $CHCl_3$ layers are washed once more with H_2O and then transferred to another conical-bottom test tube, dried under N_2 stream, and dried once again *in vacuo* over P_2O_5. At this stage, aliquots can be removed (in $CHCl_3$) for direct-probe mass spectrometry and/or sialic acid linkage analysis.

Hydrolysis and Derivatization for Analysis as Partially Methylated
   Alditol Acetates (PMAAs)

To the dry permethylated glycolipid is added 100 μl of 0.5 N H_2SO_4 in 90% HAc. After flushing with N_2 and sealing, the tube is heated for 6–7 hr at 80° in an oven, to avoid microdistillation of the contents that occurs

[22] H. Rauvala, *Carbohydr. Res.* **72**, 257 (1979).
[23] J. Finne, T. Krusius, and H. Rauvala, *Carbohydr. Res.* **80**, 336 (1980).
[24] L. R. Phillips and B. A. Fraser, *Carbohydr. Res.* **90**, 149 (1981).
[25] J. Paz Parente, P. Cardon, J. Montreuil, B. Fournet, and G. Ricart, *Carbohydr. Res.* **141**, 41 (1985).

with use of a heating block. After cooling in an ice bath, the reaction mixture is neutralized using 130 μl of 0.5 N NaOH. Absolute EtOH is added, and the contents are dried under N_2 stream at ~35°, with additional EtOH as necessary to evaporate H_2O by azeotropic effect.

After drying *in vacuo* over NaOH (under N_2), the methylated sugars are reduced with aqueous $NaBD_4$ (300 μl of a 10 mg/ml solution in 0.01 N NaOH); after standing loosely capped at 4° overnight, the excess $NaBD_4$ is destroyed by addition of a few drops of glacial acetic acid. After evaporation under N_2 stream with addition of EtOH as needed, borate is removed by repeated (5×) addition of methanol (1 ml containing 1–2% acetic acid) and evaporation under N_2 stream at ~35°. The contents are dried *in vacuo* over P_2O_5 overnight.

Acetic anhydride (300 μl) is added and the test tube flushed with N_2, sealed, and heated in a block at 100° for 2 hr, with occasional sonication to disperse the salt residue, which catalyzes the acetylation. After cooling to room temperature, excess reagent is removed by flushing with N_2 (~35°) with repeated addition of toluene (1–2 ml) to evaporate acetic anhydride by azeotropic effect. (At this stage, care should be taken not to overdry the residue, since the more highly methylated neutral sugar derivatives are volatile enough to be lost, and the salts, following this last step, are light and nonadherent to the glass walls and may be blown out.) The residue is partitioned between $CHCl_3$ and H_2O and is treated the same as the methylated intact glycolipid. The $CHCl_3$ layers are transferred to a conical-bottom test tube and again dried (carefully) under N_2 stream, and taken up in 10–20 μl of hexane for injection onto the GC column.

Methanolysis and Derivatization for Analysis of Methylated Neuraminic Acid Methyl Ester Methylglycosides[26–28]

An aliquot (5–10 μg) of a permethylated ganglioside (sialic acid-containing glycolipid) is degraded by heating in 500 μl of 0.5 N methanolic HCl (under N_2) at 80° overnight. (In any case where an *internal* 2 → 8NeuNGcα2 → linkage is suspected to be present, it is advisable to use much milder methanolysis conditions, i.e., 0.05 N HCl, 80°, 1 hr,[29] since it is known that although the permethylated terminal NeuNGc residue is stable to N-deacylation, the internal residue with a free 8-hydroxy group is N-deacylated nearly quantitatively under the former condition.[28,30]) After cooling to room temperature, the solution is extracted

[26] H. Rauvala and J. Kärkkäinen, *Carbohydr. Res.* **56,** 1 (1977).
[27] H. van Halbeek, J. Haverkamp, J. P. Kamerling, J. F. G. Vliegenthart, C. Versluis, and R. Schauer, *Carbohydr. Res.* **60,** 51 (1978).
[28] S. Inoue and G. Matsumura, *Carbohydr. Res.* **74,** 361 (1979).
[29] R. K. Yu and R. W. Ledeen, *J. Lipid Res.* **11,** 506 (1970).
[30] S. Inoue and G. Matsumura, *FEBS Lett.* **121,** 33 (1980).

(3×) with hexane (1 ml). Ag_2CO_3 is added to neutralize the HCl, and methanol is removed to another test tube following centrifugation. The salts are washed 2 times with MeOH (1 ml), and the combined methanol extracts are evaporated under N_2, followed by drying *in vacuo* over P_2O_5. Acetylation is carried out using 200 μl each of dry pyridine and acetic anhydride, and the tube sealed under N_2 and heated at 80° for a half-hour. After cooling, the reagents are removed under N_2 with the addition of toluene. The dried residue is taken up in $CHCl_3$ for GC-MS.

Equipment for GC-MS

Satisfactory results are obtained on a Finnigan 3300 GC-MS/6110 data system that has been modified for use with fused silica capillary columns by installation of a glass-lined open/split injector and direct deactivated fused silica capillary interface into the ion source. In configuring a system, it is essential to make certain that the PMAA sample does not come in contact with heated metal or contaminated surfaces at any point prior to entry to the ion source. If the path from the oven is long, a deactivated fused silica capillary interface to the ion source employing a zero dead volume butt connector to the column is recommended. A siliconized glass injector liner should be employed, cleaned and resurfaced periodically. The yield of hexosamine derivatives is markedly decreased by contact with metal surfaces or contaminants in the injector. Columns used are 30 m DB-5 and DB-225 (J & W Scientific, Rancho Cordova, CA) bonded-phase fused silica capillaries, with outer diameter of 0.32 mm and a 0.25-μm coating thickness.

Conditions for Analysis

Splitless injection is used for all gas chromatography; injector temperature, 230°; transfer line, 230 240°; splitter is closed with oven at 50° for 40 sec after injection, then opened, and oven temperature is raised as rapidly as possible to program starting temperature (see below).

The chemical ionization source (methane, 300 μm) is maintained at 100°; electron energy, 120 V; emission current, 0.5 mA; collector, 35 eV.

PMAAs on DB-5. Head pressure, 16 psi helium. Temperature program is 140–250° at 4°/min with 110-sec delay to reach starting temperature from 50° (2.5 min following injection). Data system (starting at 2 min after injection) is set to collect sets of four ions (cycle time, 1 sec) in the SIM mode. Newer versions of SIM programs do not have such a severe limitation on the number of ions monitored. Because of the limitations of the

SIM version used, ion groups monitored may have to be chosen carefully according to derivatives expected and/or a second run performed under identical conditions but with different ions selected. In practice, however, a single run may be sufficient (see conditions given in Figs. 2, 3, and 5).

PMAAs on DB-225. Head pressure, 12 psi helium. Temperature program, 130° or 140° to 220° at 4°/min with 110 sec allowed to reach starting temperature. Data collection is the same as for DB-5. Because N-acetylhexosamine derivatives are irreversibly adsorbed on this column, it is simpler to select groups of four ions to monitor (see Fig. 4).

Neuraminic Acid Derivatives on DB-5. Head pressure, 16 psi helium. Oven temperature 230° isothermal, reached 6 min, 20 sec after injection.

Identification of Derivatives

The identification of PMAAs is made on the basis of characteristic ions produced at the proper retention times, and in some cases, on the basis of ion ratios.

In methane chemical ionization (CI), the most useful ions for deoxyhexose and hexose derivatives appear to be $(MH - 32)^+$ and $(MH - 60)^+$, produced by loss of neutral MeOH and HAc, respectively, from the parent pseudomolecular ion. The spectra of these compounds in methane and isobutane CI have been amply discussed.[10,11] For N-acetylhexosamines, MH^+ and $(MH - 60)^+$ are the most abundant.[11] The nominal masses for these fragments are listed in Table I. More ions may be monitored if software allows. Given the limitations of four ions per cluster, we have found the programs given with the accompanying figures to be most useful. The neuraminic acid derivatives give abundant MH^+, $(MH - 32)^+$, and $(MH - 64)^+$ fragments from loss of one and two molecules of MeOH from the pseudomolecular ion. These are listed, along with retention data on DB-5, in Table II. Spectra for the terminal and 8-O-substituted NeuNAc derivatives in methane CI have been published.[31]

It is outside the scope of this article to give extensive listings of retention data for PMAAs. Given the variation in columns, conditions, and temperature programs used by different laboratories, there is no pressing need to publish here yet another tabulation of retention indices. Readers are referred to tables previously published,[1-14] and should keep in mind that, in most cases, the *order* of elution will be identical to those published for similar phases, with few exceptions (see discussion of Fig. 4). The best policy is for workers to establish their own tables based on experience

---

[31] E. Nudelman, S. Hakomori, R. Kannagi, S. Levery, M.-Y. Yeh, K. E. Hellström, and I. Hellström, *J. Biol. Chem.* **257**, 12752 (1982).

TABLE I
NOMINAL MASSES (m/z) OF MH⁺, (MH − 32)⁺, AND (MH − 60)⁺ IONS OBTAINED
IN METHANE CHEMICAL IONIZATION MASS SPECTROMETRY OF MONODEUTERATED
PARTIALLY METHYLATED ALDITOL ACETATES

Parent sugar	Substituents	MH⁺	(MH − 32)⁺	(MH − 60)⁺
Deoxyhexose	0	294	262	234
	1	322	290	262
	2	350	318	290
	3	378	346	318
Hexose	0	324	292	264
	1	352	320	292
	2	380	348	320
	3	408	376	348
	4	436	404	376
N-Acetylhexosamine	0	365	333[a]	305
	1	393	361[a]	333
	2	421	389[a]	361
	3	449	417[a]	389

[a] Produced in very low abundance; (MH − 42)⁺ may be more useful.

TABLE II
NOMINAL MASSES (m/z) OF MH⁺, (MH − 32)⁺, AND (MH − 64)⁺ IONS OBTAINED IN
METHANE CHEMICAL IONIZATION MASS SPECTROMETRY OF ACETYLATED METHYLATED
N,N-ACYL,METHYL-NEURAMINIC ACID METHYL ESTER METHYLGLYCOSIDES WITH
RETENTION TIMES ON DB-5[a]

NeuNAcNMe- methyl ester- methylglycosides	Ret. index[b]		MH⁺	(MH − 32)⁺	(MH − 64)⁺
	α	β			
NeuNAcNMe- 4,7,8,9-Me₄-O-	0.93[c]	1.00	408	376	344[e]
NeuNAcNMe-8-Ac-O- 4,7,9-Me₃-O-	1.03[c]	1.11	436	404	372[e]
NeuNMeGcNMe- 4,7,8,9-Me₄-O-	1.08[c]	1.18	438	406	374[e]
NeuNMeGcNMe-8-Ac-O- 4,7,9-Me₃-O-	N.D.[d]	1.32	466	434	402[e]

[a] 230° isothermal (see text).

[b] Relative to 4,7,8,9-Me-O-NeuNAcNMe-methyl ester-β-methylglycoside; retention
times measured from opening of splitter.

[c] Minor component.

[d] N.D., Not determined.

[e] Most abundant ion.

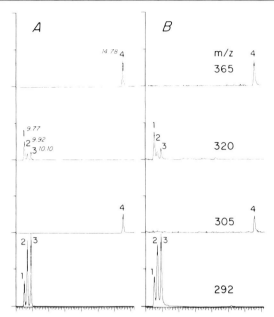

FIG. 2. GC-MS on DB-5 of PMAAs from 15 μg (10 nmol) of globoside (Gb₄Cer). (A) Injection of 1 μl/20 μl of sample, with electron multiplier set at 1200 V; (B) injection of 1 μl/ 200 μl, electron multiplier at 2900 V. Ordinate, intensity of each ion at mass number indicated. Abscissa, retention time. The italicized numbers are retention times in minutes, measured from the point at which the splitter is opened. Peaks identified are (1) 2,3,6-tri-O-Me-Gal; (2) 2,3,6-tri-O-Me-Glc; (3) 2,4,6-tri-O-Me-Gal; and (4) 3,4,6-tri-O-Me-GalNAcMe.

with standard compounds analyzed on individual systems. Identifications can be confirmed by coinjection with appropriate standards (see, for example, Fig. 5), and by analysis on a second column.[14]

The ratios of the major fragments can be useful in distinguishing certain isomeric derivatives, as established by McNeil and Albersheim for isobutane CI spectra.[10] These differences can also be of use in methane CI, for example in distinguishing 3,4,6- from 2,4,6-tri-O-Me-Gal (see discussion of Fig. 3).

Illustrative Examples

The analysis of PMAAs from 10 nmol of globoside (Gb₄Cer, Fig. 2) demonstrates that a sample of this size presents no difficulty in either derivatization or detection (even for a 10-year-old quadrupole instrument not originally built for use with capillary column GC). In Fig. 2A, the injection of PMAAs at 500 pmol gives a flat baseline at low electron

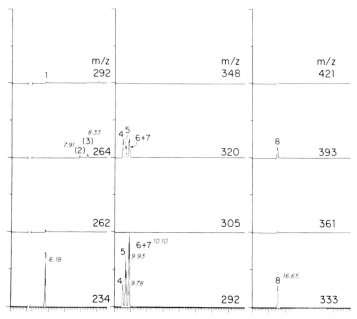

FIG. 3. GC-MS on DB-5 of PMAAs from Fucα1 → 2Galβ1 → 3Gb₄Cer (MBr1 antigen) obtained from a breast cancer cell line.[34] Peaks identified are (1) 2,3,4-tri-O-Me-Fuc; (2) 2,3,4,6-tetra-O-Me-Glc (retention standard); (3) 2,3,4,6-tetra-O-Me-Gal (from impurity of Galβ1 → 3Gb₄Cer); (4) 2,3,6-tri-O-Me-Gal; (5) 2,3,6-tri-O-Me-Glc; (6) 2,4,6-tri-O-Me-Gal; (7) 3,4,6-tri-O-Me-Gal; and (8) 4,6-di-O-Me-GalNAcMe. The coelution of (6) and (7) is normal with this phase.

multiplier (EM) voltage. In Fig. 2B, with the EM near maximum, levels of PMAAs at 50 pmol are still detected with only a small increase in signal-to-noise level, even though a significant degree of tailing occurred during this analysis (i.e., under less than optimum conditions).

The analysis of a more complex glycolipid is illustrated in Figs. 3 and 4. The glycolipid Fucα1 → 2Galβ1 → 3GalNAcβ1 → 3Galα1 → 4Galβ1 → 4Glcβ1 → 1Cer, or globo-H, was first isolated from teratocarcinoma cells[16] as part of an investigation of structures reactive with the stage-specific antibody α-SSEA-3,[32,33] and subsequently found in carcinoma and normal epithelia of human breast (identified as the MBr1 antigen[34]) and in normal erythrocytes.[35] Although the complete structural identification re-

[32] L. H. Shevinsky, B. B. Knowles, I. Damjanov, and D. Solter, *Cell (Cambridge, Mass.)* **30**, 697 (1982).

[33] P. W. Andrews, P. N. Goodfellow, L. H. Shevinsky, D. L. Bronson, and B. B. Knowles, *Int. J. Cancer* **29**, 523 (1982).

[34] E. G. Bremer, S. B. Levery, S. Sonnino, R. Ghidoni, S. Canevari, R. Kannagi, and S. Hakomori, *J. Biol. Chem.* **259**, 14773 (1984).

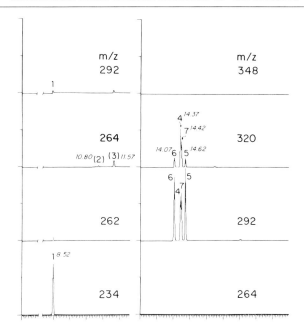

FIG. 4. GC-MS on DB-225 of PMAAs from Fucα1 → 2Galβ1 → 3Gb₄Cer sample used in Fig. 3. Peaks are identified with the same numbers as in Fig. 3. Temperature program was 140/4/220°.

lied in large part on ^{1}H NMR, enzyme degradation, HPTLC immunostaining, and direct probe MS of the permethylated compound, GC-MS was at the time crucial in determining unambiguously the linkage positions of the saccharide units comprising the carbohydrate chain. Compared with Gb₄Cer, this glycolipid produces 2,3,4-tri-O-Me-Fuc, 3,4,6-tri-O-Me-Gal, and 4,6-di-O-Me-GalNAcMe in place of 3,4,6-tri-O-Me-GalNAcMe, arising from the additional H determinant linked to GalNAc. The characteristic production of 3,4,6-tri-O-Me-Gal, which coelutes with the 2,4,6 derivative on DB-5 (Fig. 3, peak 6+7), illustrates a case for which the use of a second column is mandatory. Although routine use of more than one column is desirable, in this analysis the presence of some component other than 2,4,6-tri-O-Me-Gal in the peak at 10.10 min is already indicated by the large ratio of m/z 320 fragment (compare peak 3, Fig. 2 or peak 6, Fig. 4). All four tri-O-Me-Hex components are resolved on the DB-225 column (Fig. 4), and their identities corroborated by the additional set of retention times and production of ions in their proper ratios for these conditions. Note that the order of elution of (4) and (7) are reversed compared with OV-225 packed column.[9]

[35] R. Kannagi, S. B. Levery, and S. Hakomori, *FEBS Lett.* **175,** 397 (1984).

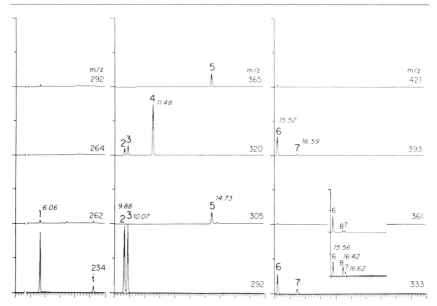

FIG. 5. GC-MS on DB-5 of PMAAs from type 3 chain A glycolipid.[36] Peaks identified are (1) 2,3,4-tri-O-Me-Fuc; (2) 2,3,6-tri-O-Me-Glc; (3) 2,4,6-tri-O-Me-Gal; (4) 4,6-di-O-Me-Gal; (5) 3,4,6-tri-O-Me-GalNAcMe; (6) 3,6-di-O-Me-GlcNAcMe; (7) 4,6-di-O-Me-GalNAcMe. The inset shows the di-O-hexosaminitols (m/z 333, 393) obtained when the sample is coinjected with standard 4,6-di-O-Me-GlcNAcMe (8), and illustrates the separation between the two 4,6-di-O-Me-HexNAcMe derivatives.

An example in which the improved resolution and sensitivity of the capillary column–chemical ionization system helped to solve a complex problem is illustrated in Fig. 5. In this case, with the use of monoclonal antibodies and improved HPTLC separation techniques, a glycolipid was isolated from A erythrocyte membranes, making up about 50% of the fraction formerly assigned the structure GalNAcα1 → 3[Fucα1 → 2]Galβ1 → 4GlcNAcβ1 → 3Galβ1 › 4GlcNAcβ1 → 3Galβ1 → 4Glcβ1 → 1Cer (A^b). Previously undetected, the new glycolipid, on the basis of its completely different antibody reactivity and distinct ^{1}H NMR spectrum, was tentatively assigned the following structure.[36]

GalNAcα1 → 3Galβ1 → 3GalNAcα1 → 3Galβ1 →
                    2   2
                    ↑   ↑
               Fucα1   Fucα1

                                   4GlcNAcβ1 → 3Galβ1 → 4Glcβ1 → 1Cer

[36] H. Clausen, S. B. Levery, E. Nudelman, S. Tsuchiya, and S. Hakomori, *Proc. Natl. Acad. Sci. U.S.A.* **82**, 1199 (1985).

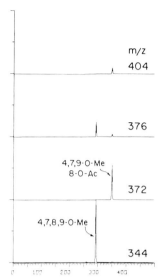

FIG. 6. GC-MS on DB-5 of acetylated, partially methylated NeuNAcMe-methyl ester-methylglycosides obtained from melanoma G_{D3} (II³NeuNAcα2 → 8NeuNAcLc₂Cer).[31]

Qualitatively, the GC-MS of this substance (Fig. 5) differs from the analysis of A^b only by the presence of 4,6-di-O-Me-GalNAcMe, which is consistently produced in poor yield by this glycolipid. In the previous analysis of the mixed A^b fractions, using packed columns and electron ionization–MS detection, the presence of this derivative had gone undetected, or at best, had been noted as an unidentifiable impurity. Given the extra assurances that the compound is indeed homogeneous with respect to the carbohydrate moiety, this analysis is fully consistent with the proposed structure; compared with the analysis of A^b run under identical conditions, the response for neutral sugars was also quantitatively correct.

Figure 6 illustrates the application of methylation analysis to glycolipids containing sialic acid. The derivative corresponding to internally linked NeuNAc was detected in this sample of G_{D3} from human melanoma[31] by its characteristic (MH − MeOH)$^+$ and (MH − 2MeOH)$^+$ ions[31] and retention time compared with the terminal NeuNAc derivative.[26,27] One particular advantage of these derivatives over the corresponding TMS compounds lies in their stability. This chromatogram was made using a sample that had been stored (at −20°) for 2½ years. Sample loading was approximately 200 pmol, electron multiplier set at 1800 V.

[3] Reductive Cleavage of Permethylated Polysaccharides

By GARY R. GRAY

Because of the compositional and structural complexity encountered in polysaccharides, no single method is available for establishing their structure. Where ample amounts of material have been available ^{1}H and ^{13}C NMR spectroscopy[1–6] have proved to be particularly valuable, yielding structural information relative to anomeric configuration, composition, and, in some cases, linkage positions and sequence. In general, these spectroscopic techniques have proved to be most useful for small oligosaccharides or for polysaccharides with relatively small repeating units. In most cases where NMR has been used in the structural characterization of complex carbohydrates, it has been used in combination with various chemical methodologies.

The most generally applicable chemical method for polysaccharide structure determination, by far, is "methylation analysis."[7–9] This technique, when used in combination with specific chemical cleavage methods such as partial acid hydrolysis,[10] Smith degradation,[11] nitrous acid deamination of 2-amino-2-deoxyaldoses,[12] acetolysis,[13] chromium trioxide oxidation,[14] β-elimination at uronic acid residues,[15] and oxidation and β-elimination,[16] provides structural information relative to sequence, linkage positions, ring forms, the location of noncarbohydrate substit-

[1] G. Kotowycz and R. U. Lemieux, *Chem. Rev.* **73**, 699 (1973).
[2] D. R. Bundle and R. U. Lemieux, *Methods Carbohydr. Chem.* **7**, 79 (1976).
[3] H. J. Jennings and I. C. P. Smith, this series, Vol. 50, p. 39.
[4] A. S. Perlin, *Methods Carbohydr. Chem.* **7**, 94 (1976).
[5] J. F. G. Vliegenthart, L. Dorland, and H. van Halbeek, *Adv. Carbohydr. Chem. Biochem.* **41**, 209 (1983).
[6] P. A. J. Gorin, *Adv. Carbohydr. Chem. Biochem.* **38**, 13 (1981).
[7] B. Lindberg, this series, Vol. 28, p. 178.
[8] B. Lindberg and J. Lönngren, this series, Vol. 50, p. 3.
[9] B. Lindberg, J. Lönngren, and S. Svensson, *Adv. Carbohydr. Chem. Biochem.* **31**, 185 (1975).
[10] J. N. BeMiller, *Adv. Carbohydr. Chem.* **22**, 25 (1967).
[11] I. J. Goldstein, G. W. Hay, B. A. Lewis, and F. Smith, *Methods Carbohydr. Chem.* **5**, 361 (1965).
[12] J. E. Shively and H. E. Conrad, *Biochemistry* **15**, 3932 (1976).
[13] R. D. Guthrie and J. F. McCarthy, *Adv. Carbohydr. Chem.* **22**, 11 (1967).
[14] S. J. Angyal and K. James, *Aust. J. Chem.* **23**, 1209 (1970).
[15] B. Lindberg, J. Lönngren, and J. L. Thompson, *Carbohydr. Res.* **28**, 351 (1973).
[16] S. Svensson, this series, Vol. 50, p. 33.

METHODS IN ENZYMOLOGY, VOL. 138

uents, and, in some cases, anomeric configuration.[14] The combination of methods based upon methylation analysis is especially valuable because the products (partially methylated alditol acetates or partially methylated oligosaccharides) are analyzed by mass spectrometry,[17-19] and the amounts of material which can be analyzed (1–5 μg)[20] are therefore much smaller than the amounts required for NMR spectroscopy. This sensitivity has proved to be of critical importance in the structural analysis of lipid- and protein-derived complex carbohydrates.[21-26]

In spite of the fact that methylation analysis has been utilized so successfully in the determination of structure of complex carbohydrates, the method still suffers two significant disadvantages: (1) ambiguous results with regard to position of linkage and ring form are obtained for aldopyranosyl and aldofuranosyl residues linked at O-4 and O-5, respectively, as well as for ketohexopyranosyl and ketohexofuranosyl residues linked at O-5 and O-6, respectively, and (2) the method is laborious.

These difficulties led us to propose a new method for carrying out methylation analysis, as shown in Scheme 1.[27] The salient feature of this method, *regiospecific* reductive cleavage of the glycosidic carbon–oxygen bond, is accomplished with triethylsilane as the reducing agent in the presence of either trimethylsilyltrifluoromethane sulfonate (TMSOTf)[28] or boron trifluoride etherate (BF$_3$·Et$_2$O).[27] Pyranosides are converted to 1,5-anhydroalditol derivatives (**1**) and furanosides are converted to 1,4-anhydroalditol derivatives (**2**), therefore establishing the linkage position(s) and ring form simultaneously. The reaction is carried out in such a way that the permethylated polysaccharide is converted to derivatives suitable for GLC-MS analysis with only one reaction workup. The reductive cleavage method therefore overcomes both of the aforementioned disadvantages of standard methylation analysis.

[17] H. Björndal, B. Lindberg, and S. Svensson, *Carbohydr. Res.* **5**, 433 (1967).
[18] M. W. Spellman, M. NcNeil, A. G. Darvill, and P. Albersheim, *Carbohydr. Res.* **122**, 137 (1983).
[19] L.-E. Franzen, W. F. Dudman, M. McNeil, A. G. Darvill, and P. Albersheim, *Carbohydr. Res.* **117**, 157 (1983).
[20] T. J. Waeghe, A. G. Darvill, M. McNeil, and P. Albersheim, *Carbohydr. Res.* **123**, 281 (1983).
[21] C. C. Sweeley and G. Dawson, *Biochem. Biophys. Res. Commum.* **37**, 6 (1969).
[22] W. J. Esselman, R. A. Laine, and C. C. Sweeley, this series, Vol. 28, p. 40.
[23] H. Rauvala, J. Finne, T. Krusius, J. Kärkkäinen, and J. Jarnefelt, *Adv. Carbohydr. Chem. Biochem.* **39**, 389 (1981).
[24] K. Watanabe and S.-I. Hakomori, *Biochemistry* **18**, 5502 (1979).
[25] K. Watanabe, T. Matsubara, and S.-I. Hakomori, *J. Biol. Chem.* **251**, 2385 (1976).
[26] W. G. Carter and S.-I. Hakomori, *Biochemistry* **18**, 730 (1979).
[27] D. Rolf and G. R. Gray, *J. Am. Chem. Soc.* **104**, 3539 (1982).
[28] D. Rolf, J. A. Bennek, and G. R. Gray, *J. Carbohydr. Chem.* **2**, 373 (1983).

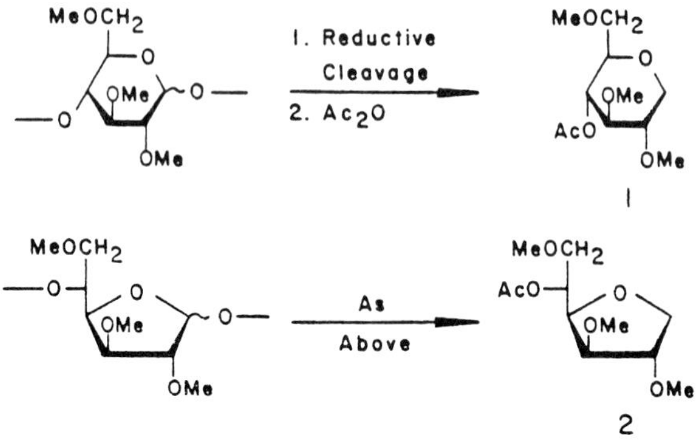

Scheme 1.

In addition to its obvious advantages over standard methylation analysis, preliminary studies have shown that the reductive cleavage method possesses another significant advantage, i.e., selectivity. By appropriate choice of the catalyst, it is possible to achieve *total* reductive cleavage (TMSOTf) or partial, selective reductive cleavage ($BF_3 \cdot Et_2O$). With $BF_3 \cdot Et_2O$ as the catalyst, for example, the glycosidic bond between 6-linked glycopyranosyl residues is relatively stable to reductive cleavage, and preliminary results indicate that β-1,3- and β-1,4-linked glucopyranosyl residues are not cleaved at all. This selectivity is likely to be very useful in sequence studies.

Experimental protocols for carrying out this reaction and the results obtained with some model permethylated polysaccharides are given in the following sections.

Experimental Procedures

Reductive Cleavage with Et_3SiH and TMSOTf. A stock solution of Et_3SiH and TMSOTf in CH_2Cl_2 was prepared in a 10-ml ampule that had previously been silylated inside by treating it with 10% Me_2SiCl_2 in toluene for 1.5 hr, and then washed with dry methanol, dried at 150°, and cooled in a desiccator. To the ampule were sequentially added CaH_2 (0.2 g), CH_2Cl_2 (9.0 ml), Et_3SiH (0.8 ml), and TMSOTf (1.0 ml). The ampule was capped with a rubber septum, and then vented through a syringe filled with Drierite for 1 hr, to allow H_2 gas to escape. The ampule was then stored, desiccated, at 4°.

Reductive cleavages were conducted by adding the per-O-methylated polysaccharide (5 mg) and a small stirring bar to a Wheaton V-vial equipped with a Teflon-lined screw top. The vial and contents were kept under high vacuum for 2 hr, and then the Et_3SiH–TMSOTf stock solution

(0.28 ml) was added (starting concentrations: ~0.1 M in glucan acetal, 0.5 M in Et_3SiH, and 0.5 M in TMSOTf). The vial was now capped, and the mixture was stirred for 20 hr at room temperature, at which time 13 μl of Ac_2O (5 equivalents per 1 equivalent of acetal) was added. Stirring was continued for 20 hr at room temperature, and the reaction was then quenched by the addition of 0.5 ml of saturated aqueous $NaHCO_3$. The biphasic reaction mixture was stirred for 1 hr (mild evolution of gas occurs), and the aqueous layer was carefully removed. Saturated aqueous $NaHCO_3$ (0.5 ml) was added, and stirring was continued for 30 min. The aqueous layer was removed, and the CH_2Cl_2 solution was used directly for GLC analysis.

Reductive Cleavage with Et_3SiH and $BF_3 \cdot Et_2O$. A 5-mg sample of the per-*O*-methylated polysaccharide and a small stirring bar were added to a V-vial, and the contents were dried as described before. Dichloromethane (0.25 ml, predried with CaH_2), Et_3SiH (20 μl), and redistilled $BF_3 \cdot Et_2O$ (15 μl) were sequentially added, and the vial was capped. After the mixture had been stirred for 20 hr at room temperature, MeOH (0.25 ml) was added, and the mixture was deionized by adding Dowex AG 501-X8-D resin, a few beads at a time, until the blue color of the resin was retained. The resin was removed by filtration and washed with MeOH (~3 ml), and the filtrate was collected in a 5-ml, round-bottom flask. Solvents were removed by evaporation under vacuum at 30°, with precaution not to allow evaporation to dryness to occur, and most of the remaining solvent was removed by purging under a stream of dry nitrogen. Subsequent acetylation of the product was accomplished by dissolving in acetic anhydride (0.75 ml) and pyridine (0.75 ml) and heating for 1 hr at 100° in a sealed ampule. After being cooled to room temperature, the solution was added to ice (20 g) contained in a separatory funnel. Dichloromethane (30 ml) was added, and the organic layer was washed successively with 30-ml portions of 2 N H_2SO_4 (twice), saturated aqueous $NaHCO_3$ (twice), and water (twice), dried (anhydrous Na_2SO_4), and concentrated to ~1 ml under vacuum at room temperature. The concentrate was transferred to a small vial, and evaporated to dryness under a stream of dry nitrogen at room temperature. The product was dissolved in a small amount of methanol (50 μl) for GLC analysis.

Results and Discussion

Saccharomyces cerevisiae D-*Mannans.*[29] The D-mannans derived from *S. cerevisiae* X2180 wild type and its mutants are particularly useful models because of their content of branch points and several types of

[29] J. U. Bowie, P. V. Trescony, and G. R. Gray, *Carbohydr. Res.* **125,** 301 (1984).

linkages. Given below are the structures of the outer-chain mannans from these yeasts and the products expected to be produced by sequential perethylation and reductive cleavage.

$$\rightarrow M_2^1 \rightarrow {}^6M_2^1 \rightarrow {}^6M_2^1 \rightarrow {}^6M_2^1 \rightarrow {}^6M_2^1 \rightarrow {}^6M \rightarrow \qquad \rightarrow M_2^1 \rightarrow {}^6M_2^1 \rightarrow {}^6M_2^1 \rightarrow {}^6M \rightarrow$$

$$\qquad\quad \uparrow \quad\ \uparrow \quad\ \uparrow \quad\ \uparrow \quad\ \uparrow \qquad\qquad\qquad\qquad \uparrow \quad\ \uparrow \quad\ \uparrow$$

$$P \ \rightarrow {}^6M_2^1 \quad M_2^1 \quad M_3^1 \quad M_2^1 \quad M^1 \qquad P \rightarrow {}^6M_2^1 \quad M_2^1 \quad M^1$$

$$\uparrow \qquad\ \uparrow \quad\ \uparrow \quad\ \uparrow \quad\ \uparrow \qquad\qquad\qquad \uparrow \quad\ \uparrow \quad\ \uparrow$$

$$M_3^1 \quad M_3^1 \quad M_3^1 \quad M^1 \quad M^1 \qquad\qquad\qquad M^1 \quad M^1 \quad M^1$$

$$\uparrow \qquad\ \uparrow \quad\ \uparrow$$

$$M^1 \quad M^1 \quad M^1$$

X2180 wild type   mnn1

$$\rightarrow M^1 \rightarrow {}^6M^1 \rightarrow {}^6M \rightarrow \qquad\qquad \rightarrow M_2^1 \rightarrow {}^6M_2^1 \rightarrow {}^6M_2^1 \rightarrow {}^6M_2^1 \rightarrow {}^6M \rightarrow$$

$$\qquad\qquad\qquad\qquad\qquad\qquad\qquad\qquad\quad \uparrow \quad\ \uparrow \quad\ \uparrow \quad\ \uparrow$$

mnn2

$$\qquad\qquad\qquad\qquad\qquad\qquad M_2^1 \quad M_3^1 \quad M_2^1 \quad M^1$$

$$\qquad\qquad\qquad\qquad\qquad\qquad \uparrow \quad\ \uparrow \quad\ \uparrow$$

$$\qquad\qquad\qquad\qquad\qquad\qquad M_3^1 \quad M^1 \quad M^1$$

$$\qquad\qquad\qquad\qquad\qquad\qquad \uparrow$$

$$\qquad\qquad\qquad\qquad\qquad\qquad M^1$$

mnn4

3 $R^2 = R^3 = R^4 = R^6 = Et$
4 $R^2 = Ac, R^3 = R^4 = R^6 = Et$
5 $R^3 = Ac, R^2 = R^4 = R^6 = Et$
6 $R^6 = Ac, R^2 = R^3 = R^4 = Et$
7 $R^2 = R^6 = Ac, R^3 = R^4 = Et$
8 $R^3 = R^6 = Ac, R^2 = R^4 = Et$

When reductive cleavage was carried out with TMSOTf as the catalyst, the expected 1,5-anhydro-D-mannitol derivatives were indeed obtained from all four mannans and they were formed in the expected molar ratios (Table I). Nonreducing (terminal) mannopyranosyl groups gave rise to 3 and 2-, 3-, and 6-linked mannopyranosyl residues gave 4, 5, and 6, respectively. Branch-point mannopyranosyl residues linked at both O-2 and O-6 gave 7, whereas 3,6-linked mannopyranosyl residues of the inner core of the mannans gave 8. When reductive cleavage was carried out with $BF_3 \cdot Et_2O$ as the catalyst, 2-, 3-, and 6-linked residues were incompletely cleaved and branch-point residues linked at both O-2 and O-6 or O-3 and O-6 were not cleaved at all (Fig. 1 and Table I). Reductive cleavage under these conditions is nicely complementary to acetolysis, which selectively cleaves 6-linked backbone residues.

D-*Fructans*.[30] The suitability of the reductive cleavage method for the analysis of linkage positions in fructofuranosyl residues was examined

[30] D. Rolf and G. R. Gray, *Carbohydr. Res.* **131,** 17 (1984).

TABLE I

MOLAR RATIOS OF 1,5-ANHYDRO-D-MANNITOL DERIVATIVES DERIVED BY
REDUCTIVE CLEAVAGE OF *Saccharomyces cerevisiae* D-MANNANS[a]

D-Mannan	Catalyst	Molar ratio[b]					
		3	4	5	6	7	8
X2180	TMSOTf	1.00	0.53	0.59	0.06	0.78	0.03
	BF$_3$	1.00	0.23	0.46	0.08	0.01	—
mnn1	TMSOTf	1.00	0.46	0.03	0.05	0.70	0.02
	BF$_3$	1.00	0.14	0.01	0.04	—	—
mnn2	TMSOTf	1.00	0.46	0.18	6.85	0.13	0.17
	BF$_3$	1.00	0.26	0.14	4.90	—	—
mnn4	TMSOTf	1.00	0.64	0.57	0.05	0.84	0.02
	BF$_3$	1.00	0.26	0.42	0.04	0.01	—

[a] Reprinted from Bowie *et al.*[29]
[b] Normalized to the basis of 3 as unity.

using sucrose, chicory root inulin, and *Aerobacter levanicum* levan as
models. As shown in Scheme 2, the glucopyranosyl group of sucrose is
expected to give rise only to **9** whereas the fructofuranosyl group can give
2,5-anhydrohexitols having both the *manno* (**10**) and *gluco* (**11**) configuration. Indeed, the expected products were formed with both catalysts, and
where care was taken to prevent volatile loss of **9** (TMSOTf-catalyzed
reaction), the combined amount of **10** and **11** formed was equal, within
experimental error, to the amount of **9** (Table II).

TABLE II

MOLE FRACTIONS OF PRODUCTS DERIVED BY REDUCTIVE CLEAVAGE OF
PERMETHYLATED SUCROSE, INULIN, AND LEVAN[a]

Fructan	Catalyst	Mole fraction							
		9	10	11	12	13	14	15	16
Sucrose	BF$_3$	0.41	0.41	0.17	—	—	—	—	—
	TMSOTf	0.49	0.42	0.08	—	—	—	—	—
Inulin	BF$_3$	0.04	0.05	0.01	0.65	0.24	—	—	—
	TMSOTf	0.05	0.04	0.01	0.67	0.22	—	—	—
Levan	BF$_3$	—	0.08	0.03	0.26	—	0.53	0.03	0.07
	TMSOTf	—	0.09	0.04	0.25	—	0.54	0.03	0.05

[a] Reprinted from Rolf and Gray.[30]

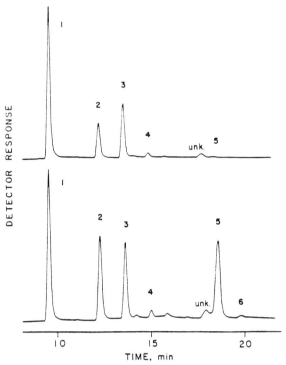

FIG. 1. Gas–liquid chromatograms on a column (3.18 mm × 3.66 m) of SP-2401, programmed from 170 to 250° at 6°/min, of the partially ethylated anhydroalditol acetates
derived by reductive cleavage of perethylated mnn4 mannan with $BF_3 \cdot Et_2O$ as the catalyst
(upper) and TMSOTf as the catalyst (lower). The numbered peaks were identified as follows:
(1) 1,5-anhydro-2,3,4,6-tetra-O-ethyl-D-mannitol, **3**; (2) 1,5-anhydro-2-O-acetyl-3,4,6-tri-O-
ethyl-D-mannitol, **4**; (3) 1,5-anhydro-3-O-acetyl-2,4,6-tri-O-ethyl-D-mannitol, **5**; (4) 1,5-an-
hydro-6-O-acetyl-2,3,4-tri-O-ethyl-D-mannitol, **6**; (5) 1,5-anhydro-2,6-di-O-acetyl-3,4-di-O-
ethyl-D-mannitol, **7**; (6) 1,5-anhydro-3,6-di-O-acetyl-2,4-di-O-ethyl-D-mannitol, **8**. Slight
differences in retention times arise from the variability of the temperature program. Reprinted from Bowie *et al.*[29]

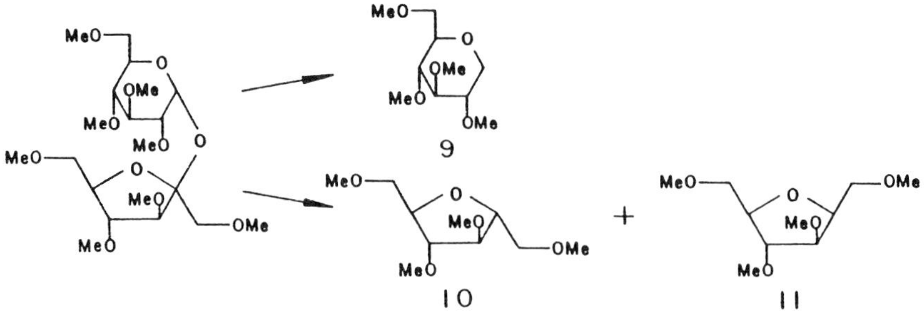

SCHEME 2.

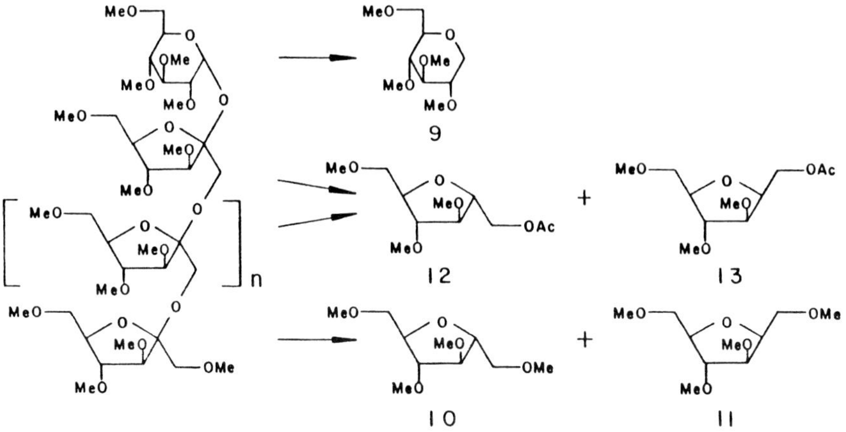

SCHEME 3.

Inulin is a well-characterized 2,1-linked fructofuranose polymer termi-nated at its "reducing end" by a sucrose residue (Scheme 3). The nonre-ducing (terminal) fructofuranosyl group is expected to give both **10** and **11**, and the glucopyranosyl group is expected to give a single product (**9**) as was observed for sucrose. The internal 2,1-linked fructofuranosyl resi-dues can also give rise to two products, namely, the 1-*O*-acetyl deriva-tives of the 2,5-anhydrohexitols having the *manno* (**12**) and *gluco* (**13**) configurations. All residues gave the expected products. For the terminal fructofuranosyl group, the product (**10**) having the *manno* configuration was again preponderant, and the total mole fraction of the two 2,5-anhy-droalditols (**10** and **11**) derived from that group was the same as the mole fraction of **9**, formed from the terminal glucopyranosyl group. The inter-nal 2,1-linked fructofuranosyl residues also gave rise to two products (**12** and **13**) and again, the product having the *manno* configuration (**12**) was preponderant. The average number of 2,1-linked fructofuranosyl residues in the chain, obtained by dividing the combined mole fraction of **12** + **13** (0.89) by the combined mole fraction of **10** + **11** (0.05), was found to be 18, in agreement with the reported value.

Levan is a high molecular weight fructan composed of a 2,6-linked fructofuranose backbone that is branched at some of the O-1 positions (Scheme 4). This polysaccharide therefore meets one of the criteria for developing the reductive cleavage method, namely the inability of stan-dard methylation analysis to distinguish between 6-linked ketohexo-furanosyl and 5-linked ketohexopyranosyl residues. All residues in the polysaccharide gave two products, as expected (Table II, Fig. 2). Nonre-ducing fructofuranosyl groups gave **10** and **11**, with the *manno* product

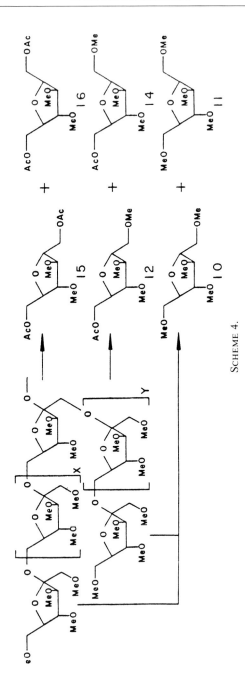

SCHEME 4.

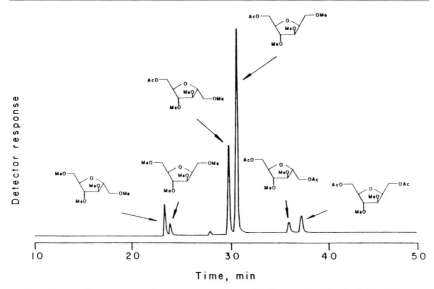

FIG. 2. Gas–liquid chromatogram on a column (3.18 mm × 3.66 m) of SP-2401, programmed from 70 to 310° at 6°/min, of the partially methylated anhydroalditol acetates derived by reductive cleavage of permethylated levan with TMSOTf as the catalyst. Reprinted from Rolf and Gray.[30]

(**10**) again predominating. 6-Linked fructofuranosyl residues also gave two products (**12** and **14**), but in this case the *gluco* product (**14**) was preponderant. Branch-point fructofuranosyl residues linked at both O-1 and O-6 also gave two products (**15** and **16**) and in this case also, the *gluco* product (**16**) was preponderant (Table II). The reason for this stereoselectivity is unknown, but it is considered to be an advantage, since, for the residues examined, the *manno/gluco* ratio was characteristic of the linkage position.

D-Glucans.[31] Reductive cleavage analysis of the D-glucans amylose, cellulose, laminarin, and pullulan gave the products whose structures are given below.

9   $R^3 = R^4 = R^6 = $ Me

17   $R^3 = $ Ac, $R^4 = R^6 = $ Me

1   $R^4 = $ Ac, $R^3 = R^6 = $ Me

18   $R^6 = $ Ac, $R^3 = R^4 = $ Me

19   $R^3 = R^6 = $ Ac, $R^4 = $ Me

2

20

TABLE III

MOLE FRACTIONS OF PRODUCTS DERIVED BY REDUCTIVE CLEAVAGE OF
PERMETHYLATED AMYLOSE, CELLULOSE, LAMINARIN, AND PULLULAN[a]

		Mole fraction[b]						
D-Glucan	Catalyst	9	17	1	18	19	2	20
Amylose	TMSOTf	Trace	—	0.94	—	—	0.04	—
	BF$_3$	Trace	—	0.91	—	—	0.06	—
Cellulose	TMSOTf	Trace	—	0.89	—	—	0.04	—
	BF$_3$	—	—	—	—	—	—	—
Laminarin	TMSOTf	0.11	0.77	—	0.01	0.07	—	—
	BF$_3$	—	—	—	—	—	—	—
Pullulan	TMSOTf	0.02	—	0.62	0.33	—	0.02	—
	BF$_3$	0.02	—	0.42	0.07	—	0.03	0.37

[a] Reprinted from Rolf et al.[31]

[b] Small amounts of unidentified products, presumably rising from incompletely methylated residues in the polysaccharides, were also observed.

Both amylose and cellulose are linear polymers composed of 1,4-linked glucopyranosyl residues, so the major product of reductive cleavage of both polymers is expected to be the 4-O-acetyl-1,5-anhydro-D-glucitol derivative (1). As seen in Table III, compound 1 was indeed the major product of reductive cleavage when TMSOTf was the catalyst. In addition, traces of 9 arising from nonreducing glucopyranosyl groups were also observed in these reactions, as expected. Unexpectedly, small amounts of 2 were also formed during reductive cleavage. A careful study of this reaction[32] demonstrated that the formation of 2 was attributable to the presence of water in the reaction mixture, which led to hydrolysis of the permethylated polysaccharide and subsequent reductive cleavage of 2,3,6-tri-O-methyl-D-glucose to produce both 1 and 2. The formation of 2 can be minimized (2–4%) by taking precautions to exclude water. Interestingly, BF$_3$ · Et$_2$O effectively catalyzed the reductive cleavage of permethylated amylose, but under the same conditions, permethylated cellulose was not cleaved. BF$_3$·Et$_2$O was also ineffective in catalyzing the reductive cleavage of permethylated laminarin, a β-1,3-linked D-glucan branched at some of the O-6 positions. The complete reductive cleavage of permethylated laminarin was readily accomplished with TMSOTf as the catalyst, however (Table III). The major product, as expected, was the 3-O-acetyl-1,5-anhydro-D-glucitol-derivative (17). Branch-point resi-

[31] D. Rolf, J. A. Bennek, and G. R. Gray, Carbohydr. Res. 137, 183 (1985).

[32] J. A. Bennek, D. Rolf, and G. R. Gray, J. Carbohydr. Chem. 2, 385 (1983).

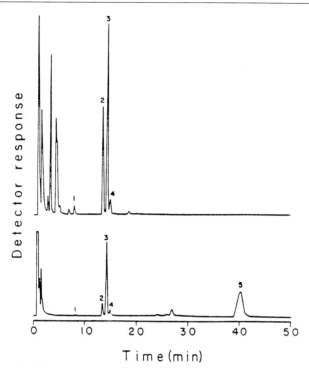

Time (min)

FIG. 3. Gas–liquid chromatograms, on a column (3.18 mm × 2.44 m) of SP-2401, pro-grammed from 110 to 220° at 6°/min, of the partially methylated anhydroalditol acetates derived by reductive cleavage of permethylated pullulan with TMSOTf as the catalyst (up-per) and BF$_3$ · Et$_2$O as the catalyst (lower). The numbered peaks were identified as follows: (1) 1,5-anhydro-2,3,4,6-tetra-O-methyl-D-glucitol, **9**; (2) 6-O-acetyl-1,5-anhydro-2,3,4-tri-O-methyl-D-glucitol, **18**; (3) 4-O-acetyl-1,5-anhydro-2,3,6-tri-O-methyl-D-glucitol, **1**; (4) 5-O-acetyl-1,4-anhydro-2,3,6-tri-O-methyl-D-glucitol, **2**; and (5) 6-O-(4-O-acetyl-2,3,6-tri-O-methyl-α-D-glucopyranosyl)-1,5-anhydro-2,3,4-tri-O-methyl-D-glucitol, **20**. Peaks eluted prior to peak 1 were present in a reagent control prepared without polysaccharide. Unnum-bered peaks were not identified. Reprinted from Rolf *et al.*[31]

dues gave the 3,6-di-O-acetyl-1,5-anhydro-D-glucitol derivative (**19**), and terminal D-glucopyranosyl groups gave **9**. Compounds **17** and **19** were produced in a ratio of 11 : 1, respectively, demonstrating that 1 of every 12 backbone residues is branched.

The results obtained for permethylated pullulan (Table III) provide a good example of *selective* reductive cleavage. This D-glucan is a linear polysaccharide composed of a trisaccharide repeating unit containing one α-1,6-linked and a two α-1,4-linked D-glucopyranosyl residues. With TMSOTf as the catalyst, all glycosidic linkages were cleaved and the expected 4-O-acetyl-1,5-anhydro-D-glucitol derivative (**1**) and 6-O-acetyl-

1,5-anhydro-D-glucitol derivative (18) were formed (Table III, Fig. 3). From the combined amounts of 1 and 2 formed relative to 18, the calculated ratio of 1,4-linked to 1,6-linked residues is 1.94 : 1.00, in good agreement with the theoretical ratio of 2 : 1. With $BF_3 \cdot Et_2O$ as the catalyst under the same reaction conditions, however, the major products were the 4-O-acetyl-1,5-anhydro-D-glucitol derivative (1) and the disaccharide anhydroalditol (20, Fig 3)! The 1,6-linkages of the permethylated polysaccharide are therefore quite stable to reductive cleavage, as was also observed in the reductive cleavage of yeast mannans.

[4] Oligosaccharide Structure by Two-Dimensional Proton Nuclear Magnetic Resonance Spectroscopy

By THEODORE A. W. KOERNER, JAMES H. PRESTEGARD, and ROBERT K. YU

The marriage of sophisticated computers and high-field superconducting magnets approximately 10 years ago created the potential for a revolution in the analytical power of nuclear magnetic resonance (NMR) spectroscopy. The full realization of this potential, however, had to await the imagination of R. R. Ernst and co-workers who, by developing pulsing programs[1-4] that have allowed increasingly more complex and structurally revealing "dialogues" between spectrometer and sample nuclei,[5] have transformed NMR spectroscopy into a method capable of the complete and independent structural analysis of biomolecules. The graphical products of these pulsing programs have become known as "two-dimensional" (2-D) NMR spectra, even though they contain three dimensions of information. Two-dimensional NMR methods and their physical basis have been reviewed.[6-8]

[1] W. P. Aue, E. Bartholdi, and R. R. Ernst, J. Chem. Phys. 64, 2229 (1976).
[2] A. A. Maudsley and R. R. Ernst, Chem. Phys. Lett. 50, 368 (1977).
[3] J. Jeener, B. H. Meier, P. Bachman, and R. R. Ernst, J. Chem. Phys. 71, 4546 (1979).
[4] S. Macura, Y. Huang, D. Suter, and R. R. Ernst, J. Magn. Reson. 43, 259 (1981).
[5] One cannot help comparing the development of these increasingly complex pulsing programs and their use in "spectrometer-nuclei dialogues" with the final scene in the movie Close Encounters of the Third Kind in which communication between terrestrial and extraterrestrial intelligence is through increasingly complex musical pulses, each sequence yielding more information than the last.
[6] L. W. Jelinski, Chem. Eng. News 62, 26 (1984); A. Bax and L. Lerner, Science 232, 960 (1986).
[7] R. Benn and H. Günther, Angew. Chem., Int. Ed. Engl. 22, 350 (1983).
[8] A. Bax, "Two-Dimensional Nuclear Magnetic Resonance in Liquids." Reidel Publishing, Boston, Massachusetts, 1982.

For no other class of molecule have 2-D methods been more useful than oligosaccharides and their conjugates, glycolipids and glycopeptides. This is because of two factors. First, the natural allocation of protons within oligosaccharides is well suited to yield all or most of their primary and secondary structure. Second, through 2-D NMR the severe resolution problem peculiar to oligosaccharides (the "hidden resonance problem") has been generally overcome. This resolution problem results from the facts that almost all resonances of oligosaccharides occur between 3 and 5 ppm, regardless of solvent, and that within this same narrow region the substantial water-derived resonance occurs. By reducing spectra into subspectra, each attributable to only one oligosaccharide residue, 2-D NMR allows overlapping envelopes of resonances to be deciphered.

The reason that oligosaccharide structures lend themselves to proton NMR analysis is that protons are spaced linearly around each oligosaccharide residue. Thus there is usually one proton only at each optically active ring carbon, which serves as a structural "reporter group" at that carbon, the C-2 of ketose residues being the only important exception. When more than one proton is present at a particular carbon it signifies the terminus of the residue or important constitutional data (e.g., the residue is a deoxysugar). Because of the linear allocation of protons around each residue ring, a vicinal J-coupled connectivity or "trail" exists for each residue such that each proton leads to the next proton all around the ring, revealing the stereochemistry and constitution at each carbon. Conveniently, one terminus of each aldose residue J connectivity, the H-1 or anomeric proton of each residue, resonates in a characteristic region (4.0–5.0 ppm) that contains few other signals. Thus the J-connectivity trail for such residues can easily be "picked up" in this region.

Just as important as these intraresidue J connectivities is the potential proximity of anomeric protons and linkage site protons across the glycosidic bonds. Thus when through-space coupling is considered, oligosaccharides manifest a continuously coupled series of proton "reporter groups" from one end of the molecule to the other, making their structures readily accessible to proton NMR analysis. However, in order to undertake such an analysis the J connectivities of each residue must be separated from each other and the through-space coupling revealed. Both of these problems have been solved via 2-D NMR.

Selection of 2-D NMR Experiments

Two general classes of 2-D NMR experiments exist at this time. They are *J-resolved* and *correlated* spectroscopy. In the first type, spectra are

TABLE I

PULSING SEQUENCES OF NMR EXPERIMENTS

Experiment	Pulsing sequence[a]
1-D (basic)	$90°-t_2 \ (t_1 = 0)$
2-D COSY	$90°-t_1-90°-t_2$
2-D SECSY	$90°-t_{1/2}-90°-t_{1/2}-t_2$
2-D NOESY	
COSY matrix	$90°-t_1-90°-\Delta t_1-90°-t_2$
SECSY matrix	$90°-t_{1/2}-90°-\Delta t_1-90°-t_{1/2}-t_2$
2-D J-resolved	$90°-t_{1/2}-180°-t_{1/2}-t_2$

[a] Symbols: 90°, ninety degree pulse; 180°, one-hundred-eighty degree pulse; t_1, evolution time; t_2, detection time during which the FID (free induction decay) signal is recorded.

characterized by one frequency axis (F_1) containing coupling (J) information and by another frequency axis (F_2) containing chemical shift information (ppm). The third dimension of resonance intensity is plotted as a contour. In the second class of 2-D experiment, spectra have both frequency axes (F_1 and F_2) containing chemical shift data and the third dimension is resonance intensity plotted as a contour. Two types of J-resolved and correlated 2-D spectra are possible, *homonuclear* and *heteronuclear,* that is those involving couplings between the *same* or *differing* nuclear types, e.g., ^{1}H and ^{1}H or ^{1}H and ^{13}C nuclei. Only proton homonuclear 2-D methods will be considered in this article, since ^{1}H–^{13}C heteronuclear 2-D methods require too much sample to be useful in studies of most oligosaccharide samples of biological origin. The pulsing sequences for the different types of homonuclear 2-D NMR experiments and their variants are shown in Table I. Though proton homonuclear J-resolved 2-D NMR is of some use in assigning the spectra of simple structures, its lack of sensitivity and inability to provide J-connectivity information render it a fairly useless procedure for the analysis of oligosaccharide structure. In contrast, proton homonuclear correlated 2-D methods have been extremely useful for such studies.

Two types of correlated 2-D NMR experiments[1-4] are possible, those involving *scalar* couplings through coherent transfer of transverse magnetization or those involving *dipole* couplings through incoherent transfer of magnetization. The significant difference between the two experiments is that scalar correlated spectra can be used to reveal *through-bond* connectivities and dipole correlated spectra can be used to reveal *through-space* connectivities. Two experimental variations of scalar correlated 2-D

NMR (Table I) are correlation spectroscopy (COSY) and spin echo correlated spectroscopy (SECSY). The dipole correlated experiment useful in oligosaccharide analysis is 2-D nuclear Overhauser effect spectroscopy (2-D NOESY). It should be pointed out that some authors have referred to scalar correlated spectra as "*J*-correlated" or "chemical shift correlated" spectra.

In 1982 investigators in New Haven[9] and Vancouver[10] reported almost simultaneously the first applications of correlated 2-D NMR methods to oligosaccharide moieties. In these early studies the full power of 2-D methods was manifest as the complete analysis of oligosaccharide primary structure, including sequence and linkage sites, was possible. Since then 2-D SECSY and NOESY have been applied to gangliosides,[11,12] their asialo derivatives,[11,12] ganglioside G_{M1} oligosaccharide,[13] and the neutral glycolipid globoside.[14,15] Two-dimensional experiments including COSY and NOESY have been performed on peracetylated globo-series glycolipids.[16a] Two-dimensional NOE[16] and COSY[17] studies have also been carried out on N-linked oligosaccharides derived from glycoproteins. Early interest in *J*-resolved 2-D NMR studies of glycolipids[18,19] and oligosaccharides[20,21] has abated for the reasons mentioned above. Two recent review articles by Yu *et al.*[15] and Sweeley and Nunez[22] that consider the application of 2-D NMR methods to glycoconjugates have been published

[9] J. H. Prestegard, T. A. W. Koerner, P. C. Demou, and R. K. Yu, *J. Am. Chem. Soc.* **104,** 4993 (1982).

[10] M. A. Bernstein and L. D. Hall, *J. Am. Chem. Soc.* **104,** 5553 (1982).

[11] T. A. W. Koerner, J. H. Prestegard, P. C. Demou, and R. K. Yu, *Biochemistry* **22,** 2676 (1983).

[12] T. A. W. Koerner, J. H. Prestegard, P. C. Demou, and R. K. Yu, *Biochemistry* **22,** 2687 (1983).

[13] R. L. Ong and R. K. Yu, *Arch. Biochem. Biophys.* **245,** 157 (1986).

[14] T. A. W. Koerner, J. N. Scarsdale, J. H. Prestegard, and R. K. Yu, *J. Carbohydr. Chem.* **3,** 565 (1984).

[15] R. K. Yu, T. A. W. Koerner, P. C. Demou, J. N. Scarsdale, and J. H. Prestegard, *Adv. Exp. Med. Biol.* **174,** 87 (1984).

[16a] S. Gasa, M. Nakamura, A. Makita, M. Ikura, and K. Hikichi, *Eur. J. Biochem.* **155,** 603 (1986).

[16] S. W. Homans, R. A. Dwek, D. L. Fernandes, and T. W. Rademacher, *FEBS Lett.* **164,** 231 (1983).

[17] R. Geyer, H. Geyer, S. Stirm, G. Hunsmann, J. Schneider, U. Dabrowski, and J. Dabrowski, *Biochemistry* **23,** 5628 (1984).

[18] A. Yamada, J. Dabrowski, P. Hanfland, and H. Egge, *Biochim. Biophys. Acta* **618,** 473 (1980).

[19] J. Dabrowski and P. Hanfland, *FEBS Lett.* **142,** 138 (1982).

[20] L. D. Hall, G. A. Morris, and S. Sukumar, *J. Am. Chem. Soc.* **102,** 1745 (1980).

[21] R. C. Bruch and M. D. Bruch, *J. Biol. Chem.* **257,** 3409 (1982).

[22] C. C. Sweeley and H. A. Nunez, *Annu. Rev. Biochem.* **54,** 756 (1985).

and the majority of one issue[23] of the new *Journal of Carbohydrate Chemistry* has been devoted to 2-D NMR spectroscopy of monosaccharides and oligosaccharides. In this article the procedures for the analysis of oligosaccharide structure by correlated 2-D proton NMR spectroscopy will be presented. The emphasis will be on the determination of primary structure; however, studies of secondary structure will be briefly considered.

Procedure for Primary Structure Determination

Preliminary Studies and An Overview

The first step in any proton NMR study of an oligosaccharide or derivative should be obtaining an integrated, one-dimensional spectrum from 0 to 10 ppm. From this study the purity and actual concentration of the sample, appropriateness of the solvent and temperature, and the structural complexity of the oligosaccharide can be assessed. In the case of spectra of simple structures or of complex structures that are identical to ones already in a well-defined database, a one-dimensional study may suffice for structure determination. Examples of databases already in existance are oligosaccharides[24,25] in deuterium oxide and glycolipids[11,26-28] in methyl sulfoxide-d_6 deuterium oxide (98:2, v/v). Particularly useful in such one-dimensional analyses are the anomeric and methyl resonances located between 4 and 5 and between 1 and 2 ppm, respectively. In the case of oligosaccharides whose one-dimensional spectrum does not resemble previously encountered structures but is technically good, then 2-D NMR analysis is indicated. Such a 2-D analysis requires at least 0.5 mg of sample to be practical at the present time. For a trisaccharide, or more complex oligosaccharide, at least a 360-MHz spectrometer is necessary.

Analysis of primary structure via 2-D proton NMR involves two types of correlated experiments. First, one of the various scalar correlated spectra[1-4] is obtained to reveal through-bond (*J*) couplings. From this first

[23] S. L. Patt, *J. Carbohydr. Chem.* **3**, 493 (1984); W. B. Wisc, P. E. Pfeffer, and P. Kovac, *ibid.* p. 513; B. Coxon, *ibid.* p. 525; R. E. Botto, *ibid.* p. 545; T. A. W. Koerner, J. N. Scarsdale, J. H. Prestegard, and R. K. Yu, *ibid.* p. 565; K. Bock and H. Pedersen, *ibid.* p. 581; A. Bax, W. Egan, and P. Kovac, *ibid.* p. 593.

[24] J. F. G. Vliegenthart, L. Dorland, and H. Van Halbeck, *Adv. Carbohydr. Chem. Biochem.* **41**, 209 (1983).

[25] J. P. Carver and J. R. Brisson, *in* "Biology of Carbohydrates" (V. Ginsburg and P. W. Robbins), Vol. 2, p. 289. Wiley, New York, 1984.

[26] J. Dabrowski, P. Hanfland, and H. Egge, this series, Vol. 83, p. 69.

[27] S. Gasa, T. Mitsuyama, and A. Makita, *J. Lipid Res.* **24**, 174 (1983).

[28] H. Clausen, S. B. Levery, J. M. McKibbin, and S. Hakomori, *Biochemistry* **24**, 3578 (1985).

experiment J connectivities are established and the spectrum is thus divided into subspectra, each of which originates from and identifies one oligosaccharide residue. From the pattern of the chemical shifts and coupling constants found in each subspectrum the identity of the residue of origin is established. The second type of 2-D NMR experiment carried out is 2-D NOESY, which reveals through-space couplings. By examination of the anomeric region of the NOESY spectrum of an oligosaccharide, linkage sites and sequence are determined for all glycosidic linkages except those involving ketose residues (see "Limitations," below).

A key concept in the 2-D NMR analysis of an oligosaccharide is the comparison of scalar correlated spectral data with NOESY data through chemical shift identities of resonances. As an example, the comparison of 2D-SECSY and 2D-NOESY spectra of an idealized trisaccharide is shown in Fig. 1. From such a comparison, information on residue type, sequence, and linkage sites can be combined to yield the complete primary structure of an oligosaccharide or derivative. Details of the use of these two types of 2-D NMR experiments are presented as follows, using the glycolipid ganglioside G_{M2} (Fig. 2) as an example.

Generation of J Connectivities via Scalar Correlated 2-D NMR

Both SECSY and COSY experiments have been performed with oligosaccharides; however, in our work we have usually chosen the SECSY experiment because it is advantageous in situations where differences in resonance frequencies of coupled resonances are a small fraction of the total spectral width. This is typically the case for oligosaccharides and glycoconjugates, where differences in chemical shifts of ring protons are always less than 2 ppm. For these cases the SECSY experiment allows contraction of the F_1 domain to half the chemical shift difference with very substantial savings in data-set size, processing time, and acquisition time. An example of the SECSY spectrum of ganglioside G_{M2} is shown in Fig. 3, together with its 1-D, integrated spectrum for comparison.

After a scalar correlated spectrum of an unknown oligosaccharide is obtained, a graphical analysis is undertaken to extract J connectivities. In the case of the SECSY experiment, contour densities above and below the central horizontal axis identify J-coupled protons. Thus, sequential construction of vertical-upward, 135° diagonal upfield, and vertical-upward lines of lengths x, $2\sqrt{2}x$, and x, respectively, leads one through the J connectivity. It is easiest to begin in the well-resolved 4.0–5.0 ppm region in which the terminal anomeric protons of aldopyranoside residues are found, then to proceed upfield for each J connectivity into the complex ring proton region (3.0–4.0 ppm). In this manner, J connectivities are

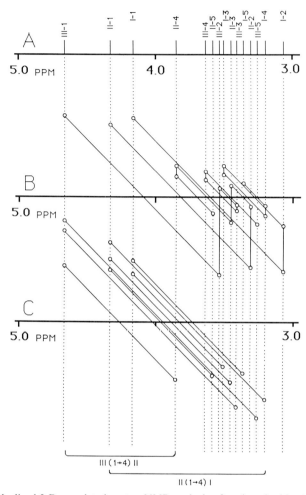

Fig. 1. Idealized 2-D correlated proton NMR analysis of a trisaccharide: (A) 1-D spectrum for comparison; (B) 2-D SECSY spectrum; and (C) 2-D NOESY spectrum (only anomeric dipole couplings are shown). Multiplicities in resonance lines of the 1-D spectrum and in the contours of the 2-D spectra have been eliminated for clarity. The trisaccharide is assumed to have an NMR-transparent aglycone, such as the tri-deutero-methyl group. Thus, resonance I-1 "sees" only the two intraresidue dipole couplings in the 2-D NOESY spectrum.

generated that are usually complete at high field (500 MHz). At lower resolution at least the first four protons of each oligosaccharide residue (H-1 → H-4) can usually be followed. Once the *J* connectivity is determined, the chemical shift and coupling constants are extracted for each

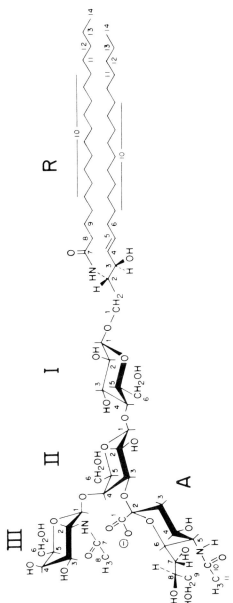

FIG. 2. The structure of ganglioside G_{M2}. Residue labels: I, β-Glc; II, β-Gal; III, β-GalNAc; A, α-NeuAc; and R, ceramide (as assigned in the text).

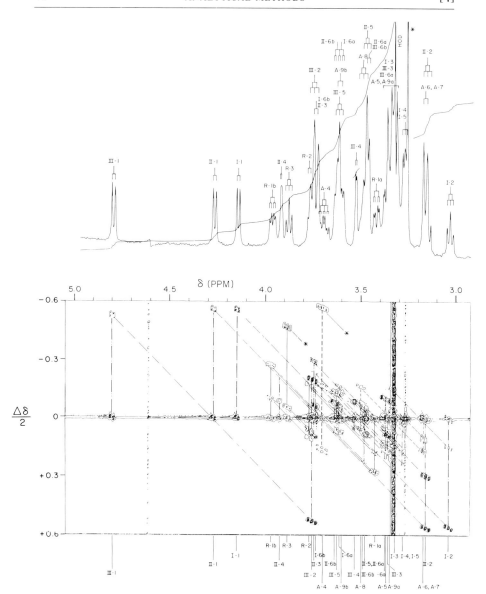

Fig. 3. *Above:* Integrated, 1-D proton NMR spectrum of ganglioside G_{M2} between 3 and 5 ppm, obtained at 500 MHz and 30°. An asterisk (*) indicates EDTA peak. *Below:* 2-D SECSY proton NMR spectrum of ganglioside G_{M2}, revealing the J connectivities of its component residues in the oligosaccharide ring proton region (3–5 ppm). J connectivities are labeled for residues as follows: I, β-Glc (- · -); II, β-Gal (- · · -); III, β-GalNAc (- · · · -); A, α-NeuAc (- - -); and R, ceramide (——). An asterisk (*) indicates J connectivities to protons outside observed window. From Koerner *et al.*[11] Reprinted with permission from *Biochemistry.* Copyright (1983) American Chemical Society.

subspectral proton either from the SECSY spectrum directly or by also using the accompanying 1-D spectrum. These data are then tabulated and each J connectivity is analyzed for its proton number, pattern, vicinal coupling constant sequence, and chemical shifts.

Identification of Residue Type via J-Connectivity Data

Probably the most important determinants of oligosaccharide residue constitution are the number and pattern of protons manifest in a residue's J connectivity. These characteristic J-connectivity numbers and patterns are shown for most oligosaccharide residues in Table II. The proton number can only be known if the complete J connectivity has been extracted; however, if only the first four protons are known, the constitution of the residue can be deduced from the pattern associated with this J-connectivity fragment. For example a linear pattern, AHMR. . . , is characteristic of an aldose. Depending on the number of couplings manifest at R of this pattern, an aldopentose can be distinguished from an aldohexose. If the entire J connectivity is known, a 6-deoxyaldohexose can be distinguished from an aldohexose. Similarly 2-deoxyaldoses are heralded by an A(HJ)M. . . pattern and sialic acid residues by (AB)HM. . . patterns.

After its constitution is established, the stereochemistry of the residue is determined by consideration of the values of the first four vicinal (^3J) coupling constants displayed in its J connectivity. The correlation of all possible vicinal coupling constant sequences with the stereochemistry of aldohexopyranose residues is shown in Table III. Similar tables may be constructed from data[29] available for aldopentoses and ketoses. Though there are 16 theoretically possible coupling constant patterns for aldohexopyranoses, four are stereochemically impossible and seven others would arise from monosaccharides not yet encountered in oligosaccharides. In determining these vicinal coupling constant sequences, J values of less than 5 Hz are considered small (S) and of more than 5 Hz are considered large (L).

The last type of data considered are the chemical shifts of protons within each J connectivity. Examples of residue chemical shifts are shown in Table IV.[30,31] As pointed out earlier, the chemical shifts of anomeric (H-1) protons are between 4.0 and 5.0 ppm. The remaining protons of each J connectivity are much less regular. The importance of

[29] C. Altona and C. A. G. Haasnoot, *Org. Magn. Reson.* **13,** 417 (1980).

[30] J. Dabrowski, P. Hanfland, and H. Egge, *Biochemistry* **19,** 5652 (1980).

[30a] V. D. Dua, B. N. N. Rao, S. Wu, V. E. Dube, and C. A. Bush, *J. Biol. Chem.* **261,** 1599 (1986).

[31] J. Dabrowski, P. Hanfland, H. Egge, and V. Dabrowski, *Arch. Biochem. Biophys.* **210,** 405 (1981).

TABLE II
OLIGOSACCHARIDE RESIDUE SPIN SYSTEMS AND THEIR CHARACTERISTICS

Number of protons	J-connectivity pattern	Symbol	Residue constitution	Example residues[a]
5		AHMRV	Aldohexuronic acid	Glucuronic acid, iduronic acid
6		AHMR(XZ)	Aldopentose	Arabinose, ribose, xylose, lyxose
7		A(HJ)MR(XZ)	2-Deoxyaldopentose	2-Deoxyribose
7		AHMRV(XZ)	Aldohexose	Glc, Gal, Man, GlcNAc, GalNAc, ManNAc
8		A(HJ)MRV(XZ)	2-Deoxyaldohexose	2-Deoxyglucose
8		AHMRVX₃	6-Deoxyaldohexose	Fuc, Rha, Qui
9		A(HJ)MRVX₃	2,6-Dideoxyaldohexose	Digitoxose
9		(AB)HMRSV(XZ)	3-Deoxy-2-nonulosonic acid	NeuAc, NeuGy
2 + 5		AB-AHM(XZ)	2-Hexulose	Fructose, tagatose, psicose, sorbose

Broken J connectivity

[a] Abbreviations used are defined in Table III; also NeuAc, N-acetylneuraminic acid; NeuGy, N-glycoylneuraminic acid.

TABLE III

THE CORRELATION OF ALL POSSIBLE VICINAL COUPLING CONSTANT PATTERNS WITH
THE STEREOCHEMISTRY OF ALDOPYRANOSYL RESIDUES OF OLIGOSACCHARIDES

Vicinal coupling constants[a]				Aldopyranose residue stereochemistry[b]		
$J_{1,2}$	$J_{2,3}$	$J_{3,4}$	$J_{4,5}$	Configuration	Conformation	Example residues[c]
L	L	L	L	β-gluco	$^4C_1$	β–D-Glc, β-D-GlcNAc, β-D-Qui
L	L	L	S	α-ido	$^1C_4$	(α-D-Idose)
L	L	S	L	Impossible stereochemistry		
L	L	S	S	β-galacto	$^4C_1$	β-D-Gal, β-D-GalNAc, β-L-Fuc, β-D-Fuc
				α-altro	$^1C_4$	(α-D-Altrose)
L	S	L	L	Impossible stereochemistry		
L	S	L	S	Impossible stereochemistry		
L	S	S	L	β-allo	$^4C_1$	(β-D-Allose)
L	S	S	S	β-gulo	$^4C_1$	(β-D-Gulose)
S	L	L	L	α-gluco	$^4C_1$	α-D-Glc, α-D-GlcNAc, α-D-Qui
S	L	L	S	β-ido	$^1C_4$	(β-D-Idose)
S	L	S	L	Impossible stereochemistry		
S	L	S	S	α-galacto	$^4C_1$	α-D-Gal, α-D-GalNAc, α-L-Fuc, α-D-Fuc
S	S	L	L	α-manno	$^4C_1$	α-D-Man, α-D-ManNAc, α-L-Rha
				β-manno	$^4C_1$	β-D-Man, β-D-ManNAc
S	S	L	S	α-gulo	$^1C_4$	(α-D-Gulose)
S	S	S	L	α or β-altro, α-allo	$^4C_1$	(α or β-D-Altrose, α-D-Allose)
S	S	S	S	α or β-ido	$^4C_1$	(α or β-D-Idose)
				α-gulo	$^4C_1$	(α-D-Gulose)
				α or β-talo	$^4C_1$	(α or β-D-Talose)

[a] The symbols L and S represent large (>5 Hz) and small (<5 Hz) vicinal proton–proton coupling constants, respectively.

[b] The absolute stereochemistry of a residue cannot usually be established by NMR. Stereochemical designators used in this table apply as well to the L form of a residue as to the D form. Thus the $^4C_1$ conformation of β-D-glucose and its mirror image, the $^4C_1$ conformation of β-L-glucose, would both manifest an L,L,L,L vicinal coupling pattern.

[c] Abbreviations used: Glc, glucopyranose; GlcNAc, N-acetyl-2-amino-2-deoxygluco-pyranose; Qui, quinovose (6-deoxyglucopyranose); Gal, galactopyranose; GalNAc, N-acetyl-2-amino-2-deoxygalactopyranose; Fuc, fucose (6-deoxygalactopyranose); Man, mannopyranose; ManNAc, N-acetyl-2-amino-2-deoxymannopyranose; and Rha, rhamnose (6-deoxymannopyranose). Residues shown in parentheses are theoretically possible, but have not yet been encountered in natural oligosaccharides.

TABLE IV

REFERENCE CHEMICAL SHIFTS OF TERMINAL, UNSUBSTITUTED OLIGOSACCHARIDE RESIDUES
FOUND IN GLYCOLIPIDS

Residue[b]	Chemical shift (δ)[a]							Example
	H-1	H-2	H-3	H-4	H-5	H-6a	H-6b	
β-Glc	4.09	2.98	3.16	3.1	3.1	3.44	3.66	Glucosylceramide[c]
β-Gal	4.20	3.40	3.3[c]	3.61	3.4[c]	3.53	3.58	Lactosylceramide[f]
β-GlcNAc	4.65	3.36	[d]	[d]	[d]	[d]	[d]	Lactotriaosylceramide[g]
β-GalNAc	4.46	3.61	3.52	3.61	3.34	3.40	3.48	Gangliotriaosylceramide[f]
α-Gal	4.81	3.65	3.56	3.76	4.05	[d]	[d]	Globotriaosylceramide[g]
α-GalNAc	4.74	4.09	3.56	3.73	3.80	[d]	[d]	Globopentaosylceramide[g]
α-L-Fuc(1 → 2)	4.99	3.51	3.51	3.51	4.07	1.04	1.04	Structure I[h]
α-L-Fuc(1 → 4)	4.78	3.50	3.59	[d]	4.58	[d]	[d]	Structure III[h]

[a] Obtained at 360 MHz or greater in methyl sulfoxide-d_6–D_2O (98 : 2, v/v), referenced to internal tetramethylsilane.
[b] Abbreviations as used in Table III.
[c] Obscured by other resonance.
[d] Not reported.
[e] Values reported at 65° by Yamada et al.[18]
[f] Values reported at 30° by Koerner et al.[11]
[g] Values reported at 65° by Dabrowski et al.[30]
[h] Values reported at 65° by Dabrowski et al.[31]

chemical shift data is that they allow fine details of oligosaccharide residue constitution to be determined, such as the presence of an acetamido group (Glc vs GlcNAc) or O-acetylation. Chemical shift data are also important in assigning the site of sialic acid attachment (below).

The combination of J-connectivity pattern, coupling constant, and chemical shift data to assign oligosaccharide residue structure is illustrated in algorithm form in Figs. 4 and 5, for the specific example of assigning the structure of a glycolipid aldohexopyranose residue of unknown structure. The use of this algorithm will be illustrated using the example of the oligosaccharide residues of ganglioside G_{M2} (Fig. 2).

Inspection of the 1-D and 2-D SECSY spectra of G_{M2} (Fig. 3) and graphical analysis of the later spectrum yields five J connectivities or subspectra whose chemical shifts and coupling constants are shown in Table V. Of the five subspectra, three are seen to be seven-spin AHMRV(XZ) systems, which are labeled I, II, and III. All three of these subspectra have a linear pattern (AHMR. . .) for their first four protons with only one proton coupled to the R proton. Furthermore, they all have an A part resonating between 4 and 5 ppm. Thus subspectra I–III arise

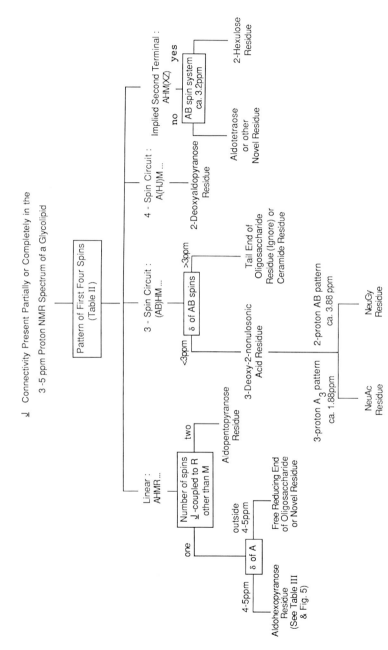

Fig. 4. An algorithm for the identification of the constitution of oligosaccharide residues present in glycolipids, based on the pattern and chemical shifts of *J* connectivities manifest in the 3- to 5-ppm region of their proton NMR spectra.

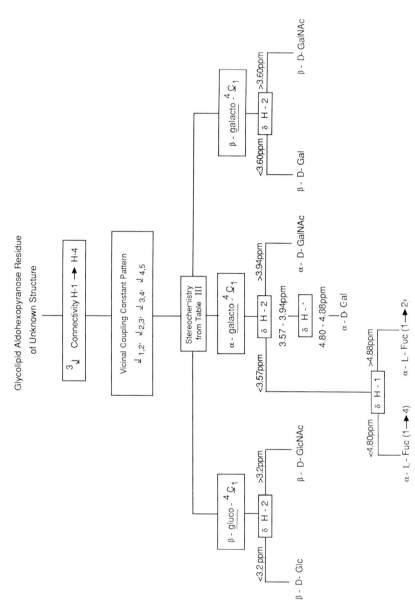

FIG. 5. An algorithm for the identification of the stereochemistry and fine constitutional structure of oligosaccharide aldopyranose residues present in glycolipids, based on vicinal coupling constant pattern and chemical shift data derived from the first four protons of their J connectivities.

TABLE V

CHEMICAL SHIFTS AND COUPLING CONSTANTS FOR GANGLIOSIDE G_{M2} OLIGOSACCHARIDE RESIDUES[a]

Proton chemical shifts, δ (ppm)

Residue	H-1	H-2	H-3	H-4	H-5	H-6a	H-6b	H-7	H-8	H-9a	H-9b
I	4.150	3.035	3.325	3.277	3.289	3.613	3.744	—	—	—	—
II	4.274	3.158	3.744	3.920	3.470	3.463	3.613	—	—	—	—
III	4.794	3.753	3.39[b]	3.524	3.642	3.39[b]	3.463	—	1.777	—	—
A[c]	—	—	2.558 (e) 1.621 (a)	3.700	3.39[b]	3.148		3.161	3.485	3.340	3.613

Proton–proton coupling constant, J (Hz)

Residue	$J_{1,2}$	$J_{2,3}$	$J_{3,4}$	$J_{4,5}$	$J_{5,6a}$	$J_{5,6b}$	$J_{6a,6b}$	$J_{6,7}$	$J_{7,8}$	$J_{8,9a}$	$J_{8,9b}$	$J_{9a,9b}$
I	7.9	8.2	9.6	9.6	6.0	1.5	−11.5	—	—	—	—	—
II	9.6	8.7	<1.5	<1.5	6.0	5.6	−12.1	—	—	—	—	—
III	8.8	9.6	1.5	<1.5	5.9	6.0	−12.0	—	—	—	—	—
A	−12.4 (3e,3a)		5.1 (3e,4) 10.6 (3a,4)	10.6	9.1		—	1.9	9.8	6.0	1.5	−10.8

[a] The estimated error for δ is ±0.001 ppm and that for J is ±0.6 Hz. Spectrum obtained in methyl sulfoxide-d_6–deuterium oxide (98 : 2, v/v) at 500 MHz and 30°; referenced to internal tetramethylsilane.
[b] Obscured by HOD resonance.
[c] A-11 methyl resonance at 1.875 ppm.

from aldohexopyranose residues, according to Fig. 4. In Fig. 5, the coupling constant sequences of each subspectra are considered. For subspectrum I, $J_{1,2}$, $J_{2,3}$, $J_{3,4}$, and $J_{4,5}$ are all large (Table V) and a β-*gluco*-$^4C_1$ stereochemistry is assigned according to Table III. Similarly, subspectra II and III manifest the coupling constant sequence L,L,S,S and are assigned β-*galacto*-$^4C_1$ stereochemistries. Continuing with the Fig. 5 algorithm, subspectrum I is noted to have an H-2 resonance at 3.035 ppm, indicating that it arises from a β-D-glucopyranosyl residue. Likewise, the fact that the H-2 resonance of II is at 3.158 ppm and of III is at 3.753 ppm indicates that these subspectra arise from β-galactopyranosyl and N-acetyl-β-galactosaminyl residues, respectively.

The fourth subspectrum of ganglioside G_{M2} is seen to have a nine-spin (AB)HMRSV(XZ) pattern which, when analyzed according to Fig. 4 (or Table II), reveals it originates from an N-acetylneuraminic acid residue. Analysis[11] of the fifth subspectrum reveals it to be the aglycone ceramide (R, Fig. 2).

In summary, through 1-D and 2-D SECSY NMR, the oligosaccharide moiety of ganglioside G_{M2} is seen to be composed of four oligosaccharide residues: β-Glc, β-Gal, β-GalNAc, and α-NeuAc. Integration data obtained from the 1-D spectrum (Fig. 3) reveal that all residues are present in equimolar concentrations relative to the aglycone ceramide. Thus the residue composition of ganglioside G_{M2} is known.

Oligosaccharide Sequence and Linkage Sites via 2-D NOESY

It is important that the 2-D NOESY experiment be carried out so that the spectrum produced can be easily compared to the scalar correlated spectrum already obtained. Since we prepared a SECSY spectrum of ganglioside G_{M2}, we next needed a 2-D NOESY spectrum prepared using a SECSY matrix. Such a 2-D NOESY spectrum is shown in Fig. 6, produced using the same axes as used for the 2-D SECSY spectrum (Fig. 3). Graphical identification of dipole-coupled protons is by finding 135°-aligned contour densities above and below the horizontal axis, as described above for the SECSY spectral analysis. Dipole couplings are observed between anomeric (H-1) and nonanomeric resonances and between pairs of nonanomeric resonances; however, only the former will be considered, since only they are essential for establishment of oligosaccharide primary structure.

Two types of dipole (or through-space) couplings are observed for anomeric protons. These are intraresidue or "across the ring" couplings, helpful in confirming resonance assignments, and interresidue couplings, which are across the glycosidic linkage and very important in identifying

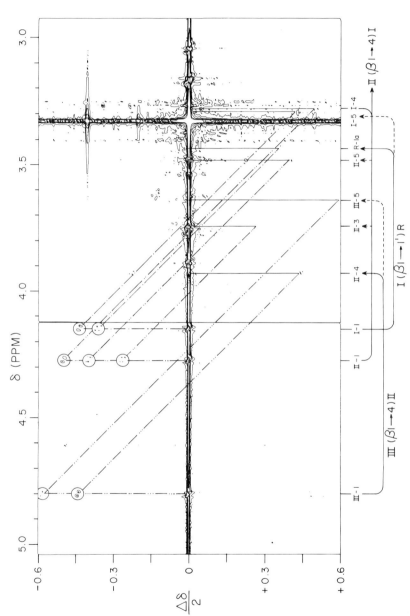

FIG. 6. 2-D NOESY proton NMR spectrum of ganglioside G_{M2} between 3 and 5 ppm, obtained at 500 MHz. Dipole couplings are labeled for the anomeric protons of the following oligosaccharide residues: I, β-Glc (- · -); II, β-Gal (- · · -); III, β-GalNAc (- · · · · -). Anomeric couplings are circled for clarity. At the bottom, interresidue couplings are labeled with solid lines and intraresidue couplings with dashed lines. The glycosidic linkages that are indicated by interresidue proton pairs are noted. From Koerner *et al.*[11] Reprinted with permission from *Biochemistry.* Copyright (1983) American Chemical Society.

linkage sites and sequence. Using the example of ganglioside G_{M2} (Fig. 6), we identify the three anomeric (H-1) resonances at 4.79, 4.27, and 4.15 ppm, arising from residues III, II, and I, respectively. The 4.79 ppm resonance (III-1) is seen to dipole-couple to two upfield signals at 3.92 and 3.64 ppm, assigned respectively to II-4 and III-5 in the 2-D SECSY study (Table V). Thus the III-1 anomeric proton has an intraresidue coupling to III-5 and an interresidue coupling to II-4 and the following glycosidic linkage exists: III-1 → II-4. Similarly II-1 (4.27 ppm) couples to II-3 (3.74), II-5 (3.47), and I-4 (3.28 ppm), the latter interresidually, and a II-1 → I-4 glycosidic linkage exists. When resonance assignments[11] for the aglycone ceramide (R) are considered, a I-1 → R-1 glycosidic linkage is seen, as well as an intraresidue coupling to I-5 (3.29 ppm).

Overlapping this glycosidic linkage information yields the following sequence for ganglioside G_{M2}: **III** (1 → 4)**II** (1 → 4)**I** (1 → 1)**R**. When the residue identities (see previous section) are substituted into this sequence the oligosaccharide primary structure is seen to be GalNAc(β1 → 4)Gal(β1 → 4)Glc(β1 → 1)ceramide. However, we still do not know the linkage site of the NeuAc residue, which is a ketose and lacks an anomeric proton.

Sialic Acid-Linkage Sites via Sialylation Shifts

In order to determine the sialic acid attachment sites(s) the chemical shifts of the ganglioside are compared with those of the actual asialoganglioside, whose structure has been deduced from the 2-D NOESY analysis. In the case of ganglioside G_{M2}, the chemical shift data[9,11] for asialo-G_{M2} are needed. It should be noted that the chemical shift data[11,26] for almost any desired asialo or core oligosaccharide structure are now available, at least in the case of glycolipids. Comparison of the chemical shifts at each possible sialylation site and calculation of the sialylation shift or chemical shift difference for ganglioside G_{M2} are shown in Table VI. Only at II-3 is Δ greater than 0.2 ppm, indicating that the sialic acid residue is attached at II-3 and that only one sialylation site is present.

Since only one α-NeuAc is present in ganglioside G_{M2} (last section), the sialic acid glycosidic linkage must be NeuAc(α2 → 3)II. Combining this information with the partial structure deduced via 2-D NOESY and it is concluded that the primary structure of ganglioside G_{M2} is NeuAc(α2 → 3)[GalNAc(β1 → 4)]Gal(β1 → 4)Glc(β1 → 1)ceramide. This 2-D NMR-derived structure is in agreement with that assigned through chemical and enzymatic methods, as previously noted.[12]

TABLE VI
SIALYLATION-INDUCED CHEMICAL SHIFT
DIFFERENCES (Δ) FOR GANGLIOSIDE G_{M2}[a]

Proton	Chemical shift (ppm)		Δ
	G_{M2}[b]	Asialo-G_{M2}[c]	
I-2	3.035	3.036	−0.00
I-3	3.325	3.335	−0.01
I-6b	3.744	3.745	−0.00
II-2	3.158	3.245	−0.09
II-3	3.744	3.519	+0.23
II-6b	3.613	3.604	+0.01
III-3	3.39	3.519	−0.13
III-4	3.524	3.614	−0.09
III-6b	3.463	3.478	−0.02

[a] Note that sialylation is impossible at I-4, II-4,
and III-2.
[b] Values taken from Table V.
[c] Values taken from Koerner et al.[11]

So far the complete primary structure of at least seven glycoli-pids[9,11,12,14,32] has been determined through correlated 2-D NMR methods similar to those discussed above. Included in this group[14,32] are oligosac-charide glycones that contain α- as well as β-glycosidic linkages and an octasaccharide[32] glycone. These examples support the generality and po-tential range of 2-D NMR methods for structural analysis.

Limitations

As mentioned already, 2-D NOESY sequencing of oligosaccharides fails for ketose residues. In studies of oligosaccharides derived from ani-mal sources, the ketoses of greatest concern are the derivatives of neuraminic acid (sialic acids). In the case of monosialo structures, such as our example ganglioside G_{M2}, the sequence and linkage site can be deter-mined using sialylation shifts; however, for di-, tri-, and oligosialo struc-tures this method breaks down. What is needed is more spectral informa-tion on the NeuAc ($\alpha2 \rightarrow 8$) NeuAc linkage, which in our hands has been difficult to obtain due to a tendency of gangliosides containing this

[32] J. N. Scarsdale, S. Ando, T. Hori, R. K. Yu, and J. H. Prestegard, *Carbohydr. Res.* (1986) (in press).

structural fragment to manifest poorly dispersed oligosaccharide spectra. Though we are not aware of any studies as yet, when fructose and other 2-hexuloses are encountered in 2-D studies of plant oligosaccharides, these ketoses will present a problem in sequence studies. Another problem encountered in studies of oligosaccharides derived from glycoproteins is the difficulty in assigning anomeric linkage to mannopyranose residues. This problem stems from the degeneracy of the vicinal coupling constant pattern manifest by both anomeric forms of this sugar (Table III). Finally, the assumption that 2-D NOESY will always detect trans-glycosidic anomeric proton couplings has been challenged.[30a] However, we have not encountered any exceptions to this assumption in our own work.

Future Prospects

Perhaps the limitations noted above will be overcome by future technical improvements in 2-D NMR methods. Already the scope of structural complexity amenable to 2-D analysis has been increased by improved experiments such as relay COSY[32,33] or RECSY, pure absorption (multiple quantum filtered) COSY,[32] and pure absorption NOESY.[32] As in the past, the imagination of Ernst and co-workers[34,35] has led the way. Another trend is toward simplification of spectra and shortening of experimental and computer processing time.[32,36,37] Some of our 2-D NOESY experiments which originally took 20 hr now only take 2 hr.

Secondary Structure Studies

2-D NOESY studies of the conformation of the glycolipid globoside[14,15] and a nonasaccharide[16] derived from the glycoprotein transferrin have been carried out. In these studies transglycosidic interprotonic distances are estimated from semiquantitative 2-D NOE measurements. When enough transglycosidic or other interprotonic distances can be determined, a unique or limited range of conformations can be defined. Such studies of globoside[15] indicate that its oligosaccharide moiety has a bent or

[33] S. W. Homans, R. A. Dwek, D. L. Fernandes, and T. W. Rademacher, *Proc. Natl. Acad. Sci. U.S.A.* **81**, 6286 (1984).

[34] G. Eich, G. Bodenhausen, and R. R. Ernst, *J. Am. Chem. Soc.* **104**, 3731 (1982).

[35] M. Rance, O. W. Sorenson, G. Bodenhausen, G. Wagner, R. R. Ernst, and K. Wüthrich, *Biochem. Biophys. Res. Commun.* **117**, 479 (1983).

[35a] B. Maggio, T. Ariga, and R. K. Yu, *Arch. Biochem. Biophys.* **241**, 14 (1985).

[36] D. G. Davis and A. Bax, *J. Am. Chem. Soc.* **107**, 7197 (1985).

[37] A. Bax and D. G. Davis, *in* "Advanced Magnetic Resonance Techniques in Systems of High Molecular Complexity" (N. Nicolai and G. Valensin, eds.). Birkhaueser, Basel (in press).

"inverted L" conformation. Though this approach has been criticized for not being as quantitative as procedures based on 1-D NOE measurements, we have found that other physical measurements such as calorimetry[35a] confirm the conformations predicted by 2-D NOE studies.

Acknowledgments

The authors wish to thank Henry Weinfeld (Tulane University School of Medicine) for help in preparing computer graphics. This work was supported by USPHS grant NS 11853.

[5] Direct Chemical Ionization Mass Spectrometry of Carbohydrates

By VERNON N. REINHOLD

Current Developments in Mass Spectrometry Instrumentation

Traditionally, mass spectrometry has been a valuable tool for the chemist, but with the availability of gas chromatography–mass spectrometry (GC-MS) instrumentation and computer data systems, this instrumental combination has proved successful in many aspects of analytical inquiry. The utility of GC-MS systems in biological laboratories has been stimulated by the advances in GC capillary column technology and the production of more practical and less expensive equipment. Biological and environmental scientists have found excellent and widespread applications in the determination of individual components in complex mixtures. Carbohydrate researchers have found direct application of GC-MS in resolving the complex mixtures generated during linkage analysis.

The broader application of MS instrumentation to many compounds of biological interest (usually very polar and of high molecular weight) has been limited by two critical factors; first, samples could only be ionized when in the gas phase; and second, conventional ion sources (electron ionization, EI) deposit an excess amount of energy with the sample during ionization. Sample vaporization causes pyrolysis while high-energy ionization induces further fragmentation. The resultant EI spectra are very complex with a preponderance of low mass fragments and little indication of molecular weight and structural detail at high mass.

These limitations have now been largely overcome by a series of desorptive techniques that allow direct sample ionization from a condensed phase, such as fast atom bombardment (FAB), or processes that impart little energy to the sample during ionization, such as chemical ionization (CI). Mass spectrometers updated with these new "soft" ionization techniques can now provide usable spectra for compounds of increased polarity and higher molecular weight. With the constraints of molecular ionization somewhat diminished, the attention of many mass spectroscopists has focused on the limits of high mass ion transmission and detection, and more detailed analysis of ion structures.

Effective application of MS to large biopolymers must, in the final analysis, be dependent on methods of partial degradation. In spite of these advances in instrumentation, it is unlikely that some form of partial depolymerization can be avoided. These procedures encompass glycosidic cleavage at specific residues and/or random depolymerization to obtain a family of overlapping oligomers, separation of the mixture, and identification of the smaller components. Constructing a biopolymer sequence from degradation data is always fraught with compromise and each polysaccharide generates its own set of qualifications. For proteins, partial degradation procedures have been very successful and may be approached chemically or enzymatically with a number of specific cleavage points. Unfortunately, polysaccharide researchers do not, as yet, share this analytical flexibility and frequently must contend with additional complications of component instability, broad variations in interresidue bond strengths, and separation of extremely complex mixtures of isomers and anomers. Current procedures lack specificity and sensitivity and much research must be initiated to overcome these very difficult problems. Mass spectrometry can aid greatly in this endeavor.

The new developments in ionization have brought molecular characterization by MS to different areas of research, but it should not be forgotten that the sustaining principles in mass spectrometry have always been gas-phase ion separation in a magnetic and/or electric field according to the mass-to-charge ratio of the compound and detection of these gaseous ions at remarkable sensitivity. These are two important characteristics which instrumentalists have yet to nurture to their fullest potential, and combined with the areas of "soft" ionization, liquid chromatography-MS, and MS-MS, indicate that the discipline of mass spectrometry is only starting to mature and will have an ever-increasing role in carbohydrate structural determination.

Chemical Ionization Mass Spectrometry

Chemical ionization MS provides a technique to maximize molecular and high mass information by replacing the highly exothermic process of electron ionization with a low-energy, chemical process involving ion–molecule reactions. The essential feature of CI is the EI of a reagent gas maintained at near-atmospheric pressures which thereby generates an ion plasma. Injection of a sample into this plasma causes ionization by one of three processes: (1) charge exchange, by using an aprotic reagent gas; (2) proton transfer, where ions of the reagent gas function as a Brönsted acid; or (3) formation of collision-stabilized adduct ions, in which the reagent ion functions as a Lewis acid. Processes (2) and (3) are thermodynamic in nature and distribution of ion products is dependent on reactant structure, their proton affinities, and reagent gas pressures. This topic has been reviewed by one of the major contributors to this area[1] and a book covering the topic has recently been published.[2]

Standard instrumental procedures for CI requires that the sample be introduced into the ion chamber as a gas, analogous to the requirement for EI. For carbohydrate materials, this means the preparation of derivatives to avoid pyrolytic decomposition during vaporization. A number of derivatization procedures for CI-MS have been used to overcome this problem, and several groups have described CI-MS of acetylated,[3–6] methylated,[6,7] and trimethylsilyl ether[8] derivatives of relatively small molecular weight samples. The reagent gases employed are most frequently methane, isobutane, or ammonia which produce the corresponding reacting species CH_5^+, $C_4H_9^+$, and NH_4^+. The protonating capability of these reagent gases decreases in the above order, and thus, fragment ions are more abundant in the methane and isobutane spectra of saccharides, whereas the ammonia CI spectra are dominated by ammonia clusters[9] and molecular weight-related ions.[3] The first carbohydrate ammonia CI study[3] reported single adduct ions for peracetylated mono- and disaccharides. The acetylated derivatives of glucose and methyl glucopyranoside showed "in addition to the $(M + NH_4)^+$ ion, a smaller peak corresponding to $(M + NH_4 - H_2O)^+$ and $(M + NH_4 - CH_3OH)^+$, respectively."[3] The combination of protecting samples by derivatization and using a weak

[1] B. Munson, *Anal. Chem.* **49**, 772A (1977).
[2] A. G. Harrison, "Chemical Ionization Mass Spectrometry." CRC Press, Boca Raton, Florida.
[3] A. M. Hogg and T. L. Nagabhushan, *Tetrahedron Lett.* p. 4827 (1972).
[4] R. C. Dougherty, J. D. Roberts, W. W. Binkley, O. S. Chizov, V. I. Kadentsev, and A. A. Solov'yov, *J. Org. Chem.* **39**, 451 (1974).
[5] D. Horton and J. D. Wander, *Anal. Biochem.* **55**, 123 (1973).
[6] M. McNeil and P. Albersheim, *Carbohydr. Res.* **56**, 239 (1977).
[7] E. G. de Jong, W. Heerma, and C. A. X. G. F. Sicherer, *Biomed. Mass Spectrom.* **6**, 242 (1979).
[8] T. Murata and S. Takahashi, *Carbohydr. Res.* **62**, 1 (1978).
[9] A. M. Hogg and P. Kebarle, *J. Chem. Phys.* **43**, 449 (1965).

protonating reagent gases is a technique that has proved to be most successful for oligosaccharides. For this latter reason further discussion in this article will only consider the use of ammonia as a reagent gas.

Carbohydrate Ion Structures Using Ammonia Chemical Ionization

The use of stable isotopes in the reagent gas or incorporated in the sample greatly aids the study of fragmentation patterns. These techniques and the observation of mass ion shifts for a series of permethylated saccharides have illustrated some important characteristics of ammonia as a reagent gas.[10] For example, the ability of the reagent gas to associate with both charged and neutral species can be followed by comparing the mass shifts between $^{14}N^{1}H_3$ and $^{14}N^{2}H_3$ in the CI-MS of permethylated butyl-α-D-glucopyranoside (Fig. 1). The base ion (at m/z 310) is 18 Da greater than the molecular weight of the compound, suggesting adduct formation with an ammonium ion (Fig. 1a). This is substantiated by a 4-Da shift to m/z 314 when $^{14}N^{2}H_3$ is used as the reagent gas (Fig. 1b). However, the ion at m/z 236 (Fig. 1a) shifts only 3 Da when $^{14}N^{2}H_3$ is used (Fig. 1b). This disparity can be accounted for by assuming reagent gas-induced glycosidic cleavage, stabilization as the pyranoxonium ion, and subsequent adduct formation with neutral ammonia to give the even-mass ion at m/z 236 (Scheme 1, pathway a).[10] These samples were introduced into the mass spectrometer from a gas chromatograph and, therefore, glycosidic cleavage must be a consequence of reaction exothermicity following adduct formation within the CI chamber. Since glycosidic cleavage represents the only major fragment, these results provide some indication that ammonia CI operated under these experimental conditions would be the most appropriate energy probe to obtain a combination of molecular weight and sequence information with larger oligosaccharides.

Ammonia CI-induced glycosidic rupture to yield the pyranoxonium ion appears most obvious (Scheme 1), but there are alternative isomeric structures that may be considered. One of these could be a consequence of direct reagent gas (NH₃) attack at the C-1 carbon and elimination of the butyl group from the molecular ion adduct to yield the protonated 1-amino-2,3,4,6-O-tetramethyl-α-D-glucose, an isomeric structure (Scheme 1, pathway b). Support for covalent nitrogen attachment can be found by a comparison of the two spectra in Fig. 1. The ion fragment m/z 236 loses 3 mol of methanol to form the ion series, m/z 204, 172, and 140 (Fig. 1a). When $^{14}N^{2}H_3$ is used as the reagent gas the ions observed are m/z 207, 174, and 142 (Fig. 1b). The reproducible losses of 32, 33, and 32 Da can be

[10] V. N. Reinhold and S. A. Carr, *Mass Spectrom. Rev.* **2**, 153 (1983).

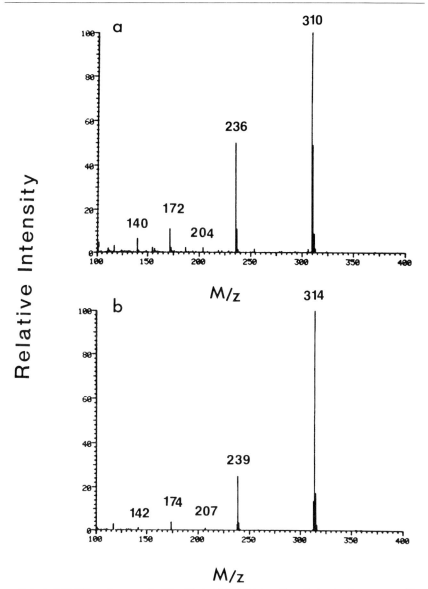

FIG. 1. GC-MS and ammonia CI of butyl-2,3,4,6-tetra-O-methyl-α-D-glucopyranoside. Reagent gas: (a) $^{14}N^{1}H_{3}$; (b) $^{14}N^{2}H_{3}$; source temperature, 125°; accelerating voltage, 3.0 kV. From Reinhold and Carr.[10] Reprinted with permission from *Mass Spectrom. Rev.* Copyright 1983, John Wiley & Sons, Inc.

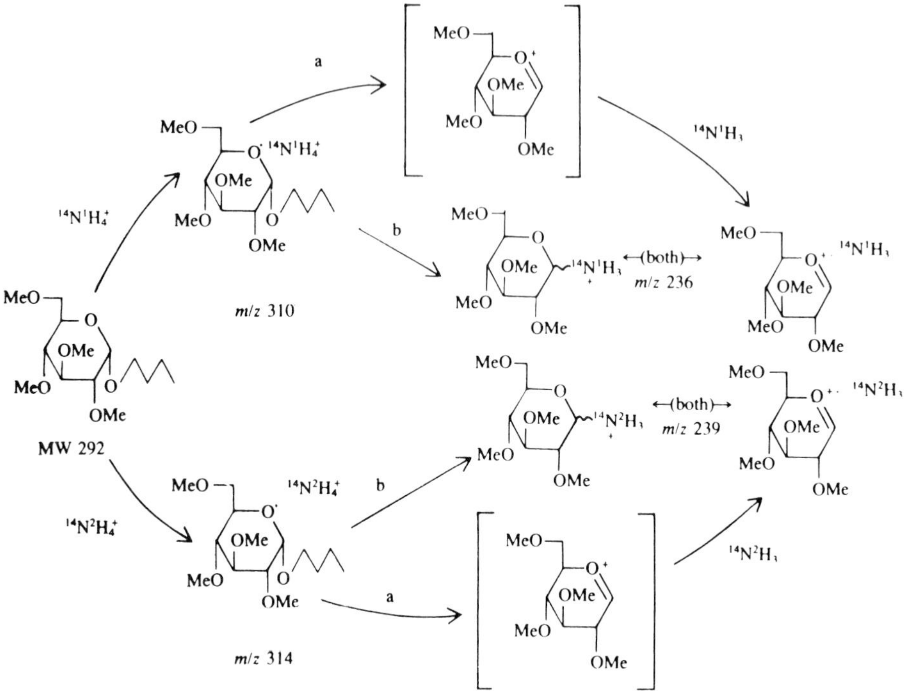

SCHEME 1.

accounted for by postulating the second mole of methanol to include a deuterium atom, CH_3O^2H, as shown in Scheme 2, pathway b. This strongly suggests gas-phase chemistry and covalent nitrogen attachment. These results are not surprising, for there have been earlier reports of ammonia reacting with aldehydes,[11] alcohols,[12-14] and ketones[15-18] under CI conditions. As discussed in the next section, ammonia DCI and gas-phase nucleophilicity may enhance glycosidic cleavage and this would account for the more pronounced sequence information observed with this reagent gas.

[11] D. F. Hunt, in "Advances in Mass Spectrometry" (A. R. West, ed.), p. 143. Applied Science Publishers, London, 1974.

[12] J. Bastard, D. K. Manh, M. Fetizon, and J. C. Tabet, J. Chem. Soc., Perkin Trans. 2 p. 1591 (1981).

[13] P. Tecon, Y. Hirano, and C. Djerassi, Org. Mass Spectrom. 17, 277 (1982).

[14] F. O. Gulacar, F. Mermoud, J. Winkler, and A. Buchs, Helv. Chim. Acta 67, 488 (1984).

[15] A. Maquestiau, R. Flammang, and L. Nielsen, Org. Mass Spectrom. 15, 376 (1980).

[16] J. C. Tabet and C. Fraisse, Org. Mass Spectrom. 16, 45 (1981).

[17] B. R. DeMark and P. D. Klein, J. Lipid Res. 22, 166 (1981).

[18] P. Rudewicz and B. Munson, Anal. Chem. 57, 786 (1985).

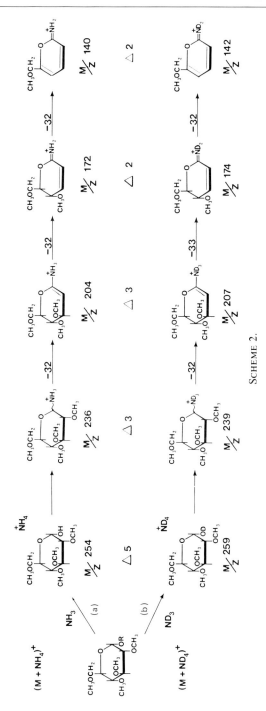

Scheme 2.

Direct Chemical Ionization Mass Spectrometry

An important modification to the CI technique has been the placement of sample material on extended probes directly within the ionization chamber. This approach has increased the structural information available from large and more polar samples by minimizing pyrolysis and can be considered comparable to the "in-beam" procedures utilized for the EI-MS of glycolipids.[19]

The development of "direct" or "in-beam" mass spectral analysis has followed a somewhat sinuous pathway since it was first discussed in 1958 as a method to enhance molecular weight information when using electron ionization.[20] In addition to the physical placement of material, two research groups[21,22] have provided a basis for much of the subsequent work that has led to the development of direct chemical ionization (DCI). There has been some confusion in the literature about the name DCI (direct or desorptive chemical ionization) and the actual placement of sample. In the discussions that follow we will use the term "direct" to mean placement of sample on an electrically resistant wire (emitter) which can be inserted directly into a confined ion plasma. The materials are "distilled" from this emitter by a programmed heating current. This should not be confused with techniques that distill sample material from a reservoir or cup outside of the ion source, techniques which unfortunately have been called direct probe chemical ionization.[23] Commercial instruments can be supplied with DCI capability; this heated emitter may also serve to introduce samples for "in-beam" electron ionization by inserting the sample-coated emitter until it rests adjacent to the electron beam (monitored by trap deflection current) and applying a small heating current. Samples analyzed by this direct method frequently provide mass spectra with enhanced high mass fragments. Simple resistance wires can replace commercial emitters for a fraction of the cost and these can be coated with a polyimide surface to augment the desorptive process.[24] A somewhat comparable process has been demonstrated with a moving belt HPLC-MS

[19] J. Angstrom, M. E. Breimer, K.-E. Falk, G. Hansson, K.-A. Karlsson, H. Leffler, and I. Pascher, *J. Biol. Chem.* **257,** 682 (1982).

[20] R. I. Reed, *J. Chem. Soc.* p. 3432 (1958).

[21] M. A. Baldwin, F. W. McLafferty, and R. J. Beuhler, *Org. Mass Spectrom.* **7,** 1111 (1973).

[22] R. J. Beuhler, E. Flanigan, L. J. Green, and L. Friedman, *J. Am. Chem. Soc.* **96,** 3990 (1976).

[23] M. J. Kreek, F. A. Bencsath, A. Fanizza, and F. H. Field, *Biomed. Mass Spectrom.* **10,** 544 (1983).

[24] V. N. Reinhold and S. A. Carr, *Anal. Chem.* **54,** 499 (1982).

interface in which the samples are heat desorbed from a polyimide surface[25-27] into a chemical ionization chamber.

Considering the overall ease of DCI operation, the ionization flexibility available by using different, and the use of isotopically labeled reagent gases for the study of molecular fragmentation, it is surprising that it has not seen widespread application. There have been preliminary studies on a variety of sample types: nucleic acids,[28,29] glycolipids,[30] peptides,[31] phospholipids,[32] glycosidic flavonoids,[33] and carbohydrates.[10,24,34-35a] The topic has been reviewed.[36]

The essential features and limitations of DCI-MS can be directly related to the heat-initiated desorptive process. Spectra acquired early during sample desorption (minimal heat) tend to show molecular weight-related ions only. Subsequent spectra, taken at higher emitter heating currents, show an increase in low mass fragments with decreasing abundance of molecular weight-related ions. This increasing pyrolytic component means that spectra cannot be compared unless they are taken at identical points during the desorption profile. Although this may be somewhat disconcerting, in practice it is highly advantageous in that it provides additional structural detail within a single sample analysis. However, as one proceeds to the analysis of higher molecular weight and more polar samples, the thermal energy needed to initiate desorption becomes excessive and the spectra exhibit a preponderance of pyrolytic fragments. At this point a DCI-MS analysis becomes less informative.

[25] P. Dobberstein, E. Korte, G. Meyerhoff, and R. Pesch, *Int. J. Mass Spectrom. Ion Phys.* **46**, 185 (1983).

[26] K. Levsen, K. H. Schafer, and P. Dobberstein, *Biomed. Mass Spectrom.* **11**, 308 (1984).

[27] D. E. Games, M. A. McDowall, K. H. Schafer, P. Dobberstein, and J. L. Grower, *Biomed. Mass Spectrom.* **11**, 87 (1984).

[28] R. Hagemann and K. Jankowski, *Biomed. Mass Spectrom.* **10**, 559 (1983).

[29] H. Virelizier, R. Hagemann, and K. Jankowski, *Biomed. Mass Spectrom.* **10**, 559 (1983).

[30] S. A. Carr and V. N. Reinhold, *Biomed. Mass Spectrom.* **11**, 633 (1984).

[31] S. A. Carr and V. N. Reinhold, *in* "Methods in Protein Sequence Analysis" (M. Elzinga, ed.), p. 263. Humana Press, Clifton, New Jersey, 1982.

[32] D. Dessort, P. Bisseret, Y. Nakatani, G. Ourisson, and M. Lates, *Chem. Phys. Lipids* **33**, 323 (1983).

[33] K. Hostettmann, J. Doumas, and M. Hardy, *Helv. Chim. Acta* **64**, 297 (1981).

[34] K. Harada, S. Ito, N. Takeda, M. Suzuki, and A. Tatematsu, *Biomed. Mass Spectrom.* **10**, 5 (1983).

[35] N. Takeda, K. Harada, M. Suzuki, A. Tatematsu, and I. Sakata, *Biomed. Mass Spectrom.* **11**, 608 (1983).

[35a] G. A. Gabel, C. E. Costello, V. N. Reinhold, L. Kurz, and S. Kornfeld, *J. Biol. Chem.* **259**, 13762 (1984).

[36] R. J. Cotter, *Anal. Chem.* **52**, 1589A (1980).

The precise mechanism of ion generation by DCI remains unclear, but we would like to discuss briefly an attractive model recently discussed for desorption from the condensed phase[37] and integrate the data of others[36,38] and observations by ourselves. Much evidence has been presented to support a process involving evaporation of polar neutrals and preformed ions from heated surfaces.[39-44] These processes appear to fall into two categories: those that are assisted by particle bombardment and those affected by superheating a condensed phase. Both processes cause disruption of the surface and ejection of droplets, clusters, or packets. In the DCI context, this surface disruption would be initiated by substrate heating, inducing bubble formation at the sample–substrate interface (superheating). Bubbles so formed would migrate to the surface and eventually spray droplets into the ion plasma of the CI source to become collisionally declustered and ionized. Mechanisms of a bursting bubble have been shown by holography to be a two-step process. As the bubble reaches the surface its top membrane ruptures spewing forth, in 100 msec, as many as a thousand "film drops," each a thousand times smaller than a raindrop. Next, the bubble turns inside out to form a single peak, which breaks off as a "jet drop." The energy resident with the sample following droplet disintegration and the thermodynamics of ion–molecule interactions are reflected in the resultant spectrum.

Heat desorption of sample from a DCI emitter results in an effective molecular weight fractionation. The distillation can be shown with a series of oligosaccharides by plotting scan number (linear programmed heating) and ion abundance for oligomer MH^+ (Fig. 2). This distillation of individual species presupposes two important factors: little sample interaction with its environment and a finite molecular depth for bubble formation. However, when the thermal energy required to disrupt the interactions between molecules is greater than that of intramolecular bonding, or when distillation continues to monolayers, other explanations must be embraced. As discussed above, it has been our observation that with samples of higher molecular weight and polarity, fragmentation increases

[37] M. L. Vestal, in "Mass Spectrometry Reviews" (G. R. Waller and O. C. Dermer, eds.), Vol. 2, p. 447. Wiley, New York, 1983.
[38] R. N. Stillwell, D. I. Carroll, J. G. Nowlin, and E. C. Horning, Anal. Chem. 55, 1313 (1983).
[39] R. Stoll and F. W. Rollgen, J. Chem. Soc., Chem. Commun. pp. 103, 3943 (1980).
[40] M. Ohashi, R. P. Barron, and W. R. Benson, J. Am. Chem. Soc. 103, 3943 (1981).
[41] R. Stoll and F. W. Rollgen, Org. Mass Spectrom. 16, 72 (1981).
[42] M. Ohashi, R. P. Barron, and W. R. Benson, J. Am. Chem. Soc. 103, 3943 (1981).
[43] R. J. Cotter and A. L. Yergey, Anal. Chem. 53, 1306 (1981).
[44] G. J. Q. van der Peyl, K. Isa, J. Haverkamp, and P. G. Kistemaker, Org. Mass Spectrom. 16, 416 (1981).

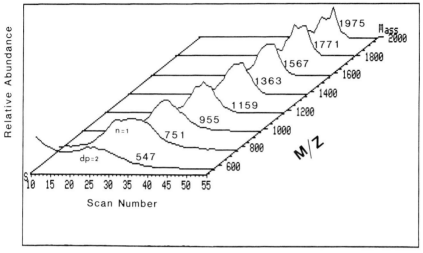

FIG. 2. Ammonia direct chemical ionization of a polysaccharide mixture obtained from corn syrup. Distillation of oligomers with increasing scan number. Elution profiles generated by ion chromatograms of protonated molecular ions for a series of fluorescently labeled materials obtained from corn syrup. Source temperature, 125°; accelerating voltage, 3.0 kV.

with a corresponding decrease in molecular weight-related ions, a result inconsistent with a cluster-only desorption process. This suggests a pyrolytic contribution which may augment the resultant spectra. Considerable effort has been undertaken to diminish DCI-related pyrolysis[21,22,45–51] and this has extended the structural information available for more polar and higher molecular weight samples. It is with these materials that the ion plasma proximity ("direct") is critical and the greater molecular stability

[45] D. F. Hunt, J. Shabanowitz, F. K. Botz, and D. Brent, *Anal. Chem.* **49,** 1160 (1977).
[46] G. Hansen and B. Munson, *Anal. Chem.* **50,** 1130 (1978).
[47] J. P. Thenot, J. Nowlin, D. I. Carroll, E. Montgomery, and E. Horning, *Anal. Chem.* **51,** 1101 (1979).
[48] R. J. Cotter and C. Fensleau, *Biomed. Mass Spectrom.* **6,** 287 (1979).
[49] E. Constantin, Y. Nakatani, G. Ourisson, and R. I. Hueber, *Tetrahedron Lett.* p. 4745 (1980).
[50] P. Bisseret, Y. Nakatani, G. Ourisson, R. Hueber, and G. Teller, *Chem. Phys. Lipids* **33,** 383 (1983).
[51] N. Takeda, K. Harada, M. Suzuki, A. Tatematsu, and I. Sakata, *Mass Spectrosc.* **33,** 59 (1985).

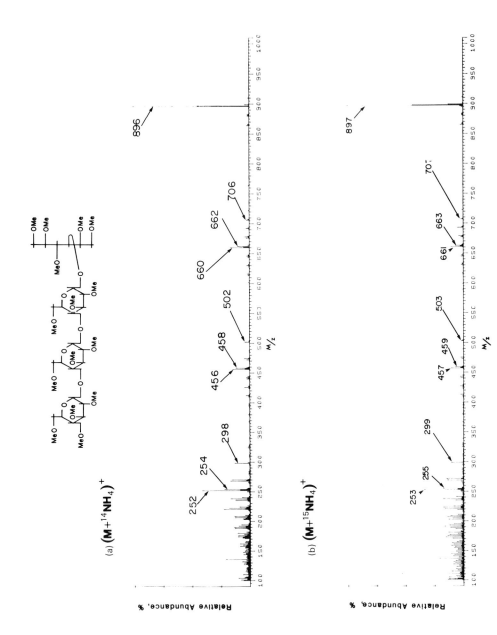

observed in DCI may simply be a more rapid sample ionization and extraction previous to the effects of pyrolytic decomposition.

Analysis of Oligosaccharides

Sequence information obtained by MS is dependent on a periodic polymer fragmentation that is characteristic of monomer array. For oligosaccharides, the major fragments from these processes can be explained by cleavage on either side of the glycosidic oxygen followed by hydrogen transfer. These cleavages are observed under chemical and FAB ionization, as well as for neutral gas collisions. In addition to the ammonia-induced glycosidic cleavage discussed above, the exceptional lability can probably be related to two additional features: that of bond energetics and the fact that the intervening ring rupture requires the cleavage of two bonds for depolymerization. Factors that enhance glycosidic rupture relative to ring fragmentation would be expected to accentuate sequence information. Thus, hydroxyl group blocking by chemical derivatization would be expected to stabilize the ring and thereby provide spectra that show greater glycosidic cleavage. These features are important in thermal processes and are demonstrated below.

This combination of ring stabilization and reagent gas selectivity for glycosidic rupture provides an ideal energy measurement for both sequence and molecular weight information. The principle features and advantages can be illustrated by comparing the spectra in Fig. 3a and b for the analysis of a reduced permethylated tetrasaccharide[10] using $^{14}N^1H_3$ and $^{15}N^1H_3$, respectively, and a corresponding heptasaccharide[24] sample (Fig. 4). Before analyzing these spectra in detail, two major features are worthy of note: (1) the stability of the molecular ion adduct, and (2) the enhanced ion abundance for rupture of each glycosidic linkage (ions differing by 2 Da which appear first at m/z 252/254 and at the glycosidic intervals of 204 Da, i.e., 456/458; 660/662; . . .). As observed in Fig. 3b, all ions shift to one higher mass when using $^{15}N^1H_3$ as the reagent gas. A detailed look at this spectrum indicates that this shift is without exception and one must conclude that two principle reactions are occurring in the CI chamber; i.e., pyrolytic or even-electron species must react with an ammonium ion and any charged, odd-mass fragment must react with ammonia to produce the even-mass array. The clustering of ammonia (ammo-

FIG. 3. Ammonia DCI mass spectrum of reduced permethylated glucose tetrasaccharide. Reagent gas: (a) $^{14}NH_3$; (b) $^{15}NH_3$. Source temperature, 125°; accelerating voltage, 3.0 kV. From Reinhold and Carr.[10] Reprinted with permission from *Mass Spectrom. Rev.* Copyright 1983, John Wiley & Sons, Inc.

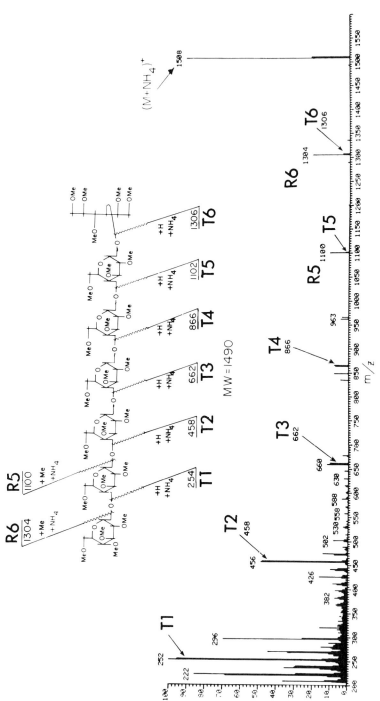

FIG. 4. Ammonia DCI mass spectrum of reduced permethylated glucoheptasaccharide isolated from corn syrup and purified by HPLC. Source temperature, 125°; accelerating voltage, 3.0 kV. From Reinhold and Carr.[24] Reprinted with permission from *Anal. Chem.* Copyright 1982. American Chemical Society.

nia/ammonium) with itself, $(NH_3)_n NH_4^+$, parent ions $(M + NH_4)^+$, and fragments $(F + NH_3)^+$, as well as neutral species, provides a situation in which a large portion of desorbed sample and its fragments are observed. This fact has been illustrated earlier[10] and is in contrast with many other MS techniques in which neutral species are undetected.

The terminal "T" and reducing "R" ends of the oligomer (Fig. 4) are observed to each produce two separate ion series which differ by 14 Da. These fragment ions can be accounted for as hydrogen or methyl transfer products. In Figs. 3 and 4, many of the R and T fragments are isomeric but can be differentiated by altering the mass of the reducing end. Reduction with NAB^2H_4 or tagging with a fluorescent group can serve this purpose and a spectrum of this latter derivative better illustrates the R fragments, (i.e., m/z 547/533 and 343/329, Fig. 5). The T fragments can be explained by cleavage on the reducing side of the glycosidic oxygen, transfer of a hydrogen (or methyl group), and association with an ammonium ion (m/z 254, 458, 662). The major R fragments can also be accounted for by rupture of the glycosidic bond, but on the terminal side of the glycosidic oxygen with transfer of a hydrogen (or a methyl group) and association with an ammonium ion. Depending on the compounds basicity, fragments may occur as protonated or ammonium adduct ions. Thus, for the oligosaccharide in Fig. 5 with a terminally located basic pyridinylamine, all R fragments are protonated while all T fragments occur as ammonium adduct ions.

From these considerations and the study of derivatives of varying structure, a fragmentation scheme can be proposed which accounts for the major ions detected (Scheme 3). Of particular interest is the repeating

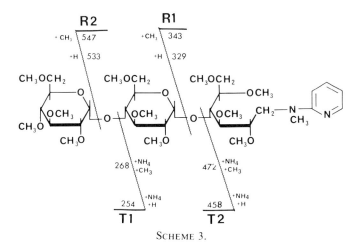

Scheme 3.

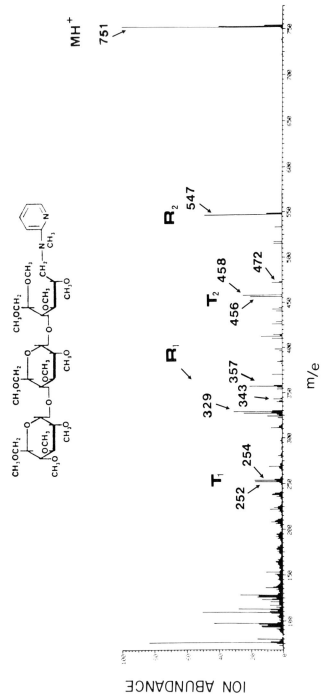

FIG. 5. Ammonia DCI mass spectrum of fluorescently labeled, permethylated glucose trisaccharide. Source temperature, 125°; accelerating voltage, 3.0 kV.

TABLE I
POSSIBLE STRUCTURES BASED ON DIFFERENT MECHANISMS
OF FORMATION

	Structure	m/z using various reagent gases		
		$^{14}N^{1}H_3$	$^{15}N^{1}H_3$	$^{14}N^{2}H_3$
(A)	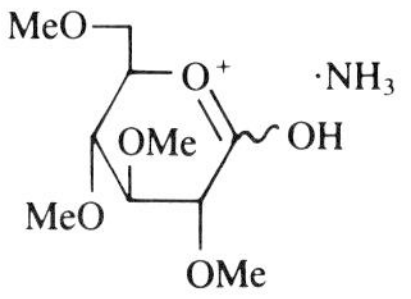 MeO— / O⁺ ·NH₃ / OMe ~OH / MeO / OMe	252	253	256
(B)	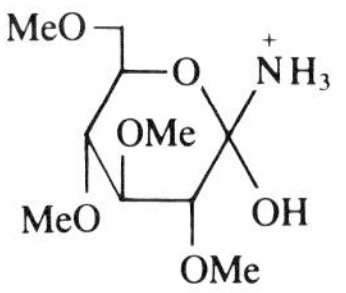 MeO— / O ⁺NH₃ / OMe / MeO OH / OMe	252	253	256
(C)	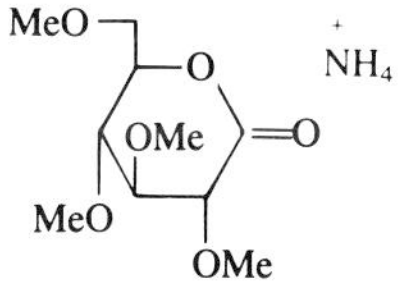 MeO— / O ⁺NH₄ / OMe =O / MeO / OMe	252	253	256

ion doublet (m/z 252/254, 456/458, . . .) that occurs at intervals equal to the intervening glycosidic residue. It is clear that the fragments both originate from the terminal side of the oligomer (Fig. 5) and the structure of the higher mass ion in each doublet is easy to justify. Isotopically labeled ammonia and perdeuteromethylation has shown that the m/z 252 fragment contains one nitrogen, three exchangeable hydrogens, and four methyl groups. Three structures are possible which could result from very different mechanisms of formation (Table I). The first of these structures (**A**) could arise by attachment of neutral ammonia to a CI fragment ion. A second possibility (**B**) could be a product of gas-phase chemical reaction involving covalent bond formation to nitrogen, and a third isomeric structure (**C**) could be derived from ammonium attachment to a neutral fragment as a product of pyrolysis. Although the C-1 linked amine structure (**B**) is not confirmed by these studies it is consistent with the above study and the deuterium work discussed earlier (Fig. 1b, Schemes 1 and 2).

The preceding study of ammonia DCI-MS, using relatively simple materials and stable isotopes, provides a basis for understanding fragmen-

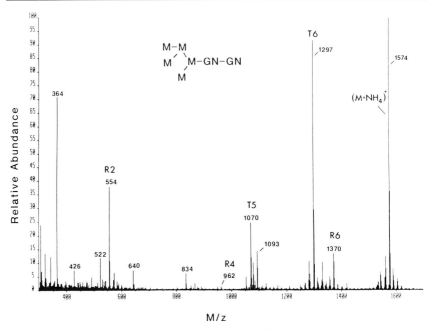

FIG. 6. Ammonia DCI mass spectrum of doubly branched and permethylated Man_5GlcN_2.[52] Source temperature, 125°; accelerating voltage, 3.0 kV.

tation for sequence and molecular weight determinations. In the following examples this understanding is expanded to N-linked oligosaccharides[35a,52] and glycolipids[30] of increasing structural complexity. These amide-containing samples are sufficiently basic to occur as protonated species at low ammonia reagent gas pressures; however, when ion sources are designed "tighter" (fewer leaks) this allows normal instrument operation with higher internal CI chamber pressures, thus pushing the equilibrium in the direction of adduct formation. Under these conditions, amides occur as ammonium adducts (Figs. 6 and 7), while amine samples with greater basicity (Fig. 5) occur as protonated molecular ions.

Ammonia DCI-MS of Man_5GlcN_2 using stable isotopes indicated the molecular weight to be 1556 Da, m/z 1574, $(M + NH_4)^+$. The fragment, m/z 1297 (Fig. 6), did not change when the sample was reduced or prepared as the fluorescent derivative, indicating it to be derived from the terminal end of the oligomer, T6. The ion abundance of this fragment suggests a

[52] S. Santikarn, V. N. Reinhold, C. Warren, R. Trimble, and R. Jeanloz, *Proc. Int. Symp. Glycoconjugates, 8th, 1985* Abstract, p. 95 (1985).

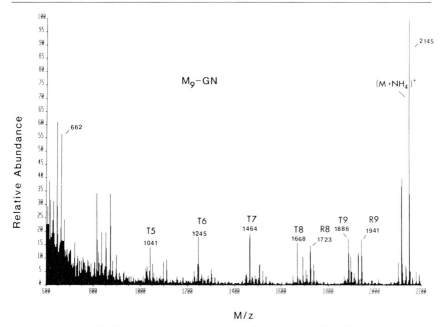

FIG. 7. Ammonia DCI mass spectrum of permethylated Man₉GlcN.[52] Source temperature, 125°; accelerating voltage, 3.0 kV.

facile elimination of the reducing terminal glycosamine residue. The doubly branched structure drawn on the figure can be rationalized from the presence of R6, T6, T5, R2 fragments and the absence of R5 and R3 fragments. These same T and R fragments can be observed on the higher homolog Man₉GlcN (Fig. 7) and a phosphorylated homolog Man₆₋₉GlcN (Fig. 8), but these materials are contaminated with structural homologs and close interpretation of these fragments for branching information may be misleading.

In addition to the sequence and molecular weight information provided by the ammonia DCI-MS technique, a methylated phosphorylmannosyl oligosaccharide sample posed an additional problem for locating an indigenous methyl group. This problem was sorted out using the techniques of oligomer perdeuteromethylation and ammonia DCI-MS.[35a] The results are summarized below. Composition analysis, HPLC, and FAB-MS had shown the sample to be a mixture of four major homopolymers related by incrementing residues of mannose, the composition consistent with $(GlcNAc)_1$ $(Man)_{6-9}$ (phosphate)$_1$. It was also determined that the phosphate group was in diester linkage and a molecular weight determination by fast atom bombardment mass spectrometry indicated this to be a

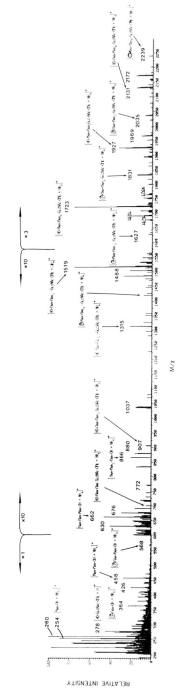

FIG. 8. Ammonia DCI mass spectrum of permethylated methylphosphomannosyl oligosaccharide. Source temperature, 125°; accelerating voltage, 3.0 kV. From Gabel et al.[35a] Copyright 1984, American Society of Biological Chemists.

methyl group of unknown location. Permethylation and ammonia DCI-MS (Fig. 8) supported earlier findings and provided a way to localize the point of methylation. This was realized from the spectra which showed a facile loss of 108 Da ($Me_2O_2PO^-$) from each of the molecular weight-related oligomers, indicating lability of the phosphate group (e.g., m/z 2239 to 2131; m/z 2035 to 1927; m/z 1831 to 1723). Perdeuteromethylating the sample and noting the mass shifts for phosphate loss (111 observed and not 114) indicated that the indigenous alkyl group was attached to the phosphate residue and not directly to a carbohydrate residue.

Analysis of Glycosphingolipids by DCI-MS

The application of ammonia DCI to glycolipid materials has proved equally successful by providing molecular weight, sequence, and lipid ester detail in a single analysis.[30] This can be illustrated in Fig. 9 for a permethylated and reduced ceramide trihexoside. Reduction of this amide fatty acid conjugate previous to analysis[53] enhances DCI elimination of the methylated alkyl amine. In addition, these fragments, which are observed at the low mass end of the spectrum, provide a separate evaluation of the fatty ester composition (i.e., m/z 342, 368, 370, 398, 400). The opportunity to compare these two groups of ions is of considerable value especially for relating structural features to the sphingosine moiety. Carbohydrate sequence ions can be observed which are identical to those in Fig. 3 (e.g., m/z 254, 458, 662) and represent the major fragments in the spectra. All of these fragments are ammonium adduct ions. A summary of the major fragments detected and their structural origin is presented in Scheme 4. Variations in structure are readily detected by DCI as exemplified below with a related sample. Extension of the carbohydrate chain by an amino sugar moiety (globotetraglycosylceramide, GL-4) provides the same sequence information with the appropriate shift in mass. Because of the basicity of this amino sugar residue, the T fragments (m/z 266, 470, 674, 878) appear as protonated fragments and not as ammonium adduct ions as in Figs. 3 and 4. However, the spectral information is not compromised and a complete structural assignment is still very easy (Fig. 10).

Ammonia DCI-MS of permethylated and reduced gangliosides can also provide considerable structural information. For ganglioside G_{M1} reduction converts the carboxyl group of NANA to a primary alcohol, in addition to the conversion of N-methylamides to tertiary amines. DCI mass spectral analysis of the resulting material (Fig. 11) follows the over-

---

[53] K.-A. Karlsson, I. Pascher, W. Pimlott, and B. E. Samuelsson, *Biomed. Mass Spectrom.* **1**, 49 (1974).

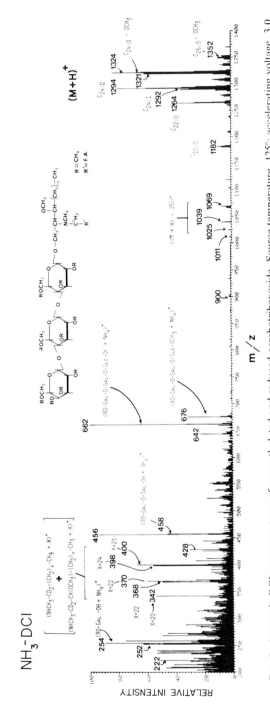

FIG. 9. Ammonia DCI mass spectrum of permethylated and reduced cerebotrihexoside. Source temperature. 125°; accelerating voltage. 3.0 kV. From Carr and Reinhold.[30] Reprinted with permission from *Biomed. Mass Spectrom.* Copyright 1984. Wiley-Haden Ltd.

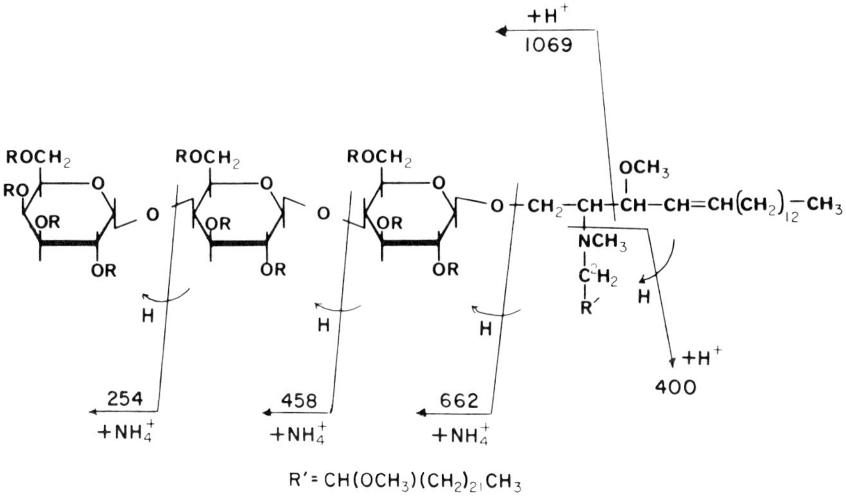

SCHEME 4.

all fragmentation patterns observed for GL-4 and G_{M3}. The protonated molecular ions at m/z 1736, 1764, 1766, 1790, and 1792 represent homologs of G_{M1} with varying fatty acid composition. These structural elements can be confirmed by observation of the alkylamine fragments at m/z 286 and 314. This is consistent with the major fatty acids being the $C_{18:0}$ and $C_{20:0}$ homologs. Lesser amounts of $C_{16:0}$, monohydroxy $C_{16:0}$ and $C_{18:1}$ can be detected at m/z 258, 288, and 312. The composition and sequence of the carbohydrate residues can be observed (m/z 254, 470) for the terminal Glc-GalN sequence, the Glc-ceramide sequence (m/z 786, 814), and the fragments related to the total carbohydrate chain (m/z 1185, 1215, 1229). The presence of a sialyl moiety is confirmed by the facile loss of this residue from the molecular ion (m/z 1397, 1427, 1455) and the occurrence of the protonated species (m/z 338).

Summary

    The mass spectra obtained under ammonia DCI conditions are quite distinct from the results obtained by either conventional EI or CI mass spectrometry. Abundant $(M + H)^+$ or $(M + NH_4)^+$ ions and sequence-specific oligosaccharides fragments are characteristic for permethylated glycoconjugates. Polar materials must be derivatized for DCI-MS and permethylation appears most satisfactory. Although this provides, in

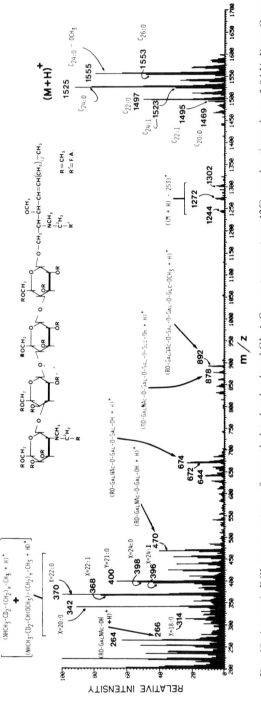

Fig. 10. Ammonia DCI mass spectrum of permethylated and reduced GL-4. Source temperature, 125°; accelerating voltage, 3.0 kV. From Carr and Reinhold.[30] Reprinted with permission from *Biomed. Mass Spectrom.* Copyright 1984, Wiley-Haden Ltd.

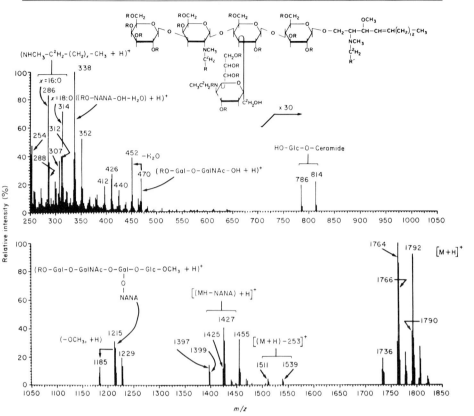

FIG. 11. Ammonia DCI mass spectrum of permethylated and reduced ganglioside G_{M1}. Source temperature, 125°; accelerating voltage, 3.0 kV. From Carr and Reinhold.[30] Reprinted with permission from *Biomed. Mass Spectrom.* Copyright 1984, Wiley-Haden Ltd.

many cases, a burdensome problem, these procedures are fully consistent with saccharide linkage analysis using GC-MS.

Preliminary evidence has been presented that indicates gas-phase ammonolysis may enhance oligosaccharide sequence information and produce amine-linked products. When compared with fast atom bombardment ionization, DCI shows greater sensitivity, more specific sequence fragmentation, and better overall signal-to-noise measurements. We have used ammonia DCI-MS for oligosaccharide materials beyond the 3000-Da range with less than satisfying results; thus this range may represent the upper limit of analysis under the current conditions due to increasing pyrolysis. It may be that this mass range could be extended by an alternative desorption technique. Interestingly, fast atom bombardment has shown marked improvement in sensitivity by desorption into a CI plasma.[54]

Acknowledgments

The author would like to express his appreciation to Dr. Sitthivet Santikarn for sample analysis and discussions relevant to this article and to the National Science Foundation and the National Institutes of Health for financial support.

[54] R. B. Freas, M. M. Ross, and J. E. Campana, *J. Am. Chem. Soc.* **107,** 6195 (1985).

[6] BioGel P-4 Column Chromatography of Oligosaccharides: Effective Size of Oligosaccharides Expressed in Glucose Units

By Akira Kobata, Katsuko Yamashita, and Seiichi Takasaki

Since we have reported the method to analyze oligosaccharides by BioGel P-4 (<400 mesh) column chromatography, it has been used as an indispensable tool to obtain sugar patterns of plasma membrane glycoproteins as well as purified glycoproteins. In a previous paper of this series (K. Yamashita, T. Mizuochi, and A. Kobata, Vol. 83 [6]), we have included the cumulative data of the effective sizes of sugars as expressed in glucose units. Since then, the values of many additional oligosaccharides have been reported. For the convenience of the researchers of glycoproteins, these data will be summarized in Table I.

TABLE I

EFFECTIVE SIZE OF SUGARS[a]

Residue	Glucose units	Reference[a]
Manα1 (3) 6 Manβ1→4GlcNAcβ1→4GlcNAc-OH GlcNAcβ1→4Manα1 (6) 3	9.0	1
Fucα1 ↓ 6 Manα1 6 Manβ1→4GlcNAcβ1→4GlcNAc-OH GlcNAcβ1→4Manα1 3	10.0	2

TABLE I (*continued*)

Residue	Glucose units	Reference[a]

```
                GlcNAcβ1
                   ↓
              Manα1    4
                 ⁶⧹  Manβ1→4GlcNAcβ1→4GlcNAc
        GlcNAcβ1→2Manα1 ³⧸                    OH
```
Glucose units: 10.2   Reference: 3

```
Galβ1→4GlcNAcβ1→4Manα1 (3)
                      ⁶⧹  Manβ1→4GlcNAcβ1→4GlcNAc
                Manα1 ³⧸                        OH
                     (6)
```
Glucose units: 10.2   Reference: 3

```
                GlcNAcβ1
                   ↓
        GlcNAcβ1→2Manα1 4
                     ⁶⧹  Manβ1→4GlcNAcβ1→4GlcNAc
                Manα1 ³⧸                        OH
```
Glucose units: 10.4   Reference: 3

```
        GlcNAcβ1
              ⁶⧹Manα1
        GlcNAcβ1 ²⧸   ⁶⧹  Manβ1→4GlcNAcβ1→4GlcNAc
                Manα1 ³⧸                        OH
```
Glucose units: 10.9   Reference: 4

```
          Manα1→3Manα1
                      ⁶⧹  Manβ1→4GlcNAcβ1→4GlcNAc
Galβ1→4GlcNAcβ1→4Manα1 ³⧸                        OT
```
Glucose units: 10.9   Reference: 5

```
        GlcNAcβ1              Fucα1
            ↓                   ↓
  GlcNAcβ1→2Manα1  4            6
               ⁶⧹  Manβ1→4GlcNAcβ1→4GlcNAc
          Manα1 ³⧸                        OH
```
Glucose units: 11.0   Reference: 6

```
                            Fucα1
                             ↓
    GlcNAcβ1                 6
           ⁶⧹Manα1
    GlcNAcβ1 ²⧸   ⁶⧹  Manβ1→4GlcNAcβ1→4GlcNAc
          Manα1 ³⧸                        OH
```
Glucose units: 11.6   Reference: 2

```
                GlcNAcβ1
                   ↓
        GlcNAcβ1→2Manα1  4
                     ⁶⧹  Manβ1→4GlcNAcβ1→4GlcNAc
        GlcNAcβ1→4Manα1 ³⧸                        OH
```
Glucose units: 11.7   Reference: 6

```
            Manα1
                ⁶⧹Manα1
            Manα1 ³⧸   ⁶⧹  Manβ1→4GlcNAcβ1→4GlcNAc
Galβ1→4GlcNAcβ1→4Manα1 ³⧸                        OH
```
Glucose units: 11.8   Reference: 5

```
                GlcNAcβ1
                   ↓
              Manα1    4
          ⁴⧹     ⁶⧹  Manβ1→4GlcNAcβ1→4GlcNAc
  GlcNAcβ1   Manα1 ³⧸                        OH
  GlcNAcβ1 ²⧸
```
Glucose units: 12.0   Reference: 3

```
    GlcNAcβ1
           ⁶⧹Manα1  (3)
    GlcNAcβ1 ²⧸   ⁶⧹  Manβ1→4GlcNAcβ1→4GlcNAc
    GlcNAcβ1→4Manα1 ³⧸                        OH
                    (6)
```
Glucose units: 12.3   Reference: 1

```
    GlcNAcβ1
           ⁶⧹Manα1
    GlcNAcβ1 ²⧸   ⁶⧹  Manβ1→4GlcNAcβ1→4GlcNAc
    GlcNAcβ1→2Manα1 ³⧸                        OH
```
Glucose units: 12.5   Reference: 7

```
        GlcNAcβ1              Fucα1
            ↓                   ↓
  GlcNAcβ1→2Manα1  4            6
               ⁶⧹  Manβ1→4GlcNAcβ1→4GlcNAc
  GlcNAcβ1→2Manα1 ³⧸                        OH
```
Glucose units: 12.7   Reference: 6

TABLE I (*continued*)

Residue	Glucose units	Reference[a]

```
             GlcNAcβ1              Fucα1
                ↓                    ↓
GlcNAcβ1→2Manα1  4                   6
              ↘6                                          12.8      6
                Manβ1→4GlcNAcβ1→4GlcNAc
              ↗3                                  OH
GlcNAcβ1→4Manα1


             GlcNAcβ1
                ↓
Galβ1→4GlcNAcβ1→2Manα1 (3) 4
                     ↘6                                   12.8      8
                       Manβ1→4GlcNAcβ1→4GlcNAc
                     ↗3                           OH
     GlcNAcβ1→2Manα1 (6)


             GlcNAcβ1              Fucα1
                ↓                    ↓
GlcNAcβ1→2Manα1  4                   6
              ↘6                                          12.8      8
                Manβ1→4GlcNAcβ1→4GlcNAc
              ↗3                                  OH
GlcNAcβ1→2Manα1


             GlcNAcβ1
                ↓
  Manα1→3Manα1  4
              ↘6                                          12.9      5
                Manβ1→4GlcNAcβ1→4GlcNAc
              ↗3                                  OH
Galβ1→4GlcNAcβ1→2Manα1


GlcNAcβ1→2Manα1
               ↘6
GlcNAcβ1          Manα1                                   13.1      9
        ↘4     ↗3 Manβ1→4GlcNAcβ1→4GlcNAc
          Manα1                                  OH
GlcNAcβ1↗2


                                Fucα1
                                  ↓
GlcNAcβ1                          6
        ↘6
          Manα1 (3)
GlcNAcβ1↗2    ↘6                                          13.3      1
                Manβ1→4GlcNAcβ1→4GlcNAc
GlcNAcβ1→4Manα1↗3 (6)             OH


             GlcNAcβ1              Fucα1
                ↓                    ↓
GlcNAcβ1→2Manα1 (3) 4                6
              ↘6
                Manβ1→4GlcNAcβ1→4GlcNAc                   13.3      10
              ↗3 (6)                             OH
Galβ1→4GlcNAcβ1→2Manα1
                     3
                     ↑
                  Fucα1


                                Fucα1
                                  ↓
GlcNAcβ1→2Manα1 (3)               6
              ↘6
                Manβ1→4GlcNAcβ1→4GlcNAc                   13.3      11
              ↗3 (6)                             OH
Galβ1→4GlcNAcβ1→2Manα1


             GlcNAcβ1
                ↓
GlcNAcβ1→2Manα1  4
              ↘6                                          13.4      6
GlcNAcβ1         Manα1↗3 Manβ1→4GlcNAcβ1→4GlcNAc
        ↘2                                       OH
          Manα1
GlcNAcβ1↗4


                  Manα1
Galβ1→4GlcNAcβ1       ↘6
              ↘4   Manα1 Manβ1→4GlcNAcβ1→4GlcNAc          13.5      12
                ↗2 ↗3                           OH
Galβ1→4GlcNAcβ1


             GlcNAcβ1
                ↓
  Manα1→3Manα1  4
Galβ1→4GlcNAcβ1    ↘6                                     13.6      5
              ↘4  Manα1 Manβ1→4GlcNAcβ1→4GlcNAc
                ↗2 ↗3                           OH
GlcNAcβ1


             GlcNAcβ1
                ↓
Galβ1→4GlcNAcβ1→2Manα1  4
                     ↘6                                   14.0      8
                       Manβ1→3GlcNAcβ1→4GlcNAc
                     ↗3                           OH
Galβ1→4GlcNAcβ1→2Manα1
```

TABLE I (*continued*)

Residue	Glucose units	Reference[a]
GlcNAcβ1 Fucα1 ↓ ↓ Galβ1→4GlcNAcβ1→2Manα1↘(3) 4 6 6Manβ1→4GlcNAcβ1→4GlcNAc$_{OH}$ GlcNAcβ1→2Manα1↗3 (6)	14.0	8
GlcNAcβ1↘6 Manα1↘ GlcNAcβ1↗2 6 Manβ1→4GlcNAcβ1→4GlcNAc$_{OH}$ GlcNAcβ1↘4 3 Manα1↗ GlcNAcβ1↗2	14.2	13
GlcNAcβ1 ↓ GlcNAcβ1→2Manα1↘(3) 4 6Manβ1→4GlcNAcβ1→4GlcNAc$_{OH}$ Galβ1→4GlcNAcβ1→2Manα1↗3 3 (6) ↑ Fucα1	14.3	10
Manα1↘6 GlcNAcβ1 Manα1↘ ↓ Manα1↗3 6 4 Galβ1→4GlcNAcβ1↘4 3Manβ1→4GlcNAcβ1→4GlcNAc$_{OH}$ Manα1↗ GlcNAcβ1↗2	14.3	5
GlcNAcβ1 Fucα1 GlcNAcβ1↘6 ↓ ↓ Manα1↘6 4 6 GlcNAcβ1↗2 Manβ1→4GlcNAcβ1→4GlcNAc$_{OH}$ GlcNAcβ1→4Manα1↗3	14.4	6
GlcNAcβ1 Fucα1 ↓ ↓ GlcNAcβ1→2Manα1↘6 4 6 GlcNAcβ1↘2 Manβ1→4GlcNAcβ1→4GlcNAc$_{OH}$ Manα1↗3 GlcNAcβ1↗4	14.4	6
GlcNAcβ1↘6 GlcNAcβ1 Manα1↘ ↓ GlcNAcβ1↗2 6 4 GlcNAcβ1↘4 Manβ1→4GlcNAcβ1→4GlcNAc$_{OH}$ Manα1↗3 GlcNAcβ1↗2	14.5	3
Fucα1 ↓ 6 Manα1↘6 Galβ1→4GlcNAcβ1↘4 Manβ1→4GlcNAcβ1→4GlcNAc$_{OH}$ Manα1↗3 Galβ1→4GlcNAcβ1↗2	14.5	12
GlcNAcβ1 ↓ GlcNAcβ1→2Manα1↘6 4 Galβ1→4 { GlcNAcβ1↘4 Manβ1→4GlcNAcβ1→4GlcNAc$_{OH}$ Manα1↗3 GlcNAcβ1↗2	14.6	8
Fucα1 ↓ GlcNAcβ1 3 ↓ Galβ1→4GlcNAcβ1→2Manα1↘6 4 Manβ1→4GlcNAcβ1→4GlcNAc$_{OH}$ Galβ1→4GlcNAcβ1→2Manα1↗3 3 ↑ Fucα1	15.1	10

(*continued*)

TABLE I (*continued*)

Residue	Glucose units	Reference[a]
GlcNAcβ1↘6 Manα1(3) ↓GlcNAcβ1↗2 ↘4 / GlcNAcβ1↘4 Manα1(6)↗3 ... GlcNAcβ1 ↓4 / Fucα1↓6 / Manβ1→4GlcNAcβ1→4GlcNAc_OH	15.4	6
Galβ1→4GlcNAcβ1→Manα1 ↘6 (3)↓4 / Galβ1→4GlcNAcβ1→Manα1↗3 (6) / GlcNAcβ1 / Fucα1 / Manβ1→4GlcNAcβ1→4GlcNAc_OH	15.5	10
Galβ1→4 { GlcNAcβ1↘4 GlcNAcβ1↗2 Manα1 ↘6 / GlcNAcβ1↘4 GlcNAcβ1↗2 Manα1↗3 / GlcNAcβ1↓4 / Manβ1→4GlcNAcβ1→4GlcNAc_OH	15.5	3
Galβ1→4 { GlcNAcβ1↘6 GlcNAcβ1↗2 Manα1 ↘6 / GlcNAcβ1↘4 GlcNAcβ1↗2 Manα1↗3 / GlcNAcβ1↓4 / Manβ1→4GlcNAcβ1→4GlcNAc_OH	15.5	3
GlcNAcβ1↘6 Manα1 / GlcNAcβ1→4Manα1↗2 ↘6 / GlcNAcβ1↘4 GlcNAcβ1↗2 Manα1↗3 / GlcNAcβ1↓4 / Manβ1→4GlcNAcβ1→4GlcNAc_OH	15.5	14
(Galβ1→4)₂ { GlcNAcβ1→2Manα1 ↘6 / GlcNAcβ1↘4 GlcNAcβ1↗2 Manα1↗3 / Fucα1↓6 / Manβ1→4GlcNAcβ1→4GlcNAc_OH	15.6	15
Galβ1→4 { GlcNAcβ1→2Manα1 ↘6 / GlcNAcβ1↘4 GlcNAcβ1↗2 Manα1↗3 / GlcNAcβ1↓4 / Fucα1↓6 / Manβ1→4GlcNAcβ1→4GlcNAc_OH	15.7	8
(Galβ1→4)₂ { GlcNAcβ1→2Manα1 ↘6 / GlcNAcβ1↘4 GlcNAcβ1↗2 Manα1↗3 / GlcNAcβ1↓4 / Manβ1→4GlcNAcβ1→4GlcNAc_OH	15.7	8
Fucα1↓3 / Galβ1→4GlcNAcβ1→2Manα1 ↘6 / Galβ1→4GlcNAcβ1→2Manα1↗3 / GlcNAcβ1↓4 / Fucα1↓6 / Manβ1→4GlcNAcβ1→4GlcNAc_OH	15.7	16
Galβ1→4GlcNAcβ1↘6 Manα1 / Galβ1→4GlcNAcβ1↗2 ↘6 / Galβ1→4GlcNAcβ1→2Manα1↗3 / Manβ1→4GlcNAcβ1→4GlcNAc_OH	16.0	4

TABLE I (*continued*)

Residue	Glucose units	Refer-ence[a]

```
    Fucα1
      ↓
      3              GlcNAcβ1              Fucα1
Galβ1→4GlcNAcβ1→2Manα1      ↓                  ↓
                        6   4                  6
Galβ1→4GlcNAcβ1→2Manα1   Manβ1→4GlcNAcβ1→4GlcNAc_OH          16.0     10
      3
      ↑
    Fucα1
```

```
                                              Fucα1
           ⎛ GlcNAcβ1→2Manα1                      ↓
           ⎜ GlcNAcβ1                             6
(Galβ1→4)₂ ⎨        4Manα1   6                                    16.1     15
           ⎜ GlcNAcβ1    2  Manβ1→4GlcNAcβ1→4GlcNAc_OH
           ⎝                3
```

```
                                              Fucα1
           ⎛ GlcNAcβ1                              ↓
           ⎜        6Manα1                         6
Galβ1→4    ⎨ GlcNAcβ1   2   6                                     16.3     15
           ⎜ GlcNAcβ1       Manβ1→4GlcNAcβ1→4GlcNAc_OH
           ⎜        4Manα1 3
           ⎝ GlcNAcβ1   2
```

```
           GlcNAcβ1    GlcNAcβ1
                   6        ↓
Galβ1→4GlcNAcβ1→4Manα1      4
           GlcNAcβ1   2  6                                        16.6      3
           GlcNAcβ1      Manβ1→4GlcNAcβ1→4GlcNAc_OH
                   4Manα1 3
           GlcNAcβ1   2
```

```
           GlcNAcβ1    GlcNAcβ1
                   6        ↓
           GlcNAcβ1→4Manα1  4
           GlcNAcβ1   2  6                                        16.6      3
Galβ1→4GlcNAcβ1      Manβ1→4GlcNAcβ1→4GlcNAc_OH
                   4Manα1 3
           GlcNAcβ1   2
```

```
           ⎛ GlcNAcβ1    GlcNAcβ1
           ⎜         6Manα1  ↓
(Galβ1→4)₂ ⎨ GlcNAcβ1   2    4                                    16.6      8
           ⎜ GlcNAcβ1      6 Manβ1→4GlcNAcβ1→4GlcNAc_OH
           ⎝         4Manα1 3
             GlcNAcβ1   2
```

```
                        GlcNAcβ1        Fucα1
           ⎛ GlcNAcβ1→2Manα1   ↓            ↓
           ⎜ GlcNAcβ1          4            6
(Galβ1→4)₃ ⎨        4Manα1   6                                    16.7      8
           ⎝ GlcNAcβ1    2  Manβ1→4GlcNAcβ1→4GlcNAc_OH
                            3
```

```
                        GlcNAcβ1
                           ↓
Galβ1→4GlcNAcβ1→2Manα1      4
Galβ1→4GlcNAcβ1        6                                          16.8      6
                   4Manα1 Manβ1→4GlcNAcβ1→4GlcNAc_OH
Galβ1→4GlcNAcβ1    2     3
```

```
                                              Fucα1
                                                 ↓
Galβ1→4GlcNAcβ1                                   6
                   6Manα1                                         16.9     15
Galβ1→4GlcNAcβ1    2     6
Galβ1→4GlcNAcβ1→2Manα1   Manβ1→4GlcNAcβ1→4GlcNAc_OH
                        3
```

(*continued*)

TABLE I (*continued*)

Residue	Glucose units	Refer-ence[a]

Structures with Glucose units and References:

```
                                                      Fucα1
Galβ1→4GlcNAcβ1                                         ↓
             6Manα1                                     6
Galβ1→4GlcNAcβ1 2      6Manβ1→4GlcNAcβ1→4GlcNAc          17.1    15
Galβ1→4GlcNAcβ1→2Manα1 3                        OH
```

```
             GlcNAcβ1
                 ↓
Galβ1→4GlcNAcβ1→2Manα1  4
                      6Manβ1→4GlcNAcβ1→4GlcNAc           17.0    8
Galβ1→4GlcNAcβ1 4Manα1 3                        OH
             2
Galβ1→4GlcNAcβ1
```

```
         GlcNAcβ1                              Fucα1
                6Manα1                          ↓
         GlcNAcβ1 2                             6
(Galβ1→4)2 GlcNAcβ1    6Manβ1→4GlcNAcβ1→4GlcNAc  17.2    15
         GlcNAcβ1 4Manα1 3                OH
                 2
```

```
Galβ1→4GlcNAcβ1→2Manα1
Galβ1→4GlcNAcβ1        6Manα1→4GlcNAcβ1→4GlcNAc          17.3    13
     3          4Manα1 3                     OH
     ↑          2
  Fucα1
Galβ1→4GlcNAcβ1
```

```
                                          Fucα1
                                           ↓
Galβ1→4GlcNAcβ1→2Manα1                      6
Galβ1→4GlcNAcβ1     6Manβ1→4GlcNAcβ1→4GlcNAc 17.5    12
             4Manα1 3                   OH
             2
Galβ1→4GlcNAcβ1
```

```
         GlcNAcβ1  GlcNAcβ1
                6Manα1  ↓
         GlcNAcβ1 2     4
(Galβ1→4)3 GlcNAcβ1    6Manβ1→4GlcNAcβ1→4GlcNAc  17.5    8
         GlcNAcβ1 4Manα1 3                OH
                 2
```

```
         GlcNAcβ1  GlcNAcβ1      Fucα1
                6Manα1  ↓         ↓
         GlcNAcβ1 2     4         6
(Galβ1→4)2 GlcNAcβ1    6Manβ1→4GlcNAcβ1→4GlcNAc  17.5    8
         GlcNAcβ1 4Manα1 3                OH
                 2
```

```
         GlcNAcβ1  GlcNAcβ1
                6    ↓
Galβ1→4GlcNAcβ1→4Manα1  4
         GlcNAcβ1 2     6Manβ1→4GlcNAcβ1→4GlcNAc  17.6    3
Galβ1→4GlcNAcβ1 4Manα1 3                   OH
             2
         GlcNAcβ1
```

```
             GlcNAcβ1      Fucα1
                 ↓          ↓
Galβ1→4GlcNAcβ1→2Manα1  4   6
Galβ1→4GlcNAcβ1     6Manβ1→4GlcNAcβ1→4GlcNAc  17.8    6
             4Manα1 3                   OH
             2
Galβ1→4GlcNAcβ1
```

```
                                    Fucα1
                                     ↓
Galβ1→4GlcNAcβ1→2Manα1 (3)           6
Galβ1→4GlcNAcβ1     6Manβ1→4GlcNAcβ1→4GlcNAc  17.8    16
     3          4Manα1 3                 OH
     ↑          2   (6)
  Fucα1
Galβ1→4GlcNAcβ1
```

TABLE I (*continued*)

Residue	Glucose units	Reference[a]

```
                        GlcNAcβ1              Fucα1
                           ↓                    ↓
(Galβ1→4GlcNAcβ1→)₁ ⎰Manα1    4                6
                    ⎱     ⁶₃Manβ1→4GlcNAcβ1→4GlcNAc
(Galβ1→4GlcNAcβ1→)₂ ⎱Manα1                              OH
                    3
                    ↑
                  Fucα1
```

18.0   10

```
                          GlcNAcβ1           Fucα1
                          (3)↓                 ↓
Galβ1→4GlcNAcβ1→2Manα1    4                  6
Galβ1→4GlcNAcβ1    ⁶₃Manβ1→4GlcNAcβ1→4GlcNAc
              3    ⁴Manα1 (6)                    OH
              ↑  ₂
            Fucα1
Galβ1→4GlcNAcβ1
```

18.2   16

```
        ⎡GlcNAcβ1                        Fucα1
        ⎢      ⁶Manα1                      ↓
(Galβ1→4)₃⎢GlcNAcβ1 ²                      6
        ⎢GlcNAcβ1    ⁶₃Manβ1→4GlcNAcβ1→4GlcNAc
        ⎣GlcNAcβ1 ⁴Manα1²                      OH
```

18.3   15

```
                        GlcNAcβ1
                           ↓
(Galβ1→4GlcNAcβ1→)₂ ⎰Manα1    4
                    ⎱     ⁶₃Manβ1→4GlcNAcβ1→4GlcNAc
(Galβ1→4GlcNAcβ1→)₁ ⎱Manα1                         OH
                    3
                    ↑
                  Fucα1
```

18.3   10

```
Galβ1→4GlcNAcβ1  GlcNAcβ1
              ⁶Manα1  ↓
Galβ1→4GlcNAcβ1²    4
Galβ1→4GlcNAcβ1    ⁶₃Manβ1→4GlcNAcβ1→4GlcNAc
              ⁴Manα1                        OH
Galβ1→4GlcNAcβ1²
```

18.5   8

```
        ⎡GlcNAcβ1   GlcNAcβ1        Fucα1
        ⎢      ⁶Manα1  ↓              ↓
(Galβ1→4)₃⎢GlcNAcβ1²    4              6
        ⎢GlcNAcβ1    ⁶₃Manβ1→4GlcNAcβ1→4GlcNAc
        ⎣GlcNAcβ1 ⁴Manα1²                  OH
```

18.5   8

```
Galβ1→4GlcNAcβ1
              ⁶Manα1
Galβ1→4GlcNAcβ1²
Galβ1→4GlcNAcβ1    ⁶Manβ1→4GlcNAcβ1→4GlcNAc
              ⁴Manα1³                       OH
Galβ1→4GlcNAcβ1²
```

18.6   13

```
              Fucα1
               ↓
               3
Galβ1→4GlcNAcβ1→2Manα1    1βGlcNAc   Fucα1
     Fucα1             ↓         ↓
      ↓            ⁶  4          6
      3            ⁵₃Manβ1→4GlcNAcβ1→4GlcNAc
Galβ1→4GlcNAcβ1 ⁴Manα1                    OH
Galβ1→4GlcNAcβ1²
             3
             ↑
           Fucα1
```

18.7   10

(*continued*)

TABLE I (*continued*)

Residue	Glucose units	Refer- ence[a]

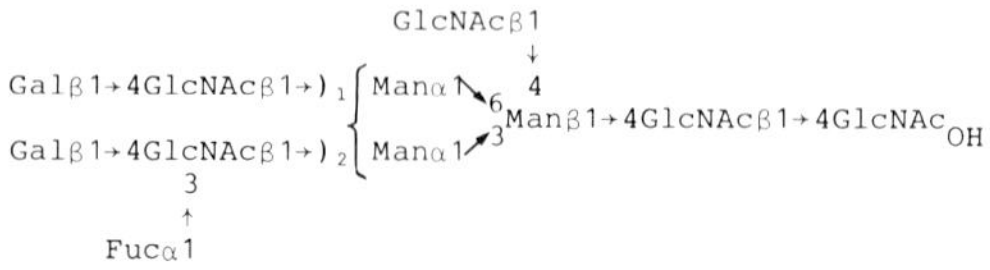

The structures in the Residue column, with their associated values:

19.0 — 10

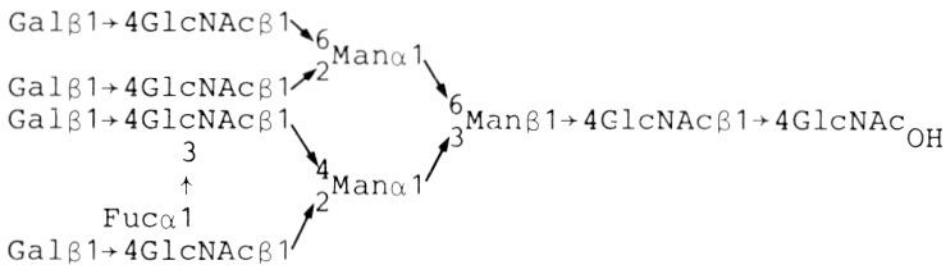

19.3 — 13

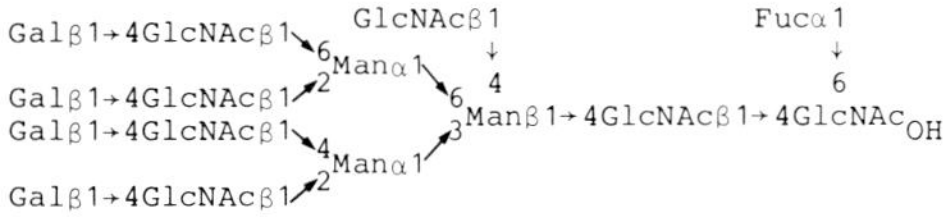

19.4 — 6

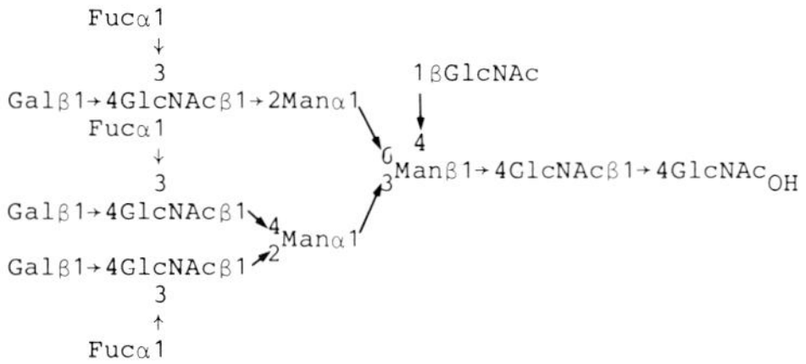

19.5 — 10

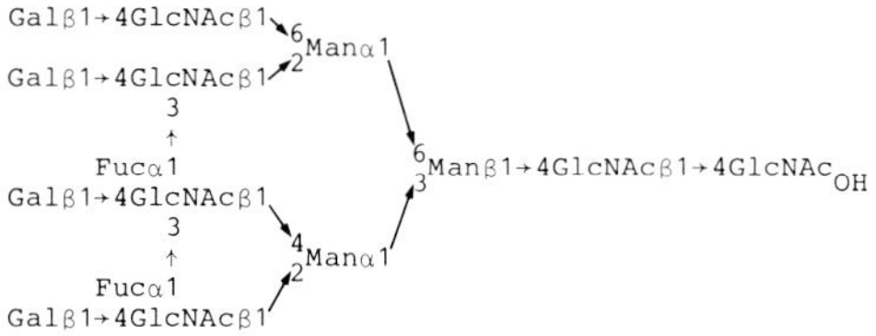

20.0 — 13

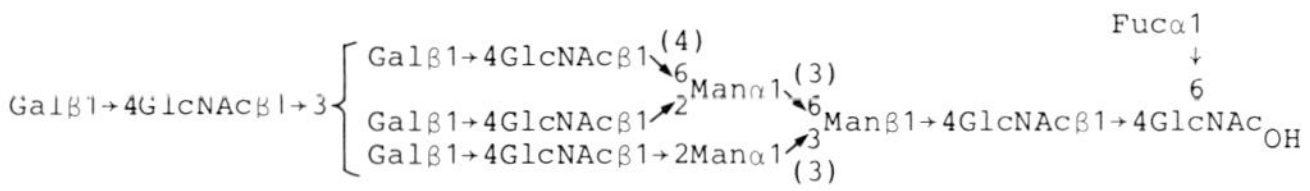

20.0 — 15

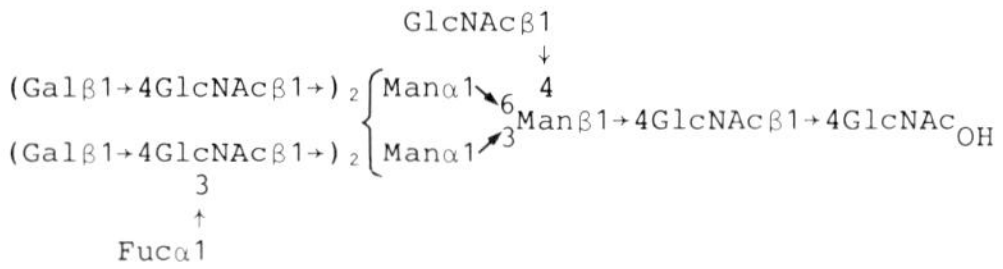

20.6 — 10

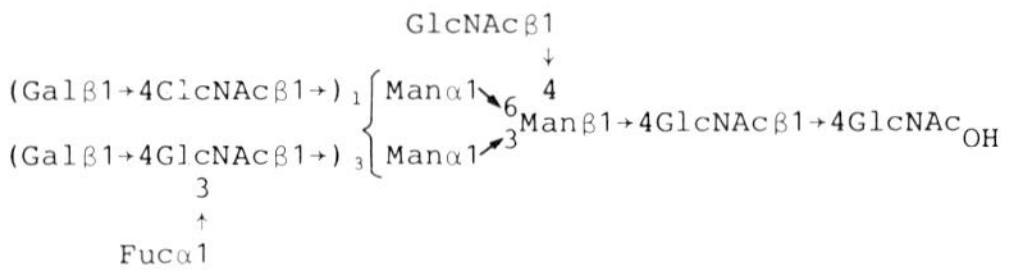

21.2 — 10

TABLE I (*continued*)

Residue	Glucose units	Reference[a]

```
                    Fucα1
                     ↓
                     3
            Galβ1→4GlcNAcβ1
                            6
                             Manα1
            Galβ1→4GlcNAcβ1
                   3
                   ↑                    1βGlcNAc
                 Fucα1                    ↓
                 Fucα1                    4
                                        6
                   ↓                      Manβ1→4GlcNAcβ1→4GlcNAc     22.1     10
                   3                    3                        OH
            Galβ1→4GlcNAcβ1
                            4
                             Manα1
            Galβ1→4GlcNAcβ1
                   3
                   ↑
                 Fucα1
```

```
            ⎡ Galβ1→4GlcNAcβ1                        Fucα1
            ⎢               6                          ↓
            ⎢                Manα1                     6
            ⎢ Galβ1→4GlcNAcβ1   2                      6
 Galβ1→4GlcNAcβ1→3 ⎨ Galβ1→4GlcNAcβ1        Manβ1→4GlcNAcβ1→4GlcNAc     22.7     15
            ⎢               4    3
            ⎢                Manα1
            ⎣ Galβ1→4GlcNAcβ1   2
```

```
            ⎡ Galβ1→4GlcNAcβ1                        Fucα1
            ⎢               6                          ↓
            ⎢                Manα1                     6
            ⎢ Galβ1→4GlcNAcβ1   2                      6
 (Galβ1→4GlcNAcβ1→3)₂ ⎨ Galβ1→4GlcNAcβ1        Manβ1→4GlcNAcβ1→4GlcNAc     25.8     15
            ⎢               4    3                              OH
            ⎢                Manα1
            ⎣ Galβ1→4GlcNAcβ1   2
```

```
            ⎡ Galβ1→4GlcNAcβ1                        Fucα1
            ⎢               6                          ↓
            ⎢                Manα1                     6
            ⎢ Galβ1→4GlcNAcβ1   2                      6
 (Galβ1→4GlcNAcβ1→3)₃ ⎨ Galβ1→4GlcNAcβ1        Manβ1→4GlcNAcβ1→4GlcNAc     28.7     15
            ⎢               4    3                              OH
            ⎢                Manα1
            ⎣ Galβ1→4GlcNAcβ1   2
```

[a] References:
1. H. Yoshima, N. Shiraishi, A. Matsumoto, S. Maeda, T. Sugiyama, and A. Kobata, *J. Biochem.* (*Tokyo*) **91**, 233 (1982).
2. K. Yamashita, T. Ohkura, Y. Tachibana, S. Takasaki, and A. Kobata, *J. Biol. Chem.* **259**, 10834 (1984).
3. K. Yamashita, J. P. Kammerling, and A. Kobata, *J. Biol. Chem.* **258**, 3099 (1983).
4. S. Takasaki, G. J. Murray, S. Furbish, R. O. Brady, J. A. Barranger, and A. Kobata, *J. Biol. Chem.* **259**, 10112 (1984).
5. K. Yamashita, Y. Tachibana, A. Hitoi, and A. Kobata, *Carbohydr. Res.* **130**, 271 (1984).
6. K. Yamaashita, Y. Tachibana, H. Shichi, and A. Kobata, *J. Biochem.* (*Tokyo*) **93**, 135 (1983).
7. T. Mizuochi, T. Taniguchi, K. Fujikawa, K. Titani, and A. Kobata, *J. Biol. Chem.* **258**, 6020 (1983).
8. K. Yamashita, unpublished data.
9. K. Yamashita, C.-J. Liang, S. Funakoshi, and A. Kobata, *J. Biol. Chem.* **256**, 1283 (1981).

(*continued*)

References to TABLE I (*continued*)

10. K. Yamashita, A. Hitoi, N. Tateishi, T. Higashi, Y. Sakamoto, and A. Kobata, *Arch. Biochem. Biophys.* **240**, 573 (1985).
11. K. Yamashita, Y. Tachibana, T. Takeuchi, and A. Kobata, *J. Biochem.* (*Tokyo*) **90**, 1281 (1981).
12. T. Mizuochi, R. Nishimura, C. Derrape, T. Taniguchi, T. Hamamoto, M. Mochizuki, and A. Kobata, *J. Biol. Chem.* **258**, 14126 (1983).
13. H. Yoshima, A. Matsumoto, T. Mizuochi, T. Kawasaki, and A. Kobata, *J. Biol. Chem.* **256**, 8476 (1981).
14. K. Yamashita, J. P. Kammerling, and A. Kobata, *J. Biol. Chem.* **257**, 12809 (1982).
15. A. Hitoi, K. Yamashita, Y. Niwata, M. Irie, N. Kochibe, and A. Kobata, *J. Biochem.* (*Tokyo*) (1986) in press.
16. A. Mizoguchi, S. Takasaki, S. Maeda, and A. Kobata, *J. Biol. Chem.* **259**, 11949 (1984)

[7] Separation of Benzoylated Oligosaccharides by Reversed-Phase High-Pressure Liquid Chromatography: Application to High-Mannose Type Oligosaccharides

By PETER F. DANIEL

Currently, amino-bonded columns are the most widely used for HPLC of oligosaccharides, employing either direct detection of amino sugars at 190–206 nm[1-4] or radioactive monitoring of the column effluent.[5-7] While UV detection is carried out in real time, radioactive monitoring usually involves collection of aliquots for later batch processing in a liquid scintillation counter; this nullifies much of the speed and resolution that makes HPLC such an attractive procedure in the first place. However, recent improvements in HPLC radioactive flow detectors have made them a viable, though expensive, alternative for on-line detection. Although HPLC analysis of labeled oligosaccharides provides much greater sensitivity than measurement of intrinsic UV absorption, comparable sensitiv-

[1] A. Boersma, G. Lamblin, P. Degand, and P. Roussel, *Carbohydr. Res.* **94**, C7 (1981).
[2] M. L. E. Bergh, P. Koppen, and D. H. van den Eijnden, *Carbohydr. Res.* **94**, 225 (1981).
[3] J. P. Parente, G. Strecker, Y. LeRoy, J. Montreuil, and B. Fournet, *J. Chromatogr.* **249**, 199 (1982).
[4] C. D. Warren, S. Sadeh, P. F. Daniel, B. Bugge, L. F. James, and R. W. Jeanloz, *FEBS Lett.* **163**, 99 (1983).
[5] S. J. Turco, *Anal. Biochem.* **118**, 278 (1981).
[6] S. J. Mellis and J. U. Baenziger, *Anal. Biochem.* **114**, 276 (1981).
[7] S. J. Mellis and J. U. Baenziger, *Anal. Biochem.* **134**, 442 (1983).

TABLE I
COMPARISON OF AMINO- AND REVERSED-PHASE COLUMNS FOR HPLC
OF OLIGOSACCHARIDES

Aminopropyl columns	Octylsilica columns
Rapid and convenient, no derivatization of the sample	Involves careful derivatization of the sample by perbenzoylation
Reduction unnecessary since anomers are not separated	Reduction advisable due to separation of anomers
Good resolution, but columns are unstable	Excellent resolution, with stable columns
Retention times vary from day to day	Reproducible retention times
Microgram sensitivity with detection at 190 nm	Nanogram sensitivity with detection at 230 nm
Relative (peak to peak) quantitation only	Easy and accurate quantitation with internal or external standard

ity can be achieved by modification of oligosaccharides at the reducing terminus, for example by reductive amination to form a UV-absorbing or fluorescent chromophore.[8-10] However, although such derivatives may improve chromatographic resolution in addition to enhancing sensitivity,[10] a serious drawback is that regeneration of the starting material is not possible. Peracylation of oligosaccharides, in contrast, permits recovery of the original oligosaccharide or alditol after preparative HPLC by treatment with mild base[11] and offers the possibility of improved resolution by reversed-phase chromatography. Reversed-phase HPLC on C_8 or C_{18} bonded columns has been used for the analysis of peracetyl,[12] peralkyl,[13] and perbenzoyl[14-16] derivatives of oligosaccharides. A comparison of the respective merits of amino- and reversed-phase columns for the analysis of native and perbenzoylated oligosaccharides is presented in Table I.

Perbenzoylation of glycoconjugates prior to HPLC analysis permits the use of gradient elution for the rapid separation of components of

[8] W. T. Wang, N. C. LeDonne, Jr., B. Ackerman, and C. C. Sweeley, *Anal. Biochem.* **141**, 366 (1984).

[9] C. Prakash and I. K. Vijay, *Anal. Biochem.* **128**, 41 (1983).

[10] E. Coles, V. N. Reinhold, and S. A. Carr, *Carbohydr. Res.* **139**, 1 (1985).

[11] S. K. Gross and R. H. McCluer, *Anal Biochem.* **102**, 429 (1980).

[12] G. B. Wells, V. Kontoyiannidou, S. J. Turco, and R. L. Lester, this series, Vol. 83, p. 132.

[13] M. McNeil, A. G. Darvill, P. Aman, L.-E. Franzen, and P. Albersheim, this series, Vol. 83, p. 3.

[14] P. F. Daniel, D. F. DeFeudis, I. T. Lott, and R. H. McCluer, *Carbohydr. Res.* **97**, 161 (1981).

[15] P. F. Daniel, I. T. Lott, and R. H. McCluer, *Chromatogr. Sci.* **18**, 363 (1981).

[16] P. F. Daniel, C. D. Warren, and L. F. James, *Biochem. J.* **221**, 601 (1984).

different size and allows their nondestructive analysis by a sensitive UV detector. Furthermore, collected peaks can be permethylated by the method of Hakomori[17] without prior debenzoylation. Previous studies from this laboratory have demonstrated the usefulness of this procedure for the microanalysis of glycosphingolipids[18] and gangliosides.[19,20] This article describes the preparation and rapid HPLC analysis of picomolar amounts of perbenzoylated oligosaccharides. The procedure has been used routinely in this laboratory for the past 5 years to separate and quantitate urinary oligosaccharides from patients with glycoprotein storage disorders, with a special emphasis on mannosidosis.[14,15,21,22] Recently, we have successfully used the method for the more challenging analysis of oligosaccharides in diseased plasmas and in cultured cells.[23-25]

Experimental

Materials

BioGel P-2, AG 50W-X8 and 50W-X2 (H$^+$ forms), and AG 1-X8 and 1-X2 (acetate forms) were purchased from Bio-Rad Laboratories. Yeast mannan, raffinose, isomaltose, isomaltotriose, bovine and porcine thyroglobulins, benzoic anhydride, and 4-dimethylaminopyridine were obtained from Sigma Chemical Company. Bond-Elut disposable extraction columns (C$_{18}$ coating, 1-ml capacity) were purchased from Analytichem International, and C$_{18}$ Sep-pak columns were obtained from Waters Associates. HPLC-grade solvents (pyridine, water, and acetonitrile) were obtained from Burdick and Jackson. Ultrasphere and Microsorb HPLC columns were obtained from Rainin Instruments. Endo-β-N-acetylglucosaminidases (Endo D and Endo H) were purchased from Miles Scientific. Bovine mannosidosis urine was supplied by Dr. R. D. Jolly, Department

[17] S. Hakomori, *J. Biochem.* (*Tokyo*) **55**, 205 (1964).
[18] M. D. Ullman and R. H. McCluer, this volume [8].
[19] E. G. Bremer, S. K. Gross, and R. H. McCluer, *J. Lipid Res.* **20**, 1028 (1979).
[20] M. D. Ullman and R. H. McCluer, *J. Lipid Res.* **26**, 501 (1985).
[21] I. T. Lott and P. F. Daniel, *Neurology* **31**, 1159 (1981).
[22] R. H. McCluer, G. A. Schwarting, P. F. Daniel, E. H. Kolodny, and J. E. Evans, *in* "Glycoconjugates" (E. A. Davidson, J. C. Williams, and N. M. DiFerrante, eds.), p. 459. Praeger, New York, 1985.
[23] P. F. Daniel, N. E. O'Neil, and E. H. Kolodny, unpublished results.
[24] P. F. Daniel, N. E. O'Neil, and R. J. Molyneux, *in* "Glycoconjugates" (E. A. Davidson, J. C. Williams, and N. M. DiFerrante, eds.), p. 200. Praeger, New York, 1985.
[25] P. F. Daniel, D. S. Newburg, N. E. O'Neil, P. W. Smith, and G. W. J. Fleet, submitted for publication.

of Veterinary Pathology, Massey University, Palmerston North, New Zealand. Urine from locoweed-intoxicated sheep was provided by Dr. L. F. James, USDA Poisonous Plant Research Laboratory, Logan, Utah. Swainsonine was provided by Dr. R. J. Molyneux, USDA Natural Products Chemistry Research Unit, Albany, California.

Per-O-Benzoylation

Samples containing <1 μmol of total saccharides in a screw-cap tube (13 × 100 mm) are dried *in vacuo* in the presence of phosphorus pentoxide and benzoylated for a minimum of 4 hr (usually overnight) at 37° with dry pyridine (0.5 ml) containing 10% (w/v) benzoic anhydride plus 5% (w/v) 4-dimethylaminopyridine as acylation catalyst.[14,15] Samples containing 1 nmol or less of total sugars are benzoylated in silanized tubes with 0.1 ml of reagent. Following benzoylation, each sample is diluted with 9 volumes of water, stirred in a vortex mixer, and applied to a C_{18} disposable extraction column. Initially we used Sep-pak columns; we have since switched to Bond-Elut columns because they are less expensive and give less particulate contamination. The column is then washed with 10% (v/v) aqueous pyridine, followed by water (2 ml of each). Benzoylated oligosaccharides are eluted with acetonitrile (2 ml), dried under N_2 and dissolved in acetonitrile for analysis by HPLC.

Reversed-Phase Chromatography

In our initial studies, HPLC was performed with a Varian model 5020 Chromatograph (Varian Associates) equipped with a Rheodyne model 7125 injector fitted with a 20-μl loop and a 5-μm octylsilica column (4.6 × 250 mm; Ultrasphere). The output from a model SF770 variable-wavelength detector (Kratos/Schoeffel) was connected in series to an Autolabs system 1 computing integrator (Spectra-Physics) and a strip-chart recorder. Over time this system was upgraded by purchase of a more sensitive Spectroflow 757 variable wavelength detector (Kratos) to replace the SF770, and an SP4270 integrator-plotter (Spectra-Physics). Also, switching to 3-μm octylsilica columns (4.6 × 100 mm; Microsorb Short-One) gave better resolution than 5-μm columns for analytical separations (cf. Fig. 3). A 0.5 μm low dead volume in-line filter was found to offer significant protection against column inlet frit clogging with 3-μm columns.

A 15-min linear gradient from acetonitrile/water (4:1, v/v) to pure acetonitrile at a flow rate of 1 ml/min is employed to elute benzoylated oligosaccharides, which are detected by their absorbance at 230 nm.

Sources of Oligosaccharides

High-Mannose Oligosaccharides from Thyroglobulins. Bovine and porcine thyroglobulins (pentuplicate and triplicate samples, respectively, 10 mg each) were digested with papain[26] and then subjected to chromatography on BioGel P-2. The glycopeptide fractions were digested with Endo H[27]; cleaved high-mannose oligosaccharides were isolated in the unbound fraction following ion-exchange chromatography on mixed bed resins, reduced, and benzoylated.

Pathological Urines. Urines from two patients with mannosidosis and from animal models of the disease (sheep poisoned with locoweed, *Astragalus lentiginosus,* and calves with genetic mannosidosis) were desalted on mixed-bed resins, further purified by adsorption chromatography on charcoal–Celite, and then subjected to a second ion-exchange chromatography to obtain a neutral oligosaccharide fraction as previously described,[28] which was reduced and benzoylated. The charcoal–Celite chromatography was shown to be an essential part of the procedure as very low yields of benzoylated oligosaccharides were obtained when it was omitted, probably owing to consumption of the reagent by nonsaccharide components of the urine. Alternatively, BioGel P-2 chromatography can be substituted for the charcoal–Celite step.[4]

Cultured Fibroblasts. Normal human fibroblasts and fibroblasts from a patient with mannosidosis were grown in the presence and absence of 100 μM swainsonine, respectively, for 7 days. Harvested cells were sonicated in distilled water and the supernatant was deionized on mixed-bed resins. The resultant neutral oligosaccharide fraction was reduced and benzoylated.

Acetolysis Cleavage Products. Yeast mannan (200 μg) was acetylated with pyridine/acetic anhydride and then acetolysis was performed at 37° for 16 hr in acetic anhydride/acetic acid/concentrated H_2SO_4 (10 : 10 : 1) as described.[29] The reaction products were deacetylated with barium methoxide in dry methanol for 1 hr at room temperature,[30] reduced with sodium borohydride, and then benzoylated. Collected oligosaccharide fractions from preparative HPLC of high-mannose oligosaccharides from pathological urines were debenzoylated with 0.6 M sodium hydroxide in methanol or with barium methoxide in methanol prior to acetylation and acetolysis.

[26] J. Finne and T. Krusius, this series, Vol. 83, p. 269.
[27] A. L. Tarentino, R. B. Trimble, and F. Maley, this series, Vol. 50, p. 574.
[28] G. Strecker and A. Lemaire-Poitau, *J. Chromatogr.* **143,** 553 (1977).
[29] T. Tai, K. Yamashita, M. Ogata-Okawa, N. Koide, T. Muramatsu, S. Iwashita, Y. Inoue, and A. Kobata, *J. Biol. Chem.* **250,** 8569 (1975).
[30] L. Rosenfeld and C. E. Ballou, *Carbohydr. Res.* **32,** 287 (1974).

Comments

Benzoylation Conditions

Perbenzoylation of oligosaccharides prior to analysis serves two important functions: (1) it produces derivatives having improved chromatographic properties on reversed-phase columns as compared to the native oligosaccharides; and (2) it imparts a large extinction coefficient that facilitates the UV detection and quantitative determination of picomolar amounts of individual oligosaccharides (see Table I). Complete benzoylation of glycoconjugates can be achieved by treatment for 16 hr at 37° with either benzoyl chloride in pyridine[23] or benzoic anhydride in pyridine with 4-dimethylaminopyridine as acylation catalyst.[11] A shift in retention time between the benzoic anhydride and benzoyl chloride derivatives of an unknown compound can be used to predict the presence of one or more N-acetylhexosamines in the sample because N-acetylhexosamines undergo N-benzoylation in addition to O-benzoylation on derivatization with benzoyl chloride, whereas benzoic anhydride yields solely the O-benzoyl derivatives.[11] Quantitative N-debenzoylation cannot be achieved because a mixture of N-benzoyl- and N acetylhexosamines is obtained upon treatment with alkali. For this reason, derivatization with benzoic anhydride is well suited for both analytical and preparative HPLC of oligosaccharides containing N-acetylhexosamine residues, such as those obtained from N-linked glycans.

It is crucial that benzoylation be complete because incomplete benzoylation gives rise to additional peaks on HPLC. Benzoylation of the nonreducing trisaccharide raffinose (11 free hyrdoxyl groups) with benzoic anhydride for only 2 min yielded ~60% underbenzoylated products, which are more polar and are therefore less strongly retained on a reversed-phase column.[14] After a 5-min benzoylation, ~10% incompletely benzoylated products were obtained, but these disappeared with longer reaction times and benzoylation was complete in 15 min (Fig. 1). In order to ensure complete benzoylation of long-chain oligosaccharides, which may be sterically hindered (e.g., 34 derivatizable hydroxyl groups on reduced $Man_9GlcNAc_2$), we have standardized the procedure to a reaction time of 16 hr. This is adequate for quantitative benzoylation of all oligosaccharides that we have tested. The presence of silica strongly interferes with benzoylation, so it is necessary to eliminate silica from oligosaccharide samples obtained by preparative TLC. We have found gel filtration on BioGel P-2 to be effective for this purpose.

Excess benzoylating reagents must be removed prior to HPLC, otherwise early eluting peaks are lost in a huge background absorbance. We have found disposable C_{18} Bond-Elut columns to be ideal for sample

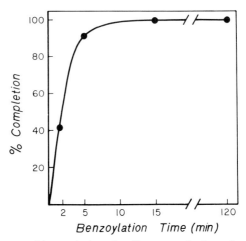

FIG. 1. Time course of benzoylation. A raffinose standard was benzoylated at 37° for various times and then subjected to HPLC analysis. The area of the fully benzoylated raffinose peak is expressed as a percentage of the total area from all the raffinose peaks. From Daniel *et al.*[14]

cleanup. Samples prepared in this manner appear to be completely stable in the absence of alkali and this has enabled us to accumulate a library of standards. We have detected no changes in such standards over several years when they are tightly capped and stored in acetonitrile at 4°.

Recovery and Sensitivity

The linearity of the detector response was investigated by injection of progressive dilutions of a benzoylated raffinose standard. As shown in Fig. 2, the response was linear over the range of 1–600 pmol with the UV detector set at maximum sensitivity (range 0.01 absorbance units full scale at 230 nm). Linearity above this value could be restored by attenuating the signal at the detector.

By adding small amounts of [^{3}H]raffinose to samples prior to benzoylation, we have established that recovery of radioactivity in the acetonitrile eluent from a Sep-pak column is 81% ($\pm$5.4) and is constant over the entire range from 100 pmol to 100 nmol, thus making the procedure eminently suitable for processing very small amounts of sample.[14] For the combined derivatization and analysis of amounts of raffinose ranging from 10 nmol to 1 μmol the overall response was >90% of maximum with excellent reproducibility (Table II). For the successful analysis of material in amounts <10 nmol, special precautions must be taken to minimize losses. These include silanization of glassware to reduce absorptive losses

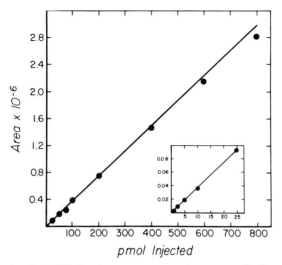

pmol Injected

FIG. 2. Linearity of the UV absorbance response to the amount of raffinose injected onto an Ultrasphere octyl column. The eluted raffinose was determined by measuring the UV absorption at 230 nm with a Schoeffel 770 detector set at maximum sensitivity. The deviation from linearity at the top of the range can be corrected by attenuating the signal at the detector. From Daniel *et al.*

TABLE II

UNIFORMITY OF RESPONSE FOR DERIVATIZATION AND ANALYSIS
OF RAFFINOSE[a]

Amount benzoylated (nmol)	Amount injected[b] (pmol)	Mean response per nmol (area $\times$ 10^{-6})[c]
0.1	10	3.38 $\pm$ 1.22 (9)
0.3	30	3.02 $\pm$ 0.90 (6)
1	100	2.69 $\pm$ 0.78 (4)
3	200	3.42 $\pm$ 1.17 (3)
10–1000[d]	200	4.15 $\pm$ 0.14 (8)

[a] Small samples of raffinose (0.1–1 nmol) were benzoylated with 100 μl of benzoic anhydride reagent; the remainder was benzoylated with 500 μl. Duplicate injections of samples containing >1 nmol differed by <2%; duplicate injections of 100 and 300 pmol samples differed by <7%. Based on recovery of counts from raffinose samples to which a small amount of [³H]raffinose had been added, recovery from the Sep-pak cartridge was 81% ($\pm$5.4) and was uniform over the range from 100 pmol to 100 nmol. From Daniel *et al.*

[b] Volume injected, 20 μl.

[c] $\pm$, Standard deviation, with number of samples given in parentheses.

[d] Two samples each of 10, 30, 100, and 1000 nmol.

and microbenzoylation in 100 μl of reagent. As shown in Table II, $>70\%$ of maximum response could be obtained for the analysis of as little as 100 pmol of raffinose, although not surprisingly the variation between samples increased significantly.

It has been shown that the molar extinction of perbenzoylated glycoconjugates is independent of the compound studied and is strictly additive as a function of the degree of substitution.[31] The absorbance increment for each additional benzoyl group is $\sim$12,600. It is usually impractical to use an internal standard because of the complexity of pathological urine samples. Quantitative determination is best achieved by determining the response per benzoyl group of a known amount of a suitable external standard that is derivatized at the same time as the experimental sample. For this purpose, a nonreducing trisaccharide such as raffinose is ideal. The detection limit for raffinose is <0.5 pmol. Thus for oligosaccharides with nine or more sugar residues a realistic detection limit is <0.2 pmol. This *increase* in sensitivity with increasing size is a most desirable feature for trace analysis, as the converse is true for many detection procedures, for example those that depend on derivatization of the reducing terminal sugar residue.

Column Selection

We initially tested a C_{18} column (Lichrosorb RP-18) for the separation of benzoylated oligosaccharides, but it proved to be too retentive since oligosaccharides with more than six sugar residues could not be eluted even with absolute acetonitrile. The choice of stronger eluotropic solvents was severely limited by the requirement for UV transparency at 230 nm. However, with a C_8 column (Lichrosorb RP-8) packed with irregular 10-μm particles we were able to resolve mannosidosis oligosaccharides containing up to 10 sugar residues in less than 25 min.[15] Much greater resolution was obtained on an Ultrasphere octylsilica column packed with spherical 5-μm particles, and a further significant increase in resolution was obtained by switching to a short Microsorb C_8 column packed with spherical 3-μm particles (Fig. 3). The resolving power of this column ($>100,000$ plates/m) was sufficient to partially resolve minor components that had escaped attention previously.

Patients with mannosidosis excrete a homologous series of oligosaccharides containing a single glucosamine residue and two to nine mannose residues. Examination of the elution profile of reduced oligosaccharides shown in Fig. 3B reveals that $Man_2GlcNAc$ and $Man_9GlcNAc$ appear as symmetrical peaks, whereas $Man_3GlcNAc–Man_8GlcNAc$ exhibit addi-

[31] F. Nachtmann, *Z. Anal. Chem.* **282**, 209 (1976).

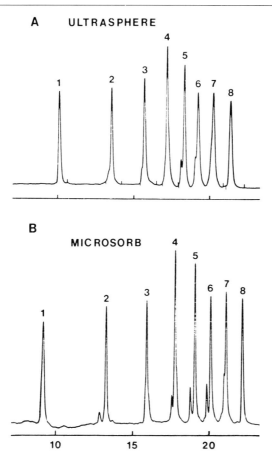

FIG. 3. Comparison of Microsorb and Ultrasphere columns for the separation of a mixture of standard oligosaccharides isolated from human mannosidosis urine by preparative HPLC. (A) Results obtained on a 5-μm Ultrasphere octyl column (4.6 × 250 mm) at a flow rate of 2 ml/min. (B) Separation obtained on a 3-μm Microsorb C_8 column (4.6 × 100 mm) at a flow rate of 1 ml/min. Peak identification: 1, M_2G; 2, M_3G; 3, M_4G; 4, M_5G; 5, M_6G; 6, M_7G; 7, M_8G; 8, M_9G (M, D-mannose; G, N-acetyl-D-glucosamine).

tional peaks or shoulders on the leading and/or trailing edges, suggesting the presence of structural isomers that are partially separated. This is consistent with reports on the structure of oligosaccharides in mannosidosis urine.[32,33] Acetolysis studies have shown that $Man_3GlcNAc–$

[32] K. Yamashita, Y. Tachibana, K. Mihara, S. Okada, H. Yabuuchi, and A. Kobata, J. Biol. Chem. 255, 5126 (1980).
[33] F. Matsuura, H. A. Nunez, G. A. Grabowski, and C. C. Sweeley, Arch. Biochem. Biophys. 207, 337 (1981).

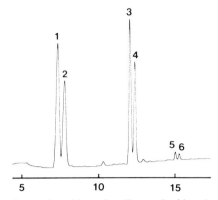

FIG. 4. Separation of nonreduced isomalto-oligosaccharides. A mixture of standard iso-maltose and isomaltotriose was benzoylated and analyzed on a Microsorb C_8 column as described in the text. Peak identification: 1, α-isomaltose; 2, β-isomaltose; 3, α-isomalto-triose; 4, β-isomaltotriose; 5, α-isomaltotetraose; 6, β-isomaltotetraose.

$Man_5GlcNAc$ each contained two isomers, and $Man_6GlcNAc$–$Man_8GlcNAc$ each consisted of three isomers, differing in their disposition of 2-linked mannose residues on the three arms of the mole-cule; $Man_2GlcNAc$ and $Man_9GlcNAc$, however, were shown to be single components.

Because of the high resolving power of 3-μm Microsorb columns, reduction of complex biological samples is a virtual necessity. The need for this is illustrated in Fig. 4, which shows essentially baseline resolution between anomeric forms of isomalto oligosaccharides. The retention times listed in Table III indicate that anomers of isomaltose, isomaltotriose and isomaltotetraose are separated by 25, 18, and 15 sec, respectively. Table III is included to illustrate quantitation by means of an external standard. The areas are first normalized to 11 benzoyl groups and then the absolute amount of each component is calculated from the measured response of a raffinose standard. The analysis shown required only 600 pmol of benzoylated oligosaccharides and this was sufficient to indicate a 3.4% contamination of isomaltotriose by the next higher homolog.

Applications

Species Differences in Thyroglobulins

Endo H treatment followed by reduction and perbenzoylation of the released oligosaccharides is a sensitive tool to fingerprint differences in

TABLE III
QUANTITATIVE DETERMINATION OF NONREDUCED ISOMALTO OLIGOSACCHARIDES[a]

Peak	Component[b]	Retention time (sec)	Area ($\times 10^{-6}$)	Number of benzoyl groups	Normalized area ($\times 10^{-6}$)[c]	Amount injected (pmol)[d]
1	α-IM$_2$	436	1.4171	8	1.9485	215
2	β-IM$_2$	461	0.9859	8	1.3556	149
3	α-IM$_3$	713	1.1033	11	1.1033	122
4	β-IM$_3$	731	0.7603	11	0.7603	84
5	α-IM$_4$	886	0.0447	14	0.0351	4
6	β-IM$_4$	901	0.0342	14	0.0269	3

[a] These data were obtained from the analysis shown in Fig. 4.
[b] IM$_2$, isomaltose; IM$_3$, isomaltotriose; etc.
[c] The areas were normalized to 11 benzoyl groups.
[d] The amount of each oligosaccharide was calculated from the response of an external standard of raffinose (11 benzoyl groups).

the content of high-mannose glycans between specific glycoproteins. The high-mannose oligosaccharides isolated from bovine and porcine thyroglobulins are compared in Fig. 5. It is evident that Man$_5$GlcNAc and Man$_9$GlcNAc are predominant in bovine thyroglobulin, as reported by Spiro and Spiro,[34] whereas porcine thyroglobulin contains more or less equal amounts of all five high-mannose glycans (Table IV).[35] Such large differences in a highly conserved glycoprotein are surprising and most likely arise from species differences in the number of high-mannose chains and their accessibility to processing enzymes: bovine thyroglobulin is reported to contain only 5–6 chains/mole,[36] whereas porcine thyroglobulin contains 8–11 chains/mole.[37] Interestingly, we observed a much higher standard deviation in the molar composition of high-mannose glycans isolated from bovine thyroglobulin than from porcine thyroglobulin, suggesting that an intrinsically greater microheterogeneity may account for why the ratios we obtained differ substantially from reported values.[34] In contrast, the values we obtained for porcine thyroglobulin agree closely with those in the literature.[35]

[34] R. G. Spiro and M. J. Spiro, *Philos. Trans. R. Soc. London, Ser. B* **300**, 117 (1982).
[35] T. Tsuji, K. Yamamoto, T. Irimura, and T. Osawa, *Biochem. J.* **195**, 691 (1981).
[36] T. Arima, M. J. Spiro, and R. G. Spiro, *J. Biol. Chem.* **247**, 1825 (1972).
[37] M. Fukuda and F. Egami, *Biochem. J.* **123**, 407 (1971).

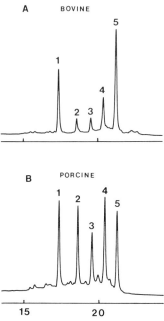

FIG. 5. HPLC of high-mannose oligosaccharides from bovine and porcine thyroglobulins. High-mannose oligosaccharides were prepared from native thyroglobulins and then reduced, benzoylated, and analyzed on a Microsorb C_8 column as described in the text. (A) Bovine thyroglobulin; (B) porcine thyroglobulin. Peak identification: 1, M_5G; 2, M_6G; 3, M_7G; 4, M_8G; 5, M_9G (M, D-mannose; G, N-acetyl-D-glucosamine).

TABLE IV

MOLAR DISTRIBUTION OF HIGH-MANNOSE OLIGOSACCHARIDES
FROM THYROGLOBULINS

	Molar ratio (% ± S.D.)		Reported values	
Oligosaccharide	Bovine ($N = 5$)	Porcine ($N = 3$)	Bovine[a]	Porcine[b]
Man$_5$GlcNAc	35.1 ± 3.3	28.6 ± 0.9	18	27
Man$_6$GlcNAc	5.6 ± 0.8	20.6 ± 0.8	10	27
Man$_7$GlcNAc	6.3 ± 1.3	13.2 ± 0.6	9	20
Man$_8$GlcNAc	13.7 ± 1.3	20.1 ± 0.8	20	17
Man$_9$GlcNAc	39.3 ± 3.9	17.7 ± 0.6	43	9

[a] Spiro and Spiro.[34]
[b] Tsuji et al.[35]

Comparison of Genetic and Induced Mannosidosis

The active component in spotted locoweed is the indolizidine alkaloid swainsonine,[38] which is a powerful inhibitor of both Golgi mannosidase II[39] and lysosomal α-mannosidase.[40] The urinary oligosaccharide profiles given by a calf with mannosidosis and a sheep after 42 days of locoweed treatment are given in Fig. 6. It is apparent that identical oligosaccharides are present in both samples but that the relative amounts differ significantly. The major oligosaccharide in bovine mannosidosis urine is $Man_2GlcNAc_2$ (45%), whereas $Man_4GlcNAc_2$ and $Man_5GlcNAc_2$ are the most abundant in locoweed-treated sheep urine. These differences in the treated sheep are consistent with the superimposition of a processing block on a mannosidosis condition.

The structures of the isolated oligosaccharides are given in Table V; note the ability of the Microsorb column to separate closely related structural isomers. Two homologous series of oligosaccharides containing one and two residues of glucosamine, respectively, were present in urine from both sources[41,42]; the chitobiosyl series was the most abundant and comprised 85% or more of the total. We found no trace of a pentasaccharide with a chitotriosyl structure previously reported to be a major component in bovine mannosidosis urine.[43]

The calf urine gave similar analyses from different specimens regardless of the age at which they were obtained, unlike the locoweed-treated sheep urine, which changed markedly over the course of the feeding period.[16,44] $Man_3GlcNAc_2$ isolated after 6 days of treatment (peak M_3G_2-I in the legend to Fig. 6) is probably derived from the trimannosyl core of complex glycoproteins synthesized prior to locoweed feeding. It is predominantly a single isomer (88% M_3G_2-Ia; see Table V for its structure) that is resistant to Endo H but is digested by Endo D.[16] A different isomer, M_3G_2-Ib, derived from the breakdown of hybrid and high-mannose gly-

---

[38] R. J. Molyneux and L. F. James, *Science* **216**, 190 (1982).

[39] D. R. P. Tulsiani, T. M. Harris, and O. Touster, *J. Biol. Chem.* **257**, 7936 (1982).

[40] P. R. Dorling, C. R. Huxtable, and S. M. Colegate, *Biochem. J.* **191**, 649 (1980).

[41] P. F. Daniel, N. E. O'Neil, C. D. Warren, J. E. Evans, and R. D. Jolly, *in* "Carbohydrates 1984" (J. F. G. Vliegenthart, J. P. Kamerling, and G. A. Veldink, eds.), p. 350. Vonk Publishers, Zeist, 1984.

[42] P. F. Daniel, C. D. Warren, L. F. James, and R. D. Jolly, *in* "Plant Toxicology" (A. A. Seawright, M. P. Hegarty, L. F. James, and R. F. Keeler, eds.), pp. 290–300. Poisonous Plants Committee, Yeerongpilly, Queensland (1985).

[43] A. Lundblad, B. Nilsson, N. E. Norden, S. Svensson, P.-A. Ockerman, and R. D. Jolly, *Eur. J. Biochem.* **59**, 601 (1975).

[44] S. Sadeh, C. D. Warren, P. F. Daniel, B. Bugge, L. F. James, and R. W. Jeanloz, *FEBS Lett.* **163**, 104 (1983).

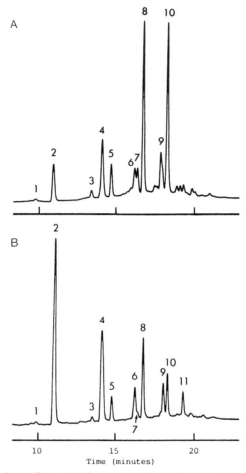

FIG. 6. Comparison of the HPLC elution profiles of (A) sheep urine after 42 days of locoweed treatment and (B) bovine mannosidosis urine. Samples were reduced, benzoylated, and analyzed on a Microsorb C_8 column as described in the text. Peak identification: 1, M_2G; 2, M_2G_2; 3, M_3G; 4, M_3G_2-I; 5, M_3G_2-II; 6, M_4G; 7, M_4G_2-I; 8, M_4G_2-II; 9, M_5G; 10, M_5G_2; 11, M_6G_2 (M, D-mannose; G, N-acetyl-D-glucosamine); structures are given in Table V.

cans predominates from day 13 on. After 51 days of locoweed feeding it comprised >90% of the M_3G_2-I peak. This isomer has a slightly longer retention time on HPLC (6 sec), is digested by Endo H but not by Endo D, and has the structure shown in Table V. A third isomer, M_3G_2-II, is resistant to cleavage by both Endo H and Endo D and elutes later than the other two isomers of $Man_3GlcNAc_2$. Endo H digestion of M_3G_2-I from

TABLE V

STRUCTURES OF OLIGOSACCHARIDES ISOLATED FROM BOVINE MANNOSIDOSIS AND
LOCOWEED-TREATED SHEEP URINE

Fraction from HPLC (Fig. 6)		Structure[a]
1.	M_2G	Manα1 $\rightarrow$ 6Manβ1 $\rightarrow$ 4GlcNAc
2.	M_2G_2	Manα1 $\rightarrow$ 6Manβ1 $\rightarrow$ 4GlcNAcβ1 $\rightarrow$ GlcNAc

3. M_3G

$$\begin{array}{c} \text{Man}\alpha 1 \\ \searrow \\ 6 \\ \text{Man}\beta 1 \rightarrow 4\text{GlcNAc} \\ 3 \\ \nearrow \\ \text{Man}\alpha 1 \end{array}$$

4a. M_3G_2-Ia

$$\begin{array}{c} \text{Man}\alpha 1 \\ \searrow \\ 6 \\ \text{Man}\beta 1 \rightarrow 4\text{GlcNAc}\beta 1 \rightarrow \text{GlcNAc} \\ 3 \\ \nearrow \\ \text{Man}\alpha 1 \end{array}$$

| 4b. | M_3G_2-Ib | Manα1 $\rightarrow$ 3Manα1 $\rightarrow$ 6Manβ1 $\rightarrow$ 4GlcNAcβ1 $\rightarrow$ 4GlcNAc |
| 5. | M_3G_2-II | Manα1 $\rightarrow$ 6Manα1 $\rightarrow$ 6Manβ1 $\rightarrow$ 4GlcNAcβ1 $\rightarrow$ 4GlcNAc |

6. M_4G

$$\begin{array}{c} \text{Man}\alpha 1 \\ \searrow \\ 6 \\ \text{Man}\beta 1 \rightarrow 4\text{GlcNAc}\beta 1 \rightarrow 4\text{GlcNAc} \\ 3 \\ \nearrow \\ \text{Man}\alpha 1 \rightarrow 2\text{Man}\alpha 1 \end{array}$$

7. M_4G_2-I

$$\begin{array}{c} \text{Man}\alpha 1 \rightarrow 3\text{Man}\alpha 1 \\ \searrow \\ 6 \\ \text{Man}\beta 1 \rightarrow 4\text{GlcNAc}\beta 1 \rightarrow 4\text{GlcNAc} \\ 3 \\ \nearrow \\ \text{Man}\alpha 1 \end{array}$$

8. M_4G_2-II

$$\begin{array}{c} \text{Man}\alpha 1 \\ \searrow \\ 6 \\ \text{Man}\alpha 1 \rightarrow 6\text{Man}\beta 1 \rightarrow 4\text{GlcNAc}\beta 1 \rightarrow 4\text{GlcNAc} \\ 3 \\ \nearrow \\ \text{Man}\alpha 1 \end{array}$$

9. M_5G

$$\begin{array}{c} \text{Man}\alpha 1 \\ \searrow \\ 6 \\ \text{Man}\beta 1 \rightarrow 4\text{GlcNAc} \\ 3 \\ \nearrow \\ \text{Man}\alpha 1 \rightarrow 2\text{Man}\alpha 1 \rightarrow 2\text{Man}\alpha 1 \end{array}$$

10. M_5G_2

$$\begin{array}{c} \text{Man}\alpha 1 \\ \searrow \\ 6 \\ \text{Man}\alpha 1 \\ 3 \qquad \searrow \\ \nearrow \qquad\quad 6 \\ \text{Man}\alpha 1 \qquad\quad \text{Man}\beta 1 \rightarrow 4\text{GlcNAc}\beta 1 \rightarrow 4\text{GlcNAc} \\ 3 \\ \nearrow \\ \text{Man}\alpha 1 \end{array}$$

[a] Structures were determined by GC-MS of permethylated alditol acetates.

bovine mannosidosis urine showed that the branched isomer made up two-thirds of the total, indicating the greater contribution from complex glycoproteins in this disease as compared to the locoweed-treated sheep.

Microanalysis of Fibroblast Oligosaccharides

Mannosidosis fibroblasts accumulated substantial amounts of oligosaccharides (44 nmol/mg cell protein), which were shown to contain two to nine mannose residues and a single glucosamine residue (Fig. 7).[24] The profile is very similar to that obtained from mannosidosis urine.[14] $Man_2GlcNAc$, $Man_3GlcNAc$, and $Man_4GlcNAc$ comprised >90% of the total oligosaccharides and were shown by micropermethylation studies[45] to be identical to the major components previously identified from mannosidosis urine[46,47]: $Man\alpha1,3Man\beta1,4GlcNAc$, $Man\alpha1,2Man\alpha1,3Man\beta1,$ $4GlcNAc$ and $Man\alpha1,2Man\alpha1,2Man\alpha1,3Man\beta1,4GlcNAc$.

No mannose-containing oligosaccharides were detected in normal fibroblasts in the absence of swainsonine, but in the presence of 100 μM swainsonine stored oligosaccharides increased from 4 nmol/mg cell protein after 2 days of treatment to 10 nmol/mg after 7 days.[24] Oligosaccharides containing three to nine mannose residues were present (Fig. 7) and $Man_3GlcNAc$ and $Man_5GlcNAc$ were the major components (67%). Surprisingly, $Man_2GlcNAc$ was only observed in swainsonine-treated control cells during the recovery period after removal of swainsonine, although it was the major oligosaccharide (60%) in untreated mannosidosis fibroblasts. Cenci di Bello et al.[48] have concluded from similar results obtained by TLC that human fibroblasts contain a lysosomal α-mannosidase that is unaffected in genetic mannosidosis but is inhibited by swainsonine.

We observed slight differences in the retention times of $Man_3GlcNAc$ from mannosidosis fibroblasts and from swainsonine-treated control fibroblasts (Fig. 7). Micropermethylation analysis of M_3G-I from control fibroblasts indicated that it had a branched structure identical to a minor component from bovine mannosidosis urine (M_3G, Table V), whereas M_3G-II from mannosidosis fibroblasts had a linear structure, as already indicated. One possible explanation for these differences, which we are currently exploring, is that $Man_3GlcNAc$ and $Man_4GlcNAc$ are synthe-

[45] R. Geyer, H. Geyer, S. Kuhnhardt, W. Mink, and S. Stirm, Anal. Biochem. **133**, 197 (1983).
[46] N. E. Norden, A. Lundblad, S. Svensson, P.-A. Ockerman, and S. Autio, J. Biol. Chem. **248**, 6210 (1973).
[47] N. E. Norden, A. Lundblad, S. Svensson, and S. Autio, Biochemistry **13**, 871 (1974).
[48] I. Cenci di Bello, P. Dorling, and B. Winchester, Biochem. J. **215**, 693 (1983).

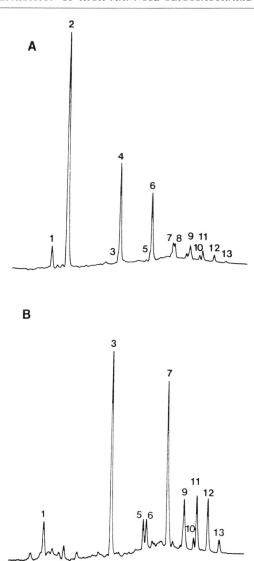

FIG. 7. HPLC of high-mannose oligosaccharides from cultured fibroblasts. (A) Oligosaccharides from untreated mannosidosis fibroblasts ($2-4 \times 10^6$ cells) after 7 days in culture ($15/1000 \mu l$ injected). (B) Oligosaccharides from swainsonine-treated control fibroblasts ($2-4 \times 10^6$ cells) after 7 days of treatment with $100 \mu M$ swainsonine ($20/300 \mu l$ injected). A neutral oligosaccharide fraction was prepared from cultured fibroblasts and then reduced, benzoylated, and analyzed as described in the text. Peak identification: 1, unknown disaccharide; 2, M_2G; 3, M_3G-I; 4, M_3G-II; 5, M_4-G-I; 6, M_4G-II; 7, M_5-G-I; 8, M_5G-II; 9, M_6G; 10, M_7G-I; 11, M_7G-II; 12, M_8G; 13, M_9G (M, D-mannose; G, N-acetyl-D-glucosamine).

TABLE VI
QUANTITATION OF YEAST MANNAN
ACETOLYSIS PRODUCTS

| | Percentage molar ratio | |
Component	This analysis	Reported values[a]
Mannose	25.4	26.0
Mannobiose	37.3	40.9
Mannotriose	25.9	19.5
Mannotetraose	11.4	13.5

[a] Cohen et al.[51]

sized in mannosidosis fibroblasts by addition of 2-linked mannose residues to $Man_2GlcNAc$, provoked by its nonphysiological accumulation. Other unorthodox products that are not known to occur in any glycoprotein have also been reported from patients with aspartylglucosaminuria[49] and from goats with β-mannosidosis.[50]

Analysis of Acetolysis Products

Acetolysis of yeast mannan at 37° for 16 hr gave the HPLC elution profile shown in Fig. 8A. Quantitative determination of this analysis is presented in Table VI, which illustrates the close correspondence of these results to the values reported by Cohen et al.[51] Analysis of acetolysis products as their perbenzoyl derivatives on a Microsorb column has two advantages over the μBondapak/carbohydrate column employed by Cohen et al.: (1) ease of quantitation and (2) separation of the mannotriose peak into two isomers (Fig. 8A). Preparative HPLC of these mannotriose fractions and GC-MS analysis of the permethylated alditol acetates indicated that M_3-I has the structure $Man\alpha 1,3Man\alpha 1,2Man$ and M_3-II has the structure $Man\alpha 1,2Man\alpha 1,2Man$.

Incomplete acetolysis of $Man_2GlcNAc_2$ from bovine mannosidosis urine at 30° for 16 hr indicated that it must contain a 6-linked mannose residue, since free mannose and $ManGlcNAc_2$ were the major cleavage products (Fig. 8B). The structure was subsequently confirmed by permethylation studies. $Man\alpha 1,6Man\beta 1,4GlcNAc\beta 1,4GlcNAc$ is the major oligosaccharide accumulating in cattle with mannosidosis and differs

[49] A. Lundblad, P. K. Masson, N. E. Norden, S. Svensson, P.-A. Ockerman, and J. Palo, *Eur. J. Biochem.* **67,** 209 (1976).
[50] F. Matsuura and M. Z. Jones, *J. Biol. Chem.* **260,** 15239 (1985).
[51] R. E. Cohen, L. Ballou, and C. E. Ballou, *J. Biol. Chem.* **255,** 7700 (1980).

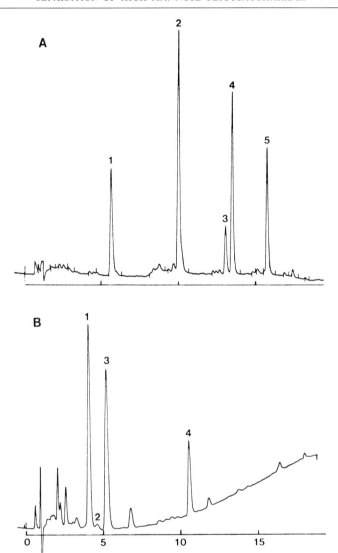

FIG. 8. HPLC of acetolysis cleavage products. Following acetolysis, samples were de-acetylated, reduced, and benzoylated and then analyzed as described in the text. (A) Acetolysis of yeast mannan (200 μg); an amount equivalent to 1.6 μg was injected on an Ultrasphere octyl column at a flow rate of 2 ml/min. Peak identification: 1, M; 2, M_2; 3, M_3-I; 4, M_3-II; 5, M_4. (B) Acetolysis (30° for 16 hr) of $Man_2GlcNAc_2$ (200 nmol) isolated from bovine mannosidosis urine; an amount equivalent to 600 pmol was injected onto a Microsorb C_8 column at 1 ml/min. Peak identification: 1, M; 2, MG; 3, MG_2; 4, M_2G_2 (M, mannose; G, N-acetyl-D-glucosamine).

markedly from the major storage product in human mannosidosis urine,[46] Manα1,3Manβ1,4GlcNAc, indicating species differences in substrate specificity of not only lysosomal α-mannosidase but also endo-β-N-acetylglucosaminidase and/or peptide : N-glycosidase.

Application to Complex Glycans

Reversed-phase HPLC is ideal for the analysis of high-mannose glycans, which usually contain only one or two hexosamine residues, but is less well suited for the analysis of complex glycans. As shown in Fig. 8, ManGlcNAc and ManGlcNAc$_2$ have retention times only slightly greater than mannose and much less than mannobiose; their elution is earlier than would be predicted on the basis of the number of sugars (and hence the number of benzoyl groups) because of reduced binding affinity to the hydrophobic column matrix due to the polar nature of the N-acetyl groups on the glucosamine residues. For this reason, complex glycans with multiple hexosamine residues elute considerably earlier than high-mannose glycans with the same number of sugars, with consequent crowding of the peaks. However, so long as appropriate standards are run, complex glycans can be analyzed on reversed-phase columns.

We have successfully used this HPLC procedure to monitor the much smaller accumulation of characteristic oligosaccharides in G$_{MI}$ fibroblasts (2–4 nmol/mg protein) as compared to mannosidosis cells (>40 nmol/mg). The profile (Fig. 9) is similar to G$_{MI}$ urine, although the relative abundance of the octasaccharide and the decasaccharide is altered. Peak identities were confirmed by means of fast atom bombardment mass spectrometry (FAB-MS) after permethylation and their probable structures are listed in Table VII.[51a]

Conclusions

Because of the rapid separation, extreme sensitivity, and ease of quantitation, the method described herein has considerable potential for the microanalysis of oligosaccharides from biological sources. The detection is nondestructive and individual compounds can easily be collected for determination of radioactivity or for further analysis. Isolated components can be permethylated for structural studies without prior debenzoylation. Clearly, this degree of sensitivity is not really necessary for analysis of the large amounts of oligosaccharides found in urines from patients with glycoprotein or glycolipid storage diseases, although it is certainly useful for the quantitative determination of minor components. This ana-

[51a] K. Yamashita, T. Ohkura, S. Okada, H. Yabuuchi, and A. Kobata, *J. Biol. Chem.* **256,** 4789 (1981).

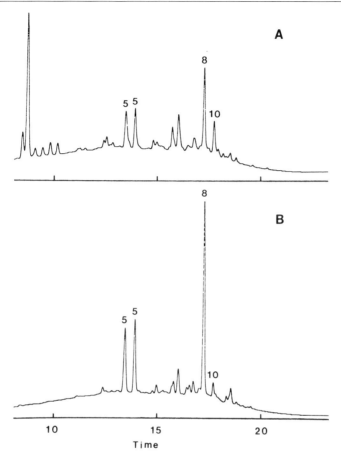

Fig. 9. HPLC elution profiles of oligosaccharides isolated from G_{M1}-gangliosidosis fibroblasts and from G_{M1} urine. (A) Oligosaccharides isolated from G_{M1} fibroblasts ($\sim 6 \times 10^6$ cells) after 21 days in culture (20/1000 μl injected). (B) Oligosaccharides isolated from G_{M1} urine as described in the text (an amount equivalent to 0.4 μl of urine was injected). The number above each peak indicates the number of sugar residues present.

lytical procedure has enabled us to measure the low level of oligosaccharides found in the serum of patients with mannosidosis[21] and G_{M1}-gangliosidosis.[23]

Potential clinical uses include the examination of stored material in cultured skin fibroblasts and tissue biopsies from patients suffering from disorders of carbohydrate metabolism and the analysis of amniotic fluids for prenatal diagnosis[52]; the method can also be applied to the analysis of

[52] T. G. Warner, A. D. Robertson, A. K. Mock, W. G. Johnson, and J. S. O'Brien, *Am. J. Hum. Genet.* **35**, 1034 (1983).

TABLE VII
STRUCTURES OF MAJOR G_{M1}-GANGLIOSIDOSIS OLIGOSACCHARIDES

Number of sugars	Probable structure[a]	Molecular weight	
		Calculated	$[MH]^{+b}$
5a	Galβ1 → 4GlcNAcβ1 → 2Manα1 → 3Manβ1 → 4GlcNAc	1164.36	1165
5b	Galβ1 → 4GlcNAcβ1 → 2Manα1 → 6Manβ1 → 4GlcNAc	1164.36	1165
8	Galβ1 → 4GlcNAcβ1 → 2Manα1 ↘ 6 Manβ1 → 4GlcNAc ↗ 3 Galβ1 → 4GlcNAcβ1 → 2Manα1	1817.55	1818
10	Galβ1 → 4GlcNAcβ1 → 2Manα1 ↘ 6 Manβ1 → 4GlcNAc Galβ1 → 4GlcNAcβ1 ↘ 4 ↗ 3 Manα1 2 ↗ Galβ1 → 4GlcNAcβ1	2470.74	ND

[a] Yamashita et al.[51a]
[b] Determined by FAB-MS after permethylation at the Mass Spectrometry Facility, MIT, by courtesy of Dr. Catherine Costello. ND, not done.

cerebrospinal[53] and ocular fluids.[54] The procedure may be applied to the rapid isolation of intermediates involved in glycoprotein processing and their acetolysis products. It could also be used to study changes in the level of high-mannose glycoproteins with the state of growth or differentiation of cells,[55,56] or structural changes in carbohydrate chains on a specific glycoprotein as a function of malignant transformation.[57]

Acknowledgments

This work was supported by NIH Grants HD 16942 and HD 05515. Technical assistance from Nancy O'Neil and collaborations with Drs. L. F. James, R. D. Jolly, R. J. Molyneux, and C. D. Warren are gratefully acknowledged.

[53] P. F. Daniel and I. T. Lott, unpublished results.
[54] C. D. Warren, J. Alroy, B. Bugge, P. F. Daniel, S. S. Raghavan, E. H. Kolodny, J. J. Lamar, and R. W. Jeanloz, FEBS Lett. 195, 247 (1986).
[55] T. Muramatsu, N. Koide, C. Ceccarini, and P. H. Atkinson, J. Biol. Chem. 251, 4673 (1976).
[56] N. Koide, T. Muramatsu, and A. Kobata, J. Biochem. (Tokyo) 85, 149 (1979).
[57] K. Yamamoto, T. Tsuji, O. Tarutani, and T. Osawa, Eur. J. Biochem. 143, 133 (1984).

[8] High-Pressure Liquid Chromatography Analysis of Neutral Glycosphingolipids: Perbenzoylated Mono-, Di-, Tri-, and Tetraglycosylceramides

By M. D. Ullman and R. H. McCluer

Quantitative HPLC analysis of glycosphingolipids becomes a useful and convenient procedure provided high-sensitivity on-line detection is possible. Neutral glycosphingolipids do not possess a characteristic chromophore that permits their quantitative UV detection where high sensitivity is required. However, they can be derivatized with benzoyl chloride or benzoic anhydride to form stable per-O,N- or per-O-benzoylated products, respectively.[1,2] These products can be quantitatively measured by their absorption of UV light at 280 nm or, with higher sensitivity, at 230 nm.

Benzoyl chloride reacts with the hydroxyl groups and amide nitrogen of nonhydroxy fatty acid (NFA) cerebrosides. When the resulting diacylamine derivatives are treated with mild alkali a mixture of parent cerebrosides and N-benzoylpsychosine is obtained.[3,4] When hydroxy fatty acid (HFA) cerebrosides are perbenzoylated the substitution of the amide hydrogen is stearically blocked so that only O-benzoylation occurs. Thus, mild alkaline methanolysis of these HFA derivatives yields only the parent compounds.

Both NFA and HFA neutral glycosphingolipids derivatized with benzoic anhydride yield only the O-benzoyl derivatives.[2] Therefore, native glycolipids are recovered after mild alkaline hydrolysis. Because sphingolipids which contain only HFA as N-acyl substituents form the same derivative with either benzoyl chloride or the benzoic anhydride reaction, they can easily be distinguished from NFA-containing sphingolipids which form different derivatives with these reagents. The NFA and HFA derivatives are separated by HPLC. The benzoyl chloride reaction is most useful for analytical purposes, because resolution of components containing HFA and/or phytosphingosine is superior to that obtained with the O-benzoates formed with benzoic anhydride.

[1] M. D. Ullman and R. H. McCluer, *J. Lipid Res.* **18**, 371 (1977).
[2] S. K. Gross and R. H. McCluer, *Anal. Biochem.* **102**, 429 (1980).
[3] R. H. McCluer and J. E. Evans, *J. Lipid Res.* **14**, 611 (1973).
[4] R. H. McCluer and J. E. Evans, *J. Lipid Res.* **17**, 412 (1976).

Quantitative Analysis of per-O,N-Benzoylated
  Neutral Glycosphingolipids

Standards

Purified neutral glycosphingolipids to be used as standards can be isolated from erythrocytes or other tissue.[1] It is critical that the standard preparations are chromatographically pure and free of closely related isomers and inorganic material such as silicic acid. We utilize the hexose or long-chain base content of standard preparations to establish and verify the peak area response factor of each glycolipid.

Derivatization

Neutral glycosphingolipids (at least 200 ng of each) and a known amount of internal standard,[5] such as *N*-acetylpsychosine, are dried under a stream of nitrogen in a 13 × 100 mm screw-capped culture tube with a Teflon-lined cap. They are per-O,N-benzoylated with 0.5 ml of 10% benzoyl chloride in pyridine (v/v) which is added directly to the bottom of the tube. The sample reaction mixture is warmed at 37° for 16 hr. The pyridine is removed with a stream of nitrogen at room temperature until the residue appears as an oil-covered solid. Then, 3 ml of hexane is added to the reaction tube. The hexane is washed three times with 1.8 ml of methanol–water (80 : 20, v/v) which is saturated with sodium carbonate (1.2 g sodium carbonate in 300 ml of methanol–water, 80 : 20). While removing the lower aqueous methanol layer a slight positive pressure is exerted on the pipet bulb as the pipet passes through the upper (hexane) layer. The hexane layer is then washed with 1.8 ml of methanol–water (80 : 20, v/v). The lower phase is withdrawn and discarded and the hexane is evaporated with a stream of nitrogen. The derivatives are dissolved in 100–500 μl of carbon tetrachloride and an aliquot is injected into the HPLC column.

The derivatives can also be isolated from the reagents and reaction by-products with a C_{18} reversed-phase cartridge (C_{18} Bond-Elut, Analytichem International, Harbor City, CA) by a procedure that has been developed for per-O,N-benzoylated gangliosides.[6] The dried reaction mixture is dissolved in 0.8 ml of methanol and transferred to a C_{18} reversed-phase cartridge which is prewashed with 2 ml of methanol. The reaction vial is rinsed with an additional 0.8 ml of methanol and the rinse solvent is transferred to the cartridge. The collected and combined eluates are

[5] M. D. Ullman and R. H. McCluer, *J. Lipid Res.* **19,** 910 (1978).
[6] M. D. Ullman and R. H. McCluer, *J. Lipid Res.* **26,** 501 (1985).

passed through the cartridge again. The cartridge is eluted with 4 ml of methanol to remove reaction by-products. Perbenzoylated neutral glyco-sphingolipids are eluted from the cartridge with 3 ml of methanol–benzene (8:2). The derivatives are collected in a 13 × 100 mm screw-capped culture tube and the solvent is removed at room temperature with a stream of nitrogen. They are then dissolved in 100–500 μl of carbon tetrachloride and an appropriate aliquot is injected into the HPLC.

The cartridge procedure is fast and the recoveries are comparable to the partition workup. However, N-acetylpsychosine cannot be used as an internal standard with this method of isolation because the partition char-acteristics of its per-O,N-benzoylated derivative on the reversed-phase cartridge are substantially different from those for the per-O,N-benzoyl-ated derivatives of neutral glycosphingolipids. The per-O,N-benzoylated products are stabile for months providing they are completely free of alkali.

Chromatography

The derivatives are separated on a pellicular (Zipax, E. I. DuPont de Nemours, Inc., Wilmington, DE) column (2.1 mm i.d. × 500 mm) (see packing procedure in "Comments" section) with a 10-min linear gradient of 2–17% water-saturated ethyl acetate in hexane and a flow rate of 2 ml/min and detected by their UV absorption at 280 nm.[1] The minimum level of detection (twice baseline noise) by this procedure is approximately 70 pmol of each neutral glycosphingolipid. The sensitivity of the procedure can be increased by detection at 230 nm, the absorption maximum of the derivatives, which requires the use of a mobile phase that is transparent at this wavelength.

To accomplish this increased sensitivity with 230 nm detection, a 13-min linear gradient of 1–20% dioxane in hexane and a flow rate of 2 ml/min is used (Fig. 1).[5] The mobile phase is directed through a preinjector flow-through reference cell to cancel the residual absorption of the diox-ane. Several UV detectors have flow-through reference cells that are pressure rated high enough to be utilized in this system. The maximum pressure rating required is about 1,000 psi. Per-O,N-benzoylated ceramides (derivatized with benzoyl chloride) and neutral glycosphingo-lipids are separated in a single chromatographic run with a 15-min linear gradient of 0.23–20% dioxane in hexane.[7]

A consistent separation of per-O,N-benzoylated glucocerebroside from per-O,N-benzoylated galactocerebroside, at all ratios, is obtained by

[7] K-H. Chou and F. B. Jungalwala, *J. Neurochem.* **36**, 394 (1981).

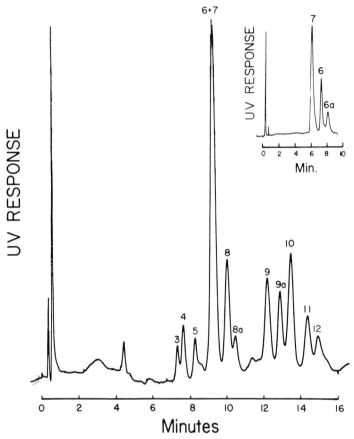

FIG. 1. HPLC of male (C57BL/6J) mouse kidney perbenzoylated glycosphingolipids on a Zipax column with detection at 230 nm. A 13-min linear gradient of 1–20% dioxane in hexane at a flow rate 2 ml/min was used as the mobile phase. Inset shows the separation of the collected peak 6 and 7, when reinjected on the same column but with a 14-min linear gradient of 0.25–1% isopropanol in hexane as the solvent. Peak 3, Glc-Sph-NFA; 4, Gal-Sph-NFA; 5, Glc-Phyto-NFA; 6, Glc-Sph-HFA + Glc-Phyto-Hfa; 6a, Gal-Sph-Hfa + Gal-Phyto-HFA; 7, GaOse$_2$-Sph-NFA; 8, GaOse$_2$-Sph-HFA + GaOse$_2$-Phyto-NFA; 8a, Lac-Sph-NFA; 9, GbOse$_3$-Sph-NFA; 9a, GbOse$_3$-Phyto-NFA, 10, GbOse$_3$-Sph-HFA; 11, GbOse$_4$-Sph-NFA; and 12, GbOse$_4$-Phyto-NFA + GbOse$_4$-Sph-HFA. Sph refers to C$_{18}$-sphingosine; Phyto, C$_{18}$-phytosphingosine; NFA, nonhydroxy fatty acid; HFA, hydroxy fatty acid. Reproduced from McCluer and Gross.[12]

HPLC of the per-O,N-benzoylated derivatives on the pellicular silica column maintained at 60° and a 10-min linear gradient of 1–20% dioxane in hexane at a flow rate of 2 ml/min.[8] This system is particularly useful for

[8] E. M. Kaye and M. D. Ullman, *Anal. Biochem.* **138**, 380 (1984).

the analysis of brain cerebrosides since the ratio of galactocerebroside to glucocerebroside is usually very high.

Quantitative Analysis of Per-O-benzoylated Neutral Glycosphingolipids

The per-O-benzoyl derivatives of neutral glycosphingolipids are produced via their reaction with 10% benzoic anhydride in pyridine with 5% 4-N-dimethylaminopyridine (DMAP) as a catalyst. The reaction is run at 37° for 4 hr.[2] Excess reagents are removed by solvent partition between hexane and aqueous alkaline methanol as described above.

The derivatives are separated by HPLC as described for the O,N-benzoyl derivatives (Fig. 2). The per-O-benzoyl derivatives of HFA cerebrosides (formed by reaction with benzoyl chloride or benzoic anhydride) elute earlier than the per-O-benzoyl derivatives of NFA cerebrosides (formed with benzoic anhydride) but later than the per-O,N-benzoyl derivatives of NFA cerebrosides (formed with benzoyl chloride).

Comments

Derivatizations

Several factors influence the benzoyl chloride derivatization. It is important that neutral glycosphingolipid samples to be perbenzoylated are relatively free of silica or silicic acid particles. These particles arise either from dissolved silica after extraction of the glycosphingolipids from TLC or from "fines" that avoid filtration.

Moisture is another factor that influences the benzoyl chloride derivatization. Water (e.g., atmospheric moisture) reacts instantaneously with benzoyl chloride to form benzoic acid and hydrochloric acid. Thus, it is important to store benzoyl chloride in dry surroundings. A 500-ml bottle of benzoyl chloride can usually be used for 3–6 months.

Furthermore, the pyridine must be dried by storage over 4A molecular sieves. Some batches of pyridine may have a very high moisture content and the molecular sieves may not trap all of the water. This situation is evident on mixing the pyridine with the benzoyl chloride because a crystalline precipitate forms almost immediately. There are other indications that moisture has been introduced into the reaction medium. For example, the solvent front of the chromatogram may be broader than usual or there may be a broad peak that elutes just after the solvent front. These anomalies are due to benzoic acid. Finally, the sample(s) must be dried *in vacuo* over phosphorus pentoxide at least 1 hr before the perbenzoylation is performed.

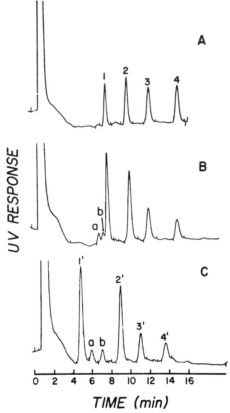

TIME (min)

FIG. 2. HPLC of per-O- and per-O,N-benzoylated standard and plasma glycosphingolipids. The derivatized glycosphingolipids were injected into a Zipax column (2.1 mm × 50 cm) and eluted with a 13-min linear gradient of 2.5–25% dioxane in hexane with detection at 230 nm. (A) Standard glycosphingolipids (GSL) per-O-benzoylated with benzoic anhydride and 4-dimethylaminopyridine (DMAP). (B) Plasma GSL per-O-benzoylated with benzoic anhydride and DMAP. (C) Plasma GSL perbenzoylated with benzoyl chloride. Glycosphingolipid peaks are identified as (1) glucosylceramide, (2) lactosylceramide, (3) galactosyllactosylceramide, (4) N-acetylgalactosaminylgalactosyllactosylceramide. Peak A is unidentified, and peak B is hydroxy fatty acid containing galactosylceramide. Reproduced from Gross and McCluer.[2]

Chromatography

In our laboratories, the formation of O,N-benzoyl derivatives is used most frequently for the quantitative analysis of neutral glycosphingolipids. Those derivatives of NFA and HFA neutral glycosphingolipids are resolved very well on a pellicular silica column. The O-benzoyl derivatives are also used analytically, especially as an adjunct procedure for the tentative identification of HFA-glycosphingolipids. The per-O-benzoyl derivatives can be used to distinguish between HFA- and NFA-glycosylceramides by comparing their elution times to the per-O,N-benzoylated derivatives. The O-benzoyl derivatives have also been used for preparative purposes since native neutral glycosphingolipids are recovered by mild alkaline methanolysis of the isolated derivatives.[9]

Pellicular columns can be inexpensively dry-packed in less than 20 min and they routinely provide very high resolution for long periods of time (many months of extensive usage). The column can be easily packed in about 15 min by the "tap–fill" method.[10] Briefly, the bottom of a 500 mm × 2.1 mm i.d. stainless-steel tube is fitted with a male end-fitting which contains a stainless-steel frit. The top of the tube is fitted with a male end-fitting which has a 4-cm piece of tubing and no frit. The tubing is filled with pellicular silica (Zipax, E. I. Dupont de Nemours, Wilmington, DE) to about 1–2 cm high. The bottom of the column is then tapped on a bench top about 80 times. During this procedure the column is also rotated. The silica is added and packed repetitively until the column is filled. The top end-fitting is replaced with one which contains a stainless-steel frit so that no dead space remains and the column is ready to be equilibrated and used.

The mobile phases for the HPLC of neutral glycosphingolipids are mixtures of relatively apolar solvents, such as hexane, with reasonably polar solvents, such as dioxane, isopropanol, or isopropanol and water. Poor mixing characteristics of the solvents which comprise the mobile phase creates baseline instabilities. A dual chamber dynamic (mechanical) mixer or two single-chambered mixers in series are required for baselines sufficiently stable to allow detection at high sensitivity such as 0.04 absorbance units full scale (AUFS).

A variety of chromatographic systems have been used to separate perbenzoylated derivatives. Mobile phases of isopropanol in hexane have been used which do not have to be directed through a preinjector flow-through reference cell since they are reasonably UV transparent at 230 nm.[11,12] These systems are used with a totally porous 5- or 10-μm silica column, but they can be used with the pellicular silica columns also.

[9] R. H. McCluer and M. D. Ullman, in "Cell Surface Glycolipids" (C. C. Sweeley, ed.), p. 1. Am. Chem. Soc., Washington, D.C., 1980.
[10] L. R. Snyder and J. J. Kirkland, "Introduction to Modern Liquid Chromatography," 2nd ed. Wiley (Interscience) New York, 1979.
[11] P. M. Strasberg, I. Warren, M. A. Skomorowski, and J. A. Lowden, Clin. Chim. Acta 132, 29 (1983).
[12] R. H. McCluer and S. K. Gross, J. Lipid Res. 26, 593 (1985).

Both dioxane/hexane and isopropanol/hexane systems have been used to separate perbenzoylated derivatives. The dioxane/hexane system provides consistent resolution of the major glycosphingolipids and in our laboratories provides a better resolution of the minor plasma and tissue neutral glycosphingolipids. Further, the selectivity and efficiency of this system are such that the neutral glycosphingolipid derivatives yield only one peak for each neutral glycosphingolipid rather than further partial separation of each neutral glycosphingolipid on the basis of its fatty acid composition. This provides more consistent automatic integration of the peaks and yields a less complex chromatogram. The disadvantage of the system is that the mobile phase, in a gradient run but not in an isocratic run, must pass through a preinjector flow-through reference cell. Many detectors will accommodate the rather modest pressure requirements and the hexane/dioxane system is the one of choice in our laboratories.

The isopropanol/hexane mobile phase used with a porous silica column produces adequate resolution of the major plasma glycosphingolipids. It also shows some selectivity for the fatty acid portion of derivatized neutral glycosphingolipid and the chromatographic peaks are often broadened or split into doublets based on the population of medium and long-chain fatty acids.

The isopropanol/hexane mobile phase can be used with a pellicular silica column but the resolution of the neutral glycosphingolipid derivatives is not as high as that obtained with either the hexane/dioxane with a pellicular column or the isopropanol/hexane system with a totally porous silica column. The isopropanol/hexane mobile phase with a pellicular column will, however, provide resolution of some hydroxy fatty acid- or phytosphingosine-containing neutral glycosphingolipids which are not resolved in the dioxane system. Sometimes combinations of systems can be used. For example, if one system does not adequately resolve peaks of interest, the peaks can be collected and separated in the other system (Fig. 1, inset). Obviously, the system of choice in any laboratory is dependent on the specific objectives.

At high sensitivities many UV detectors are vulnerable to ambient temperature fluctuations. These fluctuations are caused by temperature-induced alterations in the refractive index of the mobile phase. Inadequate insulation around the tubing which leads from the column outlet to the detector inlet will lead to unstable baselines. This piece of tubing should be well insulated. A piece of rubber tubing slit from end to end and slipped over the tubing provides excellent insulation. Some manufacturers have incorporated heat exchangers into their detectors. An easy method to detect ambient temperature problems is to simply place your finger on the tubing in question for a few seconds. If there is a baseline disturbance it is an indication there is an ambient temperature problem.

Finally, for analytical procedures, it is important to know the absolute recoveries of the glycosphingolipids. There are several examples in the literature in which losses of neutral glycosphingolipid that occur during the isolation procedures are not taken into consideration. It is documented that neutral glycosphingolipid recoveries from silicic acid columns decrease with increasing length of neutral glycosphingolipid carbohydrate chains.[13] It is important to add, if possible, a high specific activity radiolabeled (or other) standard that can be perbenzoylated and used to determine the total recovery, and, consequently, the absolute lipid concentrations of the tissue source. N-Acetylpsychosine has been used as an internal standard[5] for neutral glycosphingolipid determinations. However, difficulty is encountered because the short-chain N-acetyl group imparts characteristics that are slightly different than those of long-chain fatty acid glycolipids. Finally, for further details concerning this procedure and other lipid HPLC methods see this series, Vol. 125.[14]

---

[13] R. E. Vance and C. C. Sweeley, *J. Lipid Res.* **8**, 621 (1967).
[14] R. H. McCluer, M. D. Ullman, and F. B. Jungalwala, this series, Vol. 125, Part M.

[9] Short-Bed, Continuous Development, Thin-Layer Chromatography of Glycosphingolipids[1]

By WILLIAM W. YOUNG, JR. and CHERYL A. BORGMAN

Thin-layer chromatography is one of the major techniques used for the analysis and purification of glycosphingolipids. In recent years high-performance thin-layer chromatography (HPTLC) plates have been widely used because of increased band resolution as compared to conventional plates. However, HPTLC plates have the disadvantage that the bands,

---

[1] This investigation was supported by Grants IM-335/BC-489A from the American Cancer Society and Grant AI-21916 from the NIH. Abbreviations: ceramide dihexoside (CDH), Galβ1 → 4GlcCer; gangliotriaosylceramide (GgOse₃Cer; asialo-G_{M2}), GalNAcβ1 → 4Galβ1 → 4GlcCer; globoside, GalNAcβ1 → 3Galα1 → 4Galβ1 → 4GlcCer; Forssman glycolipid, GalNAcα1 → 3GalNAcβ1 → 3Galα1 → 4Galβ1 → 4GlcCer; N-glycolylhematoside (G_{M3}), NeuNGcα2 → 3Galβ1 → 4GlcCer; G_{M1}, Galβ1 → 3GalNAcβ1 → 4 [NeuNAcα2 → 3]Galβ1 → 4GlcCer; G_{M1b}, NeuNAcα2 → 3Galβ1 → 3GalNAcβ1 → 4Galβ1 → 4 GlcCer; and G_{T1}, NeuNAcα2 → 3Galβ1 → 3GalNAcβ1 → 4[NeuNACα2 → 8NeuNAcα2 → 3]Galβ1 → 4GlcCer.

METHODS IN ENZYMOLOGY, VOL. 138

although clearly resolved, may be very closely spaced. Thus, scraping of individual bands from HPTLC plates may be difficult. As described in detail by Perry,[2] the separation by TLC of any two components can be increased in almost all cases by decreasing the solvent strength, which for practical purposes is equivalent to decreasing solvent polarity. The limitation of this theory is that a decrease of solvent strength is accompanied by an exponential decrease in R_f.[2] This difficulty can be overcome by the high solvent velocities achieved by continuous development in a short-bed chamber. The inverse relationship between solvent velocity and bed length is illustrated elsewhere.[3] The following examples describe the use of continuous development (CD-HPTLC) for increasing the separation of several types of glycosphingolipids. Our original description of this information was presented elsewhere.[4]

Materials

The short-bed continuous development chamber was obtained from Regis Chemical Co. (Morton Grove, IL), glass-backed and aluminum-backed HPTLC plates from E. Merck (Federal Republic of Germany), and Enhance spray from New England Nuclear (Boston, MA). Glycolipid standards were purified from the following sources using established procedures[5]: ceramide dihexoside (CDH) and globoside from human erythrocytes, N-glycolylhematoside (G_{M3}) from horse erythrocytes, gangliotriaosylceramide (GgOse$_3$Cer) from guinea pig erythrocytes, gangliosides G_{M1} and G_T from bovine brain, and Forssman ceramide pentasaccharide from sheep erythrocytes. The neutral glycolipids from human meconium were partially purified by Iatrobead chromatography as previously described.[6] Mouse L5178Y lymphoma cells were metabolically labeled with [^{3}H]galactose and the glycolipids purified as previously described.[7]

Chromatography Conditions

Continuous development HPTLC was performed by applying samples at the origin 1.5 cm from the bottom of the plate. The short-bed continu-

[2] J. A. Perry, *J. Chromatogr.* **165,** 117 (1979).

[3] Regis Technical Manual, "Short Bed/Continuous Development." Regis Chem. Co., Morton Grove, Illinois, 1979.

[4] W. W. Young, Jr., and C. A. Borgman, *J. Lipid Res.* **27,** 120 (1986).

[5] R. A. Laine, K. Stellner, and S. Hakomori, *in* "Methods of Membrane Biology" (E. D. Korn, ed.), p. 205. Plenum, New York, 1974.

[6] W. W. Young, Jr., H. S. Johnson, Y. Tamura, K. Karlsson, G. Larson, J. M. R. Parker, D. P. Khare, U. Spohr, D. A. Baker, O. Hindsgaul, and R. U. Lemieux, *J. Biol. Chem.* **258,** 4890 (1983).

[7] W. W. Young, Jr., C. A. Borgman, and D. M. Wolock, *J. Biol. Chem.* **261,** 2279 (1986).

ous development tank was filled with either 50 ml of solvent B or D or 30 ml of solvent F. Plates were placed in position 4 (next to the most horizontal position) for solvents B and D and position 3 for solvent F; see Fig. 1 of Ref. 3 which shows the plate positions. In CD-HPTLC the solvent reached the top of the tank in about 10–15 min, and development was allowed to continue for the times indicated below. For conventional HPTLC, plates were removed from tanks when the solvent reached the top of the plate (8.5 cm bed length). Solvents and development times: A, chloroform–methanol–water (C–M–W; 62 : 30 : 6, v/v), 30 min; B, C–M–W (75 : 18 : 2.5), 2.5 hr; C, C–M–0.25% $CaCl_2$ in H_2O (60 : 40 : 9), 37 min; D, C–M–0.25% $CaCl_2$ (62.5 : 30 : 6), 2 hr; E, C–M–W (55 : 40 : 11), 47 min; and F, C–M–W (60 : 35 : 8), 45 min.

Separation of Neutral Glycolipids

Gangliotriaosylceramide (asialo-G_{M2}), the major neutral glycolipid of mouse L5178Y lymphoma cells, is present in the cells in three forms separable by HPTLC.[7] The fastest migrating form was found previously to contain mainly C_{24} fatty acids, the middle band C_{16} fatty acid, and the slowest band C_{16} α-hydroxy fatty acid.[8,9] Although the three bands were clearly resolved by conventional HPTLC (Fig. 1, lane 1), the bands were too closely spaced (5 mm between bands a and c) to allow them to be scraped separately in order to quantitate incorporation of radiolabeled precursors into individual bands. However, chromatography in a continuous development chamber resulted in a threefold increase in separation of the three bands (Fig. 1, lane 2), thus making scraping of separate bands a straightforward process. Even though chromatography proceeded for 2.5 hr, the bands were nearly as sharp as those obtained by conventional HPTLC.

Separation of Gangliosides

Mouse L5178Y lymphoma cells contain a major ganglioside that was previously identified as G_{M1b}.[10] Chromatography in a conventional tank produced a set of four bands (Fig. 2, lane 1), which were too tightly spaced to permit individual bands to be scraped separately. In contrast, continuous development increased band separation threefold (Fig. 2, lane 2) and made the scraping of each band feasible.

[8] S. Hakomori, D. Urdal, M. Yokota, and W. W. Young, Jr., *Proc. Jpn. Soc. Med. Mass Spectrom.* **5**, 3 (1980).
[9] D. L. Urdal and S. Hakomori, *J. Biol. Chem.* **258**, 6869 (1983).
[10] R. Kannagi, R. Stroup, N. A. Cochran, D. L. Urdal, W. W. Young, Jr., and S. Hakomori, *Cancer Res.* **43**, 4997 (1983).

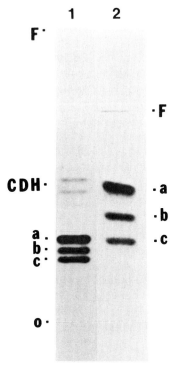

FIG. 1. Comparison of gangliotriaosylceramide bands separated by conventional versus CD-HPTLC. Lymphoma cells were metabolically labeled with [³H]galactose, extracted, and the lower phase glycolipids purified by the acetylation procedure. Aliquots (5000 cpm) were chromatographed on glass-backed HPTLC plates as follows: lane 1, regular chamber using solvent A; lane 2, continuous development chamber (plate at position 4) using solvent B. After spraying the plates with Enhance spray, fluorography was performed by exposing the plates to X-ray film at −70° for 2 days (lane 1) or 3 days (lane 2). F, solvent front; O, origin. CDH, ceramide dihexoside. R_f of glycolipid standards: lane 1, CDH 0.42, GgOse₃Cer (from guinea pig erythrocytes) 0.26, and globoside 0.19; lane 2, CDH 1.0, GgOse₃Cer 0.65, and globoside 0.29. Reprinted with permission from Young and Borgman.[4]

Separation of Acetylated Neutral Glycolipids

The following example was chosen to illustrate the improved separation of acetylated glycolipids that is possible using CD-HPTLC. Acetylated gangliotriaosylceramide and acetylated globoside comigrate when chromatographed in a conventional tank using the solvent dichloroethane–acetone (55 : 45, v/v). Initial attempts with CD-HPTLC utilized less polar mixtures of dichloroethane and acetone. The optimal conditions of dichloroethane–acetone (80 : 20) for 3 hr, bed length 6.2 cm produced a separation of bands of only 3 mm. However, diluting the solvent with hexane produced a dramatic increase in band separation. Continuous

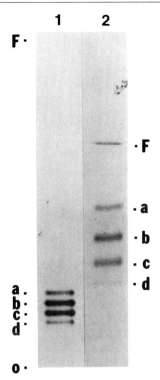

FIG. 2. Comparison of ganglioside G_{M1b} bands separated by conventional versus CD-HPTLC. Lymphoma cells were metabolically labeled with [^{3}H]galactose, extracted, and upper phase lipids purified on Bond-Elut C_{18} reversed-phase columns. Aliquots (10,000 cpm) were chromatographed on glass-backed HPTLC plates as follows: lane 1, regular chamber using solvent C; lane 2, continuous development chamber (plate at position 4) using solvent D. Fluorography was performed as described in Fig. 1, using an exposure time of 3 days at $-70°$ for both lanes 1 and 2. R_f of ganglioside standards: lane 1, N-glycolylhematoside (G_{M3}) 0.33, G_{M1} 0.19, and G_T 0.01; lane 2, G_{M3} 1.0, G_{M1} 0.77, and G_T 0.04. Reprinted with permission from Young and Borgman.[4]

development in the solvent dichloroethane–acetone–hexane (20 : 30 : 50) for 1.5 hr separated the bands by 1.5 cm (bed length 5.8 cm, acetylated GgOse$_3$Cer R_f 0.81, acetylated globoside R_f 0.55).

TLC-Immunostaining

The technique developed by Magnani et al.[11] for staining glycolipids on TLC plates with monoclonal antibodies has been of great value in

[11] J. L. Magnani, B. Nilsson, M. Brockhaus, D. Zopf, Z. Steplewski, H. Koprowski, and V. Ginsburg, J. Biol. Chem. **257,** 14365.

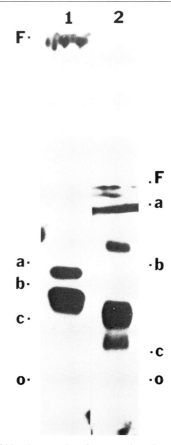

FIG. 3. Comparison of blood group Lewis a reactive glycolipids separated by conventional versus CD-HPTLC and visualized by TLC-immunostaining. Meconium neutral glycolipids were chromatographed on aluminum-backed HPTLC plates as follows: lane 1, regular chamber using solvent E; lane 2, continuous development chamber (plate at position 3) using solvent F. TLC-immunostaining was performed as described in Materials and Methods. Glycolipid standards: a, Forssman; b, G_{M1} ganglioside; and c, G_T ganglioside. The dark area at the front of the chromatogram in lane 1 and the dark line below the front in lane 2 represent nonspecific staining. Reprinted with permission from Young and Borgman.[4]

identifying new glycolipid species. The following example was chosen to indicate the increased resolution of TLC-immunostaining that can be achieved using CD-HPTLC. Antibody staining of TLC plates was performed according to the method of Magnani et al.[11] Briefly, aliquots of human meconium neutral glycolipids were applied to aluminum-backed HPTLC plates and chromatographed as described in Fig. 3. After drying, the sheets were dipped in hexane containing 0.05% poly(isobutyl methacrylate) (Polysciences, Warrington, PA), blocked with 5% bovine serum albumin in phosphate-buffered saline, and then incubated in succession with anti-Lewis a antibody CF4C4 and iodinated staphylococcal protein A. After washing and drying, the sheets were exposed at −70° for 1–2 hr to Kodak XAR-5 film utilizing an intensifying screen. Human meconium contains several glycolipids that bear the Lewis a human blood group determinant (Galβ1 → 3[Fucα1 → 4]GlcNAc) as we described previously.[6] By conventional HPTLC these reactive species appear as two bands, a faster band that comigrates with authentic Lewis a ceramide pentasaccharide and a broad slower band (Fig. 3, lane 1). CD-HPTLC in solvent F not only increased the separation between the two main bands but also separated the slower main band into two components (Fig. 3, lane 2). An alternative solvent for separating complex glycolipids consists of n-propanol–H_2O (80 : 20), which produced excellent separation in the G_{MI} to G_T area when plates were run in position 3 for 1.5–2 hr. (data not shown).

Comments

The short-bed continuous development HPTLC method should be of use for the analysis and preparative isolation of glycosphingolipids. One of the drawbacks to conventional HPTLC analysis is the difficulty of separating complex glycolipids in chloroform–methanol systems.[12] Several approaches have been taken to improve the separation of these complex glycolipids: (1) some poorly resolved glycolipids can be well separated after acetylation; (2) certain glycolipid mixtures can be separated on borate-impregnated plates[13]; and (3) a conventional plate can be developed multiple times with intermediate drying to achieve greater separation than that possible by a single development.[14] Continuous development using a short-bed tank accomplishes the same goal of the latter approach but in a much more efficient fashion.

The four examples listed above should serve as a starting point for determining ideal CD-HPTLC conditions for a given application. The initial choice of solvent should be one which is less polar than that which is optimal for separation of the compounds in question by conventional HPTLC. Generally this means preparing a less polar mixture of the standard components (such as chloroform, methanol, and water); only with dichloroethane–acetone systems used for acetylated glycolipids did we find the technique of dilution with hexane[3] to be of practical use. For analytical purposes the shortest bed lengths (i.e., the most vertical plate

[12] J. N. Kanfer and S. Hakomori, *Handb. Lipid Res.* **3**, 1 (1983).
[13] C. G. Gahmberg and S. Hakomori, *J. Biol. Chem.* **250**, 2438 (1975).
[14] K. Stellner, K. Watanabe, and S. Hakomori, *Biochemistry* **12**, 656 (1973).

positions) may be preferred because highest solvent velocities can be achieved with minimal diffusion of bands. However, for preparative purposes a longer bed length will be required to obtain the necessary distance between bands (Figs. 1 and 2). In such preparative cases the autoradiographic intensity of CD-HPTLC bands in general will be less than that obtained by conventional HPTLC due in part to continuous development separation of species that are superimposed in conventional HPTLC.

[10] Analysis of Sialic Acids

By ROLAND SCHAUER

Since 1978, when the last review on the analysis of sialic acids appeared in this series[1] interest in these compounds has increased, as their involvement in the biological functions of many different molecules and cells became evident.[2-5] For example, sialic acids play a role in cell biology by their negative charge, by influencing the conformation of (glyco-) proteins, by acting as receptors for microorganisms, toxins, and hormones, and by masking receptors and other, e.g., immunological, recognition sites of molecules and cells. The latter function attracts more interest as it seems to be of outstanding importance.[4,5] In these functions sometimes sialic acids other than N-acetylneuraminic acid are involved. For instance, N-glycolylneuraminic acid as component of gangliosides has been discussed as a cancer-specific antigen (Hanganutziu–Deicher antigen) in man.[6,7] The ganglioside G_{D3} containing N-acetyl-9-O-acetylneuraminic acid has been demonstrated to be a specific antigen for human melanoma cells.[8] O-Acetyl groups seem to be also involved in the regulation of sialic acid catabolism,[9] and have been shown to bind specifically

[1] This series, Vol. 50 [64].
[2] R. Schauer, Adv. Carbohydr. Chem. Biochem. 40, 131 (1982).
[3] W. Reutter, E. Köttgen, C. Bauer, and W. Gerok, Cell Biol. Monogr. 10, 263 (1982).
[4] R. Schauer, Trends Biochem. Sci. 10, 357 (1985).
[5] R. Schauer, A. K. Shukla, C. Schröder, and E. Müller, Pure Appl. Chem. 56, 907(1984).
[6] R. Schauer, Biochem. Soc. Trans. 11, 270 (1983).
[7] H. Higashi, Y. Fukui, S. Ueda, S. Kato, Y. Hirabayashi, M. Matsumoto, and M. Naiki, J. Biochem. (Tokyo) 95, 1517 (1984).
[8] D. A. Cheresh, R. A. Reisfeld, and A. P. Varki, Science 225, 844 (1984).
[9] A. P. Corfield and R. Schauer, Cell Biol. Monogr. 10, 195 (1982).

influenza C viruses[10] or interfere with the attachment of other influenza virus strains.[11]

Sialic acids have also gained increasing attention in pathological processes, e.g., sialidosis, in which sialoglycoconjugates accumulate due to a lack of sialidase.[12] Several pathological states are characterized by the excretion of large amounts of free N-acetylneuraminic acid in urine.[13,14] Much interest is focused on malignancies, which are often accompanied by an elevated level of sialic acids in serum. Therefore, determination of these sugars may be useful for the diagnosis and prognosis of cancer.[15] It was also observed that many cancer cell types contain more membrane sialic acids than the corresponding normal cells.[1,3,16]

From this follows that sensitive and reliable techniques are necessary for the qualitative and quantitative analysis of sialic acids, which are present in biological materials only in low amounts. They are components of glycoconjugates (glycoproteins and gangliosides) as well as of oligo- and polysaccharides of some microorganisms and of higher animals from the echinoderms upward. In these substances sialic acids are α-glycosidically linked to galactose, N-acetylgalactosamine, N-acetylglucosamine, or sialic acid residues. They usually represent terminal sugars which are most frequently linked to the terminal, nonreducing positions of oligosaccharide chains, but sometimes they are bound to internal carbohydrates. More rarely, sialic acids also occur within oligosaccharide chains. The occurrence and chemistry of these compounds have been reviewed elsewhere.[17,18] Natural β-linkages of sialic acids are only known from their CMP-glycosides.[19,20]

The analytical progress made in the carbohydrate field in general and with sialic acids in particular during the past 10 years has led to the discovery of several new sialic acids in natural sources. Thus, the number

[10] G. Herrler, R. Rott, H.-D. Klenk, H.-P. Müller, A. K. Shukla, and R. Schauer, *EMBO J.* **4,** 1503 (1985).

[11] J. C. Paulson, G. N. Rogers, S. M. Carroll, H. H. Higa, T. Pritchett, G. Milks, and S. Sabesan, *Pure Appl. Chem.* **56,** 797 (1984).

[12] M. Cantz, *Cell Biol. Monogr.* **10,** 307 (1982).

[13] J. P. Kamerling, J. F. G. Vliegenthart, R. Schauer, G. Strecker, and J. Montreuil, *Eur. J. Biochem.* **56,** 253 (1975).

[14] M. Renlund, P. Aula, K. O. Raivio, S. Autio, K. Sainio, J. Rapola, and S.-L. Koskela, *Neurology* **33,** 57 (1983).

[15] K. M. Erbil, J. D. Jones, and G. G. Klee, *Cancer (Philadelphia)* **55,** 404 (1985).

[16] J. A. Alhadeff and R. T. Holzinger, *Int. J. Biochem.* **14,** 119 (1982).

[17] A. P. Corfield and R. Schauer, *Cell Biol. Monogr.* **10,** 5 (1982).

[18] J. Montreuil, *Adv. Carbohydr. Chem. Biochem.* **37,** 157 (1980).

[19] J. Haverkamp, T. Spoormaker, L. Dorland, J. F. G. Vliegenthart, and R. Schauer, *J. Am. Chem. Soc.* **101,** 4851 (1979).

[20] H. H. Higa and J. C. Paulson, *J. Biol. Chem.* **260,** 8838 (1985).

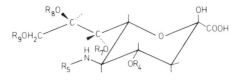

R_5	$R_{4,7,8,9}$
$-\overset{\overset{\displaystyle O}{\|\|}}{C}-CH_3$	$-H$ (4,7,8,9)
$-\underset{\underset{\displaystyle OH}{\|}}{\overset{\overset{\displaystyle O}{\|\|}}{C}}-CH_2$	$-\overset{\overset{\displaystyle O}{\|\|}}{C}-CH_3$ (4,7,8,9)
	$-\underset{\underset{\displaystyle OH}{\|}}{\overset{\overset{\displaystyle O}{\|\|}}{C}}-CH-CH_3$ (9)
	$-CH_3$ (8)
	$-SO_3H$ (8)
	$-PO_3H_2$ (9)

FIG. 1. Structure of natural sialic acids. N- and O-substituents together with their position (in parentheses) in the neuraminic acid molecule are indicated. Several combinations of these substituents are known and up to three O-acetyl residues have been found in one sialic acid molecule.[2,17] N-Acetylneuraminic acid 9-phosphate is an intermediate in sialic acid biosynthesis and has been found so far only as free sugar.[2,9]

of sialic acids increased to 23,[2,17,21] not including the six unsaturated sialic acids (N- and O-derivatives of 2-deoxy-2,3-didehydroneuraminic acid)[13,21,22] and the three 2,7-anhydrosialic acids[21,23] found in the urine of man and animals or in human cerumen.[24] All these sialic acids represent derivatives of neuraminic acid (systematic name 5-amino-3,5-dideoxy-D-glycero-D-galactononulosonic acid) with acetyl or glycolyl groups at the amino function and often acetyl, lactyl, sulfate, phosphate, or methyl groups at one or several of the hydroxyl groups (Fig. 1). Combinations of these residues are possible, as in N-glycolyl-8-O-methyl-9-O-acetyl-neuraminic acid of the starfish *Asterias rubens*.[23,25] Highly O-acetylated

[21] R. Schauer, C. Schröder, and A. K. Shukla, *Adv. Exp. Med. Biol.* **174,** 75 (1984).

[22] A. K. Shukla, C. Schröder, U. Nöhle, and R. Schauer, *Proc. Int. Symp. Glycoconjugates, 7th, 1983,* p. 155 (1983).

[23] C. Schröder, U. Nöhle, A. K. Shukla, and R. Schauer, *Proc. Int. Symp. Glycoconjugates, 7th, 1983,* p. 162 (1983).

[24] M. Suzuki, A. Suzuki, T. Yamakawa, and E. Matsunaga, *J. Biochem. (Tokyo)* **97,** 509 (1985).

[25] R. Schauer and M. Wember, *Proc. Int. Symp. Glycoconjugates, 8th, 1985* Vol. 1, p. 264 (1985).

sialic acids are also known, e.g., *N*-acetyl-7,8,9-tri-*O*-acetylneuraminic acid from bovine submandibular gland glycoprotein. The structures and systematic names of the sialic acids known are listed elsewhere.[2,17,21]

In this article some new techniques for the analysis of sialic acids will be described together with improvements of older methods. Other well established assays have been reviewed or described in detail earlier in this series.[1]

Isolation and Purification of Sialic Acids

As sialic acids occur in a great diversity of molecules and often in minute amounts together with various other anionic compounds, special care must be taken in their quantitative purification. Furthermore, sialic acids often carry labile *O*-acetyl groups, which easily undergo hydrolysis or intramolecular migration[26,27] during the isolation procedure. The following procedures are therefore prerequisites to obtaining exact quantitative and qualitative sialic acid analyses.

Before liberation of sialic acids from crude biological materials, e.g., lyophilized cells or tissue homogenates, low molecular weight substances, which may interfere with sialic acid analysis, should be removed by 24-hr dialysis against water in the cold. Before dialysis, however, it is further recommended to inactivate enzymes by either heating for 5 min at 60° or addition of 50% ethanol and keeping for 1 hr at room temperature. In this way sialidase and esterase activities are destroyed which otherwise could lead to loss of sialic acids or their *O*-acetyl groups (see below). As sometimes free sialic acids occur in larger quantities, the dialysates should be checked for the presence of diffusible sialic acids.

Release from Glycosidic Linkage

Sialic acids may be released from glycosidic linkages by either dilute acids (0.1 *N* hydrochloric or sulfuric acid, 80°, 50–60 min) or sialidases (EC 3.2.1.18) as described in this series.[1] The conditions for acid hydrolysis must be milder (0.01 *N* hydrochloric acid or sulfuric acid, or formic acid of pH 2.2; 70°, 60 min) in the case of *O*-acylated sialic acids, to avoid hydrolysis of more than 30–50% of *O*-acetyl groups. Because under the milder conditions not all sialic acids are liberated, hydrolysis should be repeated in 0.1 *N* mineral acid after dialysis, for optimum recovery of sialic acids. The use of sulfuric acid is recommended, as it can be removed by barium hydroxide from the solution after hydrolysis, and in this way further damage of sialic acids by other acids is avoided. Varki and

[26] J. P. Kamerling, H. van Halbeek, J. F. G. Vliegenthart, R. Pfeil, A. K. Shukla, and R. Schauer, *Proc. Int. Symp. Glycoconjugates, 7th, 1983*, p. 160 (1983).
[27] A. Varki and S. Diaz, *Anal. Biochem.* **137**, 236 (1984).

Diaz[27] found that the hydrolytic release of sialic acids is quantitative and the loss and migration of O-acetyl groups in the sialic acid side chain reach a minimum by the use of 2 M acetic acid at 80° for 3 hr. In the case of isolation and analysis of 7-O-acetylated sialic acids, the pH after hydrolysis should not deviate from 4–6, since the migration rate of this O-acetyl group to O-9 is lowest under these conditions (Kamerling et al., unpublished).[27]

Acid hydrolysis may be preferred to the use of sialidases, in view of the facts that sialic acids O-acetylated at the side chain are released from their glycosidic linkage only at reduced rates (20–50%) by these enzymes when compared with N-acetylneuraminic acid, and sialic acids with an O-acetyl residue at C-4 are completely resistant toward the action of all sialidases tested so far.[2,9,28] Furthermore, sialic acids bound to internal galactose residues to which N-acetylgalactosamine is linked by its glycosidic group, as in gangliosides and some glycoproteins, are almost or completely resistant to the action of most sialidases.[9,29] The addition of detergents (e.g., sodium cholate) may improve the hydrolysis of this "internal" sialic acid, as was studied in detail with the ganglioside G_{M1} and Clostridium perfringens sialidase.[29] Exceptions are sialidases from Arthrobacter ureafaciens[9,28] and Sendai virus,[30] which were found to release this sialic acid residue easily already in the absence of detergents. Data obtained by 500 MHz proton NMR spectroscopy may explain the resistance of the "internal" sialic acid residue of G_{M1} or the corresponding oligosaccharide insofar as the glycosidic bond of this residue has less conformational freedom than the terminal sialic acid of gangliosides.[29,29a]

When using sialidases it should be considered that $\alpha(2-3)$ linkages of sialic acids are hydrolyzed faster (2–5 times in the case of bacterial and mammalian enzymes) than $\alpha(2-6)$ linkages, with the exception of the A. ureafaciens enzyme, which shows almost double the hydrolysis rate of $\alpha(2-6)$ bonds.[9,28] Because sialidases from influenza viruses act extremely slow on $\alpha(2-6)$ linkages when compared with $\alpha(2-3)$ bonds, they can be used for tentative analysis of the type of sialic acid linkage present in an oligosaccharide or glycoconjugate.[28,31,32] For example, Newcastle disease

[28] A. P. Corfield, J.-C. Michalski, and R. Schauer, in "Sialidases and Sialidoses, Perspectives in Inherited Metabolic Diseases" (G. Tettamanti, P. Durand, and S. Di Donato, eds.), Vol. 4, p. 3. Ermes, Milan, 1981.

[29] A. P. Corfield, R. Schauer, L. Dorland, J. F. G. Vliegenthart, and H. Wiegandt, J. Biochem. (Tokyo) 97, 449 (1985).

[29a] L. Dorland, H. van Halbeek, J. F. G. Vliegenthart, R. Schauer, and H. Wiegandt, Carbohydr. Res. 151, 233 (1986).

[30] Y. Suzuki, T. Morioka, and M. Matsumoto, Biochim. Biophys. Acta 619, 632 (1980).

[31] A. P. Corfield, M. Wember, R. Schauer, and R. Rott, Eur. J. Biochem. 124, 521 (1982).

[32] A. P. Corfield, H. Higa, J. C. Paulson, and R. Schauer, Biochim. Biophys. Acta 744, 121 (1983).

virus sialidase hydrolyzes $\alpha(2-3)$ linkages almost 500-fold faster than $\alpha(2-6)$ linkages.[31] $\alpha(2-8)$ bonds usually are cleaved at intermediate rates by the viral, bacterial, and mammalian sialidases tested. The natures of the penultimate sugar and of the total oligosaccharide, glycoprotein, or ganglioside are of great influence on the hydrolysis rate, too.[2,9,28-32]

Usually bacterial sialidases are used for enzymatic hydrolysis of sialic acids.[33] Due to the frequent occurrence of sialic acid $\alpha(2-6)$ bonds and "internal" sialic acid residues, use of the enzyme from *A. ureafaciens* is recommended. Samples containing bound sialic acids are dissolved in 0.1 *M* acetate buffer, pH 5 and incubated with 50 mU/ml of this enzyme for 1–2 hr at 37°. The sialic acids liberated are either determined directly by colorimetric means[1] or dialyzed for qualitative analysis (see below). To check whether all sialic acids were liberated by sialidase, acid hydrolysis of the incubation mixture after dialysis may be carried out.

In some cases the use of immobilized sialidase, which can be removed from the incubation mixture by centrifugation, without, e.g., heating of the material to inactivate soluble enzyme, is advantageous.[34] A closed circuit has been developed allowing continuous pumping of the substrate over the sialidase beads and removal of sialic acids liberated by dialysis. By this method complete and gentle desialylation is possible with substrates, e.g., mucins or cells, from which total removal of sialic acids is difficult by common techniques.

Purification

The procedures applied, including several precautions and the necessity to delipidify sialic acid solutions after hydrolysis and dialysis with diethyl ether or *n*-hexane to obtain quantitative yields and qualitative analyses of sialic acids without errors, have been described in this series.[1] In the case of high sialic acid concentrations and the absence of lipid material, sialic acids can directly, without dialysis, be applied to ion-exchange columns after hydrolysis and removal of sulfuric acid by barium hydroxide. Regarding the anion-exhange step, two improvements were made. (1) With the use of Dowex 2X8 resin and a flat formic acid gradient (0–0.6 *M*), prefractionation of a sialic acid mixture is possible and contaminants are more efficiently removed.[34] For example, four sialic acid pools differing in their composition were obtained from bovine submandibular gland glycoproteins, which were further fractionated on cellulose using 1-butanol–1-propanol–water (1:2:1) as solvent.[1] This modifica-

[33] R. Schauer and U. Nöhle, *in* "Methods of Enzymatic Analysis" (H. U. Bergmeyer, ed.), 3rd ed., Vol. 4, p. 195. Verlag Chemie, Weinheim, 1984.
[34] A. P. Corfield, J.-M. Beau, and R. Schauer, *Hoppe-Seyler's Z. Physiol. Chem.* **359**, 1335 (1978).

tion, instead of using the earlier gradient of 0–2 M formic acid,[1] led to the discovery of a variety of new sialic acids.[35] (2) Elution of sialic acids from the anion exchanger can also be carried out by pyridine–acetic acid mixtures of pH 5.4. For gradient elution concentrations from 0.05 to 0.15 M may be chosen, and for routine batch elutions 0.1 M solutions are used. This method is advantageous because too-acidic conditions, which may lead to loss of O-acetyl groups and to their intramolecular migration,[35a] can be avoided.

The column chromatography procedures originally designed for the isolation and fractionation of sialic acids in larger quantities can easily be scaled down for analytical purposes. Often sialic acids cannot be purified by the described procedures to an extent sufficient for further definite analyses including thin-layer chromatography or gas–liquid chromatography. In these cases rinsing of sialic acid samples after the ion-exchange steps over cellulose in 1-butanol–1-propanol–water (1 : 2 : 1) is recommended (30–50 μg sialic acid and 10 ml column volume). Final purification may also be achieved by gel filtration in water on Sephadex G-10 or G-25, or in pyridinium acetate (0.1 M, pH 5.4) on BioGel P-2.

Colorimetric Analysis of Sialic Acids

The methods for colorimetric analysis of sialic acids have been described and discussed in the last carbohydrate issue of this series with regard to their sensitivity, specificity, and applicability to free, bound or differently substituted sialic acids.[1] Of the diphenol assays, the orcinol/ Fe^{3+}/HCl test and, of the periodic acid/thiobarbituric acid assays, the procedure according to Aminoff have been described.[1] The application of these colorimetric reactions in micro scales has been emphasized, as they enable the easy determination of minimum amounts of 1–3 μg sialic acid in the diphenol reactions and 0.5 μg in the periodic acid/thiobarbituric acid assays. The molar extinction coefficients of some sialic acids exhibited in the most frequently used assays have been summarized.[2,36]

Periodate/Acetylacetone Reaction

Several sensitive fluorimetric assays are available for sialic acid determination. Two methods[37,38] were described in this series.[1] However, to

[35] G. Reuter, R. Pfeil, S. Stoll, R. Schauer, J. P. Kamerling, C. Versluis, and J. F. G. Vliegenthart, *Eur. J. Biochem.* **134**, 139 (1983).
[35a] J. P. Kamerling, R. Schauer, A. K. Shukla, S. Stoll, H. van Halbeek, and J. F. G. Vliegenthart, *Eur. J. Biochem.*, in press.
[36] R. Schauer and A. P. Corfield, *Cell Biol. Monogr.* **10**, 77 (1982).
[37] H. H. Hess and E. Rolde, *J. Biol. Chem.* **239**, 3215 (1964).
[38] K. S. Hammond and D. S. Papermaster, *Anal. Biochem.* **74**, 292 (1976).

the author's knowledge, they are not widely used. A third test has been added[39] which is based on methods described by Massamiri et al.[40] and Brearly and Weiss.[41] It allows discrimination between sialic acids with or without O-acyl groups at C-8 or C-9 of the glycerol side chain. Sialic acids are oxidized by periodate, and the formaldehyde produced from C-9 of sialic acid reacts with acetylacetone in the presence of ammonium acetate leading to a fluorogen. The reaction scheme and the formula are shown in Fig. 2. The procedure is as follows.

Reagents

Phosphate buffer (PBS), pH 7.2: 8 g sodium chloride; 0.2 g potassium chloride; 1.4 g disodium hydrogen phosphate dihydrate and 0.2 g potassium dihydrogen phosphate are dissolved to a final volume of 1 liter with bidistilled water.

Acetylacetone reagent: 200 μl acetylacetone, 300 μl glacial acetic acid, and 15 g ammonium acetate are dissolved to a final volume of 100 ml with bidistilled water. The reagent is stable for 1 week at 4° in the dark.

Sodium periodate, 2.5 mM: 53 mg are dissolved in 100 ml of the buffer. Sodium arsenite solution, 0.154 M: 2 g sodium arsenite are dissolved in 100 ml of 0.5 M HCl.

N-Acetylneuraminic acid or N-acetyl-9-O-acetylneuraminic acid solutions, 0.3 mM, in water. These standard solutions are stable for at least 1 month at −70°.

Procedure. The analyses (reagent blanks and assays with or without de-O-acetylation) were run in triplicate. Samples containing 2–5 μg of sialic acids either free or bound to soluble complex carbohydrates, cell membranes, or intact cells are dissolved or suspended with buffer to a final volume of 100 μl. To the probes of one series 100 μl of 0.2 M NaOH is added and the samples are kept at 4° for 45 min to saponify O-acetyl groups. Then 100 μl of 0.2 M HCl is added for neutralization. To the samples of the other series, where no de-O-acetylation is carried out, including the blanks, 200 μl of 0.1 M NaCl is added. All samples are oxidized with the aid of 200 μl sodium periodate and keeping the mixtures at 4° for 15 min in the dark. After centrifugation at 12,000 g for 5 min in case of the presence of insoluble material, 400 μl of the supernatants are mixed with 100 μl sodium arsenite and 400 μl acetylacetone solutions.

[39] A. K. Shukla and R. Schauer, *Hoppe-Seyler's Z. Physiol. Chem.* **363**, 255 (1982).
[40] J. Massamiri, G. Durand, A. Richard, J. Féger, and J. Agneray, *Anal. Biochem.* **97**, 346 (1979).
[41] G. M. Brearly and J. B. Weiss, *Biochem. J.* **110**, 413 (1968).

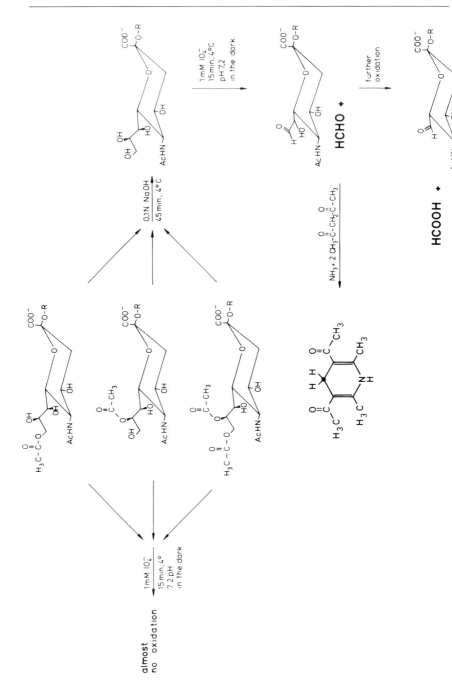

The mixtures are incubated at 60° for 10 min, and samples of 0.7 ml are diluted with 1.8 ml of bidistilled water. The sialic acid samples give a light yellow color. The relative fluorescence [λ_{max} (excitation) = 410 nm, λ_{max} (emission) = 510 nm] of each sample is measured against the reagent blank.

With this method a minimum amount of 200 ng (0.6 nmol) of sialic acids can be determined. Sialic acids being unsubstituted at the glycerol side chain including N-acyl-4-O-acetylneuraminic acids give equal yields of fluorescence, as oxidation with periodate results in the production of equal amounts of formaldehyde (1 mol/mol sialic acid, Fig. 2). In contrast, 9-O-acylated sialic acids do not yield formaldehyde and are therefore almost negative in this test. Similarly, 8-O-substituted sialic acids cannot directly be determined by the acetylacetone assay. After saponification of the O-acyl groups, the increase in formaldehyde and fluorogen, respectively, corresponds to the amount of 9-O-acylated sialic acids present in the sample, as the 9 position is by far the most frequently acylated group in the sialic acid side chain. The relative amount can thus be calculated from the difference in fluorescence between the assays run with or without the saponification step described.

This method proved to be especially useful for estimation of the content of unsubstituted or 9(8)-O-acetylated sialic acids in cells, best investigated with erythrocyte membranes, which frequently contain O-acetylated sialic acids.[39] The main advantage of this method is that sialic acids need not be removed from cells before analysis, which may be incomplete and lead to loss of sialic acids and their O-acetyl groups.

Although the test is quick and sensitive, a variety of compounds interfere. Disturbance is caused by all substances leading to formaldehyde under the influence of periodate, e.g., glycerol. It is therefore recommended to apply the test to purified glycoconjugates and to cells, which, after rupture and inactivation of sialidase and esterase by 50% ethanol, should be freed from small molecules by dialysis. With erythrocytes the best results were obtained by analysis of isolated membranes. The values are too low with intact erythrocytes and do not correspond to those obtained with other colorimetric methods, as these cells have been shown to consume some of the formaldehyde produced by periodate.[42] No inter-

---

[42] D. J. Capaldi and K. E. Taylor, *Microchem. J.* **31**, 275 (1985).

FIG. 2. Chemical reactions involved in formation of the fluorogen 3,5-diacetyl-1,4-dihydro-2,6-dimethylpyridine in the periodate/acetylacetone assay of sialic acids.[39] Carbon-4 of the pyridine derivative is derived from sialic acid-borne formaldehyde. Sialic acids substituted at C-8 or C-9 do not give fluorescence.

ference with the test was observed with desialylated cell membranes or glycerol fatty acid esters, even under the conditions of saponification of O-acetyl groups. The assay, however, cannot be carried out in Tris buffer. With the precautions discussed, the acetylacetone test is rather specific for sialic acids and useful for the investigation of these sugars in a variety of systems including cell surfaces. This is important especially with regard to 9-O-acetylated sialic acids, which attract more and more attention due to their wide occurrence, lability, and biological roles.[2,4]

Other Colorimetric Assays

Determination of the total amount of O-acetyl groups in sialic acids using hydroxylamine has been described in this series.[1] This assay is, however, less sensitive than the periodate/acetylacetone test in that it requires at least 0.5 μmol ester groups for exact analysis. On the other hand, the acetylacetone assay is restricted to O-acetyl groups at C-8 or C-9. Because these colorimetric assays indicate the number and in some cases the position of ester groups only indirectly, for exact structural analysis of sialic acids with O-acetyl groups or other residues, chromatographic methods including mass spectrometry or NMR spectroscopy are necessary. The determination of O-lactyl groups in sialic acids with the aid of L-lactate dehydrogenase after saponification has been described.[1]

Further colorimetric sialic acid assays are based on the formation of pyruvate by the action of N-acetylneuraminate pyruvate-lyase (EC 4.1.3.3, N-acetylneuraminate lyase) on free sialic acid.[1] Pyruvate is converted to lactate by L-lactate dehydrogenase (EC 1.1.1.27). The consumption of NADH for pyruvate reduction is followed at 366 nm. It is proportional to the amount of free sialic acids.[1] This assay is sensitive and specific, allowing the determination of sialic acid quantities in the micromole range. As this test is limited to free sialic acids, these sugars can first be released by treatment of sialoglycoconjugates with sialidase.

In another combined assay pyruvate derived from sialic acids by the lyase reaction is converted to acetyl phosphate, carbon dioxide, and hydrogen peroxide by pyruvate oxidase (EC 1.2.3.3). Hydrogen peroxide, which is formed in amounts equivalent to sialic acid, leads to a chromophor with λ_{max} at 505 nm wavelength from 4-aminoantipyrin and p-chlorophenol under the influence of peroxidase (EC 1.11.1.7).[43] Substitution of the latter compound by N-ethyl-N-2-hydroxyethyl-3-toluidine results in a red chromophor (λ_{max} 550) of a sialic acid test combination commercially available. These enzymatic assays are specific for sialic acids, due to the strong specificity of the lyase. However, pyruvate,

---

[43] K. Sugahara, K. Sugimoto, O. Nomura, and T. Usui, *Clin. Chim. Acta* **108**, 493 (1980).

which might naturally be present in the assay mixture, must separately be determined as one of the blanks without the lyase and subtracted from the value obtained in the presence of the lyase. Furthermore, the limitations of the use of sialidase for sialic acid release have been described above and have to be considered for evaluation of the results.

Chromatographic Analyses of Sialic Acids

Thin-Layer Chromatography

Application of this tool in one or two dimensions, including saponification of O-acyl groups, for sialic acid analysis has extensively been described and discussed in this series.[1] Principally the same systems are still in use, but chromatography on cellulose in 1-butanol–1-propanol–0.1 N HCl (1 : 2 : 1, v/v/v) has proved to be the most suitable. It shows the best and most reproducible separation of different sialic acids and is less sensitive to interfering substances when compared with the other systems. With the new materials (e.g., cellulose sheets from Merck, Darmstadt) and prerunning of the plates in the solvent, better separations of the O-acetylated sialic acid isomers can be obtained (Table I)[35,36] when compared with the data given earlier in this series.[1] For example, it is now possible to separate sialic acids mono-O-acetylated at different positions. The use of high-performance thin-layer chromatography (HPTLC) silica gel plates and 1-propanol–water (7 : 3, v/v) as solvent is also recommended. For the separation of sialylated oligosaccharides from colostrum a system on silica gel using ethanol–1-butanol–pyridine–water–acetic acid (100 : 10 : 10 : 30 : 3, v/v/v/v/v) was developed.[44] Sialyl-α(2-3)- and α(2-6)-lactoses differing in either N-acetyl- or N-glycolylneuraminic acid are separated in this system. Furthermore, chromatography in two dimensions with ammonia treatment between the first and second run, as applied originally to free sialic acids,[1] enables the detection of O-acetylated oligosaccharides, such as in milk and urine.[17] Similarly, with two-dimensional chromatography of gangliosides including intermediary saponification of ester groups, the identification of gangliosides with O-acetylated sialic acids is possible even in a complex mixture.[45-47] Such gangliosides

[44] R. W. Veh, J.-C. Michalski, A. P. Corfield, M. Sander, D. Gies, and R. Schauer, J. Chromatogr. 212, 313 (1981).

[45] R. W. Veh, M. Sander, J. Haverkamp, and R. Schauer, in "Glycoconjugate Research" (J. D. Gregory and R. W. Jeanloz, eds.), Vol. 1, p. 557. Academic Press, New York, 1979.

[46] D. C. Gowda, G. Reuter, A. K. Shukla, and R. Schauer, Hoppe-Seyler's Z. Physiol. Chem. 365, 1247 (1984).

[47] S. Sonnino, R. Ghidoni, V. Chigorno, M. Masserini, and G. Tettamanti, Anal. Biochem. 128, 104 (1983).

TABLE I
THIN-LAYER CHROMATOGRAPHIC MIGRATION RATES (R_f VALUES)
OF SIALIC ACIDS[a]

	R_f	
Compound	Cellulose	Silica gel
N-Acetylneuraminic acid	0.45	0.39
N-Acetyl-4-O-acetylneuraminic acid	0.60	0.61
N-Acetyl-7-O-acetylneuraminic acid	0.54	0.61
N-Acetyl-9-O-acetylneuraminic acid	0.63	0.61
N-Acetyl-7,9-di-O-acetylneuraminic acid	0.70	0.73
N-Acetyl-8,9-di-O-acetylneuraminic acid	0.75	—[b]
N-Acetyl-7,8,9-tri-O-acetylneuraminic acid	0.80	—[b]
N-Acetyl-9-O-L-lactylneuraminic acid	0.56	0.61
N-Acetyl-2-deoxy-2,3-didehydroneuraminic acid	0.55	0.70
N-Glycolylneuraminic acid	0.35	0.39
N-Glycolyl-4-O-acetylneuraminic acid	0.65	0.61
N-Glycolyl-9-O-acetylneuraminic acid	0.55	0.61
N-Glycolyl-7,9-di-O-acetylneuraminic acid	0.70	—[b]
N-Glycolyl-2-deoxy-2,3-didehydroneuraminic acid	0.45	—[b]

[a] On 0.1 mm cellulose using 1-butanol–1-propanol–0.1 M HCl (1:2:1, v/v/v) or on 0.2 mm silica gel with 1-propanol–water (7:3, v/v).[2,35,36]
[b] Not determined.

frequently occur in mammalian brain,[17,47] but have also been found in erythrocytes[46] and human melanoma cells.[8]

Spraying with the orcinol reagent for sialic acid staining[1] is still in use on account of its high specificity, giving the gray-lilac color typical for sialic acids. The color is orange when the carboxylic group of sialic acid is reduced to an alcohol residue.[5]

High-Performance Liquid Chromatography

Sialic acid analyses by the chromatographic systems used so far (e.g., thin-layer chromatography, gas-liquid chromatography) are hampered by the necessity to extensively purify these sugars by tedious methods before analysis. Even for colorimetric analyses purification is often necessary, as several natural compounds interfere with the assays. This is one of the reasons why the isolation and characterization of enzymes of sialic acid metabolism progressed so slowly. These difficulties have largely been overcome by the development of sialic acid analysis by HPLC.

Sialic acids are quantified as a whole on cation-exchange[48] or anion-exchange[49] resins. Separation of N-acetylneuraminic acid, N-glycolylneuraminic acid, and O-acetylated derivatives was possible for the first time as borate complexes on a sugar analyzer using a strong basic anion-exchange resin.[50]

A further improvement is the analysis of underivatized sialic acid mixtures, CMP-sialic acid, and sialyllactose isomers on the strong basic anion-exchange resin Aminex A-28 or better A-29, using stainless-steel columns of 40 × 4.6 mm.[51,52] Before the column is filled, the resin is converted to the desired ionic form and suspended in the solution in which chromatography is performed. In this way excessive back pressure is avoided. After a few hundred runs, the resin is taken out and regenerated by treatment with 3 M HCl at 60° for 30 min, washing with water, 0.25 M sodium sulfate solution, water, and finally suspension in elution buffer. For analysis, samples of 10 μl containing 0.05–1 μg of sialic acids dissolved in water or buffers are applied to the column. Sialic acids are eluted isocratically by 0.75 mM sodium sulfate solution. A gradient from 0 to 1 mM sodium sulfate may be used alternatively. The solutions are degassed by helium. The flow rates are 0.3–1 ml/min at 5–30 bar. Eluted sialic acids are monitored at wavelengths between 195 and 215 nm. Detector sensitivity is 0.04 absorbance units full scale (a.u.f.s.).

Sialic acids eluting from the column can further be analyzed colorimetrically and by thin-layer chromatography or gas–liquid chromatography in combination with mass spectrometry. For these purposes corresponding sialic acid peak areas from several runs are collected. For chromatographic analyses, salts are removed by ion-exchange chromatography.[1] Direct mass spectrometric analysis of the eluting sialic acids will be described below.

By this method, baseline separation of N-acetylneuraminic acid and N-glycolylneuraminic acid is possible. These sialic acids elute earlier than their O-acetylated derivatives (Table II).[51] Most of the mono-, di-, and tri-O-acetylated sialic acids can be separated from each other. In sialic acid mixtures, however, complete separation of N-glycolylneuraminic acid and N-acetyl-9-O-acetylneuraminic acid is not possible. In this case, a

[48] H. K. B. Silver, K. A. Karim, M. J. Gray, and F. A. Salinas, *J. Chromatogr.* **224,** 381 (1981).

[49] M. J. Krantz and Y. C. Lee, *Anal. Biochem.* **63,** 464 (1975).

[50] A. K. Shukla, N. Scholz, E. H. Reimerdes, and R. Schauer, *Anal. Biochem.* **123,** 78 (1982).

[51] A. K. Shukla and R. Schauer, *J. Chromatogr.* **244,** 81 (1982).

[52] A. K. Shukla, R. Schauer, F. M. Unger, U. Zähringer, E. T. Rietschel, and H. Brade, *Carbohydr. Res.* **140,** 1 (1985).

TABLE II
RETENTION TIMES OF DIFFERENT N,O-SUBSTITUTED SIALIC
ACIDS, METHYL- AND CMP-GLYCOSIDES OF
N-ACETYLNEURAMINIC ACID, VARIOUS
SIALYLOLIGOSACCHARIDES AND N-ACYLMANNOSAMINES ON
ANION-EXCHANGE (AMINEX A-28) HPLC[a]

Compound	Retention time (sec)
N-Acetylmannosamine	43
N-Glycolylmannosamine	49
Di-N-acetylneuraminyllactose	99
CMP-N-acetylneuraminic acid	105
N-Acetylneuraminyl-α(2-6)-N-acetyllactosamine	115
N-Acetylneuraminyl-α(2-6)-lactose	137
N-Acetylneuraminyl-α(2-3)-lactose	173
N-Acetyl-9-O-acetylneuraminyl-α(2-3)-lactose	205
N-Glycolylneuraminyl-α(2-3)-lactose	210
N-Acetylneuraminic acid-β-methylglycoside	265
N-Acetyl-7-O-acetylneuraminic acid	316
N-Glycolyl-8-O-methylneuraminic acid	321
N-Acetylneuraminic acid	361
N-Acetylneuraminic acid-α-methylglycoside	440
N-Glycolylneuraminic acid	480
N-Acetyl-4-O-acetylneuraminic acid	510
N-Acetyl-9-O-acetylneuraminic acid	530
N-Acetyl-2-deoxy-2,3-didehydroneuraminic acid	605
N-Glycolyl-4-O-acetylneuraminic acid	610
N-Glycolyl-9-O-acetylneuraminic acid	615
N-Acetyl-7,9-di-O-acetylneuraminic acid	636
N-Glycolyl-7,9-di-O-acetylneuraminic acid	730
N-Acetyl-7,8,9-tri-O-acetylneuraminic acid	740
N-Glycolyl-7,8,9-tri-O-acetylneuraminic acid	910

[a] Using 0.75 mM sodium sulfate as mobile phase. The flow rate was 0.5 ml/min at 15 bar using a 40 × 4.6 mm column.[51]

second run with saponification of the sialic acid mixture yielding N-acetylneuraminic acid from the O-acetylated species may help. Both 2-deoxy-2,3-didehydro-N-acetylneuraminic acid and CMP-N-acetylneuraminic acid can clearly be distinguished from N-acetylneuraminic acid. Different sialyllactoses including a 9-O-acetylated derivative also exhibit distinct retention times. The values given in Table II depend on the ionic strength and the nature of anions present in the mobile phase. Sodium sulfate (0.75 mM) proved to be most suitable for the separation of sialic acids. With further increases in the concentration of sodium sulfate the separation becomes poorer but sialic acids elute faster.

The peak areas of the sialic acids strongly depend on the wavelength; at 195 nm they show a 15 times higher sensitivity than at 215 nm. This dependence is not as pronounced with 2-deoxy-2,3-didehydro-N-acetylneuraminic acid, but at 215 nm its sensitivity is about 30 times higher when compared to N-acetylneuraminic acid. We analyze sialic acids routinely at 200 nm.

Minimum amounts of 30 ng (0.1 nmol) of saturated sialic acids and of 1 ng of 2-deoxy-2,3-didehydro-N-acetylneuraminic acid can be measured. When choosing the more sensitive absorption conditions at 195 nm, 0.01 a.u.f.s., a minimum amount of 6 ng (20 pmol) can be determined. An even higher increase in sensitivity has been reported.[52]

Monitoring at the wavelengths described is not specific for sialic acids, as impurities present in the sialic acid preparations may interfere. The specificity for sialic acids can be increased by the postcolumn reactions mentioned above, which, however, require sufficient sialic acid quantities. Therefore, it appears easier to increase the specificity with the aid of, e.g., N-acetylneuraminate pyruvate-lyase and to compare the elution profiles before and after this enzyme treatment. The decrease or disappearance of a peak area indicates the presence of a sialic acid (Fig. 3). Furthermore, the formation of acylmannosamines can be monitored in the same run.[53] The specificity for O-acetylated sialic acids is increased by hydrolysis of O-acetyl groups before a parallel HPLC analysis. This is achieved either by saponification under mild alkaline conditions (0.1 M NaOH, 0°, 45 min) or, in a more specific way, by the use of sialate O-acetylesterase (EC 3.1.1.53) (Fig. 3). Esterase activity specific for sialic acids has been detected in various tissues including liver[54] and, as the only enzyme, in influenza C virus.[10]

The analyses are not disturbed by the presence of small amounts of proteins, e.g., enzymes. However, if the probes are contaminated by too high concentrations of impurities, especially those absorbing UV light of this energy, at least partial purification of sialic acids is necessary before HPLC analysis. In spite of this limitation, the HPLC method described is the fastest and most sensitive technique for the analysis of different sialic acids and does not require their derivatization or extensive purification.

These facts proved useful not only for the analysis of sialoglycoconjugates but also for kinetic measurements of the enzymes of sialic acid metabolism, such as sialidase, N-acetylneuraminate pyruvate-lyase[54a], sialate O-acetylesterase, CMP-sialate synthase (EC 2.7.7.43, acylneuraminate cytidylyltransferase) and CMP-sialate hydrolase (EC

[53] A. K. Shukla and R. Schauer, *Falk Symp.* **34,** 565 (1983).
[54] A. K. Shukla, S. Stoll, and R. Schauer, *Hoppe-Seyler's Z. Physiol. Chem.* **365,** 1065 (1984).
[54a] A. K. Shukla and R. Schauer, *Anal. Biochem.* **158** (1986), in press.

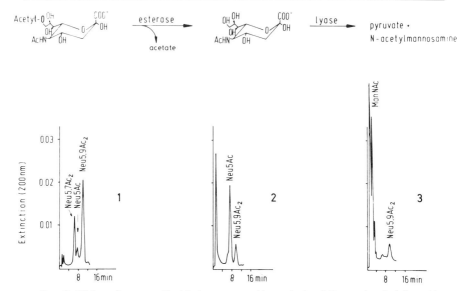

FIG. 3. High-performance liquid chromatographic analysis of O-acetylated sialic acids. To increase specificity of the analysis, after the first run (1), 9-O-acetyl groups are removed from sialic acids with the aid of sialate-O-acetylesterase resulting in N-acetylneuraminic acid (2). 7-O-Acetyl groups are cleaved only after migration to position 9.[26,54] N-Acetyl-neuraminic acid is then degraded by acylneuraminate pyruvate-lyase leading to the appearance of a N-acetylmannosamine (ManNAc) peak (3).[51,53,54] Neu5,7Ac₂, N-acetyl-7-O-acetyl-neuraminic acid; Neu5Ac, N-acetylneuraminic acid; Neu5,9Ac₂, N-acetyl-9-O-acetyl-neuraminic acid.

3.1.4.40, CMP-N-acylneuraminate phosphodiesterase).[53] Besides the quickness, HPLC analysis of these and other enzyme reactions makes it possible to analyze reaction products and sometimes also substrates (e.g., sialyllactose) directly and in one run. It is thus possible to determine the substrate specificity of an enzyme even within a mixture of sialic acids (see for example Fig. 4). The HPLC method described also allows qualitative and quantitative analysis of other acidic sugars such as glucuronic acid and 3-deoxy-D-*manno*-2-octulosonic acid (KDO) including KDO glycosides.[52]

Isocratic HPLC analysis of underivatized sialic acids in the nanomole range is also possible on amino phases using acetonitrile–phosphate buffer mixtures as mobile phase.[55,56] Reversed-phase HPLC of *p*-nitrophenylhydrazone derivatives of sialic acids has been reported.[57] Anal-

[55] H. Fiedler and H. Faillard, *Chromatographia* **20**, 231 (1985).
[56] A. Varki and S. Diaz, *J. Biol. Chem.* **260**, 6600 (1985).
[57] P. A. McNicholas, M. Batley, and J. W. Redmond, *J. Chromatogr.* **315**, 451 (1984).

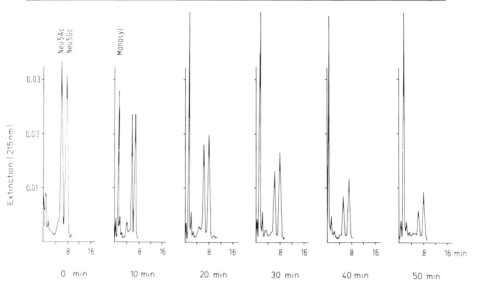

FIG. 4. High-performance liquid chromatographic analysis of substrates and products (*N*-acylmannosamines, Manacyl) of *N*-acetylneuraminate pyruvate-lyase reaction in dependence on time. *N*-Acetylneuraminic acid (Neu5Ac) is more rapidly cleaved by the enzyme than *N*-glycolylneuraminic acid (Neu5Gc).[51,53]

ysis of sialic acids as *N*-acylmannosamines after cleavage by *N*-acetyl-neuraminate pyruvate-lyase was described by Honda and Suzuki.[58]

Gas–Liquid Chromatography

Quantitative and qualitative analyses of sialic acids as trimethylsilyl esters, per-*O*-trimethylsilyl ethers (using trimethylsilylimidazole or pyridine, hexamethyldisilazane, and chlorotrimethylsilane), or as methyl esters, per-*O*-trimethylsilyl ethers (esterification by diazomethane and trimethylsilylation by pyridine, hexamethyldisilazane, and chlorotrimethylsilane, or by trimethylsilylimidazole) on packed columns was described in this series[1] and by Kamerling and Vliegenthart.[59]

Capillary gas–liquid chromatography of these sugars is now possible, for example on 25 m fused silica columns (OV-101 or OV-17, Macherey Nagel & Co. Düren, FRG).[60,61] For this purpose 100 μg of freeze-dried,

[58] S. Honda and S. Suzuki, *Anal. Biochem.* **142,** 167 (1984).
[59] J. P. Kamerling and J. F. G. Vliegenthart, *Cell Biol. Monogr.* **10,** 95 (1982).
[60] C. Schröder and R. Schauer, *Fresenius Z. Anal. Chem.* **311,** 385 (1982).
[61] G. Reuter, C. Schröder, and R. Schauer, *Proc. Int. Symp. Glycoconjugates, 7th, 1983,* p. 158 (1983).

purified sialic acids are dissolved in 30 μl of dry pyridine and silylated by 50 μl N-methyl-N-trimethylsilyltrifluoroacetamide for 30–60 min at room temperature. Aliquots of this mixture are chromatographed using a temperature program of 150–280°, 2°/min. Depending on the grade of purity and the nature of N-substituents, trimethylsilyl esters, per-O-trimethylsilyl ethers and/or trimethylsilyl esters, per-O-trimethylsilyl ethers, N-trimethylsilyl derivatives of sialic acids are formed. These derivatives are stable for at least 2 weeks at room temperature, and the solution remains clear. The pertrimethylsilylated sialic acids are well separated from each other, as can be seen from the retention times on OV-101 of a variety of sialic acids as well as of some of their natural unsaturated and anhydro derivatives, summarized in Table III. The detection limit of this method is 0.5 ng sialic acid when using a flame ionization detector. This corresponds to a sensitivity at least a hundred times higher when compared with packed columns. Application of this method in combination with mass spectrometry (see below) enabled the discovery of several new neuraminic acid derivatives, partially occurring in only minute quantities.[2,17,23,61] In addition, N-glycolylneuraminic acid could unequivocally be identified as a natural contaminant in the sialic acid fraction from human glycoproteins.[23] The improvement of the isolation procedure of sialic acids described above also contributed to the progress made in sialic acid analysis.

A disadvantage of the method described using N-methyl-N-trimethylsilyltrifluoroacetamide is the fact that two derivatives and correspondingly two peaks in the chromatograms result from one sialic acid, due to incomplete N-trimethylsilylation. For N-glycolylated sialic acids the degree of N-trimethylsilylation is always less than for N-acetylated sialic acids. Alternatively, sialic acids may therefore be trimethylsilylated by trimethylsilylimidazole or hexamethyldisilazane and chlorotrimethylsilane in pyridine for capillary gas–liquid chromatography. These procedures do not lead to N-trimethylsilylated products and result in stable derivatives, but they have other disadvantages: the samples with trimethylsilylimidazole show severe tailing of the solvent peak and those containing hexamethyldisilazane and chlorotrimethylsilane are turbid and should therefore not be used for autosampler injection.[61]

Gas–Liquid Chromatography–Mass Spectrometry

Electron impact mass spectrometry of the methyl esters, per-O-trimethylsilyl ethers of sialic acids, which led to the structural elucidation of a variety of sialic acids especially with regard to the nature, number, and

TABLE III

Relative Gas–Liquid Chromatographic Retention Times (R_t Values) of Trimethylsilyl Ester,
Per-O-Trimethylsilyl Ether Derivatives of Sialic Acids and Characteristic Fragment Ions from
Electron Impact Mass Spectrometry[a]

Compound	R_t	Characteristic ions at m/z						
		A	B	C	D	E	F	G
N-Acetylneuraminic acid	1.00	726	624	536	356	375	205	173
N-Acetyl-4-O-acetylneuraminic acid	1.09	696	594	506	356	—	205	143
N-Acetyl-9-O-acetylneuraminic acid	1.06	696	594	536	356	375	175	173
N-Acetyl-7,9-di-O-acetylneuraminic acid	n.d.[b]	666	564	—	—	375	175	173
N-Acetyl-2-deoxy-2,3-didehydroneuraminic acid	1.02	636	—	446	356	285	205	—
N-Acetyl-2,7-anhydroneuraminic acid	0.89	564	462	374	—	—	205	173
N-Glycolylneuraminic acid	1.17	814	712	624	444	375	205	261
N-Glycolyl-4-O-acetylneuraminic acid	n.d.	784	682	594	444	—	205	231
N-Glycolyl-9-O-acetylneuraminic acid	1.23	784	682	624	444	375	175	261
N-Glycolyl-2-deoxy-2,3-didehydroneuraminic acid	1.17	724	—	534	444	285	205	—

[a] GLC on an OV-101 capillary column obtained under conditions described in the text. Mass spectrometric data from Refs. 23, 60, 61.
[b] n.d., not determined.

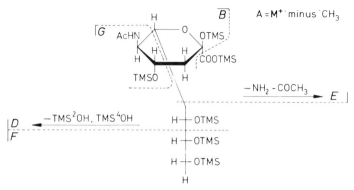

FIG. 5. Formation of the characteristic fragment ions A–G used for mass spectrometric analysis of the trimethylsilyl (TMS) esters, per-*O*-trimethylsilyl ethers of *N*-acetyl-neuraminic acid.[61] For *m/z* values see Table III.

position of *N*- and *O*-acyl groups or other residues,[35,59,62] has been described in this series.[1] The high resolution power of capillary gas–liquid chromatography and direct trimethylsilylation of sialic acids brought a further progress in sialic acid analysis, especially for the detection of minor quantities.[23] For example, this technique resulted in the identification of seven neuraminic acid derivatives in rat urine, of which four were new.[21,22] This analysis was facilitated by the use of mass chromatography.[22,63]

The mass spectrometric fragmentation scheme, which has been set up for the methyl esters, per-*O*-trimethylsilyl ethers,[57,62] is also valid for the trimethylsilyl esters, per-*O*-trimethylsilyl ethers and the corresponding *N*-trimethylsilyl derivatives, thus establishing its general applicability for sialic acids (Fig. 5). The characteristic fragment ions of these derivatives are summarized for some sialic acids in Table III. When compared to the methyl ester, per-*O*-trimethylsilyl ether of *N*-acetylneuraminic acid, the mass spectrum of the trimethylsilyl ester, per-*O*-trimethylsilyl ether of this sialic acid shows only minor alterations. A shift of +58 mass units occurs for all fragments containing the C-1 part of the molecule. The intensity of fragment B increases relative to A. Instead of fragment E the ion E minus $(CH_3)_3SiOH$ (*m/z* 285) becomes more important.[61] The mass spectrum of the trimethylsilyl ester, per-*O*-trimethylsilyl ether, *N*-trimethylsilyl derivative reveals some changes. A general shift of +72 mass

[62] J. P. Kamerling, J. F. G. Vliegenthart, C. Versluis, and R. Schauer, *Carbohydr. Res.* **41**, 7 (1975).

[63] U. Nöhle, A. K. Shukla, C. Schröder, G. Reuter, R. Schauer, J. P. Kamerling, and J. F. G. Vliegenthart, *Eur. J. Biochem.* **151**, 1–5 (1985).

units can be observed for all fragments containing the nitrogen substituent. In addition to the fragments A, B, C, and D the corresponding ions without the N-trimethylsilyl group are found (A', B', C', and D'). Instead of fragment D, fragment G becomes the base peak in the region from m/z 100 to m/z 950. Fragment H[59] is not present. E minus $(CH_3)_3SiOH$ is more intense than E. Fragment B has a relatively higher intensity than fragment A. In some cases, especially for N-glycolylated derivatives, the molecular ion was observed, which has never been found for the two other types of derivatives.[61]

Chemical ionization (CI) mass spectrometry of pertrimethylsilylated sialic acids was carried out after capillary gas–liquid chromatography on a quadrupole mass spectrometer at 220 eV using isobutane as reagent gas.[64] As could be expected, the fragment ions at lower m/z values are less intense than in the EI spectra. However, the number and position of O-acetyl groups and the type of the N-substituent can be clearly recognized. The base peak at m/z values higher than 300 is formed by elimination of the substitutents at C-2 and C-4 from the $[M + 1]^+$ ion.

High-Performance Liquid Chromatography–Mass Spectrometry

This method enables mass spectrometric analysis of underivatized sialic acids eluting from the HPLC column.[65] As only volatile salts can be used in this system, the sodium sulfate solution described above for sialic acid elution is replaced by a water–acetonitrile (4:1, v/v) mixture containing 20 mM ammonium formate, pH 6. The solid phase is Aminex A-29 and the flow rate 0.5 ml/min at 15–20 bars. One part of the effluent is monitored at 200 nm and the other (3%) is transferred to the mass spectrometer source using a direct liquid inlet technique. At the source temperature (250°) the buffer components are volatile and act as chemical ionization reagent gas. It is also possible to apply pure sialic acids dissolved in water (5–10 μg per 10 μl) directly without column, on-line to the mass spectrometer.

The mass fragments of N-acetylneuraminic acid and its 2-deoxy-2,3-didehydro derivative, of N-glycolylneuraminic acid, and of some O-acetylated sialic acids are given in Table IV. All spectra show a molecular ion peak (fragment a). In the spectra of saturated sialic acids the prominent peak is fragment h, which is the base peak, allowing discrimination between different sialic acids and determination of the degree of their O-

[64] G. Reuter and R. Schauer, *Anal. Biochem.* **157**, 39 (1986).
[65] A. K. Shukla, R. Schauer, U. Schade, H. Moll, and E. T. Rietschel, *J. Chromatogr.* **337**, 231 (1985).

TABLE IV

MASS FRAGMENTATION OF UNDERIVATIZED SIALIC ACIDS ON COMBINED HIGH-PERFORMANCE LIQUID CHROMATOGRAPHY–MASS SPECTROMETRY[a]

Compound	Characteristic ions (a–j)[b] at m/z									
	j	i	h	g	f	e	d	c	b	a
N-Acetylneuraminic acid	168	186	204	222	246	256	264	274	292	310
N-Glycolylneuraminic acid	184	202	220	238	262	272	280	290	308	326
N-Acetyl-4-O-acetylneuraminic acid ⎫ N-Acetyl-7-O-acetylneuraminic acid ⎬ N-Acetyl-9-O-acetylneuraminic acid ⎭	210	228	246	264	—	—	306	316	334	352
N-Acetyl-7,9-di-O-acetylneuraminic acid	—	—	288	306	—	—	348	358	376	394
N-Acetyl-7,8,9-tri-O-acetylneuraminic acid	—	—	330	348	—	—	390	400	418	436
N-Acetyl-2-deoxy-2,3-didehydroneuraminic acid	—	—	—	—	—	—	—	—	274	292

[a] See Fig. 6 for chemical nature of the fragments from N-acetylneuraminic acid.[65]

[b] Ion a corresponds to $[M + H]^+$ and fragments b–j are derived from ion a by loss of water (18 a.m.u.), carbon monoxide (28 a.m.u.) and/or ketene (42 a.m.u.).

acetylation. Thus, the m/z values of fragment h for N-acetylneuraminic acid and its derivatives O-acetylated 1-, 2-, or 3-fold on the side chain differ by 42 units each, corresponding to the difference of one O-acetyl residue each. Similar differences are exhibited by the other fragments of these sialic acids. The fragmentation patterns of N-acetylneuraminic acid derivatives mono-O-acetylated at C-4, C-7, or C-9 are qualitatively identical and only slight differences in the intensities of the peaks are seen. Therefore, the position of the O-acetyl groups cannot be elucidated by this method. However, different O-acetylated sialic acids separate on HPLC, thus allowing the localization of the position of O-acetyl groups indirectly by the use of standards. N-Acetylneuraminic acid and N-glycolylneuraminic acid can readily be discriminated by their mass fragmentation patterns, all m/z values differing by 16 mass units. The simplicity of the mass spectrum of 2-deoxy-2,3-didehydro-N-acetylneuraminic acid showing only two ions at $m/z = 292$ $[M + H]^+$ and 274 (292–18) is remarkable.[65]

The fragmentation reactions of sialic acids induced by chemical ionization with the solvent are assumed to lead to the fragments shown in Fig. 6.[65] The scheme is mainly based on the observation that most fragments of one sialic acid differ by 18 mass units, which corresponds to the loss of several water molecules. Most of the prominent fragments are assumed to be derived from the open-chain configuration of sialic acids. The simple spectrum of 2-deoxy-2,3-didehydro-N-acetylneuraminic acid (Table IV) may be derived from the ring form by the release of two water molecules.

This method allows the analysis of 1–10 μg of sialic acids. It may be useful for the rapid analysis of microquantities of sialic acid mixtures, isolated from biological material or originating from enzyme reactions metabolizing sialic acids. The technique can be also applied to direct analysis of, e.g., sialyllactose. The spectrum of this compound shows a fragmentation pattern typical for N-acetylneuraminic acid and the molecule ions $[M + H]^+$, etc., of lactose, galactose, and glucose.[65]

Fast Atom Bombardment Mass Spectrometry

This method can be used for investigation of the number, position, and nature of sialic acids in complex carbohydrates. Native gangliosides with up to five sialic acid residues were analyzed by this technique either in the negative ion mode or, after permethylation, in the positive ion mode. The sites of attachment and the number of N-acetylneuraminic acid residues present in the gangliosides could be deduced directly from the fragmenta-

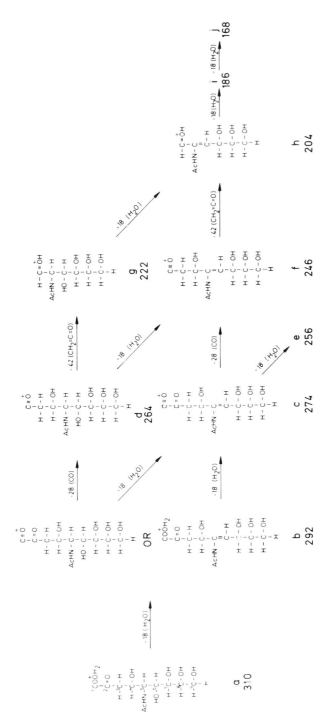

FIG. 6. High-performance liquid chromatographic–mass spectrometric analysis of underivatized N-acetylneuraminic acid (Neu5Ac).[65] The proposed fragmentation pattern for the open chain form of Neu5Ac is given; for the m/z values of corresponding fragments of other sialic acids see Table IV.

tion pattern.[66,67] The presence and localization of other sialic acids, e.g., N-glycolylneuraminic acid (unpublished results) and N-acetyl-9-O-acetyl-neuraminic acid (in native form),[68] can be identified from the mass spectra, too.

Nuclear Magnetic Resonance Spectroscopy

Since the last article on sialic acid analysis was written in this series,[1] great progress in the determination of free or glycosidically bound sialic acids was made by using 360 or even 500 MHz proton NMR spectroscopy. By this technique insight into the position and type of glycosidic linkages of sialic acids in sialylated oligosaccharides, glycopeptides,[69,70] or glycolipids[71] was obtained. These studies allowed the unequivocal determination of the number and position of sialic acids in unbranched or branched oligosaccharides. For the complete structural elucidation of many compounds only few signals of each sugar residue were necessary, so-called "structural reporter groups." Advanced two-dimensional NMR techniques make assignments of protons possible which could not be done by one-dimensional measurements.[72] The nature of sialic acids, i.e., the occurrence of a N-acetyl or N-glycolyl group or of O-acetyl groups in different positions of the neuraminic acid molecule, can be determined by NMR spectroscopy, too.[69,73]

This method also allows kinetic measurements. Thus, the nonenzymatic migration rate of O-acetyl groups from O-7 to O-9 in the sialic acid side chain was followed by 360 MHz proton NMR spectroscopy.[26] As examples of enzyme reactions, detailed insight into substrate specificity

[66] H. Egge, J. Peter-Katalinić, G. Reuter, R. Schauer, R. Ghidoni, S. Sonnino, and G. Tettamanti, *Chem. Phys. Lipids* **37**, 127 (1985).

[67] H. Rösner, H. Rahmann, G. Reuter, R. Schauer, J. Peter-Katalinić, and H. Egge, *Biol. Chem. Hoppe-Seyler* **366**, 1177 (1985).

[68] J. Thurin, M. Herlyn, O. Hindsgaul, K.-A. Karlsson, N. Strömberg, D. Elder, Z. Steplewski, and H. Koprowski, *Proc. Int. Symp. Glycoconjugates, 8th, 1985*, p. 632 (1985).

[69] J. F. G. Vliegenthart, L. Dorland, H. van Halbeek, and J. Haverkamp, *Cell Biol. Monogr.* **10**, 127 (1982).

[70] J. F. G. Vliegenthart, L. Dorland, and H. van Halbeek, *Adv. Carbohydr. Chem. Biochem.* **41**, 209, 1983.

[71] T. A. W. Koerner, J. H. Prestegard, P. C. Demou, and R. K. Yu, *Biochemistry* **22**, 2676, 2687 (1983).

[72] H. van Halbeek, *in* "Abstracts of the Third European Symposium on Carbohydrates" (J. Defaye, ed.), p. 40. Imprimerie Bibliothèque Interuniversitaire, Grenoble, 1985.

[73] J. Haverkamp, H. van Halbeek, L. Dorland, J. F. G. Vliegenthart, R. Pfeil, and R. Schauer, *Eur. J. Biochem.* **122**, 305 (1982).

of the sialyltransferase from bovine colostrum[74] and of viral or bacterial sialidases[75] was obtained. It was possible in these studies to follow the preferential incorporation of sialic acid into distinct branches of N-glycosidic oligosaccharides and to discriminate between the rates of hydrolysis of sialic acids bound in different linkages to one or to different substrate molecules. Furthermore, NMR analysis gave a possible explanation for the resistance of the "internal" sialic acid residue in G_{M1}. The glycosidic linkage of this sialic acid was demonstrated to have less conformational freedom when compared to the terminal sialic acid of the ganglioside G_{D1a}.[29,29a]

Histochemical Analysis

Several methods are available for sialic acid staining in tissues and cells. Due to their biological importance special interest is focused on membrane-associated sialic acids.[2–4] Staining of these residues is possible on the basis of their relatively high acidity or their specific reactivity with periodate or lectins.[2,76] In the first group sialic acids can be visualized with, for example, Alcian Blue, colloidal iron, cationized ferritin, or ruthenium red. After selective oxidation of the side chain with periodate the aldehyde groups formed react with p-dimethylaminobenzaldehyde (Schiff's reagent) or, in a very sensitive way, with the fluorescent reagents dansylhydrazine, rhodamine, or fluorescein-containing hydrazides. For electron microscopy oxidized sialic acids are made visible by reaction with biotin and interaction with ferritin-conjugated avidin. These procedures have been reviewed.[2,76] A recent histochemical study for the ultrastructural staining of sialoglycoconjugates provides several additional substances which bind either to the carboxylic group or to an oxidation-derived aldehyde group of sialic acids.[77] When periodate is used in these staining procedures the conditions must be mild (1–10 mM periodate, 10 min, 0° in the dark), to specifically oxidize and stain sialic acids and not other sugar residues.[2,76]

Since O-acetylated sialic acids are widespread and becoming more important,[4] interest in their cellular localization is increasing. Histochemical demonstration of sialic acids O-acetylated in the glycerol side chain is possible, because substituents at O-8 and O-9 of the neuraminic acid

---

[74] D. H. Joziasse, W. E. C. M. Schiphorst, D. H. van den Eijnden, J. A. van Kuik, H. van Halbeek, and J. F. G. Vliegenthart, *J. Biol. Chem.* **260**, 714 (1985).

[75] H. Friebolin, W. Baumann, G. Keilich, D. Ziegler, R. Brossmer, and H. von Nicolai, *Hoppe-Seyler's Z. Physiol. Chem.* **362**, 1455 (1981).

[76] C. F. A. Culling and P. E. Reid, *Cell Biol. Monogr.* **10**, 173 (1982).

[77] G. Menghi, A. M. Bondi, L. Vitaioli, and E. Baldoni, *Acta Histochem.* **72**, 101 (1983).

molecule have been shown to hinder oxidation with periodate.[78] Correspondingly, removal of these ester groups by alkaline treatment (0.5% KOH in 70% ethanol)[79] may increase the staining reaction of sialic acid, as was shown in human and bovine intestinal tissues.[2,76,79–81] By this method most of the O-acetylated sialic acids can be distinguished from the non-O-acetylated species, since their ester groups predominantly occur at O-9. Techniques for the localization of O-acetyl groups in sialic acid molecules as well as tentative determination of the presence of several O-acetyl groups are also available by application of periodate oxidation, saponification, borohydride reduction, and different staining steps in various combinations.[2,76,80–81a]

Sialic acids bound in α(2-3) or α(2-6) linkages to N-acetylgalactosamine of mucins can be distinguished histochemically in the following way[82]: Periodate (50 mM, 30 min, room temperature) oxidizes sialylated N-acetylgalactosamine at C-3 and C-4 only if the sialic acid is bound at C-6. After reduction with borohydride of the aldehyde groups formed and enzymatic release of sialic acid, the decomposed N-acetylgalactosamine residue no longer reacts with lectins, e.g., those from *Ricinus communis*, conjugated with a fluorescent dye. In the case of α(2-3) linkages, N-acetylgalactosamine remains intact by this procedure and develops full fluorescence.

Sialic acid-specific autoradiography of tissues and cells can be carried out by mild oxidation with periodate and reduction with borotritide according to Gahmberg and Andersson.[83]

A useful and specific tool for the histochemical analysis of carbohydrates, which does not require oxidation of sialic acids, is the use of lectins. For this purpose lectins may be conjugated, e.g., with gold particles,[84] peroxidase,[85–87] and rhodamine or fluorescein isothiocyanate.[87]

[78] J. Haverkamp, R. Schauer, M. Wember, J. P. Kamerling, and J. F. G. Vliegenthart, *Hoppe-Seyler's Z. Physiol. Chem.* **356,** 1575 (1975).

[79] C. F. A. Culling and P. E. Reid, *J. Microsc. (Oxford)* **119,** 415 (1980).

[80] R. W. Veh, A. P. Corfield, R. Schauer, and K. H. Andres, *Proc. Int. Symp. Glycoconjugates, 5th, 1979,* p. 191 (1979).

[81] R. W. Veh, S. Dürscheidt, D. Meessen, H. D. Kuntz, and B. May, *in* "Interdisziplinäre Onkologie am Beispiel des Magenkarzinoms" (W. Kozuschek, ed.), p. 97. Springer-Verlag, Berlin and New York, 1985.

[81a] R. W. Veh *et al.,* to be published.

[82] U. A. Zimmer, R. W. Veh, W. Kozuschek, and B. May, *in* "Interdisziplinäre Onkologie am Beispiel des Magenkarzinoms" (W. Kozuschek, ed.), p. 117. Springer-Verlag, Berlin and New York, 1985.

[83] C. G. Gahmberg and L. C. Andersson, *J. Biol. Chem.* **252,** 5888 (1977).

[84] J. Roth, *J. Histochem. Cytochem.* **31,** 987 (1983).

[85] K. Yamada and S. Shimizu, *Histochem. J.* **11,** 457 (1979).

[86] B. A. Schulte, S. S. Spicer, and R. L. Miller, *Histochem. J.* **16,** 1125 (1984).

[87] P. K. Schick and W. G. Filmyer, *Blood* **65,** 1120 (1985).

These ligands are suitable for light, fluorescence, or electron microscopy. Peroxidase binding is visualized on light microscopy by reaction, for example, with 3,3'-diaminobiphenyl[85] or diaminobenzidine.[87] This technique can now be applied also to sialic acids, as a variety of lectins specific for these sugars have been discovered. These are limulin from the American horseshoe crab *Limulus polyphemus*,[88,89] carcinosporin from the Indian horseshoe crab *Carcinoscorpius rotunda cauda*,[90] and lectins from American spiders of the genus *Aphonopelma*,[91] from the "whip scorpion" *Mastigoproctus giganteus*,[92] from the scorpions *Vaejovis spinigerus*[93] and *Heterometrus granulomanus*,[93a] from the lobster *Homarus americanus*,[94] from the African giant snail *Achatina fulica*,[95] from the snail *Cepea hortensis*,[96] from the slug *Limax flavus*,[86,97,98] and from the Pacific oyster *Cranostrea gigas*.[99] The plant lectin wheat germ agglutinin also binds to sialic acids, as has been extensively studied.[100,101]

While some of these lectins bind to both *N*-acetyl- and *N*-glycolyl-neuraminic acid, such as limulin[88] and the lectin from the slug *L. flavus*,[97] others such as the lectin from the African giant snail[95] or wheat germ agglutinin[2,101] are specific for *N*-acetylneuraminic acid only. Sialic acid *O*-acetyl groups may also influence the binding strength of the lectin. For example, an acetyl group at O-4 of neuraminic acid hinders binding of

---

[88] A.-C. Roche, R. Schauer, and M. Monsigny, *FEBS Lett.* **57**, 245 (1975).
[89] E. Cohen, G. R. Vasta, W. Korytnyk, C. R. Petrie, III, and M. Sharma, *in* "Recognition Proteins, Receptors, and Probes: Invertebrates" (E. Cohen, ed.), p. 55. Alan R. Liss, Inc., New York, 1984.
[90] D. T. Dorai, S. Mohan, S. Srimal, and B. K. Bachhawat, *FEBS Lett.* **148**, 98 (1982).
[91] G. R. Vasta and E. Cohen, *Dev. Comp. Immunol.* **8**, 515 (1984).
[92] G. R. Vasta and E. Cohen, *J. Invertebr. Pathol.* **43**, 333 (1984).
[93] G. R. Vasta and E. Cohen, *Experientia* **40**, 485 (1984).
[93a] H. Ahmed, B. P. Chatterjee, S. Kelm, and R. Schauer, *Biol. Chem. Hoppe-Seyler* **367**, 501 (1986).
[94] C. A. Abel, P. A. Campbell, J. Vanderwall, and A. L. Hartman, *in* "Recognition Proteins, Receptors, and Probes: Invertebrates" (E. Cohen, ed.), p. 103. Alan R. Liss, Inc., New York, 1984.
[95] S. M. M. Iguchi, T. Momoi, K. Egawa, and J. J. Matsumoto, *Comp. Biochem. Physiol. B* **81B**, 897 (1985).
[96] S. E. Holm, A.-M. Bergholm, B. Wagner, and M. Wagner, *J. Med. Microbiol.* **19**, 317 (1985).
[97] R. L. Miller, J. F. Collawn, Jr., and W. W. Fish, *J. Biol. Chem.* **257**, 7574 (1982).
[98] R. L. Miller and J. D. Cannon, Jr., *in* Recognition Proteins, Receptors, and Probes: Invertebrates" (E. Cohen, ed.), p. 31. Alan R. Liss, Inc., New York, 1984.
[99] S. W. Hardy, P. T. Grant, and T. C. Fletcher, *Experientia* **33**, 767 (1977).
[100] K. A. Kronis and J. P. Carver, *Biochemistry* **24**, 826, 834 (1985).
[101] M. Monsigny, A.-C. Roche, C. Sene, R. Maget-Dana, and F. Delmotte, *Eur. J. Biochem.* **104**, 147 (1980).

limulin completely.[102] In contrast, a lectin with high specificity for sialic acids acetylated at O-9 or O-4 was purified from the hemolymph of the California coastal crab *Cancer antennarius*.[103] These specificities should be kept in mind when using lectins for histochemical purposes. Limulin[89,104] and *L. flavus* lectin[105] are available now in purified form and have proved to be excellent tools for sialic acid histochemistry.

Concluding Remarks

In this article a variety of tools for the analysis of sialic acids has been described, ranging from assays which can easily be handled and are in wide use, to techniques which require expensive equipment and specialists for the interpretation of analytical data. The latter methods, as NMR spectroscopy and mass spectrometry, are necessary for the unequivocal elucidation of the structure of free or bound sialic acids. With the other methods, e.g., colorimetry and thin-layer chromatography, applicable in each laboratory, the everyday problems of sialic acid biochemistry can be solved, i.e., the occurrence, amount, and tentative structure of sialic acids. However, to get conclusive and reproducible results with these analyses, sialic acids must carefully be purified by the removal of lipids and ion-exchange chromatography. If these precautions, together with the influence of substituents on sialic acid analysis and the problems involved in the liberation of sialic acids from glycosidic bonds by acids or sialidases, are taken into consideration, knowledge about the occurrence, metabolism, and biological role of the members of the great sialic acid family will increase further.

Acknowledgments

The author thanks Margret Wember and Sabine Stoll for excellent technical assistance and Dr. Gerd Reuter for valuable advice and critical reading of the manuscript. Financial support from Deutsche Forschungsgemeinschaft and Fonds der Chemischen Industrie is acknowledged.

[102] R. Maget-Dana, R. W. Veh, M. Sander, A.-C. Roche, R. Schauer, and M. Monsigny, *Eur. J. Biochem.* **114,** 11 (1981).
[103] M. H. Ravindranath, H. H. Higa, E. L. Copper, and J. C. Paulson, *J. Biol. Chem.* **260,** 8850 (1985).
[104] V. Muresan, M. P. Sarras, Jr., and J. D. Jamieson, *J. Histochem. Cytochem.* **30,** 947 (1982).
[105] J. Roth, J. M. Lucocq, and P. M. Charest, *J. Histochem. Cytochem.* **32,** 1167 (1984).

[11] Resialylated Erythrocytes for Assessment of the Specificity of Sialyloligosaccharide Binding Proteins

By JAMES C. PAULSON and GARY N. ROGERS

Sialyloligosaccharide binding proteins have been demonstrated to mediate a variety of biological processes.[1] For example, viral attachment proteins of influenza viruses, paramyxoviruses, and polyoma virus allow adsorption of virus to host cells by binding to sialyloligosaccharides as receptor determinants.[2,3] Sialyloligosaccharides have also been demonstrated or implicated to be cell surface receptor determinants for mycoplasma,[4,5] for recirculating lymphocytes seeking capillary sites of entry to the lymph system,[6] for blood group and tumor-specific antibodies,[7,8] for cholera toxin and other bacterial toxins,[9] and for a variety of plant, animal, and bacterial lectins.[10,11] In several cases in which sufficient information is known, the sialic acid binding proteins involved in the recognition event exhibit remarkable specificity for both the terminal sialic acid residue and the penultimate oligosaccharide sequence to which it is attached. This article describes a simple method to study the binding specificity of multivalent sialyloligosaccharide binding proteins which are capable of agglutination of, or adsorption to, erythrocytes. The application to the assessment of influenza virus receptor specificity is shown as an example.

[1] Abbreviations: Sia, sialic acid; NeuAc, N-acetylneuraminic acid; NeuGc, N-glycolylneuraminic acid; 9-O-Ac-NeuAc, 9-O-acetyl-N-acetylneuraminic acid.

[2] C. Howe and L. T. Lee, *Adv. Virus Res.* **17**, 1 (1972).

[3] J. C. Paulson, in "The Receptors" (M. Conn, ed.), Vol. 2, p. 131. Academic Press, New York, 1985.

[4] L. R. Glasgow and R. L. Hill, *Infect. Immun.* **30**, 353 (1980).

[5] L. M. Loomes, K. Uemura, R. A. Childs, J. C. Paulson, G. N. Rogers, P. R. Scudder, J.-C. Michalski, E. F. Hounsell, D. Taylor-Robinson, and T. Feizi, *Nature (London)* **307**, 560 (1984).

[6] S. D. Rosen, M. S. Singer, T. A. Yednock, and L. M. Stoolman, *Science* **228**, 1005 (1985).

[7] S. Hakomori, *Annu. Rev. Immun.* **2**, 103 (1984).

[8] S. Hakomori, in "Monoclonal Antibodies and Functional Cell Lines" (R. H. Kennett, K. B. Bechtol, and T. J. McKern, eds.), p. 67. Plenum, New York, 1984.

[9] W. Ruetter, E. Kottgen, C. Bauer, and W. Gerok, *Cell Biol. Monogr.* **10**, 264 (1982).

[10] G. I. Pardoe and G. Uhlenbruck, *J. Med. Lab. Technol.* **27**, 249 (1970).

[11] R. W. Jeanloz and J. F. Codington, in "Biological Roles of Sialic Acid" (A. Rosenberg and C.-L. Schengrund, eds.), p. 201. Plenum, New York, 1978.

Principle

Erythrocytes are enzymatically modified first by removal of sialic acid with a bacterial sialidase and then by restoration of sialic acid in defined sequence on the glycoprotein carbohydrate groups using a highly purified mammalian sialyltransferase.[12,13] The sequence restored is determined by the strict specificity of the sialyltransferase used.[14] Thus, by using several sialyltransferases, erythrocyte preparations containing different sequences may be obtained. The specificity of a sialyloligosaccharide binding protein may then be deduced by testing its ability to agglutinate and/or adsorb to the resialylated cells.

Preparation of Resialylated Erythrocytes[12,15]

Sialyltransferases

Sialyltransferases are purified from porcine submaxillary glands[16] or rat liver[17] as described previously. The reaction for each enzyme with its preferred acceptor substrate is given below. Assays for the sialyltransferases have been described.[16,18]

A. Galβ1,4GlcNAcα2,6-sialyltransferase (EC 2.4.99.1)

CMP-Sia[1] +Galβ1,4GlcNAc-R → Siaα2,6Galβ1,4GlcNAc-R + CMP

B. Galβ1,3(4)GlcNAcα2,3-sialyltransferase (EC 2.4.99.5)

CMP-Sia +Galβ1,3(4)GlcNAc-R → Siaα2,3Galβ1,3(4)GlcNAc-R + CMP

C. Galβ1,3GalNAcα2,3-sialyltransferase (EC 2.4.99.4)

CMP-Sia +Galβ1,3GalNAcαThr/Ser → Siaα2,
$$3Galβ1,3GalNAcαThr/Ser + CMP$$

D. GalNAcα2,6-sialyltransferase (EC 2.4.99.3)

CMP-Sia + Galβ1,3GalNAcαThr/Ser →

$$\begin{array}{c} Galβ1,3 \\ \diagdown \\ GalNAcαThr/Ser + CMP \\ \diagup \\ Siaα2,6 \end{array}$$

[12] J. E. Sadler, J. C. Paulson, and R. L. Hill, *J. Biol. Chem.* **254**, 2112 (1979).

[13] J. C. Paulson, J. E. Sadler, and R. L. Hill, *J. Biol. Chem.* **254**, 2120 (1979).

[14] T. A. Beyer, J. E. Sadler, J. I. Rearick, J. C. Paulson, and R. L. Hill, *Adv. Enzymol.* **52**, 24 (1981).

[15] G. N. Rogers and J. C. Paulson, *Virology* **127**, 361 (1983).

[16] J. E. Sadler, T. A. Beyer, C. L. Oppenheimer, J. C. Paulson, J.-P. Prieels, J. I. Rearick, and R. L. Hill, this series, Vol. 83, p. 458.

[17] J. Weinstein, U. de Souza-e-Silva, and J. C. Paulson, *J. Biol. Chem.* **257**, 13835 (1982).

[18] J. Weinstein, U. de Souza-e-Silva, and J. C. Paulson, *J. Biol. Chem.* **257**, 13845 (1982).

Buffers and Other Materials

PBS (phosphate-buffered saline): 10 mM sodium phosphate, pH 7.0, 0.15 M NaCl. BBS (barbitol-buffered saline): 10 mM sodium diethyl barbiturate, pH 7.0, 0.15 M NaCl.

AMAB (modified Alsever's medium-containing antibiotics), 30 mM sodium citrate, pH 6.1, 77 mM NaCl, 114 mM glucose, 100 μg/ml of neomycin sulfate, 330 μg/ml of chloramphenicol.

CIB (cell incubation buffer): 25 mM sodium cacodylate, pH 6.5, 75 mM NaCl, 100 mM glucose, and 10 mg bovine serum albumin, 100 μg neomycin sulfate, 330 μg chloramphenicol per milliliter. This buffer must be prepared fresh and used within 1–2 weeks.

CMP-Sia: CMP-[^{14}C]NeuAc (5000 cpm/nmol; New England Nuclear). CMP-N-Glycolylneuraminic acid (CMP-NeuGc)[19] or CMP-9-O-acetyl-N-acetylneuraminic acid (CMP-9-O-Ac-NeuAc)[19,20] may also be used. To monitor incorporation of unlabeled CMP-Sia, a trace amount of CMP-[^{14}C]NeuAc (250 mCi/mmol; New England Nuclear) can be added to give a final specific activity of 5000 cpm/nmol.

Sialidase: *Vibrio cholera* sialidase (1 unit/ml; Grand Island Biochemical Company).

Erythrocytes: Human erythrocytes collected by venipuncture are washed by centrifugation and suspension 4 times in BBS, and stored as a 50% suspension in AMAB. Throughout these procedures the volume of the cell preparations is determined by hematocrit.

Preparation of Resialylated Erythrocytes[12,15,21]

Preparation of Sialidase-Treated Cells. Erythrocytes are washed by centrifugation and suspension 3 times in BBS, and then incubated for 3 hr at 37° in a suspension (38% v/v) with BBS containing 4 mM CaCl$_2$, bovine serum albumin at 5 mg/ml, and 150 milliunits *V. cholera* sialidase per milliliter packed cells. The asialo-cells are then washed in BBS and stored as a 50% suspension in AMAB. This procedure releases about 90% of the total sialic acid.[21]

Preparation of Sialyltransferases. The purified sialyltransferases are generally stored in buffers containing 50% glycerol and 0–2% Triton X-100 or Triton CF-54, both of which must be removed before reaction of

[19] R. Schauer, *Adv. Carbohydr. Chem. Biochem.* **40**, 131 (1982).
[20] H. H. Higa and J. C. Paulson, *J. Biol. Chem.* **260**, 8838 (1985).
[21] S. C. Carroll, H. H. Higa, and J. C. Paulson, *J. Biol. Chem.* **256**, 8357 (1981).

the enzymes with erythrocytes. This is easily accomplished by mixing 20–150 μl of the sialyltransferase (10–200 milliunits) with sufficient CIB to give a final volume of 400 μl, followed by chromatography on a freshly prepared column (0.7 × 13 cm) of Sephadex G-50 (fine) equilibrated and developed in CIB. Fractions (0.3 ml) are collected and assayed for sialyltransferase activity. Active fractions are pooled (1.1–1.2 ml) and added to a tube containing sufficient lyophilized CMP-Sia to give a final concentration of 1.1–1.3 mM. Recovery of enzyme is usually 80–90%. At this step the sialyltransferases may be stored for 24 hr at 4° with no loss of activity. It has been empirically determined that an unused column of Sephadex G-50 will remove Triton detergents by adsorption if the initial concentration in the 400-μl sample is less than 0.14%, but will not be removed if much higher than this level. If necessary, residual Triton may be removed by adsorption to Bio-Beads SM-2 (Bio-Rad) prior to addition of the CMP-Sia,[21,22] except for the Galβ1,3(4)GlcNAcα2,3-sialyltransferase, which is inactivated by this treatment.[15]

Resialylation of Sialidase-Treated Erythrocytes. Sialyltransferase in CIB with 1.1–1.3 mM CMP-Sia is mixed with an equal volume of packed sialidase-treated cells previously washed with CIB. Reaction is allowed to proceed for 3–4 hr at 37°. The cells are then washed several times in BBS and suspended in an equal volume of AMAB for storage. Quantitation of radiolabeled sialic acid[12] is done by lysis of 10–20 μl of the cells in 1 ml of 20 mM sodium phosphate, pH 7, collection of the membranes on a glass fiber filter (Whatman GF-C) by vacuum filtration, placing the filter in a minivial containing 2.0 ml 50% ethanol and 2.5 ml scintillation fluid (3a70b, Research Products International), and counting in a liquid scintillation counter.

The amount of sialic acid incorporated into the cells can be varied by the amount of enzyme used. Typically, for experiments examining influenza virus receptor specificity, the amount of sialyltransferase used per milliliter of packed cells is 100 milliunits of the Galβ1,4GlcNAcα2,6-sialyltransferase, 80 milliunits of the Galβ1,3(4)GlcNAcα2,3-sialyltransferase, 20 milliunits of the Galβ1,3GalNAcα2,3-sialyltransferase, and 180 milliunits of the GalNAcα2,6-sialyltransferase. Using CMP-NeuAc as the donor substrate, this amount of enzyme gives levels of incorporation of 35–45 nmol/ml packed cells in the NeuAcα2,6Galβ1,4GlcNAc sequence, 70–75 nmol/ml in the NeuAcα2,3Galβ1,3(4)GlcNAc sequence, 90–125 nmol/ml in the NeuAcα2,3Galβ1,3GalNAc sequence and 60–65 nmol/ml in the NeuAcα2,6GalNAc sequence, respectively. Using a higher level of sialyltransferase gives a maximum of twice the level of incorporation.[12]

[22] P. W. Holloway, *Anal. Biochem.* **53,** 304 (1973).

TABLE I

HEMAGGLUTINATION OF RESIALYLATED ERYTHROCYTES BY SELECTED
INFLUENZA VIRUSES[a]

Cell preparation[b]	HA titer for selected viruses[b]					
	1	2	3	4	5	6
Native	256	256	512	256	1024	1024
Asialo	0	0	0	0	0	0
Resialylated						
NeuAcα2,6Gal (A)	256	256	512	8	0	1024
NeuGcα2,6Gal (A)	0	0	512	0	0	0
9-O-Ac-NeuAcα2,6Gal (A)	0	0	0	0	0	0
NeuAcα2,3Gal (B)	0	0	256	256	1024	n.d.
NeuAcα2,3,Gal (C)	0	0	0	256	1024	1024
NeuGcα2,3Gal (C)	0	0	0	0	0	512
9-O-Ac-NeuAcα2,3Gal (C)	0	0	0	0	0	0
NeuAcα2,6GalNAc (D)	0	0	0	0	0	n.d.

[a] Compiled in part from Rogers and Paulson[15] and Higa et al.[23]

[b] Sequences found on resialylated cells are defined by the specificity of the sialyl-transferases used (A–D) as described under Experimental Procedures.

[c] Hemagglutination (HA) titers are expressed as the highest 2 fold serial dilution of virus that causes hemagglutination of the derivatized erythrocytes; n.d. means not determined. Viruses used are (1) A/Aichi/2/68; (2) A/swine/Iowa/1976/31; (3) A/Victoria/3/75; (4) A/duck/Ukraine/1/63; (5) A/equine/Miami/1/63; (6) A/duck/Memphis/546/78.

Lower levels of incorporation are conveniently obtained by shorter incubation times (0.3–1.5 hr). For CMP-NeuGc or CMP-9-O-Ac-NeuAc as donor substrates, either more or less sialyltransferase may be required to achieve equivalent rates of incorporation obtained with CMP-NeuAc.[20,23]

Assessment of Sialyloligosaccharide Binding Protein Specificity

Specificity of Influenza Viruses for Sialyloligosaccharide Receptor Determinants. Influenza viruses are membrane-enveloped viruses which attach to cells by means of the major surface glycoprotein, the hemagglutinin. Examples of differential agglutination of resialylated erythrocytes by selected influenza viruses are shown in Table I. This analysis is particularly good for detecting differences in the specificity of the viral hemagglutinins for their sialyloligosaccharide receptor determinants. Indeed, differences in the ability of the viruses to agglutinate cells containing

[23] H. H. Higa, G. N. Rogers, and J. C. Paulson, *Virology* **144,** 279 (1985).

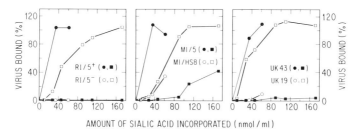

AMOUNT OF SIALIC ACID INCORPORATED (nmol / ml)

Fig. 1. Differential dependence on sialic acid incorporation of viral adsorption to resialylated erythrocytes. Each panel shows results from a pair of viruses isolated as receptor
variants or subpopulations from single influenza isolates; A/RI/5/57 (left), A/Memphis/102-72
(middle), or A/duck/Ukraine/1/63 (right). Viral adsorption is expressed as the amount of
virus bound relative to that adsorbed to 5% native erythrocytes, as measured using the viral
sialidase as an enzymatic marker.[15] Solid symbols indicate viruses specific for hemagglutination of resialylated erythrocytes containing the NeuAcα2,6Galβ1,4GlcNAc sequence (at 30–
40 nmol/ml packed cells) and open symbols indicate viruses specific for hemagglutination of
cells containing the NeuAcα2,3Galβ1,3GalNAc sequence (at 90–110 nmol/ml packed cells).
In each case adsorption to resialylated erythrocytes containing increasing amounts of the
NeuAcα2,6Gal (○,●) or the NeuAcα2,3Gal (□,■) linkages is examined.

different linkages and different sialic acids in the same linkage are readily
apparent.

It is important to consider that the maximum amount of sialic acid
incorporated in the resialylated cells usually ranges from 5 to 20% that of
native cells, and that viruses which fail to bind a sialyloligosaccharide
sequence in the routine analysis may bind at higher levels of incorporation. This is illustrated in Fig. 1 which compares three pairs of viruses
with contrasting receptor binding properties, each pair representing sub-
strains or selected variants from a single parent virus. For the initial
analysis done as in Table I, all three pairs appeared identical with one
virus agglutinating cells containing the NeuAcα2,3Gal linkage (at 100
nmol/ml packed cells) and the other agglutinating cells containing
NeuAcα2,6Gal linkage (at 35 nmol/ml packed cells). However, as shown
in the figure, when these same virus pairs are compared for their adsorption to cells resialylated in the same linkages but to varying extents,
differences in their degree of specificity and in their avidity can be readily
detected. Thus, conclusions regarding the types of sialyloligosaccharide
sequences which mediate adsorption to native cells should take into account that native cells may contain higher levels of sialic acid than the
resialylated cells.

Screening influenza virus isolates from a variety of animal species has
provided some insights into the biological significance of receptor specificity. Indeed, influenza isolates with closely related hemagglutinins (H3

serotype) were found to exhibit different receptor specificities, based on species of origin.[15] While human isolates preferentially bound erythrocytes containing the NeuAcα2,6Gal linkage, avian and equine isolates bound cells containing the NeuAcα2,3Gal linkage. Subsequently, receptor variants were selected from human viruses which exhibit the receptor specificity of avian and equine viruses and also from avian viruses which exhibit the receptor specificity of the human viruses. In each case sequence analysis of the hemagglutinins revealed that the differences in receptor specificity are due largely to a single amino acid change at residue 226,[24] which is in the proposed receptor-binding pocket of the three-dimensional structure of the hemagglutinin reported by Wilson et al.[25] These observations suggest that different species may exert selective pressures, resulting in a hemagglutinin with a receptor specificity best suited for growth in that host.[15]

Application to Other Sialyloligosaccharide Binding Proteins. In addition to influenza virus hemagglutinins, several other sialyloligosaccharide binding proteins and microorganisms have been examined for their binding specificity using resialylated erythrocytes. These include the viral attachment proteins of paramyxoviruses[13] and polyoma virus,[26] the cell attachment factor of *Mycoplasma pneumoniae*,[5] sialic acid-specific lectins from the horseshoe crab (*Limulus polyphemus*), the slug (*Limax flavus*), and an *O*-acetylsialic acid-specific lectin from the marine crab (*Cancer antennarius*).[27,28] While somewhat more difficult, receptor specificity experiments can also be performed with living tissue culture cells. This has been shown for restoration of functional receptors leading to infection for Sendai virus, polyoma virus, and influenza virus.[29-31] For this application both the resialylation and the recognition event, in this case adsorption of virus, must be done within 30 min to avoid problems due to endogenous replacement of cell surface sialic acid.

Acknowledgment

Work of the authors was supported by grants from the National Institutes of Health (GM-27904 and AI-16165) and the American Cancer Society (FRG-224).

[24] G. N. Rogers, J. C. Paulson, R. S. Daniels, J. J. Skehel, I. A. Wilson, and D. C. Wiley, *Nature (London)* **304**, 76 (1983).
[25] I. A. Wilson, J. J. Skehel, and D. C. Wiley, *Nature (London)* **289**, 368 (1981).
[26] L. D. Cahan, R. Singh, and J. C. Paulson, *Virology* **130**, 281 (1983).
[27] M. H. Ravindranath, H. H. Higa, and J. C. Paulson, *J. Biol. Chem.* **260**, 8850 (1985).
[28] M. H. Ravindranath and J. C. Paulson, this volume [44].
[29] M. A. K. Markwell and J. C. Paulson, *Proc. Natl. Acad. Sci. U.S.A.* **77**, 5693 (1980).
[30] H. Fried, L. D. Cahan, and J. C. Paulson, *Virology* **109**, 188 (1981).
[31] S. C. Carroll and J. C. Paulson, *Virus Res.* **3**, 65 (1985).

[12] Detection of Polysialosyl-Containing Glycoproteins in Brain Using Prokaryotic-Derived Probes[1]

By Frederic A. Troy, Patrick C. Hallenbeck,
Ronald D. McCoy, and Eric R. Vimr

A class of high molecular weight surface sialoglycoproteins that function in neural cell adhesion and brain development has been described by several laboratories.[2-6] These glycoproteins, designated neural cell adhesion molecule (N-CAM),[7] D2-cell adhesion molecule (D2-CAM), or brain cell surface protein-2 (BSP-2), appear to be the same antigen.[5,8,9] The embryonic (E) form of N-CAM contains a high sialic acid content that undergoes a postnatal conversion to a low sialic acid content in the adult (A).[10-12] The developmental reduction in sialic acid is postulated to modulate cell–cell adhesive properties of neuronal cells and to mediate their specific organization into adult brain tissue.[13,14]

Sialic acid rarely exists internally to form polysialosyl structures.[15] A well-characterized polysialic acid is the K1 capsular antigen of *Escherichia coli* K1 serotypes which contain at least 200 sialyl residues joined by

[1] This study was supported by Public Health Service Research Grant AI-09352 (to FAT) from the National Institutes of Health.

[2] E. Bock, Z. Yavin, O. S. Jørgensen, and E. Yavin, *J. Neurochem.* **35**, 1297 (1980).

[3] O. S. Jørgensen, *J. Neurochem.* **37**, 939 (1981).

[4] G. M. Edelman, S. Hoffman, C.-M. Choung, J.-P. Thiery, R. Brackenbury, W. J. Gallin, M. Grumet, M. E. Greenberg, J. J. Hemperly, C. Cohen, and B. A. Cunningham, *Mol. Neurobiol.* **47**, 515 (1983).

[5] C. Goridis, H. Deagostini-Bazin, M. Hirn, M.-R. Hirsch, G. Rougon, R. Sadoul, O. K. Langley, G. Gombos, and J. Finne, *Mol. Neurobiol.* **47**, 527 (1983).

[6] U. Rutishauser, *Mol. Neurobiol.* **47**, 501 (1983).

[7] Abbreviations used: N-CAM, neural cell adhesion molecule (E, embryonic; A, adult); Endo-N, endoneuraminidase specific for polysialic acid; NeuNAc, *N*-acetylneuraminic acid (sialic acid); SDS–PAGE, sodium dodecyl sulfate–polyacrylamide gel electrophoresis; DP, degree of polymerization; PMSF, phenylmethylsulfonyl fluoride.

[8] O. S. Jørgensen, A. Delouvee, J.-P. Thiery, and G. Edelman, *FEBS Lett.* **111**, 39 (1980).

[9] J. M. Lyles, D. Linnemann, and E. Bock, *J. Cell Biol.* **99**, 2082 (1984).

[10] O. S. Jørgensen and M. Moller, *Brain Res.* **194**, 419 (1980).

[11] J. Finne, *J. Biol. Chem.* **257**, 11966 (1982).

[12] J. B. Rothbard, R. Brackenbury, B. A. Cunningham, and G. M. Edelman, *J. Biol. Chem.* **257**, 11064 (1982).

[13] G. M. Edelman, *Annu. Rev. Neurosci.* **7**, 339 (1984).

[14] U. Rutishauser, *Nature (London)* **310**, 549 (1984).

[15] R. Schauer, *Adv. Carbohydr. Chem. Biochem.* **40**, 131 (1982).

α-2,8-ketosidic linkages.[16] The K1 capsule can confer on cells the ability to invade and colonize the meninges of neonates.[17-19] Preliminary studies suggested that the E form of N-CAM may contain internal sialic acid residues on N-linked oligosaccharides.[11,20] Developmentally associated changes in the protein moiety of N-CAM,[21] including posttranslational glycosylation, sulfation, phosphorylation, and intracellular transport reactions, have been described.[22] In contrast, little is known about the biosynthesis or temporal expression of the polysialosyl units on N-CAM or the mechanism that mediates the developmentally regulated reduction of polysialic acid during the embryonic to adult conversion. Three prokaryotic-derived probes that detect polysialosyl-containing glycoproteins in brain were developed to study these processes.[22a]

Prokaryotic-Derived Reagents for the Analysis and Characterization of Polysialosyl Glycoproteins

The probes that recognize or synthesize polysialic acid containing α-2,8-ketosidic linkages are (1) a soluble form of a bacteriophage-derived endo-N-acetylneuraminidase, designated Endo-N, produced by infection of E. coli K1 with a lytic bacteriophage; (2) an equine polyclonal IgM antibody, designated H.46, prepared by immunizing a horse with Neisseria meningitidis serogroup B; the polysialic acid capsule of N. meningitidis B is structurally identical to the E. coli K1 capsule; (3) a membranous sialyltransferase from E. coli K1 that can transfer sialic acid (NeuNAc) from CMP-NeuNAc to exogenous acceptors of oligo- or polysialic acid. The bacterial sialyltransferase will recognize embryonic but not adult rat brain membranes as exogenous acceptors.

These probes have been used in combination with structural studies to (1) prove the presence of polysialosyl carbohydrate units in membranes of embryonic chick and rat brains[22a]; (2) quantitate the amount of [14C]NeuNAc incorporated specifically into α-2,8-linked polysialosyl units in N-CAM by a CMP-NeuNAc : poly-α-2,8-sialosyl sialyltransferase[23]; and (3) to show that the polysialic acid on N-CAM has a direct

[16] T. E. Rohr and F. A. Troy, J. Biol. Chem. 255, 2332 (1980).
[17] M. S. Schiffer, E. Oliveira, M. P. Glode, G. H. McCracken, Jr., L. M. Sarff, and J. B. Robbins, Pediatr. Res. 10, 82 (1976).
[18] F. A. Troy, Annu. Rev. Microbiol. 33, 519 (1979).
[19] K. Jann and B. Jann, Prog. Allergy 33, 53 (1983).
[20] J. Finne, U. Finne, H. Deagostini-Bazin, and C. Gordis, Biochem. Biophys. Res. Commun. 112, 482 (1983).
[21] K. L. Crossin, G. M. Edelman, and B. A. Cunningham, J. Cell Biol. 99, 1848 (1984).
[22] J. M. Lyles, D. Linnemann, and E. Bock, J. Cell Biol. 99, 2082 (1984).
[22a] E. R. Vimr, R. D. McCoy, H. F. Vollger, N. C. Wilkison, and F. A. Troy, Proc. Natl. Acad. Sci. U.S.A. 81, 1971 (1984).
[23] R. D. McCoy, E. R. Vimr, and F. A. Troy, J. Biol. Chem. 260, 12695 (1985).

effect on N-CAM mediated adhesion and that the amount of this carbohydrate is important for normal development of nerve tissue.[24] This article will review the development and use of these probes to detect polysialosyl carbohydrate units in brain and bacterial glycoconjugates.[22a,23,24a]

Bacteriophage-Derived Endo-N-Acetylneuraminidase

$$[\rightarrow 8]\text{-}\alpha\text{-NeuNAc-}(2\rightarrow]_n \rightarrow X \longrightarrow [\rightarrow 8]\text{-}\alpha\text{-NeuNAc-}(2\rightarrow]_{2-4}$$

The bacteriophage-induced endo-N-acetylneuraminidase (Endo-N) is specific for hydrolyzing oligo- or poly-α-2,8-sialosyl carbohydrate units in sources as disparate as rat brain and bacteria. In the polysialic acid capsule in *E. coli* K1, n is ~200 sialyl units and X represents endogenous acceptor molecules.[16,25–27] In N-CAM, n is at least 12–20 sialyl units and X is an N-asparaginyl-linked oligosaccharide chain of the complex type.[1,11] The enzyme can also hydrolyze the α-2,8 linkage in bacterial polysialic acids containing alternating α-2 $\rightarrow$ 8 and α-2 $\rightarrow$ 9-ketosidic linkages.[24a] The soluble form of the enzyme reported here is derived from a bacteriophage (K1F) lysate of *E. coli* K1. The K1F phage was isolated from University of California, Davis, sewage after enrichment on *E. coli* K-235 by standard procedures.[28] K1F encodes for the synthesis of both a soluble form and phage-bound form of the enzyme. Properties of a similar endoneuraminidase associated with phage particles are described by Stirm and colleagues.[29] Another phage-associated Endo-N with somewhat different properties has been reported.[30] Tomlinson and Taylor also recently described the purification of a soluble form of a K1 depolymerase that has substantially different properties than the K1F-derived enzyme.[31]

Assay Method

Principle. Oligo- or polysialic acid from *E. coli* K1 is depolymerized by either the soluble or phage-bound Endo-N. The reaction is followed by quantitating either loss of substrate or ([^{14}C]-sialyl oligomers from [^{14}C]-polysialic acid) or by measuring the amount of product ([^{14}C]-sialyl oligo-

[24] U. Rutishauser, M. Watanabe, J. Silver, F. A. Troy, and E. R. Vimr, *J. Cell Biol.* **101,** 1842 (1985).

[24a] P. C. Hallenbeck, E. R. Vimr, F. Yu, B. L. Bassler, and F. A. Troy, *J. Biol. Chem.* (1987), in press.

[25] F. A. Troy and M. A. McCloskey, *J. Biol. Chem.* **254,** 7377 (1979).

[26] E. C. Gotschlich, B. A. Fraser, O. Nishimura, J. B. Robbins, and T.-Y. Liu, *J. Biol. Chem.* **256,** 8915 (1981).

[27] M. A. Schmidt and K. Jann, *FEMS Microbiol. Lett.* **14,** 69 (1982).

[28] R. J. Gross, T. Cheasty, and B. Rone, *J. Clin. Microbiol.* **6,** 548 (1977).

[29] B. Kwiatkowski, B. Boschek, H. Thiele, and S. Stirm, *J. Virol.* **43,** 697 (1982).

[30] J. Finne and P. H. Mäkelä, *J. Biol. Chem.* **260,** 1265 (1985).

[31] S. Tomlinson and P. W. Taylor, *J. Virol.* **55,** 374 (1985).

mers) formed. Reaction products vary with depolymerases derived from different K1 specific phages[24a,29-31] and with different oligo- or polysialyl substrates. There are no known cofactors for Endo-N and the pH optimum is ~7.4. Suitable substrates include unlabeled or radiolabeled oligomers with degrees of polymerization (DP) 5 or higher and polysialic acid (DP ~200). Polysialic acid may be synthesized *in vivo* or *in vitro*.[16,25] Sialyl oligomers, such as colominic acid, can be generated by either mild acid hydrolysis[25] or Endo-N depolymerization of polysialic acid.[24a]

Procedures for Preparing Oligo- and Polysialic Acid. Unlabeled or [3]H- or U-^{14}C-labeled sialyl polymers are isolated from *E. coli* culture filtrates by a modification of our previous method.[25,32] Cells are grown in 1 liter of modified M9 medium containing the following components (grams per liter): Na_2HPO_4, 6; KH_2PO_4, 3; NaCl, 0.5; NH_4Cl, 1; casamino acids, 4; glucose, 4. After autoclaving, 10 ml of 10 mM $CaCl_2$ and 1 ml of 1 M $MgSO_4$ are added. For preparing radiolabeled polymers, one mCi of U-^{14}C-labeled glucose (or 2 mCi of [^{3}H]glucose) in 1.0 ml of 90% ethanol is added directly to the sterile medium. Cells are grown at 37° on a rotary shaker. Sialyl polymers are removed from the surface of late log phase cells by treating the entire culture in a Polytron Model PCU-2 (3–5 times for 40 sec each). Cells are removed by centrifugation (16,000 g for 25 min) and the polymers precipitated by cetyltrimethylammonium bromide (CETAB; 0.3% final concentration). After 24 hr at 4°, the sialyl polymers are extracted from the CETAB complex with 1 M $CaCl_2$ and fractionally purified by ethanol precipitation.[33] These sialyl polymers are sufficiently pure for use in Endo-N assays and may be further purified by published procedures.[34,35] Alternatively, *in vitro* synthesized [3]H- or ^{14}C-labeled polysialic acid, prepared as previously described[16] may be used. Unlabeled sialyl oligomers are commercially available (Colominic acid, Sigma Chemical Co.). These preparations contain variable degrees of sialyl oligomers that can be resolved by anion-exchange chromatography[22a,24a,30,36] before use. Unlabeled colominic acid can be labeled either in the reducing terminus by reduction with KB[3]H_4, as described previously,[25] or in the nonreducing terminus by sequential periodate oxidation and KB[3]H_4 reduction.[16,37]

[32] F. A. Troy, I. K. Vijay, M. A. McCloskey, and T. E. Rohr, this series, Vol. 83, p. 540.
[33] W. F. Vann and K. Jann, *Infect. Immun.* **25**, 85 (1979).
[34] E. C. Gotschlich, T.-Y. Liu, and M. S. Artenstein, *J. Exp. Med.* **129**, 1349 (1969).
[35] R. Schneerson, M. Bradshaw, J. K. Whisnant, R. C. Myerowitz, J. C. Parke, and J. B. Robbins, *J. Immunol.* **108**, 1551 (1972).
[36] H. Nomoto, M. Iwasaki, T. Endo, S. Inoue, Y. Inoue, and G. Matsumara, *Arch. Biochem. Biophys.* **218**, 335 (1982).
[37] L. Van Lenten and G. Ashwell, *J. Biol. Chem.* **246**, 1889 (1971).

Assay of Endo-N Activity by Substrate Depletion. Endo-N activity is approximated by measuring the depolymerization of radiolabeled polysialic acid as follows. Standard incubation mixtures (25 μl) containing 67 μg ^{14}C-labeled polysialic acid ($\sim$92 × 10^3 dpm/mg polymer) in 20 mM Tris–HCl, pH 7.4 are incubated at 37° with varying dilutions of enzyme. After 15 min, 20 μl are spotted on Whatman No. 3 MM paper and chromatographed in ethanol–1 M ammonium acetate, pH 7.5 (7 : 3), as described.[32] ^{14}C-labeled sialyl polymers remaining at the origin are quantitated by scintillation counting. Endo-N activity is calculated from the highest dilution of enzyme that gives at least a 20% loss in radioactivity from the origin substrate. In principle, this method can be used to measure Endo-N activity by quantitating oligosialic acid formation but this requires radiochromatographic scanning or the tedium of cutting and counting multiple strips. Moreover, this method does not yield initial reaction rates since Endo-N preferentially depolymerizes higher molecular weight polymers (Table II) and the chromatography assay favors mobility of limit digestion products. For this reason, substrate depletion underestimates Endo-N activity. In this assay, Endo-N activity is expressed as a function of ^{14}C-labeled sialyl oligomers released from the polymer. One unit is defined as the amount of enzyme required to mobilize 50,000 cpm of sialyl oligomers per minute. This is equivalent to about 1 μmol of sialic acid reducing terminus released per minute.

Assay of Endo-N Activity by Product Formation. Endo-N activity can be determined by measuring the number of α-2,8 linkages hydrolyzed. Since each cleavage results in the formation of a new reducing terminus which in the acyclic form is sensitive to periodate oxidation between C-6 and C-7, Endo-N activity can be quantitated by the thiobarbituric acid (TBA) method.[38] The 2-acetamido-4-deoxyhexos-5-uluronic acid resulting from periodate oxidation readily undergoes an aldol cleavage to form β-formyl pyruvate which condenses with TBA.[39] An advantage of this method is that it can be carried out using unlabeled sialyl polymers. Controls (no enzyme) are essential to determine the background of free reducing ends in the undegraded substrate, which is a more serious problem the smaller the DP of the substrate. This background can be decreased by prior reduction of the sialyl oligomers with borohydride.[25] After reduction, the oligomers no longer react with TBA. A potential disadvantage of this method for following Endo-N activity through the purification scheme is interference in formation of the TBA chromophore, particularly by salts, e.g., $(NH_4)_2SO_4$. Standard 1.0-ml incubation mix-

[38] L. Skoza and S. Mohos, *Biochem. J.* **159,** 457 (1976).
[39] R. Kuhn and P. Lutz, *Biochem. Z.* **338,** 554 (1963).

TABLE I
PURIFICATION OF THE SOLUBLE FORM OF ENDO-N-ACETYLNEURAMINIDASE (ENDO-N)[a]

Step	Total activity (units)[b]	Yield (%)	Total protein (mg)	Specific activity (units/mg)	Fold purification (×)
(1) Phage lysate	9,450	100	22,500	0.4	1
(2) 50% (NH₄)₂SO₄ pellet	5,652	42	2,290	2.5	6
(3) High-speed supernatant	3,955	42	2,840	1.4	3.4
(4) Heat treatment (60°C, 20 min)	2,482	26	2,227	1.1	2.7
(5) 40% (NH₄)₂SO₄ pellet	3,109	33	436	7.1	17.3
(6) Octyl sepharose column	2,712	29	235	11.5	28
(7) Hydroxy-apatite-DEAE chromatography	18,437	195	4.4	4219	10,290

[a] The data shown are for a representative purification. Due to limitations in the substrate depletion assay, the absolute values in any purification may vary.

[b] Endo-N activity was determined by substrate depletion by measuring the depolymerization of radiolabeled polysialic acid. One unit is defined as the amount of enzyme required to mobilize 50,000 cpm of sialyl oligomers per min. This is equivalent to about 1 μmole of sialic acid released per minute.

tures containing 2–6 mg oligo- or polysialic acid in 20 mM Tris–HCl, pH 7.4, are incubated at 37° with appropriately diluted enzyme solutions. The reaction is terminated after 1.0, 2.5, 5.0, 7.5, 10, and 15 min by the addition of ethanol to 50%. The number of reducing ends are quantitated by the TBA assay,[39] omitting the initial acid hydrolysis step. The reaction is linear for at least 10 min at 0.2–4 mg sialyl polymer/ml and 2 milliunits of enzyme. For determining initial kinetics, the OD$_{540\,nm}$ characteristically changes from ~0.02 to >0.3–0.6. Under these conditions, less than 10% of the sialyl polymer linkages are cleaved. One unit of Endo-N forms 1 μmol of product (as TBA reactive material) per minute per milligram protein at 37°.

Purification Procedure for Endo-N

A summary of the purification of Endo-N from 6 liters of concentrated media is given in Table I.

Bacterial Strains and Media. While the K1 human isolate *E. coli* K-235 can be used for phage propagation, we use an *E. coli* K-12/K1 hybrid strain, designated EV1. This strain was constructed by introducing the *kps* genetic locus necessary for biosynthesis of the K1 capsule into *E. coli* K-12 by conjugation.[40] Luria broth is used for strain storage, phage titers, and precultures.

Growth Conditions. Cultures are grown in 6 liters of concentrated media (66 g Bacto-tryptone, 135 g Bacto-yeast extract, pH 7.6), phosphate buffer (80 g K_2HPO_4, 19.2 g KH_2PO_4) and 120 g glucose in an ISCO High Density Fermentor. One liter of concentrated media is inoculated with 20 ml of a preculture obtained from a single colony isolate of EV1 and incubated in a rotary shaker at 37° overnight. The 1-liter preculture is added to 5 liters of media (plus phosphate buffer and glucose) and grown at 37° in the fermentor until mid-late log phase. pH is maintained at pH 6.9–7.2 by periodic additions of 10 N NaOH, at which time additional glucose is added. Aeration is initially at 16 ft^3 O_2/min which is increased to 20 ft^3 O_2/min at mid log phase. When the culture reaches a density of 10^{10} colony forming units per milliliter, concentrated K1 phage is added to a multiplicity of infection of 0.1–0.2. Two milliliters of a 30% solution of antifoam C is added. After 1.5 hr, lysis is complete. The contents are placed at 4°, and the lysate further treated as described below (Table I; Step 1).

Purification of Endoneuraminidase. (Table I, Steps 2–7). Step 2. Precipitation with 50% Ammonium Sulfate. The phage lysate is brought to 50% saturation with ammonium sulfate [$(NH_4)_2SO_4$] by the addition of solid $(NH_4)_2SO_4$ and left stirring at 4° overnight. The pellet is collected by centrifuging at 6,200 g for 30 min, resuspended in 350 ml of 50 mM Tris buffer (pH 7.4) containing 0.5 mM EDTA, 0.2 mM phenylmethylsulfonyl fluoride (PMSF), and sedimented at 24,000 g for 20 min. The supernatant is saved and treated as described below. The pellet is extracted with 150 ml of the above buffer and centrifuged at 24,000 g for 20 min. The pellet is re-extracted due to substantial Endo-N activity being entrained in the pellet. This step effectively concentrates Endo-N activity and removes cellular debris.

Step 3. High-Speed Centrifugation. The supernatants are combined and centrifuged at 190,000 g for 85 min. The pellet is saved for purification of the phage (see below) and the supernatant treated as described in Step 4. The purpose of the high-speed centrifugation is to recover K1F phage, which is why both the total activity and the specific activity decrease at this stage. This step can be eliminated if only the soluble Endo-N activity is desired.

Step 4. Heat Treatment. The combined supernatants are heated in

[40] E. R. Vimr and F. A. Troy, *J. Bacteriol.* **164,** 854 (1985).

250-ml centrifuge bottles at 55–60° for 25 min and then immediately centrifuged at 24,000 g for 20 min (at 4°). This results in a substantial loss of protein without appreciable loss of Endo-N activity.

Step 5. Ammonium Sulfate Fractionation. The supernatant fraction from Step 4 is brought to 25% saturation with $(NH_4)_2SO_4$, stirred for 15 min on ice, and centrifuged at 24,000 g for 15 min. The resulting supernatant is then brought to 40% saturation with $(NH_4)_2SO_4$ and centrifuged as above. The pellet is redissolved in a minimal volume of 20 mM Tris, pH 7.5, containing 0.5 mM EDTA and dialyzed overnight against the same buffer. Fractional precipitation with $(NH_4)_2SO_4$ results in about a 7-fold purification. It also concentrates Endo-N so that after dialysis it can be stored at $-20°$ for at least several months without appreciable loss in activity.

Step 6. Hydrophobic Chromatography. The retentate from Step 5 is brought to 2 M NaCl and loaded on a 3.6 × 26.5 cm column of octyl-Sepharose previously equilibrated with 20 mM Tris (pH 7.5) and 2 M NaCl. The column is washed with 300 ml of 2 M NaCl (containing 20 mM Tris, pH 7.5), followed by a decreasing linear NaCl gradient formed from 300 ml of 2 M NaCl and 300 ml of distilled water, both of which contain 20 mM Tris (pH 7.5). The active fractions are dialyzed overnight against 10 mM sodium phosphate (pH 7.5).

Step 7. Hydroxylapatite–DEAE Double-Column Chromatography. The retentate from Step 6 is applied to a "double column" consisting of a 3.6 × 18 cm column of hydroxylapatite connected in series with a 1.4 × 29 cm column of DEAE–Tris-acryl, previously equilibrated with 10 mM phosphate buffer (pH 7.5). After the sample has been applied and the columns extensively washed with equilibrating buffer, the hydroxylapatite column is disconnected, and the DEAE–Tris-acryl column developed with a linear gradient formed from 200 ml of 20 mM Tris (pH 7.5) and 200 ml of 20 mM Tris (pH 7.5), 0.25 M NaCl. A single peak of activity is eluted at about 0.15 M NaCl.

Purification of Phage Particles Containing Endoneuraminidase

The pellet from the high-speed centrifugation (190,000 g for 85 min) is resuspended in a minimal volume of 0.1 M Tris (pH 7.4), 0.1% NH$_4$Cl, 0.5% NaCl with the aid of a Teflon-glass homogenizer, layered on top of a CsCl step gradient consisting of 8 ml of 1.6 g/ml CsCl, 14 ml 1.3 g/ml CsCl, and 16 ml of the resuspended high-speed pellet, and centrifuged in a SW 27 rotor at 20,000 rpm for 120 min. The opalescent layer at the interface of the 1.3/1.6 g/ml CsCl is removed and dialyzed overnight against 0.1 M Tris (pH 7.4), 0.5% NH$_4$Cl, 0.5% NaCl and stored at 4° over a few drops of chloroform.

Properties of the Endo-N-acetylneuraminidase

Molecular Weight, Subunit Composition, and Purity. Endo-N obtained from the final hydroxylapatite column is purified ~10,000-fold and appears homogeneous by SDS–PAGE.[24a] The enzyme has a subunit molecular weight of 105,000 estimated by SDS–PAGE. This corresponds to one of the higher molecular weight phage proteins which comprises 7.5% (by weight) of the total phage protein. The holoenzyme has an extrapolated molecular weight by gel filtration on Sephacryl S-200 of 328,000, suggesting that the active enzyme is a trimer. These properties differ from those reported by Tomlinson and Taylor.[31] They reported a molecular weight of 208,000 for the holoenzyme and proposed that their *E. coli* K1-derived endoneuraminidase was a trimer that consisted of two subunits of 78,000 and one of 38,500. Whether these polypeptides were phage proteins was not addressed.

Substrate Specificity. Endo-N is specific for catalyzing the hydrolysis of α-2,8-sialosyl linkages. It will depolymerize sialyl oligomers and polymers with this linkage from bacterial capsular polysaccharides and neural membranes.[22,22a] It will also cleave the α-2,8 linkage in the *E. coli* N67 capsule which consists of sialic acid joined by alternating α-2,8 and α-2,9 linkages.[22a] It will not depolymerize the *N. meningitidis* serogroup C capsule which is a homopolymer of α-2,9-sialosyl units. The soluble and K1F-associated Endo-N both require a minimum of five sialyl residues for activity,[41] whereas Finne and Mäkelä reported a requirement for 8 sialyl residues for the PK1A phage-bound Endo-N.[30] No major difference in the products of limit digestion has been found between the two forms of the K1F-derived enzyme. The limit digest products from the *E. coli* K1 capsule are DP4 with some DP1,2 and DP3. Endo-N digestion of polysialic acid on the embryonic form of N-CAM yields sialyl oligomers with DP3 and 4.[1,22] The presence of a terminal sialitol changes both the distribution of limit digestion products and the apparent minimum substrate size from DP5 to DP6 ([NeuNAc]$_5$-NeuNAc-OH).[24a]

Stability. The soluble form of Endo-N is stable to heating at 55–60° for 25 min and can withstand multiple freezing and thawing. In the crude or partially purified state, Endo-N can be frozen at −20° for at least several months and lyophilized with reasonable recovery of activity. The purified enzyme is unstable to lyophilization but can be stabilized by addition of 0.4% bovine serum albumin. There are no metal ion requirements or cofactors and the activity is unaffected by 1 mM EDTA. Endo-N is inhibited by polyanions such as DNA, poly-γ-D-glutamic acid, and chondroitin sulfate.[24a]

[41] P. C. Hallenbeck, F. Yu, and F. A. Troy, unpublished results.

TABLE II
KINETIC PARAMETERS OF SOLUBLE AND BACTERIOPHAGE
K1F-ASSOCIATED ENDONEURAMINIDASE

Substrate	DP	Soluble Endo-N		Phage-bound Endo-N	
		K_m	V_{max}[a]	K_m	V_{max}[a]
Poly-α-2,8[b]	150–200	51 μM	36	71 μM	19
Oligo-α-2,8[c]	10–20	1.6 mM	72	1.2 mM	18
Poly-α-2,8–α-2,9[d]	150–200	6.6 μM	25	7.1 μM	13

[a] V_{max} expressed as micromoles NeuNAc reducing equivalent per minute per milligram protein.
[b] Polysialic acid capsule isolated from *E. coli* K1 (DP ~150–200).
[c] Oligosialic acid (colominic acid) isolated from *E. coli* K1 (DP ~10 < 20).
[d] Polysialic acid capsule isolated from *E. coli* N-67 (DP ~150).

Kinetic Constants and Mechanism. The kinetic constants for the purified soluble and phage-associated Endo-N are summarized in Table II. Little significant difference has been found between the phage-bound and free form of the enzyme. The apparent K_m values for the low and high molecular weight α-2,8-linked substrates and the α-2,8-α-2,9-linked sialyl polymers are within a factor of two for both forms. The high molecular weight α-2,8-linked sialyl polymers (DP ~200) are substantially better substrates (K_m 50–70 μM) than the sialyl oligomers (K_m 1.2 mM). The total number of cleavage sites is about the same at these concentrations of oligo- and polysialic acid. Analyses of the initial rate products did not reveal smaller sialyl oligomers characteristic of the products of limit digestion. These results indicate that the Endo-N likely binds and cleaves at random sites on the polysialosyl chains, in contrast to initiating cleavage at one end and depolymerizing processively. Unexpected is the observation that the alternating α-2,8-α-2,9-ketosidically linked sialyl polymer from *E. coli* N67 appears to be a better substrate than the α-2,8-linked polymer from *E. coli* K1.[24a]

Equine Polyclonal Antibody for Detecting Poly-α-2,8-Linked Sialosyl Carbohydrate Units

Anticapsular antiserum containing polyclonal IgM antibodies against α-2,8-linked polysialic acid was prepared by multiple iv injections of a horse (No. 46) with formalin-fixed *N. meningitidis* serogroup B strain B-

11.[42] The antiserum, designated H.46, was prepared by Dr. John B. Robbins of the National Institutes of Health, Bethesda, Maryland, who has graciously supplied it to many investigators. H.46 will recognize α-2,8-linked sialyl units with DP about 10 or higher.[1,30,43] It can therefore be used to detect these carbohydrate units in polysialosyl-containing glycoproteins in brain and in the polysialic acid capsules of invasive *E. coli* K1 and group B meningococci. A thorough analysis of the specificity of H.46 has recently appeared.[43] A monoclonal IgG antibody against *E. coli* K1 and the meningococcus group B capsules has also been described.[44]

Procedure for Using Anti-Polysialosyl Antibody (H.46) for Detecting Polysialic Acid. The presence of polysialyl residues associated with either N-CAM or bacterial membranes is determined by Western blot analysis using the H.46 antibody.[22a,23] Rabbit IgG anti-horse IgM is iodinated by the chloramine-T method[45] using carrier-free Na[125]I.[22a] Electrotransfer of antigens from polyacrylamide gels to nitrocellulose paper is carried out at 200 mA for 12–16 hr.[46]

Procedure for Preparing Bacterial and Neural Membranes for Immune Blotting. Eight-day-old postnatal rats (male Sprague–Dawley) or 8-day embryonic chicks (White Leghorn) are killed by severing the spinal cord. Brains and other tissues are dissected immediately into ice-cold phosphate-buffered saline (PBS) containing PMSF at 1 mg/ml. Homogenates are prepared by disruption of cells in a Dounce homogenizer. Membranes are sedimented at 27,000 g for 30 min (4°), resuspended to 20–40 mg protein/ml in 10 mM Tris–HCl (pH 7.6). For SDS–PAGE, samples are resuspended to 1–2 mg protein/ml in Laemmli sample buffer[47] and incubated at 30° for 2–4 hr (caution: do not boil). Aliquots containing 100–200 μg protein per well are electrophoresed through gradient (5–15%) SDS–PA gels on the Hoefer No. 520 vertical slab gel apparatus at 4°. After electrophoresis, proteins are electrotransferred to nitrocellulose paper as described above. The nitrocellulose blots are first treated with 3% fetal bovine serum albumin (BSA) in 10 mM Tris–HCl (pH 7.4) containing 0.9% NaCl and 0.1% NaN$_3$ for 1 hr. They are then washed with blot wash buffer (BWB; 10 mM Tris, pH 7.4, containing 0.2% SDS, 0.5% Triton X-100, 0.5% BSA, 0.9% NaCl, and 0.01% sodium azide) and treated for 3 hr

[42] L. D. Sarff, G. H. McCracken, M. S. Schiffer, M. P. Glode, J. B. Robbins, I. Orskov, and F. Orskov, *Lancet* **1,** 1099 (1975).
[43] H. J. Jennings, R. Roy, and F. Michon, *J. Immunol.* **134,** 2651 (1985).
[44] M. Frosch, I. Gorgen, G. J. Boulnois, K. N. Timmis, and D. B. Suermann, *Proc. Natl. Acad. Sci. U.S.A.* **82,** 1194 (1985).
[45] F. C. Greenwood, W. M. Hunter, and J. S. Glover, *Biochem. J.* **89,** 114 (1963).
[46] H. T. Towbin, T. Staehelin, and J. Gordon, *Proc. Natl. Acad. Sci. U.S.A.* **76,** 4350 (1979).
[47] U. K. Laemmli, *Nature (London)* **227,** 680 (1970).

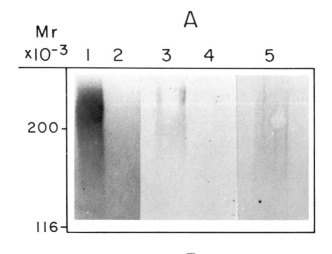

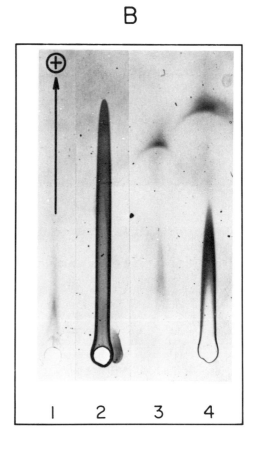

with H.46 antibodies diluted 1 : 25 or 1 : 50 in BWB. Unbound antibodies are removed by washing 3 times with BWB. Bound antibodies from H.46 are detected by incubating the blot with rabbit ^{125}I-labeled anti-horse IgM for 3 hr. Immunoblots are washed 3 times with blot wash buffer to remove unbound antibodies, dried, and placed on Kodak X-Omat AR2 or SB-5 film for autoradiography. Bacterial membranes are prepared as described.[22a,32] Alternatively, as shown in Fig. 1, intact cells can be resuspended directly in Laemmli sample buffer and an aliquot electrophoresed directly.

Preparation of the *E. coli* K1 Sialyltransferase to Detect Brain Polysialylglycoproteins as Exogenous Acceptors

The membrane-associated sialyltransferase complex from *E. coli* K1 serotypes is prepared from cells in the late logarithmic phase of growth as described previously.[16] Incorporation of [14C]NeuNAc from CMP-[14C]NeuNAc into polymeric products is also carried out as described.[16,25,32] *Escherichia coli* strain EV11, a spontaneous K1F-phage resistant mutant, is derived from a hybrid of *E. coli* K-12 and a K1 antigen expressing strain.[40] Membranes from EV11 are defective *in vitro* in the transfer of NeuNAc from CMP-NeuNAc to endogenous acceptors. These membranes can transfer NeuNAc to exogenous sialyl acceptors if they contain oligo- or polysialic acid in α-2,8-ketosidic linkages. Details of the isolation and characterization of these mutant strains have been published.[40]

Fig. 1. Detection of polysialic acid in brain and *E. coli* K-235 based on immune reactivity to anti-polysialosyl antibodies. (A) Immunoblotting using H.46 antibodies is performed as described in the text. Brain homogenate of 8-day postnatal rat (lane 1) is treated with Endo-N for 2 hr at 28° prior to immunoblotting (lane 2). Brain homogenate of 8-day-old chick embryo (lane 3) is shown after treatment with Endo-N (lane 4). Samples of protein (100 μg) are electrophoresed in lanes 1 to 4. Approximately 2×10^{10} cells of *E. coli* K-235 are suspended in 1 ml of Laemmli sample buffer; 50 μl are electrophoresed (lane 5). (B) Qualitative rocket immune electrophoresis is carried out on lantern slides coated with 11.6 ml of 1% agarose containing 1% Triton X-100, 0.02% NaN₃, and 46 μl of H.46 serum in 25 m*M* sodium barbital buffer (pH 8.6). Samples are prepared in dilution buffer (1% Triton X-100 plus NaN₃ and barbital buffer, as above), sonicated for 5 sec, loaded into wells, and electrophoresed for 2 hr at 6 V/cm. After electrophoresis, immune precipitates are identified by staining with Coomassie Blue R-250 as described. Lanes 1 and 2 are the results obtained with *E. coli* strains EV38 and K-235, respectively. The membrane fraction (21 mg protein/ml) from 8-day postnatal rat brain is treated as above with Endo-N and resuspended in dilution buffer to 2.1 mg/ml prior to electrophoresis (lane 3). A control sample not treated with Endo-N is shown in lane 4.

Use of the Prokaryotic-Derived Probes to Detect Polysialosyl-Containing
Glycoprotein in Brain

*Antibodies Specific for Bacterial Polysialic Acid Detect Similar
Antigenic Species in Neuronal Tissues*

Experiments using H.46 antibodies for immune blots of detergent-
solubilized 8-day postnatal rat brains show a major immune reactivity at
molecular weight 180,000 to 240,000. Figure 1A (lane 1) shows that this
immune reactivity exists as a broad band, suggestive of apparent molecular weight heterogeneity in the components detected by this method.
Since H.46 antibodies have immune specificity toward α-2,8-linked polysialic acids, these data indicate that rat brains contain multimers of sialic
acid. Qualitatively identical immune reactivity is also observed in 8-day
embryonic chick brain (Fig. 1A, lane 3), chick retina, and spinal cord, but
not in liver, spleen, thymus, or bone marrow (data not shown). The immune reactivity is not detected in adult rat brain, suggesting its decrease
during embryonic development is important during early development
and differentiation. Further evidence that H.46 antibodies recognize neuronal polysialic acid is provided by the finding that immune reactivity is
abolished by prior treatment of brain extracts with Endo-N (Fig. 1A, lanes
2 and 4). Additional confirmation comes from comparison with an immune blot of bacterial K1 antigen. Figure 1A (lane 5) shows the immune
blot obtained using detergent-solubilized *E. coli* K-235. The similarity in
apparent high molecular weight and band polydispersity to brain-derived
antigen is striking and shows that H.46 antibodies recognize molecular
components in bacteria and in brain that contains polysialic acid in α-2,8-
ketosidic linkage.

H.46 immune reactivity can also be detected by rocket immune electrophoresis. Figure 1B (lane 2) shows a rocket obtained using *E. coli*
K-235 as antigen. The ability to detect immune precipitates is dependent on
the presence of polysialic acid as antigen since a mutant derivative of *E.
coli* K-235 unable to synthesize sialyl polymer (*E. coli* EV38) shows no
reactivity (Fig. 1B, lane 1). Rat brain membranes produce an immune
precipitate qualitatively similar to that observed with the bacterial polymer (Fig. 1B, lane 4). The ability of brain samples to form immune precipitates is abolished by pretreatment with Endo-N (Fig. 1B, lane 3). These
results corroborate previous conclusions based on immune blotting (Fig.
1A), and provides evidence that H.46 antibodies are able to form immune
precipitin complexes with polysialic acid from sources as disparate as rat
brain and bacteria. Using K1 antigen purified from *E. coli* K-235 and
quantitative rocket immune electrophoresis, this method will detect as

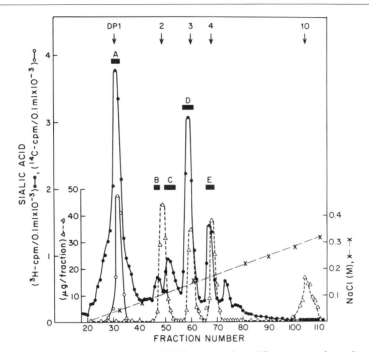

FIG. 2. DEAE chromatography of sialyl oligomers released from neuronal membranes by Endo-N. Sialyl oligomers (223 μg total sialic acid) released after 15 min incubation with Endo-N are reduced with 10 mCi KB³H₄, desalted, and mixed with a tracer amount of [¹⁴C]NeuNAc and 100–200 μg each of unlabeled sialyl oligomer with DP of 2, 3, 4, and 10. The mixture is fractionated with a 300-ml salt gradient at 0.4 ml/min; 3-ml fractions are collected and analyzed as indicated.

little as 5 ng of polysialic acid.[22a] This technique is therefore useful for quantitation of low levels of polysialic acid in neuronal tissue.

Endo-N Treatment of Brain Membranes Releases Sialyl Oligomers

Direct evidence that the products of Endo-N treatment of brain membranes are oligomers of sialic acid comes from chromatographic and structural analyses of the released material. Endo-N solubilized oligomers from a 2-hr digest of brain membranes are fractionated by DEAE chromatography.[22a] As shown in Fig. 2, greater than 80% of the total sialic acid measured by TBA elutes from the DEAE column with a DP of 3–5 NeuNAc residues. This result shows that the H.46-reactive, Endo-N-sensitive brain material consists of multimers of sialic acid. Proof of the oligomeric nature of this material comes from demonstrating its sensitivity to exoneuraminidase, after reduction of the oligomers with KB³H₄.[22a]

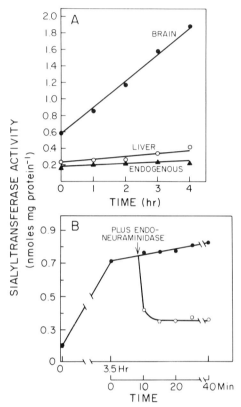

FIG. 3. Polysialylated rat brain membrane functions as an exogenous acceptor for *E. coli* K1 sialyltransferase. Frozen brains from 8-day-old rats are thawed and homogenized in 2 ml PBS with five strokes in a Dounce homogenizer. The homogenate is centrifuged for 45 min at 20,000 *g* and the membrane pellet resuspended in 0.5 ml of 50 m*M* Tris–HCl (pH 8.0), containing 3.5 m*M* magnesium acetate and 2 m*M* dithiothreitol. The membranous fraction from liver is prepared in an identical manner. (A) Sialyl polymer synthesis in *E. coli* EV11 membranes is stimulated by the addition of 8-day-old rat brain membranes as an exogenous sialyl acceptor (●). No restoration of sialyl polymer synthesis is seen with 8-day-old rat liver membranes (○). No NeuNAc is incorporated into endogenous acceptor molecules in EV11 membranes (▲). (B) Sensitivity of the sialyl polymers synthesized in (A) to Endo-N. Brain-dependent incorporation of NeuNAc by EV11 membranes is allowed to continue for 3.5 hr in incubation mixture containing 4–6 mg rat brain membrane protein. After 3.5 hr, two 120-μl aliquots are removed from the incubation mixture. To one sample, 20 μl of 100 m*M* Tris–HCl (pH 7.0) buffer is added (control, ●). To the other sample, 20 μl of Endo-N is added (○). After addition of buffer or Endo-N, 20-μl aliquots are removed at 2, 5, 10, 15, and 30 min and assayed for the amount of [14C]NeuNAc remaining in polysialic acid by chromatography on Whatman No. 3 MM paper, as described.[22a,32]

Polysialylated Brain Membrane Functions as an Exogenous Acceptor for E. coli Sialyltransferase

A membranous sialyltransferase complex from *E. coli* K1 catalyzes the synthesis of linear sialylpolymers containing ~200 sialyl residues.[16] The polysialic acid is composed exclusively of α-2,8-ketosidic linkages. Sialyl polymer synthesis can be stimulated by addition of exogenous acceptors containing oligo- or polysialic acid to the enzyme complex.[25,48] The ability of the sialyltransferase to transfer sialyl residues to exogenous acceptor is highly specific for acceptors containing α-2,8-linked sialic acid.[25] Because of this specificity, the bacterial sialyltransferase can be used to assay for the presence of α-2,8-linked polysialic acid in neuronal tissue.[22a] Sialyl polymer synthesis in membranes from wild-type *E. coli* K-235 is stimulated ~2.5-fold by embryonic rat brain membranes.[22a] This indicates the presence of α-2,8-linked polysialosyl residues in 8-day-old rat brains. This conclusion is confirmed using a mutant *E. coli* strain EV11.[40] As shown in Fig. 3, membranes from EV11 are defective in the transfer of NeuNAc from CMP-NeuNAc to endogenous acceptor molecules. Reconstitution studies show that EV11 can transfer NeuNAc to exogenous acceptors, e.g., colominic acid (results not shown). Importantly, sialyl polymer synthesis is restored in EV11 membranes when 8-day-old rat brain membranes are added as exogenous acceptors (Fig. 3A). Sialyl polymer formation is proportional to the amount of brain protein added over an 8-fold range. Figure 3A also shows that membranes similarly prepared from the liver fail to restore sialyl polymer synthesis, presumably because they lack polysialosyl chains. This would be consistent with the lack of immunoreactivity of liver membranes to H.46 antibodies. Confirmation that the brain-dependent incorporation of NeuNAc by EV11 membranes is into α-2,8-linked polysialic acid is shown by demonstrating the sensitivity of the product to Endo-N digestion. As shown in Fig. 3B, nearly all of the sialyl polymer synthesized after 3.5 hr by the exogenous addition of brain membrane is sensitive to Endo-N. This assay can detect polysialosyl units in as little as 0.22 mg of unfractionated rat brain membrane protein.

Acknowledgments

We acknowledge the contributions of H. F. Vollger, N. C. Wilkison, and B. L. Bassler for preliminary studies on immunoblotting and Endo-N purification. We thank F. Yu for contributions to the purification and kinetic analysis of Endo-N. We wish to thank Dr. J. B. Robbins for the generous gift of the anti-polysialosyl antibody. We thank L. A. Troy for expert editorial and secretarial assistance.

[48] C. Whitfield, D. A. Adams, and F. A. Troy, *J. Biol. Chem.* **259**, 12769 (1984).

[13] Inositol-Containing Sphingolipids

By ROGER A. LAINE and THOMAS C.-Y. HSIEH

Introduction

Glycophosphosphingolipids (GPS's) are membrane glycolipids containing a phosphodiester linkage between inositol and ceramide. The following structure constitutes a common core: inositol (1-*O*)-phosphoryl-(*O*-1)ceramide. Higher homologs bear extended sugar chains which are attached to the inositol. Thus far, these substances have been found in plants, yeast, and fungi, but not in any of the examined animal phyla. Early work on GPS compounds was initiated by Professor H. E. Carter, who had earlier characterized sphingosine at the University of Illinois. He showed that seeds from cotton, peanut, corn, and soybeans contained glycolipids containing phytosphingosine, inositol, and phosphate.[1-6] Substances with similar chemical properties could be isolated from the seeds and tissues of several plants, and were termed "phytoglycolipids." Plants also contain small amounts of glycosphingolipids which have sugars glycosidically attached directly to the ceramide,[7] as well as glycerolipid and steroid-based glycolipids. Therefore the term glycophosphosphingolipid ("GPS") is used here for this specific class of plant glycolipid. Carter's work culminated in 1969 with the characterization of the major component GPS in corn and in various plant seeds. In more recent work, primarily by a collaboration between the authors and Professor R. L. Lester's group at the University of Kentucky, three GPS's from tobacco and five from yeast and fungi have been characterized. As negatively charged glycosphingolipids, these ubiquitous compounds in the plant, yeast, and fungal kingdoms may be structurally analogous to, and may have as many idiotypic varieties as, the sialic acid-containing compounds known as

[1] H. E. Carter, W. D. Celmer, D. S. Galanos, R. H. Gigg, W. E. M. Lands, J. H. Law, K. L. Mueller, T. Nakayama, H. H. Tomizawa, and E. Weber, *J. Am. Oil Chem. Soc.* **35**, 335 (1958).

[2] H. E. Carter, S. Brooks, R. H. Gigg, D. R. Strobach, and T. Suami, *J. Biol. Chem.* **239**, 743 (1964).

[3] H. E. Carter, P. Johnson, and E. J. Weber, *Annu. Rev. Biochem.* **34**, 109 (1965).

[4] H. E. Carter, D. R. Strobach, and J. N. Hawthorne, *Biochemistry* **8**, 383 (1969).

[5] H. E. Carter and J. L. Koob, *J. Lipid Res.* **10**, 363 (1969).

[6] H. E. Carter and A. Kisic, *J. Lipid Res.* **10**, 356 (1969).

[7] R. A. Laine and O. Renkonen, *Biochemistry* **13**, 2837 (1974).

"gangliosides" found in animals. More than 100 kinds of gangliosides are known in animals.

In plants and fungi, little is known about the chemistry or subcellular distribution of the negatively charged membrane glycophosphosphingolipids. Less is known about their mode of biosynthesis or of their biological activity. Only four absolute structures have been published in the literature for these compounds in plants, one from corn and three from tobacco. Five other compounds have been characterized from yeast and the human pathogen *Histoplasma capsulatum*.

When the last work was performed on corn and soybeans by Carter, in 1969, no technology existed for separation of the natural mixture of GPS compounds, and the project was abandoned. New methods using high-performance liquid chromatography and high-resolution thin-layer chromatography are now available for purification of the individual oligosaccharides from GPS's. New structural data will bring knowledge to bear on metabolic studies of these complex molecules. Once information on the chemistry and fundamental biosynthesis is advanced on these compounds, further studies on their biological activities can be contemplated. As of yet, the field is neglected, but is ripe for fresh progress. The GPS's of known structure are shown in Table I.

Carter *et al.* found that on carbon–Celite and anion-exchange chromatography the corn seed GPS fraction contained 41% of tetrasaccharide **I** in Table I, 9% of a trisaccharide (lacking the mannose), and 42% penta- and higher oligosaccharides. They also reported on GPS compounds from *Phaseolus vulgaris*[5] and on amino sugar-lacking GPS compounds from other plant seeds.[6]

During this time, a GPS compound in yeast was reported by Wagner and Zofcsik to have the preliminary structure **II** as shown in Table I.[8]

GPS compounds from peanuts were also investigated by Wagner,[9] and a partial structure of one component was proposed as component **III**, Table I.

This same group reported the isolation of a GPS fraction which had glucose and glucuronic acid from the alga *Scenedesmus obliquus*.[10] This was the first report to include algae in the organisms which synthesize sphingosine. In yeast and fungi, Lester's group [11-13] confirmed the pres-

[8] H. Wagner and W. Zofcsik, *Biochem. Z.* **346**, 333 (1966).
[9] H. Wagner, W. Zofcsik, and I. Heng, *Z. Naturforsch.* **24**, 922 (1969).
[10] H. Wagner, P. Pohl, and A. Munzing, *Z. Naturforsch.* **24**, 360 (1969).
[11] S. Steiner, S. Smith, C. J. Waechter, and R. L. Lester, *Proc. Natl. Acad. Sci. U.S.A.* **64**, 1042 (1969).
[12] S. W. Smith and R. L. Lester, *J. Biol. Chem.* **249**, 3395 (1974).
[13] R. L. Lester, S. W. Smith, G. B. Wells, D. C. Rees, and W. W. Angus, *J. Biol. Chem.* **249**, 3388 (1974).

TABLE I

KNOWN GLYCOPHOSPHOSPHINGOLIPIDS

Structure number	Structure
I	Man(α1-2) Inos(1-O-phosphoryl)ceramide GlcNAc(α1-4)GlcUA(α1-6)[a]
II	Man-Inos-phosphorylceramide (MIPC)
III	Ceramide-phosphoryl-Inos(?-4)GlcUA(?1-3)GlcNH$_2$(1-?)(Gal, Ara, Man)
IV	Inos(1-O)phosphorylceramide (IPC)
V	(Inos-phosphoryl)Inos-phosphorylceramide (IP)2C
VI	(Inos-phosphoryl)Man-Inos-phosphorylceramide M(IP)2C
VII	(Man,Gal)-Inos-phosphorylceramide
VIII	(Man,Gal,Glc)-Inos-phosphorylceramide
IX	(Man$_2$,Gal$_3$)-Inos-phosphorylceramide
X	GlcNAcp(α1-4)GlcUAp(α1-2)Inos-(1-O-phosphoryl)ceramide
XI	Galp(β1-4)GlcNAcp(α1-4)GlcUAp(α1-2)Inos
XII	Galp(?1-6)Galp(?1-4)GlcNAcp(α1-4)GlcUAp(α1-2)Inos
XIII	Araf(?1-6)Galp(?1-4)GlcNAcp(α1-4)GlcUAp(α1-2)Inos
XIV	R(1-6) R = Gal, Ara Gal(β1-4)GlcNAc(α1-4)GlcUA(α1-2)Inos R'(1-3) R' = Gal, Ara
XV	Manp(α1-3)Manp(α1-2 or 6)Inos(1-O-phosphoryl)ceramide(OH)
XVI	Manp(α1-3) Manp(α1-2 or 6)Inos(1-O-phosphoryl)ceramide(OH) Galf(α1-6)
XVII	Manp(α1-3) Manp(α1-2 or 6)Inos(1-O-phosphoryl)ceramide(OH) Galp(β1-4)

[a] Carter's proposed structure contained GlcNH$_2$.

ence of MIPC (**II**, above), and also found and characterized compound **IV** and partially characterized compounds **V** and **VI** (Table I).

Brennan *et al.* reported three partially characterized GPS's (compounds **VII**, **VIII**, and **IX**) containing mannose, galactose, and glucose from *Aspergillus niger*.[14,15] These compounds are listed in Table I.

Kaul and Lester resumed research on plant GPS's by development[16,17] of a mild extraction method for obtaining a GPS concentrate from fresh

[14] P. J. Brennan and J. Roe, *Biochem. J.* **147**, 179 (1975).
[15] P. J. Brennan and D. M. Losel, *Adv. Microb. Physiol.* **17**, 47 (1978).
[16] K. Kaul and R. L. Lester, *Plant Physiol.* **55**, 120 (1975).
[17] K. Kaul and R. L. Lester, *Biochemistry* **17**, 3569 (1978).

tobacco leaves. Carter's and Wagner's laboratories had used much harsher chemical extraction procedures (acid or alkali), which could have chemically altered the GPS compounds. Numerous bands were seen on thin-layer chromatography of this GPS fraction, and the two major compounds were found to have the following compositions: (1) GlcNH$_2$-GlcUA-Inos-Phos-Ceramide, and (2) its N-acetyl analog. Interestingly, the total GPS fraction was found nearly equally divided between compounds which had a free amino group on the glucosamine and compounds with an N-acetylglucosamine. This fact complicates the separation and purification problems with GPS's. In tobacco leaves, a trisaccharide-containing, 50% N-acetylated GPS constituted about 40% of the GPS fraction (whereas in corn, Carter had found only 9% trisaccharide form). From the composition, it was possible to assign tentative sugar ratios for a series of six fractions which had, in addition to GlcNAc, GlcUA, and Inos, additional arabinose and galactose moieties.[17]

The detailed structure of the tobacco trisaccharide from the GPS fraction was determined by Hsieh et al.[18] to be as depicted in compound **X** in Table I. This structure differed from the core structure of the compounds found in corn in having the glucuronic acid connected to the 2-position of the myo-inositol, rather than the 6-position as reported by Carter et al.[4] Chemical ionization mass chromatography[19,20] was used to detect the deuterium-labeled products of periodate oxidation to determine the position of linkage on the inositol.[18]

An advance in the chromatography of oligosaccharides was developed by Wells and Lester[21] utilizing reversed-phase HPLC of the reduced acetyl derivatives of malto oligosaccharides as model compounds.[21a] This enabled the purification of some of the GPS oligosaccharides by repetitive high-resolution chromatography on low-capacity reversed-phase columns.[22] These inositol-containing oligosaccharides are nonreducing. Therefore the steps used to prepare the oligosaccharides for HPLC were (1) ammonolysis to release the phosphorylceramide, (2) reduction of the glucuronic acid carboxyl (to prevent isomerization and decarboxylation), and (3) acetylation. The major tetrasaccharide obtained by this procedure was completely characterized by Hsieh et al.[22] to have the structure as shown as **XI** (Table I). A minor tetrasaccharide, only partially purified, appeared to have the same composition as Carter's major tetrasaccharide

[18] T. C.-Y. Hsieh, R. A. Laine, and R. L. Lester, *Biochemistry* **17**, 3575 (1978).
[19] R. A. Laine, *Int. Congr. Pure Appl. Chem.* **27**, 193 (1980).
[20] R. A. Laine, *Anal. Biochem.* **116**, 383 (1981).
[21] G. B. Wells and R. L. Lester, *Anal. Biochem.* **97**, 184 (1979).
[21a] This series, Vol. 83, p. 132.
[22] T. C.-Y. Hsieh, R. L. Lester, and R. A. Laine, *J. Biol. Chem.* **256**, 7747 (1981).

from corn.[4] According to methylation analysis,[23] the next major member of this series appeared to have a galactopyranose (1-6)-linked to the terminal galactose of the major tetrasaccharide to give structure **XII** as shown in Table I.

A minor component appeared on methylation analysis[23] to contain a terminal arabinofuranose (1-6)-linked to the terminal galactose of tetrasaccharide **XI** (described above), giving structure **XIII** in Table I.

Both of these pentasaccharide fractions appeared to contain also one or more hexasaccharides, evidenced by the presence of 5–10 mol% of 2,4-di-O-methylgalactose which would indicate a 3,6-disubstituted galactose. About 5–10 mol% of terminal galactose was found in the pentasaccharide **XIII** sample and a similar small amount of terminal arabinose in the pentasaccharide **XII** sample. The branched hexasaccharides could originate from galactoses disubstituted with either or both arabinose and galactose. A tentative structure for these hexasaccharides (which cochromatographed with the pentasaccharide fractions) was proposed as structure **XIV** in Table I.[23] In tobacco, at least 20 members of this class of compounds remain to be identified.

Recently, Barr et al.[24] isolated and partially characterized from the yeast phase of *H. capsulatum* three novel GPS's, shown as structures **XV**, **XVI**, and **XVII** in Table I.

The separation and isolation of these compounds are still formidable tasks; however, methods for resolution of oligosaccharides in underivatized, reduced form by normal-phase chromatography[25] show promise. This may form a powerful combination with the already-mentioned chromatography of the acetyl derivatives by reversed-phase chromatography.[21] Detection methods useful for these separations include derivatization of the amino group of the glucosamine with [³H]acetate, chromogenic, or fluorescent compounds.

Procedures

Isolation of a Glycophosphosphingolipid Fraction from Plants

Fresh leaves are harvested and placed in iced water until homogenization, which should take place within 6 hr of harvest. Plants are homogenized and extracted according to the mild method of Kaul and Lester.[16] Briefly, for 1 kg of leaves, chopped into 1- to 2-in. squares, 65 ml of 100% (w/v) trichloroacetic acid and 200 ml of 95% ethanol are used for homoge-

[23] R. A. Laine, T. C.-Y. Hsieh, and R. L. Lester, *ACS Symp. Ser.* **128**, 65 (1980).
[24] K. Barr, R. A. Laine, and R. L. Lester, *Biochemistry* **23**, 5589 (1984).
[25] S. Turco, *Anal. Biochem.* **118**, 278 (1981).

nization in a 1-gallon Waring blendor for 1 min. After 30 min, the pH is brought to 7 with concentrated NH$_4$OH. This homogenate can be stored at $-20°$. To each liter of leaf homogenate, 649 ml 95% ethanol, 268 ml of diethyl ether, 53.6 ml pyridine, and 5 ml of concentrated NH$_4$OH is added, the latter used to titrate the pH to 8.5. After refluxing the mixture at 60° for 1 hr, stirred, the extract is filtered through four layers of cheese-cloth while still warm. The mixture is cooled in an ice bath and the pH adjusted to 5 with glacial acetic acid. The desired precipitate will form upon storage at 5° for 1 week. To collect the precipitate, the supernatant fraction is siphoned off and discarded, leaving a mixture of 10% precipi-tate and 90% supernatant. This mixture is slurried and mixed with one-tenth volume of Hyflo Super Cel (Johns-Manville Co.) and one-fourth volume of acetone. The mixture is filtered on a Buchner funnel with vacuum, the filter cake is resuspended in another one-fourth volume (as above) of acetone and refiltered. This process is repeated twice, reserving the filter cake. After air-drying the filter cake is slurried in Solvent A (chloroform–methanol–water, 16 : 16 : 5, v/v), poured into a glass chroma-tography column, and eluted with the same solvent containing 50 mM ammonium acetate until sphingolipids no longer appear as judged by thin-layer chromatography. About 300 ml of this eluate results from processing each kilogram of leaves. One-half volume of methanol is added to the eluate and a precipitate forms in 2 days at 5°. An additional precipitate forms from the supernatant fraction at $-20°$ for 2 days and the two precip-itates are combined to form a *glycophosphosphingolipid concentrate* which is air-dried and stored in a desiccator at 5°. From tobacco, each kilogram of fresh leaves gave a concentrate which contained 100 μmol of phosphorus.

Subfractions can be obtained by chromatography on silica gel and DEAE–cellulose according to protocol established by Kaul and Les-ter.[16,17] Fractions are monitored for phosphate by a colorimetric method[17] and for lipid and sugars by gas–liquid chromatography of suitable deriva-tives as described elsewhere.[25a]

Production of a Carboxyl-Reduced Oligosaccharide Mixture from the Glycosphosphoceramide Concentrate

By procedures used in published work,[17] the carboxyl of the uronic acid in the core oligosaccharide structure is reduced by formation of an internal lactone by dehydration with a water-soluble carbodiimide fol-lowed by reduction with sodium borohydride. The reduced, base-stable oligosaccharide is then released from the lipid and phosphate by ammo-

[25a] This series, Vol. 28, p. 159.

nolysis and purified as detailed below. Briefly, the sodium form of GPS is converted to the acid form by passing a 6 mM aqueous solution (10 ml) through a 30-ml Dowex 50 (H+) column. The acid form GPS is converted to the lactone by adding 26 mg of 1-ethyl-3-(3-dimethylaminopropyl)carbodiimide (EDAC) per micromole of GPS per milliliter of aqueous solution and stirring for 2 hr at 22°, with a pH of 4.8 maintained by addition of HCl. For reduction, the pH is raised to 8.0 by addition of ammonium hydroxide, 0.6% of 1-octanol (v/v) is added to prevent foaming, and 75 mg sodium borohydride/ml is added with incubation at 55° for 18 hr. This is followed by neutralization of the reaction mixture with 1.5 × the volume of Dowex 50 (H+) resin, which is filtered and washed with 4 volumes of water. The pH of the filtrate is adjusted to 6 with ammonium hydroxide and the mixture is chromatographed batchwise on a 15-ml column of DEAE–cellulose (acetate form) in water. The column is washed with 2 column volumes each of water–methanol (1 : 1), water–methanol (1 : 3), and chloroform–methanol–water (16 : 16 : 5). The carboxyl-reduced GPS can be removed with chloroform–methanol–water (16 : 16 : 5) containing 0.5 M ammonium acetate. Recovery is near 90% of original phosphorus, and the yield of reduced product is 85%. Products can be precipitated from the column eluate with 0.5 volume of acetone, dried, and stored in a desiccator.

To cleave the ceramide moiety from the glycolipid, 20 μmol GPS is hydrolyzed in 4 ml of aqueous 0.5 N KOH at 100° for 9 hr. After acidification with 0.5 ml of 5 N acetic acid, the mixture is partitioned after adding 1.5 ml water, 6 ml chloroform, 4.5 ml methanol, and 1.5 ml of toluene. The aqueous phase which contains the carboxyl-reduced oligosaccharides and phosphooligosaccharides is collected and dried under nitrogen. After redissolution in 2 ml of water, the pH is adjusted to 8 with ammonium hydroxide and the phosphate is removed from the phosphooligosaccharides with E. coli alkaline phosphatase by incubation at 22° overnight for a quantitative yield. At this point, the glucosamine contains a free amino group which can be exploited for chromogenic, fluorogenic, or radioactive labeling. N-Acetylation is performed, if desired, by three 20-min incubations with acetic anhydride in an aqueous solution of sodium carbonate.[26]

Purification of Oligosaccharides Released from the
  Glycophosphosphingolipids

To isolate major fractions of the oligosaccharides, gel permeation chromatography on Bio-Rad P gels is used, effectively separating oligo-

---

[26] S. Roseman and I. Daffner, *Anal. Chem.* **28,** 1743 (1956).

saccharide fractions up to 10 sugars in length.[26a] These fractions will still probably contain mixtures, since in the work on tobacco[22] there were several tetrasaccharides, three to six pentasaccharides, and mixtures of the higher oligomers. Sugar chain length may approach 20 units or more. Several methods in high-performance liquid chromatography (HPLC) can be utilized for further purification of the oligosaccharides, including reversed-phase separation of the acetylated derivatives[21] and amino-derivatized normal-phase chromatography on silica gel.[21a,25] Detection methods include benzoylation of the amino group on the glucosamine, use of fluorescamine or o-phthaldehyde to synthesize a fluorophore on the glucosamine in glucosamine-containing GPS's for fluorescence detection of the molecules, use of tritiated acetic anhydride for radiolabeling of the aminosugar, among others. High-performance thin-layer chromatography can complement these methods.

Carbohydrate Analysis to Solve Unique Structural Problems Peculiar to GPS Oligosaccharides

Intersugar linkage positions are determined by methylation analysis, described elsewhere in these volumes.[26b] A modified method using chemical ionization mass spectrometry in the detection of the gas chromatographically separated derivatives gives higher sensitivity and selectivity.[19,20] Anomeric configuration can be determined by CrO_3 oxidation,[27,28] glycosidase digestion, or ^{1}H NMR (see Koerner, this volume [4]). Sequences of sugars can be partially determined by fast atom bombardment mass spectrometry of the intact oligosaccharide and by partial hydrolysis followed by linkage determination of the fragments. When possible, sequential glycosidase digestion is very useful for direct sequence determination. Branching points can be determined by location of disubstituted sugars in the methylation analysis and by direct mass spectrometry, using fast atom bombardment ionization.

In the classically applied methylation analysis, one cannot distinguish between a 4-linked aldohexopyranose residue and a 5-linked aldohexofuranose residue, since both give the same partially methylated sugar (2,3,6-tri-O-methylhexose) after hydrolysis of the glycosidic linkage. The same problem exists in the case of 4- and 5-linked aldopentose residues. To solve these problems, two approaches can be used.

(1) Reductive cleavage of the glycosidic linkages of the permethylated

[26a] This series, Vol. 83, p. 105.
[26b] This series, Vol. 28, p. 178.
[27] J. Hoffman, B. Lindberg, and S. Svensson, *Acta Chem. Scand.* **26**, 661 (1972).
[28] R. A. Laine and O. Renkonen, *J. Lipid Res.* **16**, 102 (1975).

oligosaccharides is carried out in triethylsilane, boron trifluoride, and trifluoroacetic acid. Subsequent acetylation would give stable cyclic anhydroalditol derivatives (with an O-acetyl group of each original linkage site) suitable for GC–MS analysis in a single step.[29,30]

(2) Treatment of an oligosaccharide with exoglycosidases to remove the sugar residues from the terminal sequentially until the particular sugar residue in question becomes terminal. Since methylation analysis can easily distinguish the ring forms of a terminal sugar residue, methylation analysis on the treated oligosaccharide and a comparison with the data from the methylation analysis on the original oligosaccharide will reveal the ring form of the sugar residue in question.

To determine the ratio of the GlcNAc- and the GlcNH$_2$-containing counterparts of a glycophosphosphingolipid, the GPS can be N-acetylated using trideuteroacetic anhydride in the presence of sodium bicarbonate. Fast atom bombardment mass spectrometry of the GPS will give ion pairs containing the GlcNAc and GlcNAc-d_3 residues (with a mass difference of 3 a.m.u. in any of these pairs of ions). The ratio of any such pair of ions can be used to determine the proportion of the GlcNAc- and the GlcNH$_2$-containing GPS's. Another approach is to use enzyme digestion to liberate the GlcNAc and the GlcNAc-d_3 residues from the GPS's and then GC–MS can be used to determine the ratio of the trimethylsilyl ether derivatives of trimethylsilyl glycosides of GlcNAc and GlcNAc-d_3.[31]

New Methods of Sequence Determination Using Fast Atom Bombardment Mass Spectrometry

Recently, a derivatization strategy has been developed using addition of phosphate to enhance sensitivity and sequencing information in using fast atom bombardment mass spectrometry for sequencing of neutral oligosaccharides.[32] Since the GPS's are intrinsically phosphorylated, they are very suitable for analyses by negative ion fast atom bombardment mass spectrometry. Molecular ion and additional fragment ions can often be detected with suitable matrices such as glycerol, thioglycerol, and triethylamine. Molecular weight and some sequence and branching information can thus be obtained readily.

[29] D. Rolf and G. R. Gray, *J. Am. Chem. Soc.* **104**, 3539 (1982).
[30] A. Van Langenhove and V. N. Reinhold, *Proc. 32nd Annu. Meet. Am. Soc. Mass Spectrom. Allied Top.* p. 675 (1984).
[31] D. C. DeJongh, T. Radford, J. D. Hribar, S. Hanessian, M. Bieber, G. Dawson, and C. C. Sweeley, *J. Am. Chem. Soc.* **91**, 1728 (1969).
[32] T. C.-Y. Hsieh and R. A. Laine, *Proc. 33rd Annu. Meet. Am. Soc. Mass Spectrom. Allied Top.* pp. 977–978 (1985).

Biosynthetic Pathway

The primary core compound of GPSs in plants is GlcUA(α1-2)Inos-phosphorylceramide. Thus, glucuronosyltransferase which specifically transfers GlcUA to the inositol would be the first glycosyltransferase in this pathway This enzyme has not yet been reported. The UDP-*N*-acetylglucosaminyl(α1-4)transferase which transfers *N*-acetylglucosamine in an α configuration to the C-4 position of the glucuronic acid is the next transferase in the plant system. One question remains unanswered from the tobacco results, where 50% of the glucosamine is unacetylated. Either a deacetylase follows the transfer of *N*-acetylglucosamine, or a unique glucosamine transferase is present. The series of GPS compounds in tobacco contains the compound Gal(β1-4)GlcNAc(α1-4)GlcUA(α1-2)Inos-P-ceramide which implicates a Gal(β1-4) transferase. An enzyme with a similar activity occurs in human milk, and two kinds seem to occur in porcine trachea, one α-lactalbumin insensitive. It will be interesting on an evolutionary basis to compare the properties, and possibly the sequence of the Gal transferase to the mammalian milk transferase (which synthesizes lactose in the presence of α-lactalbumin). Nurminen and Suomalainen[33] reported the presence of compound **II** (Table I) in both the plasma membrane and cell wall of *Saccharomyces cerevisiae*. However, subcellular localization of the GPS's in plants has not yet been reported, so it is uncertain whether they are components of only plasma membrane, or intracellular membranes and organelles.

[33] T. Nurminen and H. Suomalainen, *Biochem. J.* **125**, 963 (1971).

[14] Antibodies against Cell Surface Carbohydrates: Determination of Antigen Structure

By JOHN L. MAGNANI, STEVEN L. SPITALNIK, and VICTOR GINSBURG

The carbohydrates on cell surfaces change during development as they are probably involved in cell recognition. These developmentally regulated changes allow some antibodies directed against carbohydrates to discriminate among various tissues, both normal and malignant. The most important glycosyl residues as far as immunological specificity is concerned are generally at nonreducing ends of the carbohydrate chains. These "immunodominant" residues determine the antigenic specificity of the molecules in which they occur. Since the terminal sequences of sugars in glycoproteins and glycolipids are sometimes identical, some antibodies directed against carbohydrates react with both glycoproteins and glycolipids; other antibodies against carbohydrates react only with glycoproteins or only with glycolipids.

To obtain cell-specific monoclonal antibodies, mice and rats have been immunized with various cell types in many laboratories. Many of these antibodies are apparently specific for certain cells and developmental stages. Some of these antibodies are directed against carbohydrates, about one-fourth in our experience.[1]

In this article we present methods for determining whether antibodies that react with cell membranes are directed against carbohydrates; for determining whether the carbohydrate antigens are in glycolipids, glycoproteins, or both; and for determining the structure of the antigens. General methods of glycolipid purification have recently been reviewed.[2]

General Procedure

A scheme for the analysis of carbohydrate antigens is outlined in Fig. 1. The procedures are described below according to number.

1. Total Lipid Extraction

This lipid extraction is based on a previous method.[3] A cell pellet or tissue is homogenized in 3 volumes of distilled H_2O on ice. The homogenate is then poured into 10 volumes of methanol followed by the addition of 5 volumes of chloroform under constant stirring for 30 min at room temperature. After centrifugation at 2000 rpm for 10 min, the clear supernatant is saved, and the pellet is resuspended in 4 volumes of distilled H_2O followed by the addition of 15 volumes of chloroform–methanol (1:2). The mixture is centrifuged again for 10 min at 2000 rpm and the supernatants are combined. The combined supernatants are dried in a rotary evaporator and resuspended in 5 volumes of chloroform–methanol–H_2O (60:35:4.5) to yield a total lipid extract containing the lipids from 0.2 mg of tissue per microliter. The delipidated protein pellet is resuspended in 3 volumes of distilled H_2O and lyophilized. The resulting delipidated powder is tested for glycoprotein antigens as described below in Section 3.

[1] V. Ginsburg, J. L. Magnani, S. L. Spitalnik, P. F. Spitalnik, D. D. Roberts, and C. Dubois, *Proc. Int. Symp. Glycoconjugates, 8th, 1985* Vol. 2, p. 550 (1985).
[2] S. Hakomori, *Handb. Lipid Res.* **3**, 1 (1983).
[3] L. Svennerholm and P. Fredman, *Biochim. Biophys. Acta* **617**, 97 (1980).

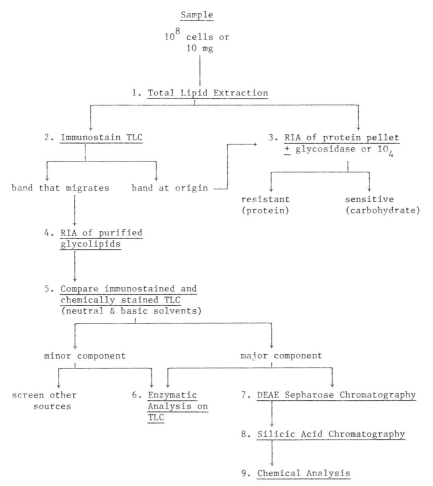

FIG. 1. Analysis of antigens.

2. Immunostaining Thin-Layer Chromatograms

Glycolipid antigens are detected on thin-layer chromatograms by auto-radiography as previously described.[4,5] The total lipid extract is chromatographed on aluminum-backed high-performance thin-layer chroma-

[4] J. L. Magnani, D. F. Smith, and V. Ginsburg, *Anal. Biochem.* **109**, 399 (1980).

[5] J. L. Magnani, B. Nilsson, M. Brockhaus, D. Zopf, Z. Steplewski, H. Koprowski, and V. Ginsburg, *J. Biol. Chem.* **257**, 14365 (1982).

tography plates (silica gel 60, E. Merck, Darmstadt, West Germany) in chloroform–methanol–0.25% KCl (5 : 4 : 1 by volume). The dried chromatogram is soaked for 1 min in a 0.1% solution of poly(isobutyl methacrylate) beads (Polysciences, Inc., Warrington, PA) dissolved in hexane. After drying in air, the chromatogram is sprayed with Tris-buffered saline (0.05 M Tris–HCl, 0.15 M NaCl, pH 7.4) containing 1% bovine serum albumin and 0.1% sodium azide (buffer A) and immediately soaked in buffer A until all of the silica gel is wet (about 15 min). The TLC plate is then removed and layed horizontally on a slightly smaller glass plate. Monoclonal antibody diluted in buffer A (2–5 μg/ml) is then layered on the chromatogram (60 μl/cm^2 of chromatogram) and incubated for 1 hr at room temperature. The chromatogram is washed by dipping in four successive changes of cold phosphate-buffered saline at 1-min intervals and overlayed with buffer A containing 2 × 10^6 cpm of ^{125}I-labeled goat anti-mouse IgM or IgG per milliliter. These second antibodies (Kirkegaard and Perry, Inc., Gaithersburg, MD) are labeled to a specific activity of 30–40 μCi/μg using Na^{125}I and Iodogen.[6] High specific activity is crucial for detection. After 1 hr at room temperature, the chromatogram is washed as before in cold phosphate-buffered saline, dried, and exposed to X-ray film (XAR-5, Eastman Kodak, Rochester, NY) for 16 hr.

As some antibodies nonspecifically adsorb to hydrophobic areas on the chromatogram, the amount of antigen should be estimated and a control experiment using only the second ^{125}I-labeled antibody should also be included. Using this technique, antibodies can detect about 0.5 ng of purified glycolipid antigen. High amounts of lipids sometimes bind antibodies nonspecifically. As shown in Fig. 2,[6a] glycolipid antigens immunostain as dark doublets, whereas nonspecific staining of faster migrating lipids are lighter and more diffuse.

3. Radioimmunoassay of Delipidated Protein Pellet

The lyophilized delipidated protein pellet obtained after total lipid extraction is dissolved at 1 mg/ml in phosphate-buffered saline, pH 7.2 (0.01 M sodium phosphate, 0.15 M NaCl), containing sodium azide (0.1%) and guanidine hydrochloride (4 M). The dissolved proteins are serially diluted in phosphate-buffered saline and used to coat the wells of a round-bottom, 96-well poly(vinyl chloride) microtiter plate. Each well contains 50 μl of dissolved proteins. The plate is covered with parafilm and incubated at 37° for 1 hr or at 4° overnight.

[6] P. J. Fraher and J. C. Speck, *Biochem. Biophys. Res. Commun.* **80**, 849 (1978).
[6a] J. L. Magnani, E. D. Ball, M. W. Fanger, S. Hakomori, and V. Ginsburg, *Arch. Biochem. Biophys.* **233**, 501 (1984).

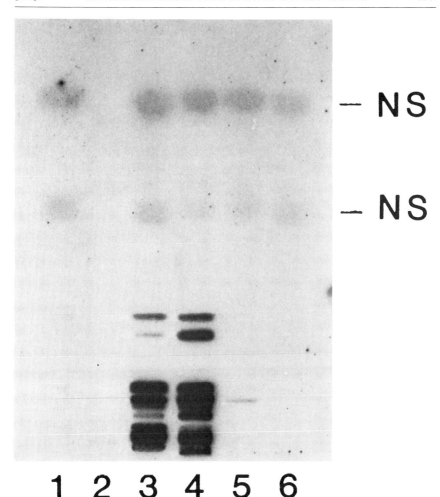

FIG. 2. Immunostained thin-layer chromatograms of total lipid extracts of cells. Monoclonal antibody PM-81, directed against laco-N-fucopentaose III,[6a] immunostained glycolipid antigens in total lipid extracts intensely. Diffuse nonspecific staining (NS) is observed on fast-migrating, hydrophobic areas in the neutral fraction of total lipid extracts. Each lane represents the lipid extract from 2 mg (wet weight) of cells. Lane 1, neutral fraction of acute myelogenous leukemia cells; lane 2, ganglioside fraction of acute myelogenous leukemia cells; lanes 3–6, total lipid extract of HL-60 cells, granulocytes, monocytes, and K560 cells, respectively.

Mild periodate oxidation cleaves vicinal hydroxyl groups on carbohydrates without altering the structure of polypeptide chains[7] and destroys most carbohydrate antigens. Therefore, the treatment of antigen-coated wells with sodium periodate is a simple and useful method for determining whether the antigen is a carbohydrate.[8]

Antigen-coated wells are washed once with sodium acetate buffer (0.05 M, pH 4.5) and then filled with the same buffer containing 10 mM sodium periodate. After incubation at room temperature for 1 hr in the dark, the wells are emptied and each is filled with buffer A. After 30 min, the wells are emptied and to each is added 50 μl of monoclonal antibody diluted in buffer A (2.5 μg/ml). The plate is covered with parafilm, incubated for 1 hr at room temperature, washed once with buffer A, and then to each is added about 100,000 cpm of ^{125}I-labeled goat anti-mouse IgG or IgM (30–40 μCi/μg). After 1 hr at room temperature, the wells are washed 6 times with cold phosphate-buffered saline, cut from the plate, and analyzed for ^{125}I in a gamma scintillation spectrometer.

The involvement of a particular terminal monosaccharide in antigenic specificity can sometimes be determined by incubating the antigen-coated wells with specific glycosidases rather than sodium periodate. For example, sialyl residues can be removed by adding to the wells 100 μl of *Clostridium perfringens* neuraminidase at 0.1 U/ml in sodium acetate buffer (0.05 M sodium acetate, 0.15 M NaCl, 0.009 M CaCl$_2$, pH 5.5). After a 16-hr incubation at room temperature, the plate is assayed with monoclonal antibodies as described above.

As glycosidases are specific for both a unique sugar and its linkage to the larger oligosaccharide, some sialic acids are not cleaved by *C. perfringens* neuraminidase. Likewise, oligosaccharides that do not contain vicinal hydroxyl groups will be resistant to mild periodate oxidation. These limitations should be kept in mind when interpreting results of treating antigen-coated wells with either sodium periodate or specific glycosidases.

4. Radioimmunoassay of Purified Glycolipids

Monoclonal antibodies that detect glycolipids on thin-layer chromatograms can be further analyzed for binding to purified glycolipids using a simple solid-phase radioimmunoassay.[9,10] Purified glycolipids dissolved in

[7] J. M. Babbit, *Adv. Carbohydr. Chem.* **11**, 1 (1956).

[8] M. P. Woodward, W. W. Young, Jr., and R. A. Bloodgood, *J. Immunol. Methods* **78**, 143 (1985).

[9] W. W. Young, Jr., E. M. S. MacDonald, R. C. Nowinski, and S. Hakomori, *J. Exp. Med.* **150**, 1008 (1979).

[10] M. Brockhaus, J. L. Magnani, M. Herlyn, M. Blaszczyk, Z. Steplewski, H. Koprowski, K. A. Karlsson, G. Larson, and V. Ginsburg, *J. Biol. Chem.* **256**, 13223 (1981).

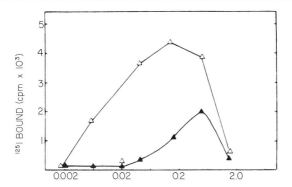

GANGLIOSIDES (mg wet weight of tissue extracted)

FIG. 3. Solid-phase radioimmunoassay of total lipid extracts of cells. Monoclonal 18B8,[10a] directed against ganglioside G_{T3},[10b] detects ganglioside antigens in serial dilutions of total lipid extracts of embryonic chicken retina ($\triangle$) or brain ($\blacktriangle$) by solid-phase radioimmunoassay. No antigen is detected in high concentrations (2 mg wet weight) of total lipid extracts by this method.

methanol (30 μl) are added to wells of a round-bottom, poly(vinyl chloride) microtiter plate and the solutions dried by evaporation. As complete evaporation of methanol is required, the plates may be subjected to a high vacuum at room temperature. The glycolipid-coated wells are then filled with buffer A and assayed as described above for glycoprotein-coated wells. This method is not used as a first step in testing the binding of a monoclonal antibody to a crude lipid extract since impurities in the lipid extract may inhibit the coating of the well by the desired glycolipid antigen. As shown in Fig. 3,[10a] no binding is detected at high concentrations of a crude lipid extract on a solid phase. In some cases, antigen may be detected after serial dilutions. With a semipurified sample, however, this assay may be used to examine the sensitivity of the antigen to glycosidases as described above for glycoproteins. Usually about 1–2 ng of purified glycolipid antigen can be detected in this assay.

5. Compare Immunostained and Chemically Stained Thin-Layer Chromatograms

Before purifying a glycolipid antigen, the relative abundance of the glycolipid in the total lipid extract must be determined. In many cases doublets observed on thin-layer chromatograms represent glycolipids that contain identical oligosaccharides but differ in the length of the ceramide.

[10a] G. B. Grunwald, P. Fredman, J. L. Magnani, D. Trisler, V. Ginsburg, and M. Nirenberg, *Proc. Natl. Acad. Sci. U.S.A.* **82**, 4008 (1985).
[10b] C. Dubois, J. L. Magnani, G. B. Grunwald, G. D. Trisler, M. Nirenberg, and V. Ginsburg, *J. Biol. Chem.* **261**, 3826 (1986).

Other bands that bind antibodies by immunostaining may represent larger oligosaccharides containing the same carbohydrate epitope. For simplicity, the fastest migrating band which contains the shortest oligosaccharide antigen is studied. Thin-layer chromatograms of glycolipid can be cut into several sections after chromatography. The location of the glycolipid antigens on one section is determined by immunostaining the plate as described above. An adjacent section is chemically stained[11] for neutral glycolipids using an orcinol spray or, for gangliosides, using a resorcinol spray.

The amount of glycolipid comigrating with the antigen is then determined by densitometry using glycolipid standards.[12] Resorcinol can detect from 0.2 to 2 μg of ganglioside, while orcinol is about 10 times less sensitive for neutral glycolipids.

As several different glycolipids may comigrate in the same band, this technique should be repeated on thin-layer chromatograms developed in several different solvents. Most gangliosides are resolved well in the neutral solvent, chloroform–methanol–0.25% KCl (5 : 4 : 1). Propanol–water (3 : 1) gives broader bands but may be used as a different solvent. Several gangliosides migrate differently in the basic solvent, chloroform–methanol–2.5 N NH$_4$OH (5 : 4 : 1). Small neutral glycolipids require a less polar solvent, such as chloroform–methanol–water (60 : 40 : 9 or 60 : 35 : 8). Many glycolipids that comigrate will separate on silica gel plates after O-acetylation. Glycolipids are dried under a stream of nitrogen and resuspended in anhydrous pyridine–acetic anhydride overnight.[13] O-Acetylated glycolipids are chromatographed on silica gel plates in 1,2-dichloroethane–methanol–water (88 : 12 : 0.1) or 1,2-dichloroethane acetone–water (55 : 45 : 5).[2] After chromatography, O-acetyl groups can be removed from glycolipids on the chromatogram by exposing the plate to concentrated vapors of ammonium hydroxide in a closed chromatography tank for 5 hr. Once deacetylated the chromatogram can be immunostained with antibodies as described above.

Some glycolipids which comigrate on normal silica gel will separate on borate-impregnated silica gel.[14] Finally, chromatographing samples on thin-layer plates by short-bed continuous development (Regis Chemical Co., Morton Grove, IL) separates glycolipids that normally comigrate by conventional chromatography.[15] Using the latter method even doublets due to differing ceramide composition are separated by at least 1 cm. If

[11] R. A. Laine, K. Stellner, and S. Hakomori, in "Methods in Membrane Biology," Vol. 2, E. Korn (Ed.), Plenum, New York.
[12] S. Ando, N. Chang, and R. K. Yu, Anal. Biochem. **89,** 437 (1978).
[13] T. Saito and S. Hakomori, J. Lipid Res. **12,** 257 (1971).
[14] E. L. Kiehn, J. Lipid Res. **7,** 449 (1966).
[15] W. W. Young, Jr. and C. A. Borgman, J. Lipid Res. **27,** 120 (1986).

the glycolipid antigen is present in low amounts and it is not feasible to obtain enough material for chemical characterization, the structure may be determined in some cases by enzymatic analysis directly on thin-layer chromatograms as described below.

6. Enzymatic Analysis on Thin-Layer Chromatograms

In some cases, glycolipids are analyzed by enzymatic treatment followed by immunostaining directly on thin-layer chromatograms.[16,17] Glycolipids are chromatographed on HPTLC plates, dried, and coated with plastic as described under section 2. After a 45-min incubation in buffer A to block nonspecific binding, the plates are washed 3 times in a buffer appropriate for the particular glycosidase. To cleave sialic acids with *C. perfringens* neuraminidase, the plates are washed in buffer B (0.05 M sodium acetate, 0.15 M NaCl, 0.009 M CaCl$_2$, pH 5.5) and then overlaid with neuraminidase (0.2 U/ml) in buffer B. A control plate is overlaid with buffer B alone. After incubation for 16 hr at room temperature, the plates are washed 5 times in phosphate-buffered saline (pH 7.4) and immunostained by monoclonal antibodies as described in section 2. Because incubation in buffer B alone has, in some cases, enhanced binding of monoclonal antibodies, an identical TLC plate incubated in the buffer alone should always be immunostained in parallel with the enzyme-treated plate.

With this technique, monoclonal antibody My-28 binds to the paragloboside structure exposed after treatment of α-2,3-sialoparagloboside (Fig. 4, lane 1) or α-2,6-sialoparagloboside (Figure 4, lane 2) with neuraminidase. Other larger gangliosides containing the paragloboside structure are also detected in the lipid extracts of human granulocytes (Fig. 4, lane 3B).

Other glycosidases can be used to reveal structures detected by monoclonal antibodies. As this method reveals both the relative migration on silica gel and the enzyme sensitivity of the glycolipids, it is possible to discover new structures in lipid extracts to which monoclonal antibodies have not yet been produced.

7. DEAE-Sepharose Chromatography

Glycolipids can be separated by charge using ion-exchange chromatography.[18,19] DEAE-Sepharose CL-6B resolves glycolipids better than

[16] S. L. Spitalnik, J. F. Schwartz, J. L. Magnani, D. D. Roberts, P. F. Spitalnik, C. I. Civin, and V. Ginsburg, *Blood* **66**, 319 (1985).

[17] M. Saito, N. Kasai, and R. K. Yu, *Anal. Biochem.* **148**, 54 (1985).

[18] G. Rousor, G. Kritchevsky, A. Yamamoto, G. Simmion, C. Galli, and A. J. Bauman, this series, Vol. 14, p. 272.

[19] R. K. Yu and R. W. Ledeen, *J. Lipid Res.* **19**, 390 (1972).

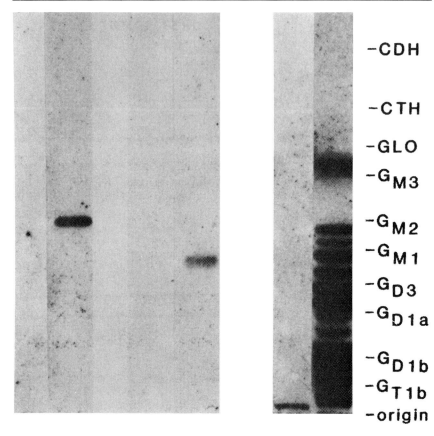

-CDH

-CTH

-GLO

-G_{M3}

-G_{M2}

-G_{M1}

-G_{D3}

-G_{D1a}

-G_{D1b}

-G_{T1b}

-origin

1A 1B   2A 2B   3A 3B

FIG. 4. Immunostained glycosidase-treated thin-layer chromatograms. Monoclonal antibody My-28, which binds lacto-N-neotetraose[16] detects the paragloboside structure exposed after neuraminidase treatment of a thin-layer chromatogram of α-2,3-sialylparagloboside (lane 1B), α-2,6-sialoparagloboside (lane 2B), and total gangliosides from granulocytes (lane 3B). No gangliosides are detected by immunostaining without prior treatment of the plate with neuraminidase (lanes 1A, 2A, and 3A).

DEAE-Sephadex. Before chromatography, a lipid extract must be desalted. For large amounts of extract, glycolipids can be safely dialyzed against water using Spectropor dialysis tubing 3000.[20] Under the critical micelle concentration of 150 μg/ml,[21] however, some glycolipids will diffuse through the membrane.

[20] H. Baumann, E. Nudelman, K. Watanabe, and S. Hakomori, *Cancer Res.* **39**, 2637 (1979).
[21] J. Kanfer and C. Spielvogel, *J. Neurochem.* **20**, 1483 (1973).

Smaller amounts of glycolipids can be desalted by chromatography on Sephadex G-25[22] or Sephadex LH-20,[23] or reversed-phase C_{18} cartridges.[24] Ten milligrams of dry total lipid extract can be desalted over 1 ml of Sephadex G-25. The gel is prewashed extensively with chloroform–methanol–water (60 : 30 : 4.5). Lipid extract dissolved in the same solvent is loaded on the column and washed with 10 column volumes of chloroform–methanol–water (60 : 30 : 4.5) followed by 5 column volumes of chloroform–methanol (2 : 1). Salts stay on the column while most glycolipids are eluted. Some large, highly charged polysialogangliosides, however, remain on the column.

For Sephadex LH-20, a total lipid extract dissolved in chloroform–methanol (25 : 10) is loaded on a column equilibrated in the same solvent. Glycolipids elute in the void volume, while sugars, amino acids, and salts are included in the gel.[20,23]

For small-scale chromatography, elution of charged lipids from the ion-exchange resin with volatile salts eliminates extensive dialysis of the separated gangliosides. Either ammonium acetate or ammonium bicarbonate dissolved in methanol can be used with DEAE-Sepharose after converting the gel to the appropriate ionic form. For ammonium acetate, the gel is soaked and washed 3 times with an excess of chloroform–methanol–0.8 M ammonium acetate in water (30 : 60 : 8) for 2 hr. The gel is then extensively washed on a sintered glass filter with chloroform–methanol–water (30 : 60 : 8) and finally stored in this solvent.[25]

A desalted lipid extract dissolved in chloroform–methanol–water (30 : 60 : 8) is loaded on a column of DEAE-Sepharose in the same solvent. The flow rate is stopped and the column incubated for 20 min at room temperature. Neutral glycolipids are then eluted with 2 column volumes of chloroform–methanol–water (30 : 60 : 8) followed by 10 column volumes of methanol. Monosialogangliosides elute with 10 column volumes of 0.01 M ammonium acetate in methanol, while simple sulfatides elute with 0.02 M ammonium acetate. At 0.04 M ammonium acetate disialogangliosides elute followed by trisialogangliosides at 0.1 M and tetra- and higher sialogangliosides at 0.5 M. Alternatively gangliosides can be eluted with a gradient of ammonium acetate in methanol. When followed by thin-layer chromatography of the separated fractions, this technique has been called "ganglioside mapping."[26] These chromatograms can also be immunostained by monoclonal antibodies to detect the presence of a glycolipid antigen. An example using monoclonal antibody 18B8 is shown in

[22] R. E. Wuthier, *J. Lipid Res.* **7,** 558 (1966).
[23] M. A. B. Maxwell and J. P. Williams, *J. Chromatogr.* **31,** 62 (1967).
[24] M. A. Williams and R. H. McCluer, *J. Neurochem.* **35,** 266 (1980).
[25] R. W. Ledeen and R. K. Yu, this series, Vol. 83, p. 139.
[26] M. Iuamori and Y. Nagai, *Biochim. Biophys. Acta* **528,** 257 (1978).

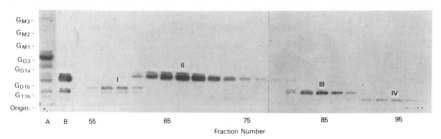

FIG. 5. Immunostained ganglioside antigens fractionated by ion-exchange chromatography. Gangliosides from 10-day chick embryonic retina are absorbed onto DEAE-Sepharose in the acetate form and eluted with a gradient of ammonium acetate in methanol (0.03–0.2 M). Aliquots of every other fraction (54–96) are immunostained directly on thin-layer chromatograms with monoclonal antibody 18B8, which is directed against G_{T3}.[10b] Lane A, unfractionated gangliosides from 10 mg (wet weight) of tissue visualized with resorcinol reagent; lane B, unfractionated gangliosides from 2 mg (wet weight) of tissue and ganglioside antigens visualized by immunostaining.

Fig. 5. Fractions containing the antigen are dried under reduced pressure by rotary evaporation. Dried samples containing high salt content may require a short dialysis against cold water followed by lyophilization to remove all salts.

8. Silicic Acid Chromatography

Final purification of glycolipids is usually obtained by chromatography on silicic acid. Either BioSil HA (−325 mesh) (Bio-Rad, Richmond, CA) or Iatrobeads[5,27,28] give good separation. Depending on the mobility of the glycolipids on thin-layer chromatograms, several different solvents may be used. Most gangliosides are immobile on silicic acid in chloroform–methanol (4:1) and will concentrate in a narrow band on the top of a column of BioSil HA (−325 mesh) packed in this solvent. Simple sulfatides and nonglycosylated lipid contaminants will separate from the immobilized gangliosides in this solvent. Glycolipids are then eluted and fractionated using a gradient of increasingly polar solvents created in an all-glass and Teflon reservoir. Smaller neutral glycolipids are fractionated by a gradient from chloroform–methanol–water (83:16:0.5) to methanol,[16] while monosialogangliosides are separated by a simple concave gradient from chloroform–methanol (4:1) to methanol–0.02% $CaCl_2$ in water (50:3).[5] Nagai and co-workers have used an interesting gradient

[27] S. Ando, M. Isobe, and Y. Nagai, *Biochim. Biophys. Acta* **424**, 98 (1976).
[28] T. Feizi, R. A. Childs, S. Hakomori, and M. E. Powell, *Biochem. J.* **173**, 245 (1978).

created by three interconnected reservoirs for separation of neutral glycolipids.[29]

Samples containing antigen determined by solid-phase radioimmunoassay as described above should be analyzed for purity by comparing thin-layer chromatograms stained by orcinol with those stained by antibodies. Several different chromatography methods described in section 5 should be used. If contaminants separate from the desired glycolipid, the purified antigen can be isolated directly from the thin-layer chromatogram. As the yield of glycolipids extracted from HPTLC plates is low, "Soft Plus" silica gel plates (Merck, Rahway, NJ) should be used. The sample is spotted as a band along the length of the thin-layer plate and developed in the appropriate solvent. After chromatography, the plate is quickly sprayed with 0.01% Primulin (Aldrich Chemical Co., Milwaukee, WI) and the glycolipids are visualized under UV light.[30] Bromphenol blue or iodine vapor may also be used for nondestructive localization of glycolipids on silica gel plates. Primulin, however is more sensitive. Bands containing the antigen are scraped from the slightly damp plate and extracted repeatedly with chloroform–methanol–water (1 : 1 : 0.1). Primulin can be eliminated by passing the sample in chloroform–methanol (4 : 1) over a small column of silica gel (1–2 ml). Primulin will pass through the column in chloroform–methanol (4 : 1). The purified glycolipid antigen can then be eluted by 5 column volumes of methanol.

Other methods of purification of glycolipids using HPLC are described elsewhere in this volume [1].

9. Chemical Analysis of Glycolipids

Techniques to chemically determine the structures of glycolipids have dramatically changed in the past few years. A current review of these methods can be found in this volume [2].

---

[29] T. Momoi, S. Ando, and Y. Nagai, *Biochim. Biophys. Acta* **441,** 488 (1976).
[30] V. P. Skipski, this series, Vol. 35, p. 396.

[15] Immunostaining Free Oligosaccharides Directly on Thin-Layer Chromatograms

By JOHN L. MAGNANI

Oligosaccharides are chromatographed on amino-bonded high-performance thin-layer chromatography silica gel plates and after chromatography the aldehydes on the reducing ends of the oligosaccharides react with the amino groups on the silica gel. Bound oligosaccharides are immunostained directly on the chromatograms by monoclonal antibodies. Oligosaccharide antigens are detected by autoradiography after a second incubation with ^{125}I-labeled goat anti-mouse immunoglobulin.[1] This method may be used to identify and characterize oligosaccharide antigens liberated from glycoconjugates by enzymatic[2-4] or chemical[5,6] means. This technique may also be used to determine the structural specificity of other carbohydrate-binding proteins such as lectins, toxins, and hormones or of bacteria and viruses that bind to cell surface glycoproteins.

Materials and Methods

The milk oligosaccharides lacto-N-difucopentaose I, lacto-N-fucopentaose III, lacto-N-neotetraose, and lactose and a mixture of neutral oligosaccharides were kindly provided by Dr. Victor Ginsburg (National Institutes of Health, NIADDK, Bethesda, MD) and Dr. David Zopf (National Institutes of Health, NCI, Bethesda, MD). Their structures are presented in Table I. Monoclonal antibodies NS10c17 and 534F8 are produced by hybridomas obtained as previously described.[7,8] Antibody NS10c17 binds the Le^b oligosaccharide, lacto-N-difucopentaose I,[9] and antibody 534F8 binds the Le^x oligosaccharide, lacto-N-fucopentaose III.[10]

[1] J. L. Magnani, *Anal. Biochem.* **150**, 13 (1985).
[2] A. L. Tarentino, T. H. Plummer, Jr., and F. Maley, *J. Biol. Chem.* **249**, 818 (1974).
[3] N. Takahashi, *Biochem. Biophys. Res. Commun.* **76**, 1194 (1977).
[4] M. N. Fukuda, K. Watanabe, and S. Hakomori, *J. Biol. Chem.* **253**, 6814 (1978).
[5] S. Takasaki, T. Mizouchi, and A. Kobata, this series, Vol. 83, p. 263.
[6] B. Nilsson, N. E. Norden, and S. Svensson, *J. Biol. Chem.* **254**, 4545 (1979).
[7] H. Koprowski, Z. Steplewski, K. Mitchell, M. Herlyn, D. Herlyn, and P. Fuhrer, *Somatic Cell Genet.* **5**, 957 (1979).
[8] F. Cuttita, S. Rosen, H. F. Gazdar, and J. D. Minna, *Proc. Natl. Acad. Sci. U.S.A.* **78**, 4591 (1981).
[9] M. Brockhaus, J. L. Magnani, M. Blaszczyk, Z. Steplewski, H. Koprowski, K.-A. Karlsson, G. Larson, and V. Ginsburg, *J. Biol. Chem.* **256**, 13223 (1981).
[10] L. Huang, M. Brockhaus, J. L. Magnani, F. Cuttita, S. Rosen, J. D. Minna, and V. Ginsburg, *Arch. Biochem. Biophys.* **220**, 317 (1983).

TABLE I
STRUCTURE OF CARBOHYDRATES

Name	Structure	Abbreviations
Lacto-*N*-difucopentaose I	Fucα1-2Galβ1-3GlcNAcβ1-3Galβ1-4Glc 4 | Fucαc1	LND I; Le[b] hapten
Lacto-*N*-fucopentaose III	Galβ1-4GlcNaβ1-3Galβ1-4Glc 3 | Fucα 1	LNF III; Le[x] hapten
Lacto-*N*-neotetraose	Galβ1-4GlcNAcβ1-3Galβ1-4Glc	LNnt

Immunostaining Oligosaccharides Directly on Thin-Layer Chromatograms. Oligosaccharides are chromatographed on amino-bonded high-performance thin-layer silica gel plates in pyridine/ethyl acetate/acetic acid/water (6:2:1:3, by volume). The dried chromatogram is laid horizontally with the silica gel side up on a hot plate regulated at 45° for 12–15 hr (Dataplate 440, Whatman, Inc., Clifton, NJ). The chromatogram is cooled and soaked for 1 min in a 0.1% solution of poly(isobutyl methacrylate) previously dissolved in hexane, allowed to air-dry, and then soaked again for an additional minute in 0.1% poly(isobutyl methacrylate). After complete drying in air, the chromatogram is sprayed with 0.05 M Tris–HCl (pH 7.8) containing 0.15 M NaCl, 1% bovine serum albumin, and 0.1% NaN_3 (buffer A) and immediately soaked in the same buffer until all of the silica gel is wet (about 10 min). The plate is then removed and laid horizontally on a slightly smaller glass plate in a large petri dish. Monoclonal antibody diluted to 20 μg/ml in buffer A is layered on the plate (60 μl/cm² chromatogram surface). After incubation at 22° for 30 min, the chromatogram is washed by dipping in three successive changes of cold phosphate-buffered saline at 1-min intervals and overlayed with buffer A containing 2 × 10⁶ cpm ¹²⁵I-labeled goat anti-mouse IgM per milliliter. After 30 min at 22°, the chromatogram is washed as before in cold phosphate-buffered saline, dried, and exposed to XAR-5 X-ray film (Eastman Kodak, Rochester, NY) for 15 hr at 22°.

Discussion

Oligosaccharides require a reducing end to bind to amino-bonded silica gel. A minimum of about 10 pmol of oligosaccharide antigen can be detected using antibodies 10c17 and 534F8 (Figs. 1 and 2). Binding is specific as antibody 10c17 binds LNDI and not LNFIII (Fig. 1), while

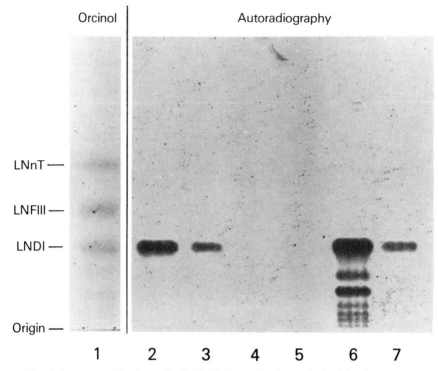

FIG. 1. Immunostaining by antibody 10c17. Lanes 2 to 5 contain the following amounts of LNDI and LNFIII: 100, 25, 5, and 1 pmol, respectively. Lanes 6 and 7 contain 0.5 and 0.1 μg of milk oligosaccharides from an Le^b donor, respectively. Lane 1 contains chemically stained oligosaccharide standards.

antibody 534F8 binds LNFIII and not LNDI (Fig. 2). Each antibody also specifically detects larger oligosaccharides which presumably contain these structures (lane 6, Fig. 1; lane 7, Fig. 2). The larger oligosaccharides are unknown minor components in the neutral oligosaccharide fraction from human milk and do not comigrate with the major oligosaccharides stained by orcinol.

Although about 10 pmol of oligosaccharide antigen can be detected, this method requires high antibody concentrations and is at least 10-fold less sensitive than immunostaining glycolipids directly on thin-layer chromatograms.[11,12]

Recently, amino-bonded silica gel has been widely used for high-pressure liquid chromatography of oligosaccharides. At low flow rates, some

[11] J. L. Magnani, D. F. Smith, and V. Ginsburg, *Anal. Biochem.* **109**, 399 (1980).
[12] J. L. Magnani, B. Nilsson, M. Brockhaus, D. Zopf, Z. Steplewski, H. Koprowski, and V. Ginsburg, *J. Biol. Chem.* **257**, 14365 (1982).

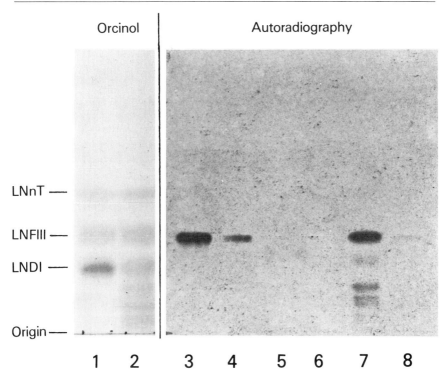

FIG. 2. Immunostaining by antibody 534F8. Lanes 3 to 6 contain the following amounts of LNDI and LNFIII: 100, 25, 5, and 1 pmol, respectively. Lanes 7 and 8 contain 0.5 and 0.1 μg of milk oligosaccharides from an Le^b donor. Lanes 1 and 2 contain chemically stained standard oligosaccharides.

sugars bind irreversibly to these columns.[13] The loss of these sugars correlates with the percentage of sugar in the aldehyde form in solution.[14] Similar results are also obtained using thin-layer silica gel plates derivatized with 3-aminopropyl groups.[15] Several investigators have suggested that Schiff bases form between these aldehydes and the amino groups on the silica gel.[16,17] Possibly this reaction may be followed by an Amadori rearrangement on TLC plates.[18] Therefore, the ability of the reducing sugar in an oligosaccharide to form a free aldehyde must be considered when using this method for quantification. The oligosaccharides used in this article are milk sugars that contain a reducing lactose core structure.

[13] B. Porsch, J. Chromatogr. 253, 49 (1982).
[14] S. M. Cantor and Q. P. Peniston, J. Am. Chem. Soc. 64, 2113 (1940).
[15] L. W. Dover, C. L. Fogel, and L. M. Biller, Carbohydr. Res. 125, 1 (1984).
[16] S. R. Abbott, J. Chromatogr. Sci. 18, 540 (1980).
[17] R. E. Majors, J. Chromatogr. Sci. 18, 488 (1980).
[18] M. Amadori, Atti Accad. Naz. Lincei Cl. Sci. Fis., Mat. Nat. 2, 337 (1925).

To determine whether an unknown oligosaccharide structure will bind to the amino-bonded silica gel plate, several micrograms of the oligosaccharide sample is spotted at the origin, allowed to react at 45° as described here, and developed in pyridine/ethyl acetate/acetic acid/water (6 : 2 : 1 : 3 by volume). After orcinol staining, any bound oligosaccharides will remain at the origin. The percentage of bound oligosaccharides can be determined by scanning densitometry.[1] To control for the reactivity of antibodies to antigens on silica gel, a small spot containing the original glycoprotein antigen should be deposited and dried at the corner of the TLC plate after chromatography. The glycoproteins remain adsorbed to the silica gel and serve as a positive control for antibody reactivity during immunostaining.

Both of the antibodies used in this article bind to the nonreducing termini of oligosaccharides. In fact, antibody 10c17 cross-reacts with the H structure.[19] Other antibodies that recognize the core structure of an oligosaccharide near the reducing end may not bind conjugated sugars. In the method described here, the reducing sugar of the oligosaccharide is bound to the amino group on the silica gel and is not available for antibody binding. In other cases, low-affinity monoclonal antibodies may only detect higher concentrations of carbohydrate antigens.[20]

As oligosaccharides can be liberated from glycoconjugates by chemical[5,6] and enzymatic[2-4] methods, this technique may be useful in defining the receptors of other proteins such as lectins, toxins, and hormones or of bacteria and viruses that bind cell surface carbohydrates.

[19] G. C. Hansson, K.-A. Karlsson, G. Larson, J. M. McKibben, M. Blascyzk, M. Herlyn, Z. Steplewski, and H. Koprowski, *J. Biol. Chem.* **258**, 4091 (1983).
[20] J. L. Magnani, E. D. Ball, M. W. Fanger, S. Hakormori, and V. Ginsburg, *Arch. Biochem. Biophys.* **233**, 501 (1984).

[16] Preparation of Total Nonacid Glycolipids for Overlay Analysis of Receptors for Bacteria and Viruses and for Other Studies

By KARL-ANDERS KARLSSON

For a relevant comparison of glycolipids of various origins it is not only necessary to isolate and structurally characterize individual glycolipids[1] but also to interpret patterns of total glycolipids. To accomplish this, a method is needed to be used with several overlay assays based on thin-

[1] S.-I. Hakomori, *in* "Sphingolipid Biochemistry" (J. N. Kanfer and S.-I. Hakomori, eds.), p. 1. Plenum, New York, 1983.

layer chromatograms of separated glycolipids, including toxins and antibodies,[2] enzymes,[3] and bacteria and viruses.[4] We have slightly modified existing isolation procedures[1] for preparing a total nonacid glycolipid fraction free of nonglycolipid contaminants. In this way it is possible to directly compare patterns after chemical detection on thin-layer chromatograms,[5] to compare chemically detected patterns from results obtained from overlay analysis, and compare chromatographic patterns with direct-inlet mass spectrometry of the same mixture.[6,7]

For acid glycolipids (sulfatides, gangliosides) ion-exchange methods in combination with specific detection reagents on thin-layer plates (e.g., resorcinol for neuraminic acid) allow an adequate comparison of total fractions.[1] These have recently been further improved through acetylation, which allows both a further removal of contaminants and a complete separation of sulfatides and gangliosides.[8] For nonacid glycolipids, however, where the more complex species are often very minor components, there has been a problem of contaminants in the final preparations. The present procedure makes use of the original acetylation procedure of Handa,[9] later modified by Saito and Hakomori.[10] The following description has been in use at our department for some years and does not include adaptations such as HPLC. However, such adaptations are easily included in the procedure.

Procedure

The different steps are defined for one transfusion unit of human plasma[11] (and in part for human small intestine) but supplementing comments are given when needed for generality. All solvents are dried and redistilled before use.[12]

[2] J. L. Magnani, M. Brockhaus, D. F. Smith, and V. Ginsburg, this series, Vol. 83, Part D, p. 235.

[3] B. E. Samuelsson, this volume [50].

[4] K.-A. Karlsson and N. Strömberg, this volume [17].

[5] M. E. Breimer, G. C. Hansson, K.-A. Karlsson, and H. Leffler, Biochem. J. 90, 589 (1981).

[6] M. E. Breimer, G. C. Hansson, K.-A. Karlsson, H. Leffler, W. Pimlott, and B. E. Samuelsson, Biomed. Mass Spectrom. 6, 231 (1979).

[7] M. E. Breimer, G. C. Hansson, K.-A. Karlsson, and H. Leffler, J. Biol. Chem. 257, 557 (1982).

[8] M. E. Breimer, G. C. Hansson, K.-A. Karlsson, and H. Leffler, Biochem. J. 93, 1473 (1983).

[9] S. Handa, Jpn. J. Exp. Med. 33, 347 (1963).

[10] T. Saito and S.-I. Hakomori, J. Lipid Res. 12, 257 (1971).

[11] K.-E. Falk, K.-A. Karlsson, and B. E. Samuelsson, FEBS Lett. 124, 173 (1981).

[12] K.-A. Karlsson, B. E. Samuelsson, and G. O. Steen, Biochim. Biophys. Acta 316, 317 (1973).

Extraction

About 250 ml of human blood plasma[11] of one transfusion unit is added to 250 ml of methanol and the mixture heated to 70° for 30 min under constant stirring in a 1-liter evaporation bottle. The extract is filtered and the residue transferred back to the extraction bottle. The procedure is repeated twice with 250 ml of chloroform–methanol 2 : 1 (v/v) and once with 250 ml of methanol. The combined extracts are evaporated to dryness with the addition of small volumes of toluene.

Small amounts of wet cells may be extracted in a similar way.[13] For a larger scale the tissue is first lyophilized in pieces and subjected to extraction in two steps in a Soxhlet apparatus.[13] In the case of human small intestine (e.g., 130 g dry weight) the first extraction is with chloroform–methanol 2 : 1 (v/v) for 24 hr (1000 ml of solvent in a 2000-ml round bottle placed in an asbestos-insulated electrical heating device). The second extraction is done with 1500 ml of chloroform–methanol 1 : 9 (v/v) for 24 hr. The combined extracts are evaporated to dryness without filtration.

Mild Alkaline Degradation and Dialysis

The dried plasma extract residue is treated with 50 ml of 0.2 M KOH in methanol for 3 hr in a bottle containing five pearls of glass for fine dispersion during occasional shaking. The KOH is neutralized with 1 ml of acetic acid and the mixture transferred to a dialysis bag with 60 ml of chloroform and 40 ml of water to produce a two-phase system. After dialysis for 4 days against running tap water the bag content is evaporated at 70° using repeated additions of toluene. The sample is finally filtered and aftereluted with chloroform–methanol 2 : 1 (v/v) and methanol. For the intestinal total extract, 500 ml of KOH in methanol is used.

Silicic Acid Chromatography

This step is to remove mainly cholesterol and fatty acid methyl esters from glycolipids and alkali-stable phospholipids including sphingomyelin. The plasma sample (about 1.4 g) is loaded on a 10-g column of silicic acid packed in chloroform. The silicic acid mostly used is from Mallinckrodt Chemical Works, St. Louis, MO, USA, and sieved to a particle size of more than 45 μm and dried.[12] However, LiChroprep Si 60 (E. Merck, Darmstadt, W. Germany) with comparable specifications has similar properties although it may leave off some silicic acid. Three fractions are eluted: 100 ml of chloroform elutes the main bulk (about 1 g) of the extract

[13] M. E. Breimer, G. C. Hansson, K.-A. Karlsson, and H. Leffler, *Exp. Cell Res.* **135,** 1 (1981).

(cholesterol and methyl esters of fatty acids); 100 ml chloroform–methanol 98 : 2 (v/v) may contain free ceramide; 100 ml of chloroform–methanol 1 : 3 (v/v) and 100 ml of methanol elutes all glycolipids and alkali-stable phospholipids (about 60 mg). In the case of the dialyzed small intestinal sample (about 40 g) 50 g of silicic acid and 500 ml of solvent is used in each step.

Ion-Exchange Chromatography[12]

The third fraction from silicic acid chromatography of the plasma sample is loaded on a column of 5 g of DEAE–cellulose (DE-23, Whatman) in acetate form packed in chloroform–methanol 2 : 1 (v/v). The loaded sample is allowed to equilibrate on the column for 1 or 2 days. Two fractions are eluted, one with 100 ml of chloroform–methanol 2 : 1 (v/v) and 100 ml of methanol eluting nonacid glycolipids and alkali-stable phospholipids, mainly sphingomyelin, and one with 50 ml of 5% (w/v) LiCl in methanol eluting acid glycolipids (sulfatides and gangliosides) and alkali-stable phospholipids. The latter fraction is dialyzed with 30 ml of chloroform and 20 ml of water against running tap water for 4 days. It may be used as a total acid glycolipid fraction or be further processed as described[8] to separate sulfatides and gangliosides. The first fraction is evaporated to dryness and acetylated. With the intestinal sample (about 2 g) 20 g of DEAE–cellulose is used.

Acetylation

Acetylation of the dry nonacid plasma sample is performed in the dark overnight in 2 ml of chloroform, 2 ml of pyridin, and 2 ml of acetic anhydride. The addition of chloroform improves solubility and assures complete reaction. Next 5 ml of methanol and 5 ml of toluene are added and the sample is evaporated in a stream of nitrogen on a heated water bath and finally subjected to vacuum suction.

Silicic Acid Chromatography of the Acetylated Nonacid Fraction

The acetylated plasma sample is loaded on a 10-g column of silicic acid packed in chloroform–methanol 98 : 2 (v/v). For this purpose particles less than 45 μm treated with methanol and dried are used.[12] Three fractions are eluted: 100 ml of chloroform–methanol 95 : 5 (v/v), 100 ml of chloroform–methanol 90 : 10 (v/v), and 100 ml of chloroform–methanol 1 : 3 (v/v) plus 100 ml methanol. The third fraction contains mainly acetylated sphingomyelin. The first two fractions, which contain acetylated glycolipids and some contaminants, are evaporated together and deacetylated. For the intestinal fraction 50 g of silicic acid is used.

Deacetylation

Deacetylation of acetylated glycolipids may be performed in two ways, one with and one without a dialysis step.

Method A. For the plasma sample 2 ml of toluene, 2 ml of methanol, and 4 ml of 0.2 M KOH in methanol are used with occasional shaking during 30 min (for the intestinal sample 5, 5, and 10 ml, respectively, are used). After addition of 0.5 ml acetic acid the sample is transferred with 10 ml of chloroform and 10 ml of water to a dialysis bag (two-phase system) and dialyzed for 4 days against running tap water. The bag content is evaporated at 70° with repeated additions of toluene.

Method B. In this case no dialysis is needed since the amount of potassium acetate formed after neutralization is very low compared to glycolipid. The reagent is composed of 1 ml of 0.2 M KOH in methanol, 14 ml of methanol, and 5 ml of toluene. Of this 0.1 ml is used for up to 2 mg of acetylated glycolipid, 0.2 ml for up to 4 mg, 0.5 ml up to 15 mg, 1 ml up to 40 mg, and so on. The sample to be deacetylated is evaporated to dryness, the appropriate amount of reagent is added, and the mixture is occasionally agitated for 2 hr after which the KOH is neutralized with acetic acid and the solvents evaporated. As an example, deacetylation of 40 mg acetylated glycolipid with 1 ml of reagent produces about 1 mg of potassium acetate remaining with the glycolipid. If needed, potassium ions may be removed by filtration through, e.g., chloroform–methanol-washed Amberlite CG-50 type I in H$^+$ form (Rohm and Haas, Philadelphia, PA, USA), where the deacetylation mixture without added acetic acid may be filtered directly. However, some irreversible adsorption of glycolipid may occur and eluted resin may contaminate. Therefore we prefer to leave potassium acetate in the sample.

Filtration through a DEAE–Cellulose Column

The deacetylated plasma sample is filtered through a column of 2 g of DEAE–cellulose packed in chloroform–methanol 2 : 1 (v/v). The loaded sample is allowed to equilibrate for 1–2 days before elution with 50 ml of chloroform–methanol 2 : 1 (v/v) and 50 ml of methanol. The purpose of this step is to remove alkali-stable, amino group-containing phospholipids which have been transferred into N-acetylated derivatives during the acetylation procedure. This makes them acid.

Silicic Acid Chromatography

This final step removes nonpolar contaminants eluted in the first two fractions. The plasma sample is loaded on a column of 5 g of silicic acid (particles less than 45 μm) packed in chloroform–methanol 98 : 2 (v/v).

After elution of two fractions each with 50 ml of chloroform–methanol 98 : 2 (v/v) the pure glycolipids are eluted with 50 ml of chloroform–methanol 1 : 3 (v/v) and 50 ml of methanol. The final yield of total nonacid glycolipids of plasma of one transfusion unit of human blood is 6–8 mg (see also below).

Thin-Layer Chromatography

The preparative steps are checked by thin-layer chromatography (silica gel 60 nanoplates; E. Merck, Darmstadt, W. Germany) preferably using chloroform–methanol–water 65 : 25 : 4 (v/v/v) for nonderivatized and chloroform–methanol 95 : 5 (v/v) for acetylated samples. Anisaldehyde[14] is preferred as a detection reagent because of characteristic colors[12]: glycolipids usually become green or blue-green, glycerophospholipids gray or violet, and sphingomyelin and ceramide blue. Some lipid bands, mainly partially degraded glycerophospholipids, remaining in the later stages of purification, give an intense red-violet color. Globotetraosylceramide, sphingomyelin, and ceramide in native and acetylated form are good reference substances. On acetylation, the glycolipid mobility changes from the polar to the nonpolar interval. Sphingomyelin remains in the polar interval but its mobility becomes slightly faster. This may be used to test the acetylation step where this phospholipid is usually still a major component. The silicic acid separation of the acetylated sample is critical for the removal of several alkali-stable phospholipids including sphingomyelin. As the silicic acid may vary in activity due to the preparation used, this step has to be carefully checked by thin-layer chromatography. Acetylated globoside is preferably used as a reference, and should elute just ahead of more complex, often minor, species, which may be difficult to detect. Also, acetylated glycolipids produce less sharp bands than the nonacetylated substances.

Comments on the Procedure

The ability to purify total nonacid glycolipids for comparative and other purposes is adequately illustrated by the present application on human plasma, a membrane-free transport medium with only 6–8 mg glycolipids of one transfusion unit of whole blood (450 ml). Figure 1 shows the patterns obtained from four donors with different blood group phenotypes. The samples look similar in the region of 1–4 sugars but present distinct differences for glycolipids with 5 and more sugars, some of them barely visible by anisaldehyde spraying (not appearing in Fig. 1)

[14] E. Stahl, "Dünnschichtchromatographie." Springer-Verlag, Berlin and New York, 1967.

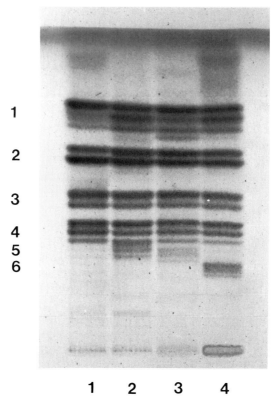

Fig. 1. Thin-layer chromatogram of total nonacid glycolipids isolated from blood plasma of the following blood group phenotypes, all of blood group O: Le(a⁻b⁻), nonsecretor (lane 1); Le(a⁻b⁻), secretor (lane 2); Le(a⁺b⁻), nonsecretor (lane 3); Le(a⁻b⁺), secretor (lane 4). About 30 μg of glycolipids were applied in each lane. The solvent was chloroform–methanol–water 60:35:8 (v/v/v) and the detection reagent was anisaldehyde. Figures to the left indicate number of sugars of the glycolipids. The structures of the triple bands in the 5- to 6-sugar interval which differ between the four lanes are in agreement with serology (see text). In addition there are slow-moving bands which differ but they are barely visible with chemical spray detection. They are, however, detectable by immunostaining.[3,15] The triple-band appearance is due to a heterogeneity of the lipophilic components.[11]

but clearly detectable by overlay analysis with blood group-specific antibodies.[3,15] The 5- and 6-sugar glycolipids seen to differ between the samples have structures[3,11,15] in agreement with the serology[3] and are as follows (see ref. 1 for complete structures): 5-sugar H glycolipid with a type 1 chain (lane 2), 5-sugar Le^a glycolipid (lane 3), and 6-sugar Le^b glycolipid (lane 4). The triple-band appearance is due to ceramide heterogeneity.[11]

[15] K.-A. Karlsson and B. E. Samuelsson, unpublished results.

The slow-moving interval of interest in Fig. 1 is virtually free of contaminants, which is notable since some glycolipids are present at a microgram level in the total preparation. In some instances with very small samples in relation to the preparation media, some minor impurities of the dialysis bags may appear in the 8- and 10-sugar region and produce a color with anisaldehyde similar to the glycolipids.

The extraction used is about 99% efficient for sphingolipids as measured[12,16] on human red cell membranes[11] and human small intestine by acid hydrolysis of the tissue residue after extraction and preparation of N-dinitrophenyl derivatives of long-chain bases.

The present procedure is only a slight modification of existing methods.[1] We have avoided Florisil chromatography[1,10] to remove phospholipids because of trailing phenomena and partial loss of complex glycolipids. If carefully handled (see also comments under "Thin-Layer Chromatography") the original silicic acid chromatography of acetylated glycolipids[9] is without this drawback. The alkaline treatment to degrade ester lipids appears to be critical for a good result. One should, however, bear in mind the possible existence of O-acetylated glycolipids[1] which are degraded in this step. Avoidance of the classical Folch partition step keeps glycolipids with from 1 up to possibly 20 or more sugars in one fraction, possible to separate from most of the alkali-stable nonglycolipids after acetylation.

Method B for deacetylation is convenient for small amounts of glycolipids. As KOH is practically not consumed during the reaction, a low amount of reagent is needed with the result that only a small part of potassium acetate remains in the sample. However, contaminants of acid nature may neutralize KOH and block the reaction. Therefore only highly purified acetylated glycolipids are recommended for this method.

The most important use for this purification procedure is for overlay analysis with bacteria and viruses.[4] Although total lipids are possible to analyze without subfractionation one obvious advantage using the total glycolipids free of nonglycolipid contaminants is the possibility to interpret positive and negative binders by a direct comparison of the overlay result with the chemically detected patterns. The preparation technique has also been used for comparative analysis of tissues and cells, including intestine of different animal species and strains,[5,7] intestine of human individuals[17,18] including meconium,[19] pancreas of human individuals,[18]

[16] K.-A. Karlsson, *Chem. Phys. Lipids* **5**, 6 (1970).

[17] K.-A. Karlsson, *Biol. Membr.* **4**, 1 (1982).

[18] S. Björk, M. E. Breimer, G. C. Hansson, K.-A. Karlsson, G. Larson, and H. Leffler, *in* "Red Cell Membrane Glycoconjugates and Related Genetic Markers" (J.-P. Cartron and C. Salmon, eds.), p. 125. Librairie Arnette, Paris, 1983.

[19] K.-A. Karlsson and G. Larson, *FEBS Lett.* **128**, 71 (1981).

separate compartments of intestine,[7,17,18] cells of different maturity,[13] normal and tumor tissue,[18,20,21] and blood of human individuals.[3,11,15,22] An important use is for mass spectrometry of derivatized mixtures,[6,7,21] where contaminants may falsify the interpretation of minor glycolipid species. The method has not been reported in detail before.

Acknowledgments

The author is indebted to Bo E. Samuelsson for the chromatogram and samples of Fig. 1 and for valuable discussions during the working out of the procedure. Karin Nilson was the constructive technician during this work, which was supported by a grant from the Swedish Medical Research Council (No. 3967).

[20] G. C. Hansson, K.-A. Karlsson, G. Larson, J. M. McKibbin, M. Blaszczyk, M. Herlyn, Z. Steplewski, and H. Koprowski, *J. Biol. Chem.* **258**, 4091 (1983).
[21] M. E. Breimer, *Cancer Res.* **40**, 897 (1980).
[22] M. E. Breimer, B. Cedergren, K.-A. Karlsson, K. Nilson, and B. E. Samuelsson, *FEBS Lett.* **118**, 209 (1980).

[17] Overlay and Solid-Phase Analysis of Glycolipid Receptors for Bacteria and Viruses

By KARL-ANDERS KARLSSON and NICKLAS STRÖMBERG

Analysis on thin-layer chromatograms of the binding of bacteria[1,2] and viruses[3] to glycolipids is rapidly proving to be a decisive tool in receptor studies. This assay has the advantage of combining the high-resolution separation of potential receptors with direct solid-phase binding, thus providing a rationalized approach to the detection and identification of carbohydrate receptors for a variety of ligands. Carbohydrates at the animal cell surface are essential binding sites for bacteria[4] at colonization and infection and for viruses[5] at invasion. The complexity of the mem-

[1] G. C. Hansson, K.-A. Karlsson, G. Larson, A. A. Lindberg, N. Strömberg, and J. Thurin, *Proc. Int. Symp. Glycoconjugates, 7th, 1983* p. 631 (1983).
[2] G. C. Hansson, K.-A. Karlsson, G. Larson, N. Strömberg, and J. Thurin, *Anal. Biochem.* **146**, 158 (1985).
[3] G. C. Hansson, K.-A. Karlsson, G. Larson, N. Strömberg, J. Thurin, C. Örvell, and E. Norrby, *FEBS Lett.* **170**, 15 (1984).
[4] G. W. Jones and R. E. Isaacson, *CRC Crit. Rev. Microbiol.* **10**, 229 (1983).
[5] N. J. Dimmock, *J. Gen. Virol.* **59**, 1 (1982).

brane-bound glycolipids[6] carries most of the known surface carbohydrate variation between animal species, individuals, and cells and therefore this overlay analysis may produce important information on the well-known tropism of both infections and normal bacterial colonizations. Interpretation of such work is further facilitated by the accessibility from a particular object of a total glycolipid fraction free of nonglycolipid contaminants,[7] containing a complex mixture with 1 up to about 15 sugars. As an estimation of the avidity of binding to an isolated glycolipid receptor, a solid-phase assay in microtiter wells is a convenient supplement.

The thin-layer overlay procedure is based on the treatment of the developed chromatogram with poly(isobutyl methacrylate)[8] and was modified from the original description of toxin and antibody overlay using poly(vinylpyrrolidone).[9–11] The handling and detection of the biological ligand and the curve assay are the essential present additions to the reported procedure.[2,3] The conditions of separate steps are, however, being continuously optimized in relation to the actual microbiological problem, including important supplementations for the cloning of genes of bacterial lectins and for clinical applications.

Overlay Assay on Thin-Layer Chromatograms

Thin-Layer Chromatography

Glycolipids are separated on silica gel coated on aluminum sheets (silica gel 60 nanoplates, E. Merck, Darmstadt, W. Germany). These sheets are more resistant to layer detachment during the many washing steps than silica gel coated on glass plates. The aluminum sheets can also be cut or bent during the handling (see below). Two identical chromatograms are developed in parallel, one for overlay analysis as described below and one for chemical spray detection, primarily with the anisaldehyde reagent.[12] This reagent produces a green or blue-green color for

[6] S.-I. Hakomori, in "Sphingolipid Biochemistry" (J. N. Kanfer and S.-I. Hakomori, eds.), p. 1. Plenum, New York, 1983.

[7] K.-A. Karlsson, this volume [16].

[8] M. Brockhaus, J. L. Magnani, M. Blaszczyk, Z. Steplewski, H. Koprowski, K.-A. Karlsson, G. Larson, and V. Ginsburg, J. Biol. Chem. 256, 13223 (1981).

[9] J. L. Magnani, D. F. Smith, and V. Ginsburg, Anal. Biochem. 109, 399 (1980).

[10] J. L. Magnani, M. Brockhaus, D. F. Smith, V. Ginsburg, M. Blaszczyk, K. F. Mitchell, Z. Steplewski, and H. Koprowski, Science 212, 55 (1981).

[11] J. L. Magnani, M. Brockhaus, D. F. Smith, and V. Ginsburg, this series, Vol. 83, Part D, p. 235.

[12] E. Stahl, "Dünnschichtchromatographie." Springer-Verlag, Berlin and New York, 1967.

glycolipids while nonglycolipids usually give other colors.[13] When using the procedure as a screening for glycolipid receptors for different ligands an efficient way is to apply mixtures of glycolipids of various animal and tissue origins,[1-3] thus including a large number of receptor candidates in one single run. The preparation of such a mixture free of nonglycolipid contaminants is described separately.[7]

For total glycolipid fractions 10–40 μg are applied to each lane. For pure glycolipid species the amount needed varies depending on the avidity of the ligand. In the case of *Escherichia coli* recognizing the sequence Galα1 $\rightarrow$ 4Gal, the detection level using autoradiography is in the picomole range, requiring only a few nanograms of glycolipid.[14] Most microbiological ligands, however, appear to be of a low-affinity type[2] and need several times more glycolipid for detection. For most of the specific interactions studied, nonreceptor glycolipids are completely negative even at a level of 10 μg or more. For practical purposes a glycolipid is considered negative if 1 μg is unable to bind.

The solvent used for nonacid glycolipids is usually chloroform–methanol–water 60 : 35 : 8 (v/v/v), solvent A, and for acid glycolipids chloroform–methanol–2.5 *M* ammonia 60 : 40 : 9 (v/v/v), solvent B, but a range of other solvents are applicable as well.[15]

Treatment of Developed Plate[2,3,8]

The dried chromatogram with separated glycolipids is dipped for 1 min in 200 ml of diethyl ether containing 0.5% (w/v) poly(isobutyl methacrylate) (Plexigum P28, Röhm GmbH, Darmstadt, W. Germany) and airdried. The plastic-treated plate placed in horizontal position is sprayed with phosphate-buffered saline (pH 7.3), containing 2% bovine serum albumin and 0.1% NaN$_3$, and is then immersed in this solution in a petri dish for 2 hr. The solution is decanted off but the plate is not allowed to dry before the overlay with ligand suspension. All steps are performed at room temperature.

In our hands treatment of the developed plate with plastic is essential to avoid unspecific binding and background staining, and this treatment is superior to use of poly(vinylpyrrolidone).[9] The hydrophobic film may

[13] K.-A. Karlsson, B. E. Samuelsson, and G. O. Steen, *Biochim. Biophys. Acta* **316**, 317 (1973).
[14] K. Bock, M. E. Breimer, A. Brignole, G. C. Hansson, K.-A. Karlsson, G. Larson, H. Leffler, B. E. Samuelsson, N. Strömberg, C. Svanborg Edén, and J. Thurin, *J. Biol. Chem.* **260**, 8545 (1985).
[15] S. K. Kundu, this series, Vol. 72, Part D, p. 185.

both optimize the albumin blocking of the silica gel (compare effect of coating in plastic wells) and induce a presentation of the amphipathic glycolipids in the assay similar to the situation in the natural cell membrane. A level of plastic too low raises the background and a level too high may coat the glycolipid and block binding. Even at the optimal 0.5% plastic 1- and 2-sugar glycolipids may incidentally show a false negative binding. We have not been able to eliminate this important drawback by changing the conditions including type of plastic or age of plastic solution. One interpretation is that a too small hydrophilic head group does not always promote a "turning up" and placement of the glycolipid on the film during the rapid evaporation of solvent and settlement of the plastic film. We have never seen this effect with larger glycolipids and detection with, e.g., blood group antibodies. Alternatively, the accessibility for a ligand to a ceramide-close binding epitope may critically vary with the precise interaction of the ceramide part with the plastic. The treatment with albumin is more even if done in the two steps described and avoids irregular "fronts" on the plates in the final shape.

Overlay with Particle Suspension

For optimal interaction of bacterial or viral particles with glycolipids locally exposed in narrow bands, a rather large amount of particles is needed using the present conditions. This step may be further modified to reduce the sample size, possibly by gentle shaking or waving of the plate covered with suspension or centrifuging to approach particle to receptor. For a 5×5 cm plate a 2–3 ml suspension is needed to just cover the plate, which may be placed on a small petri dish in the humid atmosphere of a slightly larger petri dish. In case of bacteria a number of cells in the order of 10^8/ml is required and for viruses about 30 μg/ml. After a 2-hr overlay the suspension is tipped off and the plate carefully washed 6 times with phosphate-buffered saline by a gentle waving in a petri dish, 1 min each time. This wash is most effective with the layer placed downward, which may be done without damage to the layer by bending the aluminum sheet corners 90° as supports.

Detection of Bound Particles

Particles remaining on the plate after the washings may be detected in various ways depending upon the experimental purpose. If the particles are labeled an autoradiography of the dried plate may be performed after these washings. Alternatively, nonlabeled ligand may be detected by antibody or erythrocytes.

Radioactive Labeling of Virus Particles. Although reported to be effective for viruses[16] the solid-phase ^{125}I-labeling introduced by Fraker and Speck[17] has not produced useful results for the present method. Although the virus particles were well labeled the activity remaining on the plates after the overlay was low or absent (tested for Sendai virus, influenza virus, and Epstein–Barr virus). Similarly, Bolton–Hunter labeled Sendai virus required a very long time for detection by autoradiography.[3] In case of viruses we therefore prefer detection methods using nonlabeled particles (see below).

Radioactive Labeling of Bacteria. Bacteria are usually efficiently labeled by either external or metabolic labeling. The external labeling with ^{125}I may be done with Iodo-Gen[2] or with the Bolton–Hunter reagent.[18] Labeling with the solid-phase reagent tetrachlorodiphenylglycoluril[17] available as Iodo-Gen (Pierce Chemical Co., Rockford, IL, USA) may be done as follows. The reagent (0.5 ml; 0.4 mg/ml in chloroform) is evaporated in the reaction vessel and the film obtained is thoroughly dried under a stream of nitrogen at room temperature. Before use the vessel is washed 3 times with borate-buffered saline.[17] Bacterial suspension (0.5 ml of borate-buffered saline with 10^8–10^9 cells) is reacted with 500 μCi Na[125]I at 0° for 10 min, followed by gentle centrifugation and two washes in phosphate-buffered saline (pH 7.3). The effective labeling is depending on the bacterial species in question; for *Propionibacterium granulosum*[2] a total of about 5×10^6 cpm was incorporated, which is 20–200 bacteria/cpm. This high labeling requires only a 2-hr autoradiography[2] (see below) and may be necessary for producing curves from binding in microtiter wells (see below).

Labeling with Bolton–Hunter reagent[19] may be done as follows,[18,19] with the final counts per minute varying with type of bacteria. The reagent (100 μCi in benzene; New England Nuclear, Boston, MA, USA) is evaporated in the reaction vessel and 10^8 bacteria in 1 ml of borate buffer (pH 8.5) are added to the dried reagent and the reaction mixture agitated for 15 min at 0°. The labeled cells are washed twice with phosphate-buffered saline and gentle centrifugation. The total labeling usually obtained is in the range of 10^5–10^8 cpm.

Conditions for metabolic labeling of bacteria vary with the bacterial species and conditions for growth. We have used [14C]glucose for labeling

[16] M. A. K. Markwell and C. F. Fox, *Biochemistry* **17**, 4807 (1978).
[17] P. J. Fraker and R. C. Speck, Jr., *Biochem. Biophys. Res. Commun.* **80**, 849 (1978).
[18] K.-A. Karlsson, A. A. Lindberg, and N. Strömberg, in preparation.
[19] A. E. Bolton and W. M. Hunter, *Biochem. J.* **133**, 529 (1973).

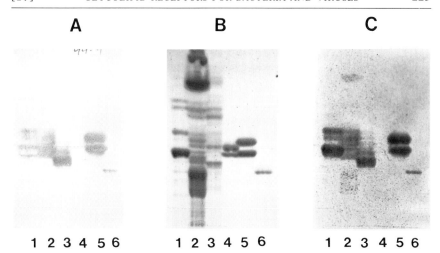

Fig. 1. Binding of bacteria to glycolipids separated by thin-layer chromatography. The bacteria used were uropathogenic *E. coli* with specificity for Galα1 → 4Gal[14] (10^9 cells/plate). (A) Detection of bound bacteria by a suspension of human erythrocytes. (B) Detection of separated glycolipids by the anisaldehyde reagent. (C) Detection of bound bacteria by autoradiography (a total of 2×10^7 cpm was overlayed and autoradiography performed for 1 hr). The following total nonacid glycolipids (20–40 μg) or pure samples (1–2 μg) were applied: human erythrocytes of blood group A (lane 1); human meconium of blood group AB (2); dog small intestine (3); dog intestine isoglobotriaosylceramide and rat intestine isoglobotetraosylceramide (4); human kidney globotriaosylceramide and human erythrocyte globotetraosylceramide (5); human erythrocyte glycolipid with blood group P1 activity, Galα1 → 4Galβ1 → 4GlcNAcβ1 → 3Galβ1 → 4GlcβCer (6). Solvent A was used. For further conditions, see text. Similar patterns are visualized with autoradiography of labeled bacteria (C) and with erythrocyte detection (A). Members of the globoseries[6] mainly with 3 and 4 sugars (lanes 1, 2, and 5) and 5 sugars (Forssman glycolipid of lane 3) do bind the bacteria and also the P1 glycolipid (lane 6). Several other glycolipids including members of the isoglobo series[6] (lane 4, containing Galα1 → 3Gal) are negative.

of *E. coli*[14] and [^{35}S]methionine for labeling of *E. coli* and several other bacteria.[20]

As a general guideline an aliquot for one standard overlay analysis (compare Fig. 1) may contain about 10^8 cells with a radioactivity of about 10^5 cpm. This requires a time for autoradiography of about 2–3 days. An example is shown in Fig. 1.

Detection of Unlabeled Bound Particles with Antibody and Labeled Anti-Antibody. Detection of bacteria or virus remaining bound to glyco-

[20] K.-A. Karlsson, N. Strömberg, and B.-E. Uhlin, in preparation.

lipids after extensive washing may be done by specific polyclonal or mono-clonal antibodies using established conditions described in the literature.

As a first example Sendai virus may be detected[3] by use of a mono-clonal antibody produced in ascites and diluted 1 : 100 with phosphate-buffered saline (pH 7.3) containing 2% bovine serum albumin and 0.1% NaN$_3$. This solution (about 2 ml) is layered over the plate (still wet after washing away unbound virus) for 2 hr, and then the plate is washed 5 times with phosphate-buffered saline. About 2 ml of ^{125}I-labeled anti-mouse immunoglobulin is overlayed for 2 hr (antiserum Z 109, Dakopatts AS, Copenhagen, Denmark). This serum is labeled with the Iodo-Gen procedure (see above), giving a specific activity in the order of 6×10^7 cpm/μg protein. This sample is diluted 2×10^4 times with phosphate-buffered saline (pH 7.5) containing 2% bovine serum albumin and 0.1% NaN$_3$, thus giving a total activity of the overlay of about 10^6 cpm. After 6 washes with phosphate-buffered saline and drying the plate is ready for autoradiography.

As a second example human rotavirus was detected with a specific rabbit hyperimmune serum[21] diluted 1 : 100. Swine anti-rabbit immuno-globulin (antiserum Z 196, Dakopatts AS, Copenhagen, Denmark) was iodinated and used with conditions as for Sendai virus detection described above.

As a third example several bacteria have been detected using both monoclonal and polyclonal antibodies and anti-antibodies and conditions as described above.

Autoradiography. The registration of labeled particles or antibody is done by autoradiography at room temperature using the dried plate and XAR-5 X-ray film (Eastman Kodak, Rochester, NY, USA) with an inten-sifying screen. The time for exposure may be as short as 2 hr (see Iodo-Gen labeling of *Propionibacterium* above) but is usually 2–3 days follow-ing the general recommendations given above for external or metabolic labeling of particles or for antibody detection.

Detection of Bound Particles by Overlay with Erythrocytes. We are at present developing the use of erythrocytes in various ways as a conven-ient way of detecting bound bacteria, viruses, and other ligands. After the washes of the plate to remove unbound particles and with the plate still wet a 4% suspension of erythrocytes in phosphate-buffered saline is over-layed for 2 hr or less. Attached erythrocytes may be seen already after a few minutes. After several washes with phosphate-buffered saline the plate is dried and may be stored or photographed. An example is shown in

---

[21] K.-A. Karlsson, N. Strömberg, and G. Wadell, in preparation.

A   B

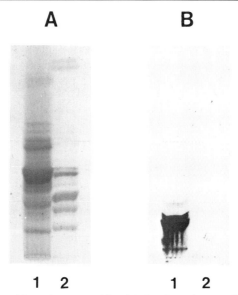

1 2   1 2

FIG. 2. Detection with erythrocytes of Sendai virus bound to gangliosides separated by thin-layer chromatography. (A) Detection of separated gangliosides by the anisaldehyde reagent. (B) Detection of bound Sendai virus by overlay with a suspension of human erythrocytes. The glycolipid samples applied were 30 μg of total gangliosides of human erythrocytes (lane 1) and 15 μg of total gangliosides of human brain (2). Solvent A was used. For further conditions, see text. There was a detectable binding only to slow-moving gangliosides of erythrocytes (lane 1). The major sialylparagloboside of erythrocytes (lane 1) and the brain gangliosides (lane 2) were negative. A similar result was earlier shown[3] after virus detection with monoclonal antibody and autoradiography.

Fig. 1 in which bound *E. coli* as detected by autoradiography of metabolically labeled cells and by erythrocytes is compared. A second example is shown in Fig. 2 for the detection of Sendai virus, and may be directly compared with published data on autoradiographic detection using the same glycolipid fractions.[3] The sensitivity of this direct detection is in the same order of magnitude as autoradiography using the standard conditions described above (see Fig. 1).

Binding Curves from Assay in Microtiter Wells

To roughly estimate the avidity of a bacterium or virus for a particular glycolipid receptor one may evaluate the detection level from dilutions of glycolipid on a thin-layer plate.[14] Alternatively, binding curves from receptor dilutions coated in microtiter wells may be applied, analogous to the binding of antibodies to glycolipids.[8] Sendai virus was earlier shown to bind to gangliosides coated on petri dishes.[22]

Microtiter wells (Cooks M 24, Nutacon, Holland) are coated with glycolipid by allowing 50 μl per well of a dilution in methanol to evaporate overnight at room temperature. Then 100 μl per well of 2% bovine serum albumin in phosphate-buffered saline is incubated for 2 hr followed by 2 rinses with 100 μl of this solution. The particle suspension is added, 50 μl in 2% albumin in phosphate-buffered saline, and plates are incubated for 4 hr, followed by 4 washes with 100 μl each of albumin in phosphate-buffered saline.

In case of nonlabeled Sendai virus we have used 1.5 μg virus in the suspension. Bound virus has been detected with mouse monoclonal antibody produced in ascites and diluted 1 : 100 in phosphate-buffered saline with 2% bovine serum albumin and 0.1% NaN$_3$. Per well, 50 μl of this suspension is incubated for 4 hr, followed by 4 washes with 100 μl each of phosphate-buffered saline. Then 50 μl of rabbit anti-mouse antiserum (4 × 10^5 cpm of ^{125}I-labeled antiserum Z 109, Dakopatts, Copenhagen, Denmark; see conditions described above) is added and plates are incubated overnight at 4°, followed by 5 washes with 100 μl each of 2% bovine serum albumin in phosphate-buffered saline and final counting in a gamma spectrometer after cutting the wells.

In the case of labeled bacteria Fig. 3 shows a curve obtained for *P. granulosum* labeled with Iodo-Gen (see above) and with lactosylceramide as specific receptor. In this case 5 × 10^6 cells/well with a radioactivity of 5 × 10^5 cpm were used.

We are at present using this binding assay to evaluate inhibition potencies of soluble receptor analogs.[23] In this case all wells are coated with, e.g., 150 ng of receptor glycolipid and the particles preincubated with a dilution series of the inhibitor in question. The inhibitor concentration at half-maximum inhibition may be used as a figure of relative inhibition potency. This approach eliminates several drawbacks of traditional inhibition assays using living target cells freshly isolated from donors.[4]

With the present conditions the wells accept only about 200 ng of glycolipid to a covering monolayer. Above this limit multilayers may form and cause inconsistencies in the curve (values may go down due to washing away of ligand bound to multilayers interacting with their polar head groups). This may sometimes mean that a weak but specific low-affinity binding detected in the overlay assay does not produce a detectable bind-

[22] J. Holmgren, L. Svennerholm, H. Elwing, P. Fredman, and Ö. Strannegård, *Proc. Natl. Acad. Sci. U.S.A.* **77**, 1947 (1980).
[23] K.-A. Karlsson and N. Strömberg, in preparation.

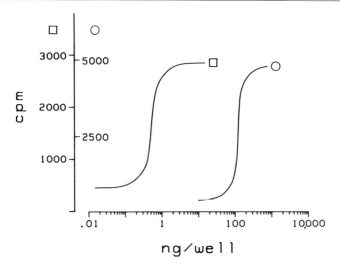

FIG. 3. Approximated curve (○) from binding of ^{125}I-labeled *Propionibacterium granulosum* (5 × 10⁶ cells/well, with a radioactivity of 5 × 10⁵ cpm) to a dilution series of dog intestinal lactosylceramide coated in microtiter wells. For further conditions, see text. For comparative purposes the location of the approximated curve from analogous binding of cholera toxin to ganglioside G_{M1} (□) is shown.

ing in the wells. In this respect the overlay assay is more sensitive than the curve assay.

Similar to the overlay detection described above, erythrocytes may be used to register bound particles in the microtiter wells.[23] Alternatively, well-established ELISA methods (see many procedures of this series) may be adapted both for the overlay assay and the curve assay.[24]

A detailed discussion of the curve assay will appear elsewhere.[23]

Comments on the Procedure

We have now used the present approach to screen for glycolipid receptors for a large number of bacterial species and many viruses, which will be discussed in separate communications. In a few instances a considerable background staining has appeared in the overlay assay due to a general stickiness of the particles. Usually this staining is negligble (see Figs. 1 and 2), allowing easy detection also of minor bands. Also, including a large number of various glycolipids in the same single run offers a reliable comparison with built-in negative controls. An important drawback of the technique is the incidental false negativity of binding in case of

[24] K.-A. Karlsson, I. Lönnroth, N. Strömberg, and S. Teneberg, unpublished.

1- and 2-sugar glycolipids, possibly caused in the plastic coating as noted above. Omitting this coating, although binding to glycolipids occurs, greatly reduces sample-to-background contrast. This problem may coincide with the known phase variation of bacteria where the investigator, apparently without possibility of cultivation control, may incidentally find a particular bacterium to be negative due to lack of expression of the lectin in question. This is especially pertinent to the large group of bacteria where the optimal receptor is lactosylceramide[1,2,18,23] (see also Fig. 3). Further work is needed to eliminate this technical problem due to the assay and the biological variation of bacteria.

Including many lanes of glycolipid mixtures of various origins in the same single overlay analysis provides a very efficient tool of receptor screening. This is facilitated by preparing total glycolipids free of nonglycolipid contaminants.[7] In this way the result from binding of the ligand may be directly compared with the glycolipid pattern as detected by the chemical spray reagent, even in the case of minor components. In many instances individual molecular species of such mixtures have been isolated and structurally characterized (e.g., intestinal epithelia of several animal species, see, for instance, Breimer et al.[25,26]). This means that a particular binding to a band of a mixture may be directly followed by more detailed studies of an already isolated component including a quantitation of the binding in the curve assay. In other cases the glycolipid of the mixture is unknown requiring isolation with high-technology methods (using the overlay assay for monitoring) and structural analysis including the use of mass spectrometry and NMR spectroscopy.[6]

One characteristic of microbiological ligands (differing from most antibodies) is the recognition of epitopes placed internally in the saccharide chain,[23] which at first sight may confuse the observer when interpreting patterns of isoreceptors with different chain extensions. One important consequence of this is the opportunity to do a detailed analysis with the present methods of the binding preferences of a ligand to these isoreceptors (in some cases the binding may be relatively hindered by the nature of the immediate substitutions of the binding epitope). A comparison of such data with results from computer-based calculations of preferred conformations of oligosaccharide chains may allow a rather detailed formulation of the binding epitope before the need of laborious organic synthesis.[14,27]

[25] M. E. Breimer, G. C. Hansson, K.-A. Karlsson, and H. Leffler, J. Biol. Chem. 257, 557 (1982).

[26] M. E. Breimer, G. C. Hansson, K.-A. Karlsson, and H. Leffler, Biochem. J. 90, 589 (1981).

[27] K. Bock, K.-A. Karlsson, N. Strömberg, and S. Teneberg, in "Protein–Carbohydrate Interactions in Biological Systems," p. 207, Academic Press, New York, 1986.

A   B

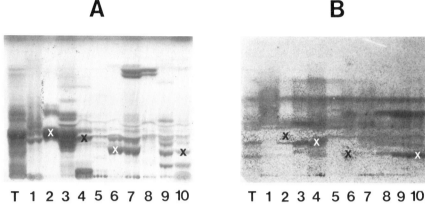

T 1 2 3 4 5 6 7 8 9 10   T 1 2 3 4 5 6 7 8 9 10

FIG. 4. Detection of novel ganglioside species by *E. coli* recognizing Galα1 → 4Gal.[14] (A) Subfractions of human erythrocytes detected by the anisaldehyde reagent. (B) Autoradiogram after overlay with ^{35}S-labeled bacteria (2×10^8 cells with a radioactivity of 5×10^7 cpm and autoradiography for 50 hr). Thirty micrograms of total erythrocyte gangliosides (T) and 10–20 μg of subfractions (1–10) after continuous elution from DEAE–Sepharose[28] were applied. Solvent A was used. For further conditions, see text. Several bands do bind the bacteria (compare Fig. 1). Only the indicated band of lane 10 has been chemically identified and is disialylglobopentaosylceramide.[29] All other positive gangliosides (some of them marked) should also contain Galα1 → 4Gal[14] and represent earlier unknown erythrocyte gangliosides.[6] The negative species in lanes 2 and 6 are sialylparagloboside and disialylparagloboside, respectively.

Another potent use of the procedure is to apply bacteria and viruses as reagents for the detection and identification of new glycolipid structures. An example is shown in Fig. 4 where *E. coli* specifically recognizing Galα1 → 4Gal[14] has been used to analyze gangliosides of human erythrocytes. The total gangliosides were subfractionated by continuous elution from a DEAE–Sepharose column.[28] Most of the bands shown to bind the bacteria must be earlier unknown gangliosides (note the negativity for the major species of NeuAcα2 → 3Galβ1→ 4GlcNAcβ1 → 3Galβ1 → 4Glcβ-Cer, sialylparagloboside).

The described procedure is being adapted for assay during the cloning of genes for bacterial lectins and for various problems of clinical interest. Also, preliminary data look promising for the logical extension of the technique to the study of adhesion phenomena of eukaryotic cells.[30]

[28] Y. Nagai and M. Iwamori, *Mol. Cell. Biochem.* **29,** 81 (1980).
[29] S. K. Kundu, B. E. Samuelsson, I. Pascher, and D. M. Marcus, *J. Biol. Chem.* **258,** 13857 (1983).
[30] R. L. Schnaar, *Anal. Biochem.* **143,** 1 (1984).

Acknowledgments

The work was supported by a grant from the Swedish Medical Research Council (No. 3967). We are indebted to A. A. Lindberg and B.-E. Uhlin for providing various bacteria and to E. Norrby, C. Örvell, L. Rymo, and G. Wadell for viruses and antibodies.

[18] Lectin Affinity Chromatography of Glycopeptides

By ROBERTA K. MERKLE and RICHARD D. CUMMINGS

Recent studies have demonstrated the vast structural heterogeneity in glycoprotein oligosaccharides from diverse sources and have broadened our understanding of the influence of glycoconjugates in many biological interactions. For these reasons it is important to purify and analyze the structures of the oligosaccharide moieties of glycoproteins under study. One method that has been used successfully in these analyses is lectin affinity chromatography, which facilitates the separation and purification of glycoprotein-derived oligosaccharides and glycopeptides by the chromatography of samples on columns of immobilized plant lectins.

The ability of plant lectins to bind to animal cell glycoconjugates and induce the agglutination of animal cells has been known for many years.[1] Until recently, the inhibition of this lectin-induced agglutination by various mono- and disaccharides was generally used to define the carbohydrate-binding specificities of lectins. However, it is increasingly apparent that lectins can distinguish more complex structural features of glycoconjugates than single sugar residues. This interesting property of lectins provides the basis for the technique of lectin affinity chromatography. Numerous studies to be cited in this article demonstrate that many immobilized lectins interact with high affinity only with certain kinds or classes of oligosaccharides.

For several reasons lectin affinity chromatography is an attractive technique for separating oligosaccharides in complex mixtures. In contrast to some other techniques, the separation of oligosaccharides by lectin affinity chromatography is not based solely on the size and/or charge of the species. The separation depends on specific interactions of oligosaccharides with immobilized lectins that recognize certain structural characteristics of the oligosaccharides. Lectin affinity chromatography, especially when used in conjunction with new techniques for radiola-

[1] H. Lis and N. Sharon, *Biol. Carbohydr.* **2**, 1 (1984).

beling oligosaccharides and glycopeptides, allows for the purification of extremely small amounts of material and facilitates the identification of minor components that might otherwise be overlooked. Because the interactions of many different oligosaccharides and immobilized lectins have been investigated, it is often possible to draw inferences about the structural characteristics of oligosaccharides under study based on their interactions with a number of different immobilized lectins. However, it must be stressed that definite conclusions about structures cannot be obtained based solely on the interactions of a glycopeptide or oligosaccharide with an immobilized lectin. Nevertheless, these inferences of structure based on the chromatographic behavior of glycoconjugates on lectin affinity columns can usually expedite subsequent detailed structural analyses. Lectin affinity chromatography is also relatively inexpensive to set up and maintain, and oligosaccharides can be fractionated reproducibly, quickly, and easily. Lectin affinity chromatography can supplement more conventional methods of separation, such as size exclusion column chromatography,[2] HPLC,[3,4] paper chromatography,[5] and paper electrophoresis[6] to provide an impressive degree of separation and purification of glycopeptides or oligosaccharides from complex mixtures. The structures of purified compounds can then be analyzed by a number of physical, chemical, and enzymatic methods.

This article describes the general procedure of serial lectin affinity chromatography of glycopeptides and oligosaccharides using several well-defined immobilized plant lectins.

The Types of Oligosaccharides Separable by Lectin Affinity Chromatography. Animal cell glycoproteins may possess Asn-linked (N-linked) or Ser/Thr-linked (O-linked) oligosaccharides or a mixture of both types.[7,8] Some of the general types of N-linked oligosaccharides and structures representative of each type are shown in Fig. 1 (structures 1–59). N-Linked oligosaccharides are generally classified as *complex, hybrid,* or *high-mannose* chains. Specific examples of these types of structures are glycopeptides **1, 55,** and **52,** respectively. In addition, the complex-type chains can be subclassified depending on the outer

[2] K. Yamashita, T. Mizuochi, and A. Kobata, this series, Vol. 83, p. 105.

[3] G. B. Wells, V. Kontoyiannidou, S. J. Turco, and R. L. Lester, this series, Vol. 83, p. 132.

[4] S. Mellis and J. U. Baenziger, *Anal. Biochem.* **114,** 276 (1981).

[5] H. Yoshima, S. Takasaki, and A. Kobata, *J. Biol. Chem.* **255,** 10793 (1980).

[6] S. Narasimhan, N. Harpaz, G. Longmore, J. P. Carver, A. A. Grey, and H. Schachter, *J. Biol. Chem.* **255,** 4876 (1980).

[7] A. Kobata, *Biol. Carbohydr.* **2,** 87 (1984).

[8] R. Kornfeld and S. Kornfeld, *in* "The Biochemistry of Glycoproteins and Proteoglycans" (W. J. Lennarz, ed.), p. 1. Plenum, New York, 1980.

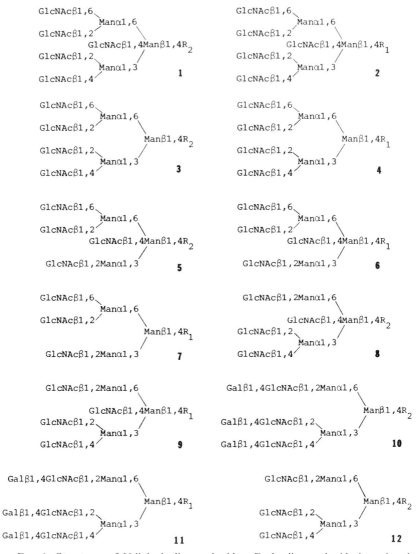

FIG. 1. Structures of N-linked oligosaccharides. Each oligosaccharide is assigned a number and is referred to by the number in the text. R_1, GlcNAcβ1, 4GlcNAcβ1Asn; R_2, GlcNAcβ1, 4Fucα1,6)GlcNAcβ1Asn.

"branching" structures. These types (and examples of each type) are biantennary (**48**), triantennary (**7** and **13**), tetraantennary (**3**), and "bisected" derivatives of these chains (**43**, **6**, and **1**), which contain a residue of GlcNAc linked β1,4 to the β-linked mannose in the core. (To simplify

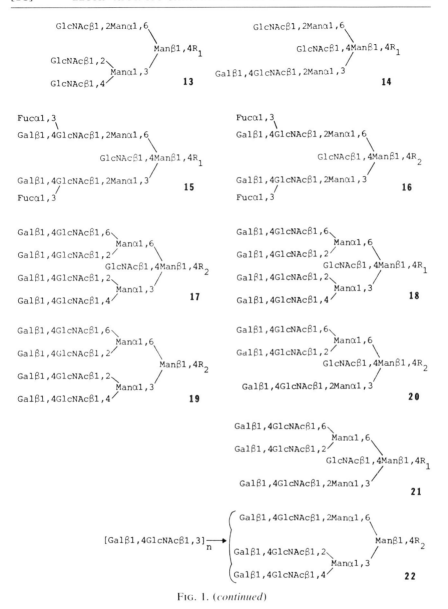

FIG. 1. (*continued*)

discussion the oligosaccharides shown in Fig. 1 do not contain sialic acid, phosphate, or sulfate residues, although these residues, especially sialic acid, are found naturally in many oligosaccharides. The influence of these residues on the interaction of glycopeptides with immobilized lectins is

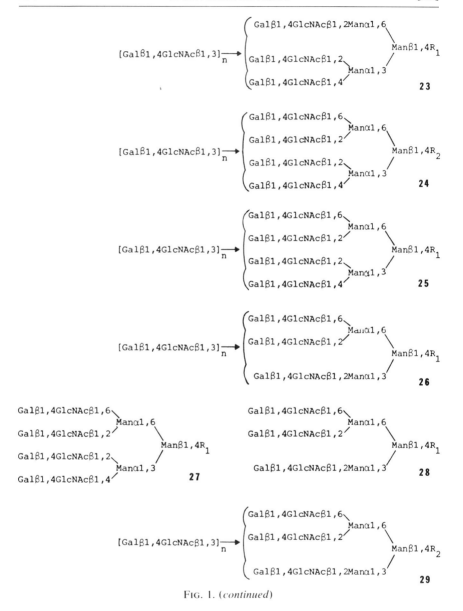

Fig. 1. (continued)

not fully documented, but where documentation is available it will be mentioned in the appropriate sections.) It is apparent from the compounds shown that N-linked oligosaccharides have many structural features in common, and that the increasing complexity of structures is manifested as modifications of a common core.

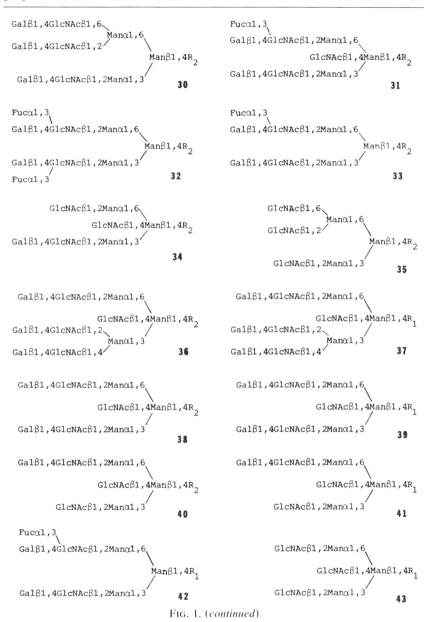

FIG. 1. (*continued*)

This article will focus on the separation of N-linked oligosaccharides by lectin affinity chromatography, since these types of structures have been fractionated successfully by this technique. There are not sufficient examples in the literature demonstrating the purification of O-linked oli-

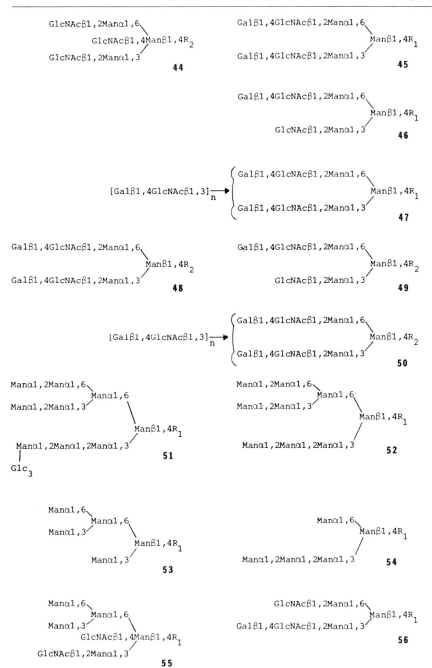

FIG. 1. (*continued*)

GlcNAcβ1,2Manα1,6
                    Manβ1,4R₁
GlcNAcβ1,2Manα1,3
                57

GlcNAcβ1,2Manα1,6
                    Manβ1,4R₂
Galβ1,4GlcNAcβ1,2Manα1,3
                58

GlcNAcβ1,2Manα1,6
                    Manβ1,4R₂
GlcNAcβ1,2Manα1,3
                59

Fig. 1. (continued)

gosaccharides by this approach to warrant including them in a discussion of this technique. Nevertheless, as the technique of lectin affinity chromatography is expanded in the future, it will most likely include lectins also capable of recognizing determinants in O-linked oligosaccharides.

Serial Lectin Affinity Chromatography. The realization of the diversity of N-linked oligosaccharides and the restricted carbohydrate-binding specificities of many different immobilized lectins has prompted the development of a general scheme for serially fractionating glycopeptides or oligosaccharides. This method has evolved from the studies of many individuals over the past 10 years.[9–16] The general procedure of serial lectin affinity chromatography is shown in Fig. 2 as a flow scheme for the separation of oligosaccharides in a complex hypothetical starting mixture composed of the glycopeptides shown in Fig. 1. The seven immobilized lectins employed in this technique are concanavalin A, erythroagglutinating phytohemagglutinin (E₄-PHA), lentil lectin, pea lectin, leukoagglutinating phytohemagglutinin (L₄-PHA), *Datura stramonium* agglutinin (DSA), and *Ricinus communis* agglutinin I (RCA I). The serial nature of the chromatographic procedure is indicated by arrows running between different schematic column profiles. Glycopeptides that interact with high affinity with the various lectins are designated by numbers which refer to the structures shown in Fig. 1. Glycopeptides or oligosaccharides bound by an immobilized lectin can be eluted by applying high concentrations of haptenic sugars. After separation of the glycopeptides from the haptenic

[9] S. Ogata, T. Muramatsu, and A. Kobata, *J. Biochem. (Tokyo)* **78**, 687 (1975).

[10] M. Tomana, W. Neidermeier, J. Mestecky, R. E. Schroehenloher, and S. Porch, *Anal. Biochem.* **72**, 389 (1976).

[11] J. Jarnefëlt, J. Finne, T. Krusius, and H. Rauvala, *Trends Biochem. Sci.* **3**, 110 (1978).

[12] J. Finne and T. Krusius, this series, Vol. 83, p. 269.

[13] H. Debray, A. Pierce-Cretel, G. Spik, and J. Montreuil, *in* "Lectins" (J. C. Bog-Hansen and G. A. Sprangler, eds.), Vol. 3, p. 335. de Gruyter, Berlin, 1983.

[14] H. Debray, D. Decout, G. Strecker, G. Spik, and J. Montreuil, *Eur. J. Biochem.* **117**, 41 (1981).

[15] D. A. Blake and I. J. Goldstein, this series, Vol. 83, p. 127.

[16] R. D. Cummings and S. Kornfeld, *J. Biol. Chem.* **257**, 11235 (1982).

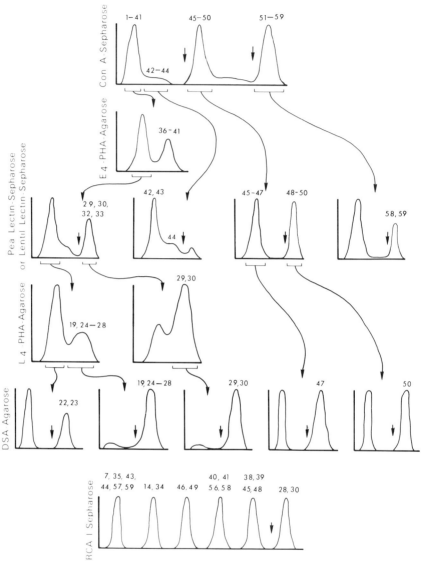

Fig. 2. Scheme for fractionating N-linked oligosaccharides by serial lectin affinity column chromatography. The N-linked oligosaccharides shown in Fig. 1 and numbered 1 through 59 can be fractionated by chromatography on a column of Con A-Sepharose, as shown at the top of this figure. Certain fractions may be pooled as indicated just below the abscissa and additional separation of glycopeptides may be achieved by passing samples over columns of other immobilized lectins, which include E₄-PHA-agarose, pea or lentil lectin-Sepharose, L₄-PHA-agarose, DSA-agarose, and RCA I-Sepharose. The elution of glycopeptides from some of the columns requires the addition of haptens which is indicated by the small arrows in the graphs. It should be noted that only those interactions of oligosaccharides and immobilized lectins which have been documented are indicated in the figure.

sugars by size exclusion chromatography, the glycopeptides can be applied to another column, and so forth. The specific interactions of each immobilized lectin with the various glycopeptides and the chromatographic techniques are described in detail below.

The scheme of serial lectin affinity chromatography is organized in the manner presented in Fig. 2 because many studies have demonstrated that this approach reduces the number of steps necessary during purification. For example, lectins that recognize features of the core structure and sugar residues substituting the core are utilized early in the fractionation to allow for the separation of oligosaccharides on the basis of the different general types. Lectins used later in the fractionation scheme recognize more peripheral sugar residues, often in addition to the core structure. With an understanding of this concept it is apparent how additional immobilized lectins having different oligosaccharide-binding specificities may be used to supplement this basic scheme.

Preparation of Samples for Lectin Affinity Chromatography. The technique of lectin affinity chromatography as discussed here is concerned with interactions of immobilized lectins with glycopeptides containing one oligosaccharide or oligosaccharides free of peptide. Glycopeptides or oligosaccharides can be prepared from purified glycoproteins, mixtures of glycoproteins, or cellular material by a number of enzymatic and chemical methods.[12] The most widely used and effective means of preparing glycopeptides is to completely digest material with Pronase, a mixture of proteases with broad specificity.[12] Exhaustive digestion usually results in glycopeptides containing one oligosaccharide chain and a very small number of amino acids. (A notable exception to this is that the peptide backbones of some glycoproteins in regions containing stretches of adjacent O-linked oligosaccharides are relatively resistant to Pronase, resulting in large-sized glycopeptides containing numerous oligosaccharide chains and amino acids.) It is also possible to prepare glycopeptides from glycoproteins purified by immunoprecipitation and polyacrylamide gel electrophoresis in sodium dodecyl sulfate. Gel slices containing the purified glycoprotein may be incubated directly with proteases to release glycopeptides.[17,18] If samples contain lipids or glycosaminoglycans it is advisable to remove them by organic solvent extraction[19] or specific precipitation[20] prior to treating the glycoproteins with proteases.

[17] R. D. Cummings, S. Kornfeld, W. J. Schneider, K. H. Hobgood, H. Tolleshaug, M. S. Brown, and J. L. Goldstein, *J. Biol. Chem.* **258**, 15261 (1983).
[18] L. Liscum, R. D. Cummings, R. G. W. Anderson, G. N. DeMartino, J. L. Goldstein, and M. S. Brown, *Proc. Natl. Acad. Sci. U.S.A.* **80**, 7165 (1983).
[19] I. Tabas and S. Kornfeld, this series, Vol. 83, p. 416.
[20] R. K. Margolis and R. U. Margolis, *J. Neurochem.* **20**, 1285 (1973).

Oligosaccharides can be prepared from glycoproteins or glycopeptides by treating samples with certain chemicals or endoglycosidases. Hydrazinolysis results in the specific release of N-linked oligosaccharides from peptide. Because the released oligosaccharides retain their reducing termini, they can be radiolabeled by reduction with NaB^3H_4.[21] O-Linked sugar chains can be released from peptide by treatment with 0.05 M NaOH containing 1 M NaBH$_4$.[22] This treatment releases little if any of the N-linked sugar chains; however, treatment with higher concentrations of base (1 M NaOH containing 1 M NaBH$_4$) can release the N-linked sugar chains, although the oligosaccharides are reduced during the treatment.[23] Several endoglycosidases are available that specifically release N-linked oligosaccharides from glycoproteins and glycopeptides. These include endo-β-N-acetylglucosaminidases D and H, which recognize and release certain high-mannose and hybrid N-linked chains[24-26] and endoglycosidase F, a mixture of endoglycosidases of broad specificity which cleaves high-mannose, hybrid, and complex N-linked chains from peptides.[27,28]

Certain immobilized lectins, notably pea and lentil lectins as discussed below, have different binding affinities for a free oligosaccharide versus the same oligosaccharide still attached to a peptide. In addition, the interaction of a free oligosaccharide with a lectin may depend on the presence or absence of a reducing terminus. These phenomena suggest that in some cases the peptide may contribute to the conformation of the attached oligosaccharide. Although most other separation techniques give the best separations using oligosaccharides instead of glycopeptides, it may be advantageous in some cases to begin fractionations with glycopeptides to take complete advantage of the technique of serial lectin affinity chromatography before preparing oligosaccharides for analysis.

Preparation of Glycopeptide Standards for Calibrating Columns of Immobilized Lectins. All columns of immobilized lectins should be standardized with glycopeptides of known structure to ensure the activity, specificity, and reproducibility of the columns. Unfortunately, purified glycopeptides are not commercially available. However, some standards

[21] S. Takasaki, T. Mizuochi, and A. Kobata, this series, Vol. 83, p. 263.
[22] R. N. Iyer and D. M. Carlson, *Arch. Biochem. Biophys.* **142**, 101 (1971).
[23] Y. C. Lee and J. R. Scocca, *J. Biol. Chem.* **247**, 5753 (1972).
[24] T. Tai, K. Yamashita, K.-A. Ogata, N. Koide, T. Muramatsu, S. Iwashita, Y. Inoue, and A. Kobata, *J. Biol. Chem.* **250**, 8569 (1975).
[25] T. Mizuochi, J. Amano, and A. Kobata, *J. Biochem. (Tokyo)* **95**, 1209 (1984).
[26] R. B. Trimble, A. L. Tarentino, T. H. Plummer, Jr., and F. Maley, *J. Biol. Chem.* **253**, 4508 (1978).
[27] J. H. Elder and S. Alexander, *Proc. Natl. Acad. Sci. U.S.A.* **79**, 4540 (1982).
[28] T. H. Plummer, Jr., J. H. Elder, S. Alexander, A. W. Phelan, and A. L. Tarentino, *J. Biol. Chem.* **259**, 10700 (1984).

can be prepared readily from a variety of purified glycoproteins that are commercially available or otherwise easily obtainable. Some of these sources are human erythrocyte glycophorin,[29,30] human immunoglobulin A,[31] bovine fetuin,[32] porcine thyroglobulin,[33,34] bovine thyroglobulin,[35] human transferrin,[36] soybean agglutinin,[37] human α_1-acid glycoprotein,[38] chicken ovalbumin,[9,39] and human ceruloplasmin.[40]

Alternatively, glycopeptide standards may be prepared from cultured cell lines. The N-linked oligosaccharides from many different cell lines have been characterized structurally, and their interactions with various immobilized lectins have been determined. For example, radioactively labeled glycopeptides can be prepared from the mouse lymphoma cell line BW5147 by growing cells in suspension culture for 2 days in normal complete medium containing either [2-^{3}H]mannose or [6-^{3}H]galactose at 0.01 mCi/ml. The procedures for metabolically radiolabeling the cells and preparing samples for analysis are described elsewhere.[16] Glycopeptides obtained after Pronase digestion can be fractionated on columns of immobilized lectins using the methods outlined here. From these cells it is possible to isolate glycopeptides that bind to Con A-Sepharose, pea and lentil lectin-Sepharose, L$_4$-PHA-agarose, DSA-agarose, and RCA I-Sepharose. (However, there are no detectable glycopeptides from these cells that interact with high affinity with E$_4$-PHA-agarose.) The structures of most of the major N-linked oligosaccharides from these cells have been determined by a number of chemical and enzymatic methods[16] and they are similar to the appropriate structures presented in Fig. 1.

Before employing columns of immobilized lectins for analyses, these types of standards should be used to demonstrate that each column contains active lectin and that the lectin is coupled to a high enough density to

[29] H. Yoshima, H. Furthmayr, and A. Kobata, *J. Biol. Chem.* **255**, 9713 (1980).

[30] T. Irimura, T. Tsuji, S. Tagami, K. Yamamoto, and T. Osawa, *Biochemistry* **20**, 560 (1981).

[31] J. Baenziger and S. Kornfeld, *J. Biol. Chem.* **249**, 7260 (1974).

[32] B. Nilsson, N. E. Nordén, and S. Svensson, *J. Biol. Chem.* **254**, 4545 (1979).

[33] T. Tsuji, K. Yamamoto, T. Irimura, and T. Osawa, *Biochem. J.* **195**, 691 (1981).

[34] K. Yamamoto, T. Tsuji, T. Irimura, and T. Osawa, *Biochem. J.* **195**, 701 (1981).

[35] R. D. Cummings and S. Kornfeld, *J. Biol. Chem.* **257**, 11230 (1982).

[36] G. Spik, B. Bayard, B. Fournet, G. Strecker, S. Bouquelet, and J. Montreuil, *FEBS Lett.* **50**, 296 (1975).

[37] L. Dorland, H. van Halbeek, J. F. G. Vliegenthart, H. Lis, and N. Sharon, *J. Biol. Chem.* **256**, 7708 (1981).

[38] B. Fournet, J. Montreuil, G. Strecker, L. Dorland, J. Haverkamp, J. F. G. Vliegenthart, J. P. Binette, and K. Schmid, *Biochemistry* **17**, 5206 (1978).

[39] T. Tai, K. Yamashita, M. Ogata-Arakawa, N. Koide, T. Muramatsu, S. Iwashita, Y. Inoue, and A. Kobata, *J. Biol. Chem.* **250**, 8569 (1975).

[40] K. Yamashita, C.-J. Liang, S. Funakoshi, and A. Kobata, *J. Biol. Chem.* **256**, 1283 (1981).

allow clear separations of the appropriate glycopeptides from other structurally different glycopeptides. It is difficult to quantify the binding capacity of each column because this can require considerable amounts of
glycopeptide standards. Therefore, during any routine analysis by lectin
affinity chromatography, glycopeptides apparently not binding to a certain immobilized lectin should be rechromatographed to ensure complete
separation.

Procedures for Separating Oligosaccharides by Lectin Affinity Chromatography

Concanavalin A-Sepharose

Specificity of Interaction. Concanavalin A (Con A) is a hemagglutinin
from the common jack bean *Canavalia ensiformis*. Con A-Sepharose interacts with high affinity with N-linked oligosaccharides in which at least
two outer mannose residues are either unsubstituted or are substituted
only at position C-2 by another sugar.[9] Thus, it can bind conventional
complex biantennary N-linked oligosaccharides, hybrid oligosaccharides,
and high-mannose oligosaccharides[9] as illustrated in Fig. 2. The lectin,
whether free or immobilized, has very low affinity for tri- and tetraantennary complex N-linked oligosaccharides and most known O-linked oligosaccharides.[40a] Krusius *et al.*[41] showed that complex biantennary glycopeptides (e.g., **45**) can be eluted from Con A by low concentrations of
hapten, whereas elution of the high-mannose glycopeptides (e.g., **52**) requires higher amounts of hapten. By differential elution with α-methylglu-
coside and α-methylmannoside, these two types of glycopeptides can be
separated from each other as well as from the tri- and tetraantennary
glycopeptides. Con A-Sepharose can also bind phosphorylated high-mannose oligosaccharides found typically in animal cell lysosomal acid hydrolases.[42] Glycopeptides containing a bisected biantennary N-linked oligosaccharide (e.g., **43** and **44**) are slightly retarded in their elution from Con
A-Sepharose and elution does not require haptenic sugars.[43-45] The presence of a core fucose in complex biantennary N-linked oligosaccharides
does not interfere with their binding to Con A-Sepharose. However, if

---

[40a] J. U. Baenziger and D. Fiete, *J. Biol. Chem.* **254**, 2400 (1979).
[41] T. Krusius, J. Finne, and H. Rauvala, *FEBS Lett.* **71**, 117 (1976).
[42] C. A. Gabel and S. Kornfeld, *J. Biol. Chem.* **257**, 10605 (1982).
[43] N. Harpaz and H. Schachter, *J. Biol. Chem.* **256**, 4894 (1980).
[44] S. Narasimhan, J. R. Wilson, E. Martin, and H. Schachter, *Can. J. Biochem.* **57**, 83 (1979).
[45] S. Narasimhan, J. C. Freed, and H. Schachter, *Biochemistry* **24**, 1694 (1985).

fucose residues are attached to the C-3 position of certain outer-branch GlcNAc residues in a biantennary chain (**42** and **15**), the affinity of the glycopeptides for Con A-Sepharose is lessened.[46] The binding affinity of Con A-Sepharose for oligosaccharides either attached or not attached to peptide does not appear to be significantly different.

Sources of Con A and Preparation of Con A-Sepharose. Con A may be purified from jack bean meal by the method of Agrawal and Goldstein[47] and then coupled to cyanogen bromide-activated Sepharose 4B. However, most recent studies utilize commercial sources of this lectin for convenience and to lessen the possibility of significant batch-to-batch variation in the amount of active lectin coupled. Pharmacia Fine Chemicals provides consistent batches of sufficiently high coupling density (10–16 mg lectin/ml gel) for the applications described here, but other sources may be found which are just as reliable. The capacity of Con A-Sepharose containing 10 mg of lectin/ml gel is approximately 100 nmol glycopeptide bound/ml gel.

For the chromatography of glycopeptides a column (0.7 × 4 cm) of Con A-Sepharose is prepared and washed with 10 ml TBS–NaN₃ (Tris-buffered saline containing 0.01 M Tris, pH 8.0, 0.15 M NaCl, 1 mM CaCl₂, 1 mM MgCl₂, and 0.02% NaN₃). New columns are prepared for each application rather than regenerating and reusing the same column. However, it is possible to use the same column for many separations over a period of several months if the column is washed with at least 20 column volumes of TBS–NaN₃ and stored at 4° between uses.

Elution of Bound Material. Chromatography is carried out at room temperature, and the glycopeptide mixture is loaded in a volume of 1 ml in TBS–NaN₃. The elution series consists of 40 ml TBS–NaN₃ followed by 40 ml of TBS–NaN₃ containing 10 mM α-methyl-D-glucoside, then 20 ml of 100 mM α-methyl-D-mannoside prewarmed to 60°. The haptens are obtained from commercial sources. Fractions of 2 ml are collected at a flow rate of approximately 1 ml/min. Recovery of glycopeptide samples from this column and all other columns described below is routinely 80–100%.

The degree of separation of many glycopeptides depends on the length of the column of Con A-Sepharose, the amount of gel, and the volume of haptenic sugar solutions used. For example, using the relatively short columns described here, small molecular weight high-mannose glycopeptides (e.g., **53** and **54**), hybrid glycopeptides (**55**), and complex bianten-

[46] K. Yamashita, Y. Tachibana, T. Nakayama, M. Kitamura, Y. Endo, and A. Kobata, *J. Biol. Chem.* **255**, 5635 (1980).

[47] B. B. L. Agrawal and I. J. Goldstein, this series, Vol. 28, p. 313.

nary glycopeptides containing terminal GlcNAc residues (57) may be partly eluted with large volumes of 10 mM α-methylglucoside. However, these glycopeptides will be eluted later from the column than the other complex biantennary chains (45 through 50) and may appear as another broad peak in the α-methylglucoside eluant.

After chromatographic separation of glycopeptides by Con A-Sepharose, the fractions containing glycopeptides of interest are pooled and dried either by evaporation under vacuum in a shaker-bath evaporator (Büchler) or by lyophilization. Glycopeptides are then separated from buffer salts and haptenic sugars by size exclusion chromatography on a column (1 × 50 cm) of Sephadex G-25-80 that is equilibrated and maintained at all times in 7% 1-propanol/water to prevent bacterial growth. For application to this column the dried samples are suspended in 1 ml of this solvent and applied to the column while collecting fractions of 1 ml. The glycopeptides are eluted in or near the void volume of the column. The fractions containing glycopeptides are pooled, dried by the methods above, and then suspended in an appropriate buffer to continue the fractionation if necessary on other columns of immobilized lectins. These general steps to separate haptenic sugars from glycopeptides are used throughout the analyses.

The detection of glycopeptides eluting from Con A-Sepharose columns or columns of other lectins can be greatly facilitated by using radioactively labeled samples. Aliquots from each fraction are simply removed and radioactivity determined by liquid scintillation counting. However, if nonradioactively labeled glycopeptides are used, chemical methods of detecting sugars must be employed.[48,49] The presence of haptenic sugars can, of course, interfere with the analyses, so it may be necessary to separate glycopeptides from haptenic sugars by the procedures above before assaying for the recovery of glycopeptide.

Pea Lectin-Sepharose

Specificity of Interaction. Pea lectin is a hemagglutinin from the common garden pea *Pisum sativum.* Immobilized pea lectin interacts with high affinity with certain complex-type N-linked oligosaccharides containing a fucose residue in the core, e.g., 29, 48, and 58, as indicated in Fig. 2.[50] It is interesting that although this fucose residue is required for high-affinity interactions of glycopeptides with the immobilized lectin, fucose is not effective as a haptenic sugar for release of bound glycopep-

[48] R. G. Spiro, this series, Vol. 8, p. 3.
[49] G. Ashwell, this series, Vol. 8, p. 85.
[50] K. Kornfeld, M. L. Reitman, and R. Kornfeld, *J. Biol. Chem.* 256, 6633 (1981).

tides.[50] The presence of outer fucose residues in biantennary N-linked oligosaccharides lacking the inner core fucose residue does not allow a high-affinity interaction with pea lectin-Sepharose.[50] Bisected biantennary N-linked oligosaccharides containing an inner core fucose residue (**44**) do not bind with high affinity but, rather, are retarded in their elution from pea lectin-Sepharose and some residual material may be eluted with α-methylglucoside.[51] Pea lectin-Sepharose has a higher affinity for glycopeptides containing terminal mannose residues in addition to the core fucose residue than for those lacking terminal mannose residues. Additionally, pea lectin-Sepharose can discriminate between isomers of triantennary Asn-linked oligosaccharides. The immobilized lectin binds with high affinity to those triantennary glycopeptides containing an inner fucose residue and one outer mannose residue substituted at position C-2, and another mannose residue substituted at both positions C-2 and C-6 by N-acetylglucosamine (**29** and **30**),[50] whereas it will not significantly interact with the isomeric triantennary glycopeptides which contain an outer mannose substituted at positions C-2 and C-4 by N-acetylglucosamine (**10** and **12**). Immobilized pea lectin does not interact with high affinity with tetraantennary glycopeptides (**3** and **19**).[16,50] Other glycopeptides, such as **5** and **20**, may interact with high affinity or may be retarded in their elution from a column of pea lectin-Sepharose, but this has not been documented.

The nature of the linkage of oligosaccharide to peptide is important for interaction of oligosaccharides with pea lectin-Sepharose. The immobilized lectin will not bind with high affinity to free oligosaccharides with or without a reducing terminus and not attached to peptide even though the free oligosaccharides may contain a residue of core fucose.[52]

Source of Pea Lectin and Preparation of Pea Lectin-Sepharose. Pea lectin is prepared according to the method of Trowbridge,[53] and the purified lectin is coupled to CNBr-activated Sepharose 4B to achieve a final coupling density of 10 mg lectin coupled/ml gel. The coupling reaction is carried out according to the directions of the manufacturer of CNBr-activated Sepharose 4B (Pharmacia Fine Chemicals) in 0.1 M NaHCO$_3$ buffer (pH 8.5), containing 50 mM α-methyl-D-glucoside and 0.5 M NaCl.[54]

Five milliliters of pea lectin-Sepharose is poured into a 1.0 $\times$ 7 cm column. The column of immobilized lectin is washed with 25 ml TBS–

[51] S. Narasimhan, *J. Biol. Chem.* **257**, 10235 (1982).

[52] K. Yamamoto, T. Tsuji, and T. Osawa, *Carbohydr. Res.* **110**, 283 (1982).

[53] I. S. Trowbridge, *J. Biol. Chem.* **249**, 6004 (1974).

[54] I. S. Trowbridge, M. Nilsen-Hamilton, R. T. Hamilton, and M. J. Bevan, *Biochem. J.* **163**, 211 (1977).

NaN$_3$ (Tris-buffered saline containing 0.01 M Tris, pH 8.0, 0.15 M NaCl, 1 mM CaCl$_2$, 1 mM MgCl$_2$, and 0.02% NaN$_3$). The capacity of the column is approximately 100 nmol of glycopeptide bound per milliliter of gel. The column is stored at 4° between uses, and is equilibrated at room temperature for 1 hr before each use. The pea lectin column is stable for at least 1 year and can be regenerated after use by washing with 5 column volumes of TBS–NaN$_3$.

Elution of Bound Material. Chromatography is carried out at room temperature, and the glycopeptides are applied to the column in a volume of 1 ml TBS–NaN$_3$. The scheme of elution is 10 ml TBS–NaN$_3$ followed by 20 ml TBS–NaN$_3$ containing 10 mM α-methyl-D-glucoside and finally 20 ml TBS–NaN$_3$ containing 0.5 M α-methyl-D-mannoside. Fractions of 2 ml are collected at a flow rate of approximately 1 ml/min.

Lentil Lectin-Sepharose

Specificity of Interaction. Lentil lectin is a hemagglutinin from the lentil bean *Lens culinaris.* Lentil lectin-Sepharose is similar to pea lectin in its binding characteristics and interacts with high affinity with complex-type Asn-linked oligosaccharides containing a fucose in the inner core (Fig. 2).[50] However, in contrast to pea lectin-Sepharose, lentil lectin-Sepharose has a higher affinity for glycopeptides with terminal GlcNAc residues rather than terminal galactose or mannose residues. The immobilized lentil lectin will similarly discriminate between the isomers of triantennary Asn-linked oligosaccharides; only those triantennary oligosaccharides containing an inner fucose residue and one outer mannose residue substituted at position C-2 and another outer mannose substituted at both positions C-2 and C-6 by GlcNAc are bound (**29** and **30**). It will not bind the alternate triantennary glycopeptide with an outer mannose residue substituted at positions C-2 and C-4 by GlcNAc (**12**). It also will not bind tetraantennary glycopeptides.[16,50]

Lentil lectin-Sepharose is somewhat different from pea-lectin-Sepharose in its ability to interact with free oligosaccharides not attached to peptide. Lentil lectin-Sepharose will interact with high affinity with free oligosaccharides if the oligosaccharides contain both the appropriate structures described above including a residue of core fucose and a reducing terminus.[52] However, lentil lectin-Sepharose will not interact with high affinity with free oligosaccharides after reduction of their reducing termini.[52]

Source of Lentil Lectin and Preparation of Lentil Lectin-Sepharose. Lentil lectin may be isolated from dried lentil beans by the method of

Howard *et al.*[55] as modified by Hayman and Crumpton.[56] Lentil lectin-Sepharose is prepared by coupling the lectin to CNBr-activated Sepharose 4B in 0.1 M NaHCO$_3$ (pH 8.4), containing 0.1 M α-methyl-D-glucoside at 4° for 18 hr[56] according to the manufacturer's directions (Pharmacia Fine Chemicals). The final coupling density of lectin should be 10 mg/ml gel.

Two milliliters of lentil lectin-Sepharose is poured into a 0.7 × 14 cm column, and the column is equilibrated by washing with 25 ml TBS–NaN$_3$ (Tris-buffered saline containing 0.01 M Tris, pH 8.0, 0.15 M NaCl, 1 mM CaCl$_2$, 1 mM MgCl$_2$, and 0.02% NaN$_3$). The capacity of this column is similar to that of pea lectin-Sepharose at equal coupling densities. The column is stored at 4° between uses, and is equilibrated at room temperature for 1 hr before each use. The lentil lectin column is stable for at least 1 year and is regenerated by washing with 5 column volumes of TBS–NaN$_3$.

Elution of Bound Material. Chromatography is carried out at room temperature. Glycopeptides are applied to the column in 1 ml of TBS–NaN$_3$ and fractions (2 ml) are collected. Bound material is eluted from the column at a flow rate of 1 ml/min, first with 40 ml TBS–NaN$_3$, then with 40 ml TBS–NaN$_3$ containing 0.5 M α-methyl-D-mannoside.

E_4-PHA-Agarose

Specificity of Interaction. E_4-PHA is a potent erythroagglutinin and leukoagglutinin derived from the red kidney bean *Phaseolus vulgaris.* E_4-PHA-agarose interacts with high affinity with bisected biantennary Asn-linked oligosaccharides containing outer galactose residues (**38** through **41**).[30,35,57,58] If E_4-PHA is coupled to agarose at low coupling densities of lectin (less than 1 mg lectin/ml gel), E_4-PHA-agarose will only retard the elution of the bisected biantennary glycopeptides and hapten is not required (see Fig. 2).[35] However, at higher coupling densities (3 mg lectin/ml gel) the glycopeptides are bound by E_4-PHA-agarose and can be eluted with either N-acetylgalactosamine or with a low pH wash.[59] High-affinity interaction with E_4-PHA-agarose is prevented if both outer galactose residues on a bisected biantennary glycopeptide are substituted at position

[55] I. K. Howard, H. J. Sage, M. D. Stein, N. M. Young, M. A. Leon, and D. F. Dyckes, *J. Biol. Chem.* **246**, 1590 (1971).
[56] M. J. Hayman and M. J. Crumpton, *Biochem. Biophys. Res. Commun.* **47**, 923 (1972).
[57] R. Kornfeld, W. T. Gregory, and S. A. Kornfeld, this series, Vol. 28, p. 344.
[58] K. Yamashita, A. Hitoi, and A. Kobata, *J. Biol. Chem.* **258**, 14755 (1983).
[59] S. J. Mellis and J. U. Baenziger, *J. Biol. Chem.* **258**, 11546 (1983).

C-6 by sialic acid[58] or if both outer GlcNAc residues are substituted by fucose at position C-3 (15).[58,60] However, if only the galactose residue on the Galβ1,4GlcNAcβ1,2Manα1,3 branch is sialylated binding is permitted.[35,45] E$_4$-PHA-agarose can also separate structural isomers of certain bisected triantennary glycopeptides. The glycopeptides containing a bisecting GlcNAc residue and one outer mannose residue substituted at positions C-2 and C-4 by Galβ1,4GlcNAc can be bound by E$_4$-PHA-agarose (36); however, the triantennary glycopeptide containing an outer mannose residue substituted at C-2 and C-6 by Galβ1,4GlcNAc (20) cannot be bound by the immobilized lectin.[58]

Source of E$_4$-PHA and Preparation of E$_4$-PHA-Agarose. E$_4$-PHA is purified from phytohemagglutinin P (PHA-P; P-L Biochemicals) or from red kidney beans following the combined procedures of Weber[61] and Leavitt *et al.*[62] as described by Cummings and Kornfeld.[35] The purified E$_4$-PHA is stored at $-10°$ in 0.05 M KH$_2$PO$_4$ (pH 6.0). To prepare the immobilized lectin, E$_4$-PHA (7 mg) is dissolved in a final volume of 6 ml of buffer containing 0.1 M NaHCO$_3$ and 0.2 M N-acetylgalactosamine (pH 8.0). A slurry of Affi-Gel 10 (10 ml) is washed as described by the manufacturer (Bio-Rad), and the moist cake is added quickly to the solution of E$_4$-PHA. The solution is mixed gently for 18 hr at 4° and the gel is then allowed to settle. Uncoupled E$_4$-PHA is recovered in the supernatant, and the coupling efficiency is estimated by determining the amount of protein remaining in the supernatant after the coupling reaction. This is accomplished by precipitating the protein in an aliquot of the supernatant with 20 volumes of ice-cold acetone and centrifuging for 5 min at 10,000 g in a high-speed table-top centrifuge. The supernatant is discarded and protein in the pellet is determined by the method of Lowry *et al.*[63] The coupling efficiency is approximately 50% with a final coupling density of 0.6 mg lectin/ml gel. The gel is resuspended in 0.1 M NaHCO$_3$ (pH 8.0), containing 0.1 M ethanolamine to block any remaining coupling sites, and the solution is mixed gently for 2 hr at 4°. A column (0.5 $\times$ 30 cm) of E$_4$-PHA-agarose is prepared at room temperature and washed with 5 column volumes of PBS–NaN$_3$ (phosphate-buffered saline: 6.7 mM KH$_2$PO$_4$, pH 7.4, 0.15 M NaCl, and 0.02% NaN$_3$). To enhance the efficiency and degree of separation of glycopeptides it is necessary to use long columns of the immobilized lectin. This point will be illustrated below in the discussion of

---

[60] U. V. Santer, M. Glick, H. van Halbeek, and J. F. G. Vliegenthart, *Carbohydr. Res.* **120**, 197 (1983).

[61] T. H. Weber, *Scand. J. Clin. Lab. Invest., Suppl.* **11**, 1 (1969).

[62] R. D. Leavitt, R. L. Felsted, and N. R. Bachur, *J. Biol. Chem.* **252**, 2961 (1977).

[63] O. H. Lowry, N. J. Rosebrough, A. L. Farr, and R. J. Randall, *J. Biol. Chem.* **193**, 265 (1951).

L_4-PHA-agarose. The capacity of the E_4-PHA-agarose column is approximately 5–10 nmol of glycopeptide retarded per milliliter of gel. Commercial preparations of E_4-PHA coupled to agarose (E-Y Laboratories, Inc.) have also been used successfully in our laboratory. The column is routinely stored in PBS–NaN$_3$ at 4°, and is equilibrated at room temperature for 1 hr before each use and the column is washed with 5 column volumes of PBS–NaN$_3$ after use. The immobilized E_4-PHA is stable and can be used repeatedly for over a year without changes in its activity.

Elution of Bound Material. Glycopeptides are dissolved in 0.5 ml of PBS–NaN$_3$ and are applied to the column of E_4-PHA-agarose at room temperature. Fractions (1 ml) are collected at a flow rate of 10 ml/hr. If E_4-PHA is coupled to agarose at higher coupling densities (3 mg/ml, as described by Mellis and Baenziger[59]), the bound oligosaccharides can be eluted with 0.4 M N-acetylgalactosamine in PBS–NaN$_3$ or with 0.05 M glycine–HCl (pH 3.5). In the latter cases the column may be regenerated by washing with at least 10 column volumes of PBS–NaN$_3$.

L_4-PHA-Agarose

Specificity of Interaction. L_4-PHA is a leukoagglutinin from the red kidney bean *Phaseolus vulgaris* and it lacks significant erythroagglutinating activity. L_4-PHA-agarose retards the elution of those tri- and tetraantennary Asn-linked oligosaccharides containing an outer mannose residue substituted at positions C-2 and C-6 by Galβ1,4GlcNAc (e.g., **24** through **30**) (Fig. 2).[35] This specificity is consistent with the finding by Hammarström *et al.*[64] that the synthetic pentasaccharide Galβ1,4Glc NAcβ1,2(Galβ1,4GlcNAcβ1,6)Man, but not the isomer Galβ1,4Glc NAcβ1,2(Galβ1,4GlcNAcβ1,4)Man, is a potent inhibitor of L_4-PHA precipitation of glycoproteins. L_4-PHA-agarose will not retard the elution of glycopeptides lacking outer galactose residues (e.g., **3**, **4**, and **7**) regardless of the substitution pattern of their outer mannose residues.[35] L_4-PHA-agarose can discriminate between the isomers of the triantennary glycopeptides, since it does not interact significantly with glycopeptides which contain galactose residues but have an outer mannose residue substituted at positions C-2 and C-4 (**10** and **11**). Many kinds of substituents on the outer branches do not interfere with the interaction of glycopeptides with L_4-PHA-agarose. For example, glycopeptides containing poly(N-acetyllactosamine) sequences (**25** and **29**) interact well with the immobilized lectin. It is not known if the presence of bisecting GlcNAc residues, as in

[64] S. Hammarström, M.-L. Hammarström, G. Sundblad, J. Arnarp, and J. Lönngren, *Proc. Natl. Acad. Sci. U.S.A.* **79**, 1611 (1982).

17, 18, 20, and **21,** interfere with the affinity of the glycopeptides with L_4-PHA-agarose. It is possible that the presence of other substituents on the outer branches, such as fucose or sialic acid, could alter the interaction of a glycopeptide with L_4-PHA-agarose, but this has not been carefully investigated.

Source of L_4-PHA and Preparation of L_4-PHA-Agarose. L_4-PHA may be purified from either red kidney beans or phytohemagglutinin P (PHA-P; P-L Biochemicals) following the procedure of Weber.[61] The final preparation of the lectin should contain negligible hemagglutinating activity at a concentration of 1 mg/ml, thus ensuring that it is not contaminated with E-PHA. The purified L_4-PHA is dialyzed against 0.05 M KH_2PO_4 (pH 6.0) and stored at $-10°$ in this buffer.

To prepare the immobilized lectin, L_4-PHA (19 mg) is dissolved in a final volume of 6 ml of $NaHCO_3$ (pH 8.0) containing 0.2 M N-acetylgalactosamine. The lectin is coupled to Affi-Gel 10 (10 ml) using the same procedure above for preparing E_4-PHA-agarose. The coupling efficiency is approximately 60% and the lectin is coupled at a density of 1.8 mg/ml gel. The gel is packed into a column (0.5 × 40 cm) at room temperature and washed with 10 column volumes of PBS–NaN_3. Commercial preparations of L_4-PHA coupled to agarose (E-Y Laboratories, Inc.) are also available and have been used successfully in our laboratory. The column is stored in PBS–NaN_3 at 4°, and is equilibrated for 1 hr at room temperature before each use. The column is regenerated after use by washing with 5 column volumes of PBS–NaN_3. The immobilized L_4-PHA is very stable and can be used for over a year with no detectable change in affinity.[35]

Elution of Bound Material. Glycopeptides are suspended in 0.5 ml of PBS–NaN_3 and are applied to the column of L_4-PHA-agarose at room temperature. Fractions (1 ml) are collected at a flow rate of 10 ml/hr. Glycopeptides interacting with the immobilized lectin are retarded in their elution and haptenic sugars are not required for their elution under these conditions.

The size and overall dimensions of the column and the amount of immobilized lectin used are important in all separations by L_4-PHA-agarose affinity chromatography. This is illustrated in Fig. 3. A small column (0.7 × 3 cm) containing approximately 1 ml of L_4-PHA-agarose (1.8 mg lectin/ml gel) was prepared. To this column was applied an approximately equal mixture of glycopeptides **10** and **19** radiolabeled with [2-³H]mannose. As can be seen in Fig. 3A, the glycopeptides passed directly through the column without apparent fractionation. However, when more of the same gel was packed in a larger column (0.5 × 40 cm) and the same mixture reapplied, an observable separation of the two glycopeptides occurred (Fig. 3B). Glycopeptide **10** was contained in frac-

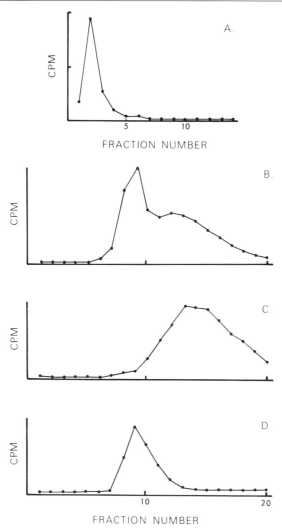

FIG. 3. Chromatography of a mixture of N-linked oligosaccharides on L$_4$-PHA-agarose and the effect of column dimensions on the degree of fractionation. (A) An approximately equal mixture of [2-^{3}H]mannose-labeled N-linked oligosaccharides having the structures of **10** and **19** was applied to a column of L$_4$-PHA-agarose with the dimensions of 0.7 × 3 cm. (B) The same mixture was applied to a column of L$_4$-PHA-agarose with the dimensions of 0.5 × 40 cm. (C) The glycopeptides in fractions 12–20 from (B) were pooled and concentrated, and the salt was removed by gel filtration. The glycopeptides were then concentrated and reapplied to the same column as in (B). (D) The glycopeptides chromatographed in (C) were reapplied to the same column in buffer containing 0.1 M GalNAc. In all figures the columns contained L$_4$-PHA coupled to agarose at a density of 1.8 mg lectin/ml gel. The eluting buffer was PBS-NaN$_3$ and the fractions were 1 ml, as described in the text.

tions 8–10 and glycopeptide **19** was retarded and contained in fractions 11–20. Glycopeptide **19** in fractions 12–20 was pooled, dried, desalted, and reapplied to the same column (Fig. 3C). Most of the glycopeptides were clearly retarded again in the elution from the column and the sample contained only a very small amount of noninteracting material representing residual glycopeptide **10**. To confirm that the separation was due to specific interaction with L_4-PHA-agarose, the same sample chromatographed in Fig. 3C was reapplied to the L_4-PHA-agarose column in the presence of 0.1 M GalNAc. In the presence of the haptenic sugar the glycopeptides were not retarded in their elution. Thus, for efficient separation of glycopeptides by L_4-PHA-agarose and any other immobilized lectin it is important to have columns of sufficient length and size.

Datura stramonium Agglutinin-Agarose

Specificity of Interaction. Seeds of the jimsonweed, *Datura stramonium,* contain a hemagglutinin and leukoagglutinin. *Datura stramonium* agglutinin (DSA)–agarose interacts with high affinity with glycopeptides containing the linear, unbranched, repeating sequence-$[Gal\beta1,4Glc$ $NAc\beta1,3]_n$, or poly(N-acetyllactosamine) sequence (Type 2 chain), e.g., **22** and **25**. The immobilized lectin will not interact with biantennary Asn-linked oligosaccharides lacking the poly(N-acetyllactosamine) sequences (**45** and **48**), but it does interact with high affinity with tri- and tetraantennary Asn-linked oligosaccharides lacking these sequences if those glycopeptides contain an outer mannose residue substituted at the C-2 and C-6 positions by N-acetyllactosamine [$Gal\beta1,4GlcNAc\beta$] (**27** and **28**).[65] This observation agrees with the studies of Crowley *et al.*,[66] who have demonstrated that the synthetic pentasaccharide $Gal\beta1,4Glc$-$NAc\beta1,2(Gal\beta1,4GlcNAc\beta1,6)Man$ is a much more potent inhibitor of hemagglutination by DSA than the isomeric pentasaccharide $Gal\beta1,4Glc$-$NAc\beta1,2(Gal\beta1,4GlcNAc\beta1,4)Man$. DSA-agarose is similar to L_4-PHA-agarose in that DSA-agarose also recognizes the branching pattern of outer mannose residues in complex-type Asn-linked oligosaccharides. It is not known whether DSA-agarose can interact with high affinity to glycopeptides which contain the branched poly(N-acetyllactosamine) sequence (I-antigen structure).

Source of DSA and Preparation of DSA-Agarose. Datura stramonium agglutinin may be purified by the method of Kilpatrick and Yeoman[67] with

[65] R. D. Cummings and S. Kornfeld, *J. Biol. Chem.* **259**, 6253 (1984).
[66] J. F. Crowley, I. J. Goldstein, J. Arnarp, and J. Lönngren, *Arch. Biochem. Biophys.* **231**, 524 (1984).
[67] D. C. Kilpatrick and M. M. Yeoman, *Biochem. J.* **175**, 1151 (1978).

an additional final step as detailed by Cummings and Kornfeld[65] and by the method of Crowley and Goldstein.[68] In our laboratory the lectin is partly purified by affinity chromatography on a column of asialofetuin-Sepharose. The partly purified lectin from this step is applied to a column of Sephacryl S-200 (3 × 60 cm) in 50 mM NaH$_2$PO$_4$ (pH 6.0), and fractions (2 ml) are collected. The fractions are assayed for protein by absorbance at 280 nm and for hemagglutinating activity.[57] Only the peak tubes of active lectin are pooled, and the material is concentrated by ultrafiltration. The purified lectin may be stored in this last buffer at −10° for at least 6 months without detectable loss of activity.

To prepare the immobilized lectin, DSA (15 mg) is suspended in a final volume of 5 ml of buffer containing 0.1 M NaHCO$_3$ and 0.1 g of a mixture of N,N'-diacetylchitobiose and N,N',N''-triacetylchitotriose (pH 8.0) and coupled to Affi-Gel 10 (5 ml), as described by the manufacturer (Bio-Rad). Approximately 75% of the lectin is coupled to the gel by this procedure with a coupling density of 1.5 mg/ml. The DSA-agarose is then washed extensively at room temperature with PBS–NaN$_3$ (phosphate-buffered saline containing 6.7 mM KH$_2$PO$_4$, pH 7.4, 0.15 M NaCl, and 0.02% NaN$_3$). Columns of DSA-agarose are prepared containing 1.0 ml of immobilized lectin in a 1-ml plastic disposable pipet plugged at the bottom with glass wool. The bed height of the gel is approximately 15 cm. (The columns are made small to decrease the overall amounts of N,N'-diacetylchitobiose and N,N',N''-triacetylchitotriose required for elution of bound glycopeptides). The column is washed with PBS–NaN$_3$ and maintained at all times in this buffer. The column is stored routinely at 4° when not in use, and is allowed to equilibrate for 1 hr at room temperature before each use. The column can be used repeatedly for several months.

Elution of Bound Material. Glycopeptides are applied to DSA-agarose in 1 ml PBS-NaN$_3$ at room temperature. The column is washed with 10 ml PBS-NaN$_3$, and then the bound glycopeptides are eluted with 5 ml PBS-NaN$_3$ containing 10 mg/ml of an approximately equal mixture of N,N'-diacetylchitobiose and N,N',N''-triacetylchitotriose. After use, the column is regenerated by washing with ten column volumes of PBS-NaN$_3$. N,N'-Diacetylchitobiose and N,N',N''-triacetylchitotriose are prepared from crustacean shells by the method of Stirling,[69] as modified by Kilpatrick and Yeoman.[67]

Glycopeptides bound by DSA-agarose and eluted by the haptenic oligosaccharides cannot be completely separated from the haptens by chromatography on Sephadex G-25-80, because small amounts of N,N',N''-

[68] J. F. Crowley and I. J. Goldstein, this series, Vol. 83, p. 368.
[69] J. L. Stirling, *FEBS Lett.* **39,** 171 (1974).

triacetylchitotriose are partly contained in fractions near the void volume of the column. Instead, eluted samples are pooled, dried, suspended in water (1 ml) and applied to a column of BioGel P-6 (2 × 50 cm), packed, and equilibrated in water. Glycopeptides elute near the void volume of this column and the haptens are clearly included.

Ricinus communis Agglutinin I-Sepharose

Specificity of Interaction. The castor bean *Ricinus communis* contains both ricin, which is not a hemagglutinin and *R. communis* agglutinin I. Immobilized *R. communis* agglutinin I (RCA I) interacts with highest affinity with bi- and triantennary Asn-linked oliogsaccharides that contain terminal galactose residues, e.g., **28, 30, 45,** and **48** (Fig. 2).[70] The lectin can also interact with derivatives of the biantennary oligosaccharides in which one or both galactose residues are sialylated. The interaction of RCA I with the sialylated derivatives is of a lower affinity than the interaction with galactose-terminating oligosaccharides.[50,70] Chromatography of glycopeptides on RCA I-Sepharose can separate biantennary Asn-linked oligosaccharides having either two terminal galactose residues (**48**) (these are retarded), one terminal galactose residue and one terminal *N*-acetylglucosamine residue (**56**) (these are less retarded), or no galactose residues (**57**) (these are not retarded) (see Fig. 2).[50] Monogalactosylated isomers of biantennary Asn-linked oligosaccharides may be separated by chromatography on the immobilized lectin. RCA I-Sepharose has a higher affinity for a glycopeptide containing a single galactose residue on the branch Galβ1,4GlcNAcβ1,2Manα3 (**56**) than for the isomer containing a single galactose on the branch Galβ1,4GlcNAcβ1,2Manα6 (**46**).[45] Narasimhan *et al.*[45] have found that the bisected, monogalactosylated isomers of biantennary Asn-linked oligosaccharides may also be separated. In this case, RCA I-Sepharose has a higher affinity for glycopeptides containing a single galactose residue on the branch Galβ1,4GlcNAcβ1,2Manα1,6 (**40**) than for the isomer containing the galactose on the Manα1,3 branch (**14**). Bisected biantennary oligosaccharides with two terminal galactoses (**38** and **39**) have also been reported to bind to[30] or to be retarded by[45] immobilized RCA I.

Source of RCA I and Preparation of RCA I-Sepharose. RCA I may be purified from castor beans by affinity chromatography on either *p*-aminophenyl-β-D-thiogalactosylsuccinimidylhydrazide polyacrylamide[70–72] or ovomucoid-Sepharose,[73] or by the method described by Olsnes.[74]

---

[70] J. U. Baenziger and D. Fiete, *J. Biol. Chem.* **254,** 9795 (1979).

[71] J. K. Inman and H. M. Dintzis, *Biochemistry* **8,** 4074 (1969).

[72] G. L. Nicolson, J. Blaustein, and M. E. Etzler, *Biochemistry* **13,** 196 (1974).

[73] W. L. Adair and S. Kornfeld, *J. Biol. Chem.* **249,** 4696 (1974).

The lectin is coupled to Sepharose-4B as described by the manufacturer (Pharmacia Fine Chemicals) at a final density of approximately 1 mg lectin/ml gel. RCA I conjugated to agarose may also be obtained from commercial sources.[45,75] The preparation of immobilized lectin is washed and equilibrated in either 0.01 M Tris–HCl (pH 7.5), 0.1 M NaCl, and 0.02% NaN$_3$ or phosphate-buffered saline (6.7 mM KH$_2$PO$_4$, pH 7.4, 0.15 M NaCl, and 0.02% NaN$_3$).

Columns (0.5 × 44 cm) of RCA I-Sepahrose are prepared as described by Narasimhan *et al.*[45] The column size may be varied[57] but it is important to use a long column in order to effect separation of the differentially retarded oligosaccharides described above and as indicated in Fig. 2.

Elution of Bound Samples. Chromatography is carried out at room temperature at a flow rate of 4.2 ml/hr and fractions (1 ml) are collected. The samples are allowed to interact with the gel for 1 hr before elution, and then the column is eluted with 90 ml of the above equilibration buffer and then 20 ml of the same buffer containing 0.1 M lactose. RCA I-agarose can be used repeatedly for 6 months with no apparent change in binding affinity.[45]

Extension of the Technique of Serial Lectin Affinity Chromatography

A number of other immobilized lectins not described in this article, including some carbohydrate-binding proteins from animal cells, can also be used for additional chromatographic separation of glycopeptides. These include immobilized wheat germ agglutinin,[73,76] pokeweed mitogen,[77,78] *Griffonia simplicifolia* A$_4$ and B$_4$ lectins,[15,79] *Vicia faba* lectin,[80] *Aleuria aurantia* lectin,[81] and many others. Additionally, the purified mannose 6-phosphate receptor has been immobilized and used to effectively separate structurally related high-mannose Asn-linked oligosaccharides differing in the amount or position of mannose 6-phosphate residues.[82,83]

---

[74] S. Olsnes, this series, Vol. 50, p. 330.
[75] P. A. Gleeson and H. Schachter, *J. Biol. Chem.* **258,** 6162 (1983).
[76] K. Yamamoto, T. Tsuji, I. Matsumoto, and T. Osawa, *Biochemistry* **20,** 5394 (1981).
[77] Y. Katagiri, K. Yamamoto, T. Tsuji, and T. Osawa, *Carbohydr. Res.* **120,** 283 (1983).
[78] T. Irimura and G. L. Nicolson, *Carbohydr. Res.* **120,** 187 (1983).
[79] D. A. Blake and I. J. Goldstein, *Anal. Biochem.* **102,** 103 (1980).
[80] Y. Katagiri, K. Yamamoto, T. Tsuji, and T. Osawa, *Carbohydr. Res.* **129,** 257 (1984).
[81] K. Yamashita, N. Kochibe, T. Ohkura, I. Ueda, and A. Kobata, *J. Biol. Chem.* **260,** 4688 (1985).
[82] A. Varki and S. Kornfeld, *J. Biol. Chem.* **258,** 2808 (1983).
[83] H. D. Fischer, K. E. Creek, and W. S. Sly, *J. Biol. Chem.* **257,** 9938 (1982).

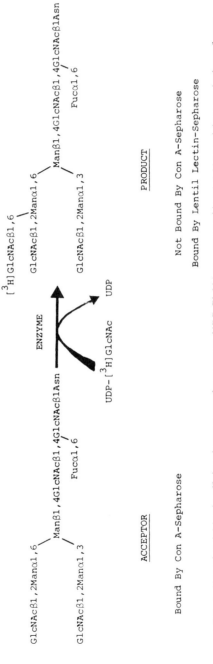

FIG. 4. Use of serial lectin affinity chromatography to assay UDP-GlcNAc: α-mannoside-β1,6-N-acetylglucosaminyltransferase.

Application of Serial Lectin Affinity Chromatography to Separate Reactants and Products in Assays of Glycosyltransferases and Glycosidases. Serial lectin affinity chromatography is obviously useful for the separation and analysis of glycopeptides in complex mixtures. However, immobilized lectins can also be utilized in the assays of many different enzymes, including glycosyltransferases and glycosidases.[43,45,51,75,84–90] An example of this approach is illustrated in Fig. 4. UDP-GlcNAc : α-mannoside-β1,6-N-acetylglucosaminyltransferase (N-acetylglucosaminyltransferase V) converts biantennary N-linked oligosaccharides to triantennary chains. This enzyme can be assayed specifically and easily by using Con A-Sepharose and lentil lectin-Sepharose to separate the reaction products.[85] The biantennary glycopeptide acceptor, containing a residue of core fucose, is bound by ConA-Sepharose, whereas the product of the reaction is not bound by this lectin, but is bound by lentil lectin-Sepharose. After incubation of a cell extract with the acceptor and UDP-[³H]GlcNAc, the entire reaction mixture is boiled and then passed over a small column of Con A-Sepharose to remove the unreacted acceptor glycopeptides. The material not bound by the lectin is passed directly over a second column containing lentil lectin-Sepharose. UDP-[³H]GlcNAc and its breakdown products do not bind to this column, but the specific reaction product does bind, as indicated in Fig. 4. The bound and radiolabeled glycopeptide can be eluted from the column directly into scintillation vials for quantifying the reaction product. Alternatively, this enzyme can be assayed by using radiolabeled glycopeptide acceptors, e.g., radioiodinated glycopeptide, and nonradiolabeled UDP-GlcNAc. The analysis is carried out in the same manner. This general method for assaying N-acetylglucosaminyltransferase V is highly specific because lentil lectin-Sepharose will not interact with other possible reaction products in the assay including structural isomers of the reaction product.

[84] S. Narasimhan, P. Stanley, and H. Schachter, *J. Biol. Chem.* **252**, 3926 (1977).

[85] R. D. Cummings, I. Trowbridge, and S. Kornfeld, *J. Biol. Chem.* **257**, 13421 (1982).

[86] M. L. Reitman, A. Varki, and S. Kornfeld, *J. Clin. Invest.* **67**, 1574 (1981).

[87] D. E. Goldberg and S. Kornfeld, *J. Biol. Chem.* **258**, 3159 (1983).

[88] M. R. Paquet, S. Narasimhan, H. Schachter, and M. A. Moscarello, *J. Biol. Chem.* **259**, 4716 (1984).

[89] D. A. Blake and I. J. Goldstein, *J. Biol. Chem.* **256**, 5387 (1981).

[90] R. D. Kilker, Jr., B. Saunier, J. S. Tkacz, and A. Herscovics, *J. Biol. Chem.* **256**, 5299 (1981).

[19] Detection of Glycosaminoglycans with ^{125}I-Labeled Cytochrome c

By Phyllis M. Sampson, Ralph Heimer, and Alfred P. Fishman

The mobility of individual glycosaminoglycans (GAG) on cellulose acetate strip electrophoresis in conjunction with selective hydrolysis by specific mucopolysaccharide lyases has been used for many years for qualitative and quantitative analyses of GAG. In this method the GAG are visualized by ortho- and metachromatic dyes.[1] The sensitivity of such dyes can be enhanced by introduction of a radiolabel as has been done with ruthenium red prepared from γ- and β-emitting [103]Ru.[2] An alternative method for visualizing GAG with a radiolabel takes advantage of the well-known ionic interactions of GAG with basic proteins. ^{125}I-Labeled cytochrome c is used to detect GAG at the 1-ng level.[3] Presumably other basic proteins may also serve this purpose.

Methods

Electrophoretic Systems. Of the many available buffer systems, three have been used.

1. pH 3.0 pyridine (0.1 M)–formic acid (0.47 M)[1]
2. pH 8.4 lithium chloride (0.05 M)–EDTA (0.01 M)[4]
3. pH 5.0 barium acetate (0.1 M)[5]

Cellulose Acetate Strips. Of the commercially available supports, two have been used.

1. Sepraphore III (Gelman)
2. Microzone (Beckman)

Preparing Cellulose Acetate Strips. The strips are first allowed to wet in distilled water and then soaked for 30 min to overnight in 30% methanol. The strips are then equilibrated in the desired buffer for 10 min with 2 changes. This technique causes no deformation of the strips.

[1] L. Rodén, J. R. Baker, J. A. Cifonelli, and M. B. Mathews, this series Vol. 28, Part B, p. 73.
[2] S. S. Carlson, *Anal. Biochem.* **122,** 364 (1982).
[3] P. Sampson, R. Heimer, and A. P. Fishman, *Anal. Biochem.* **151,** 304–308 (1985).
[4] E. H. Schuchman and R. J. Desnick, *Anal. Biochem.* **117,** 419 (1981).
[5] E. Wessler, *Anal. Biochem.* **26,** 439 (1968).

METHODS IN ENZYMOLOGY, VOL. 138

Drying Cellulose Acetate Strips. We have used GelBond, an agarose gel support obtained from BioProducts, Rockland, ME.

Autoradiography. Kodak Ortho-G X-ray film and Kodak X-Omatic Cassette with Kodak Lanex Regular Screens.

Reagent. Three hundred micrograms of horse heart Type III cytochrome *c* is dialyzed in 0.1 *M* borate buffer (pH 8.0) and then labeled in the presence of 60 μg IodoGen (Pierce) with 300 μCi Na^{125}I using the method of Fraker and Speck.[6] The protein ($\sim$0.1 μCi/μg) is separated from low molecular weight material on Sephadex G-25 PD 10 columns (Pharmacia). The preparation is diluted before each staining experiment with cold cytochrome *c* so that it contains 100 μg protein and 4×10^6 dpm in 10 ml buffer (see below), but it is recommended that the optimal amount of cold carrier be determined for different batches of cellulose acetate and ^{125}I-labeled cytochrome *c*. This can be done by dot blotting cellulose acetate strips with GAG samples.

Staining Procedure. The authors have used a Beckman Microzone Electrophoretic System which includes a 0.25-μl applicator. However, the staining procedure can be used with other cellulose acetate electrophoretic systems.

At the end of the electrophoresis, GAG are fixed by immersion of the strips in absolute ethanol–1% Tween 20 for 10 sec. The strips are then air-dried on the hydrophilic side of GelBond until they lose their translucent appearance. The GelBond support prevents deformation of the cellulose acetate during drying, but excessive drying should be avoided.

The dried strips are immersed for 5 min with gentle shaking in 10 ml 0.01 *M* ammonium formate (pH 3.0) containing 2.5% polyethylene glycol 6000, 0.05% Tween 20, and 4×10^6 dpm ^{125}I-labeled cytochrome *c* (specific activity of 0.02 μCi/μg). The strips are washed in 0.01 *M* ammonium formate–0.05% Tween 20 buffer, pH 3.0 (3 times, 5 min each) and then in 6 *M* urea dissolved in the latter buffer and then adjusted to pH 3.4. Washing with 6 *M* urea is crucial for efficient removal of background. The conditions for the urea wash may be varied from three 30-min washes to washing overnight with no appreciable differences in the relative intensity of sample to background. After further rinsing with 0.01 *M* ammonium formate buffer–0.05% Tween 20, pH 3.0, the strips are dried again on the hydrophilic side of GelBond and then covered with another sheet of GelBond for autoradiography. They are exposed to Kodak Ortho-G X-ray film in Lanex intensifying screens at $-70°$. Depending on the background, exposure times can be varied for optimal results. However, in as little as 6 hr exposure time, interpretable autoradiographs can be produced. The film is developed as directed by the manufacturer.

[6] P. J. Fraker and J. C. Speck, Jr., *Biochem. Biophys. Res. Commun.* **80**, 849 (1978).

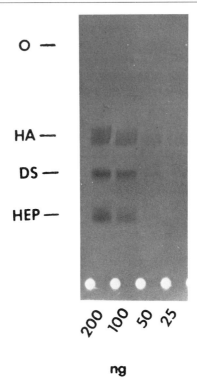

O —

HA —

DS —

HEP —

$\overset{200}{\diagdown} \quad \overset{100}{\diagdown} \quad \overset{50}{\diagdown} \quad \overset{25}{\diagdown}$

ng

FIG. 1. Electrophoretogram of hyaluronic acid (HA), dermatan sulfate (DS), and heparin (HEP), developed in 0.1 M pyridine–formate buffer, pH 3.0, for 25 min at 170 V. Strip stained with Alcian Blue. The origin is signified by 0.

A comparison of Fig. 1 with Fig. 2 illustrates the increased sensitivity that is obtained with ^{125}I-labeled cytochrome c over Alcian blue in visualizing GAG after electrophoresis in pyridine–formate buffer. Representative GAG are seen at a minimum of 50 ng using Alcian Blue (Fig. 1) while the same GAG can be seen at 0.5 ng using ^{125}I-labeled cytochrome c (Fig. 2). When lithium chloride–EDTA is used for electrophoresis, all GAG except hyaluronic acid are visible at the 1-ng level, although keratan sulfate is very faint (Fig. 3). Hyaluronic acid can be seen at the level of 2.5 ng in this system. Detection of GAG in the barium acetate system is less sensitive than using lithium chloride–EDTA. Sulfated GAG are readily seen at 5 ng but 2.5 ng seems to be the limit of detectability (Fig. 4). Hyaluronic acid is not seen at 5 ng. When using the barium acetate buffer, best results are obtained when the strip is immersed in sodium acetate-saturated absolute ethanol containing 1% Tween 20 followed by immersion in

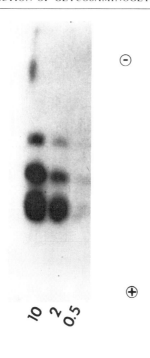

ng

FIG. 2. Electrophoretogram of HA, DS, and HEP developed as in Fig. 1, but strip stained with ^{125}I-labeled cytochrome *c* followed by autoradiography for 6 hr.

absolute ethanol–1% Tween 20 and then 70% ethanol–1% Tween 20. The strips are then dried on GelBond and the remainder of the staining procedure is the same.

Quantification. The autoradiographs can be evaluated by densitometry. The authors have obtained the results seen in Fig. 5 using an Ortec densitometer. In this experiment the pyridine–formic acid electrophoretic system (pH 3.0) was used with 1- to 20-ng amounts of GAG standards obtained from Dr. Martin B. Matthews.[1] The graphs in Fig. 5 represent plots of ng GAG vs arbitrary optical densities for four separately run sets of GAG. For each set the absorbance was obtained by calculating the areas under the densitometric curve of each GAG, using the half-height times width method. A comparison of the absorbance curves of the GAG within each set (Fig. 5A, C, and D) illustrates the different response of individual GAG to ^{125}I-labeled cytochrome *c*. This response appears to be consistent with differences in charge densities of individual GAG. Com-

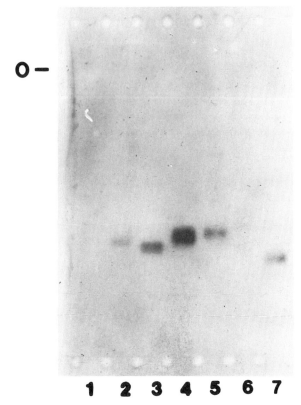

O —

1 2 3 4 5 6 7

Fig. 3. Electrophoretogram. Lane 1: HA. Lane 2: chondroitin 4-sulfate (C4S). Lane 3:
DS. Lane 4: chondroitin 6-sulfate, (C6S). Lane 5: heparan sulfate (HS). Lane 6: keratan
sulfate (KS). Lane 7: HEP. All GAG applied at 1 ng each. The origin is signified by 0. The
strip is developed in LiCl (0.05 M), EDTA (0.01 M), pH 8.4, for 15 min and stained with [125]I-
labeled cytochrome c followed by autoradiography for 12 hr.

parisons among the panels cannot be made due to variation in the time of
autoradiography. With the possible exceptions of keratan sulfate and
heparan sulfate the plots appear linear up to 5 ng and possibly 10 ng,
indicating that such nanogram ranges are optimal for quantification under
the experimental conditions.

Discussion

The interaction between cytochrome c and GAG is electrostatic[7] and
hence the potential exists for reaction with nucleic acids, another class of

---

[7] U. Lindahl and M. Höök, *Annu. Rev. Biochem.* **47**, 385 (1978).

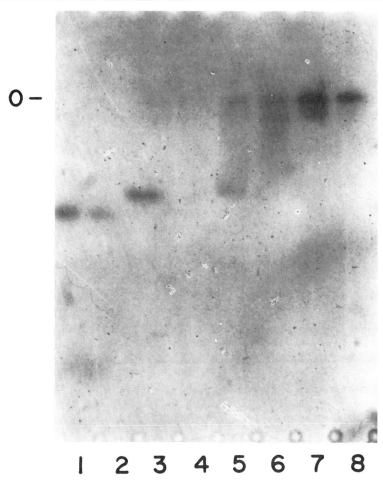

O −

I 2 3 4 5 6 7 8

FIG. 4. Electrophoretogram. Lanes 1 and 2, C6S; lanes 3 and 4, DS; lanes 5 and 6, HS; lanes 7 and 8, HEP. Samples are applied at 5 ng (first lane) and 2.5 ng (second lane) each. The origin is signified by 0. The strip is developed in barium acetate (0.1 M), pH 5.0, for a total of 2 hr at 15 mA, with replacement by a new solution after 1 hr. The strip is stained with ^{125}I-labeled cytochrome c followed by autoradiography for 8 hr.

polyanions. However, at the 5-ng level, DNA and RNA are not seen in the lithium chloride–EDTA buffer. Such amounts stain weakly in pyridine–formate buffer, but they do not interfere, as nucleic acids migrate slower than the GAG in this system. The interference of DNA and RNA in barium acetate has not be explored.

Individual GAG obtained from different sources may have differing molecular weights and charge densities, and hence give a variable re-

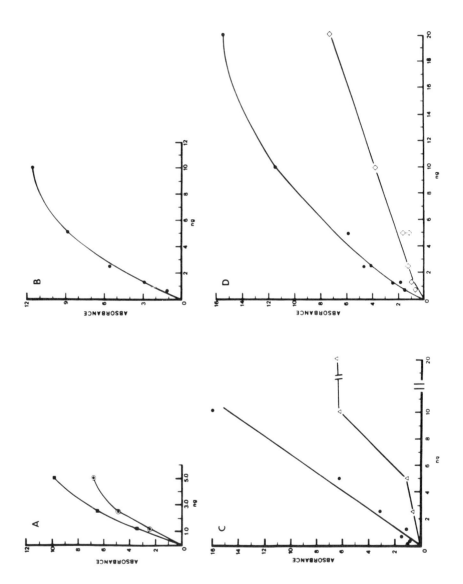

FIG. 5. Plots of concentration of GAG vs arbitrary optical density units obtained by the half-height times width method after scanning the autoradiographs in an Ortec 4310 densitometer. GAG are subjected to electrophoresis in 0.1 M pyridine–formate buffer, pH 3.0, for 25 min at 110 V in three pairings, and HA is run alone. (A) HEP, ■; DS, ⊙. (B) HA, ●. (C) C4S, ●; KS, △. (D) C6S, ●; HS, ◇.

sponse to cytochrome c. Thus when quantifying, this technique requires the same judicious use of standards as other less sensitive staining techniques.[8]

We are applying currently the electrophoretic and detection techniques described here to the analysis of GAG in plasma. Similar to other procedures for detecting plasma GAG by electrophoresis on cellulose acetate, there is interference by proteins which requires their removal by extensive proteolysis.[9] After proteolysis GAG are seen clearly by ^{125}I-labeled cytochrome c on electrophoresis of 0.5-μl aliquots of a 40-μl digest derived from 3 ml plasma. The small volume required for detection of GAG permits several electrophoretic runs for quantification and further identification using mucopolysaccharide lyases.

The assay described here has the potential of being converted to fluorescent or enzyme-linked assays, but because cytochrome c has native "peroxidase activity," coupling to an enzyme may not be necessary. The peroxidase activity of cytochrome c has been exploited in a different context in ultrastructural cytochemical techniques.[10] Other support systems such as polypropylene wells commonly used in RIA and ELISA may also have potential use in analysis of GAG by cytochrome c. Thus the interaction of GAG with cytochrome c or other basic proteins offers new opportunities for detection of GAG which are not available when GAG are visualized by ortho- and metachromatic dyes.

[8] K. Curwen and S. C. Smith, *Anal. Biochem.* **79**, 291 (1977).
[9] R. Hata, S. Ohkawa, and S. Nagai, *Biochim. Biophys. Acta* **543**, 156 (1978).
[10] M. J. Karnovsky and D. F. Rice, *J. Histochem. Cytochem.* **17**, 751 (1969).

[20] Electrophoretic Separation of Lipopolysaccharide Monomers Differing in Polysaccharide Length

By R. C. GOLDMAN and L. LEIVE

Lipopolysaccharide (LPS) is a ubiquitous component of the outer membrane of gram-negative bacteria and is important for both membrane barrier function and bacterial pathogenesis. Each LPS molecule can contain three regions: lipid A, core oligosaccharide, and O-antigen polysaccharide, and considerable variation exists among various species and strains in the precise structure of each of these regions.[1] The most variable region in the coliform bacteria *Escherichia coli* and *Salmonella* sp. is the O antigen, with more than 100 antigenic types having been described. A given O-antigen type, consisting of 3–5 sugar residues, is synthesized by assembly of its characteristic sugars on a lipid intermediate (C_{55}-polyisoprenoid), followed by polymerization of monomer units into O-antigen chains. Subsequent transfer of O antigen to core oligosaccharide, already assembled by stepwise addition of nucleotide sugars to lipid A, yields complete LPS. The following method using polyacrylamide gel electrophoresis in the presence of sodium dodecyl sulfate (SDS) allows resolution of LPS molecules based on the amount of polysaccharide attached to lipid A, and thus the number of O-antigen units attached to each lipid A–core unit. Surprisingly, current information shows that LPS isolated from a pure culture of a given O-antigen type can contain up to 50 or more sizes of molecules, which vary in the number of O-antigen units per molecule of lipid A–core oligosaccharide.[2,3]

Sample Preparation

From Whole Cells

Method A. Purification of LPS. This method enables the best subsequent resolution of complex LPS preparations. The classical phenol–water extraction uses cells suspended in H_2O at ≤60 mg dry wt/ml and extraction with an equal volume of phenol–H_2O (9 : 1, v/v).[4] Following mixing at 65–68° for 10–15 min, the sample is cooled to 10° and separated into two phases by centrifugation at 5000 *g* for 30 min. The upper aqueous phase is removed, and the lower phenol phase plus interface is reextracted with an equal volume of H_2O. When preparing LPS for gel electrophoresis,[2] the combined aqueous phases are freed of phenol by chromatography on Sephadex G-25 equilibrated with 0.05 *M* ammonium acetate (pH 8.1); LPS elutes in the void volume free of phenol (premade PD10 columns from Pharmacia are adequate for samples derived from 20–100 ml of cell culture). Subsequent treatment with pancreatic RNase, 25 μg/ml at 37° for 1 hr, is advisable to remove contaminating ribosomal and transfer RNAs which are present at this stage; this purification step is

[1] O. Luderitz, M. A. Freudenberg, C. Galanos, V. Lehmann, E. T. Rietschel, and D. H. Shaw, *Curr. Top. Membr. Transp.* **17**, 79 (1982).
[2] R. C. Goldman and L. Leive, *Eur. J. Biochem.* **107**, 145 (1980).
[3] T. Palva and H. Mäkelä, *Eur. J. Biochem.* **107**, 137 (1980).
[4] O. Westphal and K. Jann, *Methods Carbohydr. Chem.* **5**, 83 (1965).

especially important when using material labeled with phosphorus (^{33}P or ^{32}P). Chromatography on Sephadex G-25, equilibrated with 0.05 M ammonium acetate (pH 8.1), removes most of the small RNA digestion products; the column void volume (containing LPS) is lyophilized, rehydrated with 1–2 ml of H_2O, and lyophilized again. The residue is now ready for solubilization and electrophoresis (see below). If the phenol–water procedure is not appropriate for extraction of the particular LPS under study, one can use the petroleum ether–chloroform–phenol method[5] or the SDS solubilization method[6] as alternatives.

Method B. Solubilization with Partial Purification of LPS. A cell pellet from 10 ml of culture is lysed by heating at 100° in 0.125 ml of 0.06 M Tris–HCl (pH 6.8), 1.0% SDS for 5 min, and then diluted to 2.5 ml with the same Tris buffer, but lacking SDS.[7] The sample is digested with pancreatic RNase (10 μg/ml final concentration) for 16–24 hr at 37°, and subsequently treated with proteinase K (10 μg/ml final concentration) for 4 hr at 37° to remove protein. Chromatography on Sephadex G-25 equilibrated with 0.05 M ammonium acetate (pH 8.1) with lyophilization as described above yields material suitable for electrophoresis. Recoveries are usually greater than 90%.

Method C. Solubilization without Purification of LPS. An even quicker method, from which the above was derived, uses solubilization of cells obtained from agar plates and suspended to 200 Klett units (blue filter) in phosphate-buffered saline (pH 7.2).[8] The pellet from 1.5 ml of cells is dissolved in 50 μl 1 M Tris–HCl (pH 6.8) containing 2% SDS, 4% 2-mercaptoethanol, and 10% glycerol at 100° for 10 min. Proteinase K is added to 25 μg/50 μl, followed by incubation at 60° for 60 min. Samples are then applied to gels. Protease K digestion degrades most of the cellular protein, and when used with a silver stain modified to detect LPS preferentially (see below) such samples allow adequate resolution of LPS.

From Outer Membranes

Rather than purifying LPS from whole cells, isolation of outer membranes by any one of several methods[9–11] yields a preparation which is virtually RNA free, thus eliminating one of the main contaminants in preparations radiolabeled with phosphorus. LPS is solubilized for electro-

[5] C. Galanos, O. Luderitz, and O. Westphal, *Eur. J. Biochem.* **9,** 245 (1969).
[6] R. Darveau and R. Hancock, *J. Bacteriol.* **155,** 831 (1983).
[7] R. C. Goldman, K. Joiner, and L. Leive, *J. Bacteriol.* **159,** 877 (1984).
[8] P. Hitchcock and T. Brown, *J. Bacteriol.* **154,** 269 (1983).
[9] M. J. Oshorin, J. E. Gander, E. Parisi, and J. Carson, *J. Biol. Chem.* **247,** 3962 (1972).
[10] D. White, W. Lennarz, and C. Schnaitman, *J. Bacteriol.* **109,** 686 (1972).
[11] B. Kusecek, H. Wloch, A. Mercer, V. Vaisanen, G. Pluschke, T. Korhonen, and M. Achtman, *Infect. Immun.* **43,** 368 (1984).

phoresis without further purification. Phospholipid travels much more rapidly than any of the LPS monomers and can easily be disregarded. Phosphoprotein[11] and loss of resolution in areas of high concentration of outer membrane proteins, if greater than 10 μg of protein is applied to the gel, can cause some problems. However, radiolabeling to high specific activity (10^5 cpm of ^{33}P per microgram of LPS) allows so few proteins to be loaded that they do not interfere, permitting quantitative recovery of LPS, and high resolution analysis on gels.

Gel Electrophoresis. Samples are dissolved in a solution of 0.06 M Tris–HCl (pH 6.8) containing 3% (w/v) SDS, 5% (v/v) 2-mercaptoethanol, and 10% (v/v) glycerol. Samples are then heated at 100° for 5–10 min (30–60 min yields identical patterns of electrophoretic separation), allowed to cool to room temperature, and loaded onto the gel.

The best resolution is obtained using purified LPS, and electrophoresis in long (20-cm) gels using a gradient of polyacrylamide (usually 7.5 to 20% linear gradients,[2] but 7.5 to 25% exponential gradients have also been used).[11] Apparently any system based on the discontinuous buffer system given by Laemmli[12] and containing a minimum of 0.1% (w/v) sodium dodecyl sulfate (SDS) is acceptable. Urea (4 M) in combination with SDS has also been used.[11] Alternatively nongradient gels of 12.5–15% acrylamide can be used, sacrificing high resolution of LPS molecules containing the longer polymers of O antigen. The most important factors contributing to high resolution are (1) pumping the gradient at no more than 1–2 ml/min, (2) maintaining the gel polymerization time between 40 and 60 min, (3) preparing both the running and stacking gels 16–24 hr prior to use (storing at room temperature until use), and (4) using small amounts of LPS. Typically less than 1 μg of a complex mixture of radiolabeled LPS is applied per lane of 0.75-mm-thick gels. Application of more LPS leads to lack of resolution of the high molecular weight LPS species (>30 O units/lipid A–core molecule).

Gels are run at room temperature at constant current (0.3 to 1.0 mA/ml of gel). When running 20-cm gradient gels, electrophoresis can be continued for 30–60 min after the tracking dye reaches the bottom of the gel. Under these conditions, the smallest LPS species remain on the gel, and the largest species show better resolution.

Detection

Radiolabeled LPS

LPS radiolabeled with ^{14}C, ^{33}P, or ^{32}P can be detected by exposure to X-ray film, following drying of the gel with or without prior fixation using

---

[12] U. Laemmli, *Nature (London)* **227,** 680 (1970).

acetic acid–methanol–H_2O (1 : 5 : 4) for 30 min at room temperature. Most species in a complex LPS preparation (30 or more components) can be detected on X-ray film in 2–5 days exposure if 1–3 × 10^5 cpm of ^{33}P or ^{14}C-labeled LPS is applied per gel lane.[2] The use of ^{32}P-labeled LPS and Cronex Lightning-Plus intensifying screens (DuPont) should reduce the total number of cpm's required. LPS labeled with ^{3}H requires fluorography[2,13] for detection in gels.

Silver Staining

LPS can also be detected by silver staining using the basic method of Tsai and Frasch.[14] Gels are first fixed overnight in a solution of 40% ethanol, 5% acetic acid, and the fixing solution is then replaced with a solution of 40% ethanol, 5% acetic acid containing 0.7% periodic acid. After 5 min treatment, gels are washed 3 times (15 min per wash) with 500–1000 ml H_2O. Water is drained off, and replaced with a staining solution prepared as follows: (1) add 2 ml concentrated NH_4OH to 28 ml of 0.1 N NaOH; (2) add 5 ml of 20% (w/v) silver nitrate while stirring; (3) add 115 ml of H_2O. This reagent should be discarded after use, as it may become explosive if allowed to dry. The gel is agitated vigorously for 10 min, and washed 3 times (15 min each) with 500–1000 ml H_2O. The water is then replaced with developer (200 ml) consisting of 50 mg citric acid and 0.5 ml of 37% formaldehyde per liter. LPS will stain dark brown in 2–5 min and staining is terminated by washing with H_2O. The following precautions are advised: (1) use of high quality deionized or glass distilled H_2O; (2) use of detergent, acid washed, glass plates for gel electrophoresis, and similarly cleaned glass staining dishes; (3) use of rinsed vinyl gloves when handling gels; and (4) adequate coverage of the gel with each solution, and maintaining agitation at each step. As little as 5 ng of a given LPS species can be detected. Obviously more material (1–10 μg) is required if the LPS is a complex mixture. When samples contain protein, a modified silver stain is available which preferentially stains LPS.[8]

Analysis of Data

LPS from bacteria which contain O antigen, as do many strains of *E. coli* and *Salmonella typhimurium,* will separate on the basis of size following gel electrophoresis (Fig. 1). LPS from *Escherichia coli* 0111 contains one galactose residue in the core and each O-antigen unit, and phosphorus only in the lipid A–core. Thus quantitative analysis of the ratio of labeled galactose (present once in the core and once in each O

[13] W. Boner and R. Laskey, *Eur. J. Biochem.* **46,** 83 (1974).
[14] C. Tsai and C. Frasch, *Anal. Biochem.* **119,** 115 (1982).

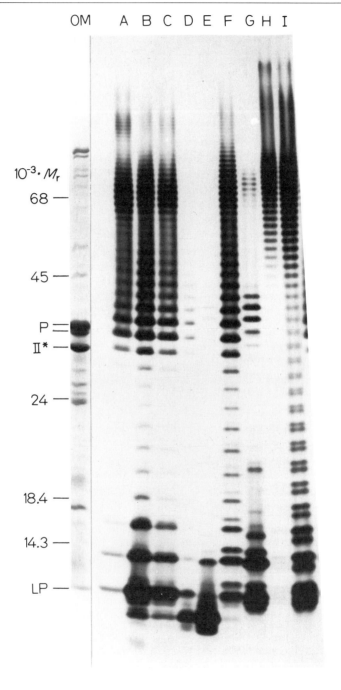

antigen unit of LPS from *E. coli* 0111) to labeled phosphate (present only in the lipid A–core region) shows that the decrease in mobility of each band reflects attachment to lipid A–core oligosaccharide of O-antigen polymers of increasing length.[2] Subjective analysis of lanes A and B of Fig. 1 (LPS labeled with galactose and phosphorus, respectively) leads to the same conclusion because phosphorus (lane B) predominately labels lower molecular weight LPS species (fewer O-antigen units/lipid A–core molecule) whereas galactose (lane A) predominantly labels higher molecular weight species (more O-antigen units/lipid A–core molecule). Some LPS preparations run as doublets (Fig. 1, lanes F and I) which stain differentially with a modified silver stain[8]; the significance of these doublets is unknown, but may reflect some additional aspect of LPS heterogeneity. One can subjectively analyze data with respect to LPS banding pattern (note the differences between lanes F and G of Fig. 1, LPS from *E. coli* O antigen groups 0111 and 086, respectively) and thus analyze clonal groups[11] or mutant phenotypes.

Quantitative analysis is best performed using LPS uniformly radiolabeled with phosphate, as a method to quantitate silver staining has not yet been described. Autoradiograms are scanned using a densitometer, and the relative contribution of each LPS species is determined. A densitometer with a small scanning aperture is required for accurate analysis of the larger LPS molecules in heterogeneous preparations. Furthermore, the entire density information of a band should be recorded for the highest accuracy, rather than a slice through the middle. Such high accuracy can only be obtained with more sophisticated densitometers such as the

Fig. 1. Banding pattern of lipopolysaccharide following electrophoresis in a 7.5–20% linear gradient of polyacrylamide. Lane OM, the relative positions of major outer membrane proteins separated in this gel system. The position of porins (P), protein II* (II*), and lipoprotein (LP) are indicated. Lane A, LPS from *E. coli* 0111 strain CL99, labeled with [³H]galactose. Lane B, as for A but labeled with [³³P]phosphate. Lane C, an equal mixture of material applied to lanes A and B. Lane D, LPS from *E. coli* 0111 strain CL99 grown in the absence of galactose and labeled with [³³P]phosphate. Lane E, as for lane D but 4-fold more sample loaded. Lane F, LPS from *E. coli* 0111, strain ATCC 29552, labeled with [³³P]phosphate. Lane G, LPS from *E. coli* 086 strain ATCC 12701, labeled with [³³P]phosphate. Lane H, LPS from *S. typhimurium* LT2, labeled with [³H]galactose. Lane I, as for lane H but labeled with [³³P]phosphate. Numbers on the left indicate the relative position of protein molecular weight markers. Total radioactivity applied was 1–3 × 10⁵ cpm per lane, except for lane D, 7 × 10⁴ cpm. Exposure time to X-ray film was 3 days. All LPS was prepared from whole cells by method A. *Escherichia coli* 0111 strain CL99 is a *galE* mutant, thus [³H]galactose is specifically incorporated into LPS as galactose. The arrow in lane D indicates the position of incomplete LPS synthesized by a *galE* mutant grown in the absence of galactose (see text).

Perkin-Elmer 1010 G microdensitometer[7,15] and data analysis by computer programs such as PIC.[7,16] However, adequate semiquantitative data can be obtained with the more standard laboratory densitometers (Ortec, Joyce Lobel, LKB, etc.).

The best method for determining precisely which band represents complete lipid A–core oligosaccharide with no O antigen is by use of mutants.[2,3] Thus the LPS synthesized by an *E. coli* mutant defective in UDPglucose 4-epimerase (EC 5.1.3.2) and grown in the absence of galactose will be truncated in the core oligosaccharide region at the point of addition of the first galactose residue. Thus the predominant species of LPS (lipid A-incomplete core) will be small in size (Fig. 1, lane D, arrow). This indicates that the next larger band [migrating opposite the lipoprotein band (LP) in lane OM] is the complete lipid A–core oligosaccharide. The next larger band (see lanes A, B, and C) would be lipid A–core oligosaccharide containing one unit of O antigen; the next higher band would contain two units of O-antigen, etc. (see Fig. 2). Mutants defective in polymerization of O-antigen units can add only one O antigen and have been used to define the band which represents lipid A–core oligosaccharide plus one O unit[3] (Fig. 2, lanes B, C, D, and E). Mutants defective in transferase cannot attach any O antigen to the lipid A–core portion of the molecule, and so define the band which represents complete lipid A–core with no O antigen (see Fig. 2, lanes F and G). The transferase mutant shown in Fig. 2, lane G, is leaky and does exhibit some transfer of O antigen to lipid A–core.

Conclusions and Recent Advances

A recent advance in methodology permits use of specific antibody to define O antigen-containing molecules. LPS is transferred to nitrocellulose filters electrophoretically analogous to Western blotting of proteins.[17–19] Such nitrocellulose sheets are amenable to antibody probing, the best resolution being attained when blots are baked at 80° for 2–4 hr prior to reaction with antibody.[19]

In conclusion, various methods are available for either rapid qualitative analysis of LPS or more detailed and precise quantitative analysis of

[15] V. Nikodem, B. Trus, and J. Rall, *Proc. Natl. Acad. Sci. U.S.A.* **78,** 4411 (1981).
[16] B. Trus and A. Stevens, *Ultramicroscopy* **6,** 383 (1981).
[17] R. Seid, D. Kopecko, J. Sadoff, H. Schneider, L. Baron, and S. Formal, *J. Biol. Chem.* **259,** 9028 (1984).
[18] M. Nurminen, E. Wahlström, M. Kleemola, M. Leinonen, P. Saikku, and H. Makela, *Infect. Immun.* **44,** 609 (1984).
[19] S. Sturm, P. Fortnagel, and K. N. Timmis, *Arch. Microbiol.* **140,** 198 (1984).

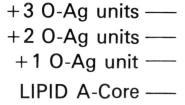

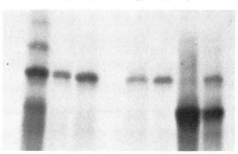

FIG. 2. Banding pattern of lipopolysaccharide from *S. typhimurium* LT2 strains with defined mutations in LPS biosynthesis. Electrophoresis was performed on a 7.5–20% linear gradient of acrylamide, and only the lower region is shown. Lane A, strain SL4917 (wild-type smooth strain). Lane B, strain SL4919 (*rfc* 985). Lane C, strain SL4920 (*rfc* 986). Lane D, strain SL901 (*rfc* 497). Lane E, strain SL1034 (*rfc* 465). Lane F, strain SL1196 (*rfb* 580). Lane G, strain SL1197 (*rfb* 604). Strains *rfc* 985, 986, 497, and 465 are polymerase mutants. Strains SL1196 and 1197 are transferase mutants which map in the *rfb* gene cluster. All bacteria were labeled with [³³P]phosphate and the LPS extracted with phenol. The dried gel was exposed to X-ray film to locate radiolabeled LPS. All LPS was prepared from whole cells by method A. O-Ag, O antigen.

complex LPS populations. These methods provide appropriate analytical techniques for studying the relationship of LPS size heterogeneity to virulence and analysis of mutants in LPS biosynthesis, as well as a general tool for detection and analysis of LPS as a constituent or contaminant in various preparations. We have recently used this method to analyze LPS synthesized by mutants which are partially defective in different stages of LPS biosynthesis. Qualitative and quantitative mathematical analysis of these data indicate that either polymerization of O-antigen units or transfer of lipid-linked polymers to lipid A–core must involve preference for O-antigen polymers of specific chain lengths, rather than complete stochastic interactions.[20]

[20] R. Goldman, F. Hunt, and L. Leive, in preparation.

Section II

Preparations

[21] Proteoglycans: Isolation and Purification from Tissue Cultures

By Masaki Yanagishita, Ronald J. Midura, and Vincent C. Hascall

Proteoglycans are macromolecules which contain a protein core with one or more covalently attached glycosaminoglycan chains. In mammalian tissues, three major classes of glycosaminoglycan are found as components of proteoglycans: chondroitin sulfate/dermatan sulfate, heparan sulfate/heparin, and keratan sulfate. The chemical properties of glycosaminoglycans and the techniques used to study them have been reviewed extensively.[1] General methods for isolating, purifying, and characterizing proteoglycans have been presented in a previous chapter of this series.[2] This chapter focuses on more recent methods for isolating and purifying proteoglycans which have proven useful for noncartilage tissues, particularly in cell culture systems.

Major developments in proteoglycan chemistry were first achieved by studying proteoglycans in cartilages, since this tissue is particularly rich in these molecules. In recent years, however, more research has been focused on the proteoglycans present in other tissues. Many of these studies involve cell cultures and metabolic labeling techniques using appropriate radioisotopic precursors. The direct application of the techniques used for studying cartilage proteoglycans to these systems is often inadequate for various reasons. For example, (1) purification using isopycnic centrifugation, one of the most powerful techniques for the cartilage proteoglycans, is not suited for proteoglycans in many tissues that have relatively low buoyant density distributions; (2) often the chemical quantities of materials are very small and losses during extraction and isolation steps occur; (3) multiple species of proteoglycans with overlapping properties are frequently present, requiring additional steps for separation of individual species; and (d) proteoglycans with strong hydrophobicity, such as cell membrane-associated proteoglycans, are not readily dissociated with the normal chaotropic 4 M guanidine–HCl extraction procedure.

This chapter describes techniques that were developed for studies of proteoglycans synthesized by rat ovarian granulosa cells in culture.[3–6]

---

[1] L. Rodén, J. R. Baker, J. A. Cifonelli, and M. B. Mathews, this series, Vol. 28 [7].
[2] V. C. Hascall and J. H. Kimura, this series, Vol. 82 [45].
[3] M. Yanagishita and V. C. Hascall, *J. Biol. Chem.* **254**, 12355 (1979).

METHODS IN ENZYMOLOGY, VOL. 138

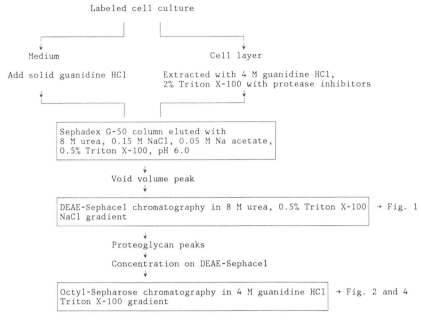

SCHEME 1.

They were designed to solve many of the technical problems associated with noncartilage proteoglycans mentioned above and are optimized for small-scale cell culture systems. However, they can be scaled up and applied to many other systems including cartilage. Scheme 1 diagrams the overall procedure.

Extraction of Proteoglycans. Detailed cell culture procedures for rat ovarian granulosa cells have been described elsewhere.[3] Typically monolayer cultures of 2–3 × 10⁶ cells are prepared in 22-mm-diameter dishes and metabolically labeled with various radioactive precursors such as [³⁵S]sulfate, [³H]glucosamine, or ³H-labeled amino acids.

After labeling, the medium is removed and treated as described below. The cell layer is extracted with 4 *M* guanidine–HCl, 0.05 *M* sodium acetate (pH 6.0), containing 2% (w/v) Triton X-100 and protease inhibitors as described in detail elsewhere.[5,7] The extraction buffer must contain deter-

[4] M. Yanagishita and V. C. Hascall, *J. Biol. Chem.* **258**, 12857 (1983).
[5] M. Yanagishita and V. C. Hascall, *J. Biol. Chem.* **259**, 10260 (1984).
[6] M. Yanagishita and V. C. Hascall, *J. Biol. Chem.* **259**, 10270 (1984).
[7] Y. Oike, K. Kimata, T. Shinomura, K. Nakazawa, and S. Suzuki, *Biochem. J.* **191**, 193 (1980).

gent to dissociate membrane-associated proteoglycans completely. Although extraction with 4 M guanidine–HCl without detergent may solubilize most of the membrane-associated proteoglycans, they remain associated with other hydrophobic macromolecules in micellar structures.[5] Such structures are prevented when sufficient amounts of detergent are included in the initial extraction solvent. Although it is possible to dissociate such structures using detergent at later purification stages, dissociation generally requires much stronger detergent conditions, and thus the process of such aggregate formation in 4 M guanidine–HCl in the absence of detergent is not entirely reversible. Other detergents which are compatible with 4 M guanidine–HCl, such as the zwitterionic detergents CHAPS[7a,8] or Zwittergent 3-12,[9] can also be used in the extraction solvent. The combination of 4 M guanidine–HCl and detergent appears to be the most effective solvent for complete dissociation, though the order of addition should be tested.[2] For the medium samples, solid guanidine–HCl is added to give an approximate concentration of 4 M.

Isolation of Radiolabeled Macromolecules by Gel Filtration. For medium and cell extracts, the sample buffer is exchanged to a solvent with 8 M urea, 0.05 M sodium acetate, 0.15 M sodium chloride (pH 6.0) containing 0.5% Triton X-100, using Sephadex G-50 (fine) chromatography. With minimal dilution of sample, this step removes unincorporated radioactive precursors and maintains the macromolecules in a denaturating solvent, but one which is compatible with the subsequent ion-exchange chromatography step. Typically, 1–1.5 ml (up to 25% of bed volume) of sample can be eluted on an 8-ml column prepared in a disposable 10-ml plastic serological pipet. Multirack plastic holders can be prepared such that 10 columns can be spaced appropriately over scintillation trays for multiple analyses. Elution is carried through the excluded volume peak, retaining the unincorporated radioisotope in the bed volume. The column can then be conveniently disposed of in radioactive waste. Proteoglycans recovered in the excluded volume of the Sephadex G-50 columns are then purified using ion-exchange chromatography (below).

Exchanging the 4 M guanidine–HCl solvent with the urea and detergent solvent by molecular sieve chromatography has several advantages over that using dialysis: (1) substantially better sample recoveries are achieved, (2) the procedure is more rapid, (3) it can easily be scaled up,

[7a] Abbreviations used: CHAPS, 3-[(3-cholamidopropyl)dimethylammonio]-1-propane sulfonate; Zwittergent 3-12, *N*-dodecyl-*N,N*-dimethyl-3-ammonio-1-propane sulfonate.

[8] S. Lohmander, J. H. Kimura, K. Kuettner, M. Yanagishita, and V. C. Hascall, *Arch. Biochem. Biophys.*, in press (1986).

[9] J. H. Kimura, E. J.-M. Thonar, V. C. Hascall, A. Reiner, and A. R. Poole, *J. Biol. Chem.* **256,** 7890 (1981).

especially in terms of number of analyses, and (4) it provides easier ways to handle radioactive waste. These features are not only convenient for ordinary isolation procedures, but are also very useful for quantitative analyses involving large number of samples such as those for pulse-chase experiments.[6]

Ion-Exchange Chromatography in Urea and Detergent. Each sample (excluded volume of Sephadex G-50 column) is applied to a column (0.7 × 4 cm) of DEAE-Sephacel which has been preequilibrated with 8 M urea, 0.15 M NaCl, 0.05 M Na acetate (pH 6.0) containing 0.5% Triton X-100. Unbound molecules are washed through the column with 2–3 volumes of this application buffer, and the column is then eluted with a NaCl gradient (0.15–1.0 M) in the same urea solvent with a total of 40–50 ml. The 0.15 M NaCl concentration in the application buffer is included so that all of the proteoglycans bind to the column whereas the bulk of other proteins do not. This NaCl concentration maximizes the washing effect of the 8 M urea solvent and minimizes carryover of other proteins into the proteoglycan fractions, which elute during the gradient at higher NaCl concentrations. Because the proteoglycans bind to the ion-exchange column, small amounts of proteoglycans in large volumes can be applied with excellent recovery. During the NaCl gradient, hyaluronic acid as well as negatively charged glycoproteins elute between 0.2–0.3 M and thus are generally well separated from proteoglycans, which elute at >0.3 M. If there is a sufficient charge difference between different proteoglycan species, then they also can be separated. As illustrated in Fig. 1 for the granulosa cell culture system, heparan sulfate proteoglycans, having an average of 0.8–0.9 sulfate groups per repeating disaccharide,[4] are separated from dermatan sulfate proteoglycans, which have 1.2–1.3 sulfate groups per repeating disaccharide.[10]

The presence of the high concentration of urea in the Sephadex G-50 void volume fractions maintains denaturing conditions while permitting the use of ion-exchange chromatography. However, when small chemical quantities of proteoglycans isolated from cell cultures are analyzed by ion-exchange chromatography, low recoveries are often encountered even in the presence of high concentrations of urea. This problem can be overcome by the use of a detergent such as 0.5% Triton X-100 or 0.5% CHAPS in the gradient buffer,[10] which probably disrupts hydrophobic interactions between the ion-exchange matrix and the macromolecules in the sample. Although most commercially available ion exchange matrices claim to have few hydrophobic sites, in practice the considerable im-

---

[10] M. Yanagishita and V. C. Hascall, *J. Biol. Chem.* **258,** 12847 (1983).

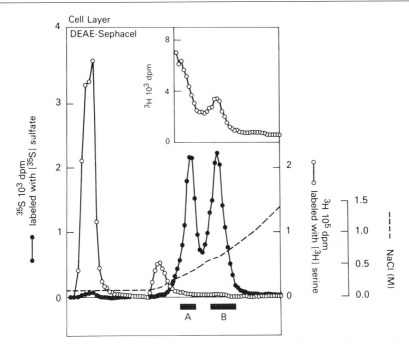

FIG. 1. DEAE-Sephacel chromatography of a rat granulosa cell extract. The cell culture was metabolically labeled with [^{35}S]sulfate and [^{3}H]serine, and the cell layer was extracted with a 4 M guanidine–HCl solvent containing 2% Triton X-100 and protease inhibitors. Labeled macromolecules were separated from unincorporated isotopes by Sephadex G-50 chromatography and analyzed by DEAE-Sephacel chromatography (see flow diagram). A column (2 ml bed volume) was eluted with a total of 46 ml of a NaCl gradient in the same solvent with a flow rate of 2.5 ml/hr at room temperature. The inset shows the profile of ^{3}H activity with an expanded scale across the two ^{35}S-labeled peaks. Peaks A and B represent heparan sulfate and dermatan sulfate proteoglycans, respectively. Data are from Yanagishita et al.[5]

provement in recovery of proteoglycans when detergents are used indicates that hydrophobic interactions can cause significant problems.[10a]

Two different detergents have been used successfully to increase the recovery of proteoglycans with DEAE-Sephacel ion-exchange chromatography: Triton X-100 (nonionic detergent) and CHAPS (zwitterionic detergent). While their effects on proteoglycan recovery is similar, they differ in their physicochemical properties, especially in their critical mi-

[10a] Similar results were observed in other procedures such as gel filtration where the presence of detergent in denaturing buffers such as 4 M guanidine–HCl significantly improves recoveries (M. Yanagishita, unpublished observation).

cellar concentrations (CMC). Triton X-100 has a relatively low CMC,[11] whereas CHAPS has a high CMC[12]; thus CHAPS is easier to remove from the sample solution by procedures such as dialysis or gel filtration. Therefore, one can choose an appropriate detergent depending on the requirements of the following procedures.

Concentration of DEAE-Sephacel Fractions and the Exchange of Detergent. After DEAE–Sephacel chromatography, fractions containing proteoglycans can be concentrated using a second DEAE-Sephacel column. First, each recovered peak is mixed with an equal volume of the urea buffer lacking NaCl so as to reduce the NaCl concentration by half. The sample is then applied onto a DEAE-Sephacel column; under these conditions, the proteoglycans bind to a relatively small volume of the ion-exchange matrix at the top of the column (depending on the chemical quantities of proteoglycans present in the sample). After all the proteoglycans are bound, they can be extracted with a small volume of 4 M guanidine–HCl with detergent. The nature of the detergent can also be changed at this step by first washing with the low salt–urea solvent containing an appropriate detergent before step elution with the guanidine–detergent solvent. An advantage of this procedure is that it avoids some potential aggregation artifacts caused by lyophilization.[12a]

Hydrophobic Chromatography of Proteoglycans. A column of octyl-Sepharose CL-4B (0.7 × 4 cm) is equilibrated with 4 M guanidine–HCl, 0.05 M sodium acetate (pH 6.0).[12b] A sample of DS-PG from the medium of granulosa cell cultures is isolated by ion-exchange chromatography similar to that illustrated in Fig. 1. This sample, free of detergent, is equilibrated in 4 M guanidine–HCl and applied to the octyl-Sepharose column. Unbound molecules are washed through the column with approximately 2–3 column volumes of the application solvent. The column is then eluted with a gradient of Triton X-100 ranging from 0 to 0.8% in the 4 M guanidine–HCl solvent in a total volume of 46 ml (Fig. 2). The 4 M guanidine–HCl solvent will maintain the core proteins in a maximally unfolded state, thereby reproducibly exposing the polypeptide backbone to the hydrophobic support and permitting interactions which might be inaccessible in the native, folded protein conformations. Approximately 15% of the ^{35}S activity does not bind to the octyl-Sepharose column; the rest binds and elutes in two major peaks at different detergent concentrations. The major peak contains 70% of the radioactivity and elutes at

[11] L. M. Hjelmeland, D. W. Nebert, and J. C. Osborn, Jr., *Anal. Biochem.* **130**, 72 (1983).

[12] A. Helenius, D. R. McCaslin, E. Fries, and C. Tanford, this series, Vol. 56 [63].

[12a] Y. Yonekura, K. Oguri, K. Nakazawa, S. Shimizu, Y. Nakanishi, and M. Okayama, *J. Biol. Chem.* **257**, 11166 (1982).

[12b] Treatment of guanadine solvent with activated charcoal removes trace impurities which adversely affect chromatography (R. J. Midura, in preparation).

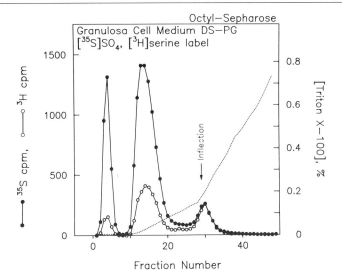

Fraction Number

FIG. 2. Octyl-Sepharose chromatography of dermatan sulfate proteoglycans in the medium fraction from a granulosa cell culture. Dermatan sulfate proteoglycans labeled with [^{35}S]sulfate and [^{3}H]serine were isolated as outlined in the flow diagram. The sample (equivalent to peak B, Fig. 1, except derived from medium compartment) was then applied onto a column of octyl-Sepharose CL-4B (bed volume 2 ml) in 4 M guanidine–HCl solvent and eluted with a Triton X-100 gradicnt (0–0.8%) in thc same solvent with a flow rate of 2.5 ml/hr at room temperature. The Triton X-100 concentration was quantitated by measuring its absorbance at 280 nm.

0.07% Triton X-100, while a small sharp peak elutes later at 0.17%. The Triton X-100 concentration was estimated by measuring UV absorbance. Initially, the concentration of Triton X-100 in the column effluent rises more slowly than expected from the concentrations generated by the gradient, indicating that some detergent binds to the hydrophobic matrix. At a concentration of 0.17%, the Triton X-100 concentration begins to rise rapidly, suggesting that binding of this detergent to available hydrophobic sites has reached saturation. It is at this inflection that all the remaining tightly bound proteoglycan molecules elute as a sharp peak, suggesting that at this point all bound material is displaced by the detergent. Recovery of proteoglycans from the octyl-Sepharose columns is quantitative.

The concentration of Triton X-100 at the inflection point depends on the concentration of guanidine–HCl in the solvent (Fig. 3).[12c] Inflection occurs at very low Triton X-100 concentrations in guanidine–HCl solvents of 0–2 M concentration and then the inflection-point concentration

[12c] Recent studies have shown that the inflection point always occurs at the CMC of Triton X-100 in the particular concentration of guanidine HCl being used (Yanagishita, M., in preparation).

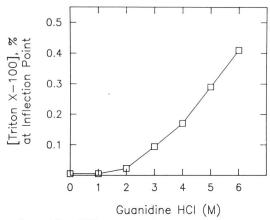

Guanidine HCl (M)

FIG. 3. Effect of guanidine–HCl concentration on the concentration of Triton X-100 which gives an inflection in the octyl-Sepharose effluent. The inflection of Triton X-100 in octyl-Sepharose effluent (see Figs. 2 and 4) was measured using the same Triton X-100 gradient formed in various concentrations of guanidine–HCl.

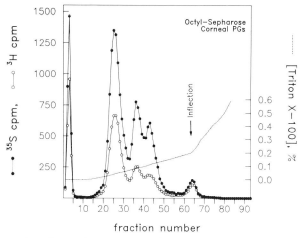

fraction number

FIG. 4. Separation of corneal stroma proteoglycans on octyl-Sepharose in the presence of 4 M guanidine–HCl. Corneal explants from day 18 embryonic chickens were metabolically labeled with [^{35}S]sulfate and [^{3}H]glucosamine. Stromal tissue was dissected from the explants and processed as described in the flow diagram for purification of proteoglycans from cell layers. A portion of the purified proteoglycans concentrated in 4 M guanidine–HCl buffer was chromatographed on octyl-Sepharose CL-4B (2 ml bed volume) in 4 M guanidine–HCl, 0.05 M Na acetate (pH 6.0). After sample application, the column was washed with an additional 2 bed volumes of the guanidine–HCl solvent followed by a gradient elution of the bound molecules using 0–0.65% Triton X-100 in the same solvent (3 ml/hr flow rate). The fraction volume was 0.7 ml, and the recovery of radioisotope was >70%.

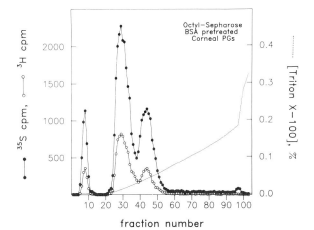

fraction number

FIG. 5. Separation of corneal stroma proteoglycans on octyl-Sepharose pretreated with bovine serum albumin (BSA) in the presence of 4 M guanidine–HCl. Corneal stroma proteoglycans were metabolically labeled and purified as described in the legend to Fig. 4. A portion of the purified proteoglycans concentrated in 4 M guanidine–HCl buffer was chromatographed on a column of octyl-Sepharose CL-4B (2 ml bed volume) that was treated with 0.2 mg of BSA in 0.2 ml of 4 M guanidine–HCl, 0.05 M sodium acetate, pH 6.0 prior to sample application. After sample application, the column was washed with an additional two bed volumes of the guanidine–HCl solvent followed by a gradient elution of the bound molecules using 0–0.5% Triton X-100 in the same solvent (4 ml/hr flow rate). The fraction volume was 0.6 ml and the recovery of radioisotope was >95%.

increases, reaching about 0.41% by 6 M guanidine–HCl. Since the best resolution of proteoglycans occurs in the gradient before the inflection point, guanidine–HCl concentrations lower than 2 M yield essentially either unbound or bound fractions without significant resolution within the column. Practically, guanidine–HCl concentrations of 4 M or greater are required for effective resolution within the gradient.

Another example of multiple proteoglycan species resolving on octyl-Sepharose is shown in Figs. 4 and 5. Shown are the octyl-Sepharose chromatograms for proteoglycans isolated from corneal stroma tissue of the embryonic chick labeled with [^{35}S]-sulfate and [^{3}H]-glucosamine. The columns were run either directly (Fig. 4) or after (Fig. 5) prior treatment with highly purified bovine serum albumin (BSA) in 4 M guanidine-HCl, 0.05 M sodium acetate, pH 6.0. Comparison of the profiles and sample recoveries in Figs. 4 and 5 indicates that the octyl-Sepharose most likely contains some high-affinity hydrophobic sites in addition to the octyl ligands. The addition of BSA to the column (0.1 mg BSA per ml of octyl-Sepharose gel) prior to sample application appears to completely block these sites while only partially reducing the number of available octyl ligands. Such pretreatment of octyl-Sepharose provides a more uniform hydrophobic matrix with higher reproducibility and sample recovery. In Fig. 5, representing the best conditions, the bound proteoglycans constitute approximately 90% of the total, and they are clearly separated into two distinct peaks within the Triton X-100 resolution area. These peaks represent a separation of dermatan sulfate (first peak) and keratan sulfate proteoglycans (second peak) of the corneal stroma.

Another detergent that has been used to elute proteoglycans from octyl-Sepharose in the presence of 4 M guanidine–HCl is CHAPS. The CHAPS concentrations required for the elution of proteoglycans were, in general, approximately one order of magnitude higher than those required by Triton X-100. The shape of the CHAPS gradient in the column effluent was smooth and did not show any inflection point as was observed in the Triton X-100 gradient. The temperature of the hydrophobic column, between 10 and 30°, only slightly affected the chromatographic profiles of proteoglycans.

Most isolation procedures used to characterize proteoglycans depend on the physicochemical properties of the glycosaminoglycan chains. For example, (1) equilibrium density gradient centrifugation resolves on the basis of buoyant density differences between glycosaminoglycan and protein; (2) ion-exchange chromatography takes advantage of high anionic charge densities of glycosaminoglycans; (3) various precipitation procedures using alcohol, cationic detergents,[13] etc. rely on differential solubility or selective cation binding; and (4) gel filtration and velocity gradient centrifugation utilize their hydrodynamic properties. However, many tissues contain multiple proteoglycan species with overlapping distributions of these properties which make it difficult to separate them by these conventional techniques.[14] In some cases,[15] isolation of specific proteoglycans was accomplished by differential degradation of one population of proteoglycans using specific enzymes, for example, purification of keratan sulfate proteoglycans after degrading chondroitin sulfate proteoglycan species with chondroitinase. However, such procedures risk the possibility of also degrading the desired proteoglycans by the frequent presence of contaminating proteases. Hydrophobic chromatography using a gradient elution with detergent can be used to resolve proteoglycans further by taking advantage of potential differences in hydrophobicity of their core proteins.[15a] Thus, it is a procedure which does not rely on

[13] J. E. Scott, *Methods Carbohydr. Chem.* **5**, 38 (1965).
[14] J. R. Hassell, D. A. Newsome, and V. C. Hascall, *J. Biol. Chem.* **254**, 12346 (1979).
[15] K. Nakazawa, D. A. Newsome, B. Nilsson, V. C. Hascall, and J. R. Hassell, *J. Biol. Chem.* **258**, 6051 (1983).
[15a] For example, the core protein preparations from chondroitinase digested dermatan sul-

properties of the glycosaminoglycan chains that do not interact with the column matrix. In other investigations, the binding of proteoglycans to hydrophobic matrices has been considered to be an indication that they are integral membrane components.[16-18] However, as indicated above, this is not necessarily the case. In appropriate solute conditions, many different species of proteoglycans can bind to hydrophobic matrices.

The hydrophobic column chromatography procedure described above is still under development and the future refinements of the conditions could make it even more useful.

fate proteoglycan and keratanase digested keratan sulfate proteoglycan of cornea elute on the octyl-Sepharose identically with the respective intact proteoglycan species, indicating that the glycosaminoglycan chains have little influence on the binding (Midura, R. J., in preparation).

[16] L. Kjellén, I. Petterson, and M. Höök, *Proc. Natl. Acad. Sci. U.S.A.* **78,** 5371 (1981).
[17] B. Norling, B. Glimelius, and Å. Wasteson, *Biochem. Biophys. Res. Commun.* **103,** 1265 (1981).
[18] G. A. Maresh, E. A. G. Chernoff, and L. A. Culp, *Arch. Biochem. Biophys.* **233,** 428 (1984).

[22] Isolation of Sialyl Oligosaccharides and Sialyl Oligosaccharide Phosphates from Bovine Colostrum and Human Urine

By JAAKKO PARKKINEN and JUKKA FINNE

Bovine colostrum and human urine are two readily available sources for the isolation of sialyl oligosaccharides and sialyl oligosaccharide phosphates. The major sialyl oligosaccharides in these two sources are tri- and tetrasaccharides,[1-4] whereas human milk and the urine of sialidosis patients contain sialyl oligosaccharides of larger molecular size.[5,6] In human pregnancy urine, the excretion of some sialyl oligosaccharides is specifi-

[1] R. W. Veh, J.-C. Michalski, A. P. Corfield, M. Sander-Wewer, D. Gries, and R. Schauer, *J. Chromatogr.* **212,** 313 (1981).
[2] J. Parkkinen and J. Finne, *Eur. J. Biochem.* **136,** 355 (1983).
[3] J. Parkkinen and J. Finne, *Eur. J. Biochem.* **140,** 427 (1984).
[4] J. Parkkinen and J. Finne, *J. Biol. Chem.* **260,** 10971–10975 (1985).
[5] D. F. Smith, D. A. Zopf, and V. Ginsburg, this series, Vol. 50, p. 221.
[6] L. Dorland, J. Haverkamp, J. F. G. Vliegenthart, G. Strecker, J.-C. Michalski, B. Fournet, G. Spik, and J. Montreuil, *Eur. J. Biochem.* **87,** 323 (1978).

TABLE I

SIALYL OLIGOSACCHARIDES OF BOVINE COLOSTRUM AND HUMAN URINE

		Approximate yield	
Abbreviation	Structure	Colostrum (μmol/liter)	Urine (μmol/48 hr)
3'SL	NeuAcα2-3Galβ1-4Glc	150	20
3'SL(NeuGc)	NeuGcα2-3Galβ1-4Glc	2	
6'SL	NeuAcα2-6Galβ1-4Glc	30	3
3'SLN	NeuAcα2-3Galβ1-4GlcNAc		8
6'SLN	NeuAcα2-6Galβ1-4GlcNAc	70	4
SGL	NeuAcα2-3GalI-3GalI-4Glc[a]	3	
SGI$_1$	NeuAcα2-3Galβ1-0-*myo*-inositol[a]		0.5
SGI$_{II}$	NeuAcα2-3Galβ1-0-*scyllo*-inositol		1
SGI$_{III}$	NeuAcα2-3Galβ1-0-*myo*-inositol[a]		1
DSL	NeuAcα2-8NeuAcα2-3Galβ1-4Glc	30	

	NeuAcα2
	        |
	        6		

DSGGN	NeuAcα2-3Galβ1-3GalNAc		3

	NeuAcα2
	        |
	        6		

DSGGN-Ser	NeuAcα2-3Galβ1-3GalNAcα1-0Ser		0.5
3'SLN-1-P	NeuAcα2-3Galβ1-4GlcNAcα-P		1
6'SLN-1-P	NeuAcα2-6Galβ1-4GlcNAcα-P	3	1
SGGN-1-P	NeuAcα2-3Galβ1-3GalNAcα-P		0.5

	P
	                |
	                6		

6'SLN-6-P	NeuAcα2-6Galβ1-4GlcNAc	1	

[a] Complete structure not reported.

cally stimulated.[7] If a large collection of model sialyl oligosaccharides is needed, several sources have to be used, since each contains a limited variety of oligosaccharides. The oligosaccharides present in bovine colostrum and human urine complete each other so that structures similar to most of the common sialyl oligosaccharides termini of glycoproteins and glycolipids can be isolated. In the following we describe a procedure for the isolation of sialyl oligosaccharides and sialyl oligosaccharide phosphates from these two sources. The oligosaccharides and their approximate yields are listed in Table I.

[7] M. Lemonnier and R. Bourillon, *Carbohydr. Res.* **51,** 99 (1976).

Preparation of the Crude Sialyl Oligosaccharide Fractions

Bovine colostrum is collected during the first day after delivery and stored frozen. One liter of thawed colostrum is shaken with 4 liters of chloroform–methanol (2 : 1, v/v) in the cold.[8] After centrifugation (1000 g, 1 hr) at 4°, the upper phase is collected, concentrated by rotary evaporation to 60 ml, and subjected to gel filtration in two aliquots as described below.

Human urine is collected from healthy individuals not subjected to dietary restrictions, phenylmercuric nitrate is added to prevent bacterial growth (30 ml of saturated solution/liter urine), and the urine is cooled to 4°, or frozen if stored longer than 2 days. A sample of pooled urine corresponding to the excretion of 2 days is filtered, 300 g of activated charcoal (Darco 60) is added, and the mixture is shaken for 4 hr at 23°. The charcoal is washed with 2 liters of distilled water on a Büchner funnel, and shaken with 800 ml of 80% (v/v) ethanol for 4 hr. The ethanol eluate is collected on the Büchner funnel, and the charcoal is eluted with an additional 1000 ml of 50% (v/v) ethanol.[9] The combined ethanol eluates are evaporated by rotary evaporation, and the residue is dissolved in 30 ml of 10 mM pyridine–acetic acid buffer (pH 5.0).[10]

The sample prepared from colostrum or urine (30 ml) is centrifuged (15000 g, 30 min) at 23° to remove insoluble material, and the supernatant is applied to a column (4 × 56 cm) of Sephadex G-25 Fine eluted with 10 mM pyridine–acetic acid buffer (pH 5.0) at 23°. The sialyl oligosaccharide-containing fractions are pooled as shown in Fig. 1 and concentrated by rotary evaporation. Care should be taken to exclude the fractions containing salt, which would interfere with the subsequent ion-exchange chromatography step. For an improved yield of the colostrum sialyl oligosaccharides the fractions immediately before and after the fractions indicated should be pooled and rerun on the gel filtration column.

Group Separation of the Sialyl Oligosaccharides According to Charge[10,11]

The crude oligosaccharide fraction of colostrum or urine (about 500 and 180 μmol sialic acid, respectively) is dissolved in 15 ml of 10 mM pyridine–acetic acid buffer (pH 5.0), and applied to a column (1.4 ×

[8] R. Öhman and O. Hygstedt, *Anal. Biochem.* **23**, 391 (1968).
[9] J. K. Huttunen, *Ann. Med. Exp. Biol. Fenn.* **44**, Suppl. 12, 1 (1966).
[10] J. Finne and T. Krusius, this series, Vol. 83, p. 269.
[11] J. Finne, *Biochim. Biophys. Acta* **412**, 317 (1975).

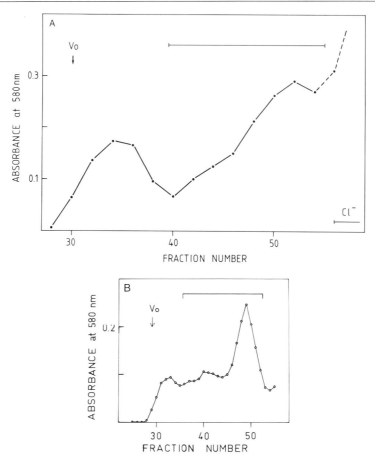

Fig. 1. Isolation of the crude sialyl oligosaccharide fractions from bovine colostrum (A) and human urine (B) by gel filtration on Sephadex G-25. Fractions of 9.2 ml (colostrum) and 9.5 ml (urine) were collected and analyzed for sialic acid (A_{580}). The oligosaccharide fractions are obtained by pooling the fractions indicated by the bars. Chloride is detected by precipitation with AgNO$_3$.

15 cm) of DEAE-Sephadex A-25 (acetate form) equilibrated with 50 mM pyridine–acetic acid buffer (pH 5.0). The column is rinsed with 10 ml of the 50 mM buffer and eluted with a linear gradient created by mixing 205 ml of 50 mM and 200 ml 1250 mM pyridine–acetic acid buffer (pH 5.0) at 23°. The monosialyl (I), disialyl (II), and sialyl oligosaccharide phosphate (III) fractions are obtained by pooling the fractions indicated in Fig. 2. The pooled fractions are concentrated by rotary evaporation below 35°, and to ensure the removal of the more slowly evaporating acetic acid,

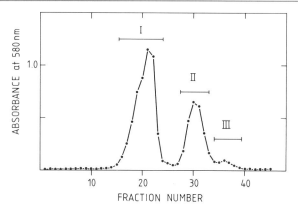

Fig. 2. Group separation of the sialyl oligosaccharides from bovine colostrum on DEAE-Sephadex. Fractions of 5.5 ml were collected. The monosialyl (I), disialyl (II), and sialyl oligosaccharide phosphate (III) fractions are obtained by pooling the fractions indicated. A similar separation is obtained for the sialyl oligosaccharides from human urine.

reevaporated a few times with distilled water. Each time, 1 ml of 1 M pyridine is added to raise the pH of the concentrated solution.

The final purification of the individual sialyl oligosaccharides is achieved by ion-exchange chromatography[12,13] on AG 1-X2 with isomolar elution. For the isolation of the urinary sialyl oligosaccharides, a preceding paper chromatography step of the fractions I, II, and III is needed. The purification of the sialyl oligosaccharides is monitored by thin-layer chromatography.[1-4]

Isolation of Monosialyl Oligosaccharides

The DEAE-Sephadex fraction I of urine is dissolved in 0.1 M pyridine (5–10 μmol sialic acid/100 μl), applied to Whatman 3 MM paper as a streak (20 μl/cm), and subjected to descending paper chromatography with the solvent ethyl acetate–pyridine–acetic acid–water (5 : 5 : 1 : 3, by volume). After 2 days, the paper is dried, and a 1-cm guide strip is cut off from the margin and stained with alkaline silver nitrate.[14] It should be noticed that not all of the components revealed (Fig. 3) correspond to sialyl oligosaccharides and, on the other hand, some major sialyl oligosaccharides are not very well visualized. The paper is cut as shown in Fig. 3, and the oligosaccharides are eluted from the unstained paper strips with

[12] M. L. Schneir and M. E. Rafelson, Jr., *Biochim. Biophys. Acta* **130**, 1 (1966).
[13] M. Koseki and K. Tsurumi, *J. Biochem. (Tokyo)* **82**, 1785 (1977).
[14] E. F. L. J. Anet and T. M. Reynolds, *Nature (London)* **174**, 930 (1956).

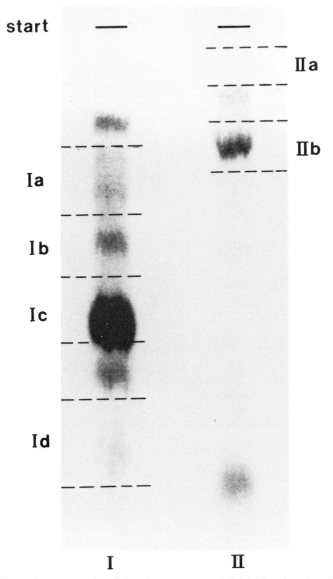

FIG. 3. Paper chromatography of the urinary mono- and disialyl fractions (I and II). The paper was developed for 2 days and stained with alkaline silver nitrate. The oligosaccharides in the zones indicated are further purified as shown in Figs. 5, 6, and 8.

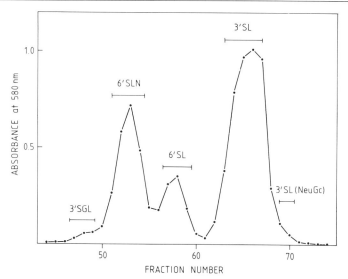

FIG. 4. Ion-exchange chromatography of the monosialyl fraction (I) of colostrum on AG 1-X2. Fractions of 9.2 ml were collected. The individual sialyl oligosaccharides are obtained by pooling the fractions indicated.

0.1 M pyridine–acetic buffer (pH 5.0),[5] and concentrated by rotary evaporation as described above, or lyophilized.

The DEAE-Sephadex fraction I of colostrum and the urinary fractions Ib, Ic, and Id are dissolved in 2 ml of 50 mM pyridine–acetic acid buffer (pH 5.0) and each fraction is run on a column (1.5 × 100 cm) of AG 1-X2 (200–400 mesh, acetate form) equilibrated, and eluted with 0.1 M pyridine–acetic acid buffer (pH 5.0) at 23°. The major monosialyl oligosaccharides obtained from bovine colostrum are the α2-3 and α2-6 isomers of sialyllactose (3'SL and 6'SL) and the α2-6 isomer of sialyl-N-acetyllactosamine (6'SLN) (Fig. 4). In addition, sialylgalactosyllactose (SGL) and a sialyllactose containing N-glycolylneuraminic acid, 3'SL(NeuGc), are obtained from colostrum (Fig. 4). 3'SL, 6'SL, and 6'SLN are also obtained from human urine (Fig. 5), although the amounts recovered are smaller than those obtained from colostrum (Table I). In contrast, the α2-3 isomer of sialyl-N-acetyllactosamine (3'SLN) is obtained only from urine. NeuAcα2-3Galβ1-3GalNAc migrates in paper chromatography between fractions Ic and Id but is not obtained in pure form by chromatography on AG 1-X2. This oligosaccharide can, however, be obtained by phosphatase treatment of the corresponding sialyl oligosaccharide phosphate (see below).

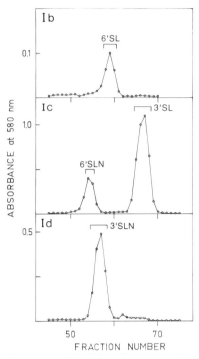

FIG. 5. Ion-exchange chromatography of the urinary paper chromatography fractions Ib, Ic, and Id on AG1-X2. Fractions of 9.2 ml were collected.

The paper chromatography function Ia of urine consists of three inositol-containing sialyloligosaccharides, which are not separated from each other on AG 1-X2. These sialylgalactosylinositols (SGI_{I-III}) are obtained in pure form by high-pressure liquid chromatography on an amino-bonded silica column (Fig. 6).[15,16] It is advisable to remove paper debris and other impurities from fraction Ia by chromatography on DEAE-Sephadex or AG 1-X2 before high-pressure liquid chromatography.

Isolation of Disialyl Oligosaccharides

The urinary DEAE-Sephadex fraction II is dissolved in 0.1 M pyridine and subjected to descending paper chromatography as described above for fraction I. The development is continued for 4 days, and fractions IIa and IIb are localized on the guide strip stained with alkaline silver nitrate

[15] J. Parkkinen, FEBS Lett. 163, 10 (1983).
[16] M. L. E. Bergh, P. Koppen, and D. H. van den Eijnden, Carbohydr. Res. 94, 225 (1981).

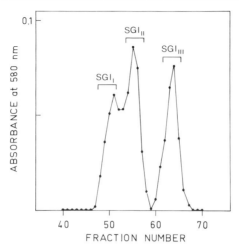

FIG. 6. High-pressure liquid chromatography of the urinary paper chromatography fraction Ia on NH$_2$-μBondapak (Waters Associates). Mobile phase: acetonitrile–15 mM potassium phosphate buffer (pH 5.2) (76:24, v/v). Flow rate 2 ml/min. Fractions of 1 ml were collected.

as shown in Fig. 3. The sialyl compounds are eluted from the unstained paper strips and concentrated by rotary evaporation as described above.

The DEAE-Sephadex fraction of colostrum and the urinary paper chromatography fractions IIa and IIb are dissolved in 1 ml of 200 mM pyridine–acetic acid buffer (pH 5.0), and applied to a column (1 × 50 m) of AG 1-X2 (200-400 mesh, acetate form) equilibrated and eluted with 0.4 M pyridine–acetic acid buffer (pH 5.0) at 23°. The major compound of the colostrum fraction is disialosyllactose (DSL), which is obtained in pure form as shown in Fig. 7. From the urinary fraction IIb, disialylgalactosyl-N-acetylgalactosamine (DSGGN) is obtained in pure form, and the same oligosaccharide chain bound to a serine residue (DSGGN-Ser) can be isolated from fraction IIa (Fig. 8).

Isolation of Sialyl Oligosaccharide Phosphates

The urinary DEAE-Sephadex fraction III is dissolved in 0.1 M pyridine and subjected to descending paper chromatography as described above for fraction I. The development is continued for 4 days. The major sialic acid-containing compounds are obtained between $R_{3'SL}$ values 0.19–0.33 (fraction IIIa) and 0.33–0.48 (fraction IIIb). These fractions are also weakly stained by alkaline silver nitrate. The fractions are eluted from the unstained paper strips and evaporated to dryness as described above.

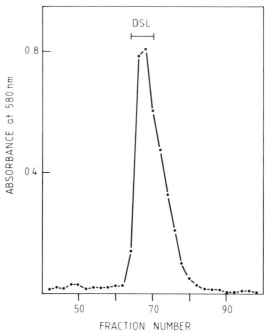

FIG. 7. Ion-exchange chromatography of the disialyl fraction (II) of colostrum on AG 1-X2. Fractions of 2.2 ml were collected.

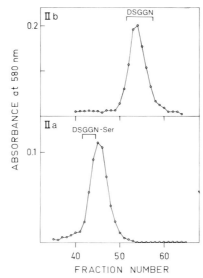

FIG. 8. Ion-exchange chromatography of the urinary paper chromatography fractions IIa and IIb on AG 1-X2. Fractions of 2.6 ml were collected.

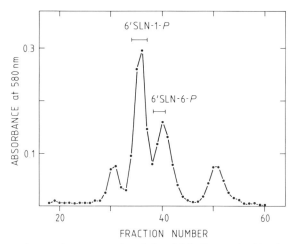

FIG. 9. Ion-exchange chromatography of the sialyl oligosaccharide phosphate fraction (III) of colostrum on AG 1-X2. Fractions of 2.5 ml were collected. 6'SLN-1-P and 6'SLN-6-P are obtained in pure form after rechromatography of the indicated fractions on AG 1-X2. The first peak eluted is due to residual DSL from DEAE-Sephadex chromatogrphy (Fig. 2).

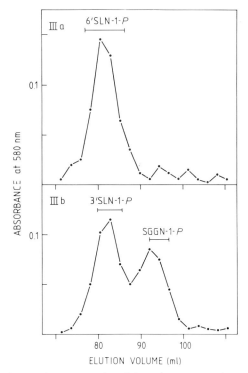

FIG. 10. Ion-exchange chromatography of the urinary paper chromatography fractions IIIa and IIIb on AG 1-X2.

The DEAE-Sephadex fraction III of colostrum and the urinary paper chromatopraphy fractions IIIa and IIIb are dissolved in 1 ml of 0.4 M pyridine–acetic acid buffer (pH 5.0) and applied to a column (1 × 50 cm) of AG 1-X2 (200-400 mesh, acetate form), equilibrated and eluted with 0.7 M pyridine–acetic acid buffer (pH 5.0) at 23°. From colostrum, two phosphorylated derivatives of sialyl-N-acetyllactosamine can be isolated, as shown in Fig. 9. One of them contains phosphate at the C-1 position (6'SLN-1-P) and the other at the C-6 position of N-acetylglucosamine (6'SLN-6-P). Three sialyl oligosaccharide 1-phosphates are obtained in pure form from the urinary fractions IIIa and IIIb as shown in Fig. 10. They are the α2-6 and α2-3 isomers of sialyl-N-acetyllactosamine 1-phosphate (6'SLN-1-P and 3'SLN-1-P) and sialylgalactosyl-N-acetylgalactosamine 1-phosphate (SGGN-1-P).

[23] Isolation and Purification of Gangliosides from Plasma

By STEPHAN LADISCH and BAIBA GILLARD

While highest concentrations of gangliosides are found in brain tissue, gangliosides of extraneural tissues and gangliosides found in the circulation have been the subject of increasing attention for several reasons. The high degree of structural diversity of this class of lipids and alterations in the cellular ganglioside complement associated with differentiation and with the neoplastic state[1] have suggested that particular gangliosides may be markers for certain tumors. The recent direct demonstrations of substantial shedding of gangliosides from the membranes of tumor cells[2,3] imply, furthermore, that certain tumor-associated gangliosides, shed into the plasma, may be useful *circulating* tumor markers.

The difficulty in the practical application of these concepts lies in obtaining highly resolved ganglioside thin-layer chromatographic patterns from small samples of plasma. Such clear patterns are necessary if one is to detect diagnostically valuable quantitative or qualitative changes in circulating gangliosides associated with disease states. It is the low concentration of circulating gangliosides (i.e., on the order of 10 nmol lipid-bound sialic acid per milliliter human plasma) that creates this difficulty.

[1] S. Hakomori and R. Kannagi, *JNCI, J. Natl. Cancer Inst.* **71**, 231 (1983).
[2] S. Ladisch, B. Gillard, C. Wong, and L. Ulsh, *Cancer Res.* **43**, 3808 (1983).
[3] G. I. Shapushnikova, N. V. Prokazova, G. A. Buznikov, N. D. Znezdina, N. A. Teplitz, and L. D. Bergelson, *Eur. J. Biochem.* **140**, 567 (1984).

The degree of purification necessary to resolve individual plasma gangliosides by thin-layer chromatography has, using available methodology, required larger volumes of plasma and sometimes is associated with significant losses during purification.

The method described here was originally developed to address the specific problem of the purification of gangliosides from small volumes of biological fluids in which the ganglioside concentration is low.[4] As described below, the procedure results in highly resolved ganglioside thin-layer chromatographic patterns using small (1.0-ml) samples of human plasma.

Principle

The overall procedure consists of three main steps. Each step in itself yields nearly quantitative recovery of gangliosides, resulting in high overall recovery when small plasma samples are processed. The first step is total lipid extraction of the plasma sample. This step markedly reduces the high plasma concentrations of protein and salt relative to gangliosides. Next, gangliosides are separated from most other lipids by partitioning the dried total lipid extract in diisopropyl ether–1-butanol–aqueous NaCl (6 : 4 : 5). Finally, the lower aqueous phase which results from this partition and which contains the gangliosides is purified by gel filtration[5] to remove low molecular weight contaminants, both inorganic (salts) and organic (polypeptides). The gangliosides then can be quantitated and qualitatively analyzed by thin-layer chromatography.

Materials and Reagents

  Chloroform and methanol, Mallinckrodt, Nanograde
  Diisopropyl ether, stabilizer-free (Aldrich, No. 18530-2)
  1-Butanol, reagent grade
  Deionized distilled water
  Sephadex G-50-150; Pharmacia
  Corning No. 8142 conical screw cap centrifuge tubes (15 and 50 ml), with Teflon-lined caps
  Silica Gel 60 high-performance TLC plates (Merck, Darmstadt, W. Germany), 10 × 10 cm.

Procedure

Total Lipid Extraction. A 1.0-ml sample of human plasma (or serum) is lyophilized in a 50-ml round-bottom glass centrifuge tube. The solid mate-

[4] S. Ladisch and B. Gillard, *Anal. Biochem.* **146,** 220 (1985).
[5] K. Ueno, S. Ando, and R. K. Yu, *J. Lipid Res.* **19,** 863 (1978).

rial is then pulverized using a glass rod and resuspended in 10 ml chloro-form–methanol (1 : 1) with bath sonication (Branson bath sonicator) to aid in dispersing the solid material. The sample is extracted for 18 hr at 4° with magnetic stirring. The tube is then centrifuged (all centrifugations are performed at 750 g) and the supernatant transferred to a 50-ml Corning No. 8142 centrifuge tube. The residue is reextracted for 4 hr with 10 ml fresh chloroform–methanol (1 : 1). The second extract is likewise clarified by centrifugation, and the two extracts combined. The combined extract is reduced to about one-fourth the original volume (i.e., 5 ml) by evapora-tion under a stream of nitrogen, and then cooled overnight at −20°. This results in further precipitation of salts and other molecules (e.g., glyco-proteins) marginally soluble in chloroform–methanol, without loss of gan-gliosides. After centrifugation, the clear supernatant, which contains the gangliosides, is transferred to a 15-ml No. 8142 tube in which the partition will be performed, is completely dried under a stream of nitrogen, and finally is subjected to oil pump vacuum to remove residual traces of the solvents.

Diisopropyl ether–1-Butanol–Aqueous NaCl Partition. Two milliliters of diisopropyl ether–1-butanol (6 : 4) is added to the dried total lipid ex-tract. The sample is vortexed and sonicated in a Branson bath sonicator to achieve fine suspension of the lipid extract. Then, 1.0 ml distilled deion-ized water[6] is added, and the tube alternately vortexed and sonicated for 2 min, and then centrifuged. The upper organic phase is carefully removed using a Pasteur pipet, and the lower aqueous (ganglioside-containing) phase repartitioned with 2.0 ml fresh diisopropyl ether–1-butanol (6 : 4). The final aqueous phase is lyophilized to remove residual organic solvents and to concentrate the sample prior to gel filtration, to increase the resolu-tion of this step.

Gel Filtration. The lyophilized sample is redissolved with bath sonica-tion in 300 μl distilled water and loaded onto the column containing Sephadex G-50-150 equilibrated in distilled water. Specifications of the column used when the gangliosides of 1 ml plasma are processed are the following: inside diameter, 10 mm; height, 10 cm; flow rate, 0.3 ml/min. (This same column can accommodate up to 2 ml plasma samples; for larger samples, larger columns would be used, as described.[4,5]) The sam-

---

[6] For the partitioning of plasma, distilled water replaces 50 mM NaCl[4] as the aqueous component of the solvent system. This is because NaCl is slightly soluble (~0.7 mg/ml) in chloroform–methanol (1 : 1), and therefore the total lipid extract of 1 ml plasma prepared as described here already contains approximately 3.5 mg salt (5 ml chloroform–methanol × 0.7 mg NaCl/ml). When redissolved in the partition step, this yields a salt concentration of 50 mM. The addition of 50 mM NaCl in the partition step would raise the total salt concentration to >100 mM, which in turn would cause increasing partition of nonpolar gangliosides, e.g., G_{M3} into the organic phase.[4]

ple is eluted with distilled water, and the eluate monitored using an LKB Uvicord UV detection monitor, at 206 nm. The gangliosides elute in the void volume peak which is collected in a 15-ml Corning No. 8142 centrifuge tube, and then lyophilized in this tube, loosely capped with its Teflon-lined cap.[7] The gangliosides are redissolved in a small volume (300–400 μl) of chloroform–methanol (1:1) and any traces of insoluble material removed by centrifugation.

Thin-Layer Chromatography. Standard thin-layer chromatographic methods are used. The solvent system we use for resolving human plasma gangliosides is chloroform–methanol–0.25% aqueous $CaCl_2 \cdot 2H_2O$ (60:40:9). Total gangliosides isolated from 1 ml plasma are spotted in a 1-cm lane, 1.5 cm from the bottom of the 10 × 10 or 10 × 20 cm TLC plate. Plates are preactivated by heating at 90° for 45 min. After development the plate is sprayed with resorcinol reagent[8] and heated to 120° for 20 min. Gangliosides are visualized as purple bands.

If direct ganglioside quantitation is also required, a 2.0-ml plasma sample is used, and one-half of the final ganglioside sample is used to quantitate total gangliosides by the resorcinol assay.[9]

Comments

Several points are critical to achieving highly resolved thin-layer chromatographic patterns when beginning with small plasma samples. At each stage of preparation, the total lipid extract must be clarified to remove insoluble impurities. The final centrifugation of the concentrated total lipid extract is particularly important because it removes a significant residual amount of proteins/polypeptides which were slightly soluble in chloroform–methanol. Similarly, centrifugation of the final ganglioside preparation obtained following the gel filtration step is essential to remove traces of residual protein which could cause smearing of the ganglioside thin-layer chromatographic pattern.

As described, the total lipid extraction step is a "dry" extraction. The elimination of water in the lipid extraction step has the significant advantage of easier and faster drying of the total lipid extract, especially important when multiple samples are being processed. While ganglioside extraction is 10% less than the close to 100% obtained with most "wet" extraction methods (e.g., chloroform–methanol–water (4:8:3)),[4] no se-

[7] We have documented that loosely capping the tube, with the Teflon-lined screw cap, eliminates the frequently substantial loss of purified gangliosides which may occur in the lyophilization step, possibly by the dispersion of the ~10 μg of lyophilized material when air reenters the container after lyophilization.

[8] L. Svennerholm, *Biochim. Biophys. Acta* **24,** 604 (1957).

[9] R. Ledeen and R. K. Yu, this series, Vol 83, p. 139.

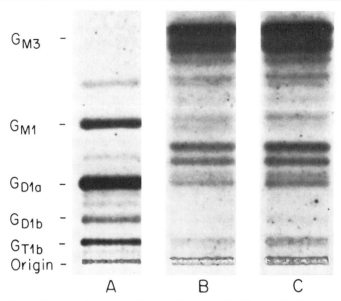

FIG. 1. Normal human plasma gangliosides. Total gangliosides purified by the procedure described herein from 1.0 ml human plasma are shown in lane B and compared to one-eighth of the total gangliosides isolated from 8 ml plasma (lane C). Lane A: Standard brain gangliosides, TLC developed and stained as described in Procedure. From Ladisch and Gillard.[4]

lective losses in plasma gangliosides have resulted from using a dry extraction method.

Applications

As described here, this partition method for ganglioside isolation and purification yields highly resolved total ganglioside patterns of human plasma. The pattern obtained using only a 1.0-ml plasma sample as starting material is identical to that of the corresponding portion of an 8-ml plasma sample which was similarly processed (Fig. 1), demonstrating the microscale applicability of the procedure. Of interest, the human plasma ganglioside pattern shows a striking complexity; at least 10 other gangliosides besides the predominant ganglioside, G_{M3},[10] are seen. This conclusion was also reached by Kundu et al.[11] in a study using a much larger plasma volume and more involved ganglioside purification preparative procedures.

[10] L. Svennerholm, J. Neurochem. 10, 613 (1963).
[11] S. K. Kundu, I. Diego, S. Osovitz, and D. M. Marcus, Arch. Biochem. Biophys. 238, 388 (1985).

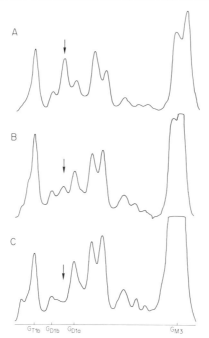

FIG. 2. Analysis of plasma gangliosides of a patient with neuroblastoma by TLC densitometry. Gangliosides were purified from 1.0 ml plasma samples obtained from a patient with metastatic neuroblastoma before (lane A) and after (lane B) a course of therapy, and from a normal donor (lane C). The TLC of these samples was developed and stained as in Fig. 1, and analyzed by scanning densitometry. Migration of standard gangliosides is shown on the abscissa, and that of the abnormal circulating ganglioside, G_{D2}, is indicated by the arrow.

The practical application of the partition method for plasma ganglioside purification is demonstrated by the studies of plasma of patients with neuroblastoma. In this work, the disialoganglioside G_{D2} was directly detected, by thin-layer chromatographic analysis of the total gangliosides of 1.0-ml plasma samples, in patient but not normal plasma samples.[12] The thin-layer chromatograms of plasma gangliosides are sufficiently resolved to allow analysis by scanning densitometry and, as shown in Fig. 2, for quantitative determination of the abnormal tumor-associated ganglioside, G_{D2}, in sequential patient samples. The detection limit under these conditions is 50 pmol LBSA/lane. In Fig. 2, the densitometric tracing of gangliosides of sequential plasma samples of a patient with neuroblastoma show

[12] S. Ladisch and Z.-L. Wu, *Lancet* **I,** 136 (1985).

the diminution of the neuroblastoma-associated G_{D2} band following therapy (lane B vs A).

The partition method is equally applicable to the study of plasma gangliosides of other species; we have recently visualized the gangliosides derived from (and shed by) the YAC-1 murine lymphoma *in vivo,* in the plasma of tumor-bearing mice.[13] The method can also be used to prepare purified plasma gangliosides for the detection of specific gangliosides by TLC immunostaining procedures[14] using monoclonal antibodies, thereby even further increasing sensitivity.

To isolate larger quantities of plasma gangliosides, the method described here for the processing of 1.0-ml plasma samples can be scaled up. In such a case, volumes used in the total lipid extraction and partitioning steps are scaled up directly, and the gel filtration conditions (column dimensions and flow rate) would be increased as described.[4]

Conversely, somewhat smaller plasma samples can also be processed by this procedure. Although overall recovery may be reduced, primarily because of increased losses in the gel filtration step, we have not observed qualitative changes in the TLC pattern (which would indicate selective losses of specific gangliosides) nor any decrease in resolution on TLC. For example, using this procedure the total ganglioside pattern of 5 ml normal human cerebrospinal fluid (0.5 nmol lipid-bound sialic acid/ml) was visualized and demonstrated to be identical to that of human brain gangliosides.[15]

Acknowledgments

This work was supported by USPHS Grant HD18171 and American Cancer Society Grant PDT-270. Dr. Ladisch is the recipient of Research Career Development Award CA00821 from the National Cancer Institute and is a Scholar of the Leukemia Society of America.

[13] H. S. Ahn and S. Ladisch, in preparation.
[14] J. L. Magnani, M. Brockhaus, D. F. Smith, and V. Ginsburg, this series, Vol. 83, p. 235.
[15] S. Ladisch, unpublished results (1985).

[24] Affinity Purification of Oligosaccharides Using Monoclonal Antibodies

By D. Zopf, K. Schroer, G. Hansson, J. Dakour, and
A. Lundblad

Monoclonal antibodies that bind carbohydrate antigens can be linked covalently or noncovalently to solid supports and used for affinity separation of oligosaccharides that contain the target antigen. The method may be generally applied to purification and analysis of naturally occurring free oligosaccharides, complex sugar chains released from glycoconjugates by chemical elimination or endohydrolytic digestion, or oligosaccharide products of specific degradative or biosynthetic reactions.

Materials. Oligosaccharides isolated from human urine[1] or human milk[2] were ^{3}H-labeled by reduction with sodium borotritide[3] (specific activity 11.6 mCi/μmol; New England Nuclear Corporation, Boston, MA). Staphylococcal protein A-Sepharose CL-4B (SPA-Seph) and cyanogen bromide-activated Sepharose (CNBr-Seph) were purchased from Pharmacia (Uppsala). Monoclonal antibodies 61.1 (IgG$_3$) and 39.5 (IgG$_{2b}$) were prepared as previously described.[4] Hybridoma NS 1116-19-9 (19-9) (IgG$_1$) was described by Koprowski *et al.*[5]

Screening of Antibodies for Use as Affinity Reagents. To be useful for solid-phase affinity purification under conditions described here, monoclonal antibodies must bind oligosaccharide ligands with $K_a \geq 10^4$ liter/mol. Usually at least 5 mg of antibody protein is required. Affinity of antibodies for radiolabeled oligosaccharides can be estimated empirically using a nitrocellulose filter assay.[6] Ascitic fluid or concentrated hybridoma culture supernant fluid containing at least 0.2 mg/ml of the monoclonal antibody is cleared by centrifugation at 10,000 g for 15 min and tested for hapten binding by incubating triplicate 10-μl aliquots of antibody with 10^6 cpm radiolabeled oligosaccharide in a total volume of 0.4 ml Tris buffer A (0.01 M Tris–HCl containing 0.14 M NaCl, 0.5 mM MgSO$_4$, 0.15 mM CaCl$_2$, pH 7.5) at 4°. In a cold room at 4°, the mixture is vacuum

[1] A. Lundblad, this series, Vol. 50 [25].
[2] A. Kobata, this series, Vol. 28 [24].
[3] A. Kobata and V. Ginsburg, *J. Biol. Chem.* **244,** 5496 (1969).
[4] A. Lundblad, K. Schroer, and D. Zopf, *J. Immunol. Methods* **68,** 217 (1984).
[5] H. Koprowski, Z. Steplewski, K. Mitchell, M. Herlyn, K. Herlyn, and J. P. Fuhrer, *Somatic Cell Genet.* **5,** 957 (1979).
[6] D. Zopf, C.-M. Tsai, and V. Ginsburg, this series, Vol. 50 [14].

TABLE I
STRUCTURES OF OLIGOSACCHARIDES

Oligosaccharide	Structure
(Glc)$_4$	Glcα1-6Glcα1-4Glcα1-4Glc
Lactosialyltetrasaccharide a (LSTa)	NeuAcα2-3Galβ1-3GlcNAcβ1-3Galβ1-4Glc
Sialyl-Le^a (SLe^a)	NeuAcα2-3Galβ1-3GlcNAcβ1-3Galβ1-4Glc
	4
	|
	Fucα1

filtered through nitrocellulose (No. BH-85, 25 mm, 0.45 μm, Schleicher and Schuell, Keene, NH) and the filter is washed by vacuum filtration with 10 ml Tris buffer A. The filter is dissolved in 2 ml dioxane and counted by liquid scintillation in Aquasol (New England Nuclear Corporation, Boston, MA). Counts unspecifically trapped by the filter are determined using an ascites blank that contains antibody of unrelated specificity. Antibodies that specifically bind at least 0.4% of added counts under these conditions can be used to construct affinity columns to separate the oligosaccharide.

Preparation of Affinity Columns

Attachment of Antibodies to a Solid Phase. Many antibodies with $K_a \geq 10^4$ liter/mol for oligosaccharide ligands belong to the IgG immunoglobulin class. These can be conveniently bound to a solid support as follows: hybridoma ascitic fluid or hybridoma culture supernatant fluid is dialyzed at least 6 hr against 50 volumes of Tris buffer A at 4°, warmed to room temperature, and then pumped at 1 ml/mm^2/min through a column of SPA-Seph. Often the column simply can be washed with the same buffer to remove unbound protein and used directly for affinity chromatography of oligosaccharides. Alternatively, IgG antibodies can be purified by pH gradient elution from SPA-Seph[7] and then bound a second time to SPA-Seph as a purified preparation or covalently bound to a chemically activated solid substrate. For example, monoclonal antibodies 61.1 and 39.5 against the glucose-containing urinary oligosaccharide Glcα1-6Glcα1-4Glcα1-4Glc bound 1 and 0.4% of added counts, respectively, in the nitrocellulose filter assay at 4° (see above). Fifteen milliliters of hybridoma ascitic fluid containing 22 mg of 61.1 (IgG$_3$) antibody was dialyzed overnight against Tris buffer A, cleared by centrifugation, and passed

[7] P. L. Ey, S. J. Prowse, and C. R. Jenkin, *Immunochemistry* **15**, 429 (1978).

through a 0.3 × 18 cm packed bed of SPA-Seph in a water-jacketed glass column at a flow rate of 0.2 ml/min. The column was washed with Tris buffer A until material absorbing at 280 nm was no longer detected. A similar column was prepared using 1 ml of a high-titer ascitic fluid from hybridoma NS 1116-19-9 (IgG$_1$).

To prepare a column with antibody covalently linked to a solid support, 61.1 antibody purified by pH gradient elution from SPA-Seph was concentrated by vacuum dialysis against 0.15 M NaCl containing 0.02 M potassium phosphate buffer (pH 8.5) and was coupled to CNBr-Seph according to the manufacturer's instructions. The final washed gel contained 1.2 mg antibody protein per milliliter packed column bed.

Temperature Effects. Monoclonal antibodies differ significantly in their dependence on temperature for oligosaccharide binding. For binding of antibody 61.1 to the tritiated alditol of (Glc)$_4$ [abbreviated ^{3}H(Glc)$_4$-ol], association constants determined by equilibrium dialysis double for each 8° downward shift in temperature from 37° (8 × 10^3 liter/mol) to 4° (1.7 × 10^5 liter/mol).[8] As a result, temperature may critically affect performance of affinity chromatography of oligosaccharides. Figure 1 shows affinity chromatography of ^{3}H(Glc)$_4$-ol on a column containing antibody 61.1 noncovalently bound to SPA-Seph at four different temperatures: at 37° ^{3}H(Glc)$_4$-ol elutes in the void volume of the column; at 15 and 10° the labeled sugar appears as retarded peaks in progressively later fractions; at 4° it is retained by the column, failing to elute even after washing with 50 column volumes of Tris buffer A. When flow is stopped, the column is warmed to 37° for 10 min, and flow is again resumed, ^{3}H(Glc)$_4$-ol elutes as a sharp peak slightly after the void volume (Fig. 1B).

A column containing antibody 39.5, which binds ^{3}H(Glc)$_4$-ol with lower affinity at a given temperature than 61.1, does not retard the reduced oligosaccharide at temperatures ≥25° but gives a retained peak well-separated from the void volume at 4° (Fig. 2). In contrast, affinity chromatography of the SLe^a oligosaccharide (Table I) on a column that contains monoclonal 19-9 bound to SPA-Seph is optimal at 32°[9] (Fig. 3). When chromatography of SLe^a is performed at room temperature or lower, the retained peak is excessively broad, similar in shape to retained peaks of ^{3}H(Glc)$_4$-ol on 61.1-SPA-Seph at temperatures between 10 and 4° (data not shown).

Flow Rates and Column Design. Flow rates of 1 ml/cm^2/min have given optimal results for the columns described above. Faster flow decreases efficiency of chromatographic separation. Choice of column di-

---

[8] A. Lundblad, K. Schroer, and D. Zopf, *J. Immunol. Methods* **68**, 227 (1984).
[9] G. C. Hansson and D. Zopf, *J. Biol. Chem.* **260**, 9388 (1985).

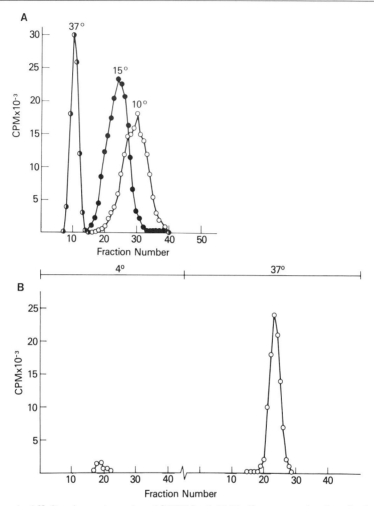

Fig. 1. Affinity chromatography of ^{3}H(Glc)$_4$-ol (Table I) on monoclonal antibody 61.1 noncovalently bound to SPA-Sepharose. (A) ^{3}H(Glc)$_4$-ol (10,000 cpm) dissolved in 50 μl Tris buffer A was chromatographed on a 0.3 × 13 cm water-jacketed column containing 22 mg monoclonal antibody 61.1. The column was eluted with Tris buffer A at 0.5 ml/min with the temperature controlled by a recirculating water bath at 10, 15, or 37°. (B) Chromatography was carried out on a 0.3 × 18 cm column at 4°. After washing with 50 ml Tris buffer A at 4°, flow was interrupted, the column was warmed to 37°, and flow was resumed.

mensions depends upon the amount of antibody available and the affinity of the antibody for oligosaccharide. The capacity of many commercial preparations of SPA-Seph for IgG is on the order of 20 mg IgG/ml packed gel. SPA-Seph columns with dimensions 0.3 × 13 to 18 cm (0.4 to 1.3 ml bed volume) were found suitable for use with antibodies 39.5 and 19-9

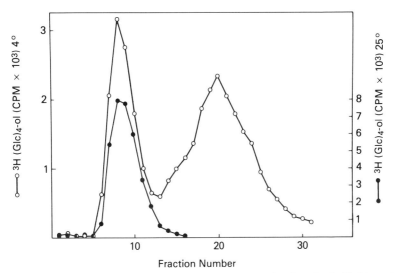

FIG. 2. Affinity chromatography of ^{3}H(Glc)$_4$-ol on monoclonal antibody 39.5 bound to SPA-Sepharose. Chromatography was carried out under the same conditions described in Fig. 1A at 4 and 25°.

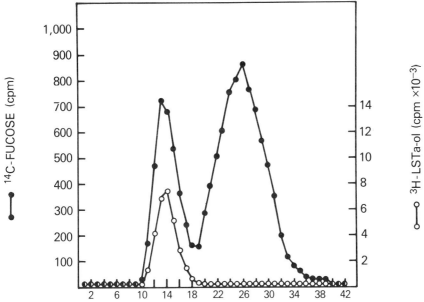

FIG. 3. Affinity chromatography of sialyl-Le^a oligosaccharide (Table I) on monoclonal antibody 19-9 bound to SPA-Sepharose. Oligosaccharides (23,000 cpm) obtained after incubating the human milk oligosaccharide acceptor LSTa (Table I) with microsomes from SW1116 cells and GDP-[^{14}C]Fuc were chromatographed on a 0.3 × 16 cm water-jacketed column maintained at 32° in Tris buffer A at a flow rate of 0.2 ml/min. Free [^{14}C]fucose and [^{3}H]LSTa-ol coelute in fractions 10–18, whereas sialyl-Le^a oligosaccharide elutes as a retarded peak in fractions 19–34.

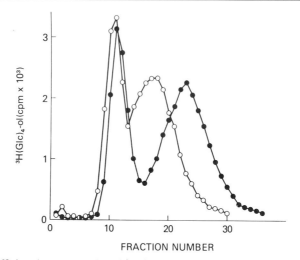

FIG. 4. Affinity chromatography of ^{3}H(Glc)$_4$-ol on a 0.3 × 18 cm packed bed column containing different amounts of monoclonal antibody 39.5 bound to SPA-Sepharose. The column was loaded from the top with 18 ml hybridoma ascitic fluid containing 5 mg of antibody, washed with Tris buffer A, and tested for affinity separation of ^{3}H(Glc)-ol as described for Fig. 1 (○). A second 18-ml aliquot of the same ascites was loaded on the same column, the column was washed, and chromatography was repeated (●). The earlier peak in both runs contains tritiated water and the retained peaks contain ^{3}H(Glc)$_4$-ol.

described above. When such antibodies are available in limited amounts and have relatively low affinities, their utilization as adsorbents is maximized by passing the antibody solution through a prepoured SPA-Seph column, thereby concentrating the antibody within a short span of the gel bed at the top column where free ligand will interact with antibody collected at the highest possible concentration in the gel. When antibody is available in larger amounts, separation of retarded peaks is generally made more efficient by adding more antibody to the top of the column, thereby saturating a longer segment of gel for use as an immunoabsorbent (Fig. 4). In contrast, batch preparation of affinity absorbents with subsaturating amounts of antibody generally results in less efficient chromatography.

Comment

Solid-phase immunoabsorbents prepared by binding IgG antibodies via their Fc portions to SPA-Seph have the advantage of greater capacity as compared with columns utilizing chemically activated linker arms.[10]

[10] D. M. Gerston and J. J. Marchalonis, *J. Immunol. Methods* **24**, 305 (1978).

For example, a column containing antibody 61.1 bound to SPA-Seph binds nearly 10 times more $(Glc)_4$-ol per milligram antibody protein bound to the gel than a column containing the same antibody covalently linked to CNBr-Seph.[8] Covalent coupling may be useful, however, for antibodies of IgM or IgA classes that are not bound by SPA-Seph and under conditions where oligosaccharides are tightly bound by antibody. Oligosaccharides can be dissociated from antibodies of high affinity by standard methods for disrupting immune complexes between immunoglobulins and carbohydrates. For example, $^3H(Glc)_4$-ol can be quantitatively eluted as a sharp peak at 4° from 61.1 covalently bound to CNBr-Seph by washing the column with 0.2 M sodium acetate buffer (pH 3.5). This method of eluting oligosaccharides might be useful for IgG-SPA-Seph columns cross-linked with dimethyl suberimidate to prevent dissociation of the antibody from protein A under acidic conditions.[10]

[25] Fluorescent Gangliosides

By SARAH SPIEGEL

Gangliosides are characteristic glycosphingolipid components of mammalian plasma membranes, where they are restricted to the outer leaflet of the bilayer with their carbohydrate residues facing the extracellular matrix.[1] Because of this orientation, gangliosides have been implicated in a variety of cell surface events, such as recognition phenomena[2] and biotransduction of membrane-mediated information.[1] Although gangliosides are receptors for bacterial toxins[3] and viruses,[4] and can function as receptor modulators for growth factor,[5] tumor promoters,[6] and neurotransmitters,[7] little is known about the physiological function of gangliosides and their influence on membrane properties.[2]

[1] P. H. Fishman and R. O. Brady, *Science* **194,** 906 (1976).

[2] S. Hakomori, *Annu. Rev. Biochem.* **50,** 733 (1981).

[3] P. H. Fishman, *J. Membr. Biol.* **69,** 85 (1982).

[4] M. A. Markwell, L. Svennerholm, and J. C. Paulson, *Proc. Natl. Acad. Sci. U.S.A.* **78,** 5406 (1981).

[5] E. G. Bremer, S. Hakomori, D. F. Bowen-Pope, E. Raines, and R. Ross, *J. Biol. Chem.* **259,** 6818 (1984).

[6] L. Srinivas, T. D. Grindhart, and N. H. Colburn, *Proc. Natl. Acad. Sci. U.S.A.* **79,** 4988 (1982).

[7] G. Dawson and E. Berry-Kravis, *Adv. Exp. Med. Biol.* **174,** 341 (1984).

To investigate the role of gangliosides in these types of cellular phenomena, fluorescent derivatives of gangliosides with an unmodified lipid moiety[8] were prepared. These derivatives can be properly inserted into the plasma membrane[10] and can be used to directly monitor the organization and dynamics of gangliosides in the plasma membrane.[10,11]

Principle

Fluorescent derivatives of gangliosides were prepared by covalent attachment of fluorophors to the sialic acid residues of gangliosides. Gangliosides were oxidized with periodate under mild conditions which generate aldehyde groups only on the sialic acid residues.[12] The aldehyde groups reacted with fluorescent dyes which contained a hydrazide group to form a Schiff base, which subsequently was stabilized by reduction (Fig. 1).

Reagents

GM1 was purified from bovine brain gangliosides after exhaustive treatment with *Vibrio cholerae* neuraminidase which hydrolyzes all of the di-, tri-, and polysialogangliosides to G_{M1} as described previously.[13]

Vibrio cholerae neuraminidase (EC 3.2.1.18) obtained from Calbiochem-Behring Corp. (La Jolla, CA).

Sodium acetate buffer, 0.1 M, pH 5.5.

Dulbecco's phosphate-buffered saline, per liter: 8.0 g of NaCl, 2.16 g of $Na_2HPO_4 \cdot 7H_2O$, 0.1 g of $MgCl_2 \cdot 6H_2O$, 0.2 g of KH_2PO_4, 0.2g of KCl, 0.1 g of $CaCl_2$ (anhydrous) final pH 7.4 (PBS).

Sodium periodate 160 mg in 5 ml distilled water (0.15 M).

50% glycerol in PBS.

---

[8] In a previous study,[9] a fluorescein-labeled analog of G_{M1} was synthesized which contained an N-fluoresceinyloctadecylamino group instead of the normal ceramide moiety. This derivative, however, has only one hydrocarbon chain and a bulky fluorescein group close to it. Thus, it may behave differently than native gangliosides or fluorescent derivatives with a normal ceramide group.

[9] J. Schlessinger, L. S. Barak, G. G. Hammes, K. M. Yamada, I. Pastan, W. W. Webb, and E. L. Elson, *Proc. Natl. Acad. Sci. U.S.A.* **74**, 2909 (1977).

[10] S. Spiegel, J. Schlessinger, and P. H. Fishman, *J. Cell Biol.* **99**, 699 (1984).

[11] S. Spiegel, S. Kassis, M. Wilchek, and P. H. Fishman, *J. Cell Biol.* **99**, 1575 (1984).

[12] R. W. Veh, A. P. Corfield, M. Sander, and R. Schauer, *Biochim. Biophys. Acta* **486**, 145 (1977).

[13] T. Pacuszka, R. O. Duffard, R. N. Nishimura, R. O. Brady, and P. H. Fishman, *J. Biol. Chem.* **253**, 5839 (1978).

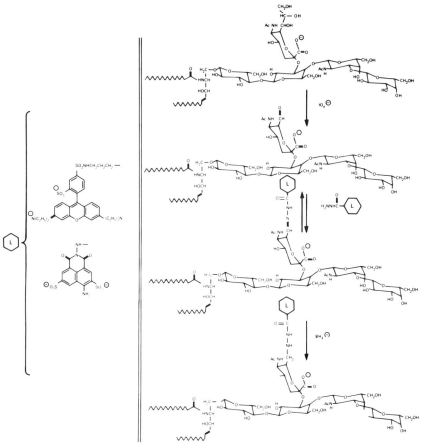

FIG. 1. The synthesis of fluorescent gangliosides. G_{M1} was chosen as a representative ganglioside. Similar reactions can occur with other gangliosides. The fluorescent dyes shown in this figure are Lissamine rhodamine-γ-aminobutyric hydrazide and lucifer yellow CH.

Lissamine rhodamine-γ-aminobutyric hydrazide was prepared as described previously.[14]

Lucifer yellow CH (from Aldrich Chemical Co.).

Potassium borohydride, 10 mM in NaOH 0.01 N.

Sodium cyanoborohydride, 100 mM in NaOH 0.01 N.

[14] M. Wilchek, S. Spiegel, and Y. Spiegel, *Biochem. Biophys. Res. Commun.* **92**, 1215 (1980).

Procedure

Preparation of Fluorescent Gangliosides. Ganglioside G_{M1}[15,16] (1 mg) dissolved in 1 ml sodium acetate buffer containing NaCl (150 mM) was oxidized with 13 μl of sodium periodate (0.15 M) for 30 min at $0°$.[12,17] The reaction was stopped by the addition of 0.1 ml of 50% glycerol and the solution was then dialyzed extensively against water and lyophilized.[18] The oxidized G_{M1} was dissolved in 1 ml of PBS and then was reacted with either Lissamine rhodamine-γ-aminobutyric hydrazide (1 mM final concentration, 5 hr) or lucifer yellow CH[19] (5 mM final concentration, 12 hr) at 37° and dialyzed extensively against PBS.[21] The rhodamine-labeled G_{M1} (Rh-G_{M1}) was reduced with 0.1 ml potassium borohydride (10 mM) for 15 min at 25°. Because the same reduction quenched the fluorescence of lucifer yellow CH, the lucifer yellow CH-labeled G_{M1} (LY-G_{M1}) was reduced with 0.1 ml sodium cyanoborohydride (100 mM) under the same conditions. The fluorescent derivatives of G_{M1} were dialyzed extensively against distilled water and lyophilized.

Analysis of Fluorescent Gangliosides. Fluorescent gangliosides were analyzed by thin-layer chromatography on silica gel 60-coated glass plates, which were developed in chloroform–methanol–0.25% $CaCl_2$ (60 : 35 : 8, v/v/v) and visualized under ultraviolet light. The fluorescent gangliosides still reacted with resorcinol reagent and could be quantified by scanning densitometry.[22,23] Two fluorescent derivatives were obtained

[15] Ganglioside G_{M1} was chosen as a representative ganglioside. Similar procedures can be applied to many other gangliosides.

[16] Ganglioside nomenclature as described by L. Svennerholm, *J. Neurochem.* **10**, 613 (1963).

[17] This treatment preferentially oxidizes the exocyclic arm of terminal unsubstituted sialic acids.[12]

[18] Alternatively, the oxidized G_{M1} can be dialyzed against water and then against a large volume of PBS.

[19] The yield of fluorescent gangliosides appeared to depend on the concentration of the fluorophore, and on the time and temperature of incubation. Other commercial fluorescent hydrazides can also be used, such as 5-({[2-(carbohydrazino)methyl]thio}acetyl)amino-fluorescein, 5-({[2-(carbohydrazino)methyl]thio}acetyl)aminoeosin,[20] eosin-5-thiosemicarbazide, dansylhydrazine, Lissamine rhodamine sulfonylhydrazide, Texas red hydrazide (from Molecular Probes, Inc., Junction City, OR). It is desirable to optimize the condition for each compound first.

[20] B. Goins, B. G. Barisas, and E. Freire, *Biophys. J.* **47**, 114a (1985).

[21] Excess of free fluorophore should be removed by dialysis against PBS containing Ca^{2+} and Mg^{2+} or by gel filtration using Sephadex G-25.

[22] P. H. Fishman, R. H. Quarles, and S. R. Max, *in* "Densitometry in Thin Layer Chromatography" (J. C. Touchstone and J. Sherma, eds.), p. 315. Wiley, New York 1979.

[23] As a large number of fluorescent derivatives were obtained from mixed bovine brain gangliosides,[11] it is desirable to compare individual species before and after the modification procedure.

from purified G_{M1}.[24] Both derivatives of Rh-G_{M1} migrated more rapidly than native G_{M1}, whereas the two derivatives of LY-G_{M1} migrated more slowly.[25] The major fluorescent derivative was separated from the minor derivative, free fluorophore, and unreacted G_{M1} by preparative thin-layer chromatography. The major fluorescent labeled G_{M1} band was scraped from the chromatograms and eluted with 10 ml of chloroform–methanol–water (10:10:3, v/v/v). Upon further chromatography, the fluorescent G_{M1} appeared as a single resorcinol-positive band which corresponded to the major fluorescent band.[26] To further assess its purity on thin-layer chromatography, cholera toxin binding was measured by the sensitive overlay technique.[27] The fluorescent band corresponded to the bound iodotoxin. There appeared to be only a trace amount of unreacted G_{M1}, which was less than 12% of the total based on densitometric scanning of the autoradiogram.[28] The fluorophores were shown to be linked to the sialyl residues as follows: (1) gangliosides G_{M1} (1 mg) were heated at 80° in 0.1 M HCl for 2 hr to hydrolyze all of the sialic acids. After dialysis against water, the material was oxidized and reacted with Lissamine rhodamine-γ-aminobutyric hydrazide and analyzed by thin-layer chromatography. No rhodaminyl glycolipid deriviatives were detected. (2) Bovine brain gangliosides (1 mg) were incubated with 0.5 U of *Vibrio cholerae* neruaminidase for 2 hr at 37°, dialyzed, oxidized, and reacted with Lissamine rhodamine-γ-aminobutyric hydrazide. Only rhodaminyl-G_{M1} was detected by thin-layer chromatography. This was expected, as neuraminidase hydrolyzes complex gangliosides to G_{M1}, which is resistant to the enzyme.

The biological activities of the fluorescent G_{M1} derivatives were com-

---

[24] The formation of the two fluorescent derivative of G_{M1} may be due to the presence of two oxidized gangliosides, one with the aldehyde at C8 of the sialic acid and the other at C-7.[12]

[25] Because the rhodamine group is relatively hydrophobic and lucifer yellow CH is both negatively charged and hydrophilic, the observed changes in mobility were predictable.

[26] S. Spiegel, *Biochemistry* **24**, 5947–5952 (1985).

[27] J. L. Magnani, D. F. Smith, and V. Ginsburg, *Anal. Biochem.* **109**, 399 (1980).

[28] The gangliosides were separated on aluminum-backed silica gel in chloroform–methanol–0.2% $CaCl_2$ (5:4:1, v/v/v). The dried chromatogram was quickly soaked twice in 0.1% poly(isobutyl methacrylate) dissolved in hexane. After drying in air, the chromatogram was sprayed with 50 mM Tris–HCl (pH 7.4) containing 150 mM NaCl and 1% bovine serum albumin. The following steps were performed rapidly. The chromatogram was soaked in the above buffer, drained, and immediately overlayed with ^{125}I-labeled choleragen (10^6 cpm/ml) in the buffer. After 30 min at 4°, the chromatogram was drained and dipped in four successive changes of the same ice-cold buffer without the bovine serum albumin. The chromatogram was air-dried and the bound toxin was detected by autoradiography using Kodak X-Omat AR-2 film and quantified by densitometic scanning of the autoradiogram.

pared with those of native G_{M1} using G_{M1}-deficient rat glioma C6 cells.[29] Exposure of the cells to either native or fluorescent G_{M1} resulted in a similar, marked increase in binding of ^{125}I-labeled cholera toxin. More importantly, the fluorescent derivatives were as effective as native G_{M1} in enhancing the responsiveness of the cells to cholera toxin and caused the same dramatic increase in toxin-stimulated cyclic AMP production.[26] It seems that the introduction of a bulky fluorophore into the sialyl residues of GM1 does not abrogate the function of the native ganglioside as the cell surface receptor for cholera toxin. Furthermore, fluorescent derivatives of gangliosides not only retain the ability to function as receptors for cholera toxin, but also appear to be as effective as native gangliosides in promoting fibronectin retention and reorganization in ganglioside-deficient mouse fibroblasts[30] and also preserve their ability to interact with organized fibrils of fibronectin.[10]

Conclusion

The ready availability of fluorescent reagents and the simplicity of the modification procedure make the preparation of these ganglioside derivatives relatively easy. The demonstration that such fluorescent gangliosides retain their biological activity should promote their use as probes for investigating the distribution, dynamics, and function of these membrane components.

Acknowledgment

The author wishes to thank Dr. Peter H. Fishman for his advice and critical review of the manuscript. This work was supported in part by the Chaim Weizmann Postdoctoral Fellowship from the Foundation of the Weizmann Institute of Science, Rehovot, Israel.

[29] P. H. Fishman, *J. Membr. Biol.* **54**, 61 (1980).
[30] S. Spiegel, K. M. Yamada, B. E. Hom, J. Moss, and P. H. Fishman, *J. Cell Biol.* **100**, 721 (1985).

[26] Lysogangliosides: Synthesis and Use in Preparing Labeled Gangliosides

By GÜNTER SCHWARZMANN and KONRAD SANDHOFF

Introduction

Gangliosides are ubiquitous in vertebrate cells and are notably abundant in neuronal plasma membranes.[1] The ceramide moiety of gangliosides is embedded in the outer leaflet of the lipid bilayer and their sialooligosaccharide residue is oriented toward the extracellular space.

Gangliosides were first discovered more than 45 years ago.[2] During the years since, gangliosides have been adequately studied for their chemical structure and cellular distribution; however, almost nothing is known about their biological function and structural organization in membranes. Nevertheless, the high concentration of gangliosides in neurons is conceivably related to membrane properties. It is the interactions of gangliosides with membrane proteins, other lipids, and themselves that have gained increasing interest in recent years.

One strategy for studying these interactions is the synthesis of gangliosides with special probes in their ceramide moiety, which are suitable to directly monitor the organization and dynamics of gangliosides in the plasma membrane. Although much effort has been put into chemical synthesis of glycolipids, a total synthesis of gangliosides has not yet been effectively achieved. This article describes the methods we have developed for the synthesis of lysogangliosides and various kinds of labeled gangliosides. These labeled gangliosides could be used as effective membrane probes.

In the first part of this article we describe two efficient methods for the synthesis of lysogangliosides, i.e., gangliosides lacking their fatty acyl residue,[3] starting from tissue gangliosides. The first procedure begins with the free sialooligosaccharide of gangliosides G_{M1} and G_{M3}, and makes possible the introduction of unnatural sphingoids or sphingoid analogs. In the second procedure lysogangliosides are synthesized which lack the fatty acyl residue but contain the natural composition of sphingoids of their parent gangliosides. Radiochemical labeling of lysogangliosides is then described.

[1] R. Ledeen, *Trends Neurosci.* **8,** 169 (1985).
[2] E. Klenk, *Hoppe-Seyler's Z. Physiol. Chem.* **273,** 76 (1942).
[3] S. Neuenhofer, G. Schwarzmann, H. Egge, and K. Sandhoff, *Biochemistry* **24,** 525 (1985).

The lysogangliosides are valuable intermediates in the economic preparation of different labeled gangliosides which are discussed in the second part of this article. Mild and selective N-acylation of lysogangliosides using various labeled fatty acid N-succinimidyl esters makes it possible to obtain different varieties of labeled gangliosides in good yields. The spin-labeled gangliosides have proved to be extremely useful in the study of ganglioside insertion[4] into membranes, and will certainly also be important in studies of membrane dynamics as well as ganglioside–protein interactions, as has been shown for spin-labeled phospholipids.[5] Gangliosides with a parinaric acid residue in their ceramide portion are conceivably valuable tools for resonance energy transfer measurements and thus for studying ganglioside–protein interactions. NBD-labeled gangliosides[5a] might prove useful for the investigation of intracellular transport and distribution of gangliosides, as has been shown for NBD-labeled phospholipids.[6]

Another group of fluorescent gangliosides, those carrying a pyrene residue in their ceramide, have also been synthesized. These pyrene gangliosides could, by means of excimer formation, be valuable tools in membrane research,[7] e.g., in the determination of lateral diffusion of gangliosides as well as their interaction with proteins. Finally, the last section of this article deals with the characterization of lysogangliosides and some of the labeled gangliosides which we have synthesized.

Materials and General Methods

Materials

Succinimidyl 12-[N-methyl-N-(7'-nitrobenz-2'oxa-1',3'-diazol-4'-yl)]-aminododecanoate (NBD-labeled fatty acid) and 1-pyrenedodecanoic acid are obtained from Molecular Probes (Junction City, OR). N,N'-Dicyclohexylcarbodiimide (DCC), di(N-succinimidyl) carbonate (DSC), 9-fluorenylmethyl chloroformate (Fmoc Cl), and sodium borohydride are from Fluka (Buchs, Switzerland). Methyl dichloroacetate, N-hydroxysuccinimide, nitromethane, diisopropylethylamine, 33% hydrogen bro-

[4] G. Schwarzmann, S. Sonderfeld, E. Conzelmann, D. Marsh, and K. Sandhoff, in "Cellular and Pathological Aspects of Glycoconjugate Metabolism" (H. Dreyfus, R. Massarelli, L. Freysz, and G. Rebel, eds.), Vol. 126, p. 195. INSERM, Paris, 1984.

[5] D. Marsh, in "Progress in Protein–Lipid Interactions" (A. Watts and J. J. H. H. M. De Pont, eds.), p. 143. Elsevier, Amsterdam, 1985.

[5a] Gangliosides containing a 12-[N-methyl-N-(7'-nitrobenz-2'-oxa-1',3'-diazol-4'-yl)]amino-dodecanoyl residue in lieu of a natural fatty acid moiety.

[6] R. G. Sleight and R. E. Pagano, J. Cell Biol. 99, 742 (1984).

[7] H.-J. Galla and W. Hartmann, Chem. Phys. Lipids 27, 199 (1980).

mide in acetic acid, and acetic anhydride are obtained from Merck-Schu-chardt (Hohenbrunn, West Germany). Anisaldehyde and ninhydrin spray reagents are from E. Merck (Darmstadt, West Germany). [^{3}H]Acetic an-hydride (500 mCi/mmol), [1-^{14}C]acetic anhydride (28 mCi/mmol), and so-dium boro[^{3}H]hydride (9.75 Ci/mmol) are obtained from Amersham Buch-ler GmbH (Braunschweig, West Germany). Sodium boro[^{3}H]hydride (5.2 Ci/mmol) is obtained from NEN Chemicals GmbH (Dreieich, West Ger-many). Silica gel LiChroprep Si 60 and Si 100 (25-40 μm), precoated thin-layer plates Kieselgel 60 and Kieselgel 60$_{F\ 254}$ (0.25 mm layer thickness) are from E. Merck (Darmstadt, West Germany). Neutral and basic alu-mina W 200 are from Woelm Pharma GmbH (Eschwege, West Germany). DEAE-Sephadex A-25 is from Pharmacia Fine Chemicals (Uppsala, Swe-den). Visking dialysis tubing type 20/32 is obtained from Serva (Heidel-berg, West Germany). All solvents are analytical grade and are obtained from either E. Merck (Darmstadt, West Germany) or Riedel-de Haën (Seelze, West Germany). Parinaric acid was a gift of Dr. R. Klein (Cam-bridge, United Kingdom).

Methods

Solvents. Some solvents are prepared dry and/or argon-saturated prior to use. Anhydrous nitromethane is obtained by distillation of the reagent-grade solvent over phosphorus pentoxide immediately prior to use. Anhy-drous toluene, acetonitrile, *N*,*N*-dimethylformamide (DMF), and ethyl acetate are obtained by purging the analytical-grade solvents over neutral alumina. Diethyleneglycol dimethyl ether (diglyme) and tetrahydrofuran (THF) are freed of peroxides by passing over basic alumina, followed by saturation with dry argon.

Gangliosides.[7a] Gangliosides are extracted from bovine brain with mixtures of chloroform–methanol–0.1% potassium chloride and parti-tioned as described by Folch *et al.*[8] prior to separation according to charge on DEAE-Sephadex A-25.[9,10] Final purification of gangliosides G_{M1}, G_{D1a}, and G_{D1b} is achieved by medium-pressure chromatography on columns (1.2 or 2.5 × 100 cm) of LiChroprep Si 100 using mixtures of chloroform–methanol–water of increasing polarity or linear gradients of 2-propanol–*n*-hexane–water (55 : 45 : 5 to 55 : 25 : 15, v/v/v). Ganglioside G_{M2} was isolated from postmortem brain of a Tay-Sachs patient and puri-

[7a] Nomenclature according to L. Svennerholm, *J. Neurochem.* **10**, 613 (1963).

[8] J. Folch, M. B. Lees, and G. H. Sloane Stanley, *J. Biol. Chem.* **226**, 497 (1957).

[9] R. W. Ledeen and R. K. Yu, this series, Vol. 83, p. 139.

[10] T. Momoi, S. Ando, and Y. Nagai, *Biochim. Biophys. Acta* **441**, 488 (1976).

fied as described.[11] Ganglioside G_{M3} was prepared from postmortem human spleen as described.[12]

Sphingosine and Sphingosine Derivatives. C_{18}-D-*erythro*sphingosine, 2-*N*-dichloroacetyl-D-*erythro*-sphingosine, and 2-*N*-dichloroacetyl-3-*O*-benzoylsphingosine are prepared essentially as described.[13] 13-(4′,4′-Dimethyl-2′-butyl-3′-oxyoxazolidin-2′-yl)tridecanoic acid (14-doxylstearic acid) and 8-(4′,4′-dimethyl-2′-nonyl-3′-oxyoxazolidin-2′-yl)octanoic acid (9-doxylstearic acid) are synthesized according to published procedures.[14]

N-Succinimidyl Ester of Labeled Fatty Acids. Fatty acid N-succinimidyl ester are made by the method of Lapidot *et al.*[15] Typically, the fatty acids (50 μmol) are dissolved in dry ethyl acetate (1 ml) and an equimolar amount of solid *N*-hydroxysuccinimide is added. After the latter has dissolved, an equimolar amount of solid *N,N′*-dicyclohexylcarbodiimide is added and the mixture is stirred under argon overnight at 25°. *N,N′*-Dicyclohexylurea is removed by centrifugation and the clear supernatant is dried in a stream of nitrogen. The residue is stored under argon at $-20°$ until use. Alternatively, the *N*-succinimidyl esters are prepared by the procedure of Ogura *et al.*[16] Briefly, a mixture of a molar equivalent of the fatty acid (50 μmol), *N,N′*-disuccinimidyl carbonate (DSC), and pyridine in acetonitrile (1.5 ml) is stirred at 25° for 8 hr. After evaporation of the solvent in a nitrogen stream, the dry residue is stored at $-20°$ under argon until use.

Notes: In these small-scale reactions both procedures gave about 70–85% yield of the activated esters. We did not find it necessary to separate the esters from the remaining free fatty acids as these are easily removed later in the purification of the labeled gangliosides. A preparation for much larger amounts of long-chain fatty acid succinimidyl ester and *trans*-parinaric acid succinimidyl ester are described by Blecher[17] and Tsai *et al.*,[18] respectively.

Thin-Layer Chromatography. Purity of the various compounds, the course of reactions, as well as column chromatographic elution profiles are routinely checked by thin-layer chromatography (TLC) using Merck

[11] L. Svennerholm, *Methods Carbohydr. Chem.* **6,** 464 (1972).

[12] G. Schwarzmann, *Biochim. Biophys. Acta* **529,** 106 (1978).

[13] F. Sarmientos, G. Schwarzmann, and K. Sandhoff, *Eur. J. Biochem.* **146,** 59 (1985).

[14] W. L. Hubbell and H. M. McConnell, *J. Am. Chem. Soc.* **93,** 314 (1971).

[15] Y. Lapidot, S. Rappaport, and Y. Wolman, *J. Lipid Res.* **8,** 142 (1967).

[16] H. Ogura, T. Kobayashi, K. Shimizu, K. Kawabe, and K. Takeda, *Tetrahedron Lett.* **49,** 4745 (1979).

[17] M. Blecher, this series, Vol. 72, p. 404.

[18] A. Tsai, B. S. Hudson, and R. D. Simoni, this series, Vol. 72, p. 483.

silica gel plates. The solvent systems used for TLC and/or column chromatography are A, toluene/acetone (1 : 1, v/v); B, toluene/acetone (8 : 2, v/v); C, toluene/acetone (2 : 8, v/v); D, chloroform/methanol/2.5 M ammonia (65 : 25 : 4, v/v/v); E, chloroform/methanol/2.5 M ammonia (60 : 35 : 8, v/v/v); F, chloroform/methanol/2.5 M ammonia (60 : 40 : 9, v/v/v); G, chloroform/methanol/15 mM calcium chloride (60 : 35 : 8, v/v/v); H, chloroform/methanol/15 mM calcium chloride (60 : 40 : 9, v/v/v); J, 2-propanol/n-hexane/water (55 : 40 : 5, v/v/v); K, 2-propanol/n-hexane/water (55 : 25 : 15, v/v/v); L, 2-propanol/n-hexane/water (55 : 35 : 10, v/v/v).

The gangliosides, the sialooligosaccharides, and their derivatives are visualized by spraying with Ehrlich's reagent[19] or anisaldehyde[20] and heating at 120° for 20 min. Deacylated gangliosides and lysogangliosides are made visible with ninhydrin spray and/or anisaldehyde spray. N-(9-Fluorenyl-methoxycarbonyl)lysogangliosides and their deacylated derivatives are visualized by fluorescence quenching followed by anisaldehyde spray. Pyrene- and NBD-labeled gangliosides are also visualized under UV light (366 nm) prior to detection by spray reagents.

Column Chromatography. Separation of reaction products and purification of the desired compounds are achieved by medium-pressure (4–20 bar) column chromatography using heavy-walled glass columns of 100-cm length of various inner diameters (0.6–1.2 cm) and having adjustable column adapters (Latek Labortechnik GmbH, Heidelberg, West Germany) and a high-pressure pump (Latek P 400) or any other suitable columns and pumps. The flow rate of solvents is usually equivalent to 2 bed volumes per hour. Fractions of 2–8 ml, depending on the column size, are collected. For column chromatography, about 50 g of silica gel LiChroprep Si is used per 50 mg of compounds. Usually separation of the products is complete. Fractions containing impure material are rechromatographed.

Analytical Assays. Glycolipid-bound Neu5Ac is measured by the method of Svennerholm[21] as modified by Miettinen and Takki-Luukkainen[22] with Neu5Ac as reference.

Radioactive samples are assayed in a Packard C 460 liquid scintillation spectrometer with Pico-Fluor (Packard, Frankfurt, West Germany) as the scintillant. ^{1}H NMR spectroscopy is performed at 25° with a Bruker WM 400 spectrometer (16K data points) in 5-mm sample tubes at 400 MHz

[19] R. A. Heacock and M. E. Mahon, *J. Chromatogr.* **17**, 338 (1965).
[20] E. Stahl and U. Kaltenbach, *J. Chromatogr.* **5**, 351 (1961).
[21] L. Svennerholm, *Biochim. Biophys. Acta* **24**, 604 (1957).
[22] T. Miettinen and J.-J. Takki-Luukkainen, *Acta Chem. Scand.* **13**, 856 (1959).

(^{1}H). Chemical shifts in the NMR spectra are given in parts per million downfield from tetramethylsilane (Me$_4$Si). For removal of exchangeable protons, each sample is repeatedly freeze-dried from D$_2$O. The lyophilized sample is dissolved in 0.6 ml of freshly prepared Me$_2$SO-d_6–D$_2$O (98:2, v/v) containing a Me$_4$Si reference (4 mM). The UV spectra are recorded at 20° in a Cary 17 spectrophotometer (Varian Instruments).

FAB mass spectrometry[23,24] is carried out with a VG analytical ZAB 1 F, reverse-geometry mass spectrometer being fitted with a FAB source and an Ion-Tech atom gun. Samples are dissolved in methanol, and 0.5–1.0 μl of the solution containing 1–10 μg of the sample is added to 2–4 μl of 1-mercapto-2,3-propanediol on the stainless-steel target. The target is bombarded with xenon atoms having 8–9 keV energy. Spectra are obtained with a hall probe controlled linear mass scan of up to 500 sec duration for a full scan from 3000 to 12 mass units. The spectra are recorded on UV-sensitive chart paper and calibrated manually by counting. Evaporations in a rotary evaporator are conducted *in vacuo,* with the bath temperature kept below 40°.

Lysogangliosides with Defined Sphingoid

Principle

This section describes the synthesis of lyso-G$_{M1}$ and lyso-G$_{M3}$, which contain a well-defined long-chain base (2S,3R,4E)-2-amino-4-octadecene-1,3-diol). Koenigs–Knorr condensation of pentadeca-O-acetyl-α-mono-sialogangliotetraosyl bromide (**III**) or deca-O-acetyl-α-monosialolactosyl bromide (**IX**) with 2-N-dichloroacetyl-3-O-benzoylsphingosine (**IV**) in the presence of mercuric cyanide affords, after removal of the protecting groups, lyso-G$_{M1}$ or lyso-G$_{M3}$, respectively, as outlined in Scheme 1 for lyso-G$_{M1}$ (**VI**).

Procedures

Monosialogangliotetraose, II^{3}Neu5AcGgOse$_4$ (**I**). This sialooligosaccharide is prepared as described.[25] G$_{M1}$ (156 mg, 0.1 mmol) is dissolved in dry methanol (30 ml) by warming. A stream of ozone (0.2 g) in oxygen,

[23] M. Barber, R. S. Bordoli, R. D. Sedgwick, and A. N. Tyler, *Chem. Commun.* p. 325 (1981).

[24] M. Barber, R. S. Bordoli, R. D. Sedgwick, and A. N. Tyler, *Nature (London)* **293,** 270 (1981).

[25] J. Sattler, G. Schwarzmann, J. Staerk, W. Ziegler, and H. Wiegandt, *Hoppe-Seyler's Z. Physiol. Chem.* **358,** 159 (1977).

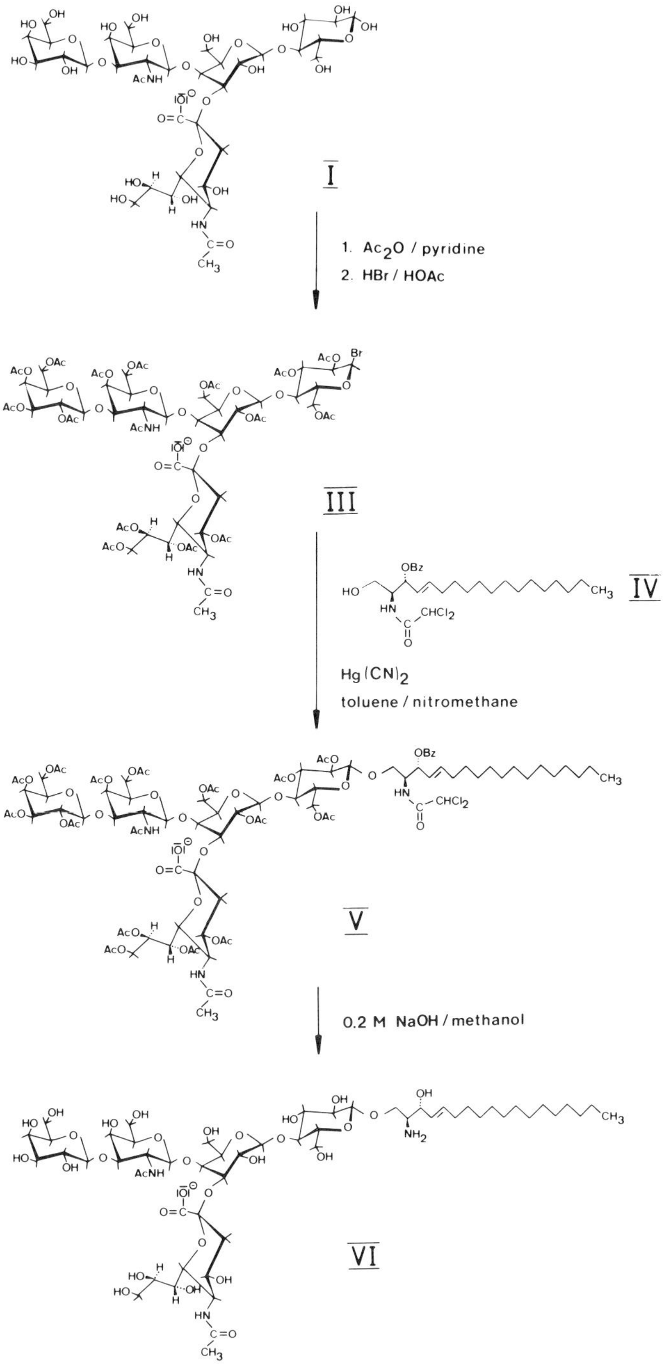

SCHEME 1.

derived from an ozone generator (Fischer, Labor and Verfahrenstechnik, Bonn West Germany) is passed through this solution within 30 min at 20°. Thereafter, methanol is promptly removed in a rotary evaporator and the residue is dissolved in deaerated 1 M sodium carbonate (15 ml) and kept under argon at 20° for 18 hr. The opaque solution is then treated with wet resin (Dowex 50 W-X2, H⁺) until no more CO_2 evolves and the mixture has reached pH 6. The resin is filtered off, and the filtrate and the rinses are passed over a small column (2 × 8 cm) of anion-exchange resin (Dowex AG 1-X2, 200-400 mesh, acetate form). After thorough washing with water, the sialooligosaccharide is eluted with 0.2 M pyridinium acetate and the elution profile is monitored by TLC in n-butanol/acetic acid/water (2 : 2 : 1, v/v/v). Fractions containing monosialogangliotetraose (R_f ~0.09) are pooled and freeze-dried. Final purification can be achieved on a BioGel P-4 column. The yield of monosialogangliotetraose is 60 mg (60%).

Hexadeca-O-acetylmonosialogangliotetraose (**II**). Monosialogangliotetraose (**I**) (50 mg, 50 μmol) is dissolved in water (0.1 ml) followed by pyridine (2 ml). This solution is mixed with acetic anhydride (1 ml) and briefly cooled in ice water. After no more heat evolves the solution is kept under argon at 50° for 3 hr before it is azeotropically dried in a rotary evaporator with several additions of toluene. The residue is now fully acetylated with acetic anhydride (2 ml) in pyridine (4 ml) at 50° under argon for 24 hr. After complete removal of pyridine, acetic acid, and acetic anhydride by azeotropic distillation with toluene in a rotary evaporator, the brownish residue is dissolved in solvent system A (2 ml) and subjected to column chromatography on LiChroprep Si 60 (1.2 × 100 cm) using a linear gradient of solvent system B and C (300 ml each). The elution profile is monitored by TLC in solvent system A. All fractions containing the pure product (R_f 0.19) are dried and lyophilized from benzene. The yield is 60 mg (72%).

Pentadeca-O-acetyl-α-monosialogangliotetraosyl bromide (**III**). Hexadeca-O-acetylmonosialogangliotetraose (**II**) (42 mg, 25 μmol) is dissolved in a 33% solution of hydrogen bromide in acetic acid (0.8 ml) in a screw-capped vial and kept at 0° for 30 min and then at 20° for 30 min. The reaction mixture is then quickly transferred to another vial containing ice-cold chloroform (4 ml). Ice water (4 ml) is added and the mixture is gently shaken for a few seconds. After separation, the upper phase is promptly withdrawn and this washing procedure is repeated twice. The chloroform layer is then immediately washed with ice-cold 0.02 M sodium bicarbonate (4 ml) and again with ice-cold water (4 ml). The chloroform layer is briefly dried over sodium sulfate and after mixing with benzene (1 ml) evaporated in a nitrogen stream with no heating. To avoid significant decomposition of the α-bromo derivative (**III**) the entire workup procedure should be done within 30 min. The dry colorless residue is dissolved in benzene (2 ml) and then freeze-dried. TLC in solvent system A shows one single major spot (R_f 0.21) with no residual starting material. The yield is 40 mg (94%). The α-bromo derivative is used without further purification.

Note: It has been observed that the addition of 2% acetic anhydride to the solution of hydrogen bromide in acetic acid 3 hr before its use was optimal for the formation of the α-bromo derivative (**III**).

Protected lyso-G$_{MI}$ [**V**, *2-N-dichloroacetyl-3-O-benzoylsphingosyl-β-(pentadeca-O-acetyl)monosialogangliotetraoside*]. A solution of the bromide (**III**) (39 mg, 23 μmol) and 2-*N*-dichloroacetyl-3-*O*-benzoyl-D-*erythro*-sphingosine (**IV**) (13 mg, 25 μmol) is stirred in 0.4 ml of a mixture of anhydrous toluene/nitromethane (1 : 4, v/v) in the presence of finely powdered anhydrous mercuric cyanide (30 mg) under argon in a screw-capped vial at 37° for 72 hr. The mixture is then diluted with benzene (3 ml) and centrifuged to remove most of the salts. The supernatant is successively washed with water containing hydrogen sulfide (3 × 2 ml), 0.05 *M* sodium bicarbonate (2 × 2 ml), and water (2 × 2 ml), dried over sodium sulfate, and evaporated in a nitrogen stream. The residue is dissolved in solvent B and chromatographed on LiChroprep Si 60 (0.9 × 100 cm) using a linear gradient of solvent system B and C (150 ml each). The elution profile is monitored by TLC in solvent A. Fractions containing the desired pure condensation product (R_f 0.31) are pooled, dried, and lyophilized from benzene. The yield of the protected lyso-G$_{MI}$ (**V**) is 20.5 mg (42%).

Lyso-G$_{MI}$ (**VI**). For deprotection, the condensation product (**V**) (20 mg, 9.4 μmol) dissolved in 0.2 *M* sodium hydroxide in methanol (1.0 ml) is warmed to 65° for 2–3 hr. When TLC in solvent system H shows a single major ninhydrin-positive spot (R_f 0.19) the mixture is neutralized with acetic acid. After evaporation of the solvent in a nitrogen stream, the residue is dissolved in water (1 ml), dialyzed at 20° for 4 hr against distilled water (10 liters), changed once, and then lyophilized. Final purification and removal of remaining salts are achieved by column chromatography on LiChroprep Si 100 (0.6 × 100 cm) in solvent system E. Fractions with the pure product are pooled and lyophilized. The yield of lyso-G$_{MI}$ is 9.6 mg (80%).

Monosialolactose, II^{3}Neu5AcLac (**VII**). G$_{M3}$ (250 mg, 0.2 mmol) is subjected to ozonolysis, alkaline fragmentation, and purification as described for the isolation of monosialogangliotetraose (**I**). The purified product shows one single spot (R_f 0.20) on TLC in *n*-butanol/acetic acid/water (2 : 1 : 1, v/v/v). The yield of monosialolactose is 76 mg (60%).

Note: Alternatively, monosialolactose can be isolated from cow colostrum and separated from its isomeric II⁶Neu5AcLac as described.[26]

Undeca-O-acetylmonosialolactose (**VIII**). Monosialolactose (**VII**) (63 mg, 0.1 mmol) is fully acetylated and purified essentially as described for the preparation of hexadeca-*O*-acetylmonosialogangliotetraose (**II**). The purified compound shows one single spot (R_f 0.42) on TLC in solvent system A. The yield of the fully acetylated monosialolactose is 82 mg (75%).

Deca-O-acetyl-α-monosialolactosyl bromide (**IX**). Undeca-*O*-acetylmonosialolactose (**VIII**) (44 mg, 40 μmol) is converted into the corresponding α-bromo derivative essentially as described for pentadeca-*O*-acetyl-α-monosialogangliotetraosyl bromide (**III**). The desired product shows one single major spot on TLC in solvent system A (R_f 0.45) and is used without further purification. The yield of the lyophilized material is 42 mg (94%).

Protected lyso-G_{M3}, 2-N-dichloroacetyl-3-O-benzoylsphingosyl-β-(deca-O-acetyl)monosialolactoside (**X**). The α-bromide (**IX**) (41 mg, 36 μmol) is subjected to Koenigs–Knorr condensation with 2-*N*-dichloroacetyl-3-*O*-benzoylsphingosine (**IV**) (21 mg, 40 μmol) in 0.5 ml of a mixture of anhydrous toluene/nitromethane (1 : 4, v/v) in the presence of mercuric cyanide (40 mg) as described above. The workup procedure is also followed as described for the synthesis of the protected lyso-G_{M1}. The condensation product (**X**) is purified by column chromatography on LiChroprep Si 60 (1.2 × 100 cm) employing a linear gradient of solvent system B and C (200 ml each). The elution profile is monitored by TLC in solvent system A. Fractions which contain the pure condensation product (R_f 0.57) are collected and lyophilized from benzene. The yield is 25 mg (45%).

Lyso-G_{M3} (**XI**). The condensation product (**X**) (25 mg, 16.1 μmol) is deprotected in 0.2 *M* methanolic sodium hydroxide as described for lyso-G_{M1} (**VI**). The residue dissolved in water (1 ml) is dialyzed at 20° for 4 hr against distilled water (10 liters), changed once, and then freeze-dried. The purification is achieved by column chromatography in solvent system D as described for lyso-G_{M1}. The elution profile is monitored by TLC in solvent system G. Fractions containing the pure product (R_f 0.28) are pooled and lyophilized from water (1 ml) following removal of the solvents. The yield of lyso-G_{M3} is generally 10–12 mg (70–80%).

[26] H. von Nicolai, H. E. Müller, and F. Zilliken, *Hoppe-Seyler's Z. Physiol. Chem.* **359**, 393 (1978).

Notes

Lyso-G_{M1} and lyso-G_{M3} with the sphingosine residue in L-threo configuration are obtained by Koenigs–Knorr condensation of 2-*N*-dichloroace-
tyl-3-*O*-benzoyl-L-*threo*-sphingosine with the appropriate fully acetylated
α-bromo sialooligosaccharides. The L-*threo*-sphingosine derivative is obtained as described.[13] These lysogangliosides with a tritium label in the 3-
position of the D-*erythro*- or L-*threo*-sphingosine moiety are prepared by
using the appropriately labeled sphingosine derivatives. The latter are
synthesized by the dichlorodicyanobenzoquinone oxidation[27] of 2-*N*-dich-
loroacetylsphingosine to its 3-keto derivative followed by reduction with
sodium boro[³H]hydride to a mixture of 2-*N*-dichloroacetyl-D-*erythro*-
[3-³H]- and L-*threo*-[3-³H]sphingosine. These two diastereomers are
separated by column chromatography[13] following removal of the *N*-
dichloroacetyl group prior to conversion into the tritium-labeled 2-*N*-
dichloroacetyl-3-*O*-benzoylsphingosines.

Lysogangliosides with the Natural Sphingoid Composition of Their
    Parent Gangliosides

Principle

Gangliosides are deacylated with potassium hydroxide in methanol at
100°.[3] By this procedure the fatty acyl as well as the acetyl group of the
sialic acid residue(s) are completely removed with very little hydrolysis of
the acetamido group of the *N*-acetylgalactosaminyl moiety of the gangliosides.[3] The sphingoid free amino group is selectively protected by reaction with 9-fluorenylmethyl chloroformate (Fmoc Cl). After re-N-acetylation of sialooligosaccharide free amino group(s) and removal of the
protective group (Fmoc), lysogangliosides are obtained in an overall yield
of usually 35%. The methodology is illustrated in Scheme 2 for the synthesis of lyso-G_{D1a} (**XIV**).

Procedure

Deacylated Gangliosides (e.g., **XII**). In a heavy-walled Teflon-lined
screw-capped flask (Pyrex) the appropriate ganglioside (50 μmol) is suspended in freshly prepared 0.8 *M* potassium hydroxide in dry methanol
(15 ml) which had been saturated with argon. After flushing with argon,
the flask is tightly sealed and the suspension is stirred at 100°. After a few

[27] M. Iwamori, H. W. Moser, and Y. Kishimoto, *J. Lipid Res.* **16**, 332 (1975).

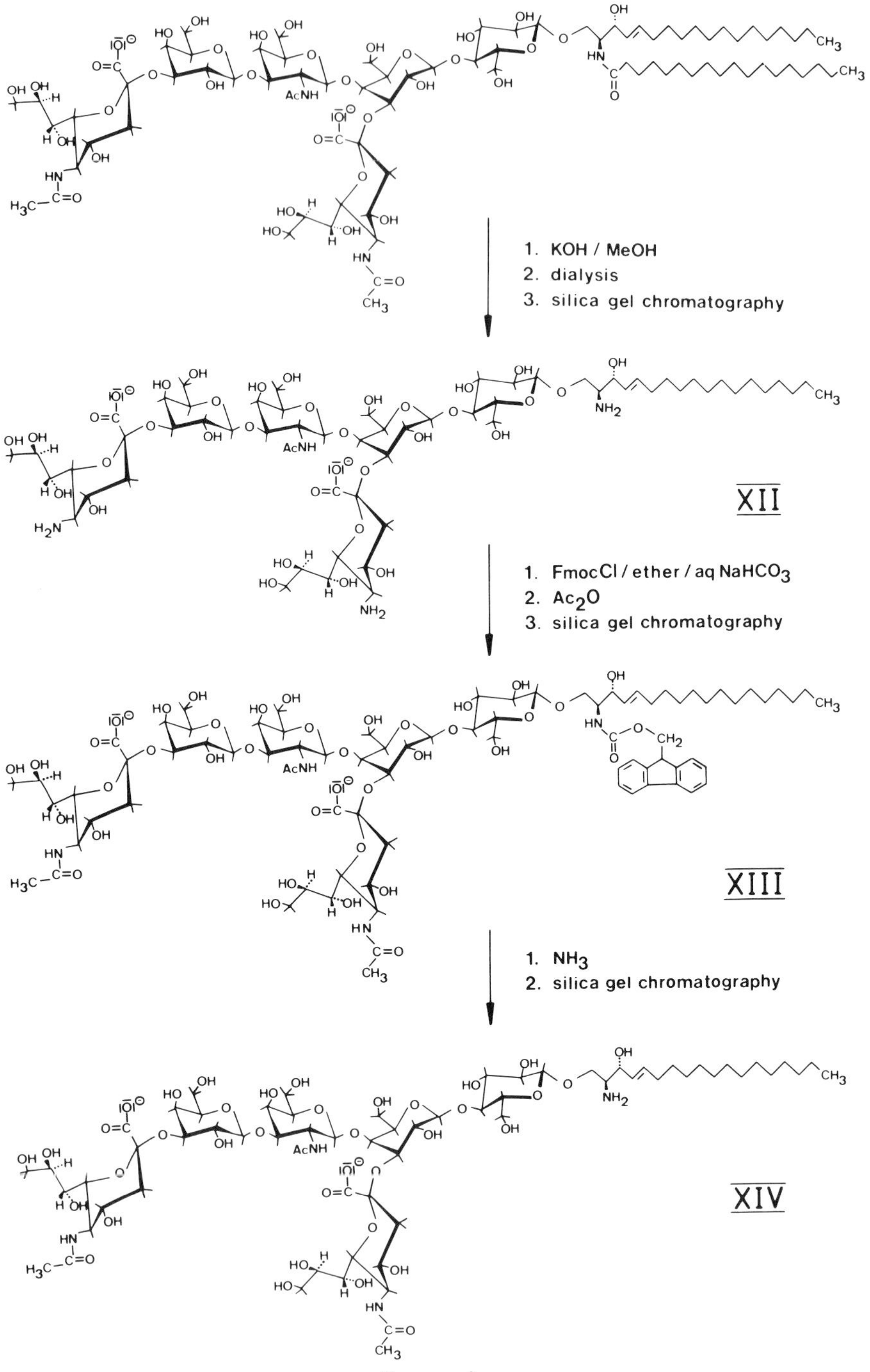

1. KOH / MeOH
2. dialysis
3. silica gel chromatography

$\overline{\text{XII}}$

1. FmocCl / ether / aq NaHCO$_3$
2. Ac$_2$O
3. silica gel chromatography

$\overline{\text{XIII}}$

1. NH$_3$
2. silica gel chromatography

$\overline{\text{XIV}}$

Scheme 2.

minutes, the ganglioside is completely dissolved and the solution turns yellow during the course of hydrolysis. After 20 hr the solution is cooled to 20° and neutralized by careful addition of acetic acid. Following removal of methanol in a nitrogen stream, the wet residue is suspended in water (3 ml) and dialyzed at 20° for 6 hr against distilled water (10 liters), with two changes. During dialysis some material sediments in the dialysis bag and is isolated by centrifugation. The pellet as well as the clear supernatant containing the deacylated ganglioside and fatty acids (in, however, different proportions) are lyophilized separately. To facilitate dissolution of the raw products the freeze-dried pellet is treated with methanol (0.8 ml) followed by chloroform (0.6 ml), whereas the freeze-dried supernatant is first treated with 2.5 M ammonia (0.18 ml) followed by chloroform (0.6 ml) under vigorous stirring. Both mixtures are then combined under vigorous stirring (Vortex) to yield an almost clear solution. The solution is then subjected to column chromatography on LiChroprep Si 100 (1.2 × 100 cm) in solvent system E (or F) for deacylated G_{M3} and G_{M2} (or G_{M1}, G_{D1a}, and G_{D1b}). The elution profile is monitored by TLC in the same solvent systems. All fractions containing a ninhydrin- as well as anisaldehyde-positive single major spot of the desired product are pooled and lyophilized from water following removal of the solvents. The yield of the deacylated gangliosides is in the range of 35–37.5 μmol (65–75%). A small amount of merely deacetylated gangliosides, i.e., gangliosides solely lacking the acetyl group(s) in their sialic acid residue(s), is usually also obtained in about 10–15% yield.

Note: An alternative method to deacylate ganglioside G_{M1} with tetramethylammonium hydroxide in aqueous n-butanol has been described by Sonnino et al.[28] In their procedure the yield of deacylated G_{M1} is lower and that of merely deacetylated G_{M1} higher than in our procedure.

N-[(9-Fluorenylmethoxy)carbonyl]lysogangliosides (c.g., **XIII**). The solution of the deacylated gangliosides (e.g., **XII**) (30 μmol) in 0.1 M sodium bicarbonate (3 ml) is mixed with diethyl ether (3 ml) and cooled in a freezer until the aqueous phase turns solid. Following the addition of 9-fluorenylmethyl chloroformate (Fmoc Cl) (7.8 mg, 30 μmol) in n-hexane (1 ml), the mixture is stirred vigorously at about 6–8° (cold room) for 24 hr. The progress of the reaction is checked by TLC in solvent systems G (or H) for N-Fmoc-deacyl-G_{M3} and -G_{M2} (or N-Fmoc-deacyl-G_{M1}, -G_{D1a}, and -G_{D1b}) using Kieselgel 60_{F254} plates. The Fmoc-protected products are visible by both fluorescence quenching and anisaldehyde spray.

If more than a negligible amount of deacylated gangliosides is still

[28] S. Sonnino, G. Kirschner, R. Ghidoni, D. Acquotti, and G. Tettamanti, *J. Lipid Res.* **26,** 248 (1985).

present, an appropriate amount of Fmoc Cl is further added to the reaction mixture and stirring is continued for another 20 hr. Thereafter, the reaction mixture is treated with small portions (10 μl) of acetic anhydride under vigorous stirring at 20° over a period of 3 hr until pII 5.5–6 is attained. Following evaporation of ether and most of the water, the salts are removed by dialysis as before and the product is freeze-dried. Final purification of N-[(9-fluorenylmethoxy)carbonyl]lysogangliosides is achieved by column chromatography on LiChroprep Si 100 (0.9 × 100 cm) using a linear gradient of solvent system J and K (or L and K) for N-Fmoc-lyso-G_{M3}, -G_{M2}, and -G_{M1} (or N-Fmoc-lyso-G_{D1a} and -G_{D1b}). The fractions giving rise to fluorescence quenching and anisaldehyde-positive spots on TLC in solvent systems G or H (see above), and which contain the pure product, are pooled and lyophilized after evaporation of the solvents. The N-protected lysogangliosides are quantified by weight and sialic acid content. The yield is usually 21 μmol (70%).

   Lysoganglioside (e.g., **XIV**). N-Fmoc-lysoganglioside (e.g., **XIII**) (20 μmol) is placed into a heavy-walled screw-capped vial (Pyrex, 16 × 100 mm) equipped with a T-piece through an open-hole cap. One end of the T-tubing is connected to an ammonia tank while the other end is fitted to a small balloon. The valve of the tank is opened a little in a way to slowly inflate the balloon with ammonia. When the balloon has reached half of its full size the vial is lowered into liquid nitrogen where ammonia condenses inside the vial. (If the condensing rate is too high the balloon starts to collapse. In this case the vial should be raised somewhat to reduce the condensing rate and to keep the balloon at a proper size.) After about 2 ml of frozen ammonia has been collected in the vial, the open-hole cap with a T-piece is quickly removed and the vial is immediately sealed with a Teflon-lined screw cap. After the vial has been kept at 20° for 2–3 hr, its content is again frozen in liquid nitrogen before the cap is unscrewed. The vial is then kept in a well-vented hood until ammonia is slowly but completely evaporated. The vial should not be placed in a metal block or water bath (not even ice water) since the energy transfer into the vial would be too fast and splashing of the vial's content would occur. The residue is purified by column chromatography on LiChroprep Si 100 in solvent system D for lyso-G_{M3}, E for lyso-G_{M2} and lyso-G_{M1}, and F for lyso-G_{D1a} and lyso-G_{D1b}. The elution profile is checked by TLC in solvent system G or H. All fractions containing ninhydrin- and Ehrlich spray-positive pure lysogangliosides (for R_f values see Table I) are pooled and freeze-dried after evaporation of the solvents. The amount of lyso-gangliosides is determined by weight and sialic acid content. General yields are 13–15 μmol (70–75%).

Lysogangliosides with Radiolabeled Sialic Acid Residue(s)

Principle

Deacylated ganglioside (e.g., **XII**) is protected at its sphingoid free amino group with the (9-fluorenylmethoxy)carbonyl (Fmoc) residue prior to the re-N-acetylation with ^{3}H- or ^{14}C-labeled acetic anhydride. The removal of the protecting group follows the same procedure as for the nonlabeled lysogangliosides.

Procedure

Tritium-Labeled Lysogangliosides. Deacylated ganglioside (5 μmol) is dissolved in 0.1 *M* sodium bicarbonate (0.5 ml) and mixed with diethyl ether (0.5 ml). After the aqueous phase is partially solidifying in a freezer, 9-fluorenylmethyl chloroformate (5 μmol, 1.3 mg) dissolved in *n*-hexane (0.2 ml) is added. The reaction mixture is stirred vigorously at 6–8° for 24 hr. If monitoring the reaction by TLC (see above) reveals only negligible amounts, if any, of unprotected deacylated gangliosides, no further addition of Fmoc Cl is necessary. After evaporation of the solvents, the residue is taken up in water (1 ml) and dialyzed against distilled water (10 liters, changed once) at 20° for 4 hr. The material retained is lyophilized and the freeze-dried material is purified by column chromatography (0.6 × 100 cm) as described for the *N*-Fmoc-lysogangliosides (see above). All fractions that on TLC in solvent systems G or H reveal one single major fluorescence-quenching and anisaldehyde-positive spot of *N*-Fmoc-deacylated gangliosides are pooled and, after evaporation of the solvent, lyophilized. The yield is usually in the range of 3 μmol (60%). Re-N-acetylation of this material is performed in DMF (0.3 ml) in the presence of diisopropylethylamine (10 μl) and [^{3}H]acetic anhydride (3.3 μmol) in toluene (16 μl). The mixture is stirred at 30° for 48 hr and the reaction is checked by TLC in solvent systems G or H. Usually the N-Fmoc-deacylated gangliosides are completely converted into labeled *N*–Fmoc-lysogangliosides with only trace amounts of their O-acetylated derivatives. The same procedure for removal of the Fmoc group and purification is followed as for the unlabeled lysogangliosides. The yield of this re-N-acetylation and *N*-Fmoc removal step is in the range of 60%.

Lysogangliosides with Tritium-Labeled Sphinganine Moieties

Principle

Lysogangliosides are hydrogenated at the sphingosine double bond with tritium gas generated *in situ* and palladium as catalyst.[12]

Procedure

³H-Labeled Lysoganglioside. Lysoganglioside (10 μmol) is dissolved in deaerated, argon-saturated water (1 ml) in a screw-capped vial. The solution is mixed with an equal volume of peroxide-free tetrahydrofuran and the mixture is kept under argon. Sodium boro[³H]hydride (10 μmol) dissolved in 50 μl of a mixture of peroxide-free diglyme and 4 M sodium hydroxide (1 : 1, v/v) is carefully layered under the lysoganglioside solution. Then palladium acetate (5 μmol) in diglyme (50 μl) is layered over the lysoganglioside solution and the vial is promptly sealed with a Teflon-lined screw cap, care being taken that any air above the solution has been replaced by argon. The mixture is now vigorously agitated (Vortex) for 24 hr at 25°. Immediately after agitating the mixture should have turned black due to finely dispersed palladium. To complete the hydrogenation solid, unlabeled borohydride (1.2 mg, 30 μmol) is added followed by the further addition of palladium acetate in diglyme (5 μmol in 50 μl), and the mixture is agitated for another 24 hr. After the addition of acetic acid (20 μl) the mixture and the rinses are transferred to a round-bottom flask and, after the addition of *n*-butanol to prevent foaming, are repeatedly dried from water in a rotary evaporator. This ensures almost complete removal of any exchangeable tritium. The residue taken up in water (2 ml) is dialyzed at 20° for 3 hr against distilled water (10 liters) (changed every hour). The material retained is freed of palladium by centrifugation and then lyophilized. Final purification and removal of the remaining salts are achieved by column chromatography on LiChroprep Si 100 (0.9 × 100 cm) as described for the unlabeled lysogangliosides. Yields are in the range of 90%.

Note: All operations should be performed in a well-vented hood. For optimal label incorporation it is mandatory to get rid of any free oxygen in the reaction vial since otherwise most of the tritium would be converted into tritiated water. Using sodium or potassium borohydride with a specific radioactivity of 5 Ci/mmol, labeled lysogangliosides with a specific radioactivity of about 1 Ci/mmol are easily obtained.

Synthesis of Labeled Gangliosides

Principle

This section describes the synthesis of gangliosides which carry labeled fatty acid residues in their ceramide portion. Lysogangliosides are selectively N-acylated with labeled fatty acid *N*-succinimidyl ester (e.g., *N*-succinimidyl pyrenedodecanoate, **XV**) in *N,N*-dimethylformamide

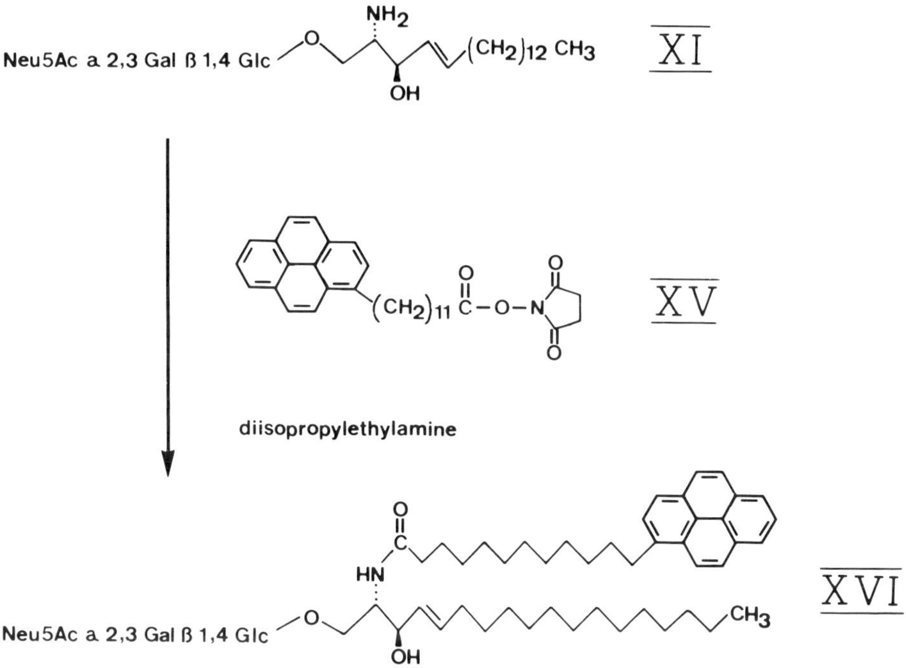

SCHEME 3.

(DMF) in the presence of diisopropylethylamine as exemplified in Scheme 3 for the synthesis of pyrene-labeled G_{M3} (**XVI**).

Procedure

Pyrene-Labeled Gangliosides. A solution of lyso-G_{M3}, -G_{M2}, -G_{M1}, -G_{D1a}, or -G_{D1b} (5 μmol) in DMF (0.4 ml) and diisopropylethylamine (20 μl) is mixed with *N*-succinimidyl pyrenedodecanoate (5 mg, 10 μmol) in DMF (0.1 ml). The clear mixture is kept under argon at 30° in the dark. After 2 days the reaction progress is monitored by TLC in solvent system G (or H) for pyrene-G_{M3}, -G_{M2}, and -G_{M1} (or -G_{D1a} and -G_{D1b}). The pyrene-labeled gangliosides have higher R_f values than their parent gangliosides and are visible both in UV light (366 nm) as well as after spraying with anisaldehyde reagent. If significant amounts of lysogangliosides should still be present in the reaction mixture, the reaction time should be prolonged. In this case it is also helpful to add more of the active ester. When TLC shows that almost all of the lysogangliosides have been converted into the labeled gangliosides the solvents are removed in a nitrogen jet and the residue is purified by column chromatography on LiChroprep Si 100 (0.6 × 90 cm). Pyrene-G_{M3}, -G_{M2}, and -G_{M1} are eluted with a linear gradient of solvent systems J and K and pyrene-G_{D1a} and -G_{D1b} are eluted with a linear gradient of solvent systems L and K. Fractions containing the pure labeled gangliosides are pooled and lyophilized. The yield is generally 4 μmol (80%) as determined by weight and sialic acid analysis.

N-Parinaroyl-lyso-G_{M1}.[28a] This compound is prepared from lyso-$_{GM1}$ (13 mg, 10 μmol) and the N-succinimidyl ester of *trans*-parinaric acid essentially as described for the synthesis of pyrene gangliosides with, however, some precautions. (1) Oxygen has to be strictly excluded at all times, and all solvents must contain 0.01% 2,6-di-*tert*-butyl-4-methylphenol (BHT). (2) The reaction and the product purification have to be performed in the dark or at least in very dim light in argon-saturated solvents.

Purification of the product is achieved by column chromatography on LiChroprep Si 100 (0.9 × 90 cm) using a linear gradient of solvent systems L and K. The N-parinaroyl-lyso-G_{M1} and G_{M1} have very similar R_f values on TLC in solvent system H. The yield of the product is in the range of 12 mg (80%). The product is stored in its column eluant under argon in screw-capped vials at $-20°$ until use.

NBD-Labeled G_{M1}.[28a] This fluorescent-labeled G_{M1} is prepared by N-acylation of lyso-G_{M1} (10 mg, 7.7 μmol) in DMF (0.48 ml) and diisopropylethylamine (20 μl) with N-succinimidyl 12-[N–methyl-N-(7'-nitrobenz-2'-oxa-1',3'-diazol-4'-yl)]aminododecanoate (12.5 mg, 25 μmol) as described for the synthesis of pyrene gangliosides. When TLC reveals a single major fluorescent and anisaldehyde-positive spot with R_f value similar to that of G_{M1} and no significant amounts of unreacted lyso-G_{M1}, the purification of the product is followed as described for N-parinaroyl-lyso-G_{M1}. In this case, however, BHT is omitted from the solvents. The yield of NBD-labeled G_{M1} is 8.4 mg (65%).

Spin-Labeled Gangliosides. 9-Doxyl-G_{M3} and -G_{M1}[4] as well as 14-doxyl-G_{M3}, -G_{M2}, -G_{M1}, and -G_{D1b} are synthesized by reacting the appropriate lysogangliosides (5 μmol) with the N-succinimidyl ester of 9-doxyl and 14-doxyl stearic acid, respectively, in DMF (0.3 ml) in the presence of diisopropylethylamine (10 μl) as described previously for the pyrene-labeled gangliosides. The reactions are monitored by TLC in solvent systems G or H. The spin-labeled gangliosides and their parent gangliosides have similar R_f values. Purification of the spin-labeled gangliosides is achieved by column chromatography on LiChroprep Si 100 (0.6 × 90 cm) using linear gradients of solvent systems J and K (or L and K) for 9-doxyl-

---

[28a] H. Stotz, Diplomarbeit, Universität Bonn (1986).

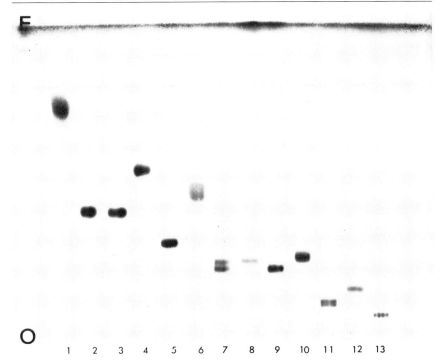

FIG. 1. Thin-layer chromatogram of lysogangliosides and their parent gangliosides. Plates were developed in solvent system H. Lane 1, G_{M3}; lane 2, lyso-G_{M3} obtained by Koenigs–Knorr condensation; lane 3, lyso-G_{M3} obtained from G_{M3}; lane 4, G_{M2}; lane 5, lyso-G_{M2}; lane 6, G_{M1}; lane 7, lyso-G_{M1} containing C_{20}- and C_{18}-sphingosine (upper and lower band, respectively); lane 8, lyso-G_{M1} (C_{20}-sphingosine); lane 9, lyso-G_{M1} (C_{18}-sphingosine); lane 10, G_{D1a}; lane 11, lyso-G_{D1a}; lane 12, G_{D1b}; lane 13, lyso-G_{D1b}. O, Origin; F, front.

G_{M3} and -G_{M1} and 14-doxyl-G_{M3}, -G_{M2}, and -G_{M1} (or 14-doxyl -G_{D1b}). The yield is in the range of 3.5 μmol (70%).

Notes: The labeled gangliosides with the exception of N-parinaroyl-lyso-G_{M1} (see above) are best stored as freeze-dried powder at −20° until use. Double-labeled gangliosides, i.e., gangliosides which in addition to the label described above contain a radiolabel in either its sialooligosaccharide or sphingoid moiety, are prepared from radiolabeled lyso-gangliosides.

Characterization of Lysogangliosides

The chromatographic behavior of the synthesized lysogangliosides and their purity are shown in Fig. 1 and their R_f values are listed in Table

TABLE I
R_f VALUES OF LYSOGANGLIOSIDES IN TLC[a]

| | R_f | |
Product	Solvent G[b]	Solvent H[b]
Lyso-G_{M3}	0.28	0.38
Lyso-G_{M2}	0.21	0.28
Lyso-G_{M1}[c]	0.17 (0.14)	0.21 (0.19)
Lyso-G_{D1a}	0.07	0.09
Lyso-G_{D1b}	0.04	0.05

[a] TLC was done at 20°. The spots were visualized by anisaldehyde spray, followed by heating for 20 min at 120°.

[b] Solvent G, chloroform–methanol–15 mM CaCl$_2$ (60:35:8, v/v/v). Solvent H, chloroform–methanol–15 mM CaCl$_2$ (60:40:9, v/v/v).

[c] Figures in parentheses are the R_f values of lyso-G_{M1} with C_{18}-sphingosine.

I. In thin-layer chromatography lysogangliosides separate partially or completely into two bands as seen most clearly for lyso-G_{M1}. This separation is due to their different sphingosine chain length. Occasionally, partial separation of these homologs have also been observed in column chromatography, Lyso-G_{M3} prepared from human spleen G_{M3} contains C_{18}-sphingosine as the only detectable base as does lyso-G_{M3} which has been synthesized from C_{18}-D-*erythro*-sphingosine by Koenigs–Knorr condensation. Lyso-G_{M2} contains mainly C_{18}-besides a little C_{20}-sphingosine.

Lysogangliosides have also been characterized by negative ion FAB mass spectrometry.[3] The high relative intensity of pseudomolecular ions (M-1) of lysogangliosides as shown in Table II confirms their composition of the respective sialooligosaccharide and sphingoid moiety. For all lysogangliosides described here, with the exception of lyso-G_{M3}, both C_{18}- and C_{20}-sphingosines were found. Beside the pseudomolecular ions, characteristic negative ions (Table II) are formed by splitting of glycosidic bonds during fast atom bombardment according to the fragmentation scheme illustrated in Fig. 2. Fragments A always contain the sphingoid residue whereas fragments B contain the terminal sugars. If fragments B contain sialic acid, they give rise to additional fragments smaller by 2 mass units.

The most diagnostic proton chemical shifts of lysogangliosides are listed in Table III. The signals of the anomeric protons observed in the spectra of lysogangliosides showed the same coupling constant as pub-

TABLE II

CHARACTERISTIC NEGATIVE IONS OF LYSOGANGLIOSIDES AND GANGLIOSIDE DERIVATIVES IN FAB MASS SPECTRA[a]

Glycolipids	M-1	M-1-NeuAc	A1	A2	A3	A4	B1	B2	B3	B4
Lyso-G_{D1a}	1569(m)	1278(m)	298(nd)	460(nd)	913(w)	1116(w)	470(m)	673(m)	1126(w)	1288(m)
	1597(s)	1306(s)	326(nd)	488(nd)	941(m)	1144(m)	468(m)	671(m)	1124(w)	1286(w)
Lyso-G_{D1b}	1569(m)	1278(m)	298(nd)	466(w)	1204(w)	1407(nd)	179	382(nd)	1126(w)	1288(w)
	1597(m)	1306(m)	326(nd)	488(w)	1232(w)	1435(nd)			1124(nd)	1286(nd)
Lyso-G_{M1}[b]	1278(nd)	987(nd)	298(nd)	460(nd)	913(nd)	1116(nd)	179	382(w)	835(w)	997(w)
	1306(s)	1015(w)	326(nd)	488(w)	941(w)	1144(w)			833(m)	995(w)
Lyso-G_{M2}	1116(s)	825(m)	298(nd)	460(w)	913(w)			220	673(m)	825(w)
	1144(w)	853(w)	326(nd)	488(nd)	941(w)				671(m)	833(w)
Lyso-G_{M3}	913(s)	622(m)	298(nd)	460(w)					470(m)	632(m)
	941(nd)	650(nd)	326(nd)	488(w)					468(m)	630(m)
14-doxyl-G_{M1}	1645(m)	1354(w)	665(w)	827(w)	1280(w)	1483(w)	179	382(nd)	835(m)	997(w)
	1673(m)	1382(nd)	693(w)	855(w)	1308(w)	1511(w)			833(m)	995(w)
14-doxyl-G_{D1b}	1936(m)	1645(nd)	665(w)	827(w)	1571(nd)	1774(nd)	179	382(nd)	1126(w)	1288(nd)
	1964(m)	1673(nd)	693(w)	855(w)	1599(nd)	1802(nd)			1124(w)	1286(nd)
G_{M1}[c]	1546(s)	1255(nd)	566(s)	728(m)	1181(w)	1384(nd)	179	382(nd)	835(m)	997(w)
	1574(s)	1283(nd)	594(s)	756(m)	1209(w)	1412(nd)			833(m)	995(w)

[a] The relative intensities of the negative ions are indicated as follows: s, strong, m, medium, w, weak; nd, not detectable. Fragments with $m/z \leq 220$ were not considered owing to the high background noise level of this region. For fragments M-1, M-1-NeuAc, A1 to A4, and B1 to B4, see text and Fig. 2.
[b] Purified lyso-G_{M1} with C_{20}-sphingosine.
[c] G_{M1} obtained upon hydrogenation of N-parinaroyl-lyso-G_{M1}.

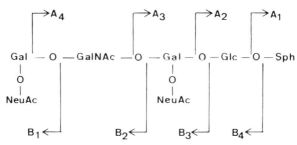

FIG. 2. Fragmentation of lysogangliosides by fast atom bombardment.

TABLE III

CHEMICAL SHIFTS[a] FOR METHYL, ANOMERIC, AND OLEFINIC PROTONS
OF LYSOGANGLIOSIDES

Protons	Lyso-G_{M3}	Lyso-G_{M2}	Lyso-G_{M1}	Lyso-G_{D1a}
Methyl				
Alkyl chain	0.86	0.86	0.85	0.86
GalNAc		1.77	1.74	1.75
Neu5Ac	1.90	1.88	1.88	1.89
Anomeric				
Glc	4.18	4.25	4.15	4.25
Gal (II)	4.25	4.28	4.27	4.25
Gal (IV)			4.20	4.25
GalNAc		4.77	4.84	4.77
Olefinic				
C_4-(sphingosine)	5.72	5.72	5.56	5.74
C_5-(sphingosine)	5.45	5.45	5.42	5.45

[a] Data in ppm. Obtained at 400 MHz at 25° in Me_2SO-d_6–D_2O (98:2,v/v);
referenced to internal Me_4Si.

lished for the respective gangliosides by Koerner et al.[29] A downfield shift
of about 0.2 and 0.1 ppm was, however, observed for the signal of the
protons at carbons 4 and 5, respectively, of the sphingosine moiety in
lysogangliosides as compared to the proton signals in the parent gangliosides that have no free amino group.

Characterization of Labeled Gangliosides

Spin-labeled gangliosides 14-doxyl-G_{M1} and -G_{D1b} have been characterized by their negative ion FAB mass spectra (Table II). Their frag-

[29] T. A. W. Koerner, J. H. Prestegard, P. C. Demou, and R. K. Yu, *Biochemistry* **22**, 2676
(1983).

ments B are the same as for their progenitor lysogangliosides, whereas fragments A and the pseudomolecular ions (M-1) which contain the doxyl-ceramide are higher by one mass unit than calculated for free radical-containing gangliosides.[30] 9-Doxyl-G_{M3} and -G_{M1} as well as 14-doxyl-G_{M3}, -G_{M2}, -G_{M1}, and -G_{D1b} gave characteristic electron spin resonance signals once incorporated into lipid bilayers. N-Parinaroyl-lyso-G_{M1} gave the same UV spectra as free *trans*-parinaric acid with λ_{max} at 285, 298, and 313 nm. After hydrogenation the expected negative ions in FAB mass spectrometry were obtained (Table II).

Acknowledgment

The authors gratefully adknowledge the excellent technical assistance of Christa Schönenberg and Annemie Laufenberg. The FAB mass spectra were recorded by Dr. Jasna Peter-Katalinić, who is gratefully acknowledged. We thank Dr. H. K. M. Yusuf for his help with the manuscript and Miss Waltraud Rau for typing it. We also thank Dr. R. Klein for his kind gift of parinaric acid. This work was supported by Grants Schw 143/5-3 and Sa 257/11-5 from the Deutsche Forschungsgemeinschaft.

[30] We have to assume that in the progress of FAB mass spectrometry a hydrogen radical combines with the nitroxide free radical.

[27] Chemical Deglycosylation of Glycoproteins

By Hakimuddin T. Sojar *and* Om P. Bahl

Introduction

Deglycosylation of glycoproteins is necessary for a number of studies such as in the structural determination of polypeptide chains, establishment of structure and function relationships of carbohydrates,[1-5] and their biosynthesis in glycoproteins. In all these studies it is important that the conditions used for chemical deglycosylation do not cause an impairment of the physicochemical integrity of the polypeptide chain. Also, the qua-

[1] N. K. Kalyan and O. P. Bahl, *J. Biol. Chem.* **258**, 67 (1983).

[2] P. Manjunath and M. R. Sairam, *J. Biol. Chem.* **257**, 7109 (1982).

[3] Y. Shimohigashi and H. C. Chen, *FEBS Lett.* **150**, 64 (1982).

[4] C. H. Pletcher, R. M. Resnick, G. J. Wei, V. A. Bloomfield, and G. L. Nelsetuen, *J. Biol. Chem.* **255**, 7433 (1980).

[5] H. C. Chen, Y. Shimohigashi, M. L. Dufau, and K. J. Catt, *J. Biol. Chem.* **257**, 14446 (1982).

ternary structure of the glycoprotein is preserved. Two chemical reagents in the anhydrous form, trifluoromethanesulfonic acid[6,7] (TFMS) and hydrogen fluoride[8] (HF), have been employed for the deglycosylation of glycoproteins. Both reagents were initially used for the deblocking of the protecting groups from synthetic peptides as well as their dissociation from the insoluble matrix in the solid-phase peptide synthesis. Both reagents hydrolyze glycosidic bonds involving neutral sugars more readily than those involving N-acetylhexosamines. The N-glycosidic bonds are less stable than the O-glycosidic bonds since the latter can be cleaved under somewhat stronger experimental conditions.[8,9] TFMS by the modified procedure described herein is definitely superior to HF. It is much more potent and sensitive and is more convenient to use than HF since it does not require any special handling. In order to prevent any secondary reactions such as the alkylation of the polypeptide chains during deprotection of the protecting group, the use of anisole was made as a scavenger. More recently dimethyl sulfide has also been used in place of anisole. However, in the deglycosylation of glycoproteins, the inclusion of anisole or dimethyl sulfide in the acid is unnecessary. This communication describes the deglycosylation of two different types of glycoproteins, fetuin and human chorionic gonadotropin (hCG), by TFMS and HF. Fetuin has a single polypeptide chain with three Asn-linked triantennary "complex" type and three Ser/Thr-linked carbohydrate units. HCG is a glycoprotein with a quaternary structure made up of two noncovalently bonded α- and β-subunits, each having two biantennary complex type carbohydrates. The β-subunit also has four Ser-linked carbohydrates.

Reagents

Anhydrous trifluoromethanesulfonic acid (Aldrich Chemical Co.), distributed in small aliquots and stored at $-20°$ in sealed ampules

Hydrogen fluoride

Pyridine, Baker's analyzed reagent grade.

Fetuin (Sigma Chemical Co.) was further purified on Sephadex G-100 using 1% NH_4HCO_3 buffer

Human chorionic gonadotropin and its α- and β-subunits (hCGα, hCGβ) prepared in the laboratory.[1]

---

[6] N. K. Kalyan and O. P. Bahl, *Biochem. Biophys. Res. Commun.* **102,** 1246 (1981).

[7] A. S. B. Edge, C. R. Faltnynek, L. Holf, L. E. Reichert, Jr., and P. Weber, *Anal. Biochem.* **118,** 131 (1981).

[8] A. J. Mort and D. T. A. Lamport, *Anal. Biochem.* **82,** 289 (1977).

[9] H. T. Sojar and O. P. Bahl, *Arch. Biochem. Biophys.* (1986) submitted.

Chemical Deglycosylation

Procedure by TFMS.[1,6] Fetuin, hCG, and the α- and β-subunits of hCG were dried under vacuum overnight at room temperature. All subsequent steps were performed at 0° or below by keeping the reaction mixture on ice. In a typical experiment 0.1 to 1 mg of fetuin, hCG, or the α- or β-subunit of hCG was treated with 15–150 μl of TFMS. The reaction mixture was incubated at 0° for 0.5–2 hr under nitrogen in a 1-ml screw-capped vial with occasional shaking. Subsequently, the reaction mixture was cooled to below −20° by placing in dry ice–ethanol contained in a small beaker, and then neutralized by the gradual addition of 60% pyridine in water, also previously cooled to below −20°. The temperature must be carefully controlled since the pyridine–TFMS salt formation is a strongly exothermic reaction. The neutralized reaction mixture was dialyzed at 4° in a Spectrapor dialysis tubing (Spectrum Medical Ind.), molecular weight cut off 2000, against several changes of 0.01% NH_4HCO_3 (pH adjusted to 7.0 by bubbling CO_2). The sample was then applied to a Sephadex G-75 (superfine) column (1 × 95 cm). The column was previously equilibrated with 0.1% NH_4HCO_3, pH 7.5 (adjusted by bubbling CO_2). The fractions containing the deglycosylated protein were pooled and lyophilized.

Procedure by HF.[2,8] Anhydrous HF causes severe burns and is hazardous and can best be handled in a closed system. The apparatus used for the experiments can be purchased from Peninsula Laboratories Inc. (San Carlos, CA) or Protein Research Foundation, Minoh (Osaka, Japan) or can be constructed in Teflon-Kel-F material.[10] The protein sample was vacuum-dried prior to treatment with HF. A 10- to 100-mg sample of fetuin or hCG was placed in the reaction vessel and the reaction was connected to the HF apparatus. After evacuating the entire HF line 10 ml of anhydrous HF was distilled over from the reservoir with stirring of the reaction vessel. The reaction is continued for 1–2 hr at 0° or at room temperature for the cleavage of O-glycosidic carbohydrate linkages. After the reaction HF was removed by a water aspirator during 15–30 min. This was done carefully and gently to avoid excessive bubbling. The last traces of HF were removed under high vacuum. Finally, the sticky reaction mixture was dissolved in 2 ml of 0.2 M NaOH to neutralize any remaining HF and the pH was readjusted to 7.5 using cold 0.2 M HCl.

The purification of the deglycosylated protein was carried out on a column of Sephadex G-100 equilibrated and eluted with 0.05 M NH_4HCO_3 at 4°. The deglycosylated protein was separated from the partially degly-

---

[10] M. P. Sanger and D. T. A. Lamport, *Anal. Biochem.* **128**, 66 (1983).

cosylated material by concanavalin A–Sepharose chromatography. A column (1.5 × 22 cm) of concanavalin A–Sepharose was equilibrated with 25 mM Tris–HCl buffer (pH 7.4) containing 100 mM $MgCl_2$, 10 mM $CaCl_2$, 0.05 NaCl, and 0.02% sodium azide. After the application of the sample the column flow was discontinued for 60 min to allow for the complete interaction of the protein with the matrix. The column was eluted with the above buffer. The deglycosylated protein eluted from the column unretarded. The fractions containing the deglycosylated material were concentrated by ultrafiltration, dialyzed, and lyophilized.

Another set of conditions[11] has also been employed for deglycosylation by HF. One microliter anisole and 10 μl 70% HF in pyridine were used for every 4 μg protein. The reaction was allowed to continue for 4–24 hr and was stopped by drying the samples *in vacuo* under NaOH trap. When these conditions were applied to erythropoietin,[12] the resulting deglycosylated hormone was inactive. It is not clear, however, from these studies whether it is the carbohydrate removal or denaturation caused by experimental conditions which is responsible for the inactivation of the hormone.

Comments

TFMS cleaves about 90% carbohydrate from fetuin, hCG, and hCG subunits as determined by the amount of hexosamines released at 0° in 0.5–1 hr. Although 1 hr reaction time is sufficient for most glycoproteins, it may be necessary to determine the length of incubation for each glycoprotein separately, particularly if the deglycosylation is incomplete in 1 hr. In the case of above proteins, even 0.5 hr incubation is adequate for the cleavage of N-glycosidic linkages. However, the O-glycosidic linkages are more stable than the N-glycoside linkages and require longer incubation period. The O-glycosidic linkages in fetuin and hCGβ are cleaved in 1 hr incubation with TFMS at 0°. The use of a base such as dimethyl sulfide or anisole is not necessary. It not only slows down the deglycosylation reaction but may cause inactivation of the protein. Its omission from the acid does not cause any detectable degradation of the polypeptide chain in a 1- to 2-hr period, as shown by polyacrylamide gel electrophoresis of fetuin and hCG after deglycosylation with TFMS with or without dimethyl sulfide (Fig. 1). It is also clear from the figure that the deglycosylation is essentially complete in 1 hr in the absence of the base. The extent of cleavage of carbohydrates from fetuin, hCG, and hCG

[11] C. Coudron, K. Ellis, L. Philipson, and N. B. Schwartz, *Biochem. Biophys. Res. Commun.* **92**, 618 (1980).

[12] M. S. Dordal, F. F. Wang, and E. Goldwasser, *Endocrinology* **116**, 2293 (1985).

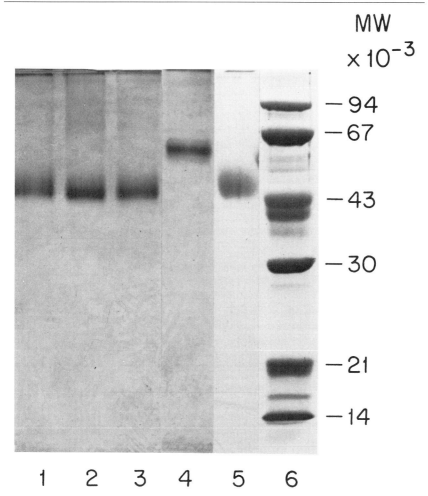

FIG. 1. SDS–polyacrylamide gel electrophoresis of fetuin before and after deglycosylation. Electrophoresis was carried out in 10% polyacrylamide gel, pH 8.3. The protein bands were visualized by silver staining. Lanes 1, 2, and 3, fetuin treated with TFMS alone for 0.5, 1, and 1.5 hr, respectively; lane 4, native fetuin; lane 5, fetuin treated with TFMS containing 25% dimethyl sulfide (v/v) for 4 hr; lane 6, molecular weight markers, phosphorylase *b* (94K), bovine serum albumin (67K), ovalbumin (43K), carbonate dehydratase (30K), soybean trypsin inhibitor (21K) and α-lactalbumin (14K).

subunits is greater by TFMS than that obtained by HF (Tables I and II). It is worth noting that in 1 hr HF does not cleave the O-glycosidic bonds in fetuin and hCG while TFMS at 0° does cleave these linkages. HF requires 3 hr incubation at room temperature for the cleavage of O-glycosidic

TABLE I

RELEASE OF HEXOSAMINES FROM FETUIN WITH TFMS AND HF

Hexosamine	TFMS[a] (% released)				HF[b] (% released)
	0.5 hr	1 hr	1.5 hr	2 hr	1 hr
GlcNAc	79.80	87.80	87.80	93.80	59.00
GalNAc	29.80	55.00	55.00	75.60	6.00

[a] Treated with TFMS at 0° for 0.5 to 2 hr.
[b] Treated with HF containing 10% anisole (v/v) at 0° for 1 hr.[8]

linkages. Similar conditions using HF have also been reported for the deglycosylation of porcine submaxillary mucin[8] which contains a large number of O-glycosidically linked carbohydrate chains.

The protein recovery by TFMS is higher than that by HF. TFMS yields homogeneous deglycosylated products (Fig. 2) and does not apparently affect the integrity of the polypeptide chains as shown by physicochemical, immunological, and biological criteria. Deglycosylated hCG (dhCG) obtained by TFMS has UV and circular dichroic spectra identical to those of the intact molecule. Figure 3 shows the circular dichroic spectra of the deglycosylated and the native α- and β-subunits of hCG. There is no loss of antibody and receptor binding activity on deglycosylation (Figs. 4, 5, and 6). In fact, the receptor binding activity of dhCG is 2-fold greater than that of native hCG (Figs. 4 and 6). Similar results have also been reported for HF.

TABLE II

HEXOSAMINE COMPOSITION OF DEGLYCOSYLATED HCG BY TFMS AND HF

Hexosamine	TFMS[a] (% released)	HF (% released)	
		I[b]	II[c]
GlcNAc	70.00	47.60	52.00
GalNAc	57.00	5.0	14.00

[a] HCG was treated with TFMS at 0° for 1 hr.
[b] I, 50 mg of hCG was treated with 10 ml of HF at 0° for 60 min.[2]
[c] II, 30 mg of hCG was treated with 8 ml of HF containing 0.5 ml of anisole at 0° for 60 min.[5]

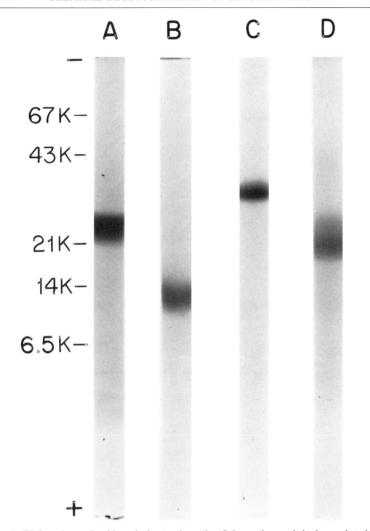

Fig. 2. SDS–polyacrylamide gel electrophoresis of the native and deglycosylated subunits. The subunits samples were run in 12.5% acrylamide gel along with the proteins of known molecular weights. A, hCGα; B, hCGdα; C, hCGβ; and D, hCGdβ where hCGdα and hCGdβ represent deglycosylated α- and β-subunits, respectively.

It is interesting to note that dhCG is an antagonist of hCG[1] *in vitro* and *in vivo*. While dhCG binds to the receptor as well as or better than the native molecule, it loses its ability to stimulate cAMP production and steroidogenesis. Furthermore it completely inhibits the hCG-induced cAMP stimulation.

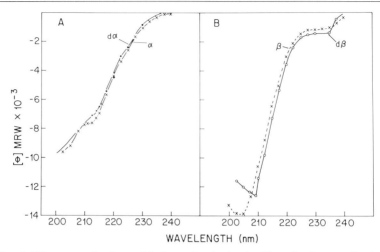

FIG. 3. CD spectra of native and deglycosylated subunits. The subunits were dissolved in the concentration of 0.3 mg/ml in 10 mM phosphate buffer, 50 mM KCl, pH 7.5. Measurement of the spectra between 240 and 200 nm was carried out in a cuvette of 1-mm path length with the spectrophotometer set at a sensititivy of 50 × 10 m°/cm, wavelength extension of 5 nm/cm, and time constant of 16S.

The protein recovery by TFMS is almost quantitative and so is the receptor binding activity of the dhCG obtained by deglycosylation of intact hCG. Thus, the modified procedure for deglycosylation with TFMS is much more effective and efficient than that reported earlier using anis-

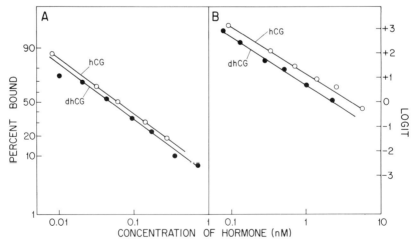

FIG. 4. Dose–response curves of the intact hormone in (A) radioimmuno and (B) radioreceptor assays. The results are expressed as logit response versus log of the dose.

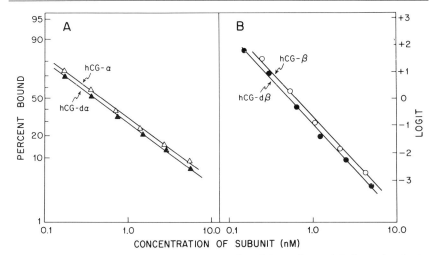

Fig. 5. Comparison of immunological properties of the native and deglycosylated subunits of hCG. A, RIA in [^{125}I]hCGα–anti-hCGα system. B, RIA in [^{125}I]hCGβ–anti-hCGβ system. HCG-dα and hCG-dβ represent deglycosylated subunits. The percentage bound, displayed as the logit transform, is plotted against the log of concentration.

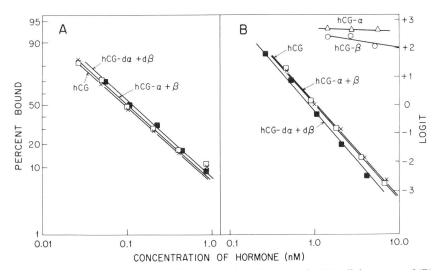

Fig. 6. Dose–response curves of the reconstituted hormone in (A) radioimmuno and (B) radioreceptor assays. The results are expressed as logit response versus log of the dose. HCG-α + β and hCG-dα + dβ represent the reconstituted hormones obtained from native and deglycosylated subunits.

ole as a scavenger. The dissociation of the hormone into subunits during the reaction is minimal in 1 hr and can be further reduced by reducing the time of incubation. The overall recovery of the active material is greater than 80%. Thus, under appropriate conditions deglycosylation of multimeric glycoproteins can be carried out with the intact molecule without causing its significant dissociation into subunits. The deglycosylation of hCG subunits can also be carried out under the above conditions. The resulting deglycosylated subunits readily reassociate to yield 90% reconstituted dhCG with receptor binding activity again twice as much as that of hCG.

In summary, TFMS provides a powerful reagent for deglycosylation of glycoproteins. The procedure described above is simple and reproducible, and does not cause any detectable alteration in the protein moiety. The reagent, therefore, should prove to be highly useful in the study of biosynthesis and function of carbohydrates in glycoproteins. It should also facilitate the amino sequence determination of highly glycosylated glycoconjugates such as mucins and proteoglycans.

Acknowledgments

This work was supported by NIH Grant HD-08766. The authors wish to thank Mrs. Ursula Brunn for typing the manuscript.

[28] Enzymatic Deglycosylation of Glycoproteins

By Nageswara R. Thotakura and Om P. Bahl

Deglycosylation of glycoproteins is important in elucidation of structure, function, and biosynthesis of biologically significant glycoproteins. For such studies, it is imperative that the integrity of the polypeptide chain is maintained during deglycosylation. In this regard, a wide variety of specific glycosidases which cleave carbohydrates[1-3] and oligosaccharide chains[4-9] have been found to be potentially useful. Enzymatic

[1] J. M. Governman, T. F. Parsons, and J. G. Pierce, *J. Biol. Chem.* **257**, 15059 (1982).

[2] W. R. Moyle, O. P. Bahl, and L. Marz, *J. Biol. Chem.* **250**, 9163 (1975).

[3] T. E. Maione and A. T. Jagendorf, *Proc. Natl. Acad. Sci. U.S.A.* **81**, 3733 (1984).

[4] A. L. Tarentino and F. Maley, *J. Biol. Chem.* **249**, 811 (1974).

[5] J. H. Elder and S. Alexander, *Proc. Natl. Acad. Sci. U.S.A.* **79**, 4540 (1982).

[6] T. H. Plummer, Jr., J. H. Elder, S. Alexander, A. W. Phelan, and A. L. Tarentino, *J. Biol. Chem.* **259**, 10700 (1984).

deglycosylation can be brought about by exoglycosidases either used sequentially or in a mixture or by endoglycosidases. A large number of highly purified exoglycosidases are available from a variety of sources[10] including *Streptococcus pneumoniae*,[11] *Clostridium perfringens*,[12] *Aspergillus niger*,[13] *Phaseolus vulgaris*,[14] Jack bean meal,[15] and hen oviduct.[16] Endoglycosidases can be classified into three major groups: (1) the endo-*N*-acetyl-β-D-glucosaminidases designated as endo D (*S. pneumoniae*),[17] endo H (*Streptomyces plicatus*),[4] and endo F (*Flavobacterium meningosepticum*)[5-8] which are specific for di-*N*-acetylchitobiose moiety in the core of Asn-linked oligosaccharide chains, (2) endo-β-D-galactosidase[18] (*S. pneumoniae*) which hydrolyzes galactosidic bonds in poly(*N*-acetyl-lactosamine) type oligosaccharide prosthetic groups, and (3) endoglyco-peptidases designated as peptide *N*-glycosidases (PNGases)[4-8] and peptide *O*-glycosidases[19] (endo-*N*-acetyl-α-D-galactosaminidase), which hydrolyze the *N*-acetylglucosaminylasparagine and *N*-acetylgalac-tosaminylserine/threonine linkages, respectively. Recently, considerable progress has been made in the isolation and characterization of several endoglycosidases. Whereas peptide *O*-glycosidase and endo-β-D-galac-tosidase (*S. pneumoniae*) are of limited interest due to their narrow substrate specificities, endo D, H, and F have been found to be efficient tools for N-deglycosylation of glycoproteins. Moreover, two other promising endoglycopeptidases, PNGase A from almond emulsin[9] and PNGase F from *F. meningosepticum*,[4-8] have been recently isolated and characterized.

Deglycosylation by Exoglycosidases

Deglycosylation of glycoproteins by exoglycosidases can be carried out by using the enzymes sequentially[2] or in a mixture.[1,3] Since both

[7] A. L. Tarentino, C. G. Gomez, and T. H. Plummer, Jr., *Biochemistry* **24**, 4665 (1985).

[8] F. K. Chu, *J. Biol. Chem.* **261**, 172 (1986).

[9] E. M. Taga, A. Waheed, and R. L. Van Etten, *Biochemistry* **23**, 815 (1984).

[10] O. P. Bahl and P. V. Wagh, *CRC Crit. Rev. Biochem.* **10**, 307 (1981).

[11] R. C. Hughes and R. W. Jeanloz, *Biochemistry* **3**, 1535 (1964).

[12] See this series, Vol. 28 [96].

[13] See this series, Vol. 28 [93–95].

[14] See this series, Vol. 28 [92].

[15] See this series, Vol. 28 [90].

[16] See this series, Vol. 28 [99] and [100].

[17] N. Koide and T. Muramatsu, *J. Biol. Chem.* **249**, 4897 (1974).

[18] M. N. Fukuda, *Biochemistry* **24**, 2154 (1985).

[19] J. Umemoto, V. P. Bhavanandan, and E. A. Davidson, *J. Biol. Chem.* **252**, 8609 (1977).

procedures call for long incubation periods for exhaustive deglycosylation, highly purified enzymes devoid of proteolytic activity should be
employed to prevent any degradation of the polypeptide chain. As a further precaution, various inhibitors of proteases such as phenylmethylsulfonyl fluoride, 6-aminocaproic acid, and p-aminobenzamidine[3] should
be included in the incubation medium with the glycosidases, or the glycosidases should be preincubated briefly prior to their addition to the substrate. For sequential degradation, it is important that the glycosidases
are free from cross-contamination. Although most of the glycosidases are
generally stable at their pH optima at room temperature and can survive
extended incubation periods, multiple additions of the enzymes are helpful. The rate of enzymatic hydrolysis depends both on the nature of the
carbohydrate and on the size of the glycoprotein substrate. The precise
conditions of enzyme concentration and time of incubation should be
determined for each unknown glycoprotein. Heat-denatured glycoproteins are better substrates than those in their native forms probably because of improved accessibility of the glycan to the enzymes on denaturation. The peripheral sugars are more susceptible to hydrolysis than those
located internally, more so than those in the proximity of the polypeptide
chain. Finally, it should be realized that while the glycoproteins are generally soluble in aqueous buffers, the recoveries of the deglycosylated analogs may be low due to their reduced solubility following deglycosylation.

Procedure

Each milligram of the α and β subunits of human choriogonadotropin
(hCG) was treated with a mixture of glycosidases from *S. pneumoniae*[1]
containing about 0.6 mU of α-L-fucosidase, 0.6 mU of β-galactosidase,
0.9 mU of α-mannosidase, 40 mU of β-N-acetylglucosaminidase, and 7
mU of α-N-acetylgalactosaminidase. One unit of each enzyme hydrolyzes
1 μmol of the corresponding p-nitrophenylglycoside per minute at 37°.
The enzyme mixture in a final volume of 250 μl of 50 mM cacodylate
buffer (pH 6.5) was preincubated with 50 μM N-ethylmaleimide for 10
min at room temperature. It was further incubated with diisopropyl fluorophosphate at 0.125% concentration for 45 min at 37°. One milligram of
the α or β subunit of hCG (final concentration 4 mg/ml) was then added
directly to the enzyme mixture and incubated for 48 hr at 37°. The resulting deglycosylated subunits were purified on a Sephadex G-100 column
using 1% ammonium bicarbonate for elution. The purification can also be
carried out by a Con A-Sepharose column.

HCG was also deglycosylated by sequential exoglycosidase digestion.[2] About 40 mg of hCG was incubated with the following enzymes in

the order and for times indicated: 100 U of *Vibrio cholerae* neuraminidase (one unit of the enzyme releases 1 μg of sialic acid in 15 min at 37°) for 18 hr, 8 U of *A. niger* β-D-galactosidase for 24 hr, 8.6 U of *A. niger* β-*N*-acetylglucosaminidase and 0.05 U of *A. niger* α-D-mannosidase for 96 hr (one unit of each of these enzymes releases 1 μmol of *p*-nitrophenol from the corresponding *p*-nitrophenylglycoside per minute at 37°). The hydrolyses were performed at 37° in 50 m*M* sodium acetate buffer (pH 4.6) except for neuraminidase for which the buffer also contained 2 m*M* calcium chloride and 0.2 m*M* EDTA. The digestions were carried out under nitrogen in the presence of toluene in a total volume of 1.2–1.5 ml. After each step, the sugar released was separated by dialysis and estimated and the carbohydrate-degraded protein was separated from the enzyme by chromatography on DEAE–cellulose (0.7 × 18 cm) using a linear gradient of 0–0.4 *M* NaCl in 0.05 *M* sodium acetate buffer (pH 5.0).

Comments

HCG is a glycoprotein hormone containing two noncovalently bonded α and β subunits. Each subunit has two biantennary Asn-linked carbohydrate units and the β subunit has additionally four Ser-linked short oligosaccharide chains.[20] The mixture of glycosidases resulted in the removal of all sugars except fucose and two N-linked *N*-acetylglucosamines from both α and β subunits of hCG.[1] The O-linked oligosaccharide chains were also cleaved as indicated by the loss of *N*-acetylgalactosamine by which the *O*-oligosaccharides are linked to serine residues in the carboxy terminus of the β subunit. The sequential[2] degradation of native hCG with exoglycosidases was not as extensive as that of the subunits with the mixture of glycosidases.[1] While sialic acid residues were removed almost quantitatively, galactose, *N*-acetylglucosamine, and mannose residues were hydrolyzed to the extent of 60–70, 50–60, and 20–30%, respectively. Apparently, the extent of hydrolysis progressively decreased from the peripheral to the internal sugar residues, probably due to the greater accessibility of the outer sugars to the enzymes. Furthermore, reduced hydrolysis of hCG carbohydrate by sequential treatment compared to that of the subunits by the mixture of glycosidases may be ascribed to the larger size of hCG and differences in the potency and specificity of the enzymes used. Thus, the amount of carbohydrate cleaved depends on the nature of the glycan substrate and the enzymes used.

The deglycosylated subunits of hCG recombined readily into deglycosylated hCG (dhCG). Both dhCG and sequentially degraded hCG analogs

[20] N. K. Kalyan and O. P. Bahl, *J. Biol. Chem.* **258**, 67 (1983).

retained their antibody and receptor binding properties, indicating that the conformational integrity of the polypeptide chains was not affected during deglycosylation and the carbohydrate does not play a role in these properties. All of the carbohydrate modified or deglycosylated derivatives of hCG were inhibitors of hCG-mediated cyclic AMP and steroid stimulation. Another example of the use of sequential exoglycosidase treatment is that of spinach chloroplast coupling factor I.[3] Treatment of this enzyme with a different set of glycosidases caused the removal of about 75% carbohydrate associated with it. The ATPase activity of the enzyme was unaltered on deglycosylation. The deglycosylated enzyme, however, was unable to catalyze photophosphorylation when recoupled with coupling factor I-depleted thylakoids.

Deglycosylation by Endoglycosidases

Recently, various endoenzymes such as endo H, endo F, PNGase A, and PNGase F have become available in highly purified form. Amino acid sequencing of endo H and the cloning of its gene in *Escherichia coli* have been accomplished.[21] Apparent molecular weights of endo H or endo F, PNGase F, and PNGase A by SDS–PAGE are 32,000, 35,500, and 89,000, respectively.

Some of the commercially available endo F preparations (Boehringer Mannheim and New England Nuclear) are contaminated with endoglycosidase PNGase F. Recently, Tarentino *et al.*[7] have been able to completely separate these two activities by a simple purification procedure. All the above endoenzymes are inactivated by SDS. However, nonionic detergents have a stabilizing effect. The digestions of the SDS denatured proteins, therefore, are carried out in the presence of a nonionic detergent such as Nonidet P-40. The detergent should be added to the enzyme before the addition of the SDS–protein mixture. Citrate completely inhibits the activity of PNGase F. The pH optimum for endo H, endo F, and PNGase A is between 4 and 5, while that for PNGase F is around 8.5. Deglycosylation with endoenzymes is carried out at 37° but it appears that temperatures higher than 23° do not accelerate the rate of deglycosylation, probably due to the heat inactivation of the enzyme PNGase F.

Denatured glycoproteins are more susceptible to hydrolysis with these enzymes than glycoproteins in their native conformation. Complete deglycosylation of glycoproteins in their native conformation can, however, be achieved by using much higher concentrations of the enzyme. PNGase F appears to be inactivated at very low concentrations of the enzyme. The

[21] P. W. Robbins, R. B. Trimble, D. F. Wirth, C. Herring, F. Maley, G. F. Maley, D. Rathin, B. W. Gibson, N. Royal, and K. Biemann, *J. Biol. Chem.* **259,** 7577 (1984).

precise conditions for hydrolysis including enzyme concentration and the incubation time vary with the substrate and, therefore, should be determined for each unknown glycoprotein separately.

Endo H and endo F preferentially cleave high-mannose carbohydrate chains but the latter can also hydrolyze biantennary complex chains at a slower rate. Unlike endo H, endo F does not seem to significantly hydrolyze oligosaccharide chains with bisect β-1,4- N-acetylglucosamine. These enzymes do not hydrolyze tri- and tetraantennary carbohydrate chains. On the other hand, PNGase A and PNGase F can readily hydrolyze high-mannose or complex multibranched oligosaccharide chains including tri- and tetraantennary structures, except when they are located on the amino or carboxy termini. The minimum carbohydrate chain required to act as a substrate for PNGase F has been shown to be di-N-acetylchitobiose core.[8] PNGase A and F have similar substrate specificities except the latter appears to be much more potent. When considering the use of N-glycosidase it should be borne in mind, particularly if the functional aspects of the glycoprotein are to be studied, that the cleavage of the N-acetylglucosaminylasparagine linkage by the enzyme results in the conversion of asparagine residue(s) to aspartic acid, thereby introducing one negative charge per glycosylation site in the protein. In the case of a glycoprotein with several glycosylation sites such as α_1-acid glycoprotein (five sites), sufficient negative charge may be introduced in the molecule to alter its conformation.

Procedure

Hydrolysis by Endo F. To 25 μg of the protein (5 mg/ml), 0.9 mU (one unit hydrolyzes 1 μmol of [^{3}H]dansylovalbumin glycopeptide per minute at 37°) of endo F from *F. meningosepticum* (Genzyme) and 3 μl of 0.1 M 1,10-phenanthroline hydrate in methanol were added in a total volume of 30 μl of 0.2 M sodium acetate buffer (pH 5.8). The incubation was performed overnight (18 hr) at 37°. The digested proteins were subjected to SDS–PAGE according to Laemmli.[22] For deglycosylation under denaturing conditions, the glycoprotein sample was treated with 0.5% SDS and 1% 2-mercaptoethanol for 3 min at 100°. Nonidet P-40 at a final concentration of 1.25% was added to the incubation mixture before the addition of the enzyme. In case of incomplete deglycosylation, the deglycosylated protein was separated from the undigested material by passing the reaction mixture through a Con A-Sepharose column. The flow-through fraction contained the deglycosylated protein with only a single N-acetylglucosamine residue linked to asparagine. If complete deglycosylation is

[22] U. K. Laemmli, *Nature* **227**, 680 (1970).

achieved as indicated by PAGE, the released oligosaccharide can be removed by filtration through a membrane (Centricon) with 5000 MW cutoff. HPLC and conventional gel filtration and ion-exchange chromatographies can also be employed depending on the molecular weight and charge differences between the native and deglycosylated proteins and the quantity of the protein being deglycosylated.

Another preparation of endo F, which has PNGase F activity as a contaminant (Boehringer), was used to determine if a mixture of these two endoenzymes would be an effective method to obtain a deglycosylated protein. Twenty-five micrograms each of hCGα and hCGβ was digested with 0.34 U (one unit hydrolyzes 1 μmol of dansyl-Asn(Glc-NAc)$_2$(Man)$_5$ in 60 min at 37° in pH 5.0 buffer—this is equivalent to 16.7 mU of endo F from the previous source) in a total volume of 30 μl of 0.2 M sodium acetate buffer (pH 5.8) at 37° overnight. Incubations were done with native as well as SDS- and 2-mercaptoethanol-denatured proteins.

Hydrolysis by PNGase F. In a final volume of 30 μl, 25 μg of hCG subunit α or β, 3 μl of 0.1 M 1,10-phenanthroline hydrate, one mU of PNGase F from *F. meningosepticum* (Genzyme) (one unit hydrolyzes 1 μmol of [³H]dansylfetuin glycopeptide per min at 37°–one mU is equivalent to one Genzyme unit) and sodium phosphate buffer (pH 8.6) (final 0.2 M) were mixed and incubated overnight at 37°. Deglycosylation under denaturing conditions was done by heating the protein sample with 0.5% SDS and 1% 2-mercaptoethanol at 100° for 3 min and Nonidet P-40 (final 1.25%) was added prior to the addition of the enzyme. Purification of the deglycosylated protein can be done by methods similar to those described for hydrolysis with endo F.

Comments

The α and β subunits of hCG were subjected to hydrolysis with endo F and PNGase F in their native conformation, after SDS denaturation and after denaturation and reduction. Under the experimental conditions, Endo F cleaved a single biantennary carbohydrate unit in hCGα, while PNGase F hydrolyzed both in the native conformation as well as after denaturation (Fig. 1A). The hydrolysis was incomplete in either case. In the case of the β subunit, both endo F and PNGase F hydrolyzed partially one of the two biantennary carbohydrate units in the native and denatured form (Fig. 1B). However, both carbohydrate units were susceptible to hydrolysis with PNGase F in the denatured–reduced hCGβ although incompletely (lane 7). The hydrolysis of a single carbohydrate unit or incomplete hydrolysis is probably due to the insufficient enzyme concentration used. It may also be due to partial inactivation of PNGase F at 37° at

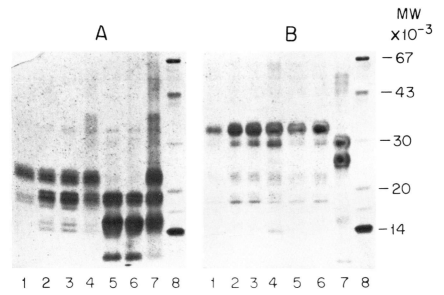

FIG. 1. SDS–PAGE of endo F- and PNGase-digested hCG subunits α (A) and β (B). The subunits were hydrolyzed in their native form (lanes 2 and 5) or after treatment with SDS alone (lanes 3 and 6) or SDS plus 2-mercaptoethanol (lanes 4 and 7), taken up directly in electrophoresis sample buffer containing 2-mercaptoethanol and electrophoresed in 12% separating gel. In each panel, lane 1 represents the untreated proteins; lanes 2–4, endo F treated; and lanes 5–7 PNGase F treated. Lane 8 shows molecular weight markers.

which temperature the incubations were carried out. When larger quantities of another preparation of endo F, which was contaminated with PNGase F, was used, both carbohydrate units in hCGα and hCGβ were susceptible to hydrolysis (Fig. 2). This may be partly due to PNGase F action but the contribution of this enzyme must be minimal since the incubations were carried out at a pH closer to the endo F pH optimum.

The concentration of the enzymes is critical in determining the extent of hydrolysis, particularly due to the fact that the endoenzymes are relatively less stable than the exoglycosidases and therefore digestion with them may require multiple additions. Tarentino et al.[7] were able to hydrolyze carbohydrate chains completely with PNGase F in the native conformation from ribonuclease B (single high-mannose chain), human transferrin (biantennary), and fetuin (triantennary). Only in the case of α_1-acid glycoprotein (tetraantennary), prior denaturation was required for complete removal of all carbohydrate units. The authors recommend the use of 0.6 to 60 mU of PNGase F for 50 μg of a glycoprotein at 37° and overnight incubation to deglycosylate most glycoproteins in their native

MW
×10⁻³

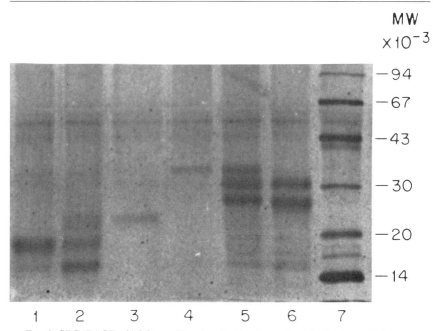

FIG. 2. SDS–PAGE of hCG α and β subunits digested with endo F (also containing some PNGase F as contaminant). After hydrolysis, the proteins were run under reducing conditions in 12% gel. Lanes 1–3, hCGα and lanes 4–6, hCGβ. The enzyme digestions were done with the native (lanes 1 and 5) and the denatured–reduced (lanes 2 and 6) subunits. Lane 7 shows molecular weight markers.

conformation. Incubations at room temperature may be more effective than at 37° as described in a subsequent communication.[8] Since glycans differ widely in their susceptibility to deglycosylation, the concentration of the endoenzymes and the time of incubation must be established for each glycoprotein. For example, denatured ribonuclease B was especially susceptible to hydrolysis with PNGase F. Less than 12 mU PNGase F/ml released 100% of the carbohydrate in 1 hr. At 0.6 mU/ml, 90% of the carbohydrate was released in 1 hr and 100% in 3 hr. In the case of α_1-acid glycoprotein, in 4 hr at 60 mU PNGase F/ml, all five carbohydrate chains were cleaved. At lower enzyme concentrations, all intermediate forms and the undigested protein were seen on SDS–PAGE.

Glycoproteins contain a wide variety of glycan structures. Since deglycosylation of a glycoprotein is necessary to study the role of carbohydrate in its function, one requires enzymes capable of cleaving all or most of the carbohydrate from a glycoprotein in its native conformation. Despite the discovery of several endoglycosidases which have greatly facilitated our investigations on the structure and biosynthesis of carbohydrates, there still remains a paucity of enzymes with broad specificity which can cleave carbohydrates from glycoproteins in their native forms. PGNases F and A offer broad specificity but these enzymes prefer denatured to the native form of the glycoprotein although some glycoproteins have been shown to be completely deglycosylated using sufficiently high concentration of the enzymes. More importantly, deglycosylation by PNGase F or A results in the conversion of carbohydrate-linked asparagines into aspartic acid residues. This can introduce an appreciable additional negative charge in the polypeptide chain and may lead to a conformational change. In this respect endo H or endo F type enzymes are better suited for deglycosylation for functional studies since the action of these enzymes is confined to the glycan moiety only. However, unlike N-glycosidases these enzymes have a narrow substrate specificity. Exoenzymes, despite long incubation times required for deglycosylation, do provide a viable alternative to the endoenzymes. In short, there is still a need to continue our efforts in search for new endoenzymes.

[29] Chemical Synthesis of Oligosaccharides

By HAROLD M. FLOWERS

General Remarks

Introduction

In my previous article in this series,[1] I discussed the biological importance of oligosaccharides and the relevance of the chemical synthesis of lower molecular weight fragments of them (disaccharides and simpler oligosaccharides). At that time, the controlled synthesis of oligosaccharides had been achieved only in limited cases and those thus prepared, whether homo- or heteropolymers, contained only a few glycosyl moieties. Some polymerization and copolymerization reactions had also been achieved. In the 7 years or so since that article was written, considerable knowledge has accrued on the structure of complex glycans. At the same time, there have also been notable advances in techniques for synthesizing oligosaccharides. Compounds containing 10 or more units have been

---

[1] H. M. Flowers, this series, Vol. 50, p. 93.

METHODS IN ENZYMOLOGY, VOL. 138

prepared and important heteroglycans have been synthesized, such as the "core region" of N-glycoproteins, the saccharide portion of O-glycoproteins and of some glycolipid and glycoprotein blood group antigens, a biologically active partially sulfated pentasaccharide from the anticoagulant heparin, and various "repeating units" of bacterial and yeast polysaccharides.

These advances have been achieved due to the introduction of new reagents and improved methods for the isolation, purification, and characterization of products, and also the mass action effect of the considerable number of laboratories engaged on this endeavour.

The basic characteristics previously pointed out[1] as required for key intermediates continue to control the synthetic steps: i.e., they should possess good protecting groups with suitable degrees of stability combined with possibilities of selective removal. They should be sufficiently stable for the necessary manipulations yet have high reactivity in the coupling step. Problems of regiospecificity accompany the preparation of these protected intermediates, while control of stereospecificity is important in the coupling reactions. The necessity for good yields at each stage becomes more critical in multistep syntheses. Steric control has always been a goal; if a particular coupling reaction is not completely stereospecific, the α- and β-anomers must be separated before the next coupling step is undertaken to prevent accumulation of the unwanted anomer as the molecular weight increases (a similar problem is encountered with racemates in polypeptide synthesis). The separation of anomers generally becomes progressively more difficult as the products increase in size. Furthermore, each separate condensation step may require different conditions since different hydroxyl groups with different protecting groups in their vicinity can be involved. Slight variations can affect the yields considerably, as well as the ratios of anomers obtained, or even the possibility of obtaining any desired product at all. These considerations may explain the poor development of solid-state oligosaccharide synthesis, in contrast to the recent advances for peptides and oligonucleotides. As new glycosyl linkages are formed, their lability may impose a limitation on reagents that can be used in subsequent stages of protection–deprotection or condensation.

In oligosaccharide synthesis, "persistent" and "temporary" blocking groups have been differentiated. Certain hydroxyl groups must be specifically protected in a different way from others so that they can be released in chosen intermediates and made available for the next condensation step to follow in the series. When the final, protected product has been attained, then the other blocking groups can be removed. The monosaccharide moieties involved in different syntheses are not necessarily identical and may vary in reactivity, so that experience derived from previous synthetic work while helpful does not obviate the necessity to consider carefully the factors involved in the particular undertaking under consideration. It should be mentioned that a number of empirical generalizations have been collected with some theoretical explanations, but surprises do crop up which apparently contradict our previous experience, and it is the unexpected which lends some of the hazard, and the piquancy, to this field of endeavour.

The importance of anhydrous conditions has already been stressed,[1] and is a factor which differentiates conditions of chemical and enzymatic synthesis. In recent years, dry boxes and reactions under nitrogen or argon have been employed and internal desiccants are very popular.

As previously, I do not intend to discuss enzyme-catalyzed glycosylation reactions nor transformations of oligosaccharides by changing the structure or stereochemistry of particular building units contained therein. Enzymatic syntheses do offer future promise with the newer techniques available for isolation and stabilization of enzymes, and some limited successes have been achieved in attempts to scale up yields to reasonable amounts. Transformations of oligosaccharides to desired products are limited in scope to commonly available precursors and a few processes since a large number of hydroxyl groups may be available per reaction, and selective protection of them is more complicated than in monosaccharides. However, the use of previously prepared derivatives of disaccharides, trisaccharides, etc. as building blocks to be coupled to other components can save much time and effort, as epitomized in some of the syntheses to be described later.

More pyranosides occur naturally than furanosides, and much of the research has applied to the former. In general terms, the fundamental problems concerned with the syntheses of the two classes are similar, although practical problems of availability of intermediates and stability of bonds do have to be taken into consideration.

Nature of the Combining Species

In stepwise syntheses of this nature, the chemistry involved can be organized essentially into preparation of suitably protected intermediates and their condensation to give products which can themselves undergo further transformations and stepwise condensations. The coupling reactions usually involve two saccharide components—an activated unit providing the glycosyl moiety (electrophile) and a hydroxyl compound providing the nucleophile. Protected glycosyl halides are still the most common glycosylating agents employed. They are generally readily available and have good reactivity in line with the leaving-group ability of the halide. Their stability varies considerably, depending on the compound involved, the anomeric configuration, and the halogen atom. Usually iodides are too unstable to be isolated but have been prepared and employed *in situ,* while fluorides have been regarded as too unreactive. Recently, some success has, however, been achieved with the latter and they may offer some advantages in certain cases. In principle, other leaving groups can replace halogen at C-1. Although acetate is a poor leaving group, it has been employed successfully in certain cases, especially where its displacement is facilitated by a suitable neighboring group. Sulfonate esters are difficult to prepare but have been made and used *in situ* with marked success. Thioglycosides can be displaced by O-nucleophiles to give O-glycosides. Glycosyl units with 1-OH have also been employed in apparently dehydrative reactions; substituted 1-imidates have been used successfully in a number of cases. Cyclic 1,2-substituents give good steric control under suitable conditions, and can be prepared from either 2-hydroxy or 2-amino sugar derivatives. Unsaturated sugars such as glycals have been employed to prepare oligosaccharides which can be reduced to deoxy compounds or hydrated with varying degrees of stereoselectivity.

The nucleophilic portion is also of importance. Usually condensation of the glycosylating agent occurs directly with a free hydroxyl group of a nucleophile. The reactivity of this hydroxyl group may be affected by vicinal groups, e.g., esters, ethers, and acetal rings, and by other stereoelectronic effects. It is often possible to perform selective glycosylations of a nucleophile bearing two free hydroxyl groups which differ considerably in their reactivity. In some cases, the hydroxyl group may be conveniently "activated" by substitution. The SH group used for synthesizing *S*-oligosaccharides is of course more nucleophilic than OH with various consequences.

Acid Acceptors

Additional reagents are required in the condensation reaction. They may be acid catalysts in some cases, but are usually salts of heavy metals used in noncatalytic, often stochiometric, amounts. Since the earlier syntheses involved displacement of halogen from the glycosyl moiety and H from the nucleophile to give HX, the salts were considered to be necessary to neutralize the acid formed and were called "acid acceptors." It was later believed that they directly facilitated removal of halogen. The selection of the salt may be quite critical. The nature of the solvent is also very important and the effect of reaction temperature may not be trivial.

Solvent polarity and temperature play a role in the complex series of equilibria available to the glycosyl halide, so that marked differences may result at different temperatures in various solvents in the anomeric configuration of the product and its yield and its possible contamination by side products.[1] Although condensation of the same protected saccharides at times produces both 1,2-*cis*- and 1,2-*trans*-glycopyranosides concomitantly, the general approach is to use different glycosylating agents for the two anomers. In this discussion, I shall therefore consider their synthesis under different headings. It will also be convenient to deal separately with 2-aminoglycosylation, and a small section will deal with sialic acid derivatives.

A vast number of publications on this topic have appeared during the last decade and it will be impossible to avoid omissions. Selection will be based on reactions of general value, newer reagents, and relation and potential applicability to products of biological interest, and will be necessarily a personal choice. Many excellent studies and some advances which could be regarded as interesting from mechanistic considerations and general carbohydrate chemistry will therefore unfortunately be omitted. The aim will be to provide the nonexpert with an understanding of the possibilities at present and in the near future of synthesizing a desired compound, and some idea of the effort required to achieve this synthesis chemically, and whether it would be worthwhile compared to other possible sources.

A number of reviews have appeared in the period from 1977 to 1984 dealing with various aspects of the subject.[2-5] The Royal Society of Chemistry Specialist Periodical Reports (Carbohydrate Chemistry), which appear annually, are an excellent source of information on carbohydrate chemistry in general and have chapters devoted to glycosides and oligosaccharides.

The Protected Glycosylating Agent

Bromides and Chlorides

The most suitable balance of properties of stability and reactivity are generally offered by the above two halides. The anomeric effect stabilizes the α-D- (or L-) halide with respect to its β-anomer,[6] and this fact may be

[2] K. Igarashi, *Adv. Carbohydr. Chem. Biochem.* **34**, 243 (1977).
[3] A. F. Bochkov and G. E. Zaikov, "Chemistry of the O-Glycoside Bond." Pergamon, Oxford, 1979.
[4] H. Paulsen, *Chem. Soc. Rev.* **13**, 15 (1984).
[5] H. Paulsen, *Angew. Chem., Int. Ed. Engl.* **21**, 155 (1982).
[6] R. U. Lemieux, *Pure Appl. Chem.* **25**, 527 (1971).

important in directing product configuration. Per-O-acylated glycosyl halides are generally more stable than per-O-alkylated derivatives, and bromides have usually been employed with the former, the acyl groups being mainly acetyl or benzoyl. Examples of their utilization in oligosaccharide synthesis will be discussed shortly. In the case of 2-O-alkylated glycosyl halides, the bromides may be inconveniently unstable and the corresponding chlorides which often do not have to be isolated anomerically pure are sufficiently reactive to provide successful condensations. Some examples of the employment of chlorides have already been provided.[1] An early synthesis of isomaltose utilized a per-O-acetylated β-D-glucopyranosyl chloride (**I**) with a protected carbohydrate (**II**) to give an α-linked disaccharide (**III**) on reaction in the presence of silver carbonate and silver perchlorate.[7]

(I) R'=Cl, R^2=ONO$_2$, R^3=OAc
(II) R'=R^2=OAc, R^3= OH

(III)

The utilization of β-chlorides and silver salts, especially together with silver perchlorate, has become quite popular in recent years for preparing 1,2-*cis*-glycosides, as will be shown later. However, it is not necessary to have β-chloride for the reaction, as was shown by an early synthesis of α,α-trehalose in 18% yield from tetra-O-benzyl-α-D-glucopyranosyl chloride and 2,3,4,6-tetra-O-benzyl-D-glucopyranose, in the presence of silver carbonate and silver perchlorate.[8]

A relatively unstable β-chloride was prepared from the disaccharide hepta-O-acetyl-α-kojibiosyl bromide and this reacted with benzyl 2,3,4-tri-O-benzyl-β-D-glucopyranoside, again in the presence of silver salts, to give a trisaccharide in 37% yield, with 1,2-cis-stereochemistry at the new glycosyl linkage. Deprotection provided O-α-Glc(1-2)-O-α-Glc(1-6)-D-glucose.[9,10] Condensation of 2,3,4-tri-O-acetyl-α-L-rhamnopyranosyl chlo-

[7] M. L. Wolfrom and I. C. Gillam, *Science* **130**, 1424 (1959).

[8] G. J. F. Chittenden, *Carbohydr. Res.* **9**, 323 (1969).

[9] V. Pozsgay, P. Nánási, and A. Neszmélyi, *Carbohydr. Res.* **75**, 310 (1979).

[10] Abbreviated glycopyranosyl moieties, if not otherwise indicated, will be of the D-configuration. Ara, Arabinose; Fuc, fucose; Gal, galactose; Glc, glucose; Man, mannose; Rha, rhamnose; Xyl, xylose.

ride with methyl 2-*O*-benzyl-4,6-*O*-benzylidene-α-D-glucopyranoside gave a 62% yield of the 3-*O*-α-linked rhamnopyranosylglucoside after 4 hr of reaction in the presence of a silver trifluoromethane sulfonate (triflate) reagent.[11] Similarly, a 1-2-*trans*-(α) configuration resulted from the condensation of a substituted α-D-mannopyranosyl chloride (**IV**), with acetate at O-2 and a protected monosaccharide in the presence of silver triflate.[12]

An α-D-mannopyranosyl chloride (**V**) with a nonparticipating C-2 substituent, used with silver carbonate, enabled the preparation of disaccharide enriched in β-anomers.[13]

It was later found[14] that silver imidazolate gave better yields than silver carbonate in this reaction and also enabled the synthesis of α-linked

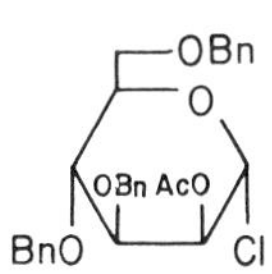

(IV)

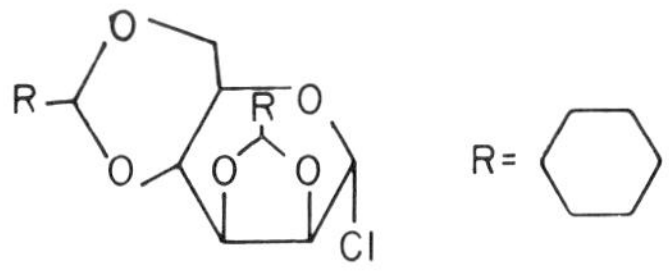

(V)

R =

(Bn = C6H5CH2)

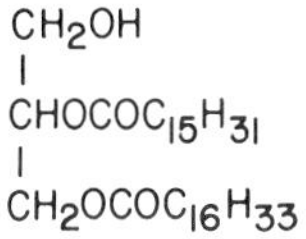

CH_2OH
|
$CHOCOC_{15}H_{31}$
|
$CH_2OCOC_{16}H_{33}$

(VII)

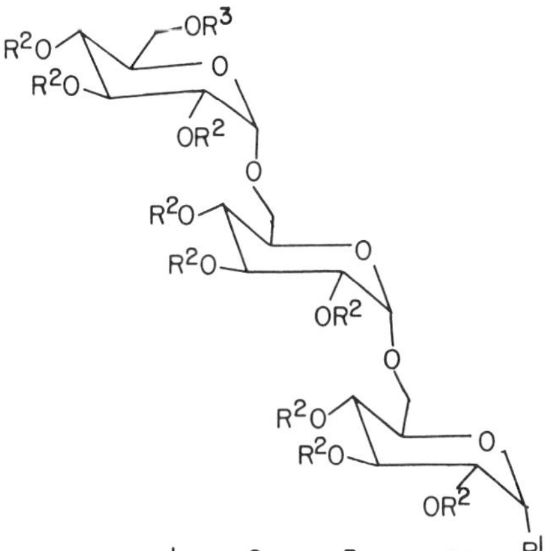

(VI) R^1 = Cl, R^2 = Bn, R^3 = CH_2COCl
(VIII) R^1 = $OCH_2CH(OC_{15}H_{31})CH_2OC_{16}H_{33}$
R^2 = H, R^3 = SO_3Na

[11] F. Konishi, S. Esaki, and S. Kamiya, *Agric. Biol. Chem.* **47**, 265 (1983).

[12] K. Bock and M. Meldal, *Acta Chem. Scand., Ser. B* **B38**, 71 (1984).

[13] P. J. Garegg and T. Iversen, *Carbohydr. Res.* **70**, C13 (1979).

[14] P. J. Garegg, R. Johansson, and B. Samuelsson, *Acta Chem. Scand., Ser. B* **B35**, 635 (1981).

galactosides and glucosides from their corresponding per-O-benzylated α-D-glycopyranosyl chlorides.[15] In the preparation of the glyceroglucolipid of human gastric secretion,[16] a substituted α-D-glucopyranosyl chloride (**VI**), prepared from a previously synthesized trisaccharide, was condensed with a diacylglycerol (**VII**) in the presence of silver triflate to give a high yield of the α-D-glucopyranoside, which was converted into the desired sulfated glycolipid (**VIII**).

In a synthesis of the human blood group P1 antigenic determinant, reaction between tetra-O-benzyl-α-D-galactopyranosyl chloride and a protected aminodisaccharide produced the desired trisaccharide with stereoselective α direction.[17] Condensation of this same chloride with the less reactive O-2 of a mannose residue in a protected disaccharide gave the α-linked trisaccharide in 76% yield, whereas no success was obtained with the corresponding bromide.[18]

The glucuronic acid derivative, methyl 2,3-di-O-benzyl-1-chloro-1-deoxy-4-O-methyl-α, β-D-glucopyranuronate (**IX**) reacted[19] with a protected xylobioside in the presence of silver perchlorate–s-collidine to give a trisaccharide derivative in 96% yield ($\alpha : \beta$ approx. 5 : 4).

(IX)   (X)   (XI)

A partially protected mannopyranoside reacted at O-2 with tetra-O-benzyl-α-D-galactopyranosyl chloride (**X**) and silver triflate to give the α-linked disaccharide in 63% yield.[20] A protected glucopyranoside, with hydroxyl groups free at C-3 and C-6, reacted with excess of X or the 3-deoxychloride (**XI**) at both available sites to give α-linked trisaccharides in reasonable yield.[21]

Some syntheses of 2-aminoglycosides have utilized glycosyl chlorides as key intermediates and will be considered later in the section dealing

[15] P. J. Garegg, C. Ortega, and B. Samuelsson, *Acta Chem. Scand., Ser. B* **B35,** 631 (1981).
[16] T. Ogawa and T. Horisaki, *Carbohydr. Res.* **123,** C1 (1983).
[17] P.-H. A. Zollo, J.-C. Jacquinet, and P. Sinaÿ, *Carbohydr. Res.* **122,** 201 (1983).
[18] P. T. Garegg, H. Hultberg, and C. Lindberg, *Carbohydr. Res.* **83,** 157 (1980).
[19] J. Hirsch, P. Kováč, J. Alföldi, and V. Mihálov, *Carbohydr. Res.* **88,** 146 (1981).
[20] T. Iversen and D. R. Bundle, *Carbohydr. Res.* **103,** 29 (1982).
[21] T. Iversen and D. R. Bundle, *Can. J. Chem.* **60,** 299 (1982).

with amino sugars. By far the majority of oligosaccharide syntheses employing glycosyl halides have used the bromides, as will appear in the various examples to be presented under the different headings later in this article.

Iodides

The addition of iodine to glycosylation reactions was sometimes found to be beneficial[22] and the effect may have been due to intermediate formation of a reactive iodide. The reaction of tetra-O-benzyl-β-D-galacto- and glucopyranosyl iodides, prepared *in situ* from the corresponding α-bromides, with various alcohols, gave mainly α-D-glycosides.[23] However, it would seem that no great advantage attends use of these compounds and they have not attained popularity. It may be added that mercury(II) iodide has been shown to provide a useful catalyst for some glycosylations.[24] Thus, reaction of 8-ethoxycarbonyloct-1-yl 2,3-di-O-benzoyl-α-L-rhamnopyranoside (**XII**) with tetra-O-acetyl-α-D-mannopyranosyl bromide and mercury(II) iodide gave an α-linked disaccharide derivative (1,2-*trans*) in 49% yield. Similarly 2-O-allyl-6-O-acetyl-3,4-di-O-benzoyl-α-D-galactopyranosyl bromide (**XIII**) reacted with 4-O-allyl-1,6-anhydro-3-O-benzyl-β-D-mannopyranoside (**XIV**) in the presence of mercury(II) iodide to give an α-disaccharide (1,2-*cis*) in 60% yield.

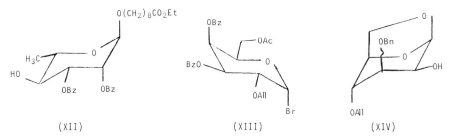

(XII)   (XIII)   (XIV)

Fluorides

Very recently, a brief report has appeared[25] on the utility of α-D-glycopyranosyl fluorides prepared from the corresponding thioglycosides under mild conditions which enable the retention of some protecting groups, such as acetals for example, which it is not easy to preserve in the preparation of bromides. Employment of a suitably protected glucosyl

[22] D. D. Reynolds and W. L. Evans, *J. Am. Chem. Soc.* **60**, 2559 (1938).
[23] F. J. Kronzer and C. Schuerch, *Carbohydr. Res.* **34**, 71 (1974).
[24] K. Bock and M. Meldal, *Acta Chem. Scand., Ser. B* **B37**, 775 (1983).
[25] K. C. Nicolaou, R. E. Dolle, D. P. Papahatjis, and J. L. Randall, *J. Am. Chem. Soc.* **106**, 4189 (1984).

fluoride together with silver perchlorate and tin(II) chloride enabled synthesis of an α(1-6)-linked hexasaccharide of D-glucose.

Glycosylsulfonates

Progress has continued on the displacement of glycosyl sulfonates[26] with special reference to steric control of the reaction. A variety of different sulfonate esters were investigated in the mannose series to optimize glycoside formation.[27] It was found that the presence of a nonparticipating 2-*O*-benzyl ether in the mannosyl sulfonate led to high yields of glycosides with a low degree of stereoselectivity. However, the presence of an acyl participating group at C-2 led to high yields of α-D-mannopyranosides (1,2-*trans*) with good steric control.[27,28] It was similarly shown in the galactose series that the presence or absence of a participating group at O-2 controlled the formation of α- or β-linked disaccharides.[29,30] These 1-sulfonates are all made from the corresponding glycopyranosyl halides, and their control of stereoselectivity does not seem to reside in the nature of the leaving group but rather in the pattern of substitution in the reacting electrophilic species (participating group at C-2), stereoelectronic effects of other substitutents). Furthermore, the high instability of these sulfonate esters demands special precautions in their preparation and employment. Their high reactivity enables excellent yields of condensation products to be obtained in many cases, but they would not generally seem to present advantages over the corresponding bromides or chlorides.

Imidates

Another good leaving group at C-1 is the *N*-methylacetimidyloxy moiety. Reaction of D-gluco- or D-galactopyranosyl halides with *N*-methylacetamide and freshly prepared silver oxide in the presence of *N*-ethyldiisopropylamine gave stable β-imidates (**XVa** and **XVb,** respectively). They react with nucleophiles in nitromethane in the presence of an acid catalyst (*p*-toluenesulfonic acid) to give α-glycosides in good yield and high stereoselectivity.[31,32] Starting from a protected derivative of *N*-ace-

[26] R. Eby and C. Schuerch, *Carbohydr. Res.* **50,** 203 (1976).
[27] E. S. Rachaman, R. Eby, and C. Schuerch, *Carbohydr. Res.* **67** 147 (1978).
[28] L. F. Awad, E. S. H. El-Ashry, and C. Schuerch, *Carbohydr. Res.* **122,** 69 (1983).
[29] H. F. Vernay, E. S. Rachaman, R. Eby, and C. Schuerch, *Carbohydr. Res.* **78,** 267 (1980).
[30] H. A. El-Shenawy and C. Schuerch, *Carbohydr. Res.* **131,** 227 (1984).
[31] P. Sinaÿ, *Pure Appl. Chem.* **50,** 1437 (1978).
[32] M.-L. Milat, P. A. Zollo, and P. Sinaÿ, *Carbohydr. Res.* **100,** 263 (1982).

α Gal (1-3)
α-L-Fuc(1-2) $\rangle\beta$ Gal (1-4) GlcNAc

(XVI)

(XVa): $R^1 = R^2 = R^3 = R^5 = OBn$, $R^4 = H$
(XVb): $R^1 = R^2 = R^4 = R^5 = OBn$, $R^3 = H$
(XVII): $R^1 = R^4 = H$, $R^2 = R^3 = R^5 = \underline{p}O_2NC_6H_4CO_2$

tyllactosamine, stepwise glycosylation with tetra-O-benzyl-1-O-(N-methylacetimidyl)-β-D-galactopyranose (**XVb**) and the corresponding β-L-fucopyranose enabled synthesis of the blood group B determinant (**XVI**).[33]

Several 2-deoxy-α-linked disaccharides were similarly prepared from the requisite β-imidate (e.g., **XVII**) in excellent yield and high stereoselectivity.[34] Again, like the 1-sulfonates, these imidates are prepared from the 1-halides and do not generally give much better yields of desired oligosaccharide products on condensation with various nucleophiles. They seem to be less reactive than the corresponding halides so that nonparticipating groups at C-2 are needed for good yields in the condensation step. N-Methylacetimidates of disaccharides do not seem to have been made so far.

A more recent development in this connection is the introduction of the 1-trichloroacetimidyl group[35] which can be substituted directly into a protected mono- or disaccharide having a free 1-OH. The 1-trichloroacetimidates produced react in high yield with suitable alcoholic components and acid catalysts to give glycosides. The steric course of the condensation does not seem to be controlled by the presence or absence of a participating group in the trichloroacetimidate, as illustrated by the production of β-linked oligosaccharides (e.g., **XX**)[36] from the trichloroacetimidates **XVIII**, **XXI**, and **XXII**. The reaction of the glycosyl trichloroacetimidate (**XVIIIa**) with 2,3,4,6-tetra-O-acetyl-α-D-glucopyranose under similar conditions to the above gave a β,β-trehalose derivative in 58% yield.[37]

[33] M.-L. Milat and P. Sinaÿ, *Carbohydr. Res.* **92**, 183 (1981).

[34] J. Szymoniak and P. Sinaÿ, *Tetrahedron Lett.* p. 545 (1979).

[35] R. R. Schmidt and J. Michel, *Angew. Chem., Int. Ed. Engl.* **19**, 731 (1980); *Tetrahedron Lett.* **25**, 821 (1984).

[36] R. R. Schmidt and J. Michel, *Angew. Chem., Int. Ed. Engl.* **21**, 72 (1983).

[37] S. J. Cook, R. Khan, and J. M. Brown, *J. Carbohydr. Chem.* **3**, 343 (1984).

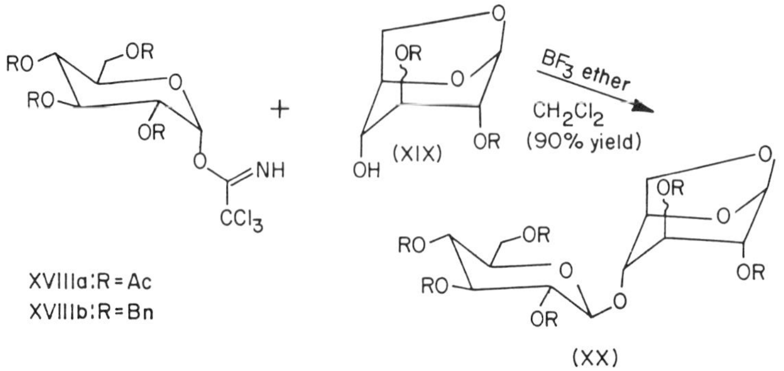

XVIIIa: R = Ac
XVIIIb: R = Bn

(XIX)

(XX)

(XIX, XX: R = Bn)

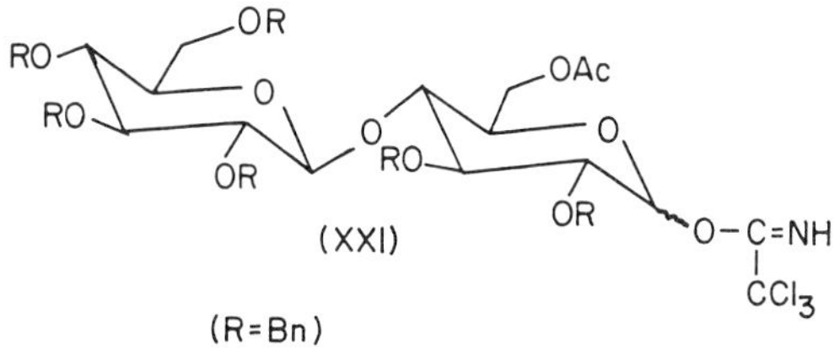

(XXI)

(R = Bn)

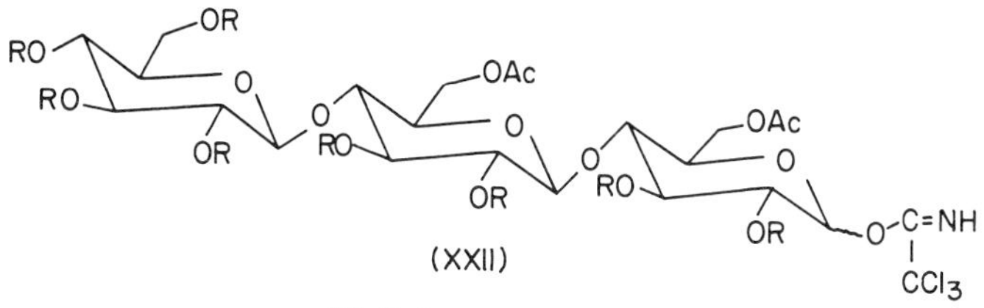

(XXII)

(R = Bn)

The trichloroacetimidates have also been utilized in 2-amino sugar glycosylations (see later).

Other C-1 Leaving Groups

Sulfur-leaving groups, such as β-thioglycosides (to prepare α-D-disaccharides),[38] have been tried with some success in a few cases, and in a recent study[39] a variety of alcohols or protected sugars were condensed with different phenylthioglycosides to give oligosaccharides, the reaction being promoted by N-bromosuccinimide in the presence of molecular sieves. Reaction was rapid and the anomeric configuration of the phenyl-

[38] S. Hanessian, C. Bacquet, and N. Lehong, *Carbohydr. Res.* **80,** C17 (1980).
[39] K. C. Nicolaou, S. P. Seitz, and D. P. Papahatjis, *J. Am. Chem. Soc.* **105,** 2430 (1983).

thioglycoside employed did not seem to be important in determining the configuration of the O-glycoside formed. However, changes in solvent did affect this stereochemistry and $\alpha : \beta$ ratios varied in certain cases from 1 : 2.3 to 9 : 1 or even 1 : 0, without any general rules being apparent.[40] More systematic investigation and a more detailed report will be necessary before a proper evaluation of the scope and possibilities of this approach can be made. The requisite thioglycosides are readily available and may be substituted with a variety of protecting groups such as acetal, carbonate ester, or dimethyl-*tert*-butylsilyl ether which are stable under the conditions of the condensation.

Substituted ammonium groups were also tried as positively charged leaving groups[41] but have not been subsequently developed to any extent.

Glycosylation and thioglycosylation were also achieved via the intermediate formation of a glycosylphosphonium chloride.[42] Tris(dimethylamino)phosphine and carbon tetrachloride at low temperature on 2,3 : 4,6-di-O-isopropylidene-α-D-mannopyranose gave an α-alkoxytris-(dimethylamino)phosphonium chloride (**XXIII**).

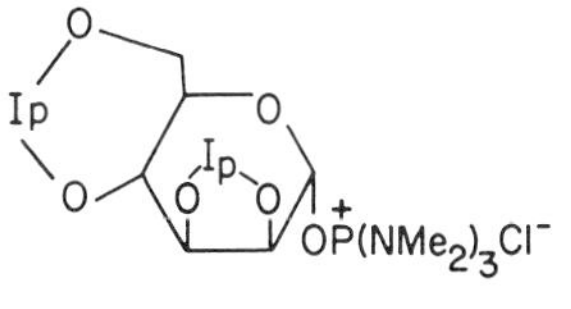

(XXIII)

$$\begin{bmatrix} \text{Ip} = (CH_3)_2 C \\ \text{Me} = CH_3 \end{bmatrix}$$

Without isolation, nucleophiles and silver p-toluene sulfonate were added. Good yields of anomeric mixtures of glycosides were obtained with simple primary alcohols (slightly enriched in α), but hydroxyl groups in protected saccharides such as 6-OH in 1,2 : 3,4-di-O-isopropylidene-α-D-galactopyranose and 3-OH in 1,2 : 5,6-di-O-isopropylidene-α-D-glucofuranose were not sufficiently reactive and gave poor yields (but apparently stereospecifically as the α-linked anomer).

Glycosyl halides are often prepared via 1-O-acetates. It has been found that the 1-O-acetate itself is a good enough leaving group under suitable conditions to enable oligosaccharide formation. The presence of a 1,2-*trans*-diacetate system is especially advantageous as participation

[40] F. J. Kronzer and C. Schuerch, *Carbohydr. Res.* **33,** 273 (1974).

[41] R. Eby and C. Schuerch, *Carbohydr. Res.* **39,** 33 (1975).

[42] F. Chretien, Y. Chapleur, B. Castro, and B. Gross, *J. Chem. Soc., Perkin Trans. I* p. 381 (1980).

by the 2-*O*-acetate facilitates removal of the *trans*-(*O*-acyl) derivative. If the 2-substituent is *N*-acetyl, participation is especially favored and an intermediate oxazoline may be formed (itself a good glycosylating agent; see later). A variety of β-glycosides have been made from different alcohols by this approach. More recent examples are the tin(IV) chloride-catalyzed preparation of 8-ethoxycarbonyloctyl-β-D-glycosides of cellobiose, lactose, and maltose,[43] and β-glycosylation of lactose with 1,2-diacylglycerol or ethanethiol catalyzed by trimethylsilyl triflate.[44] Boron trifluoride etherate was used as a catalyst in the preparation of β-2-bromethylglycosides of mono- and oligosaccharides.[45] The method has also been applied to disaccharide synthesis. Substituted 2-acetamido-2-deoxy-β-D-glucopyranosyl acetates reacted with alcohols or OH of protected sugars, including primary and secondary OH, in the presence of iron (III) chloride to give good yields of β-linked glycosides.[46]

Free 1-OH

In principle, the final result of glycosylation is the condensation of two hydroxylated species to form a new compound with removal of water, i.e., dehydration. In practice, direct dehydrations of this kind would seem to require extreme conditions and not be practical in most cases. Analogous reactions do occur in the presence of acid catalysts, e.g., the Fischer method for glycosylation.[47] This method involves treatment of a free sugar, usually monosaccharide, with large excess of an alcohol, usually a liquid alcohol which also acts as the solvent. An acid catalyst is also present and an equilibrium is set up, in which thermodynamic factors determine the structures of the final products. Complex mixtures often result, but the preferred stability of α-D-glycopyranosides in most cases enables their isolation in reasonable yield. This approach is usually not suitable for glycosides of disaccharides, since alcoholysis of the interglycosidic linkage competes with glycosylation of the reducing OH. Attachment of another sugar unit by these means in a controlled way is therefore not feasible. However, it may sometimes occur inadvertantly, e.g., the "acid reversion" encountered on acid hydrolysis of polysaccharides with formation of unexpected products, especially if the hydrolyzing solution

[43] J. Banoub and D. R. Bundle, *Can. J. Chem.* **57**, 2085 (1979).

[44] T. Ogawa, K. Beppu, and S. Nakabayashi, *Carbohydr. Res.* **93**, C6 (1981).

[45] J. Dahmén, T. Frejd, G. Magnusson, G. Noori, and A.-S. Carlström, *Carbohydr. Res.* **127**, 27 (1984).

[46] M. Kiso and L. Anderson, *Carbohydr. Res.* **72**, C15 (1979); see also *ibid.* **136**, 309 (1985), where there is evidence for intermediate oxazoline formation.

[47] E. Fischer, *Chem. Ber.* **26**, 2400 (1893).

is allowed to become too concentrated. A complex mixture of disaccharides was thus obtained, and separated as peracetates after heating D-glucose in acid solution.[48] These are again equilibrium reactions so that the thermodynamically most stable products will predominate.

It is possible, nevertheless, to activate 1-OH in the presence of the nucleophile required for the condensation step, so that by direct admixture of protected carbohydrate with free 1-OH, nucleophile, and activating agent, an oligosaccharide results. Suitable protecting groups can ensure regiospecificity and stereoselectivity. Processes of this kind presumably involve transient, *in situ* formation of an activated C-1 derivative which reacts with the nucleophile present to give the required product.

Several α-D-glucopyranosides, including disaccharides, were prepared in good yield by reaction of 2,3,4,6-tetra-O-benzyl-D-glucose with ROH in the presence of trifluoromethanesulfonic or methanesulfonic anhydride, halide ion, and *s*-collidine in a "one-pot" reaction.[49] It was presumed that glycosyl halide was formed *in situ*. A number of acid-sensitive protecting groups could be employed, such as acetals and O-trityl ethers. Further studies[50] employing ^{19}F NMR spectroscopy established the necessity for the presence of free triflic acid in the system. The very efficient trapping by triflic anhydride of the water formed in the reaction displaces the equilibrium toward glycoside formation. Although α-glycosylation is favored because of the nonparticipating C-2 substituent, considerable proportions of β-anomers may ensue. The same perbenzylated glucose also reacts with protected sugars in the presence of *p*-nitrobenzenesulfonyl chloride, silver triflate, and triethylamine to give β-linked disaccharides.[51] Modification of the conditions by the addition of N,N-dimethylacetamide and replacement of the dichloromethane solvent by N,N-dimethylformamide led to a much higher proportion of the α-anomer.[52]

1,2-Cyclic Intermediates—Orthoesters and Oxazolines

1,2-Cyclic orthoesters[53] continue to be used for preparing some 1,2-trans-linked oligosaccharides. There are technical problems and not al-

[48] A. Thompson, K. Anno, M. L. Wolfrom, and M. Inatome, *J. Am. Chem. Soc.* **76**, 1309 (1954).

[49] J. Leroux and A. S. Perlin, *Carbohydr. Res.* **67**, 163 (1978).

[50] J. A. Lacombe, A. A. Pavia, and J. M. Rockeville, *Can. J. Chem.* **59**, 472 (1981); A. A. Pavia and S. N. Ung-Chhun, *ibid.* p. 482.

[51] S. Koto, S. Inada, T. Yoshida, M. Toyama, and S. Zen, *Can J. Chem.* **59**, 255 (1981).

[52] S. Koto, N. Morishima, M. Owa, and S. Zen, *Carbohydr. Res.* **130**, 73 (1984).

[53] N. Kochetkov and A. F. Bochkov, *Methods Carbohydr. Chem.* **6**, 480 (1972).

ways does the expected stereochemistry result,[1,54] but a large number of these compounds have been prepared and used for glycosylations, including polymerization reactions.[55]

More recently, 1,2-cyano- and 1,2-thioesters have been introduced.[56,57] They react with activated nucleophiles, specifically O-trityl ethers in the presence of tritylium ions, to give high yields of disaccharides, often with very good β-anomeric specificity. However, unexpected steric results may occur. Thus, on reaction of the xylopyranose cyanoethylidene derivative (**XXIV**) with the O-trityl ether (**XXV**), a mixture of α- and β-disaccharides was obtained "as a result of anomerization."[58] It was found that the specificity for β-disaccharide formation decreased in the series $TrBF_4 > TrClO_4 > TrClO_4-Bu_4NClO_4 > Tr$ triflate used as catalysts.[59]

$$\left[Tr = (C_6H_5)_3C \right]$$

Oxazolines are employed in the synthesis of 2-aminoglycosyl oligosaccharides and will be considered under that heading.

Glycals

The use of glycals as intermediates in the preparation of glycosyl halides for synthesis of 2-amino sugar oligosaccharides has already been discussed.[1] However, glycals have been used directly as glycosylating agents, since addition to the glycal by a nucleophile can occur (with migration of the double bond) catalyzed by Lewis acids. Thus, the reaction of 3,4,6-tri-O-acetyl-D-glucal (**XXVIa**) with 1,2:3,4-di-O-isopropyli-

[54] N. K. Kochetkov and A. F. Bochkov, *Recent Dev. Chem. Nat. Carbon. Compd.* **4**, 77 (1971).

[55] N. K. Kochetkov, *Pure Appl. Chem.* **56**, 923 (1984).

[56] V. I. Betaneli, M. V. Ovchinnikov, L. V. Backinowsky, and N. K. Kochetkov, *Carbohydr. Res.* **76**, 252 (1979).

[57] L. V. Bakinowsky, Y. E. Tsvetkov, N. F. Balan, N. E. Byramova, and N. K. Kocketkov, *Carbohydr. Res.* **85**, 209 (1980).

[58] L. V. Bakinovskii, N. E. Nifant'ev, and N. K. Kochetkov, *Bioorg. Khim.* **9**, 1089 (1983); *Chem. Abstr.* **100**, 86014a (1984).

[59] L. V. Bakinovskii, N. E. Nifant'ev, and N. K. Kochetkov, *Bioorg. Khim.* **10**, 226 (1984); *Chem. Abstr.* **101**, 38739q (1984).

dene-α-D-galactopyranose, in the presence of boron trifluoride etherate, gave the 2:3-ene (**XXVII**) which could be hydroxylated, but not stereo-specifically.[60]

Triacetyl-D-glucal (**XXVIa**) or tetraacetyl-2-hydroxy-D-glucal (**XXVIb**) react similarly with an epoxyglycoside (**XXVIII**) to give an unsaturated α-linked epoxyglycoside (**XXIX**) which can be reduced to a deoxy-disaccharide.[61]

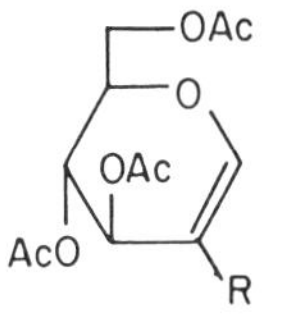

(XXVIa);R = H
(XXVIb);R =OAc

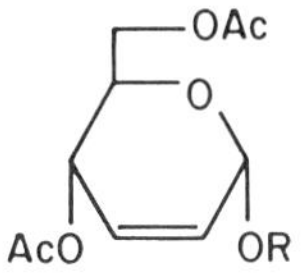

(XXVII)

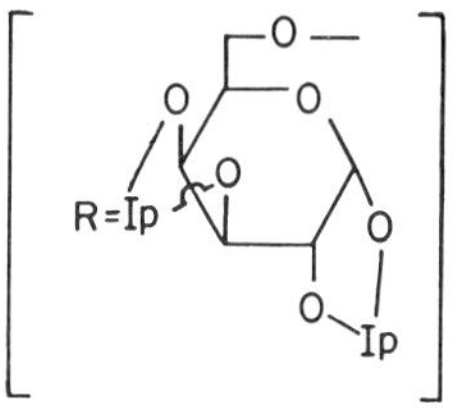

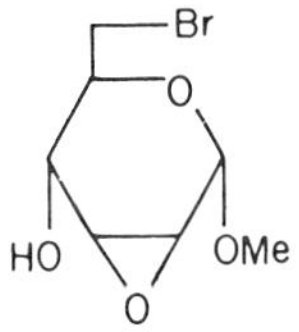

(XXVIII)

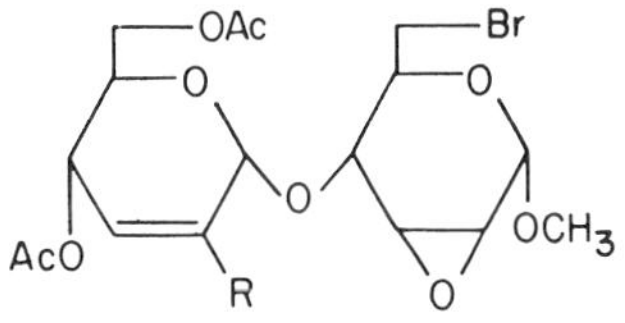

(XXIXa);R = H
(XXIXb);R =OAc (α:β= 4:1)

Unsaturated disaccharides were also obtained by condensation of a 4:5-glycal with methyl 6-deoxy-2,3-epoxy-α-D-allopyranoside, catalyzed by *N*-iodosuccinimide.[62]

The Nucleophile

Differential Reactivity

The differential reactivity of different hydroxyl groups toward various reagents is well established.[1,63] It is possible at times to exploit the selective reactivities of free hydroxyl groups so as to obtain regioselective glycosylations under competitive conditions.[64] Differences in reactivity of

[60] R. J. Ferrier and N. Prasad, *J. Chem. Soc. C* pp. 570, 575 (1969).
[61] J. Thiem, J. Schwentner, E. Schüttpelz, and J. Kopf, *Chem. Ber.* **112**, 1023 (1979).
[62] J. Thiem, P. Ossowski, and J. Schwentner, *Chem. Ber.* **113**, 955 (1980).
[63] A. H. Haines, *Adv. Carbohydr. Chem. Biochem.* **33**, 11 (1976).
[64] D. Beith-Halahmi, H. M. Flowers, and D. Shapiro, *Carbohydr. Res.* **5**, 25 (1967).

hydroxyl groups have been explained as due to stereoelectronic effects, primary versus secondary, equatorial versus axial, etc.[63] The acidity of the hydroxyl group may be important in some cases.[65] A very recent discovery is the fact that regioselectivity may be determined by the "acid acceptor" used in a glycosylation reaction.[66] Thus, in a reaction between lactose derivatives having free 3'- and 4'-hydroxyl groups (**XXX**), the site of reaction of different glycosylating agents (**XXXI, XXXII**) derived from mono- or disaccharides was determined by the nature of the "catalyst." Under heterogeneous conditions, using silver carbonate or silver silicate, β-glycosylation occurred selectively at the less reactive 4'-OH. With trimethylsilyl triflate or soluble mercury salts, however, it was almost exclusively at the more reactive 3'-OH group.

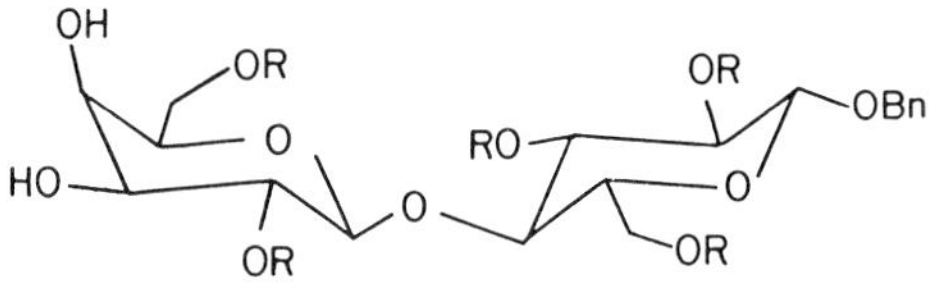

(XXXa): R = Bn
(XXXb): R = Ac

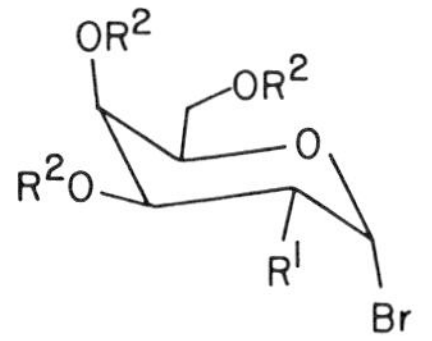

(XXXIa): R¹=N₃, R²=Ac
(XXXIb): R¹=NPh, R²=Ac
(XXXIc): R¹=OBn, R²=Bn

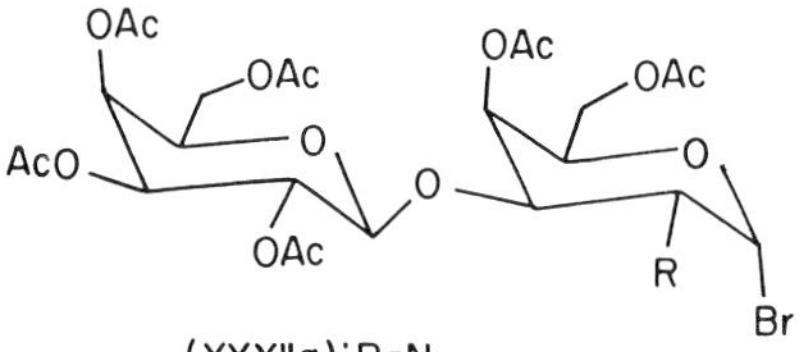

(XXXIIa): R=N₃
(XXXIIb): R=NPh

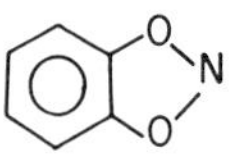

Activation

Trityl ethers have continued to be used as activated nucleophiles, especially in reactions with 1,2-orthoesters (thioesters or 1,2-cyanoethylidene derivatives),[56–59] and improvements have been made in tritylation procedures involving secondary hydroxyl groups. However, the application of these trityl ethers has been limited to a few laboratories so far.

Solvent Effects in Glycosylation

Solvent polarity can influence the reaction mechanism and product stereochemistry.[1,2,4,5] For example, in reaction of 3,4,6-tri-O-acetyl-2-O-

[65] H. M. Flowers, *Carbohydr. Res.* **99**, 170 (1982).
[66] H. Paulsen and M. Paal, *Carbohydr. Res.* **137**, 39 (1985).

benzyl-α- and β-D-glucopyranosyl chlorides with isopropyl alcohol and with methyl trityl ether in the presence of silver carbonate–silver perchlorate, α-D-glucopyranoside formation was favored in nonpolar solvents.[67,68] Ether was more effective than toluene and it was suggested that the free electrons of the ether oxygen may have contributed to this stereoselectivity. In the more polar nitromethane, the ratio of $\alpha : \beta$ anomers was 1 : 1. In glycosylation of cholesterol with tetra-O-acetyl-α-D-glucopyranosyl bromide and insoluble silver salts, 1,2-*trans*-glycosides (β) were obtained in ether solution but 1,2-orthoesters in tetrahydrofuran.[69] In dichloromethane as solvent, the addition of tetrahydrofuran raised the proportion of the α-anomers formed, whereas use of a homogeneous system in tetrahydrofuran containing s-collidine as acid acceptor led to high stereoselectivity and α-D-glycoside formation.

Mechanistic Considerations

It would be extremely helpful to be able to define *a priori* which glycosylating agent to use together with a certain acid acceptor and solvent so as to determine the steric cause of a condensation with a specific nucleophile. Attempts have been made.[1,4,5,70] to analyze possible reaction mechanisms and to offer guidelines for a rational approach to new syntheses. In recent reviews[4,5] certain conclusions were drawn on the effect on the anomeric specificity of the product by the acid acceptor used and the reactivity of the hydroxyl group undergoing reaction. A list was drawn up of the relative reactivities of various acid acceptors (so-called catalysts); it was postulated that tetraethylammonium bromide (used in "bromide-ion catalyzed reactions") was the least reactive "catalyst" and silver triflate the most. Similarly, the nature of the leaving group at C-1 and the relative reactivities of different hydroxyl groups were correlated with the product to be expected. In the substituted glycosyl halide involved in the reaction, the nature of the protecting groups is also important. Thus, tetra-O-acetyl- or benzoyl-α-D-glucopyranosyl halides have similar reactivities, but replacement of O-acyl by O-benzyl ethers increases the reactivity. The 2-O-substituent is the most critical since its participation can control stereochemistry to a considerable extent, and most approaches to 1,2-*cis*-glycosides start from halides with nonparticipating C-2 substituents. With such C-2 substituents, it was postulated that α-selectivity is improved if the catalyst employed is not too effective, since highly reactive catalysts may facilitate anomerization of the reacting halide (and any

[67] K. Igarashi, T. Honma, and J. Irisawa, *Carbohydr. Res.* **15**, 329 (1970).
[68] K. Igarashi, J. Irisawa, and T. Honma, *Carbohydr. Res.* **39**, 213 (1975).
[69] G. Wulff and G. Röhle, *Angew. Chem., Int. Ed. Engl.* **13**, 157 (1974).
[70] G. Wulff, U. Schröder, and J. Wichelhaus, *Carbohydr. Res.* **72**, 280 (1979).

intermediate ion pairs formed). It therefore becomes a matter of careful selection: more reactive nucleophiles with better catalysts will increase the reaction rate and the extent of glycosylation; at the same time, stereoselectivity may decline. Reactive halides should be used with moderately reactive hydroxyl groups to get α-glycosides with high selectivity, the catalyst being chosen to suit the reactivity of the hydroxyl group. If the hydroxyl group is especially reactive, a less effective catalyst will be necessary.

Insoluble salts used in heterogeneous reactions and soluble reagents may give different results since different reaction pathways might ensue, e.g., S_N2 with direct displacement, or S_N1 and variations thereof with intermediate formation of glycosyl cations which can react either directly or in various ways involving ion pairs, solvated cations, intermediates with the catalyst, etc., before final condensation with the nucleophile.

It should be pointed out that these generalizations must be taken with extreme caution as exceptions are not rare. For example, with most catalysts, in homogeneous or heterogeneous media, per-O-acetyl-α-glycopyranosyl bromides of D-glucose and D-galactose and 6-deoxy-L-galactose (fucose) will produce β-glycosides of protected carbohydrates stereoselectively or stereospecifically. However, a number of cases have been reported of stereoselective or even stereospecific α-glycosylations on 2-OH of protected carbohydrates on reaction with these peracetylated halides which have a C-2-participating group.[71-74]

1,2-Cis-*Glycosides*

It was recognized that replacement of a C-2 acetate group in the glycosyl halide by an ether or some other nonparticipating group could lead to α-glycoside formation. Thus, a synthesis of isomaltose was attained[7] using the 2-nitrate of 3,4,6-tri-O-acetyl-β-D-glucopyranosyl choride. Syntheses of maltose utilized the 2-O-benzyl ether[75] or the 2-trichloroacetate[76] of this same chloride. In these syntheses, the β-chloride, but not the corresponding bromide, was sufficiently stable to be isolated. Very often, the β-halide is not isolated but is prepared *in situ* by anomerization of the more stable α-halide. In the equilibrium between the two halides, the anomeric effect favors the α-anomer.[6] However, the β-halide is much the

[71] B. Helferich and J. Zirner, *Chem. Ber.* **95,** 2604 (1962).

[72] H. M. Flowers, A. Levy, and N. Sharon, *Carbohydr. Res.* **4,** 189 (1967).

[73] A. Levy, H. M. Flowers, and N. Sharon, *Carbohydr. Res.* **3,** 305 (1967).

[74] H. M. Flowers, *Carbohydr. Res.* **74,** 177 (1979).

[75] K. Igarashi, T. Irisawa, and T. Honma, *Carbohydr. Res.* **39,** 341 (1975).

[76] B. Helferich, W. M. Müller, and S. Karbach, *Liebigs Ann. Chem.* p. 1514 (1974).

more reactive, so that a process leading to rapid anomerization of the halide in the presence of a carbohydrate nucleophile which can rapidly and selectively take up the β-halide as it is formed can lead to efficient glycosylation. Displacement of the β-halogen without complete separation of glycosyl cation and halogen will lead to high α-stereoselectivity. Tetraalkylammonium salts were found to be most efficient catalysts for this anomerization reaction owing to their good solubility in the solvents of low polarity used for the reaction, and ethyldiisopropylamine or molecular sieves served as useful condensing agents in solvents such as dichloromethane.[77] The reactions were termed "halide catalyzed glycosidations."[77] The approach continues to be very useful, especially for more reactive halides, e.g., perbenzylfucosyl bromides.[78–80] Condensation of the free 4-OH of a perbenzylated lactose derivative (**XXXIIIa**) with tri-*O*-acetyl-α-L-fucopyranosyl bromide (**XXXIVa**) gave a trisaccharide containing β : α-linked fucose at 2 : 1. Analogous reaction of (**XXXIVa**) with the free 6-OH of a similar compound (**XXXIIIb**) gave β-L-fucoside in 64% yield. Halide ion fucosylation of this same 6-OH in dichloromethane-*N*,*N*-dimethylformamide with tri-*O*-benzyl-α-L-fucopyranosyl bromide (**XXXIVb**) led to α-fucoside in 52% yield.[79] However, intermediate halide anomerization is not essential to ensure 1,2-cis-glycosylation. Thus, disaccharide (**XXXV**) reacts at 2-OH with **XXXIVb** in a halide-catalyzed reaction to give an α-fucosylated product in only 30% yield. However, the use of mercury(II) cyanide increased the yield to 40%, while mercury(II) bromide increased the yield to 80%.[81] In the latter case it could be suggested that the mercury(II) bromide was supplying halide ions for the $\alpha\rightarrow\beta$-anomerization of **XXXIVb**, but this explanation would not apply to the case when mercury(II) cyanide was employed.

With less reactive hydroxyl groups then, more active catalysts can be useful. Reaction of a trisaccharide (**XXXVI**) at 3-OH of the glucopyranosyl moiety with 2,3,4,6-tetra-*O*-benzyl-α-D-galactopyranosyl bromide (**XXXIc**) gives 80% yield of an α-linked product (**XXXVII**) in the presence of mercury(II) cyanide and molecular sieve, and a 77% yield of XXVII with silver triflate–silver carbonate. However, if the catalyst is tetraethylammonium chloride (halide-catalyzed reaction), negligible tetrasaccharide is formed.[78] Care must be taken to keep the temperature low

[77] R. U. Lemiueux, K. B. Hendricks, R. V. Stick, and K. James, *J. Am. Chem. Soc.* **97**, 4056 (1975).

[78] H. Paulsen and C. Kolář, *Chem. Ber.* **112**, 3190 (1979).

[79] T. Chiba and S. Tejima, *Chem. Pharm. Bull.* **31**, 75 (1983).

[80] J. Dahmén, T. Frejd, G. Magnusson, G. Noori, and A.-S. Carlström, *Carbohydr. Res.* **125**, 237 (1984).

[81] H. Paulsen and C. Kolář, *Chem. Ber.* **114**, 306 (1981).

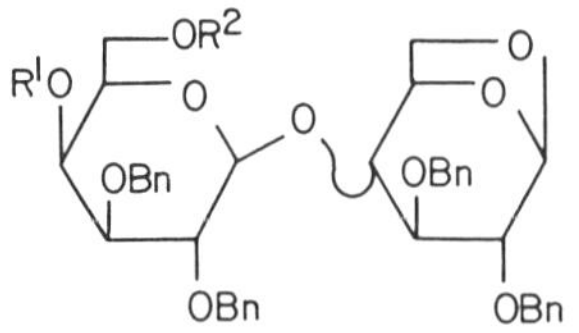

(XXXIIIa): R¹=H, R²=Bn
(XXXIIIb): R¹=Bn, R²=H

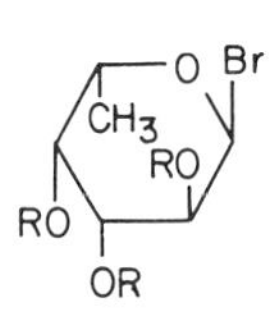

(XXXIVa): R=Ac
(XXXIVb): R=Bn

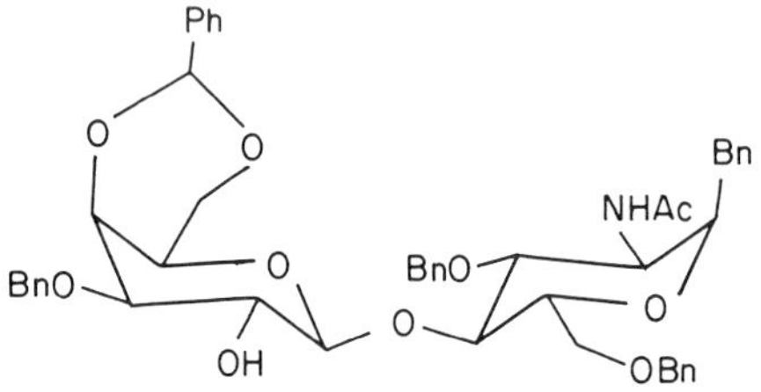

(XXXV)

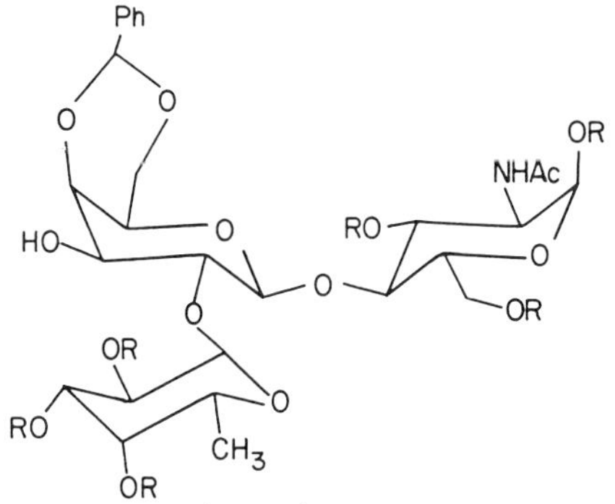

(XXXVI); R=Bn

($-25°$) for the reaction using silver salts in order to preserve good ste-reoselectivity. Increasing the polarity of the solvent, by replacing the dichloromethane or toluene with nitromethane–benzene, reduces ste-reoselectivity.

The effect of the reactivities of halide and hydroxyl components on stereochemistry were investigated[82] in coupling reactions of variously substituted galactopyranosyl bromides (**XXXVIII**) with the quite reactive 4-OH of protected rhamnosides (**XXXIX**) using mercury(II) bromide. The reactivity of the bromides (**XXXVIII**) increases more or less going down the list. In reaction with the benzyl rhamnoside (**XXXIXa**), the least reactive bromides (**XXXVIIIa,b**) gave the α-linked disaccharide stereo-

[82] H. Paulsen and O. Lockhoff, *Chem. Ber.* **114**, 3079 (1981).

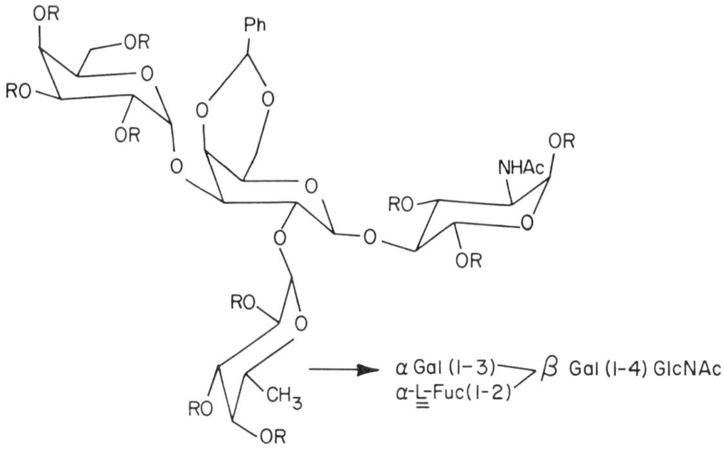

(XXXVII); R = Bn

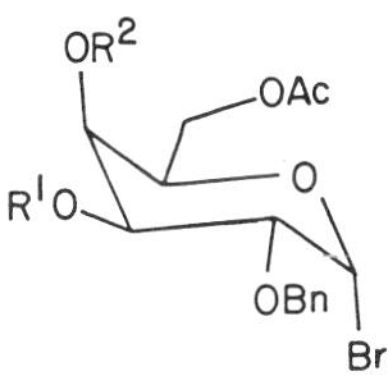

α Gal (1-3)→ β Gal (1-4) GlcNAc
α-L-Fuc(1-2)↗

(XXXVIII):(a) R^1=R^2=Ac
(b) R^1=Ac,R^2=β Man
(c) R^1=Ac,R^2=Bn
(d) R^1=βGlcNAc,R^2=Bn
(e) R^1=R^2=Bn

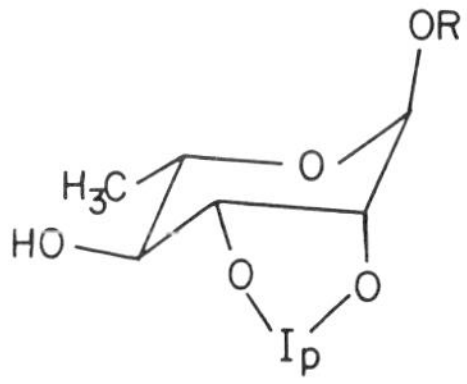

(XXXIX); (a) R=Bn
(b) R=CH$_2$CCl$_3$

specifically. The most reactive bromides (**XXXVIIId,e**) gave disaccharides with $\alpha:\beta$ ratios of approx. 1:5 and 1:4, respectively, while **XXX-VIIIc** gave $\alpha:\beta$ ratios of approx. 7:1. It will be seen from these figures that changes in substitution patterns in the bromides produce stereoelectronic effects whose results are not easily predictable and much more detailed study would be necessary to verify the possibility of devising rules based on energy calculations to predict the steric course of such condensations.

Replacement of R = Bn in **XXXIX** with a trichloroethyl group depresses the reactivity of 4-OH. In this case, reaction (of **XXXIXb**) with **XXXVIIIc** now gives disaccharides in a ratio of $\alpha:\beta$ appx. 4:1. Replacement of the mercury(II) bromide catalyst by the less effective tetraethylammonium bromide leads to a lower yield in the reaction of **XXXVIIIe** with **XXXIXa** but the product is almost exclusively the α-linked disaccharide.

By condensation of the α-halide of a protected oligosaccharide with a suitable nucleophile, higher oligosaccharides can be more readily built up than by the slower stepwise process of adding one unit at a time. Thus, a trisaccharide bromide (**XL**) can be synthesized and coupled with a disaccharide unit (**XLI**) to give a pentasaccharide derivative.[83] Use of silver perchlorate–silver carbonate as catalyst gave stereoselective α-glycosylation and the yield was quite good in spite of the lower reactivity of 4-OH in **XLI**. Deprotection of the product led to the pentasaccharide chain of the Forssman antigen (**XLII**).

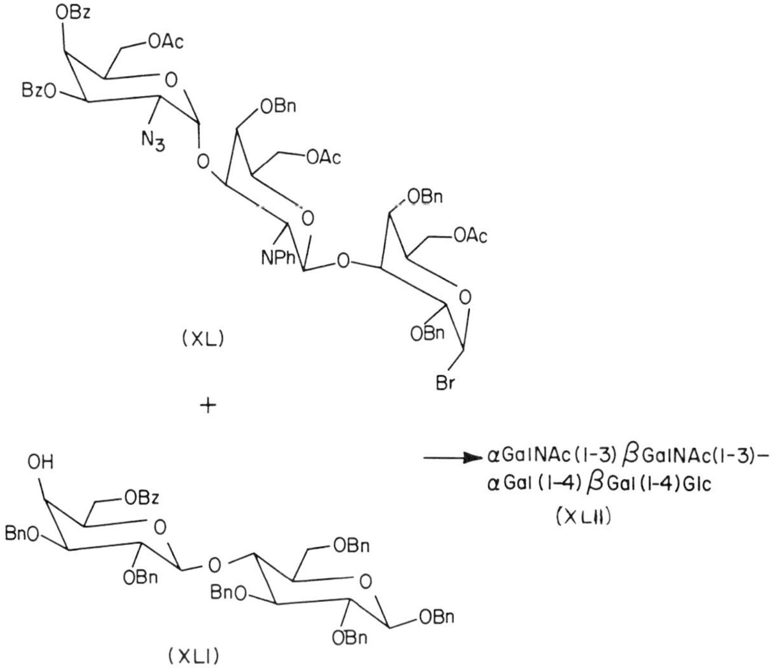

The nonahexaosyl unit of a complex type of glycan chain of a glycoprotein was synthesized by coupling (in 59% yield) a trisaccharide unit consisting of a substituted α-D-mannopyranosyl bromide with a protected trisaccharide containing free 3-OH and 6-OH. α-Linkages were obtained stereoselectively by use of silver triflate and molecular sieves.[84] It was

[83] H. Paulsen and A. Bünsch, *Liebigs Ann. Chem.* p. 2204 (1981).
[84] T. Ogawa, T. Kitajima, and T. Nukada, *Carbohydr. Res.* **123**, C5 (1983).

established that a benzylated α-D-mannopyranosyl bromide gave α-linkages with good stereoselectivity in the presence of silver triflate. However, when insoluble silver silicate was used, considerable β-anomer resulted and could exceed the amount of α-anomer formed.

Partially Acylated α-Halides

It has already been shown[1] that α-D-glycopyranosyl and α-L-fucopyranosyl bromides substituted especially at 4-OH and 6-OH with acyl groups (e.g., p-nitrobenzoate) can be condensed with carbohydrate hydroxyl groups in the presence of mercury(II) cyanide to give α-linked disaccharides with high stereoselectivity, or even stereospecifically. The work was extended[74] and a tetrasaccharide was synthesized (**XLIII**) containing α- and β-L-fucopyranosyl linkages[85] in which the α-linkage was made by using 2-O-benzyl-3,4-di-O-p-nitrobenzoyl-α-L-fucopyranosyl bromide (**XLIV**) and the β-linkage using 2,3,4-tri-O-acetyl-α-L-fucopyranosyl bromide (**XXXIVa**), both types of condensation being made in the presence of mercury(II) cyanide in 1:1 nitromethane–benzene solution.

β Gal (I-4) β-$\underline{L}$-Fuc (I-4) α-$\underline{L}$-Fuc (I-3) Glc

(XLIII)

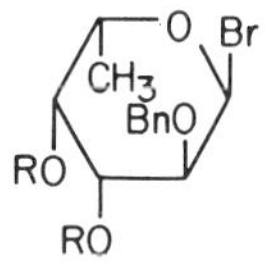

R = $\underline{P}$-O$_2$NC$_6$H$_4$CO

(XLIV)

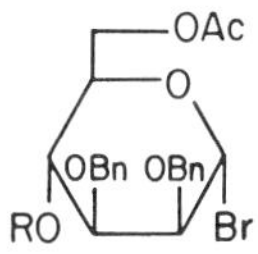

(XLVI): (a) R = ClCH$_2$CO
(b) R = Ac
(c) R = allyl (All)
(d) R = Bn

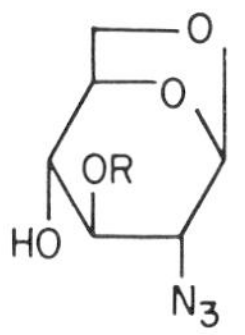

(XLV): (a) R = Ac
(b) R = Bn

Substituent effects were also found in glycosylations using insoluble reagents—silver silicate and silver zeolite.[86] The glycosyl halides were variously substituted α-D-gluco- and α-D-mannopyranosyl bromides

[85] H. M. Flowers, *Carbohydr. Res.* **119,** 75 (1983).
[86] C. A. A. van Boeckel, T. Beetz, and S. F. Van Aelst, *Tetrahedron* **40,** 4097 (1984).

which reacted with 1,6-anhydro-2-azido-2-deoxy-3-O-substituted β-D-glucopyranoses (**XLV**). In both *gluco*- and *manno*-halides, it was found that 4-O-acyl groups increase the $\beta : \alpha$ ratio compared with 4-O-alkyl while 3-O- and 6-O-acyl groups decrease this ratio. These reactions probably involved S_N2 processes since they occurred in apolar solvents such as dichloromethane and toluene and the catalysts were insoluble.[69] The mechanism is thus basically different from that followed in the α-L-fucosylations described above.[85] A 6-O-acetyl-2,3,4-tri-O-alkylated glucopyranosyl bromide gave less stereoselectivity than did the corresponding 3,6-di-O-acetate, which would presumably also be consistent with the argument that the more reactive halide (more alkylated) gave less stereoselectivity.[4,5] However, it was interesting to observe that higher selectivity and strong favoring of β-linkages were obtained in the mannose series. Employment of the α-bromides (**XLVIa,b**) with 4-O-acyl substituents gave high $\beta : \alpha$ ratios (>10 or 9.1, respectively). The 4-O-allyl substituent (**XLVIc**) reduced this ratio to approx. 5 : 1 and the 4-O-benzyl (**XLVId**) to 6 : 1.

The important factors in all these substituted mannopyranosyl bromides, then, seems to be the 6-O-acetyl substituent, since strong changes in electronic effects at C-4 seem to have little effect on the $\beta : \alpha$ ratios. However, it should be noted that replacement of the 3-O-benzyl substituent in **XLVId** by an acyl group (acetyl or trichloroacetyl) completely removes stereoselectivity and the $\beta : \alpha$ ratio becomes approx. 1.

Imidates

β-Glycopyranosyl imidates can be prepared and isolated in pure form. They condense with suitably protected carbohydrates in the presence of acid catalysts, e.g., p-toluenesulfonic acid or boron trifluoride, to give α-linked oligosaccharides in good yield and with high stereoselectivity (see p. 368).

Other Leaving Groups

The preparation of α-linked oligosaccharides, using other C-1 leaving groups such as sulfonate esters and thioglycosides, has already been discussed (pp. 368, 370). We have also mentioned the stereoselective dehydrative α-glucosylation reaction.

2-Amino Sugars (Gluco and Galacto Series)

The necessity for a nonparticipating C-2 substituent to enable 1,2-*cis*-glycoside formation applies even more strongly for amino sugars, since participation of acylamido groups is even more facile, and there is an even greater tendency to form 1,2-cyclic intermediates containing nitrogen than 1,2-orthoesters. Oxazolines are readily formed and their reaction with ROH will give β-linked products (see later). Various applications of nonparticipating C-2 substituents have been described. Since the previous article,[1] some antibiotic glycosides, e.g., neamine,[87] have been synthesized utilizing a 2-N-methoxybenzylidene substituent [**XLVII**, R = pC$_6$H$_4$(OCH$_3$)—CH=N] to produce an α-linked disaccharide.

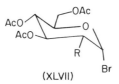

(XLVII)

2-Azido Sugars

A very successful approach at present to α-linked 2-amino sugars is by use of the corresponding azido halides which can be condensed with the requisite nucleophiles under suitable conditions to produce α-linked glycopyranosides with high stereoselectivity.[1,4,5] The 2-azido group in the product is readily converted to 2-amino or 2-acylamido functions. 2-Azido-2-deoxy-β-D-gluco- and galactopyranosyl chlorides can be prepared and isolated as unstable intermediates. These β-halides react with nucleophiles and silver perchlorate–silver carbonate in dichloromethane to give α-glycosides in high yields and with good stereoselectivity in S$_N$2-type reactions.[88] Good stereoselectivity is also obtained with reactive primary hydroxyl groups and the method is preferred[4,5] for these cases to the alternative of *in situ* anomerization since it is more stereoselective. This is rather surprising since it might be expected that the silver catalyst would be less, rather than more, selective than *in situ* anomerization in which a less active catalyst (tetraethylammonium chloride) is employed. When the other substituent groups on the 2-azido halide are acyl groups, its relatively low reactivity necessitates the employment of a very active catalyst for successful condensation with less reactive, secondary carbohydrate hydroxyl groups. Then, silver triflate-silver carbonate is the choice. Reaction of the α-bromide (**XXXIa**), for example, with 3'-OH of the protected trisaccharide (**XXXVI**) and silver triflate–silver carbonate at $-20°$ gave the desired product stereoselectively, and it could be con-

[87] A. Harayama, T. Tsuchiya, and S. Umezawa, *Bull. Chem. Soc. Jpn.* **52**, 3626 (1979).

[88] H. Paulsen and W. Stanzel, *Chem. Ber.* **111**, 2334 (1978).

verted to the tetrasaccharide blood group A determinant (**XLVIII**). Care must be taken with the temperature, since stereoselectivity decreases considerably at room temperature.

Utilization of both the *in situ* anomerization approach and application of insoluble silver salts in heterogeneous phase is illustrated in a synthesis of a branched pentasaccharide[89]—the repeating unit of the O-specific chain of *Shigella dysenteriae* lipopolysaccharide. Reaction of the 2-azido-α-glucosylbromide (**XLIX**) with 4-OH of the anhydrogalactose derivative (**L**) occurred in the presence of mercury(II) bromide to give the α-linked disaccharide which could be converted to a product with a free 3-OH in the anhydrogalactose moiety. This product was condensed with a protected 2-azido-α-galactosylbromide (**LIa**), using mercury(II) cyanide-mercury(II) bromide, again to give α-linked glycoside stereoselectively. This product was transformed into a trisaccharide α-bromide and the "block" condensed with 4'-OH of the disaccharide derivative (**LII**). Silver perchlorate–silver carbonate was the preferred catalyst in this latter reaction.

[89] H. Paulsen and H. Bünsch, *Chem. Ber.* **114**, 3126 (1981).

An illustration of the delicacy of choice in such reactions comes from later work in the same laboratory in which a very similar bromide to **LIa** (**LIb**, replacement of C-3 ester by benzyl ether which would presumably raise its reactivity somewhat) reacted with a protected trisaccharide of L-rhamnose in the presence of silver silicate in dichloromethane to give a tetrasaccharide with β-linked 2-azido-2-deoxygalactose with high stereoselectivity.[90]

A different laboratory, on the basis of studies with a model compound, cyclohexanol, concluded[91] that the most efficient reagent for synthesizing 2-acetamido-2-deoxy-α-D-galactopyranosyl oligosaccharides was the benzylated chloride (**LIII**). They found that mercury(II) salts showed high β-selectivity and silver triflate–s-collidine favored the α-anomer, but silver perchlorate–silver carbonate showed high α-selectivity. Compound **LIII** reacted with a protected trisaccharide, on 3-OH of a galactose moiety, to give 82% yield of a blood group A-active tetrasaccharide after 30 min at room temperature. In the same study, they achieved α-fucosylation by use of **XLIV**[85] on a protected disaccharide together with mercury(II) cyanide–mercury(II) bromide in acetonitrile.

Some of the 2-azido-α-bromides of mono- and oligosaccharides have been recently utilized in syntheses of the linkage region of carbohydrates to L-serine and L-threonine as it occurs in O-glycoprotein chains.[5,92,93] T_N glycopeptide related to human glycophorin A[M] was synthesized[93] as L-seryl-O-(α-GalNAc)-L-seryl-O-(α-GalNAc)-L-threonyl-O-(α-GalNAc)-L-threonylglycine trifluoroacetate.

Nitrosochlorides. An earlier approach to α-glycosides of 2-amino sugars using a nonparticipating nitrogenous function was that of nitrosochlorides, reductive processes converted the resulting oligosaccharide to the desired acetamido compound.[1,94]

As was mentioned previously,[1] problems arise with this method in the steric control of the reduction of the intermediate hydroxyaminoglycoside to the required amino compound. Thus, condensation of dimeric tri-O-acetyl-2-deoxy-2-nitroso-α-D-galactopyranosyl chloride with 1,2:5,6-di-O-isopropylidene-α-D-galactofuranose gives a product which is converted into a mixture of epimers[95] (Gal:Tal 5:7). Reduction of the corresponding derivatives derived from the D-glucosyl chloride goes more stereoselectively and does offer a reasonable route to 2-deoxy-2-amino-α-D-gluco-

[90] H. Paulsen and W. Kutschker, *Liebigs Ann. Chem.* p. 557 (1983).
[91] N. V. Bovin, S. E. Zurabyan, and A. Ya. Khorlin, *Carbohydr. Res.* **112,** 23 (1983).
[92] V. V. Bencomo, J.-C. Jacquinet, and P. Sinaÿ, *Carbohydr. Res.* **110,** C9 (1982).
[93] B. Ferrari and A. A. Pavia, *Int. J. Pept. Protein Res.* **22,** 549 (1983).
[94] R. U. Lemieux, K. James, and T. L. Nagabhushan, *Can. J. Chem.* **51,** 48 (1973).
[95] R. U. Lemieux and R. V. Stick, *Aus. J. Chem.* **31,** 901 (1978).

pyranosides. However, this method is not superior to the 2-azido method, especially with the recent introduction of a more convenient preparation of the 2-azido compound starting from the corresponding glycal[96,97] in place of the much longer process previously required which involved successive substitutions and displacements on anhydropyranose sugars.

β-D-*Mannopyranosides*

In mannopyranosides (and the related 6-deoxy compounds, rhamno-pyranosides) the 2-substituent is epimeric to that in glucose and galactose. Since the stable halide is still the α-D-anomer, this results in it being in a 1,2-trans configuration to the C-2 substituent. Participating groups at C-2 will therefore tend to produce *trans*-1,2-glycosides as before, which in this case will be the α-D-anomers. *In situ* anomerization processes would lead to unstable β-halides which would again tend to give α-glycosides on reaction with nucleophiles. Some circuitous methods have been devised for β-D-mannopyranosides depending on synthesis of the corresponding glucoside and inversion at C-2 to the required configuration.[98]

It was reasoned[4,5] that an insoluble silver catalyst could promote β-glycoside formation from an α-D-mannopyranosyl halide without a C-2 participating group by an S_N2-type process. We have already mentioned findings with some of these halides (**XLVI**) using silver silicate and silver zeolite as catalysts[86] and the discovery that the halides with lower reactivity (4-*O*-acyl versus 4-*O*-alkyl) gave higher proportions of β-anomers and better stereoselectivity.

The dibenzyl derivative (**LIV**) reacted with 4-OH of a protected anhydroglucose (**LV**) to give the desired β-linked disaccharide stereoselectively (with silver silicate, β : α ratio approx. 5 : 1).

A different laboratory constructed β-D-mannopyranosyl linkages from the tribenzylated bromide (**XLVId**) using insoluble silver salts as promoters and thus prepared[99] *p*-trifluoroacetamidophenyl *O*-β-Man(1-4)-*O*-α-L-Rha(1-3)-α-Gal and *p*-trifluoroacetamidophenyl *O*-β-Gal(1-6)-*O*-β-Man(1-4)-α-L-Rha which are related to O-antigenic polysaccharides of *Salmonella* serogroup E. Stereoselectivity was not particularly high. In the presence of molecular sieves, silver zeolite gave an α : β ratio of 1 : 3 while silver silicate gave α : β approx. 1 : 1 (yields were approx. 80%). A

[96] R. U. Lemieux and R. M. Ratcliffe, *Can. J. Chem.* **57,** 1244 (1979).

[97] T. C. Wong and R. U. Lemieux, *Can J. Chem.* **62,** 1207 (1984).

[98] C. Augé, C. D. Warren, R. W. Jeanloz, M. Kiso, and L. Anderson, *Carbohydr. Res.* **82,** 85 (1980).

[99] P. J. Garegg, C. Henrichson, T. Norberg, and P. Ossowski, *Carbohydr. Res.* **119,** 95 (1983).

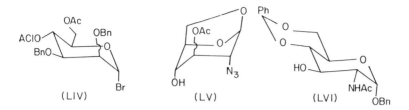

(LIV)   (LV)   (LVI)

β-L-rhamnopyranoside was similarly obtained by using 2,3,4-tri-O–ben-zyl-α-L-rhamnopyranosyl bromide with silver silicate and molecular sieves.[100] An interesting approach to α- and β-D-mannopyrano- and furan-osides was recently described[101] using alcohol triflates with either 2,3,4-tri-O-benzyl-D-mannopyranose or 2,3 : 5,6-di-O-isopropylidene-α-D-man-nofuranose, respectively. Condensation in the presence of sodium hydride led to α-glycosides, while sodium hydride together with crown ether favored the β-anomer.

1,2-*Trans*-Glycopyranosides

Possible mechanisms of reaction of acylglycosyl halides bearing par-ticipating acetyl or benzoyl groups at 2-OH with nucleophiles were dis-cussed previously.[1] It would seem that the two main possibilities, S_N2 type or S_N1 type with variations, can be roughly predicted on the basis of the acid acceptor used: insoluble salts would favor S_N2 reactions espe-cially in nonpolar solvents. By using silver carbonate, oxide, silicate, or zeolite, then it should be possible to determine the configuration of the product by selection of the halide taken for the reaction; inversion of 1,2-*cis*-halides (the most stable for gluco and galacto series) will lead to 1,2-*trans*- (i.e., β) glycosides. With mannose derivatives, on the other hand, the more stable 1,2-*trans*-halide should give 1,2-*cis*- (i.e., β) glycosides. However, the possibility of participation of the 2-substituent must also be considered. This applies especially strongly to the manno series and thereby precludes the preparation of β-D-mannopyranosides by S_N2 reac-tions from 2-O-acetyl- (or benzoyl)mannopyranosyl bromides under nor-mal circumstances, although, as I have already shown, it does offer this possibility when the 2-substituent is nonparticipating.

Galacto and Gluco Derivatives

Early employment of insoluble silver catalysts (oxide and carbonate) and, occasionally, other insoluble inorganic salts gave disaccharides in

[100] H. Paulsen and W. Kutschker, *Carbohydr. Res.* **120,** 25 (1983).
[101] R. R. Schmidt, U. Moering, and M. Reichrath, *Chem. Ber.* **115,** 39 (1982).

reasonable yield if the hydroxyl component was fairly reactive, especially 6-OH. The addition of an internal desiccant to take up the water formed as a result of the reaction of the salt with the acid liberated in the reaction was generally beneficial and increased yields considerably.

Employment of soluble mercury(II) salts (cyanide, bromide) in nonpolar or semipolar media was considered to direct the reaction mechanism via ionization of the halide. The subsequent equilibria in which glycosyl cation, halide ion, catalyst, and solvent can all be involved determined the nature of the final product(s).

Mercury(II) cyanide[102] has proved to be a most effective catalyst for many condensations with a variety of nucleophiles; the side product, hydrogen cyanide, escapes from the system and does not affect the subsequent equilibria. Occasionally cyano compounds may be formed as unwanted byproducts and can themselves react under certain circumstances.[103] Usually, the reaction takes place at ambient temperature and a good solvent system is 1 : 1 nitromethane–benzene. Care must be taken to ensure dryness of reagents and then the presence of an internal desiccant is not necessary. The acylglycopyranosyl halide dissolves readily in the solvent and can be used either crystalline or as a syrup in cases where it is not so obtainable. It is generally beneficial to use excess of halide added in portions, and long reaction times are required, several hours to several days, depending on the reactivity of the nucleophile. The temperature can be raised if necessary to aid in solution of the nucleophile without any apparently deleterious effects. Thus, reaction of tetra-O-acetyl-α-D-galactopyranosyl bromide (**XXXI**, R^1 = OAc; R^2 = Ac) with benzyl 2-acetamido-2-deoxy-4,6-O-benzylidene-α-D-glucopyranoside (**LVI**),[104] or the corresponding galacto epimer[105] at 3-OH at 60° in 1 : 1 nitromethane–benzene promoted by mercury(II) cyanide produced the β-linked disaccharide in 80% yield.

A variety of fucopyranosides have also been synthesized from tri-O-acetyl-α-L-fucopyranosyl bromide with mercury(II) cyanide but sometimes, depending on the nucleophile, 1,2-*cis*- (α) rather than 1,2-*trans*-glycosides have resulted.[72–74,85,106] A very recent publication[107] describes the preparation in yields of >90% of trisaccharide derivatives containing

[102] B. Helferich and K. Weis, *Chem. Ber.* **89**, 314 (1956).
[103] B. Coxon and H. G. Fletcher, *J. Am. Chem. Soc.* **86**, 922 (1964); M. Bobek and J. Farkaš, *Collect. Czech. Chem. Commun.* **34**, 247 (1969).
[104] H. M. Flowers and R. W. Jeanloz, *J. Org. Chem.* **28**, 1377 (1963).
[105] H. M. Flowers and D. Shapiro, *J. Org. Chem.* **30**, 204 (1965).
[106] K. L. Matta, S. S. Rana, C. F. Piskorz, and J. J. Barlow, *Carbohydr. Res.* **112**, 213 (1982).
[107] J. Hirsch, E. Petráková, and J. Schraml, *Carbohydr. Res.* **131**, 219 (1984).

arabinofuranose by condensation of 2,3,5-tri-O-benzoyl-α-L-arabino-furanosyl bromide (**LVII**) with 3-OH and 3'-OH of a protected xylo-bioside. Mercury(II) cyanide was the reagent used, in benzene solution, and no trace of any 1,2-cis (i.e., β) anomer was seen. The directing effect of the nucleophile was also shown in two recent publications from the same laboratory using mercury(II) cyanide-promoted condensations. Ga-lactosylation with **XXXI** (R¹ = OAc; R² = Ac) on 6-OH of a protected p-nitrophenyl-2-acetamidodeoxy-β-D-galactopyranoside gave β-D-galacto-pyranosylation, apparently stereospecifically.[108] However, reaction of the same halide and mercury(II) cyanide under similar conditions with 2-OH of a similar glycoside led to lower stereoselectivity (β : α ratio 9 : 2).[109] A substituted galactopyranosyl bromide with a nonparticipating fuco-pyranosyl substituent at C-2 (**LVIII**), reacted with 8-(methoxycarbon-yl)octanol in benzene together with mercury(II) cyanide to give α- and β-glycosides in a ratio of 13 : 4. Similarly, reaction of **LVIII** with 3-OH of various protected 2-acetamido-2-deoxyglucopyranosides gave mixtures enriched in α-anomers[110] (α : β, approx. 3 or 4 : 1).

Condensation of **XLVII** (R = OAc) with perbenzyl ethers of benzyl β-D-glucopyranoside and cellobioside at 4-OH and 4'-OH, respectively, us-ing mercury(II) cyanide in 1 : 1 benzene–nitromethane led to high yields of the β-D-glucopyranosides (di- and trisaccharides) stereospecifically.[111] Methyl 2,3,6-tri-O-benzyl-α-D-glucopyranoside with **XXXI** (R¹ = R² = OAc) gave a convenient synthesis of lactose.[111]

Another excellent promoter for β-glycoside synthesis in homogeneous media is silver triflate[112] which enables rapid reactions at low tempera-tures. Tetramethylurea[80,112] has generally been used as an acid acceptor,

[108] K. L. Matta, S. S. Rana, and S. A. Abbas, *Carbohydr. Res.* **131**, 265 (1984).

[109] S. A. Abbas, J. J. Barlow, and K. L. Matta, *Carbohydr. Res.* **106**, 59 (1982).

[110] K. L. Matta, S. S. Rana, C. F. Piskorz, and S. A. Abbas, *Carbohydr. Res.* **131**, 247 (1984).

[111] K. Takeo, K. Okushio, K. Fukuyama, and T. Kuge, *Carbohydr. Res.* **121**, 163 (1983).

[112] S. Hanessian and J. Banoub, *Carbohydr. Res.* **53**, C13 (1977).

(LIX)

β Glc(l-6)βGlc(l-6)βGlc(l-6)βGlc(l-6)βGlc
  3   3
  ↑   ↑
  ı   ı
 βGlc βGlc
                              3
(LX)   ↑
                              ı
                              R

sometimes in the presence of molecular sieves. *s*-Collidine has occasionally replaced the tetramethylurea[17] but more frequently for 2-aminoglycosides (see later). By use of silver triflate and molecular sieves in dichloromethane, tetra-*O*-benzoyl-α-D-glucopyranosyl bromide reacted with a protected disaccharide (**LIX**), having free 2′-OII and 3′-OH groups, regio- and stereospecifically at 3′-OH to give a trisaccharide derivative in 90% yield. Stepwise and block condensations enabled synthesis of a glucoheptaose (**LX**, R = H) and a glucooctaose (**LX**, R = D-Glc*p*) eliciting phytoalexin accumulation in soybean.[113]

Cellotriose derivatives were obtained in reasonable yield (40–50%) by condensing **XLVII** (R = OAc) with 4′-OH of a protected cellobioside or hepta-*O*-acetyl-α-D-cellobiosyl bromide with 4-OH of a protected glucopyranoside in the presence of silver triflate–*s*-collidine.[114] It will be observed that these yields, using acetates as protecting groups in the nucleophiles and silver triflate as promoter, were much less than those obtained a little later in the same laboratory[111] using benzyl ethers for protection and mercury(II) cyanide as promoter.

Trimethylsilyl triflate can act as a promoter for β-glycoside synthesis from β-1-(*O*-acetates).[43,66,115]

The *orthoester method* for β-glycoside formation has already been discussed (p. 373). A protected trisaccharide with a 1,2-cyanoethylidene and a 6″-*O*-trityl group was polymerized with triphenylcarbenium perchlorate to give a polymer of β-Man(1-4)-α-Rham(1-3)-β-Gal(1-6)-containing about 10 trisaccharide units per molecule.[55] This trisaccharide is the

[113] P. Ossowski. A. Pilotti, P. J. Garegg, and B. Lindberg, *J. Biol. Chem.* **259,** 11337 (1984).
[114] K. Takeo, T. Yasato, and T. Kuge, *Carbohydr. Res.* **93,** 148 (1981).
[115] H. Paulsen and M. Paal, *Carbohydr. Res.* **135,** 53 (1984).

repeating unit of the O-antigenic lipopolysaccharide of *Salmonella ne-wington*.

2-Amino-2-Deoxy-β-D-Galacto- and Glucopyranosides

The reduced stability of acetylated pyranosyl bromides of 2-acetamido sugars led to attempts to utilize the corresponding chlorides for glycosylations. In this way, disaccharides have been prepared with more reactive nucleophiles (primary hydroxyl) using mercury(II) cyanide.[116] With 2-dichloroacetamido halides, glycosylation is better, and again β-glycosides are formed although yields decline considerably if secondary OH groups react.[117] More recently, the oxazoline and N-phthalimido methods have been found extremely useful for β-glycosylation of 2-galacto and gluco derivatives, and the 2-azido halides can also be used as β-glycosylating agents under suitable conditions.

Oxazolines. Similarly to orthoesters, oxazolines[118] (2-methyl or 2-phenyl) can be utilized for specific direction of nucleophiles and afford a good approach to 1,2-*trans*-glycosides of 2-amino sugars. In the presence of acid catalysts, especially *p*-toluenesulfonic acid, the oxazoline ring is opened with stereospecific production of the glycoside or oligosaccharide.[119] The utility of the method has been extended by the discovery

of easier ways of preparing oxazolines, e.g., by the action of iron (III) chloride on β-1-*O*-acetates[120] or mercury(II) chloride on propenylglycosides,[121] which are readily available from allylglycosides. A phenyloxazoline has also been prepared from D-glucosamine and coupled with 1,2 : 3,4-di-*O*-isopropylidene-α-D-galactopyranose at 6-OH in 80% yield.[122] In general, very good yields of disaccharides are obtained from oxazolines by reaction with primary OH groups and the yields with secondary hydroxyls, although reduced, are still often quite reasonable. The reaction is usually performed for several hours at elevated temperatures

[116] M. Inage, H. Chaki, S. Kusomoto, and T. Shiba, *Tetrahedron Lett.* **21**, 3889 (1980).
[117] D. Shapiro, Y. Rabinsohn, A. J. Acher, and A. Diver-Haber, *J. Org. Chem.* **35**, 1464 (1970).
[118] F. Micheel and H. Köchling, *Chem. Ber.* **90**, 1597 (1957).
[119] S. E. Zurabyan, T. S. Antonenko, and A. Ya. Khorlin, *Carbohydr. Res.* **15**, 21 (1970).
[120] K. L. Matta and O. P. Bahl, *Carbohydr. Res.* **21**, 460 (1972).
[121] M. A. Nashed, C. W. Slife, M. Kiso, and L. Anderson, *Carbohydr. Res.* **82**, 237 (1980).
[122] R. Gigg and R. Conant, *Carbohydr. Res.* **100**, C1 (1982).

α Man(1-3)β Man(1-4)β GlcNAc(1-4)GlcNAc

(LXIII)

in weakly polar solvents (e.g., benzene–nitromethane). A significant finding, for example, was the reaction of the trisaccharide oxazoline (**LXI**) with 4-OH of a protected glucosamine derivative (**LXII**) to give a tetrasaccharide (**LXIII**) in 22% yield.[123]

Coupling of two different disaccharide oxazolines (**LXIV, LXVII**) sequentially on O-3 of a protected galactopyranoside (**LXV**) and O-6 of the galactopyranoside moiety of a trisaccharide derivative (**LXVIb**) enabled the synthesis of a branched pentasaccharide (**LXIX**), the coupling steps with the oxazolines proceeding with yields of 50–70%.[124]

The disaccharide oxazoline (**LXIV**) has also been condensed in toluene–nitromethane at 95° with 3′-OH of a protected 1,6-anhydrolactose derivative in 79% yield to give, after removal of the protecting groups, the milk tetrasaccharide lacto-N-tetraose.[125]

Phthalimides. A C-2 phthalimido group in the galacto and gluco series favors the β-halide due to steric reasons. Steric effects, and also some possible neighboring group activity, causes this C-2 substituent to have a strongly β-directing effect in glycosylation reactions.[126] Silver triflate–s-collidine is the preferred catalyst in dichloromethane solution at low temperatures. The phthalimido group is subsequently removed by hydrazinolysis. Methods making available a free 2-amino group in oligosaccharides enable the preparation of the necessary 2-phthalimido derivatives and their conversion into the 1-halides used for coupling reactions. The phthalimido-lactosamine (**LXXa**) reacts with 3-OH of the protected lactoside (**LXXI**) in a silver triflate–s-collidine-catalyzed reaction, to give the tetrasaccharide **LXXII** (coupling yield, 83%)[127]; an oxazoline coupling gave a yield of 79%.[125]

[123] C. Warren, R. W. Jeanloz, and G. Strecker, *Carbohydr. Res.* **71,** C5 (1979).

[124] C. Augé, S. David, and A. Veyrières, *Nouv. J. Chim.* **3,** 491 (1979).

[125] T. Takamura, T. Chiba, H. Ishihara, and S. Tejima, *Chem. Pharm. Bull.* **28,** 1804 (1980).

[126] R. U. Lemieux, M. T. Takeda, and B. Y. Chung, *ACS Symp. Ser.* **39,** 90 (1976).

[127] M. Ponpipom, R. L. Bugianesi, and T. Y. Shen, *Tetrahedron Lett.* p. 1717 (1978).

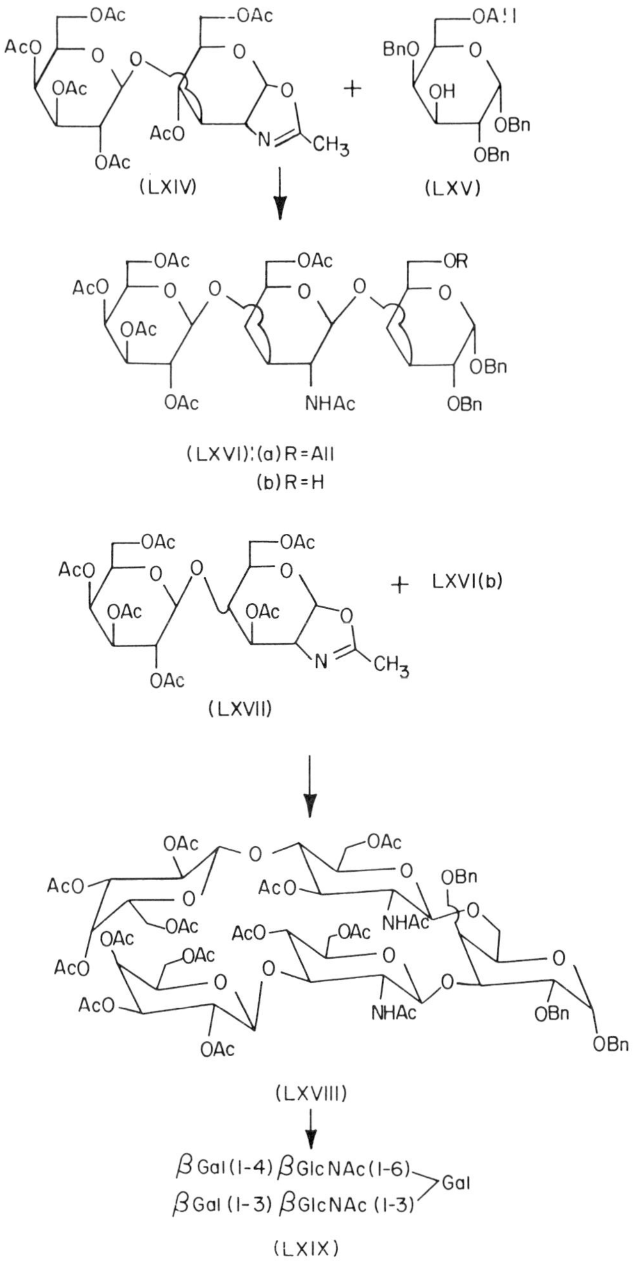

(LXIV)   (LXV)

(LXVI): (a) R = All
         (b) R = H

(LXVII)   + LXVI(b)

(LXVIII)

β Gal (1-4) β Glc NAc (1-6)
                                    Gal
β Gal (1-3) β Glc NAc (1-3)

(LXIX)

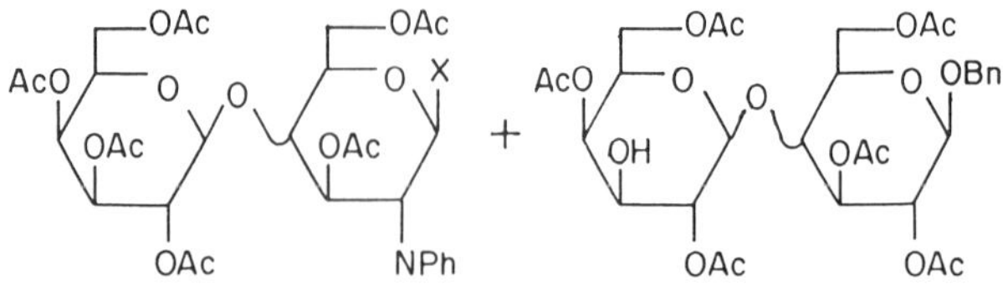

```
      OAc        OAc             OAc        OAc
  AcO   O    O      O  X      AcO   O    O      O  OBn
      OAc        OAc    +          OH        OAc
      OAc        NPh             OAc        OAc

  (LXX): (a) X = Cl              (LXXI)
         (b) X = Br
```

βGal(1-4)βGlcNAc(1-3)βGal(1-4)Glc
(LXXII)

```
R(1-4)αMan(1-3)                  R(1-2)
R(1-2)           Man            R(1-4)  αMan(1-3)
R(1-6) αMan(1-6)                                   Man
                                R(1-2)
                                R(1-4)  αMan(1-6)

  (LXXIII)                        (LXXIV)
```

(R = βGal(1-4)βGlcNAc-)

βGal(1-4)βGlcNAc(1-2)αMan(1-6)
                                 βMan(1-4)βGlcNAc(1-4)GlcNAc
βGal(1-4)βGlcNAc(1 2)αMan(1-3)

(LXXV)

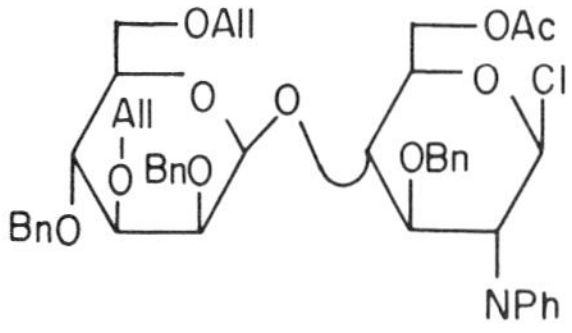

```
      OAll       OAc              OBz
  All  O     O      O  Cl     AcO   O    Br
   BnO   BnO        OBn             OBz
  BnO
             NPh              NPh
  (LXXVI)                     (LXXVII)
```

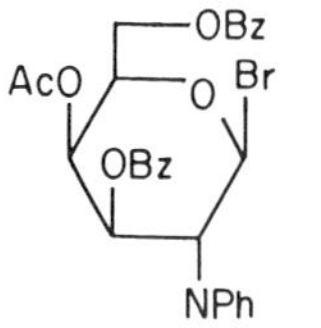

Recently, in a remarkable achievement,[128] **LXXb** was condensed with a suitably protected trimannoside using silver triflate–*s*-collidine to give nona- (**LXXIII**) and undecasaccharides (**LXXIV**) of types found in *N*-glycoprotein structures. A different laboratory recently reported[84,129] a synthesis of the nonahexaosyl unit of a complex glycan chain of *N*-glyco-

[128] H. Lönn and T. Lönngren, *Carbohydr. Res.* **120**, 17 (1983).
[129] T. Ogawa, T. Kitajima, and T. Nukada, *Carbohydr. Res.* **123**, C8 (1983).

proteins (**LXXV**). A critical coupling reaction was performed using the *N*-phthalimido-β-chloride (**LXXa**) with silver triflate-*S*-collidine.

A phthalimido disaccharide β-chloride (**LXXVI**) has been used with silver triflate–*s*-collidine in a synthesis[130] of part of the pentasaccharide *O*-α-Man(1-3)-[*O*-α-Man(1-6)]-*O*-β-Man(1-4)-*O*-β-GlcNAc(1-4)-GlcNAc, which is the core region of the carbohydrate portion of *N*-glycoproteins.

A recent finding of interest concerns the effect of nucleophile reactivity upon condensations with a phthalimido halide reagent.[131] The phthalimido-β-galactopyranosyl bromide (**LXXVII**) was prepared for reaction at 4'-OH (axial) of perbenzylated methyl β-lactosides. It was found that no desired product could be isolated when silver triflate–*s*-collidine was used. The possibility of partial hydrolysis of acyl functions in **LXXVII** was not discussed by the authors. Silver perchlorate–silver carbonate gave a reasonable yield of desired disaccharide when the 6'-OH of the lactoside was benzylated. In the presence of a 6'-*O*-benzoate in the lactoside, however, the yield of trisaccharide with **LXXVII** was much reduced. Compound **LXXa** was also used in a synthesis of a hexasaccharide hapten of i-blood group antigen[132]; this is a β(1-3)-linked trimer of *N*-acetyllactosamine.

2-Azido Halides. The use of α-halides of 2-azido sugars can be adopted to α- or β-glycoside formation depending on the reaction conditions. The synthesis of α-linked disaccharides by this method has already been discussed. Conditions tending to favor ionization and separation of charges (more elevated temperatures and more polar solvents, for example) will decrease the stereoselectivity of α-glycoside formation. At the same time, careful preservation of S_N2-like conditions can ensure a good yield of product with configuration the inverse of that of the starting halide, i.e., α-glycoside from β-halide, but β-glycoside from α-halide. Thus, the less reactive but more stable 2-azido-α-D-galacto- or glucopyranosyl halides will react with alcohols in the presence of insoluble catalysts in the heterogeneous phase with inversion of configuration. As with the synthesis of β-D-mannopyranosides, active silver catalysts are the most effective, e.g., silver silicate. The reacting components should be reasonably reactive as low temperatures in nonpolar media are preferable to minimize anomerizations. Thus, condensation of **LIa** with a protected anhydrogalactose at 3-OH (**LXXVIII**) in dichloromethane at $-20°$ gives a disaccharide in 73% yield.[83]

[130] H. Paulsen and R. Lebuhn, *Carbohydr. Res.* **130,** 85 (1984).
[131] H.-P. Wessel, T. Iversen, and D. R. Bundle, *Carbohydr. Res.* **130,** 5 (1984).
[132] J. Alais and A. Veyrières, *Tetrahedron Lett.* **24,** 5223 (1983).

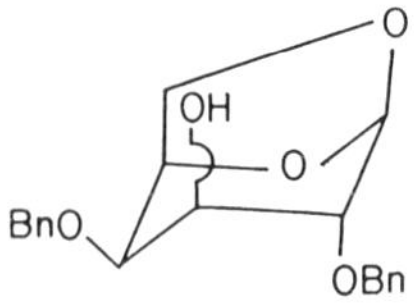

(LXXVIII)

2-Acetamido 1-O-Acetates. 1-Halido sugars are often made via their 1-O-acetates. It is not always necessary to pass through this transformation since β-1-O-acetates can themselves be converted directly into β-D-glycosides if the hydroxyl group is sufficiently reactive, as has already been described for the analogous 2-hydroxyl sugars (see p. 371). In the reaction[46] between ROH and 2-acetamido-1,3,4,6-tetra-O-acetyl-2-deoxy-β-D-glucopyranose in dichloromethane with the addition of iron(III) chloride and tetramethylurea, the yield of disaccharide varied from 80% (primary OH) to 40–60% for secondary OH, the yield declining as the reactivity of the OH fell.

In a most recent publication, the trichloroacetimidate procedure has been applied to the synthesis of β-linked D-glucosaminyl oligosaccharides.[133] N-Phthalyl derivatives of N-acetylglucosamine with free 1-OH were converted into β-trichloroacetimidates (**LXXIX**) by reaction with trichloroacetonitrile and sodium hydride. These intermediates reacted with 3-OH of a protected galactopyranoside (**LXXX**) and 6-OH of a protected disaccharide derivative (**LXXXI**) in dichloromethane, catalyzed by BF$_3$–ether with β-specificity.

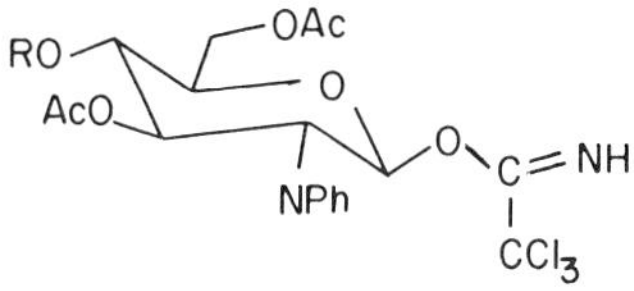

(LXXIX): (a) R=Ac
(b) R=2,3,4,6-tetra-O-Ac-β Gal

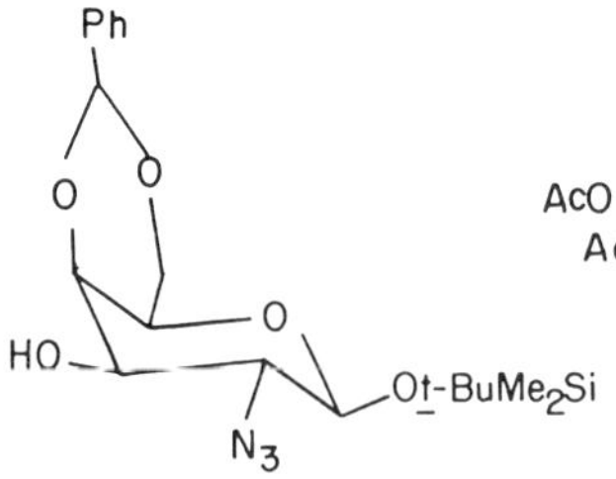

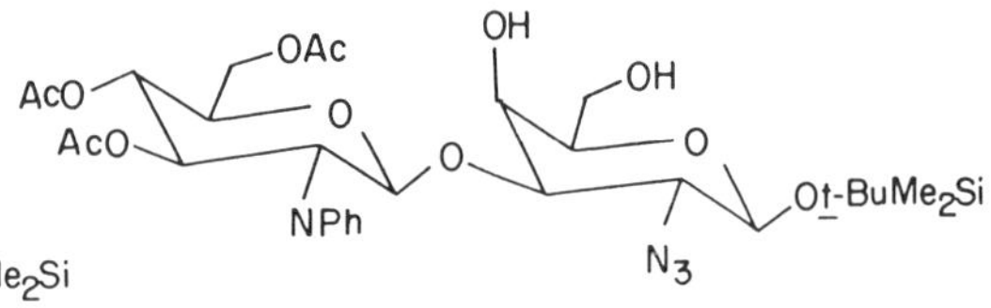

(LXXX)

(LXXXI)

[133] G. Grundler and R. R. Schmidt, *Carbohydr. Res.* **135**, 203 (1985).

In the reaction of **LXXIXa** with **LXXX**, a yield of 69% was obtained. **LXXIXb** reacted with **LXXX** to give trisaccharide in 67% yield, and with **LXXI**, a tetrasaccharide resulted in a reported yield of 94%. This latter product resulted from regiospecific condensation with 6-OH in **LXXI**, and no reaction occurred at the less reactive, axial, 4-OH.

α-D-*Mannopyranosides and* α-L-*Rhamnopyranosides*

Much of the reasoning applied to 1,2-*trans*-glycoside formation with glucose and galactose (i.e., β-D-glycosides) will also hold for the synthesis of α-D-manno- and α-L-rhamnopyranosides. An additional driving force for *trans*-glycoside formation with the above two sugars is the 1,2-*trans*-configuration of C-2 substituent and halide in the stable α-halides. When the C-2 substituent is an acyl group, participation can occur readily even without prior ionization of the halide, i.e., in S_N2 reactions as well as S_N1. This should therefore facilitate reactions with these halides in both homogeneous and heterogeneous conditions and explain the great difficulty to produce good β-stereoselectivity when required. In homogeneous media, mercury(II) cyanide[134] or 1:1 mercury(II) cyanide–mercury(II) bromide[135] have proved to be efficient promoters of α-D-mannosylation. A recent synthesis[136] of two tetrasaccharides (**LXXXII, LXXXIII**) describes the block condensation of peracetylated dimannopyranosyl bromides with suitable disaccharide acceptors using these promoters.

α Man(1–3) α Man(1–6)β Man(1–4)GlcNAc

(LXXXII)

α Man(1–2)αMan(1–3)βMan(1–4)GlcNAc

(LXXXIII)

αMan(1–2)αMan(1–6)⟩αMan-OCH₃   βGlcNAc(1–2)αMan(1–6)⟩αMan-OCH₃
αMan(1–2)αMan(1–3)   βGlcNAc(1–2)αMan(1–3)

(LXXXIV)   (LXXXV)

For more complex structures, silver triflate-tetramethylurea seems to be somewhat preferable. In syntheses[137] of the branching pentasaccharides **LXXXIV** and **LXXXV** present as part of *N*-glycoprotein chains, α-mannosylation was accomplished by use of silver triflate–tetramethyl-

[134] K. L. Matta, R. H. Shah, and O. P. Bahl, *Carbohydr. Res.* **77**, 255 (1979).
[135] J. Arnarp and J. Lönngren, *Acta Chem. Scand., Ser. B.* **B32**, 696 (1978).
[136] Y. Itoh and S. Tejima, *Chem. Pharm. Bull.* **32**, 957 (1984).
[137] T. Ogawa, K. Katano, K. Sasajima, and M. Matsui, *Tetrahedron* **37**, 2779 (1981).

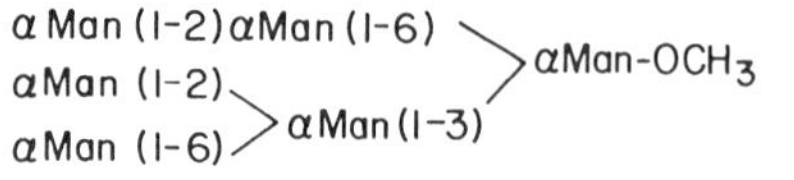

α Man (1-2)αMan (1-6)
αMan (1-2)
αMan (1-6) αMan (1-3) αMan-OCH$_3$

(LXXXVI)

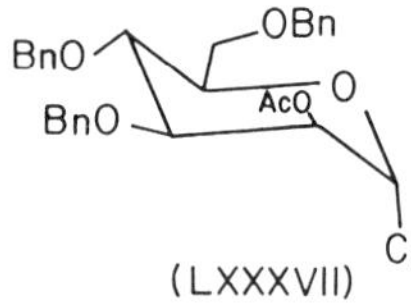

(LXXXVII)

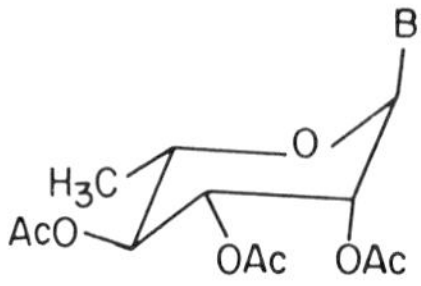

(LXXXVIII)

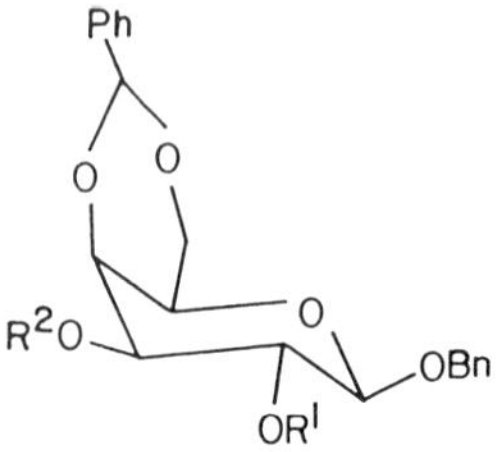

(LXXXIX):(a)R^1=H,R^2=Bz
(b)R^1=Bz,R^2=H

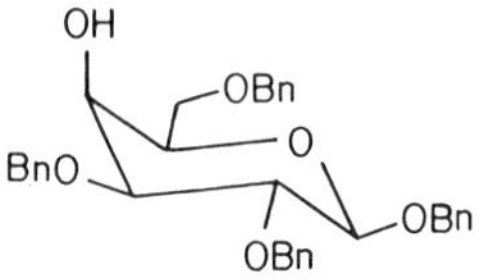

(XC)

α L-Rha(1-2)-α-L-Rha(1-3)-α-L-Rha (1-3)GlcNAc

(XCI)

urea. Glycosylation by the 2-amino sugar was achieved by use of the β-phthalimido chloride with silver triflate–s-collidine.

Linear 1,2-trans-linked mannotetraoses and -hexaoses were synthesized utilizing 2-O-acetyl-3,4,6-tri-O-benzyl-α-D-mannopyranosyl chloride and silver triflate molecular sieves. Stereospecific α-mannosylation was obtained at each step in high yield.[138]

A branched hexasaccharide (LXXXVI) was also synthesized by the silver triflate method.[139] A key step in this reaction was the preparation of a protected trisaccharide with three free OH groups (two at C-2 and one at C-6). Simultaneous reaction of these OH groups with excess α-mannosyl chloride (LXXXVII) enabled the synthesis of LXXXVI with high stereoselectivity.

Mercury(II) cyanide-promotion has also been used to prepare α-L-rhamnopyranosides from the tri-O-acetyl rhamnopyranosyl bromide (LXXXVIII), and block syntheses using disaccharide bromides have also been accomplished successfully.[140] In a study of the reaction of LXXXVIII with various protected galactopyranosides (LXXXIX, XC) at C-2, -3, and -4, the expected α-L-rhamnopyranosylation was observed.[141]

The reactivity of 4-OH in XC was very good (94% yield) in spite of its axial configuration, and this was attributed to the vicinal benzyl ethers. Reaction was generally stereospecific except for the case of LXXXIXb

[138] T. Ogawa and H. Yamamoto, *Carbohydr. Res.* **104**, 271 (1982).
[139] T. Ogawa and S. Sasajima, *Carbohydr. Res.* **93**, 53 (1981).
[140] A. Lipták, P. Nánási, A. Neszmély, I. Reiss-Maurer, and H. Wagner, *Carbohydr. Res.* **93**, 43 (1981).
[141] A. Lipták, Z. Szurmai, P. Nánási, and A. Neszmélyi, *Carbohydr. Res.* **99**, 13 (1982).

where the $\alpha:\beta$ ratio in the product was approx. $9:1$. Mercury(II) cyanide in $1:1$ nitromethane–benzene was used as the promoter in all cases.

Employment of tetra-O-acetyl-α-D-mannopyranosyl bromide labeled at C-1 with ^{13}C enabled synthesis of ^{13}C-labeled di- and trimannosides which could be useful for ^{13}C NMR spectroscopic investigations of natural products.[142] Mercury(II) bromide in acetronitrile was employed for the condensations.

In a synthesis of a tetrasaccharide hapten (**XCI**), the O-antigen repeating unit of *Shigella flexneri,* silver triflate–tetramethylurea in dichloromethane was employed for 2-rhamnosylation.[143] The silver triflate method has also been used[144] to synthesize branched tetrasaccharides related to the O-specific determinants of *Salmonella* serogroups A, B, and D1 (**XCII**).

$$\alpha Gal(1\text{-}2)\alpha Man(1\text{-}4)\alpha\text{-}\underline{L}\text{-}Rha$$
$$\underset{\alpha-Ddh}{\overset{\mid}{}}$$

$\left[\text{Ddh=3,6-dideoxy-D-}\textit{ribo}\text{-hexopyranosyl (paratose)},\text{-}\textit{xylo}\text{-hexopyranosyl (abequose), or }\textit{arabino}\text{-hexopyranosyl (tyvelose)}\right]$

(XCII)

In a synthesis of 3-O-α-Man-Man it was claimed that the best yield was obtained not by the use of the peracctylated α-D-mannopyranosyl bromide with mercury(II) salts, but rather via an intermediate 1-O-p-tosylate, analogous to the approach described previously for preparing α-D-glucosides and α-D-mannosides.[28,29] In this case, the glycosyl donor was a mannose derivative with a C-2-participating group (benzoate).

In a recent lecture[145] summarizing the work of his laboratory during the last few years on synthetic approaches to glycan chains in glycoproteins and proteoglycans, Ogawa stressed the convenient regioselective protection of hydroxyl groups achieved through stannylation[146] and the efficient glycosidation attained by use of silver salts together with molecular sieves. He described a rational design of oligosaccharide glycosyl acceptors and donors based on retrosynthetic considerations for the purpose of stereoselective convergent-type syntheses.

[142] F. M. Winnick, J. P. Carver, and J. J. Krepinsky, *J. Labelled Comp. Radiopharm.* **20**, 983 (1983).

[143] D. R. Bundle and S. Josephson, *Carbohydr. Res.* **80**, 75 (1980).

[144] K. Bock and M. Meldal, *Acta Chem. Scand., Ser. B* **B38**, 255 (1984).

[145] T. Ogawa, H. Yamamoto, T. Nukada, T. Kitajima, and M. Sugimoto, *Pure Appl. Chem.* **56**, 779 (1984).

[146] T. Ogawa, K. Katano, and M. Matsui, *Carbohydr. Res.* **64**, C3 (1978).

(XCIII)   (XCIV)

Sialyl Oligosaccharides

Early attempts to produce sialyl oligosaccharides[1] utilized a chloro derivative (**XCIII**), but yields on coupling were very low even with more reactive hydroxyl groups.

Some more recent work[147] has shown a preference for mercury(II) cyanide–mercury(II) bromide as the promoter for the condensation of **XCIII** with a protected disaccharide (**XCIV**) having free OH groups at C-4 and C-6 to give a trisaccharide in about 50% yield ($\alpha:\beta$ anomeric ratio approx. 1:1). The 4-OH group in **XCIV** did not react.

Sialic acids pose particular problems because of the expense of starting materials and the lack of stability and low reactivity of the glycosyl donors so far available. It will be observed that this most recent oligosaccharide synthesis is not stereoselective in spite of the only moderate yield obtained with a rather reactive hydroxyl group. A novel approach to such compounds has recently been devised based on an intramolecular oximercuration–demercuration scheme to synthesize sialyl-α- and β(2-6)-D-glucose.[148] By this method, a suitably protected open chain form of an 8-carbon dithioketal was prepared and converted to the 1-aldehyde. Reaction of this aldehyde with a phosphorane substituted at C-6 of a protected glucopyranoside in a Wittig reaction produced two different enol ethers in about 90% yield (ratio 2:1) which could be separated by chromatography. Further high-yielding transformations converted the major product into the "unnatural" β-linked disaccharide. The minor product was similarly converted into sialyl-α(1-6)-D-glucose.

Derivatives of 3-Deoxy-D-manno-2-octulosonic Acid (KDO)

The arrangements around C-1–C-3 in KDO and sialic acid are similar. Both are 2-keto sugars with a neighboring 3-deoxy position and thus pose some similar problems in oligosaccharide synthesis. Recent reports[149] on the reaction of a protected α-bromide of KDO with 3-OH of a protected

[147] H. Paulsen and H. Tietz, *Carbohydr. Res.* **125,** 47 (1984).

[148] F. Paquet and P. Sinaÿ, *Tetrahedron Lett.* **25,** 3071 (1984).

[149] H. Paulsen, Y. Hayauchi, and F. M. Unger, *Liebigs Ann. Chem.*, pp. 1270, 1288 (1984).

monosaccharide of N-acetylglucosamine showed that mercury(II) salts could again be used as promoters. In dichloromethane, the $\alpha:\beta$ ratio obtained was 25:1 (yield 46%), but in nitromethane, although the yield increased to 59%, the $\alpha:\beta$ ratio was now 4:1. In dichloromethane–toluene, a total yield of 43% was obtained of the α-anomer stereospecifically. Silver perchlorate–silver carbonate in dichloromethane gave a 43% yield of disaccharide ($\alpha:\beta$ approx. 12:1) but the yield with silver triflate was extremely low.

A trisaccharide was similarly obtained by reaction of the same bromide of KDO with 3'-OH of a protected derivative of β-GlcNAc(1-6)GlcNAc. In this case, condensation was promoted by mercury(II) cyanide in nitromethane at 60°. A yield of 61% was obtained ($\alpha:\beta$ approx. 10:1).

Thiodisaccharides

The increased nucleophilicity of sulfur compared with oxygen should facilitate S-glycosylation. Thiodisaccharides have been suggested for use as substrate analogs and inducers for enzymes. Thiogentiobiose was synthesized both by condensation of the sodium salt of a suitably protected 6-thio-D-glucose derivative with **XLVII** (R = OAc) and by preparing the sodium salt of 1-thio-β-D-glucose and coupling this in an S_N2 process with an appropriate D-glucose derivative having a free 6-OH.[150]

Thiolactose was prepared analogously from methyl 4-thio-α-D-glucopyranoside and **XXXI** (R^1 = OAc; R^2 = Ac) in hexamethylphosphoric triamide. A complex mixture resulted and the final yield of thiolactose obtained was only about 30%. The approach starting from 1-thio-β-D-galactopyranose was unsuccessful.[151]

A similarly low yield was also obtained in an S_N2 displacement of a 2-O-trityl ether of a substituted β-D-mannopyranoside with the sodium salt of 2,3,4,6-tetra-O-acetyl-1-thio-β-D-glucose to give thiosophorose. Thiocellobiose was obtained, however, in 75% yield by displacement of a 4-O-trityl ether from suitably protected galactopyranoside.[152]

A Lewis-acid catalyzed condensation between 2,3,4,6-tetra-O-acetyl-1-thio-β-D-glucose and 1,3,4,6-tetra-O-benzyl-β-D-fructose gave 1-thiosucrose and its α-fructofuranosyl anomer in yields of 15 and 22%, respectively.[153] It will be observed then that yields are only moderate in reactions with less reactive hydroxyl groups and optimum conditions

[150] D. H. Hutson, J. Chem. Soc. C, p. 442 (1967).
[151] L. A. Reed, III and L. Goodman, Carbohydr. Res. **94**, 91 (1981).
[152] K. Hamacher, Carbohydr. Res. **128**, 291 (1984).
[153] J. Defaye, H. Driguez, S. Poncet, R. Chambert, and M. F. Petit-Glatron, Carbohydr. Res. **130**, 299 (1984).

have not yet been apparently derived for the synthesis of thio analogs of oligosaccharides.

Concluding Remarks

There have been considerable advances in recent years in the field of oligosaccharide synthesis, and the preparation of complex structures of many saccharide units is no longer a distant dream. Such an endeavor is clearly more feasible when repeating units occur, but there have also been remarkable successes in synthesizing heterooligosaccharides containing four or five different units. On the basis of previous experience some reaction mechanisms have been derived which facilitate rational approaches to new compounds. However, the complex equilibria possible in the condensation step prevent clear-cut predictions in many cases. A variety of conditions and reagents will often have to be tried in order to obtain optimal yields and desired stereochemistry.

The vast amount of interesting glycosylated compounds which have been identified and the current discussions on the biological role of their saccharide portions heighten the synthetic interest in these compounds. Analogs and homologs can be of significant value to biologists and biochemists so that their chemical synthesis is not simply a specialized, intellectual interest. The need for such compounds has led to the marked increase in synthetic studies in the field which has been observed in recent years. Many more successes will surely be achieved in the next few years based on the recent advances in experience and understanding.

Acknowledgment

The author is the Jack and Simon Djanogly Professor of Carbohydrate Biochemistry.

[30] Nondegradable Carbohydrate Ligands

By Y. C. LEE, C. A. HOPPE, K. KAWAGUCHI, and P. LIN

Most carbohydrate ligands used in the studies of endocytosis are either natural glycoproteins or chemically or enzymatically modified proteins, and are readily degraded in the lysosomes after internalization. The lysosomal degradation of the internalized ligands can be prevented by various chemicals. Alternatively, nondegradable carbohydrate ligands can be synthesized, which can persist intracellularly even after reaching lysosomes. One successful group of nondegradable carbohydrate ligands synthesized by us is based on poly(D-lysines), which are modified with thioglycoside derivatives. The advantage of this type of ligand is that poly(D-lysine) and poly(L-lysine) derivatives of comparable sizes can be modified with the same sugars so that comparison of endocytic behavior can be made. Also, the same chemical scheme can be used to modify certain polylysines with different sugars for the ease of sugar specificity studies. The use of thioglycosides, as described here, does render glycosidic bonds more resistant to enzymatic hydrolysis.

Materials. The following materials are obtained from the indicated sources: 6-bromohexanoyl chloride (Aldrich Chem. Co.); trinitrobenzenesulfonic acid (TNBS), *p*-nitrophenyl Z-L-tyrosinate, and poly (D-lysines) (Sigma Chem. Co.).

Preparation of Pseudothiourea (PTU) Derivative of Sugars. Sugars are first converted to "acetobromo" derivatives (or "acetochloro" for hexosamines). Some examples have been given in this series earlier.[1,2] Conversion of "acetohalo" sugars to their respective pseudothiourea derivatives (PTU sugars) also have been described.[1]

Preparation of Methyl 6-Bromohexanoate. 6-Bromohexanoyl chloride (11.6 g, 54 mmol) is added to 50 ml of dry methanol, kept in an ice bath, and the mixture is magnetically stirred to homogeneity. After 5 min, triethylamine (10 ml, 71.7 mmol) is added and the mixture kept in an ice bath for 15 min, then at room temperature for 1 hr. The reaction mixture is evaporated to dryness, and shaken with a mixture of 50 ml each of water and dichloromethane. The organic layer is separated, washed with water, dried with anhydrous sodium sulfate, and evaporated to a heavy syrup (usually >90% yield), which is sufficiently pure to be used in the next step.

Coupling of 1-Thiogalactose Acetate to Methyl 6-Bromohexanoate.[3] The method of Lee *et al.*[4] is used to generate 1-thiogalactose. To a solution of 11.5 g (23.5 mmol) of PTU-D-galactose tetraacetate [2-S-(2,3,4,6-tetra-O-acetyl-β-D-galactopyranosyl)-2-thiopseudourea hydrobromide] in water (40 ml) are added 4.13 g (30 mmol) of potassium carbonate, 2.23 g (11.8 mmol) of sodium metabisulfite, and 40 mL of chloroform. The reaction is monitored by TLC on silica gel using 1:1 (v/v) toluene–acetone. After 30–45 min, when the reaction is judged complete, the chloroform

[1] C. P. Stowell and Y. C. Lee, this series, Vol. 83, p. 278.

[2] R. T. Lee and Y. C. Lee, this volume [34].

[3] K. Kawaguchi, M. Kuhlenschmidt, S. Roseman, and Y. C. Lee, *J. Biol. Chem.* **256,** 2230 (1981).

[4] R. T. Lee, S. Cascio, and Y. C. Lee, *Anal. Biochem.* **95,** 260 (1979).

layer is separated, and the aqueous layer washed with 40 ml of chloroform which is combined to the main chloroform solution. The combined chloroform solution is washed with an equal volume of 1 M NaCl, and dried with anhydrous sodium sulfate, filtered, and evaporated to syrupy 1-thio-D-galactose acetate.

The syrupy 1-thio-D-galactose acetate dissolved in 10 ml of dry methanol is mixed with 5.6 g (26.7 mmol) of methyl 6-bromohexanoate, 5.6 g (40.5 mmol) of potassium carbonate, and 4.47 g (23.5 mmol) of sodium metabisulfite. The heterogeneous mixture is stirred for 8 hr at room temperature, when most of the 1-thio sugar is consumed (as judged by TLC in 2 : 1, v/v, toluene–diethyl ether). The mixture is diluted with 100 ml of dry methanol, filtered through a layer of Celite, evaporated to dryness, and dissolved in 150 ml of chloroform. The chloroform solution is washed with 150 ml of cold 0.06 M sulfuric acid, then twice with 150 ml of cold water, dried with anhydrous sodium sulfate, filtered, and evaporated to produce 13.7 g of the crude product. Further purification is accomplished by gel filtration on a column of Sephadex LH-20 (5 × 190 cm) in 95% ethanol, followed by chromatography with a Jobin–Yvon Chromatospac Prop 100 (5-cm diameter), using 200 g of silica gel 60 (E. Merck) in 2 : 1 (v/v) toluene–diethyl ether. The desired product is obtained in pure solid form (8.1 g).

Conversion to Hydrazide. The ester is first de-O-acetylated before conversion to acyl hydrazide. The ester (8.1 g, 16.5 mmol) is dissolved in dry methanol (40 ml) and 0.4 ml of sodium methoxide (4.0 M) in dry methanol added. After standing for 3 hr at room temperature, crystalline product is formed, which is filtered and washed with methanol. Yield 5.19 g (16.0 mmol), 68% overall from the PTU sugar.

The de-O-acetylated product (5.19 g) is suspended in 40 ml of dry methanol to which is added 4.78 ml of 95% hydrazine (143 mmol). The mixture becomes clear immediately, and after 7 hr at room temperature, a copious amount of crystals are formed, which are collected by filtration, thoroughly washed with methanol, and recrystallized from hot methanol. Yield 4.6 g (14.2 mmol, 89%), m.p. 182–183°.

Coupling of 1-Thio-D-mannopyranose Tetra-O-acetate to Methyl 6-Bromohexanoate. The 1-thio sugar acetates of glucose and N-acetyl-D-glucosamine are coupled to methyl 6-bromohexanoate as described for their galactose counterpart. However, a different procedure gave better results for the mannose derivative. The 1-thio-D-mannopyranose tetra-O-acetate was generated under the conditions described above, but using only 15 min for the reaction. The subsequent extractions, drying, filtration, and evaporation are carried out quickly to minimize side-reaction product.

The syrupy 1-thio-D-mannopyranose tetra-O-acetate (7.6 g, 20.9 mmol) generated from 11.5 g (23.6 mmol) of PTU-mannose acetate is dissolved in 10 ml of dry methanol and mixed with 15.3 g (73.2 mmol) methyl 6-bromohexanoate. Triethylamine (34 ml, 244 mmol) is added to the mixture while being stirred. After 30 min at room temperature, when most of the 1-thio sugar is consumed, as determined by TLC in 2 : 1 (v/v) toluene–diethyl ether, the precipitated triethylamine bromide is removed by filtration and the filtrate is evaporated to a clear syrup. The residual amine is removed by 3 cycles of dissolution of syrup in methanol followed by evaporation. The product is purified by gel filtration through a column (5 × 190 cm) of Sephadex LH-20 in 95% ethanol, effluent fractions being analyzed with a phenol–sulfuric acid method. The isolated product, 5.4 g (11 mmol), is pure by TLC in 2 : 1 (v/v) toluene–diethyl ether.

The ester 5.4 g is dissolved in 27 ml of dry methanol and de-O-acetylated by addition of sodium methoxide to bring about a final concentration of 40 mM. Sodium ion is removed from the clear reaction mixture by passing through a column (0.5 × 5 cm) of Dowex 50 × 4 (hydrogen form) previously equilibrated with methanol. The effluent is evaporated and methyl acetate is removed from the syrup by 3 cycles of dissolution of the residue in methanol and evaporation.

Hydrazinolysis of the de-O-acetylated product is done as described for the galactose derivative, but no crystals are formed during the reaction. The bulk of excess hydrazine is removed by azeotropic distillation with toluene. The final purification is achieved by passing through a column (2 × 40 cm) of Bio-Rex 70 (hydrogen form), eluting with water. Analysis of the effluent located the peak containing the product (248–680 ml), which upon evaporation yields crystals 3.6 g (11 mmol), m.p. 127–128° (47% overall from psuedothiourea derivative).

Coupling to Poly (D-lysine). The acyl hydrazide is converted to acyl azide by nitrous acid treatment.[3] The hydrazide (24.3 mg, 75 μmol) dissolved in 500 μl of 0.5 M acetic acid is mixed with 70 μl of 4 N HCl and sodium nitrate (6.9 mg, 100 μmol). After 10 min at room temperature, sulfamic acid (9.17 mg, 100 μmol) is added and the reaction mixture gently shaken for 5 min to quench the remaining nitrous acid. The acyl azide solution thus produced is immediately added to a solution of poly(D-lysine) (d.p. ~270, 100–300 mg) in 0.4 M sodium borate (pH 10). After stirring at room temperature for 1 hr, neutralize the mixture with glacial acetic acid and purify on a column (1.5 × 145 cm) of Sephadex G-25 in 1 M acetic acid, collecting 4.5 ml fractions. Analysis of the effluent fractions with the phenol–sulfuric acid method[5] locates the Gal-modified

[5] J. McKelvy and Y. C. Lee, *Arch. Biochem. Biophys.* **132,** 99 (1969).

poly(D-lysine), well separated from the carbohydrate containing by-products. Lyophilization of the combined fractions gives the desired product in solid form.

The extent of substitution can be determined by measuring the remaining amino groups and the sugar content. Samples of poly(D-lysine) derivatives containing 0.05–0.3 μEq amino groups in 0.5 ml are mixed with 0.5 ml of 0.2 M sodium borate buffer (pH 8.5) and 0.5 ml 0.2% (w/v) TNBS solution, and incubated at 55° for 30 min. After addition of 1.5 ml of 0.5 M HCl, absorbance at 340 nm is read on a spectrophotometer[5] (N^α-Z-lysine or 6-aminohexanoic acid can be used as a reference compound). The sugar content can be determined with the phenol–sulfuric acid method,[6] with the thiogalactoside as a standard or with mercuric ion-catalyzed hydrolysis of thioglycoside first, followed by analysis of liberated sugar on a sugar analyzer.[7]

Further Modification. For the purpose of radioiodination of the sugar-containing polylysine derivative, a tyrosyl residue can be attached to the available amino group. The polylysine derivative (5 mg) is dissolved in dry dimethyl sulfoxide (2 ml, or more, if necessary), and *p*-nitrophenyl-Z-L-tyrosinate (1.2-fold molar excess) and triethylamine (1.5-fold molar excess of the available amino groups) are added. As the reaction proceeds, a yellow color (*p*-nitrophenolate) develops. Iodination can be performed by the chloramine T method.[8] Alternately, a reagent of Bolton–Hunter type[9] can be used for the same purpose. The product can be isolated by first evaporating dimethyl sulfoxide at 40–45° under reduced pressure, followed by passing through a column of Sephadex G-25 column (1 × 20 cm) in 3 M acetic acid.

It is often desirable to mask all the residual amino groups in the polylysine derivatives for certain biological experiments. This is readily accomplished by *N*-acetylation. The polylysine derivative is dissolved in saturated sodium hydrogen carbonate, and 20% excess (over amino groups) of acetic anhydride (dissolved in 10-fold volume of acetone) is added, and the mixture is stirred at room temperature for 1 hr. Completeness of N-acetylation can be tested by the TNBS method.

Application. Man-derivatized poly(L-lysine) and poly(D-lysine) (d.p. ~250) are equally avidly taken up by rabbit alveolar macrophages.[10] However, while the poly(L-lysine) derivative is degraded in a fashion similar to Man-BSA,[10] the poly(D-lysine) derivative continues to accumulate in the

[6] M. J. Krantz and Y. C. Lee, *Anal. Biochem.* **71**, 318 (1976).
[7] Y. C. Lee, this series, Vol. 28, p. 63.
[8] F. C. Greenwood, W. M. Hunter, and J. S. Glover, *Biochem. J.* **89**, 114 (1963).
[9] A. E. Bolton and W. M. Hunter, *Biochem. J.* **133**, 529 (1973).
[10] C. A. Hoppe and Y. C. Lee, *Biochemistry* **23**, 1723 (1984).

cells. Similarly, Gal-derivatized poly(D-lysine) and poly(L-lysine) both actively bind and are taken up by mammalian hepatocytes, but only the poly(L-lysine) derivatives are degraded. In both cases, the amount of poly(D-lysine) derivatives taken up by the cells far exceeds the total amount of receptor molecules, and thus renders support to the notion of receptor recycling, and that receptor recycling does not require degradation of ligands.

[31] Attachment of Oligosaccharide-Asparagine Derivatives to Proteins: Activation of Asparagine with Ninhydrin and Coupling to Protein by Reductive Amination

By ARLENE J. MENCKE, DAVID T. CHEUNG, and FINN WOLD

It is frequently most convenient to liberate naturally occurring N-linked oligosaccharides from glycoproteins by exhaustive digestion of the protein with proteases, producing oligosaccharides with a single asparagine residue still attached. Although there are glycoproteins for which the peptide bonds adjacent to the glycosylated Asn are quite resistant to proteolysis, and for which the product is a glycosylated di- or tripeptide, the glycosylated Asn appears to be a likely product if an effort is made to make the proteolytic digestion go to completion. Since the Asn moiety represents a convenient site for chemical manipulations which will not alter the oligosaccharide structure, such glycosyl-Asn derivatives are attractive starting materials for the preparation of neoglycoproteins, and some methods have been devised for derivatizing the Asn as a means of incorporating the oligosaccharide unit into proteins.[1-3] Surprisingly, however, this does not appear to be a common strategy.

One such method, based on the direct activation of the Asn moiety by oxidative deamination–decarboxylation with ninhydrin, is the subject of this article. The reaction is illustrated in Scheme 1, which also shows the subsequent reaction of the resulting aldehyde with protein amino groups and the NaCNBH$_3$ reduction of the Schiff base to the stable secondary amine link in the neoglycoprotein derivative. This method of activation and coupling was used successfully to incorporate six different ovalbumin

[1] J. C. Rogers and S. Kornfeld, *Biochem. Biophys. Res. Commun.* **45**, 622 (1971).
[2] S.-C. B. Yan, this volume [32].
[3] V. A. Chen, this volume [33].

METHODS IN ENZYMOLOGY, VOL. 138

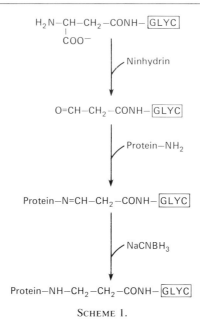

SCHEME 1.

glycosyl-Asn derivatives into serum albumin.[4] Since the desired product was albumin containing a single oligosaccharide, the procedures described involve a relatively large excess of protein in the coupling step. In fact, the proteon concentration is so high that we assume that the rate of the coupling reaction is zero order with respect to protein. If this is correct, lower protein to oligosaccharide ratios should be acceptable, but this has not been explored in a systematic manner. The product of the activation step, presumed to be the N-glycosylated malonamide semialdehyde, has not been characterized, but the rapid loss of coupling activity as a function of pH and temperature, the well-established properties of the ninhydrin reaction,[5] and the successful use of the product in the reductive amination with the protein are all consistent with the reactions outlined in Scheme I.

Materials. The different oligosaccharide-Asn derivatives were prepared from ovalbumin by the method of Huang *et al.*[6] with modifications suggested by other workers.[7,8] After fractionation of the oligosaccharide-

[4] A. J. Mencke and F. Wold, *J. Biol. Chem.* **257,** 14799 (1982).
[5] M. Friedman and L. D. Williams, *Bioorg. Chem.* **3,** 267 (1974).
[6] C. C. Huang, H. E. Mayer, and R. Montgomery, *Carbohydr. Res.* **13,** 127 (1970).
[7] K. Yamashita, Y. Tachibana, and A. Kobata, *J. Biol. Chem.* **253,** 3862 (1978).
[8] J. Conchie and I. Strachan *Carbohydr. Res.* **63,** 193 (1978).

Asn mixture by chromatography on Dowex 50H$^+$, each fraction was further purified by affinity chromatography on Sepharose-concanavalin A (Sigma). Only the fractions established to bind to Con A and to subsequently be eluted with α-methylmannoside were used in subsequent reactions. Chemical reagents, resins, and adsorbants used were all obtained from commercial sources.

Methods

Because of the instability of the product of the ninhydrin reaction, a rather rigid protocol for the reaction was developed. The main variables tested were temperature and pH in both the activation and the coupling reactions. The following procedure represents the empirical derivative of these tests. The methods are described for the activation and coupling of 1 μmol of oligosaccharide to 5 μmol of serum albumin.

Prereaction Procedures. (1) Preparation of gel filtration column for the separation of activated oligosaccharide and ninhydrin. A 12 ml (1 × 16 cm) column of BioGel P-2 was poured and equilibrated with 0.17 M potassium phosphate buffer (pH 7.5) in the cold room (4°). The column was carefully calibrated to establish accurately the elution volumes of the oligosaccharide, ninhydrin, and its reaction product, Ruheman's purple. We have found that the appropriate oligosaccharide-Asn derivative labeled at the α-amino group with either radioactive propionate (using the hydroxysuccinimide ester) or acetate (using acetic anhydride) makes a convenient reference compound for this purpose. A mixture of ninhydrin and Ruheman's purple was produced by reacting a 2 : 1 mixture of ninhydrin and asparagine at 100°, pH 6.0, for 30 min. The oligosaccharide elutes well ahead of ninhydrin. After calibration and reequilibration with buffer, the column was made ready for use by clamping it off just as the buffer meniscus reached the top of the packing. (2) Preparation of the serum albumin suspension for the coupling reaction. Serum albumin (350 mg) was weighed into a screw-capped vial and 500 μl 0.17 M phosphate buffer (pH 7.5) was added to uniformly wet and suspend the protein. Later, during the activation reaction, 100 μl of a solution containing 1.3 mmol NaCNBH$_3$/ml in the same phosphate buffer was added to the protein suspension, and gentle agitation was applied to achieve a homogeneous mixture.

Activation and Coupling. The glycosyl-Asn derivative, 1 μmol, was added to a small reaction vial containing 200 μl of a solution containing 50 μmol ninhydrin/ml of 0.1 M acetate buffer, pH 6 (the buffer was prepared by titrating a 0.1 M sodium acetate solution to pH 6.0 with acetic acid). The mixture was rapidly mixed, and the vial was immediately placed in a boiling water bath. After exactly 5 min in the water bath, the vial was transferred to an ice bath and agitated manually to bring the temperature down to about 5° as rapidly as possible. The reaction mixture was next transferred to the prepared BioGel P-2 column, the vial was rinsed with 200 μl of the pH 7.5 phosphate buffer, and the rinse was added to and gently mixed with the reaction mixture before the elution was started using the pH 7.5 buffer. Based on the data from the column calibration, the first 160–200 drops, corresponding to the void volume of the column were discarded, and the next 180–200 drops (5–6 ml), corresponding to the total elution volume of the oligosaccharide, were collected manually with continuous agitation directly into the reaction vial containing the protein–NaCNBH₃ mixture. At this time the elution was continued with a fraction collector to permit subsequent monitoring of the remaining fractions to ensure, if necessary, that all the monsaccharide had been collected in the reaction vial.

Purification of Neoglycoprotein. The homogeneous reaction mixture was left overnight at room temperature, and then applied directly to a 140-ml Sepharose-Con A column, equilibrated and eluted with a 0.01 M Tris–0.15 M NaCl buffer (pH 7.5). The unreacted albumin was carefully removed by extensive washing with the equilibration buffer before the bound fraction was eluted with 0.1 M α-methylmannoside in the same buffer. The bound fractions absorbing at 280 nm were pooled, concentrated if needed, and subjected to gel filtration on a 2.5 × 30 cm column of BioGel P-10, equilibrated and eluted with 0.01 M ammonium acetate. On this column, the neoglycoprotein fractions (phenol–sulfuric acid positive and 280 nm absorbing) were well separated from unreacted oligosaccharide (phenol–sulfuric acid positive), and the excluded fractions could be pooled, lyophilized, and characterized as the final neoglycoprotein product. It may be of interest to note that the order of the purification steps does not appear to be important. We have obtained equivalent results if the gel filtration step precedes the affinity chromatography step.

Conclusions

The direct activation of the asparagine moiety of glycoprotein-derived oligosaccharide-Asn derivatives can be accomplished by treatment with ninhydrin at pH 6.0 at 100° for 5 min. The activated product can be coupled to proteins by reductive amination with NaCNBH₃. The activated intermediate appears to be quite unstable, and it is clear that various side reactions compete effectively with the desired reaction. Under the procedures described it was possible to consistently incorporate 15–

20% of the starting oligosaccharide into serum albumin in the reaction of six different ovalbumin-derived oligosaccharide-Asn derivatives. The resulting neoglycoproteins showed several characteristic features of stable glycoproteins in binding to the appropriate lectin and being preferentially cleared from circulation of rats.

[32] Covalent Attachment of Oligosaccharide-Asparagine Derivatives: Incorporation into Glutamine Residues with the Enzyme Transglutaminase

By SAU-CHI BETTY YAN

Introduction[1]

The information encoded in oligosaccharides lies in the primary sequence, the chemistry of the linkage of the monosaccharide units, and the degree of branching. For oligosaccharides in glycoproteins, additional information may be encoded in the number and topological arrangements of the oligosaccharide units on the three-dimensional matrix of the polypeptide. Moreover, as a result of the interaction of the protein with the carbohydrate, information encoded in the oligosaccharide(s) can be modulated. The encoded information is decoded and translated into biological signals by lectins or processing enzymes interacting with the sugar moieties.

Unfortunately, it is difficult to study the biological functions of oligosaccharide with naturally occurring glycoproteins due to the high degree of heterogeneity of the oligosaccharides in a glycoprotein. To remedy this problem, one approach is to prepare synthetic glycoprotein (neoglycoprotein) with a specific number of homogeneous oligosaccharides in specific positions on the polypeptide backbone. To accomplish this chemically is a monumental task, because it is difficult to control the extent and sites of reaction in chemical modification of proteins. The reaction specificity required to prepare the desired neoglycoprotein, however, fits the descriptions of the very nature of enzymatic reactions. This article describes the use of an enzyme, guinea pig liver transglutaminase (protein-glutamine γ-glutamyltransferase), to prepare neoglycoproteins.

[1] Work done in Professor Finn Wold's laboratory, Department of Biochemistry and Molecular Biology, University of Texas Medical School, Houston, Texas. Supported by U.S. Public Health Science Grant GM31305 and by a grant from Robert A. Welch Foundation.

Transglutaminase catalyzes the reaction

$$R\text{-}CONH_2 + R'\text{-}NH_2 \xrightarrow{Ca^{2+}} R\text{-}CONHR' + NH_3$$

in which $R\text{-}CONH_2$ represents the acceptor, a protein-bound glutamine, and $R'\text{-}NH_2$, the donor, an alkylamine (see a recent review by Folk[2]). *In vivo*, the common amine donor is probably a protein-bound lysine side chain. *In vitro*, transglutaminase can recognize a wide variety of alkylamines as donors, showing high affinity for straight-chain aliphatic amines of six carbons. Taking advantage of this nonspecificity for donor, we have shown that oligosaccharide can be extended chemically with an alkylamine[3] and can serve as donor substrate for transglutaminase. Transglutaminase, on the other hand, has very stringent requirements for the acceptor site, that is, not all protein bound glutamines are acceptors. The specificity determinants[3] for the acceptor sites are as follows: (1) the glutamine cannot be in an α-helical or β-sheet structure; (2) glutamine with an adjacent hydroxyamino acid is highly preferred; (3) a valine (or possibly other branched chain amino acid) that is three residues C-terminal to the glutamine is also important.

Methods

Preparation of Proteins as Acceptor

Bovine β-casein and bovine pancreatic ribonuclease A are good candidates as acceptor proteins because they are devoid of oligosaccharide and the acceptor sites are known. Glutamines-56, -79, -167, -175, and -194 are primary acceptors (fully modified) and glutamines-54 and -182 are secondary acceptors (partially modified) recognized by transglutaminase in β-casein. For bovine pancreatic ribonuclease A, glutamines-28, -60, -69, -74, and -101 are the acceptors.[4] β-Casein can be easily purified from a mixture of bovine caseins[5] and ribonuclease A is commercially available. To prevent protein cross-linking during the transglutaminase reaction, all the lysine side chains of β-casein are blocked by ethyl acetimidate[6] and those of ribonuclease with succinic anhydride.[7] The disulfides in ribonuclease are oxidized by performic acid[8]; without this treatment, none of the seven glutamines in ribonuclease are used as acceptors.

[2] J. E. Folk, *Adv. Enzymol. Relat. Areas Mol. Biol.* **54,** 1 (1983).
[3] S. C. B. Yan and F. Wold, *Biochemistry* **23,** 3759 (1984).
[4] S. C. B. Yan, *J. Biol. Chem.* (submitted for publication).
[5] H. A. McKenzie, *Adv. Protein Chem.* **22,** 55 (1967).
[6] L. Wofsy and S. J. Singer, *Biochemistry* **2,** 104 (1963).
[7] F. S. Chu, E. Crary, and M. S. Bergdoll, *Biochemistry* **8,** 2890 (1969).
[8] S. Moore, *J. Biol. Chem.* **238,** 235 (1963).

Preparation of Oligosaccharide as Donor

Oligosaccharide-Asn. Oligosaccharide-Asn can be isolated in general by exhaustive pronase digestion of a glycoprotein. One readily available source of oligosaccharide-Asn is from ovalbumin and is isolated according to Huang *et al.*,[9] with modifications suggested by other workers.[10,11] The procedure of preparing neoglycoprotein described in this article has been successfully applied to either the hybrid or high-mannose oligosaccharides from ovalbumin. The identity and purity of the oligosaccharide-Asn can be monitored by fast atom bombardment mass spectrometry and ¹HNMR.

The Asn-oligosaccharide is transformed into a donor with the amino terminus of the asparaginyl oligosaccharide modified with 6-aminohexanoic acid by the following steps. The amino group of 6-aminohexanoic acid is first protected by 9-fluorenylmethylchloroformate (Fmoc-Cl) (from Pierce) according to a method described by Chang *et al.*[12] Five millimoles of 6-aminohexanoic acid is added to 10 ml of 10% Na_2CO_3 and 5 ml of dioxane and cooled on an ice bath. Next, 5.2 mmol of Fmoc-Cl in 7.5 ml of dioxane is added dropwise to the mixture over an hour with vigorous stirring. The mixture is then stirred further overnight at room temperature. After 150 ml of ice-cold water is added, the reaction mixture is extracted 3 times (40 ml each) with ether. The aqueous fraction (at 4°) is adjusted to pH 2 with HCl. The precipitate (product) is collected, washed with 200 ml of cold 0.1 *N* HCl, 100 ml of cold water, and then dried under vacuum. Then the carboxyl group of the above product, 6-[[[(fluorenyl-methyl)oxy]carbonyl]amino]hexanoic acid is activated by *N*-hydroxy-succinimide according to the procedure of Anderson *et al.*[13] Then 0.5 mmol of 6-[[[(fluorenylmethyl)oxy]carbonyl]amino]hexanoic acid, 0.55 mmol of *N*-hydroxysuccinimide, and 0.6 mmol of *N,N'*-dicyclohexylcar-bodiimide are mixed in 2 ml of dioxane and placed on ice immediately. After standing at 4° overnight, the precipitate formed is filtered and discarded. The supernatant is dried under vacuum to an oily solid. The *N*-hydroxysuccinimide ester is extracted from the oily solid with 3 ml of ether. After evaporating the ether under a stream of N_2, the ester is used to couple to the oligosaccharide-Asn the same day.

[9] C. C. Huang, H. E. Mayer, and R. Montgomery, *Carbohydr. Res.* **13,** 127 (1970).
[10] T. Tai, K. Yamashita, M. Ogata-Arakawa, N. Koide, T. Muramatsu, S. Iwashita, Y. Inoue, and A. Kobata, *J. Biol. Chem.* **250,** 8569–8575 (1975).
[11] J. Conchie and I. Strachan, *Carbohydr. Res.* **63,** 193 (1978).
[12] C. D. Chang, M. Waki, M. Ahmad, J. Meienhofer, E. O. Lundell, and J. D. Haug, *Int. J. Pept. Protein Res.* **15,** 59 (1980).
[13] G. D. Anderson, J. E. Zimmerman, and F. M. Callahan, *J. Am. Chem. Soc.* **86,** 1839 (1964).

The amino group of the oligosaccharide-Asn (4 μmol in 200 μl of 30 mM NaHCO$_3$) is acylated with 11.5 μmol of the N-hydroxysuccinimide ester (in 200 μl dioxane) at room temperature. After 20 hr, the [6-[[[(fluorenylmethyl)oxy]carbonyl]amino]hexanoyl]asparaginyl oligosaccharide is extracted with 2 ml of water and the 9-fluorenylmethylformate blocking group is removed with piperidine–dimethylformamide (1 : 1) at room temperature for 30 min. The mixture is dried under vacuum and the product H$_2$N-(CH$_2$)$_5$-CO-Asn-oligosaccharide is extracted with water and further purified by gel filtration chromatography on a Sephadex G-25 column. The yield is 40%. 6-[[[(Fluorenylmethyl)oxy]carbonyl]amino]-hexanoic acid and its N-hydroxysuccinimide ester migrated on silica gel G6OF$_{254}$ TLC plates with R_f of 0.42 and 0.29, respectively. Ether is the ascending solvent. The coupling of the 6-aminohexanoic acid to the oligosaccharide-Asn is confirmed by amino acid analysis after 6 N HCl hydrolysis for 24 hr at 110°. Since glycosidic bonds are sensitive to acidic conditions, acid-labile amino-protecting groups such as carbobenzoxy chloride and di-*tert*-butyl dicarbonate are avoided for the above reaction scheme. Fmoc-Cl is chosen because the protected amino group can be easily deblocked by mild alkali.

Oligosaccharide with Reducing Terminus. Any oligosaccharide with a reducing terminus (for example, obtained by endoglycosidase H digestion of glycoprotein) can be transformed into a donor for transglutaminase simply by reductive amination[3,14] with excess cadaverine. Reductive amination is a well-documented procedure and will not be described here.

Purification of Guinea Pig Liver Transglutaminase

Transglutaminase is purified from fresh guinea pig liver (from Pel Freez) according to Connellan *et al.*[15] and is more than 95% pure according to SDS–polyacrylamide gel electrophoresis. The enzyme is stored at −20° in small aliquots with no loss of activity after 1 year.

Preparation of Neoglycoprotein by the Transglutaminase Reaction

A typical reaction mixture of 400 μl contains the following: 50 mM H$_2$N-(CH$_2$)$_5$-CO-Asn-oligosaccharide, 0.75 mM amidinated β-casein, 5 mM CaCl$_2$, 5 mM dithiothreitol, 100 mM Tris (pH 7.6), and 100 μg transglutaminase added in four aliquots over a period of 20 hr at room temperature. The excess oligosaccharide can be removed from the glyco-

[14] G. R. Gray, *Arch. Biochem. Biophys.* **163,** 426 (1974).
[15] J. M. Connellan, S. I. Chung, N. K. Whetzel, L. M. Bradley, and J. E. Folk, *J. Biol. Chem.* **246,** 1093 (1971).

sylated β-casein by Sephadex G-50 gel filtration chromatography. The glycosylated β-casein is usually a mixture of β-casein containing one to five units of the oligosaccharide. The extent of glycosylation can be monitored by Lammeli SDS–PAGE on a 9 to 14% gradient slab gel. Glycosylated β-casein migrate as discrete bands of increasing molecular size with increasing number of units of oligosaccharide covalently attached.[3] For reasons that are not clear at this time, conditions to achieve complete reaction, that is, a single product of β-casein containing 5 units of oligosaccharides, have not been found. However, β-casein with different units of oligosaccharide covalently attached can be fractionated by lectin–affinity column. In the case where the oligosaccharide is Asn-(GlcNac)$_2$-(Man)$_6$ from ovalbumin, the mixture of glycosylated β-casein can be separated by a concanavalin A-Sepharose column.[16] The column is equilibrated with 20 mM triethanolamine (pH 8.0) containing 50 mM NaCl, 1 mM CaCl$_2$, 1 mM MgCl$_2$, and 0.02% sodium azide. A stepwise gradient of 1 mM, 5 mM, 50 mM, 100 mM α-methylmannoside in the column buffer will elute β-casein with one, two, three, four, and more units of oligosaccharide, respectively, from the concanavalin A-sepharose column. The eluate can be monitored by the SDS–PAGE on 9 to 14% gradient gel mentioned above, stained by Coomassie blue.

Concluding Remarks

Performic acid-oxidized and succinylated ribonuclease A is very poorly stained by Coomassie blue, silver, or Kodavue stain (Kodak). Thus, even though ribonuclease has five acceptor sites for transglutaminase, it is a less than ideal acceptor protein for neoglycoprotein synthesis. Amidinated β-casein has no such problem and is the preferred polypeptide for neoglycoprotein synthesis by the transglutaminase reaction.

Using transglutaminase to prepare neoglycoprotein has one distinct drawback. Since the lysine side chains of the protein acceptor must be blocked, the protein is at least partially denatured. Therefore, neoglycoproteins prepared by this method cannot be used as models to study the information encoded in the topological arrangement of the oligosaccharides. Neoglycoprotein models for topological studies can be prepared according to Chen.[17]

There are several advantages of the neoglycoproteins prepared by the transglutaminase reaction. (1) The glycoproteins are stable because the

[16] V. J. Chen, S. C. B. Yan, and F. Wold, *Microbiology (Washington D.C.) 1986*, 297 (1986).
[17] V. J. Chen, this volume [33].

oligosaccharides are covalently linked to the protein matrix. (2) The neoglycoproteins are constant with respect to the protein and the oligosaccharides, and the only variable is the number of oligosaccharides on the protein. The neoglycoproteins are unique in this respect and are ideal for studying the information encoded in the number of oligosaccharide chains (valence hypothesis) on a protein. (3) With digestion by the appropriate endoglycosidase, the oligosaccharides on the neoglycoprotein can be released.[16]

Using the method described in this article, with the oligosaccharide $H_2H-(CH_2)_5-CO-Asn-(GlcNac)_2(Man)_5$, i.e., (G), we have prepared neoglycoprotein derivatives[16] of (1) performic acid-oxidized and succinylated ribonuclease A as ribonuclease-$(G)_1$, ribonuclease-$(G)_2$, and ribonuclease-$(G)_3$; (2) amidinated β-casein as β-casein-$(G)_1$, β-casein-$(G)_2$, β-casein-$(G)_3$, and β-casein-$(G)_{4-5}$.

[33] Noncovalent Attachment of Oligosaccharide-Asparagine Derivatives to Proteins

By Victor J. Chen

Introduction[1]

There are four structural aspects to the interaction between a glycoprotein and a lectin (receptor) or a glycosyl-processing enzyme. These include (1) the composition of the oligosaccharide together with the nature of the linkages of the monosaccharides involved; (2) the number of chains; (3) the location of the chains relative to one another on the surface of the glycoprotein; and (4) the intramolecular interaction of the oligosaccharide with the polypeptide chain as a result of the tertiary structure of the glycoprotein.

To study the role of the oligosaccharides mediating glycoprotein–protein interaction, attempts have been made to construct glycoprotein models which allow selective variation of one of the parameters mentioned above. This article describes the concept of building a type of neoglycoprotein that can accommodate attachment of different naturally

[1] This work was done in the laboratory of Professor Finn Wold at the Department of Biochemistry and Molecular Biology, University of Texas Medical School at Houston, Houston, Texas, and was supported by U.S. Public Health Service Grant (GM 31305) and a grant from the Robert A. Welch Foundation (AU916).

occurring oligosaccharides to the same protein matrix at a defined number and location and with the possibility of altering intramolecular saccharide–peptide interaction. The linkage between the oligosaccharide and the protein is noncovalent and is formed by compounds having a high affinity for specific sites on that protein. Thus, a coenzyme is coupled to the α-amino group of an asparagine oligosaccharide and the resulting oligosaccharide–coenzyme derivative is bound to a protein containing sites for that coenzyme. Experiments have been done with high-mannose asparagine oligosaccharides from ovalbumin together with two coenzyme–protein systems: (1) pyridoxamine 5′-phosphate (PMP) and the apoenzyme of aspartate aminotransferase[2] (AAT, homodimer of molecular weight 92,000[3]) and (2) biotin and two forms of biotin-binding proteins (BBP), avidin and its microbial analog, streptavidin[4] (both are tetrameric, and of molecular weight 64,000; the former is naturally glycosylated while the latter is not[5]). In each case, there is one coenzyme site per subunit; for AAT, the two sites are related to each other by a 2-fold symmetry axis and can be considered anti[6] while for BBP, pairs of coenzyme sites are anti to each other.[5] The dissociation constants of PMP and biotin are in the 10^{-7} M and 10^{-15} M range, respectively.

Materials

Asparagine oligosaccharides (Man$_5$GlcNAc$_2$Asn and Man$_6$GlcNAc$_2$Asn) are isolated from ovalbumin according to the method of Huang *et al.*[7] N^α-(6-Amino)hexanoyl derivative of asparagine oligosaccharide is prepared according to the method of Yan.[8] NaBH$_4$ is from Aldrich and NaB^{3}H$_4$ (5 Ci/mmol) is from Amersham. Dimethyl formamide from Aldrich has been redistilled and stored over a 3-Å molecular sieve. Pyridoxal 5′-phosphate is from Sigma. Aspartate aminotransferase (EC 4.6.1.1) is prepared according to the procedure outlined by Yang and Metzler,[9] or purchased from Sigma in which case the enzyme is first allowed to pass through a concanavalin A affinity column before use.[2]

[2] V. J. Chen and F. Wold, *Biochemistry* **23**, 3306 (1984).
[3] A. E. Braunstein, *in* "The Enzymes" (P. D. Boyer, ed.), 3rd ed., Vol. 9, p. 379. Academic Press, New York, 1973.
[4] V. J. Chen and F. Wold, *Biochemistry* **25**, 939 (1986).
[5] N. M. Green, *Adv. Protein Chem.* **29**, 85 (1975).
[6] A. Arnone, P. Briley, P. H. Rogers, C. C. Hyde, and C. M. Metzler, *in* "Molecular Structure and Biological Activities" (J. F. Griffin and W. L. Duax, eds.), p. 56. Am. Elsevier, 1984.
[7] C. C. Huang, H. E. Mayer, and R. Montgomery, *Carbohydr. Res.* **13**, 127 (1970).
[8] S. B. Yan, this volume [32].
[9] B. I. Yang and D. E. Metzler, this series, Vol. 62, p. 528.

The N-hydroxysuccinimide ester of biotin is from Sigma. Its tritiated form is prepared using tritiated biotin from New England Nuclear and N-hydroxysuccinimide from Aldrich and dicyclohexylcarbodiimide from Pierce according to the method of Bayer *et al.*[10] Avidin is from Sigma, and streptavidin is a gift from Louis Chaiet of Merck, Sharp and Dohme Research Laboratories. Dimethyl sulfoxide is from Burdick and Jackson and is stored over a 3-Å molecular sieve.

Preparation of the Phosphopyridoxylated Asparagine Oligosaccharide (PAO) and Phosphopyridoxylated N^{α}-(6-Amino)hexanoylasparagine Oligosaccharide (PC6AO)

Asparagine oligosaccharide or its N^{α}-(6-amino)hexanoyl derivative (2 μmol) and pyridoxal 5'-phosphate (10 μmol) are dissolved in 0.1 M phosphate, pH 9 (1 ml). After standing at room temperature for 15 min, the yellow solution is placed in an ice bath. Solid $NaBH_4$, or NaB^3H_4 in dimethylformamide, is added until the yellow color is bleached (sometimes a trace of color persists). The unreacted borohydride is destroyed by dropwise addition of 1 M acetic acid until the pH reaches 5. To avoid bubble formation in the column chromatography that follows, the solvent is removed by lyophilization. The dried residue is dissolved in a minimum amount of water and is allowed to pass through a G-25-80 column (1.5 × 96 cm) which is eluted with 2 mM acetic acid. Eluent is monitored for pyridoxyl derivative at 310 nm and for carbohydrate by reaction with phenol and sulfuric acid.[11] PAO and PC6AO are positive by both criteria, and fractions containing the product are pooled and lyophilized. The yield of PAO, or C6PAO, is 50–75% based on the absorbancy at 327 nm in 0.02 M acetate buffer (pH 6), using 8300 for the value of molar absorptivity by analogy to PMP.[12] PMP derivatives are light sensitive, and should be stored at −20° in the dark. However, without direct exposure to sunlight, the spectra of PAO and PC6AO remain unchanged for 2 days under fluorescent room light.

Preparation of Apo Form of Aspartate Aminotransferase (AAT)

The holoenzyme of AAT contains one molecule of pyridoxal 5'-phosphate per subunit covalently bound as a Schiff base to a lysine side chain (λ_{max} 368 nm at pH 8). For resolution, this form of coenzyme must be

[10] E. A. Bayer, E. Skutelsky, and M. Wilchek, this series, Vol. 62, p. 308.
[11] M. Dubois, K. A. Gilles, J. K. Hamilton, P. H. Rebers, and F. Smith, *Anal. Chem.* **28,** 350 (1956).
[12] E. A. Peterson and H. A. Sober, *J. Am. Chem. Soc.* **76,** 169 (1954).

converted to PMP form (λ_{max} 330 nm) by transamination with an amino acid before it can be displaced from the protein by phosphate at pH 5. Thus, to a solution of holoAAT (10^{-5}–10^{-4} M, estimated by using 6.55×10^4 as the value of molar absorptivity at 280 nm for subunit of AAT) in 0.02 M triethanolamine hydrochloride (pH 8) is added cysteinesulfinate monosodium salt to a final concentration of 5 mM. After standing at room temperature for 10 min, an equal volume of 1 M phosphate (pH 4.8) is added and the mixture is incubated at 30° for 30 min. Then the mixture is cooled to 4° and 3 volumes of chilled saturated ammonium sulfate are mixed in. The precipitated protein, collected by centrifugation, is redissolved in 0.02 M triethanolamine hydrochloride (pH 8), and dialyzed against the same buffer at 4° until sulfate cannot be detected visually by precipitation with BaCl$_2$. The ratio of absorbancies at 280 nm to 330 nm should be greater than 40, and protein recovery should be 90%–95% (estimated using 6.46×10^4 as the value for molar absorptivity at 280 nm for apoAAT subunit). If needed, incubation with phosphate at pH 5 may be repeated.

Formation of AAT–Oligosaccharide Complex

Each PMP site of the apoenzyme contains one phosphate which is slowly displaced upon reconstitution with PMP or its derivatives. Thus, apoAAT (10^{-4} M in terms of subunits) and 1.2 equivalent of PAO (or PC6AO) in 0.02 M triethanolamine hydrochloride (pH 8) is incubated overnight at room temperature. The excess unbound ligand is removed by dialyzing the protein at the same concentration against the same buffer at 4° for 2 days.

Preparation of Biotinylated Asparagine Oligosaccharide (BAO) and Biotinylated N^α-(6-Amino)hexanoylasparagine Oligosaccharide (BC6AO)

Dried and desalted asparagine oligosaccharide or its N^α-(6-amino)hexanoyl derivative (2 μmol) and the N-hydroxylsuccinimide ester of biotin (5 μmol) are separately dissolved in dimethyl sulfoxide (0.25 ml). The homogeneous solutions are mixed together with a drop (40 μl) of triethylamine. The reaction mixture is stoppered and allowed to stand at room temperature overnight. The solvent is removed by lyophilization and the dried residue treated with 1 M triethanolamine hydrochloride, pH 8 (1 ml), for 4 hr to hydrolyze any unreacted reagent. Biotinylated product is purified by passage through a G-25-80 column (1.5 × 96 cm) which is eluted with water. Both BAO and BC6AO elute just behind the void volume and test positively for the biotinyl moiety using the methods of Green[13] or of McCormick and Roth,[14] as well as for carbohydrate by reaction with phenol and sulfuric acid.[11] The yield of BAO or BC6AO is 60–80%.

Formation of BBP–Oligosaccharide Complexes

BBP (70–300 nmol, concentration estimated using 24,000 and 56,000 as the values of molar absorptivity at 282 nm for the subunit of avidin and streptavidin respectively[5]) and BAO, or BC6AO, (1.2 equivalents) are incubated in 0.5 M phosphate (pH 7) buffer (1 ml) overnight at room temperature. The excess ligand is removed by dialysis against 0.1 M phosphate (pH 7) at 4° for 1 week.

Stoichiometry and Stability

Radiolabeled ligands are used in determining the stoichiometry of oligosaccharide conjugated. The number of PAO bound per dimeric AAT is typically 1.6 to 1.8.[2] The results for the two BBP complexes are consistently in the range of 2.4 to 3.2[4] despite four biotin sites in each BBP as reported in the literature and the fact that we have been able to demonstrate that biotin and BAO are taken up by both BBP to similar extent.

The stability of the complexes is assessed by assaying the amount of radiolabeled ligand released. After the complex is incubated under a desired test condition, a small aliquot (less than 0.5 ml) of the incubation mixture is passed through a YMT membrane in a Centrifree micropartition system (Amicon, product No. 4104) spun at 2000 g. The ratio of radioactivity in a volume of filtrate compared to that in the same volume of incubation mixture gives a measure of the fraction of ligand dissociated.

Using the ultrafiltration method, it is found that the AAT complex can maintain the stoichiometry mentioned above for a few days when it is kept concentrated at above 10^{-4} M in 0.02 M triethanolamine hydrochloride buffer (pH 8) 4° and in the absence of divalent anion, phosphoryl compounds, or other vitamin B_6 analogs. It becomes increasingly unstable at lower pH and under denaturing conditions. Bound PAO is quantitatively recovered when the complex is heated in boiling water for 2 min.

---

[13] N. M. Green, this series, Vol. 18A, p. 418.
[14] D. B. McCormick and J. A. Roth, this series, Vol. 18A, p. 383.

The BBP complexes are more stable and can be dialyzed for days at neutral pH. In this case, the presence of biotinylated compounds can cause the release of bound ligand. In fact, to quantitatively recover bound BAO, the complex is heated in boiling water for 10 min in the presence of 1 mM biotin.

Discussion

In a model system using noncovalent neoglycoproteins, their stability must be assessed under the desired experimental conditions with respect to the rate and extent of dissociation. At high concentration where little ligand release occurs, AAT-PAO could bind to concanavalin A-Sepharose and subsequently be recovered with little change of the stoichiometry of oligosaccharide bound to AAT.[2] By contrast, in studies of the interation with glycosyl processing enzymes, the noncovalent nature of these neoglycoproteins is an advantage which permits ready and quantitative recovery of the oligosaccharide for analysis following modification.[2,4,15] Since the coenzyme binding sites in both AAT and the BBP are deeply buried, the attached oligosaccharide in these complexes can be rendered differentially exposed with the C-6 spacer arm. These are excellent models for studying intramolecular effects of the polypeptide on processing of a maturing glycoprotein. In parallel reactions, free PAO and AAT-PAO were compared in their suceptibility toward the action of α-mannosidase. Under condition where 90% of free PAO had all five α-mannosyl residues cleaved, 70% of AAT-PAO retained concanavalin A-binding ability. Gel permeation analysis of all PAO-derivative recovered from AAT revealed that 70% had lost only one α-mannosyl residue while the remaining 30% had lost all five. Since independent evidence showed that there was 30% complex dissociation, under the acidic reaction condition it was concluded that only one mannosyl residue in AAT-bound PAO was exposed to α-mannosidase. The BBP complexes were more stable and less than 5% dissociation was detected. In a similar parallel reaction when 90% of free BAO was fully digested by α-mannosidase, a BAO derivative recovered from BBP was a heterogeneous mixture with the number of mannosyl residues ranging from two to four. Analogous experiments were performed on the biotinylated complexes with and without the spacer arm using low levels of endo-N-acetylglucosaminidase H. BAO and BC6AO were digested with a half-life of 70 min while BBP-BC6AO had a half-life of 10 hr; BBP-BAO, however, was resistant. All these experiments dem-

[15] V. J. Chen, S. B. Yan, and F. Wold, *Microbiol. Rev.* (1986).

onstrate that the protein can shield the oligosaccharide from enzymatic action as expected and as previously proposed by several investigators.

In conclusion, formation of noncovalent complexes between an oligosaccharide and a protein is a practical concept for preparation of neoglycoproteins. Depending on the ligand and the protein chosen, different modes of oligosaccharide attachment and stoichiometry of incorporation can be obtained. This together with the mild condition under which the complex is prepared and the ease of subsequent removal of the oligosaccharide for analysis make these noncovalent neoglycoproteins useful tools in studying glycoprotein biochemistry especially with reference to the role of the carbohydrate attached.

[34] Cluster Glycosides

By REIKO T. LEE and Y. C. LEE

Many animal lectins bind clustered glycosides much more tightly than can be accounted for by the arithmetic sum of the glycosyl residues in the ligand. For example, neoglycoproteins[1] derived from bovine serum albumin show logarithmic increase in binding affinity when the sugar density on the protein is increased linearly.[2] Although a similar situation also exists for some plant lectins,[3] the "cluster effect" is not as dramatic as shown by the animal lectins. Cluster glycosides can be in the form of branched oligosaccharides frequently encountered in natural glycoconjugates.[4] However, cluster glycosides of much simpler design can be readily prepared, and some of them are found to be quite potent ligands for animal lectins.[5]

Materials

The following chemicals are obtained from the indicated sources: Drierite (J. T. Baker Chemicals, NJ); 2-amino-2-hydroxymethyl-1,3-propanediol ("Tris") (Sigma Chemical Co.); 6-aminohexanol, 6-aminohexanoic acid (Aldrich Chem. Co.); 2-ethyl-N-ethoxycarbonyl-1,2-dihydro-

[1] C. P. Stowell and Y. C. Lee, Adv. Carbohydr. Chem. Biochem. 37, 225 (1980).

[2] Y. C. Lee and R. T. Lee, in "The Glycoconjugates" (M. Horowitz, ed.), Vol. 4, p. 57. Academic Press, New York, 1982.

[3] J. F. Crowley, I. L. Goldstein, J. Arnarp, and J. Lönngren, Arch. Biochem. Biophys. 231, 524 (1984).

[4] Y. C. Lee, R. R. Townsend, M. R. Hardy, J. Lönngren, J. Arnarp, M. Haraldsson, and H. Lönn, J. Biol. Chem. 258, 199 (1983).

[5] R. T. Lee, P. Lin, and Y. C. Lee, Biochemistry 23, 4255 (1984).

quinoline (EEDQ) (Aldrich Chem. Co.); N-benzyloxycarbonyl-L-aspartic acid (Z-Asp) (Sigma Chem. Co.); trifluoroacetic anhydride, ethyl trifluoroacetate (Aldrich Chem. Co.).

Glycosides of Tris(hydroxymethyl)aminomethane[6]

This type of cluster glycosides is easiest to prepare. An example of trislactoside is described below.

Preparation of 6-(Trifluoroacetamido)hexanoic Acid.[7] Trifluoroacetic anhydride (40 ml, 270 mmol) is slowly added to 6-aminohexanoic acid (13.1 g, 100 mmol) in a 100-ml round-bottom flask. To prevent evaporation of the anhydride, the reaction mixture is periodically cooled with tap water. After all the anhydride has been added, the mixture is kept for 1 hr at room temperature and evaporated to a thick syrup, which was then mixed with 100 ml of cold water and stirred overnight in the cold. The crystalline product formed is filtered off, washed with cold water, and recrystallized from diethyl ether to yield (61%) pure product (m.p. 88–89°).

Modification of Tris(hydroxymethyl)aminomethane.[6] The amino group of Tris is relatively unreactive. Attachment of functionalized "spacer arm" usually facilitates coupling of cluster glycosides to proteins or solid matrices. An example of derivatization with 6-(trifluoroacetamido)hexanoic acid is shown below:

$$CF_3CONH(CH_2)_5COOH + NH_2C(CH_2OH)_3$$

$$EEDQ \downarrow Reflux\ (ethanol)$$

$$CF_3CONH(CH_2)_5CONHC(CH_2OH)_3 \quad (FAT)$$

In a 1000-ml round-bottom flask, 12.5 g (55 mmol) of 6-(trifluoroacetamido)hexanoic acid, 6.06 g (50 mmol) of Tris, and 14.8 g (60 mmol) of EEDQ[8] are suspended to absolute ethanol, and refluxed for 5 hr. After cooling to room temperature, the reaction mixture is evaporated under vacuum to a syrup. Addition of anhydrous ether (250 ml) produces crystals, which upon recrystallization from ethyl acetate (250 ml) gives pure product, [6-(trifluoroacetamido)hexanamido]tris(hydroxymethyl)methane (FAT) (m.p. 97–99°) in 87% yield.

Preparation of Acetobromolactose.[9] Lactose octaacetate (20 g, 29.5 mmol) is mixed with 100 ml of cold, 15% (w/v) HBr in glacial acetic acid in a capped Erlenmeyer flask. The mixture is stirred at room temperature

[6] Y. C. Lee, *Carbohydr. Res.* **67,** 509 (1978).

[7] R. T. Lee and Y. C. Lee, *Carbohydr. Res.* **34,** 151 (1974).

[8] B. Belleau and G. Malek, *J. Am. Chem. Soc.* **90,** 1651 (1969).

[9] P. H. Weigel, M. Naoi, S. Roseman, and Y. C. Lee, *Carbohydr. Res.* **70,** 83 (1979).

until lactose acetate is dissolved completely. After further stirring for 45 min, 100 ml of cold chloroform is added, and the mixture is rapidly poured into a 2-liter separatory funnel containing 1.2 liters of ice water and vigorously shaken, and the chloroform layer is rapidly separated. The chloroform layer is washed twice with half-volume of cold water, twice with cold saturated sodium bicarbonate, and finally with cold 1 M NaCl. The chloroform layer is drained into an Erlenmeyer flask containing ~20 g of anhydrous sodium sulfate, is shaken until the solution becomes clear, and is suction filtered. The filtrate is concentrated by rotary evaporation under vacuum at room temperature to ~20 ml, and diluted with 50 ml of anhydrous diethyl ether. Petroleum ether (b.p. 30–60°) is added until incipient turbidity occurs. Storage of the mixture in the cold overnight produces a crystalline product, which is filtered and washed with diethyl ether–petroleum ether (yield 80–85%). The acetobromolactose thus prepared can be stored at −20° for at least several months.

Glycosylation and Fractionation of Products. Controlled glycosylation of FAT with acetobromolactose produces mono-, bis-, and trislactosides which can be readily fractionated. The proportion of the products can be varied by varying ratios of the acetobromolactose to FAT. In the following example, approximately equal amounts of bis- and trisglycosides are formed.

FAT (1.98 g, 6 mmol) and Drierite (1 g) are stirred in 55 ml of dry nitromethane for 30 min. To this suspension are added 55 ml of dry toluene, 3.8 g (15 mmol) of $Hg(CN)_2$, and 6 g (8.6 mmol) of acetobromolactose. After stirring for 2 hr, additional 4.5 g (6.4 mmol) of acetobromolactose is introduced, and the stirring is continued overnight. If monitoring of the progress of the reaction is desirable, 10–20 μl of the reaction mixture is taken into a small test tube containing ~200 μl each of chloroform and water, and the chloroform solution is analyzed by TLC on a silica gel plate developed with 4 : 1 (v/v) ethyl acetate–toluene. Carbohydrate-containing components are visualized by spraying the plate with 15% sulfuric acid in 50% ethanol followed by heating at 130–140°. The reaction is stopped when the acetobromolactose (R_f ~0.67) is spent.

After removal of insoluble materials, the solution is evaporated to a syrup, which is dissolved in 250 ml of chloroform, and washed with 100 ml of 0.5 M KBr, and dried with anhydrous sodium sulfate as mentioned above. The filtered solution is evaporated to a syrup, dissolved in 95% ethanol, and fractionated in two batches on a Sephadex LH-20 column (5 × 195 cm) equilibrated in 95% ethanol (18–20 ml per fraction) to achieve separation as shown in Fig. 1. Analysis of the effluent is by a phenol–sulfuric acid method.[10] Fractions 86–89, containing both tris- and

[10] J. F. McKelvy and Y. C. Lee, *Arch. Biochem. Biophys.* **132,** 99 (1969).

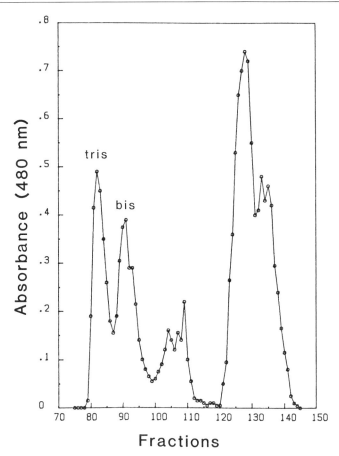

FIG. 1. Separation of glycosides.

bisglycosides are combined and chromatographed again to increase the yield of isolated products. Fractions containing pure tris- (80–85) and bis- (90–96) glycosides are combined and evaporated. The yield of tris- and bisglycosides is 1.9 g (0.87 mmol) and 1.75 g (1.12 mmol), respectively.

Deprotection of the glycosides are carried out in two steps. (1) De-O-acetylation. The per-O-acetylated trisglycoside isolated from the LH-20 column is dissolved in 40 ml of absolute alcohol, to which 20 ml of dry toluene is added, and the mixture evaporated under reduced pressure, and the azeotropic evaporation is repeated once more. The syrup is dissolved in 20 ml dry methanol containing 0.01 M sodium methoxide, and stored at room temperature overnight. The reaction mixture is decationized by shaking with 0.5 g of Dowex 50-X8 (H$^+$ form) with a small amount of water, until the solution becomes neutral or slightly acidic. The resin is filtered and washed, and the filtrate evaporated to dryness. (2) De-N-trifluoroacetylation. The resulting residue from the previous step is taken up in 10 ml of water and stirred with 2–3 g of Dowex 1-X8 (OH⁻ form) overnight at room temperature. The resin is filtered off and washed with 50% ethanol, and the filtrate evaporated. Yield: trisglycoside (AHT-tris-Lac), 0.94 g (0.78 mmol); bisglycoside, 0.95 g (1.08 mmol).

Cluster Glycosides Based on Peptide Backbone[5]

Glycosides containing ω-aminoaglycon, such as aminohexyl-β-lactoside and AHT-tris-Lac mentioned above, can be coupled to Z-Asp using EEDQ as the coupling agent or methyl chloroformate as carboxyl activator[11] to yield cluster glycosides containing two to six residues of lactosyl residues.

Preparation of 6-Aminohexylglycosides.[9] An example of 6-amino-hexyl-β-lactoside will be described. 6-Aminohexanol (3.0 g, 25.6 mmol) is mixed with ethyl trifluoroacetate (3.6 ml, 30 mmol), stirred for 5 hr at room temperature, poured into water (50 ml), and stirred overnight at 4°. Crystalline 6-(trifluoroacetamido)hexanol thus obtained is filtered and washed with cold water (3.82 g, 70% yield). De-N-trifluoroacetylation is carried out as described above to yield 6-aminohexyl lactoside in 63% yield.

Coupling of Amino-Containing Glycosides to Z-Asp. Glycosides containing ω-aminoaglycon can be coupled to the two carboxyl groups of Z-Asp to yield "bis" cluster glycosides. In the example below, the amino-containing glycoside, AHT-tris-Lac, is coupled to Z-Asp to produce a product containing six lactosyl residues.

$$
\begin{array}{c}
CH_2COOH \\
| \\
ZNHCHCOOH + NH_2(CH_2)_5COHNC(CH_2O\text{-}Lac)_3
\end{array}
$$

$$\text{EEDQ} \downarrow \ (\text{AHT-Tris-Lac})$$

$$
\begin{array}{c}
CH_2CONH(CH_2)_5CONHC(CH_2O\text{-}Lac)_3 \\
| \\
ZNHCHCONH(CH_2)_5CONHC(CH_2O\text{-}Lac)_3
\end{array}
$$

To a solution of Z-Asp (20 mg, 75 μmol) and AHT-tris-Lac (0.15 g, 120 μmol) in 3.5 ml of water are added EEDQ (250 mg, 1 mmol) and 0.5 ml of 95% ethanol. The suspension is stirred for several hours, during which time 95% ethanol is added twice more to make a final concentration of ethanol 40%. After stirring for 2 days at room temperature, the mixture is

[11] R. Barker, C. K. Chiang, I. P. Trayer, and R. L. Hill, this series, Vol. 34, p. 317.

evaporated, suspended in 2 ml of water and centrifuged to remove precipitate. The precipitate is washed twice with 2 ml of water and the combined aqueous solutions are extracted with toluene (2 ml). The aqueous solution is evaporated to ~2 ml and passed through a column (2 × 145 cm) of Sephadex G-25 with 0.1 M ammonium hydroxide as eluent (collecting 4-ml fractions). The effluent fractions are assayed for carbohydrate by the phenol–sulfuric acid method[10] using 10-μl samples, and also by TLC on silica gel using 5 : 5 : 1 : 3 (v/v) ethyl acetate–pyridine–acetic acid–water. The diamide derivative (fractions 37–45) has an R_f of 0.12, while the monoamide (41–47) has an R_f of 0.2–0.3. Fractions containing both mono- and diamide are combined and rechromatographed to increase the yield (total yield: 40% for diamide, 45% for monoamide). The concentration of these derivatives can be determined by UV absorbance ($\varepsilon_{258\ nm} = 186$) or by determining the content of lactose by the phenol–sulfuric acid method.[10]

Removal of the Z-group is readily accomplished by catalytic hydrogenolysis under atmospheric pressure (using a Brown hydrogenator)[12] in 10% acetic acid using 10% palladium on charcoal (~20% weight of the Z-compound). Normally only several hours are required for complete hydrogenolysis. The reaction mixture is filtered and the filtrate evaporated. The residue is left in a vacuum desiccator over NaOH pellets overnight to remove residual acetic acid, producing white solid in quantitative yield.

Other Examples

Preparations of other types of cluster glycosides have been described.[5]

[12] H. C. Brown, K. Sivasankaran, and C. A. Brown, *J. Am. Chem. Soc.* **28**, 214 (1963).

[35] Labeling Glycoconjugates with Hydrazide Reagents

By Meir Wilchek and Edward A. Bayer

Owing to the ubiquitous distribution of glycoconjugates and polysaccharides in nature, a variety of methods have been devised for their detection, analysis, and isolation. One approach, which involves the interaction of lectins (usually labeled with a detectable probe) with specific sugar residues on glycoconjugates, has been the subject of numerous reviews[1-3] and will not be further discussed here. An alternative approach entails the initial chemical or enzymatic modification of sugars followed by subsequent interaction of the modified saccharide(s) with an appropriate probe. This approach has been used extensively for a variety of purposes, particularly when a general labeling of glycoconjugates is desired.

Due to the chemical similarity of the various native carbohydrate groups (the major accessible chemically reactive group in most cases is of course the hydroxyl group, whereas the major biological difference is expressed by different configurational or conformational states of the various monosaccharides), the availability of suitable chemical reactions which would differentiate between the various saccharides is limited. In practice, only one enzymatic reaction and only one mild chemical reaction are currently in use, both of which generate aldehyde derivatives of the given sugar (Fig. 1). The chemical modification involves the periodate oxidation of vicinal hydroxyls of carbohydrates in general[4] which, if performed under controlled conditions, can be rendered selective for sialic acids.[5] In the enzymatic reaction, galactose oxidase is used to form the C-6 aldehyde on terminal D-galactose or N-acetyl-D-galactosamine residues.[6] In addition, the same enzyme in combination with another enzyme, neuraminidase, may be used for the modification of galactosyl residues penultimate to sialic acids (see Fig. 1).

The existing methodology is therefore restricted in scope around these two basic reactions, the major difference being the type of ligand or reagent subsequently coupled to the above-mentioned sugars.

In this article we will describe the use of these two fundamental aldehyde-generating reactions, followed by the interaction with various hydrazides. The hydrazide reagents and probes used in these reactions will be divided into the following categories: radioactive, fluorescent, target, and polymeric (including enzyme) reagents. Figure 2 illustrates several of the hydrazide reagents which will be discussed.

Radioactive Hydrazides

Radioactive hydrazides which have been used to date have contained either the ^3H, ^{35}S, or ^{125}I isotopes. The tritiated hydrazides are not particu-

---

[1] I. E. Liener, N. Sharon, and I. J. Goldstein, eds., "The Lectins: Properties, Functions, and Applications in Biology and Medicine." Academic Press, New York, 1986.
[2] N. Sharon, Adv. Immunol. **34**, 213 (1983).
[3] N. Sharon, Sci. Am. **236** (6), 108 (1977).
[4] J. A. Rothfus and E. L. Smith, J. Biol. Chem. **238**, 1402 (1963).
[5] L. Van Lenten and G. Ashwell, J. Biol. Chem. **246**, 1889 (1971).
[6] E. Avigad, D. Amaral, C. Asensio, and B. L. Horecker, J. Biol. Chem. **237**, 2736 (1962).

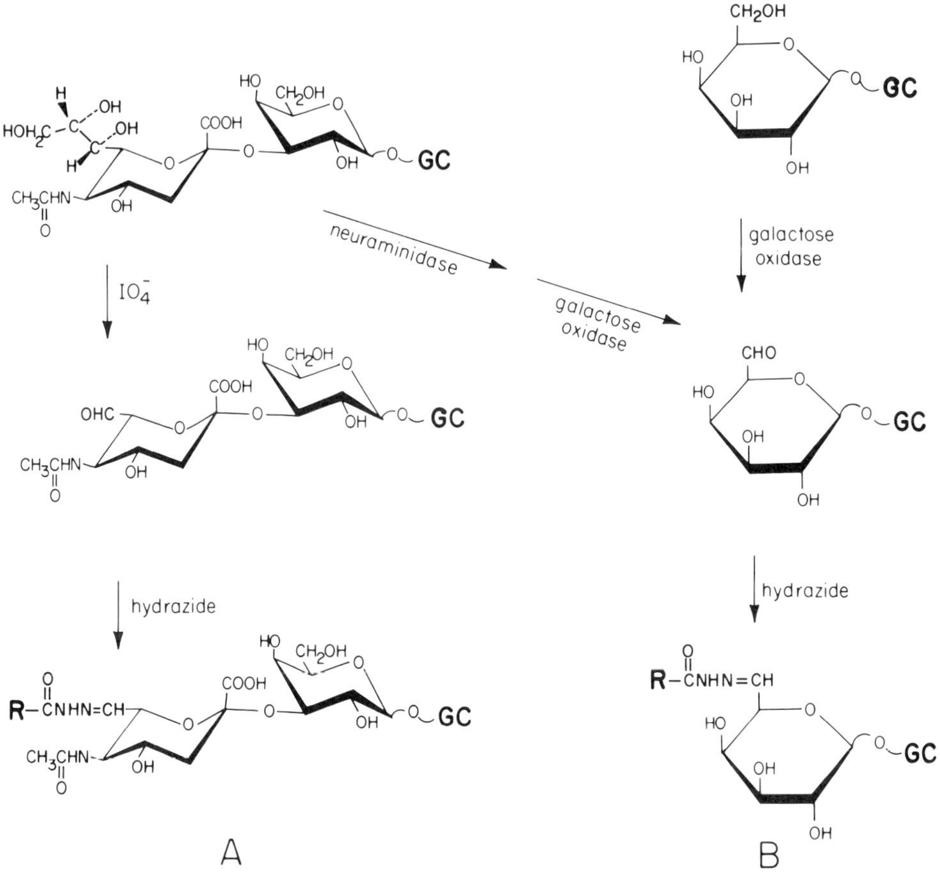

FIG. 1. Two major reaction schemes for coupling hydrazide reagents to (oxidized) sugar residues. In (A), sialyl groups of glycoconjugates are oxidized chemically with periodate and the resultant aldehyde may be reacted with an appropriate hydrazide. In (B), galactose (or N-acetylgalactosamine) residues are oxidized enzymatically using galactose oxidase to the respective C-6 aldehydo derivative which undergoes subsequent interaction with an appropriate hydrazide. In cases where galactose (or N-acetylgalactosamine) residues appear penultimate to terminal sialyl residues, hydrazides may be coupled to the sialoglycoconjugate by the combined enzymatic action of neuraminidase and galactose oxidase as shown.

larly useful for labeling since either the compound is not hot enough or tritiated borohydride can be used instead to directly reduce the aldehyde.[7-9] Perhaps the most useful case for radioactive hydrazides in the

[7] C. G. Gahmberg and S. Hakomori, *J. Biol. Chem.* **248**, 4311 (1973).
[8] C. G. Gahmberg and S. Hakomori, *Biomembranes* **8**, 131 (1976).
[9] C. G. Gahmberg and L. C. Andersson, *J. Biol. Chem.* **252**, 5888 (1977).

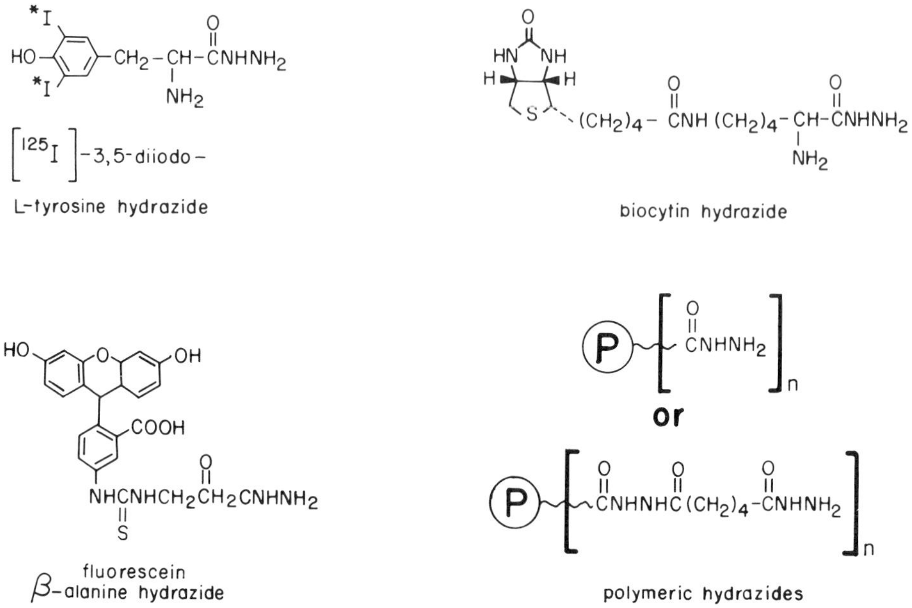

FIG. 2. Some of the hydrazide reagents described in the text.

future will be their employment as tritiated or ^{14}C-labeled hapten hydrazides (see target reagents). The combination of radioactivity with a carbohydrate-specific target reagent would add an additional dimension, particularly for the labeling of glycoconjugates for subsequent affinity-based purification procedures.

The ^{35}S-containing reagent,[10] radioactive methionine sulfone hydrazide, has been used only sparingly. One of the original authors himself claims no significant advantage for this reagent over the use of $B^3H_4^-$.[11]

The diiodotyrosine derivative, introduced recently,[12] has to our knowledge not been used since the original publication. The reason may be the laborious procedure for its synthesis which requires the exchange of radioactive ^{125}I with the precursor, followed by separation of the radioactive derivative prior to hydrazinolysis. In our laboratory, it was found that this reagent can be prepared directly from diiodotyrosine (synthesis described below). This route is much easier to perform than the original

[10] K. Itaya, C. G. Gahmberg, and S. Hakomori, *Biochem. Biophys. Res. Commun.* **64,** 1028 (1975).

[11] C. G. Gahmberg, *in* "Dynamic Aspects of Cell Surface Organization" (G. Poste and G. L. Nicolson, eds.), p. 371. Elsevier/North-Holland Biomedical Press, Amsterdam, 1977.

[12] A. Rotman, S. Linder, and V. Pribluda, *FEBS Lett.* **120,** 85 (1980).

method, since extensive purification procedures are not required. Alternatively, an iodinated hydrazide has been prepared from Bolton–Hunter reagent.[13] Although these reagents have not been used extensively to date, they may eventually find use where glycoconjugate derivatives of exceptionally high specific radioactivity are required, such as for the isolation of glycoconjugate receptors.

Preparation of 3,5-Diiodotyrosine Hydrazide

[^{125}I]Diiodotyrosine is prepared as described previously.[14] The compound (1 nmol) is suspended in dry methanol (100 μl), and the solution is cooled to about $-5°$ in an acetone–ice or NaCl–ice water bath. Thionyl chloride (25 μl) is added. The reaction mixture is left overnight at 25°, and the suspension is evaporated to dryness. Fresh solutions of methanol are added and the solvent evaporated (this procedure is repeated several times), in order to remove residual amounts of thionyl chloride. The product is dissolved in 100 μl methanol, and hydrazine hydrate (25 μl) is added. The reaction is carried out either overnight at 25° or heated at 65° for 3 hr. The solvent is removed under reduced pressure, and the free hydrazine is removed by subjecting the residue to high vacuum for prolonged periods in a desiccator using sulfuric acid as desiccant. The product is dissolved in water (0.5 ml) and applied to a Porapak Type Q column (0.5 ml, Waters Assoc. Inc., Mitford, MA). After washing the column extensively with water, the iodinated product is eluted with 80% aqueous methanol. The compound is concentrated to dryness and used directly.

Fluorescent Hydrazides

There has clearly been an increase in the use of fluorescent hydrazides compared to that of their radioactive analogs. Whereas the radioactive reagents are used mainly for detection and isolation of minute quantities of materials, the fluorescent reagents can be prepared in large amounts, and different fluorophores can be used to study the role of the carbohydrate moiety in the action of the glycoconjugate molecule.

Since a major direction of study is the function of glycoconjugates on cell surfaces, fluorescent reagents have an advantage in that they can be used efficiently in the cell sorter to separate different cell subpopulations.[15] In fact, this may be one of the major future uses of fluorescent hydrazides, since in many cell systems, the character of the carbohy-

---

[13] K. Randerath, *Anal. Biochem.* **115**, 391 (1981).
[14] K. Sorimachi and H. J. Cahnmann, *Endocrinology (Baltimore)* **101**, 1276 (1977).
[15] T. M. Jovin and D. J. Arndt-Jovin, *Trends Biochem. Sci.* **5**, 214 (1980).

drates on the cell surface reflects the respective cell type. In addition, the fluorescent reagent can be used either for analysis of cell-bound glycoconjugates in the fluorescence microscope, or for the evaluation of their distribution on SDS–PAGE, blots, TLC (for glycolipids), etc.

Fluorescent hydrazides have been used to label glycoproteins, such as thyroglobulin,[16] and ATPase,[17] as well as for glycolipids.[18] The labeled glycolipids have been used to examine the function of gangliosides in membranes by exogenous incorporation of the fluorescent gangliosides into the cell membrane.[19]

The number of fluorescent derivatives possible is virtually unlimited, since various derivatives of a variety of fluorescent compounds [including fluorescein, rhodamine, the dansyl (dimethylaminoaphthalene) group, and Lucifer yellow] can be prepared. We will describe the synthesis of two of these which bear extended side arms, the latter of which enables improved interaction with the aldehyde generated from the glycoconjugate(s) of interest.

Synthesis of Fluorescein–β-Alanine Hydrazide[16]

Z-β-Ala-NHNH-tBoc (I). To a solution of N-carbobenzoxy-β-alanine (Sigma, 2.23 g in 20 ml chloroform) is added tert-butyl carbazate (Sigma, 1.32 g in 10 ml chloroform). To this solution, dicyclohexylcarbodiimide (2.1 g in 10 ml chloroform) is added. The reaction is allowed to proceed overnight at 25°, the dicyclohexylurea is filtered, and the filtrate is concentrated to dryness. The residue is dissolved in ethyl acetate and washed successively with water, 10% citric acid, water, 0.1 M NaHCO$_3$ and water. The solvent is dried over sodium sulfate crystals, filtered, and concentrated to dryness. The residue is dissolved in a minimal volume of ether and allowed to crystallize overnight at 4°.

β-Ala-NHNH-tBoc (II). Compound I (1.65 g) is dissolved in 25 ml methanol and palladium on charcoal (100 mg) is added. Reduction is carried out in a Parr apparatus for 3 hr. The palladium is removed by filtration, and the filtrate is concentrated. The thick oily material is difficult to crystallize and is used immediately for further synthesis.

Fluorescein–β-Ala-NHNH$_2$. Fluorescein isothiocyanate (500 mg) is dissolved in dioxane (20 ml) and added to 2 equivalents of Compound II (500 mg, dissolved in a mixture of 20 ml dioxane containing 1 equivalent

[16] M. Wilchek, S. Spiegel, and Y. Spiegel, Biochem. Biophys. Res. Commun. 92, 1215 (1980).
[17] J. A. Lee and P. A. G. Fortes, Biochemistry 24, 322 (1985).
[18] S. Spiegel, M. Wilchek, and P. H. Fishman, Biochem. Biophys. Res. Commun. 112, 872 (1983).
[19] S. Spiegel, S. Kassis, M. Wilchek, and P. H. Fishman, J. Cell Biol. 99, 1575 (1984).

bicarbonate). The reaction is allowed to proceed overnight at 25°. The solution is acidified with 10% citric acid, and the product precipitated. The precipitate is filtered, washed thoroughly with water, and dried over P_2O_5. The R_f is 0.86 on TLC plates (silica gel) using methanol–ethyl acetate (1 : 1, v/v) as the solvent system. The product formed is dissolved in dioxane and treated with a solution of 2 M HCl in dioxane. After 15 min, the product precipitates. The precipitate is washed with dry ether and dried over P_2O_5 under reduced pressure. Using the above TLC system, the R_f is 0.21. Upon reaction with acetone, the R_f changes to 0.68, indicating the presence of hydrazide.

Rhodamine–β-Ala-NHNH₂

This compound is prepared from rhodamine isothiocyanate and Compound **II** by a protocol essentially analogous to that of the fluorescein derivative. TLC using methanol as solvent generates the following R_f values: 0.52 for the tBoc derivative, 0.1 for the hydrazide, and 0.32 for the acetone reaction product.

Other Fluorescent Hydrazides

Many other fluorescent hydrazides are either commercially available, e.g., Lucifer yellow (Aldrich), or can be easily prepared.[16,17]

Target Hydrazides

A target hydrazide does not necessarily comprise a chromaphore or other detectable function, but must have the capacity to bind specifically to another molecule, such as a protein, to which different probes can be attached. The target hydrazides used to date include sugar hydrazides which interact with lectins (e.g., mannose hydrazide and Con A),[20] biotin hydrazide which can interact either with the egg white glycoprotein avidin or with the bacterial protein streptavidin,[21-23] and any hapten hydrazide which can interact with appropriate monoclonal or polyclonal antibodies (e.g., DNP/anti-DNP, arsenate/antiarsenate, NIP/anti-NIP, and biotin/ antibiotin).[24,25] The target hydrazides have been used mostly to study glycoconjugates on cell surfaces where fluorescent reagents are insuffi-

[20] R. R. Rando, G. A. Orr, and F. W. Bangerter, *J. Biol. Chem.* **254,** 8318 (1979).
[21] E. A. Bayer, E. Skutelsky, and M. Wilchek, *FEBS Lett.* **68,** 240 (1976).
[22] E. Skutelsky, D. Danon, M. Wilchek, and E. A. Bayer, *J. Ultrastruct. Res.* **61,** 325 (1977).
[23] E. Skutelsky and E. A. Bayer, *J. Cell Biol.* **96,** 184 (1983).
[24] D. Wynne, M. Wilchek, and A. Novogrodsky, *Biochim. Biophys. Acta* **68,** 730 (1976).
[25] A. Ravid, A. Novogrodsky, and M. Wilchek, *Eur. J. Immunol.* **8,** 294 (1978).

cient or inappropriate for detection. By using fluoresceinated antibodies or binding proteins, the signal can be enhanced. Alternatively, any immobilization procedure (e.g., SDS–PAGE followed by blotting onto a solid matrix) is essentially analogous to labeling on cell surfaces, and target hydrazides may be used accordingly. This approach appears to be less useful for labeling glycoconjugates in solution, since introduction of a second macromolecule would significantly alter the physical characteristics of the glycoconjugates. On the other hand, this approach may be very useful for noncovalent cross-linking of two macromolecules.

Like their fluorescent analogs, the number of target hydrazides is also unlimited, since any hapten–antibody pair can be used if an appropriate hydrazide derivative of the corresponding hapten can be synthesized. Even though an extra step (using the antibody probe) is required, the target hydrazide can be the most economic and versatile system in the long run, since a variety of antibody markers (radioactive, fluorescent, enzyme-conjugated, electron dense, etc.) can be used to investigate the same hapten-derivatized specimen. In the case of biotin hydrazide, a great many avidin probes are possible, many of which are commercially available through dozens of commercial sources.[26]

In choosing the hapten or ligand portion of the target hydrazide, several considerations should be taken, one of the most important of which is the affinity between the ligand and the binding protein. Ideally, for most purposes, the higher the affinity constant the better, and it is not recommended to employ ligand–protein pairs which exhibit affinity constants less than $10^7 \ M^{-1}$. It is interesting that although the affinity constant of most lectins for sugars in solution is very low ($K_d \simeq 10^{-4} \ M$), the affinity increases markedly when the sugar is bound to cell surfaces. Sugars may therefore be good candidates for target hydrazides, provided that an uncommon sugar–lectin pair which is foreign to the experimental system is employed.

The most prevalent target system used until now has been biotin hydrazide–avidin, which has been applied for the localization of cell surface sialic acids, galactose, and other saccharides.[27] This system boasts an unprecedented ($K_d \ 10^{-15} \ M$) affinity constant, which has been the impetus for its widespread usage as a general tool in the biological sciences.[26,28,29]

We have found that the covalent bond formed between the biotin hydrazide and the aldehydosaccharide is not always stable to interaction with avidin, and the bond may be split due to physical forces imparted by

[26] M. Wilchek and E. A. Bayer, *Immunol. Today* **5,** 39 (1984).
[27] E. A. Bayer, E. Skutelsky, and M. Wilchek, this series, Vol. 83, p. 195.
[28] E. A. Bayer and M. Wilchek, *Trends Biochem. Sci.* **3,** N257 (1978).
[29] E. A. Bayer and M. Wilchek, *Methods Biochem. Anal.* **26,** 1 (1980).

the strong interaction with the deep biotin-binding pocket within the active site.[30] We therefore promote the use of biocytin hydrazide which increases the distance between the biotin moiety from the avidin pocket. From our studies, it has also been shown that the presence of the α-amino group adjacent to the hydrazide moiety serves to enhance both the extent and stability of binding to the aldehyde.[31]

The preparations of both biotin hydrazide[32] and N-γ-DNP-diaminobutyrylhydrazide[33] have been described in this series previously. In the following we will provide a newly described synthesis of biocytin hydrazide which contains a long spacer (lysine) which connects the hydrazide function to the biotin moiety.

Preparation of Biocytin Hydrazide[31]

Biocytin. Cupric carbonate (1.2 g) is added to an aqueous solution (10 ml) containing lysine hydrochloride (350 mg). The solution is heated at 100° for 5 min in a beaker, and the precipitate is filtered by gravity. The resultant blue filtrate is cooled in an ice bath, and a solution of BNHS (645 mg, 1.9 mmol in 5 ml dimethylformamide) is added. Solid sodium bicarbonate (400 mg) is introduced, and the reaction is allowed to proceed in an ice bath for 4 hr. The blue precipitate is centrifuged, and the supernatant is removed by decantation. The product is washed successively with water and ethanol, and then dried with ether. Yield, 670 mg (87%).

The blue powder is dissolved in 0.1 N HCl (10 ml) and H_2S is bubbled in until no further precipitation occurs. The mixture is heated in a hood in order to "coagulate" the CuS solid. The precipitate is filtered, and the residual H_2S is removed by evaporation. The remaining clear solution is lyophilized. Yield, 580 mg (75%).

Amino acid analysis of the hydrolyzed product yields an equivalent amount of lysine. TLC on silica gel gives one spot identical to commercially available biocytin [(Sigma) R_f 0.2, butanol–acetic acid–water (4 : 1 : 1)] with no visible contaminants using ninhydrin or dimethylaminocinnamaldehyde spray.[34]

Biocytin Methyl Ester. To a solution of methanol (5 ml) is added 0.5 ml thionyl chloride in an ice water–acetone bath. Biocytin (200 mg) is then added. The reaction mixture is left overnight with constant stirring. The solvent is reduced to minimal volume and dry ether is added. The precipitate is filtered, washed with dry ether, and dried over NaOH under re-

[30] M. Wilchek and E. A. Bayer, unpublished data.
[31] M. Wilchek and E. A. Bayer, submitted for publication.
[32] E. A. Bayer, E. Skutelsky, and M. Wilchek, this series, Vol. 62, p. 308.
[33] D. Givol and M. Wilchek, this series, Vol. 46, p. 503.
[34] D. M. McCormick and J. A. Roth, this series, Vol. 18A, p. 383.

duced pressure. The R_f of the product is 0.48 using the above TLC solvent system.

Biocytin Hydrazide. The methyl ester of biocytin (100 mg) is dissolved in 2 ml methanol, and hydrazine hydrate (0.1 ml) is added. After 48 hr at 25°, the solvent is concentrated to dryness and dried further in a desiccator under reduced pressure in the presence of H_2SO_4 until no smell of hydrazine can be detected. The product is dissolved in water (1 ml) and passed through a column containing Porapak Type Q (4 × 1 cm). The column is eluted with water and fractions (2 ml) are collected. The first several fractions usually contain residual hydrazine hydrochloride.[35] The fractions containing biocytin hydrazide are detected by dimethylaminocinnamaldehyde[34] and ninhydrin sprays.[36] To hasten elution of the hydrazide, the column is eluted with 80% aqueous methanol following the hydrazine peak. The fractions are then eluted as a sharp peak. The fractions containing the product are pooled, concentrated to dryness, dissolved in a minimal amount of methanol, and precipitated with ether. Biocytin hydrazide can be crystallized from hot ethanol. Upon reaction with acetone, three reaction products are obtained (TLC using methanol as solvent) with R_f values of 0.24, 0.55, and 0.65, representing the anticipated reaction products.

We have also prepared the same compound from t-Boc biocytin[37] and *tert*-butyl carbazate in the presence of dicyclohexylcarbodiimide in a dimethylformamide solution. The protecting groups are removed with 4 N HCl in dioxane.

Polymeric Hydrazides

The polymeric hydrazides include any polymer containing multiple hydrazide functions. Polyhydrazides can be used for a variety of purposes: they can be functional (e.g., enzyme hydrazides),[38] structural (e.g., latex hydrazide of various sizes, Sepharose hydrazide),[39,40] chromaphoric (e.g., ferritin hydrazide,[41] hydrazide-derivatized fluorescent microbeads[42]) or other hydrazides (e.g., avidin hydrazide, streptavidin hydra-

---

[35] Care should be taken to remove all hydrazine since the latter reacts well with aldehydes, thereby blocking efficient coupling of the hydrazide reagent.

[36] $R_f = 0.11$ as detected by TLC (methanol as solvent) or 0.2 using the above solvent system.

[37] E. Bayer and M. Wilchek, this series, Vol. 34, p. 265.

[38] J. M. Gershoni, E. A. Bayer, and M. Wilchek, *Anal. Biochem.* **146,** 59 (1985).

[39] Z. Malik and Y. Langzam, *Cell Differ.* **11,** 161 (1982).

[40] T. Miron and M. Wilchek, *J. Chromatogr.* **215,** 55 (1981).

[41] E. Roffman, Y. Spiegel, and M. Wilchek, *Biochem. Biophys. Res. Commun.* **97,** 1192 (1980).

[42] M. R. Kaplan, E. Calef, T. Bercovici, and C. Gitler, *Biochim. Biophys. Acta* **728,** 112 (1983).

zide, BSA–hydrazide, dextran hydrazide, or polyglutamylhydrazide[43]) which, although by themselves may not provide a specific function, may serve to enhance the number of hydrazides on a second polymer (e.g., enzyme–BSA hydrazide conjugates).

One of the advantages of polyhydrazides is that these membrane-impermeant reagents can be used to label cell surface carbohydrates directly. The polyhydrazides can be prepared in radioactive or fluorescent form, as well as electron dense for transmission electron microscopy and as large particles for scanning electron microscopy. Enzyme hydrazides also belong to the polyhydrazide category. Many enzymes have recently been introduced for enzyme immunoassay, for blotting, and for other histochemical and cytochemical purposes. Thus, the same enzyme hydrazide can be used for the microscopic analysis of cell surface glycoconjugates and as a stain for the same glycoconjugates separated from extracts by SDS–PAGE and transferred to blots. The above-described reagents have been employed for the labeling of different glycoconjugates, e.g., glycoproteins and glycolipids, both on membranes as well as for the isolated systems.

Insoluble forms of hydrazides, such as Sepharose hydrazides or hydrazide-derivatized beads, can be used for immobilization of glycoconjugates via their sugars.[29] The insolubilized glycoconjugate can then be used for affinity chromatography, for immuno- or histochemical assay, or in enzyme reactors.

Sepharose hydrazide has been described previously in this series.[44] The following provides syntheses of some other representative polyhydrazides.

Preparation of Polyhydrazides

Ferritin Hydrazide.[41,45] Adipic dihydrazide (40 mg, Sigma) is dissolved by heating into 0.9 ml double-distilled water. The pH is adjusted to 4.5 with a solution of 0.5 M HCl, and 0.1 ml ferritin (100 mg/ml, Polysciences) is added. Water-soluble carbodiimide (30 mg, WSC[46]) is added as the solid. The test tube is tapped gently and the reaction is allowed to proceed at 25° for 6 hr with periodic adjustment of the pH (between 4.5 and 5.0). The reaction is then left overnight at 4°, diluted to 10 ml with phosphate-buffered saline (pH 7.4), dialyzed exhaustively against the same buffer,

[43] E. Hurwitz, M. Wilchek, and J. Pitha, *J. Appl. Biochem.* **2,** 25 (1980).
[44] M. Wilchek and T. Miron, this series, Vol. 22, p. 72.
[45] Other proteins, in particular binding proteins such as avidin, streptavidin, and the various lectins, can be hydrazide modified by using the same protocol.
[46] 1-Ethyl-3-(3-dimethylaminopropyl)carbodiimide-HCl (Sigma).

sterilized by passage through a Millipore filter (HA 0.2 μm), and stored at 4°.

Enzyme Hydrazide.[38] An aqueous solution containing 0.5 M adipic dihydrazide is brought to pH 5 with 1 N HCl, and an aliquot (0.5 ml) is added to a solution (0.45 ml) of alkaline phosphatase (2500 units; 2.25 mg enzyme; Sigma Type VII-NT).[47] The solution is brought to 2.0 ml with saline, and 0.5 ml of an aqueous solution containing 0.1 M WSC[46] is added.[48] During the course of the reaction, the solution is maintained at pH 5 by the periodic addition of 0.5 N HCl. After 6 hr at 25°, the reaction mixture is dialyzed first against saline, and then against 100 mM Tris–HCl buffer (pH 8), containing 1 mM MgCl$_2$ and 0.02% sodium azide. The enzyme hydrazide can be stored for long periods of time at -20°, or at 4° (preferably after sterilization by passage through a Millipore filter).

Dextran Hydrazide.[43] Dextran T40 (5 g, Pharmacia) is dissolved in water (5 ml). Concentrated NaOH (40%, 38 ml) and chloroacetic acid (27 g) are added and the suspension is stirred for 12 hr at 25°. This step is repeated twice. The solution is then dialyzed exhaustively against water, and lyophilized, yielding 9.4 g carboxymethyl dextran.

Carboxymethyl-dextran (5 g) is dissolved in water (20 ml). A solution (10 ml) of aqueous hydrazine or adipic dihydrazide (10 M)[49] is then added. WSC (1 g)[46] is then added and the pH is maintained in the range of 4.5 to 5.8 for 50 min. This process is repeated twice. The solution is dialyzed exhaustively against water and cleared by centrifugation, and the product is then lyophilized. The resulting powder (derived from aqueous hydrazine) contains 7.4% nitrogen, i.e., 2.6 μmol hydrazide/mg. The powder cannot be fully dissolved in water directly, but can be dissolved first in dilute alkali followed by water dialysis.

Poly-Glu-NHNH$_2$.[43] Poly(γ-benzyl L-glutamate) (200 mg, Sigma) is suspended in 10 ml hydrazine hydrate and stirred at room temperature until dissolved.[50] The solution is then diluted 5 times and dialyzed exhaustively against distilled water. The product can be lyophilized and retains good water solubility. The content of hydrazide group can be measured either colorimetrically using β-naphthoquinone 4-sulfonate[51] or by nitrogen analysis. The product contains 4 μmol of accessible hydrazide/mg.

---

[47] The is only one example of many suitable enzymes which can be hydrazide modified by this method.

[48] The amount of WSC used in the reaction appears to be critical, since either increasing or decreasing its concentration tends to reduce the efficacy of the resultant enzyme hydrazide.

[49] The solution is adjusted to pH 5.8 by concentrated HCl.

[50] Around 3–4 hr.

[51] E. L. Pratt, *Anal. Chem.* **25**, 814 (1953).

Illustrative Labeling Procedures

Labeling Cell Surface Sialic Acids. Cells (about 10^8/ml) are washed and resuspended in an appropriate buffer.[52] A solution of sodium metaperiodate is added to a final concentration of 1 mM.[53] The treatment is carried out for 30 min in an ice bath. The cells are then washed and resuspended in a solution containing the desired hydrazide.[52,54,55] The reaction between the aldehydosaccharide and the hydrazide reagent is usually carried out for at least 1 hr at 25°. The cells are then washed and processed further for the required purpose (detection, quantification, localization, immobilization, etc.).

Labeling Cell Surface Galactose. A 5% cell suspension[56] is incubated in an agitating water bath for 60 min at 37° with a buffered solution[57] containing *Vibrio cholerae* neuraminidase[58] (0.05 U/ml, Boehringwerke, Marburg, Germany), galactose oxidase (5 U/ml, Sigma), and the desired hydrazide.[52,54] The cells are then washed twice and resuspended in the buffer of choice pending further processing.

Labeling Glycoproteins in Solution. Periodate oxidation is performed essentially as described previously.[29,59] The glycoprotein (10 mg in 5 ml of 0.1 M sodium acetate, pH 4.5, containing 5 mM sodium metaperiodate) is oxidized for 1 hr at 4° in the dark.[60] After oxidation, excess periodate is

---

[52] Tris, glycine, or other amino-containing buffers should be avoided so as to prevent their interaction with the aldehydo sugar formed.

[53] This particular concentration has been shown to be relatively specific for the oxidation of sialic acid. If the labeling of other sugar moieties is desired for the general detection of sugars, then the cell suspension should be brought to 10 mM in periodate.

[54] Optimal concentrations for the aldehyde–hydrazide interaction vary and depend upon many factors, including the hydrazide type (small molecule or macromolecule), the disposition and microenvironment of the aldehydo glycoconjugate, etc. For cell surface labeling, as much as 5 mg biotin hydrazide/ml, for example, can be used; in other systems, e.g., detection of glycoconjugates on blots, microgram quantities of enzyme hydrazide per milliliter are optimal. As a rule, it is advised to optimize each individual experimental system by titration with the desired hydrazide.

[55] Care should be taken at this point to remove excess periodate and iodide before interaction with the hydrazide reagent, since these compounds oxidize the hydrazide function resulting in coupling to amino groups (M. Wilchek and E. A. Bayer, unpublished results, 1985).

[56] The cells may be pretreated with sodium borohydride (2 mM final concentration) in order to quench endogenous oxidized membrane components.

[57] Phosphate-buffered saline, pH 7, containing 3 mM CaCl$_2$.

[58] For cells bearing exposed surface galactosyl residues, such as rabbit erythroid cells, neuraminidase can be excluded from the reaction solution.

[59] H. Debray, M. Cacan, R. Cacan, and N. Sharon, *J. Biol. Chem.* **250,** 1955 (1975).

[60] Other buffer systems, which are more suitable for the desired glycoprotein, may also be used, provided that they do not contain primary amines or other aldehyde-interacting molecules.

removed by dialysis or column chromatography on Sephadex G-25.[54] The dialyzed protein is then coupled to the appropriate hydrazide.[52,54] Excess hydrazide can be removed either by dialysis (for low molecular weight hydrazides), by gel filtration, or by affinity methods.

Labeling of Gangliosides.[61] Bovine brain gangliosides (1 mg) are suspended in 1 ml PBS.[57] The solution (10 μl) of 0.2 M sodium periodate is added and the reaction is allowed to proceed for 30 min at 4°. The reaction mixture is dialyzed extensively against the buffer,[54] after which biotin hydrazide (2.5 mg dissolved in 0.5 ml PBS containing 1 mM MnCl$_2$), fluorescein–β-Ala-NHNH$_2$ (0.5 mg/ml), or any other hydrazide[52,54] is added. The reaction is carried out for 1 hr at 37°, the modified gangliosides are dialyzed and reduced with potassium borohydride (1 mM) for 10 min at room temperature and extensively dialyzed against PBS.

Labeling Glycoconjugates on Blots.[38] Nitrocellulose membranes or any other immobilizing matrix (e.g., positively charged nylon), upon which glycoconjugates have been transferred, are quenched for 1 hr at 37° in 2% (w/v) BSA.[62] The blots are then incubated in a freshly prepared aqueous solution of sodium periodate (1 or 10 mM)[53] at room temperature for 30 min.[63] After a brief rinse in PBS, the blots are incubated for 1 hr in 2% BSA in PBS containing enzyme hydrazide (e.g., alkaline phosphatase hydrazide, 1 μg/ml),[52,54] and washed again (2 × 10 min). The blot is then treated with an appropriate substrate solution. For alkaline phosphate hydrazide, the substrate solution consists of 10 mg naphthol-AS-MX phosphate (Sigma, solubilized in 200 μl dimethylformamide) and 30 mg fast red (Sigma)[64] in 100 ml of 100 mM Tris–HCl buffer (pH 8.4). The signal, a red precipitate, appears within 30 min incubation of the blot. The stained material is extremely stable; over 6 months of exposure to light causes no appreciable fading of the signal.

---

[61] S. Spiegal, E. Skutelsky, E. A. Bayer, and M. Wilchek, *Biochim. Biophys. Acta* **687,** 27 (1982).

[62] If high levels of background are experienced, it is advisable to pretreat the BSA (which may contain detectable amounts of γ-globulins) with 10 mM periodate followed by dialysis against 25 mM Tris–glycine (pH 8.3). In most cases untreated BSA is suitable for quenching, but the background is somewhat better when the periodate-treated BSA is used.

[63] In all reactions with blots, 0.5 to 1 ml of the appropriate reagent per square centimeter of filter is used.

[64] Other "fast" colors (blue, green, violet, scarlet, etc.) may be substituted for fast red.

[36] Biochemical Characterization of Animal Cell Glycosylation Mutants

By PAMELA STANLEY

Introduction

A wide variety of selection protocols give rise to mutants with a primary defect in glycosylation. These include selections for resistance to plant lectins, reduced incorporation of ^{3}H-labeled sugars, altered cellular adhesive properties, and selections against the functional activity or the presence at the cell surface of specific membrane molecules.[1] The reason that glycosylation mutants are obtained comparatively frequently (even when they are not the desired result of a selection experiment) is presumably because carbohydrate synthesis is complex, involving molecules from almost every metabolic pathway and, in addition, because mature carbohydrates are distributed among molecules of very different structural, functional, and compartmentalization properties.

Many of the glycosylation changes expressed by different mutants have been identified, and several enzymes have been directly implicated in providing the biochemical basis of particular mutant types (Table I). The broad range of biosynthetic steps whose alteration leads to a change in glycosylation is apparent from the different phenotypes described in Table I. In addition, it can be seen that a number of different defects (e.g., nucleotide-sugar synthesis, glycosyltransferase deficiency, glycosidase deficiency) can give rise to similar carbohydrate structural changes. For example, lack of fucose residues on complex carbohydrates might be due to the loss of a fucosyltransferase, the inability to synthesize GDP-Fuc (e.g., PL^{R}1.3, Lec13, and Lec13A mutants; Table I), lack of availability of the appropriate substrate (e.g., Lec1, 15B, and Ric^{R}14 mutants; Table I), or altered compartmentalization of any one of the required substrates. On the other hand, increased fucosylation of glycoconjugates might be due to the acquisition of a new fucosyltransferase activity (e.g., LEC11 and LEC12 CHO mutants; Table I), the availability of a new substrate, increased production of GDP-Fuc, or altered compartmentalization. Therefore, the search for the biochemical basis of a glycosylation mutation must be wide ranging and, whenever possible, should be complemented by somatic cell genetic and eventually by molecular genetic studies. This

---

[1] P. Stanley, *Annu. Rev. Genet.* **18**, 525 (1984).

TABLE I

BIOCHEMICAL ALTERATIONS CORRELATED WITH GLYCOSYLATION MUTATIONS[a]

General lesion	Glycosylation mutants	Altered biosynthetic step	Probable enzymatic basis
Formation of activated sugars[b]	$PL^R1.3$ (BW5147)[c] Lec13 (CHO)[d] Lec13A (CHO)[d]	GDP-Man $\uparrow\!\!\downarrow$ GDP-Fuc	↓ GDP-Man-4,6-dehydratase (EC. 4.2.1.47)
	Thy1E (BW5147)[e] B4-2-1 (CHO)[f] (Lec15)	Dol-P $\uparrow\!\!\downarrow$ Dol-P-Man	↓ Dol-P-Man synthetase
	AD6 (3T3)[g]	Glucosamine-6-P $\uparrow\!\!\downarrow$ GlcNAc-6-P	↓ Acetyltransferase
Maturation of dolichyl-oligosaccharide[h]	B211 (CHO)[i] (Lec5)	↑ M_9Gn_2-P-P-Dol	?
	Lec1.Lec6 (CHO)[j]	↑ M_7Gn_2-P-P-Dol	?
	Lec9 (CHO)[k]	↑ M_5Gn_2-P-P-Dol	?
Glycosyltransferase or glycosidase activity reduced[l]	Lec1 (CHO)[m] 15 B (CHO)[n] Ric^R14 (BHK)[o]	M_5Gn_2Asn $\uparrow\!\!\downarrow$ $Gn(\beta1,2)M_5Gn_2Asn$	↓ GlcNAc-TI
	Ric^R21 (BHK)[p,q]	GnM_3Gn_2Asn $\uparrow\!\!\downarrow$ $Gn(\beta1,2)GnM_3Gn_2Asn$	↓ GlcNAc-TII
	$Pha^R2.1$ (BW5147)[r] Lec4 (CHO)[s]	$Gn_2M_3Gn_2Asn$ $\uparrow\!\!\downarrow$ $Gn(\beta1,6)Gn_2M_3Gn_2Asn$	↓ GlcNAc-TV
	$Pha^R2.7$ (BW5147)[t]	$Glc_3M_9Gn_2$-P-P-Dol $\uparrow\!\!\downarrow$ $Glc_2M_9Gn_2$-P-P-Dol	↓ α-glucosidase II
	CL6 (L)[u]	M_8Gn_2Asn $\uparrow\!\!\downarrow$ M_7Gn_2Asn	?
	VV6 (CTL)[v,w] (B6.1 SF.1)	(SA)GalHxSer $\uparrow\!\!\downarrow$ GalNAcβ(1,4)GalHxSer	↓ β(1,4)GalNac-T
Glycosyltransferase activity increased[x]	Wa4 (B16)[y]	+α(1,3)-fucose	↑ α(1,3)Fuc-T
	LEC11 (CHO)[z,aa]	+α(1,3)-fucose	↑ α(1,3)Fuc-TI
	LEC12 (CHO)[z,aa]	+α(1,3)-fucose	↑ α(1,3)Fuc-TII
	LEC10 (CHO)[bb]	+Bisecting GlcNAc	↑ GlcNAc-TIII
Nucleotide-sugar transport[cc]	Lec2 (CHO)[dd] 1021 (CHO)[dd]	CMP-SA $\uparrow\!\!\downarrow$ GP or GL	↓ Transport CMP-SA into Golgi
	Lec8 (CHO)[ee] 13 (CHO)[ee]	UDP-Gal $\uparrow\!\!\downarrow$ GP or GL	↓ Transport UDP-Gal into Golgi[ff]

Footnotes to TABLE I

[a] Summary of biochemical lesions associated with extensively characterized glycoslyation mutants from different parental cell lines (given in parentheses after mutant name). M or Man, Mannose; Fuc, fucose; Gal, galactose; Gn or GlcNAc, N-acetylglucosamine; GalNAc, N-acetylgalactosamine; Dol-P, dolichol phosphate; T, transferase enzyme; SA, sialic acid; GP, glycoproteins; GL, glycolipids.

[b] Other mutants which might belong to this group include ConA RII and L6C12VI from L6 myoblasts and a MOPC-315 nonsecretory variant with increased nucleotide pyrophosphatase activity. [Refs: Con A RII, G. A. Cates, A. M. Brickenden, and B. D. Sanwal, J. Biol. Chem 259, 2646 (1984). L6C12VI, C. L. J. Parfett, J. C. Jamieson, and J. A. Wright, Exp. Cell. Res. 144, 405 (1983). MOPC-315, S. Hickman, Y.-P. Wong-Yip, N. F. Rebbe, and J. M. Greco, J. Biol. Chem. 260, 6098 (1985).]

[c] M. L. Reitman, I. S. Trowbridge, and S. Kornfeld, J. Biol. Chem. 255, 9900 (1980).

[d] J. Ripka, A. Adamany, and P. Stanley, Arch. Biochem. Biophys. 249, 533 (1986).

[e] A. Chapman, K. Fujimoto, and S. Kornfeld, J. Biol. Chem. 255, 4441 (1980).

[f] J. Stoll, A. R. Robbins, and S. S. Krag, Proc. Natl. Acad. Sci. U.S.A. 79, 2296 (1982).

[g] E. J. Neufeld and I. Pastan, Arch. Biochem. Biophys. 188, 323 (1978).

[h] Another mutant which might belong to this group includes C^R-7 from CHO cells [J. A. Wright, J. C. Jamieson, and H. Ceri, Exp. Cell Res. 121, 1 (1979)].

[i] S. S. Krag, J. Biol. Chem. 254, 9167 (1979).

[j] L. A. Hunt, Cell (Cambridge, Mass.) 21, 407 (1980).

[k] A. Rosenwald, S. Krag, and P. Stanley, in preparation.

[l] Another mutant shown to belong to the GlcNAc TI defective group is 62.1 from CHO cells [C. B. Hirschberg, M. Perez, M. Snider, W. L. Hanneman, J. Esko, C. R. H. Raetz, J. Cell Physiol. 111, 255 (1982)].

[m] P. Stanley, S. Narasimhan, L. Siminovitch, and H. Schachter, Proc. Natl. Acad. Sci. U.S.A. 72, 3323 (1975).

[n] C. Gottlieb, J. Baenziger, and S. Kornfeld, J. Biol. Chem. 250, 3303 (1975).

[o] A. Meager, A. Ungkitchanukit, R. Nairn, and R. C. Hughes, Nature (London) 257, 137 (1975).

[p] P. A. Gleeson, J. Feeney, and R. C. Hughes, Biochemistry 24, 493 (1985).

[q] S. Narasimhan, S. Allen, R. C. Hughes, and H. Schachter, Glycoconjugates J. 1, 51 (1984).

[r] R. D. Cummings, I. S. Trowbridge, and S. Kornfeld, J. Biol. Chem. 257, 13421 (1982).

[s] P. Stanley, G. Vivona, and P. H. Atkinson, Arch. Biochem. Biophys. 230, 363 (1984).

[t] M. L. Reitman, I. S. Trowbridge, and S. Kornfeld, J. Biol. Chem. 257, 10357 (1982).

[u] I. Tabas and S. Kornfeld, J. Biol. Chem. 253, 7779 (1978).

[v] A. Conzelmann and S. Kornfeld, J. Biol. Chem. 259, 12528 (1984).

[w] A. Conzelmann and S. Kornfeld, J. Biol. Chem. 259, 12536 (1984).

[x] These enzyme changes might be due to activation of the respective transferase genes in the CHO mutants since no transferase activity or carbohydrate product is detected in parental CHO cells.

[y] J. Finne, M. M. Burger, and J.-P. Prieels, J. Cell Biol. 92, 277 (1982).

[z] C. Campbell and P. Stanley, Cell (Cambridge, Mass.) 35, 303 (1983).

[aa] C. Campbell and P. Stanley, J. Biol. Chem. 259, 11208 (1984).

[bb] C. Campbell and P. Stanley, J. Biol. Chem. 261, 13370 (1984).

[cc] Lec3 CHO mutants exhibit a similar sialic acid-deficient phenotype to Lec2 CHO but belong to a different complementation group [P. Stanley, T. Sudo, and J. P. Carver, J. Cell Biol. 85, 60 (1980)].

(continued)

Footnotes to TABLE I (continued)

[dd] S. L. Deutscher, N. Nuwayhid, P. Stanley, E. B. Briles, and C. B. Hirschberg, Cell (Cambridge, Mass.) **39**, 295 (1984).
[ee] S. L. Deutscher and C. B. Hirschberg, J. Biol. Chem. **261**, 96 (1986).
[ff] Several other mutants exhibit similar glycosylation phenotypes: Pha[R]1.8 from BW5147 cells, BSI.B4 from 3T3 cells, Abr[R] from CHO cells, MDW4 from MDAY-D2 cells, and Wga[RI] from L6 myoblasts. [Refs: Pha[R]1.8, I. S. Trowbridge, R. Hyman, T. Ferson, and C. Mazauskas, Eur. J. Immunol. **8,** 716 (1978). BSI.B4, W. S. Stanley, B. P. Peters, D. A. Blake, D. Yep, E. H. Y. Chu, and I. J. Goldstein, Proc. Natl. Acad. Sci. U.S.A. **76,** 303 (1979). Abr[R], I.-C. Li, D. A. Blake, I. J. Goldstein, and E. H. Y. Chu, Exp. Cell Res. **129**, 351 (1980). MDW4, J. W. Dennis, J. P. Carver, and H. Schachter, J. Cell Biol. **99**, 1034 (1984). Wga[RI], B. M. Gilfix and B. D. Sanwal, Can J. Biochem. Cell Biol. **62**, 60 (1984).]

is because an alteration in a functional activity such as a glycosyltransferase or transport protein might be due to a mutation in the gene coding for that activity or to a mutation in another gene whose product affects the first activity. Therefore, defining the actual molecular bases of glycosylation mutations will ultimately involve proof of structural alteration(s) in the genes that code for molecules involved in carbohydrate biosynthesis. At this time, however, the analysis of glycosylation genes has barely begun and so biochemical and genetic characterization must suffice.

The aim of this article is to outline general strategies for the biochemical characterization of animal cell glycosylation mutants. Emphasis will be given to approaches used in characterizing Chinese hamster ovary (CHO) cell mutants, since this laboratory has studied those mutants extensively. Experimental methods are referred to in the text but are not given in detail because of the large number of different techniques involved and because essentially all of them are now standard. The previous volume in this series describes methods for many of the approaches presented in this article.[2]

Strategic Summary

An outline of the strategies applied to the characterization of CHO glycosylation mutants is summarized below. The advantages and disadvantages of each approach are discussed in the following section.

1. Determine lectin resistance (Lec[R]) properties and/or ability to bind lectins at 4° (Lec[B] phenotype).

[2] This series, Vol. 83.

2. Determine dominance properties in somatic cell hybrids. If recessive, perform complementation analyses with related phenotypes. Isolate independent mutant(s) of identical phenotype and show that they belong to the same complementation group.

3. Infect with vesicular stomatitis virus (VSV) in the presence of different ^{3}H or ^{14}C-labeled sugars. (VSV encodes a single glycoprotein, G, that has two N-linked glycosylation sites[3,4]; infection inhibits host cell synthesis so that sugars are incorporated mainly into G). Monitor sugars incorporated into G and apparent molecular weight of G by SDS–PAGE.

4. Prepare Pronase glycopeptides of labeled G. Compare profiles on lectin affinity columns and BioGel columns before and after glycosidase treatments to identify glycopeptides for structural characterization.

5. Analyze by ^{1}H NMR spectroscopy glycopeptides missing from or acquired by G from mutant cells.

6. Determine whether carbohydrate structural change(s) observed in N-linked carbohydrates of G affect spectrum of dolichyl-oligosaccharides or other classes of glycoconjugates (glycolipids, proteoglycans, O-linked carbohydrates).

7. If indicated by structural studies, determine whether growth in a simple sugar can phenotypically revert the mutant phenotype.

8. Develop assays for appropriate enzymes or transport activities that might provide the molecular basis of the mutant phenotype.

Discussion of Each Strategy

1. Lec^R and Lec^B Phenotype. Most mutants expressing a change in glycosylation pattern are resistant to the toxicity of certain plant lectins or show reduced binding of these lectins at the cell surface.[5,6] This is because a single alteration in carbohydrate biosynthesis (even one affecting the earliest steps in dolichyl-oligosaccharide biosynthesis) usually alters the array of mature carbohydrate structures expressed at the cell surface and thereby changes the ability of the cells to bind different plant lectins. The simplest way to decide whether a mutant is likely to have arisen from a primary defect in glycosylation is to determine its resistance to a panel of lectins of different carbohydrate-binding specificities.[6,7] The method for performing this cytotoxicity test has been described in detail.[7] We use five lectins in initial phenotype tests (P-tests): wheat germ agglutinin (WGA,

[3] J. S. Robertson, J. R. Etchison, and D. F. Summers, *J. Virol.* **19,** 871 (1976).
[4] J. K. Rose and C. J. Gallione, *J. Virol.* **39,** 519 (1981).
[5] P. Stanley, V. Caillibot, and L. Siminovitch, *Cell (Cambridge, Mass.)* **6,** 121 (1975).
[6] P. Stanley, *Mol. Cell. Biol.* **5,** 923 (1985).
[7] P. Stanley, this series, Vol. 96, p. 157.

binds sialic acid and GlcNAc moieties[8-10]); concanavalin A (Con A, binds biantennary, N-linked moieties[11,12]); the leukoagglutinin from *Phasedus vulgaris* [L-PHA, binds branched, N-linked moieties with a $\beta(1,6)$-linked GlcNAc residue[13,14]]; ricin, the toxin from *Ricinus communis* (RIC, binds carbohydrates with terminal Gal and GalNAc residues[15]); and lentil lectin [LCA, binds biantennary and certain triantennary N-linked moieties with an $\alpha(1,6)$-linked fucose residue in the core region[16,17]]. If a cell line is resistant or hypersensitive to more than one of these lectins, it is almost certainly a glycosylation mutant. If it is resistant to only one lectin in this group, other lectins of different carbohydrate-binding specificity[18] should be tested. Finally, the lectin-binding properties of the cells should be examined directly. Decreased lectin binding, even of only one lectin, if observed at 4° under conditions where internalization is inhibited, is highly indicative of a cell surface glycosylation change.

The advantages of determining the Lec^R and/or Lec^B properties of a glycosylation mutant are greater than merely obtaining indirect evidence for altered glycosylation. Many CHO glycosylation mutants express characteristic patterns of cross-resistance and hypersensitivity and can be classified on that basis.[5,6] This information is critical to deciding whether a novel glycosylation phenotype is being studied or whether a previously characterized mutant has been reisolated.[6] It is also helpful in planning selection protocols specifically designed to reisolate new mutants.[7] The latter is necessary so that biochemical characterization of independently derived mutants of identical phenotype can be correlated.

2. Complementation Analysis. Even if a glycosylation mutant appears novel in its Lec^R or Lec^B properties, it is important to confirm this by complementation analysis. Novel phenotypes might arise because the mutant carries more than one mutation, as recently reported for the CHO mutant C1 which carries both Lec1 and Lec2 mutations.[6] Biochemical characterization of this line would have been complicated and even a

[8] A. K. Allen, A. Neuberger, and N. Sharon, *Biochem. J.* **131**, 155 (1973).

[9] C. S. Wright, *J. Mol. Biol.* **141**, 267 (1980).

[10] K. A. Kronis and J. P. Carver, *Biochemistry* **21**, 3050 (1982).

[11] S. Ogata, T. Muramatsu, and A. Kobata, *J. Biochem. (Tokyo)* **78**, 687 (1975).

[12] J. U. Baenziger and D. Fiete, *J. Biol. Chem.* **254**, 2400 (1979).

[13] R. D. Cummings and S. Kornfeld, *J. Biol. Chem.* **257**, 11230 (1982).

[14] S. Hammarström, M.-L. Hammarström, G. Sundbland, J. Arnarp, and J. Lonngren, *Proc. Natl. Acad. Sci. U.S.A.* **79**, 1611 (1982).

[15] J. U. Baenziger and D. Fiete, *J. Biol. Chem.* **254**, 9795 (1979).

[16] H. Debray, D. Decout, G. Strecker, G. Spik, and J. Montreuil, *Eur. J. Biochem.* **117**, 41 (1981).

[17] K. Kornfeld, M. L. Reitman, and R. Kornfeld, *J. Biol. Chem.* **256**, 6633 (1981).

[18] I. J. Goldstein and C. E. Hayes, *Adv. Carbohydr. Chem. Biochem.* **35**, 127 (1978).

waste of time, if it was being performed in the belief that a new glycosylation lesion was to be uncovered.

The methods for performing complementation analysis of CHO glycosylation mutants have been described in detail.[7] Briefly, the new isolate is fused with parental cells and the Lec[R] phenotype of the hybrids determined. If the hybrids behave like parental cells, the mutation is designated recessive. The cell line is subsequently fused with all other recessive CHO mutants of related phenotype. Only if it complements all of them does it represent a new recessive glycosylation mutation.

Mutations that behave dominantly may also be identified by the fact that hybrids formed between parental cells and the mutant behave like the mutant. Dominant phenotypes cannot, however, be classified into complementation groups. Knowledge of the dominance properties of a glycosylation phenotype is helpful in deciding whether the mutant arose from the loss of a functional activity (predicted to behave recessively) or the acquisition of a new activity (predicted to behave dominantly). For example, LEC10, LEC11, and LEC12 CHO mutants behave dominantly in somatic cell hybrids[5,19] and all appear to be due to the expression of novel glycosyltransferases not detected in parental CHO (Table I). Lec9 CHO cells possess an apparently dominant property in that they synthesize an increased proportion of branched carbohydrates (Table I). However, the Lec9 phenotype is known to be recessive,[19] and therefore most probably arises from an enzymatic loss. The dolichyl-oligosaccharides synthesized by Lec9 cells revealed an increased proportion of Man_5Gn_2-P-P-Dol with no concomitant reduction in Dol-P-Man synthetase, indicating that another glycosylation enzyme might be defective in Lec9 cells. Presumably this defect is the primary cause of the increased proportion of branched carbohydrates expressed on Lec9/VSV. Knowledge of the dominance properties of a phenotype is necessary, therefore, to ensure that the search for the primary enzymatic basis of a glycosylation change is broad enough and that the final outcome conforms with the dominance behavior of the phenotype.

3. Initial Studies with VSV. Once it seems clear that a novel glycosylation mutant is in hand, it is usually best to determine the precise change(s) in carbohydrate structure expressed by the mutant. This information identifies the sugar(s) affected by the mutation and provides a focus for the search for an enzyme defect.

Our initial studies of a putative glycosylation mutant involve a comparison by SDS–polyacrylamide gel electrophoresis (SDS–PAGE) of radioactively labeled vesicular stomatitis virus (VSV) grown in mutant and

[19] P. Stanley, *Somatic Cell Genet.* **9**, 593 (1983).

parental cells. Among the proteins encoded by VSV is a single glycoprotein termed G that has two Asn-X-Thr glycosylation sites for N-linked complex carbohydrates.[3,4] When VSV is grown in CHO cells in suspension culture, the carbohydrates associated with G are mainly biantennary, complex structures ($\sim$70–80%) and a triantennary structure with a $\beta(1,6)$-linked GlcNAc branch.[20] G from VSV grown in different CHO glycosylation mutants has been shown to exhibit changes in molecular weight as well as changes in the spectrum of radiolabeled sugars incorporated. For example, G from VSV grown in Lec1 CHO mutants carries the pentamannosyl, N-linked, processing intermediate (Man$_5$Gn$_2$Asn) instead of complex carbohydrates.[21] It consequently migrates faster in SDS gels (due to reduced molecular weight) and, when grown in radioactive sugars, does not incorporate Gal or Fuc but does incorporate Man and GlcNAc into mature G. Thus, from a few simple labeling experiments using only SDS–PAGE analysis, it could be deduced that Lec1 cells synthesized truncated carbohydrates lacking the terminal sugars typical of complex moieties.

Although all glycosylation mutants studied to date synthesize altered N-linked carbohydrates that are expressed on viral glycoproteins, these changes do not always result in detectable changes in the molecular weight of glycoproteins on SDS gels nor in significantly reduced incorporation of specific sugars. For example, Lec1A CHO cells are partially defective in GlcNac-TI activity and consequently do not complete the processing of Man$_5$Gn$_2$Asn to complex moieties.[22] Although G from Lec1A/VSV carries oligomannosyl carbohydrates as well as complex moieties, it does not migrate significantly faster in 10% SDS gels and incorporates the full spectrum of radioactive sugars.[22] However, because Lec1A mutants were known to fall into complementation group 1[19] and to exhibit reduced GlcNAc-TI activity,[22] the glycopeptides from Lec1A/VSV G were investigated. Gradient elution from Con A–Sepharose allowed detection of the expected carbohydrate structural changes.[22] Clearly, a negative result in the initial comparison of viral glycoproteins by SDS–PAGE provides no proof for the *lack* of a carbohydrate change and should not preclude further investigation of the carbohydrates by other techniques.

The reason for choosing VSV to monitor glycosylation changes is that it expresses only a small subset of the carbohydrates synthesized by CHO cells. This is both an advantage and a disadvantage—the small number of structures to be characterized enables rapid identification of changes in

[20] P. Stanley, G. Vivona, and P. H. Atkinson, *Arch. Biochem. Biophys.* **230**, 363 (1984).
[21] M. A. Robertson, J. R. Etchison, J. S. Robertson, D. F. Summers, and P. Stanley, *Cell* (*Cambridge, Mass.*) **13**, 515 (1978).
[22] P. Stanley and W. Chaney, *Mol. Cell. Biol.* **5**, 1204 (1985).

N-linked carbohydrate biosynthesis but the analysis is confined to the N-linked pathway. Changes specific to O-linked structures, glycolipids, or proteoglycans will not be detected using the virus. However, changes affecting multiple pathways (e.g., nucleotide-sugar biosynthesis or transport) will be picked up by the VSV approach. Other viruses may also be useful in initial studies of a glycosylation mutant. Sindbis virus possesses two glycoproteins (E1 and E2) which both carry N-linked carbohydrates[23] at various stages of maturation depending on the host cell.[24] Coronavirus[25] and herpesvirus[26] are of particular interest because they possess glycoprotein(s) with O-linked carbohydrates as well as those with N-linked moieties. Both Sindbis[27,28] and herpesvirus[29,30] have been used in analyzing structural carbohydrate changes expressed by glycosylation mutants.

As yet, no virus-induced glycosylation artifacts leading to erroneous conclusions of a glycosylation defect have been reported. However, this is a possibility which must always be borne in mind. Therefore, carbohydrate changes identified using viral glycoproteins should be generally confirmed for cellular glycoproteins. In fact, several laboratories prefer to investigate the mass of cellular glycoproteins in initial, indirect carbohydrate structural studies.[31-35] Although this approach has proved successful, the complexity of the array of carbohydrates associated with total cellular carbohydrates is such that several fractionation steps are required to separate the different classes of carbohydrates (complex, oligomannosyl, simple, proteoglycans, glycolipids) before any single class can be studied. Li and Kornfeld[36-38] have described some of the carbohydrates of CHO cell glycoproteins and it is apparent that the total number of micro-

[23] J. Hakimi, J. Carver, and P. H. Atkinson, *Biochemistry* **20**, 7314 (1981).
[24] P. Hsieh, M. R. Rosner, and P. W. Robbins, *J. Biol. Chem.* **258**, 2548 (1983).
[25] H. Niemann and H. D. Klenk, *J. Mol. Biol.* **153**, 993 (1981).
[26] S. Oloffson, J. Blomberg, and E. Lycke, *Arch. Virol.* **70**, 321 (1981).
[27] S. Schlesinger, C. Gottlieb, P. Feil, N. Gelb, and S. Kornfeld, *J. Virol.* **17**, 239 (1976).
[28] C. Gottlieb, S. Kornfeld, and S. Schlesinger, *J. Virol.* **29**, 344 (1979).
[29] G. Campadelli-Fiume, L. Poletti, F. Dall'Olio, and F. Serafini-Cessi, *J. Virol.* **43**, 1061 (1982).
[30] F. Serafini-Cessi, F. Dall'Olio, M. Scannavini, and G. Campadelli-Fiume, *Virology* **131**, 59 (1983).
[31] M. L. Reitman, I. S. Trowbridge, and S. Kornfeld, *J. Biol. Chem.* **257**, 10357 (1982).
[32] R. D. Cummings, I. S. Trowbridge, and S. Kornfeld, *J. Biol. Chem.* **257**, 13421 (1982).
[33] J. W. Dennis, J. P. Carver, and H. Schachter, *J. Cell Biol.* **99**, 1034 (1984).
[34] R. D. Cummings and S. Kornfeld, *J. Biol. Chem.* **259**, 6253 (1984).
[35] R. C. Hughes, G. Mills, and D. Stojanovic, *Carbohydr. Res.* **120**, 215 (1983).
[36] E. Li and S. Kornfeld, *J. Biol. Chem.* **253**, 6426 (1978).
[37] E. Li and S. Kornfeld, *J. Biol. Chem.* **254**, 1600 (1979).
[38] E. Li, R. Gibson, and S. Kornfeld, *Arch. Biochem. Biophys.* **199**, 393 (1980).

heterogeneous structures present in Pronase digests of CHO cells might number in the hundreds. The virus approach selects a comparatively small subset of these, goes a long way toward purifying them and, in effect, amplifies them because of the large amount of virus which is produced by an infected cell. Since all of the CHO mutants we have studied so far express their respective glycosylation defects on N-linked glycoproteins[1] (Table I), we have been able to use the VSV G glycoprotein to localize carbohydrate structural changes in each case.

4. Pronase Glycopeptides. To obtain detailed carbohydrate structural information, G from VSV grown in mutant cells is exhaustively digested with Pronase and the radiolabeled glycopeptides are analyzed on lectin affinity columns and on BioGel columns before and after selective glycosidase treatments. The carbohydrates at each of the two glycosylation sites may be investigated separately by preparing tryptic peptides and fractionating them by DEAE-ion exchange chromatography[3] or HPLC.[24]

Lectins are exquisitely specific carbohydrate-binding proteins.[18] The best characterized lectins are from plants, and knowledge about their carbohydrate structural requirements for high-affinity binding is continually evolving as different carbohydrates are investigated. At this time, it is clear that carbohydrates of defined and complex structure exhibit characteristic binding properties for Sepharose-conjugated lectins and that unknown carbohydrates may be partially characterized on the basis of their elution properties from different lectin columns. This approach was first exploited by Ogata *et al.*[11] and Krusius[39] to identify the requirements for glycopeptide binding to Con A. Subsequently it has been expanded by many laboratories to the point where a battery of lectin affinity columns can now be used to obtain a large amount of indirect structural information and to fractionate complex mixtures.[40] Although this approach has been extremely helpful in identifying the altered carbohydrates synthesized by glycosylation mutants, data must be interpreted with caution. For example, the presence of a bisecting GlcNAc residue in a biantennary carbohydrate will affect its binding to Con A very differently depending on the degree of completion of the $\beta(1,2)$-linked GlcNAc branches and on the commercial source of Con A.[41–43] The relative affinities of particular structures for specific lectins may alter with changes in column volume, column length, the temperature of elution, and the flow rate, as well as the

[39] T. Krusius, *FEBS Lett.* **66,** 86 (1976).
[40] R. D. Cummings and S. Kornfeld, *J. Biol. Chem.* **257,** 11235 (1982).
[41] S. Narasimhan, *J. Biol. Chem.* **257,** 10235 (1982).
[42] P. A. Gleeson and H. Schachter, *J. Biol. Chem.* **258,** 6162 (1983).
[43] C. Campbell and P. Stanley, *J. Biol. Chem.* **261,** 13370 (1984).

source of the lectin conjugate[40–45] (unpublished observations). Therefore, although general guidelines may be drawn concerning the binding behavior of certain structures for different lectin columns,[40] these are subject to experimental conditions and therefore, lectin column binding properties do not suffice as proof of carbohydrate structure.

Lectin affinity chromatography is, however, an excellent empirical approach for the separation of carbohydrates on the basis of their fine structure. Species missing from or acquired by glycosylation mutants are readily identified and a strategy for their simple, essentially single-step, purification is developed. Comparisons of CHO glycosylation mutant- and parent-derived G glycopeptide profiles on Con A-sepharose,[20,22,43,45,46] PSA-agarose,[20,46] L-PHA-agarose,[22] RCA$_{II}$-agarose,[43] or E-PHA-agarose[43] have identified novel structures. Mutant profiles vary from an altered proportion of radioactivity in a particular peak to the complete absence of a peak or the appearance of a new peak. Because the behavior of CHO/VSV G glycopeptides on several lectin columns is well characterized and the carbohydrate structures of CHO/VSV G have been determined by ^{1}H NMR spectroscopy,[20] even small changes in elution are significant and worthy of pursuit.[46] The basic strategy is to continue chromatographing individual species sequentially on lectin columns of different carbohydrate binding specificities until no further fractionation is achieved. Once a glycopeptide has been purified by lectin affinity chromatography, it is usually pure enough for structural analysis by ^{1}H NMR spectroscopy.

Chromatography on BioGel columns has also been useful in obtaining indirect structural information on the carbohydrates synthesized by CHO glycosylation mutants. High-resolution columns of BioGel-P4 separate the complex sialylated glycopeptides from each other and from oligomannosyl oligosaccharides (released by endoglycosidase H treatment). Sequential exoglycosidase treatments can be used to determine the terminal sugars on each glycopeptide. With this approach, a subtle decrease in the synthesis of branched sialylated moieties was observed for Lec4/VSV glycopeptides[46] and the appearance of Man$_5$Gn$_2$Asn among the glycopeptides of Lec1A/VSV was documented.[22] In addition, sequential glycosidase digestions were used to structurally characterize the Man$_5$Gn oligosaccharide released by endoglycosidase H from Lec1/VSV.[21] However,

[44] H. Schachter, S. Narasimhan, P. Gleeson, and G. Vella, *Can. J. Biochem.* **61,** 1049 (1983).

[45] C. Campbell and P. Stanley, *Cell (Cambridge, Mass.)* **35,** 303 (1983).

[46] P. Stanley, *Arch. Biochem. Biophys.* **219,** 128 (1982).

because this approach is indirect, more direct structural information is sought by ^{1}H NMR spectroscopy.

5. *^{1}H NMR Spectroscopy.* The development of ^{1}H NMR spectroscopy for the determination of complex carbohydrate structures has occurred over the last 10 years.[47-49] More than a hundred different carbohydrate structures have been assigned and provide well-characterized model compounds for comparison with unknowns. Provided that about 200 μg (more if possible) of a purified carbohydrate is available, ^{1}H NMR spectroscopy at 500 MHz can provide the composition, molar quantities, and linkages of most, if not all, sugars in the carbohydrate and, based on this information, a complete structure can often be deduced. If all of the spectral parameters of an unknown are not accounted for by published information for model compounds, most of the structure is usually deducible and remaining resonances can be assigned by selective glycosidase treatments. Using ^{1}H NMR spectroscopy of VSV G glycopeptides from CHO glycosylation mutants we have shown that Lec4 CHO mutants do not synthesize triantennary N-linked moieties with a $\beta(1,6)$-GlcNAc branch[20]; LEC10 mutants attach a bisecting GlcNAc to biantennary carbohydrates[43]; Lec1A mutants express parental-type biantennary moieties but, in addition, exhibit significant quantities of the Man_5Gn_2Asn processing intermediate[22]; and LEC11 and LEC12 mutants add $\alpha(1,3)$-linked fucose residues to biantennary carbohydrates (P. Stanley, unpublished observations). A major advantage of ^{1}H NMR spectroscopy is that it is nondestructive and therefore the characterized carbohydrate is available for further studies. A disadvantage is that reasonable amounts ($\sim$0.05 μmol) of carbohydrate are required to obtain a good spectrum.

Alternative procedures for direct structural determinations of complex carbohydrates such as fast atom bombardment (FAB) in combination with mass spectroscopy are being developed.[50] This approach, although destructive, requires much less carbohydrate and therefore is an attractive alternative to NMR for carbohydrates in short supply. Complete structural information can also be obtained by detailed analysis of radiolabeled carbohydrates using highly specific glycosidase enzymes and separation of glycopeptides or oligosaccharides on high-resolution columns of BioGel P-4 calibrated with markers of known structure. This approach has been developed to a high level of sophistication by Kobata and col-

[47] J. P. Carver and A. A. Grey, *Biochemistry* **20,** 6607 (1981).

[48] J. F. G. Vliegenthart, L. Dorland, and H. van Halbeek, *Adv. Carbohydr. Chem. Biochem.* **41,** 209 (1983).

[49] E. F. Hounsell, D. J. Wright, A. S. R. Donald, and J. Feeney, *Biochem. J.* **223,** 129 (1984).

[50] M. N. Fukuda, A. Dell, J. E. Oates, P. Wu, J. C. Klock, and M. Fukuda, *J. Biol. Chem.* **260,** 1067 (1985).

leagues[51] and is, in many respects, the most practical way to characterize glycopeptides present in minute quantities on cellular glycoconjugates.[52] Briefly, N-linked carbohydrates are released from glycoproteins by hydrazinolysis and labeled to high specific activity by NaB^3H_4. Acidic and neutral species are separated by paper electrophoresis and each peak of radioactivity is analyzed on BioGel P-4 before and after sequential exoglycosidase treatments. The number of sugars removed and their specific linkages are deduced from the specificity of the exoglycosidase and the degree to which it changes the elution behavior of the oligosaccharide on BioGel P-4. Although the technology involved is very simple, the problem for the novitiate is acquiring the battery of exo- and endoglycosidases for sequential digestion of oligosaccharides and the authentic oligosaccharide markers of known structure for calibrating BioGel P-4 columns and characterizing the specificity of glycosidase enzymes. Although many glycosidases are now available commercially, there are several which are not, and these must be purified from appropriate sources and subsequently characterized.

6. *Extent of a Glycosylation Lesion.* The extent of a carbohydrate structural defect must be determined by examining the degree to which the carbohydrate change(s) observed for N-linked moieties affect different glycoconjugates. For example, Lec2[53] and Lec8[54] CHO mutants express truncated glycolipids as well as glycoproteins[53] because they are defective in the transport of CMP-SA (Lec2)[55] and UDP-Gal (Lec8)[56] into the Golgi. Presumably their O-linked carbohydrates and proteoglycans are similarly affected by the reduced availability of nucleotide sugars. Experiments with clones 1021 and 13 (which belong to the same complementation groups as Lec2 and Lec8, respectively[6]) showed no reduction in the levels of nucleotide sugars in these mutants and a compartmentalization defect was suggested to explain the phenotypes.[57]

Ideally, all of the glycoconjugates synthesized by the cell should be checked to determine which classes are affected. This is an arduous task, however, and usually not necessary. In practice, mutations that affect glycosylation reactions involved in several biosynthetic pathways can be

[51] K. Yamashita, T. Mizuochi, and A. Kobata, this series, Vol. 83, p. 105.
[52] K. Yamashita, T. Ohkura, Y. Tachibana, S. Takasachi, and A. Kobata, *J. Biol. Chem.* **259,** 10834 (1984).
[53] P. Stanley, T. Sudo, and J. P. Carver, *J. Cell Biol.* **85,** 60 (1980).
[54] P. Stanley, *ACS Symp. Ser.* **128,** 213 (1980).
[55] S. L. Deutscher, N. Nuwayhid, P. Stanley, E. B. Briles, and C. B. Hirschberg, *Cell (Cambridge, Mass.)* **39,** 295 (1984).
[56] S. L. Deutscher and C. B. Hirschberg, *J. Biol. Chem.* **261,** 96 (1986).
[57] E. B. Briles, E. Li, and S. Kornfeld, *J. Biol. Chem.* **252,** 1107 (1977).

detected by examining, in addition to N-linked carbohydrates, the spectrum of glycolipids synthesized by the cell. The major glycolipid of CHO cells is G_{M3}[53,54,57,58] so mutations affecting the addition of sialic acid, galactose, or glucose will affect the amounts of G_{M3} synthesized. By investigating the N-linked carbohydrates of G and the spectrum of glycolipids, lesions affecting the addition of all sugars except N-acetylgalactosamine (GalNAc) should be detected. To monitor GalNAc addition, it would be necessary to characterize O-linked carbohydrates as recently described by Conzelman and Kornfeld.[59] Proteoglycans can be labeled and partially characterized by chromatography on DEAE-Sepharose before and after treatment with specific glycosidases.[60,61]

7. Correction of a Mutant Phenotype. The mutants PL^{R}1.3, Lec13, and Lec13A do not add normal amounts of fucose to N-linked carbohydrates (and presumably to other glycoconjugates) because they cannot synthesize GDP-Fuc from GDP-Man (Table I). Since this is a synthetic and not a compartmentalization problem, the fucose deficiency can be corrected by the addition of fucose to the medium. Fucose is converted to GDP-Fuc by the fucose salvage pathway[62] and thereby "reverts" the mutant phenotype. Since this experiment is very easy to do, it is worth surveying simple sugars for their ability to correct a mutant phenotype whenever a general glycosylation defect is suspected. In fact, investigation of this point at an early stage (e.g., during characterization of lectin resistance) is probably useful since growth in the appropriate sugar might be all that is required to "revert" a variety of lesions that inhibit the biosynthesis of nucleotide sugars.

8. Development of Glycosylation Assays. Once a particular enzyme or transport reaction is identified as the putative basis of a glycosylation phenotype, an assay to compare the properties of parent and mutant cells for that activity must be established. It can be seen from Table I that the pertinent assay might be to measure nucleotide-sugar formation or localization, or the maturation of dolichyl-oligosaccharides, or glycosyltransferase or glycosidase specific activities. Other lesions from pathways more distantly related to carbohydrate biosynthesis might also be the cause of an altered glycosylation phenotype. For example, mutations affecting nucleotide metabolism might alter the availability of nucleotide sugars, mutations affecting lipid metabolism might alter the location or

[58] G. Yogeeswaran, R. K. Murray, and J. A. Wright, *Biochem. Biophys. Res. Commun.* **56**, 1010 (1974).

[59] A. Conzelmann and S. Kornfeld, *J. Biol. Chem.* **259**, 12528 (1984).

[60] A. G. Atherly, B. J. Barnhart, and P. M. Kraemer, *J. Cell. Physiol.* **89**, 375 (1977).

[61] B. J. Barnhart, S. H. Cox, and P. M. Kraemer, *Exp. Cell Res.* **119**, 327 (1979).

[62] P. D. Yurchenco, C. Ceccarini, and P. H. Atkinson, this series, Vol. 50, p. 175.

activity of membrane-bound enzymes or lipid-linked intermediates, mutations affecting intracellular trafficking might affect the compartmentalization of glycosyltransferase or glycosidase enzymes, and mutations affecting protein synthesis might alter the availability of glycosylation enzymes. The methods to be used will depend, therefore, on the specific sugar under study and all of the reactions in which that sugar might be involved. It is beyond the scope of this article to decribe all these possible approaches. In general, an empirical approach is required that takes into account all the preliminary information on the mutant and the published literature describing the successful characterization of previously isolated glycosylation-defective types.

Alternative Strategies

Alternative approaches to deciding whether a new isolate is glycosylation defective include determination of the sugar composition of crude cellular membranes[57,63] although it is unlikely that subtle changes would be detected by this approach; labeling of available surface carbohydrates by galactose oxidase and NaB^3H_4 before and after neuraminidase treatment and analysis on SDS–PAGE to detect gross changes in galactyosylation and sialylation of membrane glycoproteins[46,53,64]; analysis of glycolipid composition by thin-layer chromatography to detect mutants making truncated glycolipids[53,54,57]; and, as mentioned previously, measurement of the incorporation of sugars into glycoproteins and dolichyl-oligosaccharides to reveal mutants with general glycosylation defects.[65–67] All of these approaches provide evidence for changes in carbohydrate biosynthesis, but often do not identify the class of glycoconjugates affected.

Summary

Glycosylation mutants are complicated to characterize because of the many different metabolic pathways involved in the biosynthesis of complex carbohydrates. After establishing that a new isolate expresses a novel phenotype consistent with a change in carbohydrate biosynthesis and that it does not belong to a previously characterized class of glycosylation mutant, it is best to determine the nature of the carbohydrate structural change(s) expressed by the mutant. Knowledge of the breadth of

[63] C. Gottlieb, A. M. Skinner, and S. Kornfeld, *Proc. Natl. Acad. Sci. U.S.A.* **71,** 1078 (1974).
[64] R. L. Juliano and P. Stanley, *Biochim. Biophys. Acta* **389,** 401 (1975).
[65] S. S. Krag, M. Cifone, P. W. Robbins, and R. M. Baker, *J. Biol. Chem.* **252,** 3561 (1977).
[66] A. Tenner, J. Zieg, and I. E. Scheffler, *J. Cell. Physiol.* **90,** 145 (1977).
[67] A. J. Tenner and I. E. Scheffler, *J. Cell. Physiol.* **98,** 251 (1979).

these changes (i.e., the number of classes of glycoconjugates affected), the ability of the mutant phenotype to be "reverted" by growth in a simple sugar, and the dominance properties of the mutant help to narrow the focus in searching for the molecular basis of the glycosylation change. Defining the latter requires developing the appropriate assay to measure a glycosyltransferase or glycosidase activity, synthesis of dolichyl-oligo-saccharide precursors of N-linked carbohydrates, synthesis of nucleotide sugars, or their transport into the correct intracellular compartment. When a defect which appears to provide the basis of a mutant phenotype has been identified, it should be confirmed in an independently derived mutant that exhibits the same general properties and belongs to the same complementation group.

Acknowledgments

The writing of this article was supported by a grant from the National Cancer Institute (R01 36434) and faculty awards from the American Cancer Society, and the Irma T. Hirschl Trust. Thanks are extended to William Chaney for comments on the manuscript. Partial support from Core Cancer Grant 3P0 CA13330 is also acknowledged.

[37] Isolation of Mutant Chinese Hamster Ovary Cells Defective in Endocytosis

By APRIL R. ROBBINS and CALVIN F. ROFF

Receptor-mediated endocytosis involves the following events: ligand (e.g., diphtheria toxin,[1] acid hydrolases,[2] epidermal growth factor,[3] trans-ferrin[4]) binds to its receptor on the cell surface; the receptor–ligand complex moves rapidly into a specialized region of the membrane—the clathrin coated pit; the coated pit seals, detaches from the plasma membrane, and plunges into the cytosol. The lumen of this endocytic vesicle (endosome) quickly becomes acidic[5] due to pumping of protons inward[6];

[1] J. H. Keen, F. R. Maxfield, M. C. Hardegree, and W. H. Habig, *Proc. Natl. Acad. Sci. U.S.A.* **79**, 2912 (1982).
[2] M. C. Willingham, I. H. Pastan, G. G. Sahagian, G. W. Jourdian, and E. F. Neufeld, *Proc. Natl. Acad. Sci. U.S.A.* **78**, 6697 (1981).
[3] H. T. Haigler, J. A. McKanna, and S. Cohen, *J. Cell Biol.* **81**, 382 (1979).
[4] M. Karin and B. Mintz, *J. Biol. Chem.* **256**, 3245 (1981).
[5] B. Tycko and F. R. Maxfield, *Cell (Cambridge, Mass.)* **28**, 643 (1982).
[6] C. J. Galloway, G. E. Dean, M. Marsh, G. Rudnick, and I. Mellman, *Proc. Natl. Acad. Sci. U.S.A.* **80**, 3334 (1983).

METHODS IN ENZYMOLOGY, VOL. 138

acidification causes some ligands to be released from their receptors (e.g., acid hydrolases[7]), others to fuse with the vesicle membrane and thus penetrate into the cytosol (e.g., diphtheria toxin,[8]) and still another (transferrin[9,10]) to release its bound iron. Following this, segregation of endosomal contents occurs[11]: free ligands and some ligand–receptor complexes (e.g., epidermal growth factor receptor) are sorted into a compartment setting off toward lysosomes; unoccupied receptors and at least one type of ligand–receptor complex (apotransferrin receptor) are sorted into a compartment leading back to the cell surface. The number and nature of intermediate compartments involved in receptor recycling and ligand delivery are unknown.

Evidence suggests that the Golgi/GERL apparatus is, among other things, an endocytic compartment. Endocytosed ligand has been found in or near the Golgi/GERL.[12] Internalized receptor has been shown to pass through the late or trans regions of this organelle on the way back to the plasma membrane.[13] In addition, newly synthesized acid hydrolases are segregated from other soluble glycoproteins in the Golgi/GERL,[14] and launched, in coated vesicles,[15] to meet with endocytosed ligands at or en route to lysosomes.

Biochemical genetics, i.e., isolation and analysis of mutants, should be a powerful tool in dissection of a process as complex as that of receptor-mediated endocytosis, and in determination of the compartments and components involved. In this article we present methods for obtaining what should eventually be a large group of endocytosis mutants with a wide variety of genetic defects. The two classes of mutants currently identified support the possibility of an intimate relationship between the Golgi/GERL and endocytosis: in both groups of mutants we observe al-

---

[7] A. Gonzalez-Noriega, J. H. Grubb, V. Talkad, and W. S. Sly, *J. Cell Biol.* **85,** 839 (1980).

[8] R. K. Draper and M. I. Simon, *J. Cell Biol.* **87,** 849 (1980).

[9] A. Dautry-Varsat, A. Ciechanover, and H. F. Lodish, *Proc. Natl. Acad. Sci. U.S.A.* **80,** 2258 (1983).

[10] R. D. Klausner, G. Ashwell, J. van Renswoude, J. B. Harford, and K. R. Bridges, *Proc. Natl. Acad. Sci. U.S.A.* **80,** 2263 (1983).

[11] H. J. Geuze, J. M. Slot, G. J. A. M. Strous, H. F. Lodish, and A. L. Schwartz, *Cell (Cambridge, Mass.)* **32,** 277 (1983).

[12] M. C. Willingham and I. H. Pastan, *J. Cell Biol.* **94,** 207 (1982).

[13] M. D. Snider and O. C. Rogers, *J. Cell Biol.* **100,** 826 (1985).

[14] H. J. Geuze, J. W. Slot, G. J. A. M. Strous, A. Hasilik, and K. von Figura, *J. Cell Biol.* **98,** 2047 (1984).

[15] C. H. Campbell, R. E. Fine, J. Squicciarini, and L. H. Rome, *J. Biol. Chem.* **258,** 2628 (1983).

terations in glycosylation activities associated with the *trans* Golgi/ GERL.[16,17]

Isolation of Mutants

To obtain mutants altered in endocytosis at steps subsequent to ligand binding, we isolate cells defective in endocytosis of two ligands, entry of which is dependent on unrelated receptors.[18] Use of a single ligand results in a predominance of mutants defective in the specific receptor. We employ as first ligand a toxin in a single round of selection. Surviving cells are grown to colonies which are screened *in situ* for decreased accumulation of radiolabeled second ligand, lysosomal hydrolases bearing the Man-6-P recognition marker. Others have isolated endocytosis mutants by simultaneous selection with ricin and *Pseudomonas* toxin,[19] or with diphtheria toxin plus modeccin,[20] and by selection with *Pseudomonas* toxin followed by screening of cloned survivors for toxin cross-resistance.[21]

Materials. Ethyl methanesulfonate (Eastman Kodak Co.); diphtheria toxin (List Biological Laboratories, Inc.); modeccin (Pierce); polyester-PeCap HD7-17 (TETKO, Inc.); Autofluor (National Diagnostics Institute, Inc.); polyethylene glycol 1000 (J. T. Baker Chemical Co. and Research Products International). Ricin and *Pseudomonas* toxin were generous gifts of Drs. Richard J. Youle and Stephen Leppla, respectively. Other reagents were from standard commercial suppliers.

Cells. Chinese hamster ovary (CHO) cells offer several advantages for biochemical genetics. They are pseudodiploid and show reasonable karyotypic stability for an immortal cell line, yet they act as "functional hemizygotes," i.e., recessive mutants are observed at high frequency 10^{-5} to 10^{-7}.[22] CHO cells grow rapidly (doubling time of the parent is 20 hr at 34°) and their efficiency of plating is >90%; thus mutants are readily cloned from single cells. The many CHO cell mutants which have been isolated and analyzed by others[23] are very useful in both genetic and phenotypic analyses of new mutants. As parental cell line we use WTB,[24]

[16] A. R. Robbins, C. Oliver, J. L. Bateman, S. S. Krag, C. J. Galloway, and I. Mellman, *J. Cell Biol.* **99**, 1296 (1984).
[17] C. F. Roff, R. Fuchs, I. Mellman, and A. R. Robbins, *J. Cell Biol.* **103** (1986).
[18] A. R. Robbins, S. S. Peng, and J. L. Marshall, *J. Cell Biol.* **96**, 1064 (1983).
[19] B. Ray and H. C. Wu, *Mol. Cell. Biol.* **2**, 535 (1982).
[20] M. H. Marnell, L. S. Mathis, M. Stookey, S.-P. Shia, D. K. Stone, and R. K. Draper, *J. Cell Biol.* **99**, 1907 (1984).
[21] J. M. Moehring and T. J. Moehring, *Infect. Immun.* **41**, 998 (1983).
[22] L. Siminovich, *Cell (Cambridge, Mass.)* **7**, 1 (1976).
[23] P. Stanley, this series, Vol. 96, p. 157.
[24] L. H. Thompson and R. M. Baker, *Methods Cell Biol.* **6**, 209 (1973).

TABLE I
CALENDAR FOR MUTANT ISOLATION

Day	Step
1	Mutagenize
2	Remove mutagen, replate
7	Add toxin for selection
8	Remove toxin, replate
9	Add polyester disks for replicas
19–23	Screen replicas
21–27	Pick colonies, amplify cells
35–45	Clone mutants
47–57	Pick clones, amplify cells

a subclone of CHO-K1,[25] which is available from the American Type Culture Collection, or WTB-111, a ouabain-resistant, thioguanine-resistant derivative of WTB.[16]

Cell Culture. Cells are grown in monolayer cultures in a humidified atmosphere at 34° under 5% CO_2 in Eagle's minimal essential medium supplemented with nonessential amino acids, penicillin, streptomycin, and 5% fetal bovine serum, unless otherwise indicated. For labeling with radioactive precursors dialyzed fetal bovine serum is substituted, and concentrations of glucose or amino acids are reduced as described in the appropriate sections. For both routine culture and experiments, cells are always kept in exponential growth. For storage, cells are harvested by trypsinization, centrifuged at 1,000 g at room temperature, resuspended to 1–2 × 10^6 cells/ml in medium containing 10% fetal bovine serum and 10% dimethyl sulfoxide, stored in aliquots in freezing vials (1 ml/vial) that are placed on ice immediately, then transferred to −70° as soon as possible. New cultures are initiated from frozen stocks every 2 months: a vial is thawed at 37°, the entire contents are added to 25 ml of medium plus 10% fetal bovine serum in a T-75 flask, then medium is replaced 2 hr later.

Mutagenesis. A calendar for the entire isolation procedure, beginning with mutagenesis, is given in Table I. Exponentially growing cultures (2 × 10^6 cells/T-75 flask) are incubated with ethyl methanesulfonate (200 μg/ml) in 15 ml of growth medium at 34°. The concentration of mutagen chosen should result in about 50% survival. After 18–24 hr (about one generation time) cells are harvested and replated. Cells are grown for 5 days between mutagenesis and selection to allow phenotypic expression. To increase the possibility of obtaining independent as opposed to sister mutants, a number of flasks are treated with mutagen and the cells from

[25] F.-T. Kao and T. T. Puck, *Genetics* **55**, 513 (1967).

the various mutageneses are kept separate throughout the rest of the procedure.

Selection. Because selection is followed by easy and efficient screening procedures (assaying 10^4 colonies requires only 4 hr of work), we can apply minimal selective pressure, a safeguard against inadvertently killing any mutants of interest. Diphtheria toxin provides an example: mutants altered at the level of translation showed much greater toxin resistance than mutants affected in toxin penetration[26]; too stringent a selection would eliminate the latter class.

Diphtheria toxin, modeccin, or *Pseudomonas* exotoxin are used as selective agents at 100, 2, or 3000 ng/ml, respectively, for 24 hr. Dose–response curves had shown that 1–10 in 10^5 cells survived these conditions; inclusion of 10 mM NH$_4$Cl completely blocked toxicity, indicating that toxicity under selective conditions results only from molecules whose entry is dependent on an acidic endocytic compartment.

Nonconditional mutants. (It should be noted that the nonconditional mutants isolated in our laboratory actually were obtained in an unsuccessful search for temperature-sensitive mutants[18]; cells were kept at 39° for 5 hr prior to and 24 hr during toxin treatment, then were returned to 34°). Cells (1×10^7/ T-150 flask) are incubated with the selective agent for 24 hr, rinsed to remove free toxin, trypsinized, and replated on 100-mm dishes to obtain ≤300 colonies /dish. Because survival varies from experiment to experiment, especially with diphtheria toxin, we routinely plate serial dilutions ranging from $\frac{1}{16}$ to $\frac{1}{2}$ T-150 on a dish.

Temperature-sensitive (ts) mutants. Cells are shifted to the nonpermissive temperature (39°) for 3 hr prior to toxin addition. After 24 hr at 39° free toxin is removed and cells are plated on 100-mm dishes as above; *but,* before cells may be shifted back to the permissive temperature (34°), cell-associated toxin must be removed or rendered inactive. We accomplish this using any of three methods:

1. Diphtheria toxin is removed from the cells by rinsing three times, trypsinizing, then washing three times by centrifugation (1000 g) and resuspending, all in Dulbecco's phosphate-buffered saline (without divalent cations) containing 10 mg inositol hexaphosphate/ml.[27]

2. Penetration of all of the toxins at the permissive temperature is prevented by including 10 mM NH$_4$Cl in the growth medium for 7 days following selection.[20] We find that replating with NH$_4$Cl reduces the number of surviving colonies to 25%; therefore, appropriate adjustments in cell number should be made. NH$_4$Cl does not affect the efficiency of replica plating.

---

[26] T. J. Moehring and J. M. Moehring, *Cell (Cambridge, Mass.)* **11,** 447 (1977).
[27] R. B. Dorland, J. L. Middlebrook, and S. H. Leppla, *J. Biol. Chem.* **254,** 11337 (1979).

3. The cells on 100-mm dishes are returned to the nonpermissive temperature for 5 days to dilute the level of toxin per cell through cell division; this last method obviously excludes mutants with lethal defects (see "Comments" below).

Replica Plating. Within 24 hr of plating the medium is replaced with medium containing 10% fetal bovine serum. Individual disks of Whatman 50 filter paper[28] or stacks of disks of polyester-PeCap HD7-17,[29] all cut to fit within the 100-mm dishes, are added and weighted down with glass beads. (The advantages—quality of replicate colonies, ability to make multiple replicas simultaneously—of the latter material are such that we have abandoned replication on paper.) Because the top disk in the stack is unusable, one extra disk should be added in making multiple replicas; five polyesters per dish yield four excellent replicas. New samples of polyester on occasion have proved toxic to CHO cells; we recommend pretreating new polyester in HCl, just as is done in recycling the disks.[29] Medium is replaced every 4–5 days.

It is very important that cells be replicated for the optimal period. If the time is too short, colonies are too small to be assayed; if too long, colonies on the replicas appear as centerless rims. However, the optimal replication period varies with the procedure (temperature, number of replicas desired, inclusion of NH_4Cl). Therefore, it is useful to set up extra replicate stacks for the sole purpose of periodically monitoring colony size.

Screening for Mutants Defective in Man-6-P Uptake. Cells on the replicate disks are screened for uptake via the Man-6-P receptor 10–16 days after plating. Prior to screening, cells kept at 39° following selection are grown for 5 days at 34°, and cells treated with NH_4Cl following selection are kept in medium without the inhibitor for 5–7 days.

As a crude but effective source of radiolabeled ligand we use ammonia-induced secretions from parental CHO cells[18]: 1×10^7 cells/T-150 flask are washed three times with medium lacking methionine, then labeled for 12–16 hr with 10 ml of medium containing 1 mCi [^{35}S]methionine, 10 μg of nonradioactive methionine, and NH_4Cl, 10 mM. The medium is clarified by centrifugation at 1000 g, protein is precipitated with ammonium sulfate (0.436 g/ml of medium), and the precipitate is dissolved in 0.5 ml of dialysis buffer and dialyzed for 36 hr against four 1-liter changes of 0.15 M NaCl, 10 mM NaHPO$_4$, pH 6.5, all at 4°. Following dialysis the secretions are clarified by centrifugation at 40,000 g for 20 min. One T-150 flask of WTB yields 1 ml of secretions containing 3×10^7

[28] J. D. Esko and C. R. H. Raetz, *Proc. Natl. Acad. Sci. U.S.A.* **75,** 1190 (1978).
[29] C. R. H. Raetz, M. M. Wermuth, T. M. McIntyre, J. D. Esko, and D. C. Wing, *Proc. Natl. Acad. Sci. U.S.A.* **79,** 3223 (1982).

cpm. About 10% of this can be internalized by recipient CHO cells, and 90–95% of this uptake is inhibited in the presence of 10 mM Man-6-P. Secretions from human diploid fibroblasts are unsatisfactory. Only 2–3% of the protein is internalized by recipient cells and more than half of this uptake is via a Man-6-P-independent route.

The replicas are removed from the master dishes (medium is replaced on the latter and they are returned to the incubator) and placed cell side up in 100-mm dishes containing growth medium. Colonies are visible both on the master dishes and as white dots on the polyester disks. In screening for nonconditional mutants the medium is simply replaced with 4 ml of medium containing 5×10^5 cpm/ml ^{35}S-labeled ammonia-induced secretions and dishes are incubated for 4 hr at 34° in the CO_2 incubator. In screening for temperature-sensitive mutants one disk from each replica stack is placed at 34°, one at 39°, for 3 hr prior to addition of ligand. Because the density of cells per colony decreases from bottom to top of the replica stack, the lower disk of a pair should be used for assay at the nonpermissive temperature. This reduces the possibility of false negatives. After incubation with the secretions the dishes are transferred to 4°, the disks are rinsed with cold serum-free medium containing bovine serum albumin (1 mg/ml), transferred to clean dishes (the less expensive bacteriological dishes may be used from here on), rinsed 3 times with cold Dulbecco's phosphate-buffered saline (without divalent cations), and then transferred to clean dishes containing cold 15% trichloroacetic acid. After 30 min the dishes are returned to room temperature, TCA is removed and colonies on the disks are stained by incubating for 1 hr in 50% methanol, 10% acetic acid, 10% TCA containing 0.5 mg Coomassie blue/ml, then destained by rinsing 3–4 times in 40% methanol, 10% acetic acid. For autoradiography the disks are rinsed 15 min in water then immersed in Autofluor for 2 hr, dried, and exposed to prefogged XR-2 film[30] for 16–24 hr at −70°. Because much of the stain is lost in Autofluor, we restain the colonies after ascertaining that the film is exposed appropriately.

Screening for Toxin Resistance. Toxin resistance is determined by assaying incorporation of [^{35}S]methionine into protein in colonies on the replicas. In screening for nonconditional mutants replicas are incubated for 2–3 hr in medium containing toxin at a concentration sufficient to inhibit protein synthesis in parental cells to 5–10% over this time period. (Generally this turns out to be the concentration used in selection. A quantitative assay for toxin sensitivity is given below.) In looking for temperature-sensitive mutants one replica is preincubated for 3 hr at 34°,

[30] R. A. Laskey and A. D. Mills, *Eur. J. Biochem.* **56,** 335 (1975).

one at 39°, prior to toxin addition. Cells are rinsed in methionine-free medium and labeled with [^{35}S]methionine (8 μCi/4 μg/4 ml/dish for 1 hr); protein is precipitated with TCA, disks are autoradiographed and stained, all as described in the preceding section.

On simultaneous screening of replicate colonies for toxin resistance and Man-6-P-dependent uptake, a mutant temperature-sensitive for endocytosis appears in all the possible combinations on autoradiography; i.e., with toxin at 39° it appears as a positive spot among negatives, at 34° as a negative among negatives; for Man-6-P uptake at 39° it appears as a negative spot among positives, at 34° as a positive among positives. Nonconditional mutants appear as positives in the first two assays and as negatives in the last two.

Purification of Mutants. Colonies of interest are detached from the master dishes using sterile, trypsin-soaked 0.5-cm disks cut from filter paper with a hole punch. (As insurance, especially for the novice, cells remaining on the master dish may be trypsinized, transferred to a T-75 flask, grown, and stored as a frozen stock. If ill befalls the colonies picked for purification, or if the wrong colonies are picked, this stock can be thawed, plated, and replicated, the replicas assayed, and a colony picked once again.) Each disk is placed in growth medium in a T-25 flask, and cells are grown and transferred to a T-75 flask. A poor yield of cells from the disk may necessitate spreading the resultant colonies in the T-25 flask by trypsinization and readdition of growth medium, so that cells will grow to a monolayer prior to transfer. Cells are grown, trypsinized, counted, diluted, and plated at 1.7 or 2.5 cells/ml (0.2 ml/well) in a 96-well cluster dish. Cells are placed in frozen stock at this time also. Wells are scored at both 6 and 10 days to identify single colonies. About 5 colonies are trypsinized and grown for testing. We find it useful to plate 200 cells on a 100-mm dish at the time we plate the 96-well tray. Cells on the dish are replicated and the replicas are screened for the original parameters. Autoradiograms from this screen are ready at the time one is picking the clones. If the wrong colony was picked the procedure can be aborted prior to picking, growing, and testing individual clones.

Frequencies. The choice of selective agent influences the frequency at which endocytosis mutants are obtained. Nonconditional endocytosis mutants were found at 1.1 × 10^{-6} with diphtheria toxin and modeccin, and 8 × 10^{-8} with *Pseudomonas* toxin. Mutants with temperature-sensitive endocytic defects are rarer than nonconditional mutants; in a selection with modeccin, the former were found at 1 × 10^{-7}, the latter at 1 × 10^{-6}. No temperature-sensitive mutants were obtained with *Pseudomonas* exotoxin. Although we have isolated a temperature-sensitive mutant using diphtheria toxin, technical problems precluded determination of an absolute frequency. Based on primary screens of the replicate colonies for toxin resistance and Man-6-P uptake at 34° and 39°, the frequency is at least that obtained with modeccin.

Quantitative Testing of Mutant Phenotype

We suggest that temperature-sensitive mutants be assayed at 41°, as well as at 34 and 39°. Although 39° was used as the nonpermissive temperature in isolating the mutants, some aspects of the mutant phenotype are not affected at this temperature.[17] Shifting to 41° for 1–4 hr prior to assay is sufficient to effect maximal alteration of all facets of the mutant phenotype; the precise time required is dependent on the cellular function under study.[17] Nonconditional mutants are assayed at 34°.

Mutants are compared to parental cells with respect to toxin resistance by measuring protein synthesis after incubation with diphtheria toxin, modeccin, *Pseudomonas* toxin, and ricin. Cells in 24-well trays (1.5 × 10^5 cells/well) are incubated with varying concentrations of toxin for 2–3 hr,[18] rinsed in methionine-free medium, and then labeled for 1 hr with [^{35}S]methionine (2 μCi/1 μg/1 ml/well). After the wells are rinsed 3 times with cold Dulbecco's phosphate-buffered saline (without divalent cations) the protein is solubilized in 75 μl of 0.1 N NaOH and neutralized with an equal volume of 0.1 N HCl. Then 25-μl aliquots are spotted on strips of Whatman 3MM paper[8]; strips are soaked in 10% TCA, rinsed with ethanol, and dried. Radioactivity is determined by liquid scintillation counting. Assessing resistance by the concentration of toxin or toxic lectin required to reduce protein synthesis to 50% of that measured in untreated controls, the most affected of the mutants are 100 times more resistant to diphtheria and *Pseudomonas* toxins, >6000 times more resistant to modeccin, and 50 times more *sensitive* to ricin than parental cells. Several mutants are highly resistant to modeccin and diphtheria toxin (and have ≤5% Man-6-P-dependent uptake) but show normal sensitivity to *Pseudomonas* toxin and only 3- to 5-fold increase in sensitivity to ricin.

Uptake via the Man-6-P receptor is compared in mutants and parent using the ammonia-induced ^{35}S-labeled secretions described above. Cells in 6-well trays (5 × 10^5 cells/well) are incubated in 1 ml of medium containing 0.5–1.0 × 10^6 cpm/ml secretions for 1 hr in the presence or absence of 10 mM Man-6-P. Cells are rinsed to remove free ligand, solubilized in 0.1 N NaOH (0.5 ml), and neutralized with HCl; then radioactivity and protein are determined. Man-6-P-dependent uptake is <5–10% of normal in the majority of mutants.

Endocytosis mutants can show qualitative differences in phenotype, yet belong to the same genetic group. Such phenotypic differences apparently reflect the relative severity of the lesion. For this reason we suggest that once a mutant is verified by quantitative assay and its spectrum of toxin resistance is determined, the mutant be analyzed first by genetic complementation (see below), and then by detailed examination of phenotype. Some aspects of the phenotype altered in the endocytosis mutants are accumulation of iron from transferrin, ATP-dependent acidification of endosomes *in vitro,* compartmentalization of newly synthesized acid hydrolases, galactosylation and sialylation of Sindbis virus, release of Sindbis virus, and sialylation of some endogenous (primarily secreted) glycoproteins.[16–18,31]

Analysis of Mutants—Genotype

Isolation of Revertants. Starting with the assumption that a defect in endocytosis would be pleiotropic, we isolate mutants exhibiting multiple phenotypic alterations; once the mutants are obtained we worry that they contain multiple genetic lesions. To ensure that all the phenotypic changes seen in a single mutant result from a single mutation, we select variants which are restored to normal for one characteristic, then screen those variants to determine whether other parameters also have reverted to normal.

The marked increase in ricin sensitivity exhibited by the most affected mutants can be exploited to select revertants; $6–8 \times 10^5$ cells/100-mm dish are incubated for 48 hr in growth medium containing 10% fetal bovine serum and 10 ng ricin/ml. Selection medium then is replaced with growth medium without ricin, and cells are fed every 4 days until colonies are large enough to be picked. The same protocol is used for isolating revertants of *ts* mutants, except that cells are shifted to 39° for 3 hr prior to and during ricin selection, following which they are returned to 34°. Viability of parental cells is >90% under these conditions. With one of the nonconditional mutants, 4 out of 1.2×10^6 cells survived ricin treatment. Two (presumably independent) of these were analyzed: one showed full restoration of all activities, i.e., endocytic and Golgi/GERL-associated functions; the other exhibited normal endocytosis, and partial (25–35% of normal) restoration of Golgi/GERL-associated activities.[16] The fact that spontaneous revertants increased in all affected activities can be isolated at a frequency of 3×10^{-6} argues that a single genetic defect is responsible for all facets of the mutant's phenotype.

[31] R. D. Klausner, J. van Renswoude, C. Kempf, K. Rao, J. L. Bateman, and A. R. Robbins, *J. Cell Biol.* **98,** 1098 (1984).

Testing the Homogeneity of the Mutant Population. Once a revertant arises it has a selective advantage among a population of nonconditional mutants, because its growth rate is faster under standard culture conditions. Assays of the total population, e.g., measurements of residual Man-6-P uptake, will reveal the presence of a normal subpopulation only when it represents about 5% of the total population. This level of contamination is unacceptable under any circumstances and is fatal in a procedure such as genetic complementation in which one blindly selects a subset of the cells (see below). To check for revertants we plate 1200–1600 mutant cells on three to four 100-mm dishes, replicate the colonies onto polyester, and then screen the replicas for Man-6-P uptake (all as above). Populations with any positive colonies are discarded.

Genetic Complementation. To determine whether the mutants represent the same or different genetic groups we analyze them for complementation. Hybrids are generated by fusing ouabain-sensitive, thioguanine-sensitive with ouabain-resistant, thioguanine-resistant cells, and then selecting for ability to grow in HAT medium (Dulbecco's modified Eagle medium containing 5% fetal bovine serum, 15 μg hypoxanthine, 0.2 μg aminopterin, and 5 μg thymidine per milliliter supplemented with 11.5 μg proline/ml) containing 2 mM ouabain. Endocytosis mutants with the appropriate markers have been generated both by sequentially selecting spontaneous ouabain-resistant then ouabain-resistant, thioguanine-resistant variants of endocytosis mutants, and by isolating endocytosis mutants from a ouabain-resistant, thioguanine-resistant variant of WTB.

Cells, plated 24 hr before fusion at 4×10^5 of each cell type/well in a 6-well tray, are rinsed 3 times with growth medium lacking both serum and Ca^{2+}.[32] One milliliter of a 1 : 1 mixture of polyethylene glycol 1000 and this medium is added gently to the well, the tray is rocked, and then the mixture is removed after 30 sec. Cells are gently rinsed 3 times in medium without Ca^{2+} or serum, and placed at 34° for 30 min in this medium supplemented with 0.2% bovine serum albumin. Medium is then replaced with standard growth medium. About 16 hr after fusion cells are harvested and plated in selection medium. After 10–12 days cells are trypsinized, grown, and cloned on 96-well dishes as described above. About 1–3% of the starting cells form viable hybrids.

Cloned hybrid cells as well as uncloned populations of hybrids are tested for toxin resistance and uptake via the Man-6-P receptor. Hybrids involving *ts* mutants are tested both at 34 and 41°. We find that all of our mutants are recessive in hybrids formed with parental cells. Fusions between mutants demonstrate the existence of two genetic groups (referred

[32] S. Schneiderman, J. L. Farber, and R. Baserga, *Somatic Cell Genet.* **5,** 263 (1979).

to as End1 and End2). There are also temperature-sensitive and nonconditional representatives of both End1 and End2.[17] End1 × End2 hybrids
show restored Golgi/GERL-associated activity; hybrids within each
group do not.

Summary and Comments

Our basic strategy for isolating mutants defective in a process as complex and uncharted as endocytosis may be summarized as follows: (1)
Isolate a wide variety of mutants by applying the least selective pressure
practicable, whenever possible screen rather than select. (2) Include temperature shifts in hopes of obtaining temperature-sensitive mutants, since
these will be most useful for identification of the primary defect. As with
the endocytosis mutants more than simple temperature shifts may be
required, but nonconditional mutants will be obtained irrespective of inclusion of the shift (see below for caveat). (3) Screen colonies *en masse*
rather than individually for desired parameters; the procedure must be
efficient if emphasis is to be placed on screening rather than selection. In
addition to toxin resistance and Man-6-P uptake, other endocytosis-re-
lated parameters measurable on the polyester replicas are uptake of [125]I-
labeled low density lipoprotein and intracellular retention of endogenous
lysosomal hydrolase, the latter assayed by indirect antibody binding.

The distribution of nine independent mutants into only two classes
suggests that our isolation procedure may bias the results. Clearly the use
of toxins as selective agents limits the target to steps occurring prior to
toxin penetration into the cytosol. Because diphtheria toxin enters the
cytosol from the endosome,[33] using diphtheria toxin one will not select
mutants altered specifically in postendosomal compartments. However,
there must be more than two steps prior to diphtheria toxin penetration.
Evidence indicates that modeccin exerts its toxic effect from an endocytic
compartment beyond the endosome,[33] and thus modeccin might be expected to select a somewhat wider spectrum of mutants than diphtheria
toxin. So far we have obtained only End2 mutants with modeccin, and
both End1 and End2 mutants with diphtheria toxin. It is unclear whether
screening for defects in Man-6-P receptor-mediated uptake introduces a
bias. However, mutants isolated solely on the basis of toxin cross-resistance were found to be defective in Man-6-P uptake also.[34] Screening the
replicas for toxin resistance is a fairly new addition to our procedure; as

[33] R. K. Draper, D. O. O'Keefe, M. Stookey, and J. Graves, *J. Biol. Chem.* **259**, 4083 (1984).
[34] W. S. Sly, J. H. Grubb, J. M. Moehring, and T. J. Moehring, *in* "Molecular Basis of
Lysosomal Storage Disorders" (R. O. Brady and J. A. Barranger, eds.), p. 163. Academic
Press, New York, 1984.

more colonies are tested for both parameters we will determine whether any cross-resistant cells show normal Man-6-P uptake.

Finding a limited variety of nonconditional mutants suggests that many defects along the pathway have lethal consequences. Isolating mutants with temperature-sensitive defects has not yet expanded our horizons. In this context it should be noted that the End1 and End2 defects are conditionally lethal: at low cell densities (<1000 cells/100-mm dish) the plating efficiencies of both temperature-sensitive and nonconditional mutants are reduced to zero at 39°. The latter are reminiscent of several classes of glycosylation mutants in which nonconditional defects result in temperature-sensitive viability.[23] If lethality is the problem, decreasing the time cells are kept at the nonpermissive temperature should help in future selections. Eliminating temperature shifts and isolating nonconditional mutants by selecting and screening at 34° also may yield new classes of defects.

Acknowledgments

We thank C. Hall, S. Krag, S. Laurie, and K. MacKay for help in translation of this manuscript.

Section III

Carbohydrate-Binding Proteins

[38] Sulfatide-Binding Proteins

By DAVID D. ROBERTS

Cell surface glycoconjugates have been extensively studied as potential receptors for cell–cell and cell–ligand interactions.[1,2] Carbohydrates occur on cell surface glycoproteins, glycolipids, and proteoglycans. As glycolipids containing a single oligosaccharide are readily purified and terminal structures on glycolipids are often identical to those found on glycoproteins, glycolipids are useful for examining the specificity of proteins which may bind to cell surface carbohydrates. Evidence for specific interactions with carbohydrates present on glycolipids has been obtained using monoclonal and polyclonal antibodies,[3] toxins,[4] lectins,[5,6] bacteria,[7] viruses,[8] and eukaryotic cells.[9] In general, these studies have focused on the role of neutral glycolipids and acidic lipids containing sialic acid as receptors.

A second class of acidic glycolipids contains sulfate esters. These lipids are found in many tissues and occur with various carbohydrate sequences. Most contain sulfate esters of galactose, but sulfation also occurs on N-acetylgalactosamine, N-acetylglucosamine, glucose, and sialic acid. Structures of reported sulfated lipids found in animals are summarized in Table I.

The simplest member of this family, sulfatide (galactosylceramide I[3]-sulfate), interacts with several proteins including factor XII,[10–12] kinino-

---

[1] W. Frazier and L. Glaser, *Annu. Rev. Biochem.* **48,** 91 (1979).

[2] S. H. Barondes, *Annu. Rev. Biochem.* **50,** 207 (1981).

[3] J. L. Magnani, this volume [14] and [39].

[4] W. E. van Heyningen, C. C. J. Carpenter, N. F. Pierce, and W. B. Greenbough, *J. Infect. Dis.* **124,** 415 (1971).

[5] S.-I. Hakomori, J. Koscielak, K. J. Block, and R. W. Jeanloz, *J. Immunol.* **98,** 31 (1967).

[6] D. F. Smith, *Biochem. Biophys. Res. Commun.* **115,** 360 (1983).

[7] G. C. Hansson, K.-A. Karlsson, G. Larson, N. Strömberg, and J. Turin, *Anal. Biochem.* **146,** 158 (1985); K.-A. Karlsson, this volume [16]; K.-A. Karlsson and N. Strömberg this volume [17].

[8] G. C. Hansson, K.-A. Karlsson, G. Larson, N. Strömberg, J. Turin, C. Örvell, and E. Norrby, *FEBS Lett.* **170,** 15 (1984); K.-A. Karlsson, this volume [16]; K.-A. Karlsson and N. Strömberg this volume [17].

[9] C. C. Blackburn and R. L. Schnaar, *J. Biol. Chem.* **258,** 1180 (1983).

[10] K. Fujikawa, R. L. Heimark, K. Kurachi, and E. W. Davie, *Biochemistry* **19,** 1322 (1980).

[11] T. Shimada, H. Kato, H. Maeda, and S. Iwanaga, *J. Biochem. (Tokyo)* **97,** 1637 (1985).

[12] M. A. Griep, K. Fujikawa, and G. L. Nelsestuen, *Biochemistry* **24,** 4124 (1985).

METHODS IN ENZYMOLOGY, VOL. 138

TABLE I
Structures of Sulfated Glycolipids

Nomenclature	Structure	Source	Ref.
Glycosphingolipids			
GalactosylCer-I^3-SO$_4$ (sulfatide)	Gal(3SO$_4$)β1-Cer	Many tissues	a–e
LactosylCer-II^3-SO$_4$	Gal(3SO$_4$)β1-4Glcβ1-Cer	Human kidney	e
GangliotriaosylCer-II^3-SO$_4$	GalNAcβ1-4Gal(3SO$_4$)β1-4Glcβ1-Cer	Rat kidney	f
GangliotriaosylCer-II^3,III^3-bisSO$_4$	GalNAc(3SO$_4$)β1-4Gal(3SO$_4$)β1-4Glcβ1-Cer	Rat kidney	g
LactotriaosylCer-III^6-SO$_4$	GlcNAc(6SO$_4$)β1-3Galβ1-4GlcCer	Hog gastric mucosa	h
TrihexosylCer-III^3-SO$_4$	Gal(3SO$_4$)1-4Gal1-4GlcCer	Hog gastric mucosa	i
GangliotetraosylCer-II^3,IV^3-bisSO$_4$	Gal(3SO$_4$)β1-3GalNAcβ1-4Gal(3SO$_4$)β1-4GlcCer	Rat kidney	j
LactoneotetraosylCer-III^6-SO$_4$	Galβ1-4GlcNAc(6SO$_4$)β1-3Galβ1-4GlcCer	Hog gastric mucosa	k
	GalNAcβ1-4[Sial(8SO$_4$)α2-3]Galβ1-4GlcCer	Bovine gastric mucosa	l
	Sial(8SO$_4$)α2-8Sialα2-3Galβ1-4GlcCer	Bovine gastric mucosa	m
	GlcUA(3SO$_4$)β1-3Galβ1-4GlcNAcβ1-3Galβ1-4GlcCer	Human peripheral nerve	n, o
	GlcUA(3SO$_4$)β1-3Galβ1-4GlcNAcβ1-3Galβ1-4GlcCer	Human peripheral nerve	n, o
	Sial(8SO$_4$)α2-6GlcCer	Sea urchin	p, q
	Sial(8SO$_4$)2-6Glc1-8Sial2-6GlcCer	Sea urchin	p, q
GangliotetraosylCer-IV^3-SO$_4$	Gal(3SO$_4$)β1-3GalNAcβ1-4Galβ1-4GlcCer	Mouse small intestine	r
Glycerolipids			
Galactosylβ1-3alkylacylGro-I^3-SO$_4$ (seminolipid)	Gal(3SO$_4$)β1-3alkylacylglycerol	Testis	s

Triglucosylalkylacylglycerol-III⁶-SO₄	Glc(6SO₄)α1-6Glcα1-6Glcα1-3alkylacylglycerol	Human saliva and gastric mucosa	t, u

[a] T. Yamakawa, N. Kiso, S. Handa, A. Makita, and S. Yokoyama, *J. Biochem. (Tokyo)* **52**, 226 (1962).

[b] P. Stoffyn and A. Stoffyn, *Biochim. Biophys. Acta* **70**, 218 (1963).

[c] N. S. Radin, *in* "Handbook of Neurochemistry" (A. Lajtha, ed.), p. 163. Plenum, New York, 1983.

[d] C. G. Hansson, K.-A. Karlsson, and B. E. Samuelsson, *J. Biochem. (Tokyo)* **83**, 813 (1978).

[e] E. Martensson, *Biochim. Biophys. Acta* **116**, 521 (1966).

[f] K. Tadano and I. Ishizuka, *J. Biol. Chem.* **257**, 1482 (1982).

[g] K. Tadano and I. Ishizuka, *J. Biol. Chem.* **257**, 9294 (1982).

[h] A. Slomiany, B. L. Slomiany, and C. Annese, *Eur. J. Biochem.* **109**, 471 (1980).

[i] B. L. Slomiany, A. Slomiany, and M. I. Horowitz, *Biochim. Biophys. Acta* **348**, 388 (1974).

[j] K. Tadano, I. Ishizuka, M. Matsuo, and S. Matsumoto, *J. Biol. Chem.* **257**, 13413 (1982).

[k] B. L. Slomiany and A. Slomiany, *J. Biol. Chem.* **253**, 3517 (1978).

[l] B. L. Slomiany, K. Kojima, Z. Banas-Gruszka, V. L. N. Murty, N. I. Galicki, and A. Slomiany, *Eur. J. Biochem.* **119**, 647 (1981).

[m] A. Slomiany, K. Kojima, Z. Banas-Gruszka, and B. L. Slomiany, *Biochem. Biophys. Res. Commun.* **100**, 778 (1981).

[n] K. H. Chou, A. A. Ilyas, J. E. Evans, R. H. Quarles, and F. B. Jungalwala, *Biochem. Biophys. Res. Commun.* **128**, 383 (1985).

[o] D. K. H. Chou, A. A. Ilyas, J. E. Evans, R. H. Quarles, and F. B. Jungalwala, *Proc. Int. Symp. Glycoconjugates, 8th, 1985* Abstract XVI-15 (1985).

[p] N. K. Kochetkov, G. P. Smirnova, and N. V. Chedareva, *Biochim. Biophys. Acta* **424**, 274 (1976).

[q] N. V. Prokazova, A. T. Mikhalilov, S. L. Kocharov, L. A. Malchenko, N. D. Zvezdina, G. Buznikov, and L. D. Bergelson, *Eur. J. Biochem.* **115**, 671 (1981).

[r] H. Leffler, G. C. Hanson, and N. Strömberg, *J. Biol. Chem.* **261**, 1440 (1986).

[s] I. Ishizuka, M. Suzuki, and T. Yamakawa, *J. Biochem. (Tokyo)* **73**, 77 (1973).

[t] B. L. Slomiany, A. Slomiany, and G. B. J. Glass, *Eur. J. Biochem.* **78**, 33 (1977).

[u] R. Narasimhan, A. Bennick, B. Palmer, and R. K. Murray, *J. Biol. Chem.* **257**, 15122 (1982).

gen,[11] and myelin basic protein.[13-15] Recently we reported that the cell adhesion molecule laminin[16] and the platelet glycoprotein thrombospondin[17] also bind specifically to sulfatides and that this binding probably accounts for the agglutination of aldehyde-fixed erythrocytes by these proteins. Three proteins from rat spermatogenic cells bind specifically to galactosylalkylacylglycerol-I[3]-sulfate, the major glycolipid of mammalian male germ cells.[17a]

Several methods have been used to study sulfatide-binding proteins. Enzymatic activation[12] and changes in fluorescence polarization[11] of factor XII and kininogen occur in the presence of sulfatide micelles. Myelin basic protein binding to sulfatides has been studied using Ouchterlony double diffusion with sulfatide micelles,[13] changes in compressibility of sulfatide monolayers at an air–water interface,[14] electron spin resonance of spin-labeled sulfatides,[15] and differential scanning calorimetry.[15] Surface-adsorbed sulfatides and sulfated galactosylglycerolipid were used to study binding of antibodies by solid phase radioimmunoassays[18] and enzyme immunoassays.[18a,18b] Covalent and noncovalent affinity columns bearing galactosylalkylacylglycerol sulfate were used to isolate binding proteins from rat testis.[17a] Recently, we characterized laminin and thrombospondin binding using solid phase assays and binding to sulfatides separated on thin-layer chromatograms.[16,17]

This article presents a generalized approach for characterizing sulfatide-binding proteins using immobilized sulfatides and outlines the criteria for defining the specificity of these interactions.

Preparation of Lipid Extracts for Detection of Sulfatides

Numerous methods for the preparation of lipid extracts from tissues[19-21] are effective for sulfated glycolipids as well. Addition of 0.8 *M*

[13] H. C. Yohe, R. I. Jacobson, and R. K. Yu, *J. Neurosci. Res.* **9**, 401 (1983).

[14] G. D. Fidelio, B. Maggio, and F. A. Cumar, *Chem. Phys. Lipids* **35**, 231 (1984).

[15] J. M. Boggs, G. Ranjaraj, M. A. Moscarello, and K. M. Koshy, *Biochim. Biophys. Acta* **816**, 208 (1985).

[16] D. D. Roberts, C. N. Rao, J. L. Magnani, S. L. Spitalnik, L. A. Liotta, and V. Ginsburg, *Proc. Natl. Acad. Sci. U.S.A.* **82**, 1306 (1985).

[17] D. D. Roberts, D. M. Haverstick, V. M. Dixit, W. A. Frazier, S. A. Santoro, and V. Ginsburg, *J. Biol. Chem.* **260**, 9405 (1985).

[17a] C. A. Lingwood, *Can. J. Biochem. Cell Biol.* **63**, 1077 (1985).

[18] W. Hofstetter, C. H. Heusser, N. Herschkowitz, and K. Blaser, *J. Immunol. Methods* **57**, 99 (1983).

[18a] E. M. Eddy, C. H. Muller, and C. A. Lingwood, *J. Immunol. Methods* **81**, 137 (1985).

[18b] C. Goujet-Zalc, A. Guerci, G. Dubois, and B. Zalc, *J. Neurochem.* **46**, 435 (1986).

[19] S.-I. Hakomori and B. Siddiqui, this series, Vol. 32, p. 345.

[20] S.-I. Hakomori, this series, Vol. 50, p. 207.

[21] L. Svennerholm and P. Fredman, *Biochim. Biophys. Acta* **617**, 97 (1980).

sodium acetate increases the yields of complex sulfated lipids extracted from some tissues.[22] Since many sulfated lipids partition significantly into the lower phase in a Folch extraction,[23] this method is not useful for removal of phospholipids and cholesterol. Phosphoglycerolipids can be degraded, however, by alkaline methanolysis (0.2 M NaOH in methanol for 30 min at 37°). The extract is neutralized with 0.9 volumes of 0.2 N HCl in methanol, reduced on a rotary evaporator to 1/4 volume, and desalted by dialysis against water or by chromatography on Sephadex G-25 in 60 : 30 : 4.5 chloroform–methanol–water[24] (2 ml gel/10 mg total lipid). Lipid extracts may be further fractionated by anion exchange chromatography on DEAE-Sepharose 4B[25,26] to remove neutral glycolipids, sphingomyelin, and cholesterol. Simple sulfatides elute between the mono- and disialoganglioside fractions.

Solid Phase Binding Assay

Sulfatides (1–500 ng) are coated on flat-bottom 96-well poly(vinyl chloride) microtiter plates (Falcon No. 3912) by drying from methanol (30 μl/well).[16] In a humid environment, efficient binding may require drying under vacuum for 30 min. Addition of a carrier lipid mixture (50 ng phosphatidylcholine and 50 ng cholesterol per well) improves reproducibility and enhances binding of laminin to adsorbed sulfatides. After drying, the wells are filled with Tris-buffered saline, pH 7.8, containing 50 mM Tris, 110 mM NaCl, 1% bovine serum albumin,[27] 5 mM CaCl$_2$, 0.1 mM phenylmethanesulfonyl fluoride,[28] and 0.02% NaN$_3$ (Tris–BSA) and incubated for 30–45 min to mask nonspecific binding sites. The buffer is removed and iodinated protein (specific activities of 0.2 to 10 μCi/μg can be used depending on the stability of the protein to iodination) in 30 μl of Tris–BSA is added. The time and temperature of incubation for optimal binding are determined empirically. For the proteins examined, 3–24 hr are re-

---

[22] K. Tadano and I. Ishizuka, *Biochem. Biophys. Res. Commun.* **103,** 1006 (1981).

[23] K. Tadano-Aritomi and I. Ishizuka, *J. Lipid Res.* **24,** 1368 (1983).

[24] K. Tadano and I. Ishizuka, *J. Biol. Chem.* **257,** 1482 (1982).

[25] R. W. Ledeen and R. K. Yu, this series, Vol. 83, p. 139.

[26] If volatile buffers are used, the bicarbonate cation should be used instead of acetate as concentration of ammonium acetate solutions under reduced pressure produces a low pH and promotes solvolysis of sulfatides.

[27] Some preparations of bovine serum albumin contain proteases which degrade laminin or thrombospondin and other impurities which inhibit binding. New lots of albumin should be tested for protease activity by incubating with iodinated protein. Degradation may be detected by sodium dodecyl sulfate gel electrophoresis or by progressive loss of binding activity. Other inhibitors can be detected by progressive loss of specific binding in the presence of increasing concentrations of albumin.

[28] Phenylmethanesulfonyl fluoride is unstable in aqueous solution. A stock solution can be prepared in ethanol and should be added to the buffer immediately before use.

quired to reach equilibrium.[16,17] Thrombospondin binds faster at 25° than at 4°, but equilibrium binding is higher at the lower temperature.[17] Because iodination of laminin abolishes sulfatide binding, the wells are first incubated with unlabeled laminin (30 μl, 10 μg/ml) for 2 hr. The wells are washed 3 times with Tris–BSA and incubated for 1 hr with specific rabbit antibodies to laminin diluted in Tris–BSA. The wells are again washed and incubated for 1 hr with iodinated protein A (200 ng/ml, 20–50 μCi/μg).[16] The wells are washed 3 times, cut out of the plate, and the bound radiolabel is quantified in a gamma counter.

Overlays of Thin-Layer Chromatograms

Purified sulfatides or desalted lipid extracts from tissues or cells are chromatographed on aluminum-backed silica gel high performance TLC plates (E. Merck) in chloroform–methanol–0.2% aqueous $CaCl_2$ (60:35:7). Reference glycolipids are spotted on lanes at one end of the same plate. After development and drying these lanes are cut off and the glycolipids visualized with orcinol–H_2SO_4 or other detection reagents. The chromatograms are air-dried, soaked for 1 min in 0.1% poly(isobutyl methacrylate) in hexane (Polysciences), air dried, sprayed with PBS (0.1 M sodium phosphate, 0.15 M NaCl, pH 7.4), and immersed in Tris–BSA for 30 min. The chromatogram is overlayed with 0.5 μg/ml ^{125}I-labeled protein in Tris–BSA (60 μl/cm^2) and incubated in a covered Petri dish under optimal conditions determined using the solid phase assay. To minimize nonspecific binding in the laminin binding assay, the ionic strength of the Tris–BSA buffer is increased to 0.22 by addition of sodium chloride. Further increase in ionic strength will also inhibit specific binding. Following incubation with unlabeled laminin (10 μg/ml) the chromatogram is washed by dipping at 1-min intervals in 4 changes of cold saline and incubated for 1 hr with anti-laminin antibodies. The chromatogram is again washed and incubated for 1 hr with iodinated protein A. After incubation with labeled reagents, the chromatograms are washed by dipping at 1-min intervals in 4 changes of cold saline, dried, and exposed to X-ray film (XAR-5, Eastman Kodak) for 8–24 hr.

The amount of lipid extract required to detect binding to sulfatides depends on both the concentration of sulfatides in the tissue and the relative affinity of the binding protein. Thrombospondin can detect less than 2 ng of sulfatide by the chromatogram overlay method,[17] whereas approximately 50 ng of sulfatide is required to detect laminin binding[16] (Fig. 1). Tissue sulfatide concentrations range from approximately 1 nmol/g wet weight of erythrocytes to about 10,000 nmol/g in brain white matter. With an untested tissue, lipid from 1–50 mg of tissue should be sufficient to detect sulfatides by this method.

Thrombospondin   Laminin

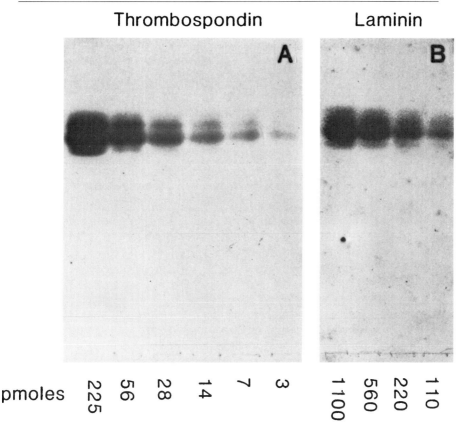

pmoles   225 56 28 14 7 3   1100 560 220 110

FIG. 1. Binding of laminin and ^{125}I-labeled thrombospondin to sulfatides separated on thin-layer chromatograms. Bovine brain sulfatides were chromatographed on aluminum-backed silica gel high-performance TLC plates (E. Merck) in chloroform–methanol–0.2% aqueous $CaCl_2$ (60 : 35 : 7). The chromatograms were air dried, soaked for 1 min in 0.1% poly(isobutyl methacrylate) in hexane, dried, sprayed with PBS, and immersed in Tris–BSA for 30 min. The chromatogram in (A) was overlayed with 0.5 μg/ml ^{125}I-labeled thrombospondin (10 μCi/μg) in Tris–BSA (60 μl/cm²) and incubated in a covered Petri dish for 3 hr at 4°. The chromatogram was washed by dipping in 5 changes of cold PBS at 1-min intervals, dried, and exposed to X-ray film (XAR-5, Eastman Kodak) for 4 hr. The chromatogram in (B) was incubated with 10 μg/ml laminin in Tris–BSA, ionic strength 0.22, for 2 hr at room temperature. The chromatogram was washed by dipping in 5 changes of saline at 1-min intervals and then incubated for 1 hr in rabbit anti-laminin diluted 1 : 1000 in Tris–BSA. The chromatogram was washed again, incubated for 1 hr in ^{125}I-labeled protein A (0.5 μg/ml), washed, dried, and exposed to X-ray film for 15 hr.

Solvolysis of Sulfatides

Sulfate esters on simple sulfatides are selectively removed by solvolysis at room temperature under anhydrous conditions using 50 m*M* HCl in

TABLE II
CRITERIA FOR SPECIFIC BINDING TO SULFATIDES

Binding is not due to low affinity ionic interactions indicated by similar binding to anionic phospholipids, gangliosides, and sulfatides.

Binding is labile to acid solvolysis but stable to base or neuraminidase treatment of lipids.

Acidic but not neutral lipid fractions are active.

Protein binds with high affinity to purified sulfatide but not to cholesterol 3-sulfate. Binding to sulfatide is saturable.

methanol.[29] Some complex sulfated glycolipids, however, are not sensitive to these conditions but can be desulfated by solvolysis in 9:1 dimethyl sulfoxide–methanol containing 4–8 mM H_2SO_4.[24] In some cases complete desulfation requires conditions which also cleave some glycosidic bonds.[24]

Criteria for Sulfatide Specificity

For characterizing lectins and antibodies which bind neutral sugars, specific inhibition by mono- or oligosaccharides suffices to determine the specificity of binding. With proteins that bind to sulfatides or other charged sugars, however, ion exchange properties of the ligand must be considered, and more extensive criteria are required to define specific binding. Table II outlines the criteria used to define the specificity of laminin and thrombospondin for sulfated glycolipids.

Hofstetter and co-workers reported that some IgM antibodies bind nonspecifically to sulfatides.[18] We have also observed low affinity nonspecific binding of several mouse monoclonal antibodies of both IgG and IgM isotypes to sulfatides and some phospholipids.[30] Thus it is important to define specificity for binding of proteins to sulfatides relative to binding obtained using unrelated anionic lipids. For example, fibronectin binds to sulfatides on thin-layer chromatograms but also binds with equal or higher affinity to anionic phospholipids such as phosphatidyl serine.[16] Fibronectin also agglutinates phospholipid vesicles nonspecifically[31] and binds weakly to various gangliosides on overlays.[16] Thus binding of fibronectin to anionic lipids may be nonspecific.

In contrast, thrombospondin binds to at least 1000-fold lower levels of sulfatide using the solid phase assay than to unrelated anionic lipids and 100-fold lower levels than to cholesterol 3-sulfate.[17] Laminin is somewhat

---

[29] P. Stoffyn and A. Stoffyn, *Biochim. Biophys. Acta* **70,** 107 (1963).

[30] D. Roberts, unpublished data (1984).

[31] J. D. Rossi and B. A. Wallace, *J. Biol. Chem.* **258,** 3327 (1983).

less specific: cholesterol 3-sulfate is approximately 10% as active as sulfatide and phosphatidyl serine is 1% as active on a per weight basis.[16] However, the concentration dependence of laminin binding indicates that binding to cholesterol 3-sulfate is very low affinity, whereas binding to sulfatide is high affinity and saturated at 20 μg/ml laminin.[16]

Examination of binding following treatment with mild acid, base, or neuraminidase is useful for differentiating classes of lipids which may account for binding activity. Small amounts of lipid extract can be subjected to each treatment and then plated at serial dilutions on microtiter plates for assay of binding. Base treatment (0.2 M NaOH in methanol for 30 min at 37°) destroys phosphoglycerolipids but not sulfated glycosphingolipids or most gangliosides. A few tissues including testis[32] contain alkali-labile sulfated glycolipids with an alkylacylglycerol tail. In tissues where this lipid predominates, binding may be base labile. Furthermore, using more vigorous base hydrolysis glycolipids which may contain 6-sulfate esters instead of the more common 3-sulfate can be degraded.[33] Solvolysis with mild acid in methanol[29] or dimethyl sulfoxide–methanol[24] selectively desulfates many sulfated glycolipids and abolishes binding without hydrolysis of gangliosides. As noted above, however, not all sulfate esters are labile under these conditions, and the effect of varying the time and temperature of solvolysis on binding activity and mobility on thin-layer chromatography should be studied. Digestion with neuraminidase can be done with an aqueous lipid suspension[34] or directly on lipids coated on microtiter plates or thin-layer chromatograms.[35] Binding to most gangliosides is destroyed, but binding to sulfatides will be stable and in some cases will increase.

Fractionation by anion-exchange chromatography is used to determine the charge of the active lipids. Laminin and thrombospondin do not bind to neutral lipids which are not retained by the column. The active sulfatides are bound and elute between the mono- and disialoganglioside fractions at approximately 0.02 M salt. Tissues such as rat kidney also contain bis-sulfated glycosphingolipids[36,37] which bind laminin and thrombospondin[38] and elute at high salt concentrations after the disialogangliosides.

[32] I. Ishizuka, M. Suzuki, and T. Yamakawa, *J. Biochem. (Tokyo)* **73,** 77 (1973).

[33] E. G. V. Percival, *Q. Rev. London* **3,** 369 (1949).

[34] R. Schauer, this series, Vol. 50, p. 67.

[35] S. L. Spitalnik, J. F. Schwartz, J. L. Magnani, D. D. Roberts, P. F. Spitalnik, C. I. Civin, and V. Ginsburg, *Blood* **66,** 319 (1985).

[36] K. Tadano and I. Ishizuka, *J. Biol. Chem.* **257,** 9294 (1982).

[37] K. Tadano and I. Ishizuka, *J. Biol. Chem.* **257,** 13413 (1982).

[38] D. D. Roberts, C. N. Rao, L. A. Liotta, H. R. Gralnick, and V. Ginsburg, *J. Biol. Chem.* **261,** 6872 (1986).

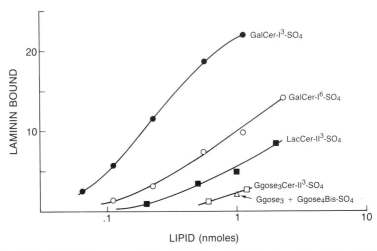

FIG. 2. Laminin binding to purified sulfated glycolipids. Binding of laminin to lipids after thin-layer chromatography was detected using rabbit antibodies to laminin and ^{125}I-labeled protein A and quantified by densitometry of autoradiograms in the linear range. Structures of the sulfated glycolipids are given in Table I.

To further define the specificity for binding to sulfated lipids and to estimate the avidity of binding, binding to purified lipids is examined using the solid phase and chromatogram overlay assays. If radiolabeled protein is used, it is essential to establish that excess unlabeled protein is able to completely inhibit binding of the labeled protein. This demonstrates that unlabeled protein is binding to the same lipids as the labeled protein and that binding is saturable. Using increasing concentrations of labeled protein, saturation should also be obtained and should require similar concentrations for half-saturation as required for 50% inhibition of binding of a trace amount of labeled protein by unlabeled protein.

The relative avidity for other lipids is then examined using a fixed subsaturating concentration of labeled protein and measuring binding to varying concentrations of immobilized lipids. By this method it was demonstrated that thrombospondin binds with high affinity to sulfatide and weakly to cholesterol 3-sulfate but not to any phospholipids, neutral glycolipids or gangliosides.[17] As mentioned previously, laminin also binds weakly to cholesterol 3-sulfate and to phosphatidylserine but not other phospholipids or glycolipids.[16] Using quantitative binding assays, the specificity of laminin for other sulfated glycolipids was also analyzed (Fig. 2). Laminin binds best to galactosylceramide I^3-sulfate and with progressively lower avidity to sulfatides with increasing carbohydrate chain length. Binding is also lower for the synthetic analog galactosylceramide I⁶-sulfate than for the natural sulfatide with a 3-sulfate ester.

Characteristics of Sulfatide Binding

Detailed analysis of the effects of divalent cations, ionic strength, and pH on binding have been reported elsewhere.[16,17,39] For the three proteins examined to date, binding is primarily ionic as high salt concentrations inhibit binding, and, in the case of thrombospondin, free galactose 3-sulfate is only a slightly better inhibitor than inorganic sulfate.[17] Binding in general is maximal at neutral pH and decreases at lower pH and is stimulated by divalent cations.[17,39] Some sulfated polysaccharides are effective inhibitors of sulfatide binding, although the relative potency of various sulfated polysaccharides is different for each protein. The high activity of heparin and certain monoclonal antibodies as inhibitors of thrombospondin binding to sulfatides suggests that a single heparin binding domain is responsible for both heparin and sulfatide binding.[17] In contrast, examination of sulfatide binding of several proteolytic fragments of laminin suggest that heparin and sulfatide binding sites are located on different regions of the laminin molecule.[16]

These results suggest that binding of some proteins to sulfatide may be due to a secondary specificity of a site which recognizes other sulfated glycoconjugates on glycoproteins or proteoglycans. Other proteins, however, may have binding sites that recognize only sulfated glycolipids. Sulfated glycolipids provide well-characterized ligands for studying interactions of proteins with carbohydrate sulfates and will complement studies of interactions with other cell-surface components including membrane-bound heparan sulfates and sulfate esters on N- and O-linked sugars of membrane glycoproteins.

---

[39] D. D. Roberts, S. B. Williams, H. R. Gralnick, and V. Ginsburg, *J. Biol. Chem.* **261**, 3306 (1986).

[39] Mouse and Rat Monoclonal Antibodies Directed against Carbohydrates

By JOHN L. MAGNANI

Many monoclonal antibodies obtained from hybridomas of mice and rats are directed against carbohydrates found in glycolipids and/or glycoproteins. This chapter provides a current list of these antibodies. Human carbohydrate-specific monoclonal antibodies are listed elsewhere in this volume [40] and are not included. Previous reviews on the subject are found in Refs. 1 and 2.

Some oligosaccharides are highly immunogenic (e.g., X-hapten, LNF III) and have been reported to bind many monoclonal antibodies. In such cases, only those antibodies which are well characterized or first published are listed. Different antibodies which bind the same oligosaccharide may, however, react with slightly different epitopes.[3] Other antibodies which react with identical epitopes but differ in binding affinities will react differently with cell surfaces. The binding of antibodies depends on antigen density. Thus, cells containing antigen below a certain threshold concentration may bind high affinity but not low affinity monoclonal antibodies directed against the same epitope.[4] Finally, differences in isotypes must be considered when comparing antibodies from this list.

Most of the antibodies presented in Table I were produced against developmentally regulated or cancer-associated antigens. Several have been used to either detect or treat human cancers. As the presence of these carbohydrates change with development and transformation, they may have interesting functions. The availability of monoclonal antibodies directed against specific oligosaccharide sequences combined with new immunological techniques will make it possible to study the function of these specific oligosaccharide sequences and their putative complementary receptor molecules.

---

[1] S. Hakomori, *in* "Monoclonal Antibodies Directed to Cell Surface Carbohydrates" (R. H. Kennett, K. B. Bechtol, and T. J. McKearn, eds.), p. 67. Plenum, New York, 1984.

[2] T. Feizi, *Nature (London)* **314**, 53 (1985).

[3] Y. Fukushi, S. Hakomori, E. Nudelman, and N. Cochran, *J. Biol. Chem.* (1984).

[4] J. L. Magnani, E. D. Ball, M. W. Fanger, S. Hakomori, and V. Ginsburg, *Arch. Biochem. Biophys.* **233**, 501 (1984).

TABLE I
MONOCLONAL ANTIBODIES DIRECTED AGAINST CARBOHYDRATES

Oligosaccharide structure	Name	Antibody	Isotype	Ref.
Galβ1-3GlcNAcβ1-3Galβ1-R	LNT	FC10.2	IgM	5
Galβ1-4GlcNAcβ1-3Galβ1-R	LNnT	My28, IB2, A5	IgM	6, 7, 8
Galβ1-4GlcNAcβ1 \ $\quad$ ^{6}Galβ1-R \ $\quad$ ^{3} \ Galβ1-4GlcNAcβ1	—	C6	IgM	8
Fucα1-2Galβ1-4GlcNAcβ1-R	H type 2	H-11 \ BE2	IgM \ IgM	9 \ 6
Galβ1-3GlcNAcβ1-R \ $\quad\quad$ 4 \ $\quad\quad$ Fucα1	Leᵃ, LNFII	CF4-C4 \ CF4-F4 \ DG4-1 \ BC9-E5 \ CO514	IgG$_{\gamma1}$ \ IgG$_{\gamma2}$ \ IgG$_{\gamma3}$ \ IgG$_{\gamma1}$ \ IgG$_{\gamma3}$	10, 11 \ \ 10
Galβ1-4GlcNAcβ1-R \ $\quad\quad$ 3 \ $\quad\quad$ Fucα1	Leˣ, LNFIII, X-hapten	SSEA-1, My1, VEP8, VEP9, VIM-D5, 534F8, PM-81, FH2, FH3	IgM	12, 13, 14, 15, 16, 17, 4, 18
Galβ1-4GlcNAcβ1-3Galβ1-4GlcNAcβ1-3Gal \ $\quad\quad$ 3 $\quad\quad\quad\quad\quad\quad$ 3 \ $\quad\quad$ Fucα1 $\quad\quad\quad\quad\quad$ Fucα1	Repeating Leˣ	FH4 \ FH5	IgG$_{\gamma3}$ \ IgM	18
Galβ1-3GlcNAcβ1-3Galβ1-4GlcNAcβ1-R \ $\quad\quad$ 4 $\quad\quad\quad\quad\quad\quad$ 3 \ $\quad\quad$ Fucα1 $\quad\quad\quad\quad\quad$ Fucα1	Leᵃ/Leˣ	C14	—	19
Fucα1-2Galβ1-3GlcNAβ1-R \ $\quad\quad\quad\quad$ 4 \ $\quad\quad\quad\quad$ Fucα1	Leᵇ, LNDI	CO431, CO294 \ 10c17	IgG$_{\gamma3}$ \ IgM	20, 11

(continued)

TABLE I (continued)

Oligosaccharide structure	Name	Antibody	Isotype	Ref.
Fucα1-2Galβ1-4GlcNAcβ1-R $\quad$3 $\quad$Fucα1	Difucosyl H type 2 (Y-hapten)	AH-6 C14/1/46/10 F-3 75.12	IgM — — —	21 22 23 24
NeuAcα2-3Galβ1-3GlcNAcβ1-3Galβ1-R	Sialylated LNT	K4	IgM	25
NeuAcα2-3Galβ1-3GlcNAcβ1-R $\quad$4 $\quad$Fucα1	Sialylated Le^a	19-9	IgG$_{\gamma 1}$	26
NeuAcα2-3Galβ1-4GlcNAcβ1-R $\quad$3 $\quad$Fucα1	Sialylated Le^x	CSLEX	IgM	27
NeuAcα2-3Galβ1-4GlcNAcβ1-4GlcNAcβ1-R $\quad$3$\qquad$3 $\quad$Fucα1$\quad$Fucα1 NeuAcα2	Sialylated difucosyl Le^x	FH6	IgM	28
NeuAcα2-3Galβ1-3GlcNAcβ1-R $\quad$4 $\quad$Fucα1 $\quad$6 NeuAcα2	Disialylated Le^a	FH7	IgG$_{\gamma 3}$	29
NeuAcα2-3Galβ1-3GlcNAcβ1-3Galβ1-R $\quad$6 NeuAcα2	Disialyl LNT	FH9	IgG$_{\gamma 2a}$	30
GalNAcα1-3Galβ1-3GlcNAcβ1-3Galβ1-R $\quad$2 $\quad$Fucα1	A (type 1)	AH21, CHL6 MASO16C	IgM	31, 32, 33

Structure	Antigen	Antibodies	Ig class	Ref.

GalNAcα1-3Galβ1-R
   2
   |
Fucα1
	A (type 1 and 2)	AH16, HT29-36
		MH2/6D4, A15/3D4,
		  TL5
			IgGγ3
			IgM
				33, 34
				35

GalNAcα1-3Galβ1-4GlcNAcβ1-R
   2
   |
Fucα1
	A (type 2)	HH4
		455
			—	33, 36

GalNAcα1-3GalNAcα1-3Galβ1-4GlcNAcβ1-3Galβ1-R
   2
   |
Fucα1
	A₁ (type 3)	TH-1	IgGγ2a	37

GalNAcα1-3Galβ1-3GlcNAcα1-3Gal-R
   2   4
   | |
Fucα1 Fucα1
	ALe^b	HH3	IgGγ2a	38

GalNAcα1-3Galβ1-4GlcNAcβ1-3Gal-R
   2   3
   | |
Fucα1 Fucα1
	ALe^y	HH2	IgGλ2a	38

Galα1-3Galβ1-4GlcNAc-R	B	2C5	IgM	39

Galα1-3Galβ1-3(4)GlcNAcβ1-R
   2
   |
Fucα1
	B (type 1 & 2)	1-3E-4, 1-3E-7, 1-1A-12
		E₂83-52
			IgM	40
				41

Galα1-3Galβ1-4GlcNAcβ1-R
   2
   |
Fucα1
	B (type 2)	E₁ 15-2, KH10, LD2,
		TC6, TD10, LA4
			IgM	41
				42

Galβ1-4Glcβ1-R	CDH	T₅A₇	—	43
Galα1-4Galβ1-4Glcβ1-R	CTH (blood group PK)	38.13	IgM	44
Galβ1-3GalNAcα1-R	T	49H.24	IgM	45
GalNAcα1-serine (threonine)	Tn	NCC-LU-35, NCC-LU-81	IgG	46
R-3GalNAcβ1-3Galα1-4R'	Globo series	SSEA-3	IgM	47

(continued)

TABLE I (*continued*)

Oligosaccharide structure	Name	Antibody	Isotype	Ref.
NeuAcα2-3Galβ1-3GalNAcβ1-R	Sialylated globo series	SSEA-4	IgG$_{γ3}$	48
Fucα1-2Galβ1-3GalNAcβ1-3Galα1-4Galβ1-R	GL-6 (globo H)	MBr1	—	49
GalNAcα1-3GalNAcβ1-3Galα1-4Galβ1-R	Forssman	M1:22:25:8	IgM	50
NeuAcα2-6Galβ1-R	LM1	1B2, 1B9	IgG$_{γ2b}$	51
NeuAcα2-3Galβ1-4Glcβ1-R	GM3	M2590	—	52
GalNAcβ1-4Galβ1-4Glcβ1-R 　　　3 　　　| 　NeuAcα1 　(NeuGc)	GM2	MoAb5-3	IgM	53
GalNAcβ1-4Galβ1-4Glcβ1-R	Asialo GM2	2D4 D11 G10	IgM, IgG$_{γ3}$	54
Fucα1-2Galβ1-3GalNAcβ1-4Galβ1-4Glcβ1-R 　　　　　　　　3 　　　　　　　| 　　　　　NeuAcα2	Fucosyl GM1	—	IgM	55
R-NeuAcα2-8NeuAcα2-3Galβ1-R 　| 　R	GD2 > GT3 > GD3	LeoMel3	IgM	56
NeuAcα2-8NeuAcα2-3Galβ1-4Glcβ1-R	GD3	R24 4.2 2B2, IF4, MG-21	IgG$_{γ3}$ IgM IgG$_{γ3}$	57 58 59
Neu5,9Acα2-8NeuAcα2-3Galβ1-4Glcβ1-R	9-O-acetyl GD3	D1.1	—	60
GalNAcβ1-4Galβ1-4Glcβ1-R 　　　3 　　| 　NeuAcα2	GD2	3F8, 2F7, 3G6, 3A7	IgM	61
Galβ1-3GalNAcβ1-4Galβ1-4Glcβ1-R 　　　　　　　3 　　　　　　| 　　　　NeuAcα2-8NeuAcα2	GD1b	—	IgM	62

Structure	Antibody	Common name	Ig class	No.
NeuAcα2-8NeuAcα2-8NeuAcα2-3Galβ1-4Glcβ1-R	18B8	GT3	IgM	63
(NeuAcα2-8NeuAcα2-8)n	2-2-B	Colominic acid	IgM	64

Galβ1-4GlcNAcβ1-3Galβ1-R
      6
      |
   SO_4^{2-}
	MB32, MB34	Keratan sulfate	IgM	65

	1/20/S-D-4		IgG$_{\gamma 1}$	66
SO_4-3GlcUAβ1-3Galβ1-4GlcNAcβ1-3Galβ1-4Glcβ1-R	HNK-1 (Leu-7)	—	IgM	67

^{a} N.R., not reported.

5. H. C. Gooi, L. K. Williams, K. Uemura, R. A. J. McIlhinney, and T. Feizi, *Mol. Immunol.* **20**, 607 (1983).
6. W. W. Young, Jr., J. Portoukalian, and S. Hakomori, *J. Biol. Chem.* **256**, 10967.
7. S. L. Spitalnik, J. F. Schwartz, J. L. Magnani, D. D. Roberts, P. F. Spitalnik, C. I. Civin, and V. Ginsburg, *Blood* **66**, 319 (1985).
8. B. A. Fenderson, C. Hahnel, and E. M. Eddy, *Devel. Biol.* **100**, 318 (1983).
9. R. W. Knowles, Y. Bai, C. L. Daniels, and W. Watkins, *J. Immunogenet.* **9**, 69 (1982).
10. W. W. Young, Jr., H. S. Johnson, Y. Tamura, K.-A. Karlsson, G. Larson, J. M. R. Parker, D. P. Khare, U. Sophr, D. A. Baker, O. Hindsgoul, and R. V. Lemieux, *J. Biol. Chem.* **258**, 4890 (1983).
11. M. Blaszczyk, G. C. Hanson, K. A. Karlsson, G. Larson, N. Stromberg, J. Thurin, M. Herlyn, Z. Steplewski, and H. Koprowski, *Arch. Biochem. Biophys.* **233**, 161 (1984).
12. S. Hakomori, E. Nudelman, S. Levery, D. Solter, and B. B. Knowles, *Biochem. Biophys. Res. Commun.* **100**, 1578 (1981).
13. M. Brockhaus, J. L. Magnani, M. Herlyn, M. Blaszczyk, Z. Steplewski, H. Koprowski, and V. Ginsburg, *Arch. Biochem. Biophys.* **217**, 647 (1982).
14. L. C. Huang, M. Brockhaus, J. L. Magnani, F. Cuttitta, S. Rosen, J. D. Minna, and V. Ginsburg, *Arch. Biochem. Biophys.* **220**, 318 (1983).
15. L. C. Huang, C. I. Civin, J. L. Magnani, J. H. Shaper, and V. Ginsburg, *Blood* **61**, 1020 (1983).
16. H. C. Gooi, S. J. Thorpe, E. F. Hounsell, H. Runysold. D. Kraft, O. Forster, and T. Feizi, *Eur. J. Immunol.* **13**, 306 (1983).
17. D. L. Urdal. T. A. Brentnall, I. D. Bernstein, and S. Hakomori, *Blood* **62**, 1022 (1983).
18. Y. Fukushi, S. Hakomori, E. Nudelman, and N. Cochran, *J. Biol. Chem.* **259**, 4681 (1984).
19. H. C. Gooi, N. J. Jones, E. F. F. Hounsell, P. Scudder, J. Hilkens, J. Hilgers, and T. Feizi, *Biochem. Biophys. Res. Commun.* **131**, 543 (1985).
20. M. Brockhaus, J. L. Magnani, M. Herlyn, M. Blaszczyk, Z. Steplewski, H. Koprowski, K. A. Karlsson, G. Larson, and V. Ginsburg, *J. Biol. Chem.* **256**, 13223 (1981).
21. K. Abe, J. M. McKibben, and S. Hakomori, *J. Biol. Chem.* **258**, 11793 (1983).

(continued)

References to TABLE I (*continued*)

22. A. Brown, T. Feizi, H. C. Gooi, M. J. Embleton, J. K. Picard, and R. W. Baldwin, *Biosci. Rep.* **3**, 163 (1983).

23. K. O. Lloyd, G. Larson, N. Strömberg, J. Thurin, and K. A. Karlsson, *Immunogenetics* **17**, 537 (1983).

24. C. Blaineau, J. LePendu, D. Arnaud, F. Connan, and P. Avner, *EMBO J.* **2**, 2217 (1983).

25. M. N. Fukuda, B. Botler, K. O. Lloyd, W. J. Rettig, P. R. Tiller, and A. Dell, *J. Biol. Chem.* **261**, 5145 (1986).

26. J. L. Magnani, B. Nilsson, M. Brockhaus, D. Zopf, Z. Steplewski, H. Koprowski, and V. Ginsburg, *J. Biol. Chem.* **257**, 14365 (1982).

27. K. Fukushima, M. Hirota, P. I. Terasaki, A. Wakisaka, H. Togashi, D. Chia, N. Suyama, Y. Fukushi, E. Nudelman, and S. Hakomori, *Can. Res.* **44**, 5279 (1984).

28. Y. Fukushi, R. Kannagi, S. Hakomori, T. Shepard, B. G. Kulander, and J. W. Singer, *Can. Res.* **45**, 3711 (1985).

29. E. Nudelman, Y. Fukashi, S. B. Levery, T. Higichi, and S. Hakomori, *J. Biol. Chem.* **261**, 5487 (1986).

30. Y. Fukushi, E. Nudelman, S. B. Levery, T. Higuchi, and S. Hakomori, *Biochem.* **25**, 2859 (1986).

31. K. Abe, S. B. Levery, and S. Hakomori, *J. Immunol.* **132**, 1951 (1984).

32. K. Furukama, H. Claussen, S. Hakomori, J. Sakamoto, K. Look, A. Lundblad, M. J. Mattes, and K. O. Lloyd, *Biochemistry* **24**, 7820 (1985).

33. H. C. Gooi, E. F. Hounsell, J. K. Picard, A. D. Lowe, D. Voak, E. S. Lennox, and T. Feizi, *J. Biol. Chem.* **260**, 13218 (1985).

34. K. Abe, S. B. Levery, and S. Hakomori, *J. Immunol.* **132**, 1951 (1984).

35. D. Voak, S. Sacks, T. Alderson, F. Takei, E. Lennox, J. Jarvis, C. Milstein, and J. Darnborough, *Vox. Sang.* **39**, 134–140 (1980).

36. H. Claussen, S. Hakomori, N. Graem, and E. Dabelsteen, *J. Immunol.* **136**, 326 (1986).

37. H. Claussen, S. B. Levery, E. Nudelman, S. Tsuchiya, and S. Hakomori, *Proc. Natl. Acad. Sci.* **82**, 1199 (1985).

38. H. Claussen, J. M. McKibben, and S. Hakomori, *Biochem.* **24**, 6190 (1985).

39. B. J. Randle, *J. Embryol. Exp. Morphol.* **70**, 261 (1982).

40. D. R. Bundle, M. A. J. Gidney, N. Kassuri, and A. F. R. Rahman, *J. Immunol.* **129**, 678 (1982).

41. G. C. Hansson, K.-A. Karlsson, G. Larson, J. M. McKibbin, M. Blaszczyk, M. Herlyn, Z. Steplewski, and H. Koprowski, *J. Biol. Chem.* **258**, 4091 (1983).

42. T. M. Jessel and J. Dodd, *Philos. Trans. R. Soc. Lond.* (*Biol.*) **308**, 271 (1985).

43. F. W. Symington, I. D. Bernstein, and S. Hakomori, *J. Biol. Chem.* **259**, 6008 (1984).

44. E. Nudelman, R. Kannagi, S. Hakomori, M. Parsons, M. Lipinoki, J. Wills, M. Fellows, and T. Tursy, *Science* **220**, 509 (1983).

45. A. F. R. Rahman and B. M. Longenecker, *J. Immunol.* **129**, 2021 (1982).

46. S. Hirohashi, H. Claussen, T. Yamada, Y. Shimosato, and S. Hakomori, *Proc. Natl. Acad. Sci. U.S.A.* **82**, 7039 (1985).

47. R. Kannagi, S. B. Levery, F. Ishigami, S. Hakomori, L. H. Shevinsky, B. B. Knowles, and D. Solter, *J. Biol. Chem.* **258**, 8934 (1983).

48. R. Kannagi, N. A. Cochran, F. Ishigami, S. Hakomori, P. W. Andrews, B. B. Knowles, and D. Solter, *EMBO J.* **2**, 2355 (1983).

49. S. B. Levery, E. G. Brenner, S. Hakomori, S. Sonnino, R. Ghidoni, and M. I. Colnaghi, *Fed. Proc., Fed. Am. Soc. Exp. Biol.* **43**, 1751 (1984).

50. K. R. Willison, R. A. Karol, A. Suzuki, S. K. Kunder, and D. M. Marcus, *J. Immunol.* **129**, 603 (1982).

51. S. Hakomori, C. M. Patterson, E. Nudelman, and K. Sekiguchi, *J. Biol. Chem.* **258**, 11819 (1983).

52. Y. Hirabayashi, A. Hamaoka, M. Matsumoto, T. Matsuara, M. Tagawa, S. Wakabagashi, and M. Taniguchi, *J. Biol. Chem.* **260**, 13328 (1985).

53. E. N. Natoli, Jr., P. O. Livingston, C. S. Pukel, K. O. Lloyd, H. Wiegandt, J. Szalay, H. F. Oettgen, and L. J. Old, *Can. Res.* **46**, 4116 (1986).

54. W. W. Young Jr., E. M. S. MacDonald, R. C. Nowinski, and S. Hakomori, *J. Exp. Med.* **150**, 1008 (1979).

55. P. Fredman, T. Brezicka, J. Holmgren, L. Lindholm, O. Nilsson, and L. Svennerholm, *Biochim. Biophys. Acta* **875**, 316 (1986).

56. S. Fukuta, J. A. Werkmeister, V. Ginsburg, and J. L. Magnani, *Fed. Proc.* **45**, 1902 (1986).

57. C. S. Pukel, K. O. Lloyd, L. R. Trabassos, N. G. Dippold, H. F. Oettgen, and L. J. Old, *J. Exp. Med.* **155**, 1133 (1982).

58. E. Nudelman, S. Hakomori, R. Kannagi, S. Levery, M. Yeh, K. E. Hellström, and I. Hellström, *J. Biol. Chem.* **257**, 12752 (1982).

59. I. Hellström, V. Brankovan, and K. E. Hellström, *Proc. Natl. Acad. Sci. U.S.A.* **82**, 1499 (1985).

60. D. A. Cheresh, A. P. Varki, N. M. Varki, W. B. Stallcup, J. Levine, and R. A. Reisfeld, *J. Biol. Chem.* **259**, 7453 (1984).

61. N. V. Cheung, V. M. Saarinen, J. E. Neely, B. Landmeier, D. Donovan, and P. T. Coccia, *Cancer Res.* **45**, 2642 (1985).

62. P. Fredman, S. Jeansson, E. Lycke, and L. Svennerholm, *FEBS Lett.* **189**, 23 (1985).

63. C. Dubois, J. L. Magnani, G. B. Grunwald, S. L. Spitalnik, G. D. Trisler, M. Nirenberg, and V. Ginsburg, *J. Biol. Chem.* **261**, 3826 (1986).

64. W. D. Zollinger and R. E. Mandrell, *Infect. Immun.* **40**, 257 (1983).

65. C. Moreno, J. Hewitt, K. Hastings, and D. Brown, *J. Gen. Microbiol.* **129**, 2451 (1983).

66. B. Caterson, J. E. Christner, and J. R. Baker, *J. Biol. Chem.* **258**, 8848 (1983).

67. K. H. Chou, A. A. Ilyas, J. E. Evans, R. H. Quarles, and F. B. Jungalwala, *Biochem. Biophys. Res. Commun.* **128**, 383 (1985).

[40] Human Monoclonal Antibodies Directed against Carbohydrates

By STEVEN L. SPITALNIK

Studies using human monoclonal antibodies have been instrumental in elucidating basic concepts in immunology and basic disease mechanisms in a variety of human disorders. Many human monoclonal antibodies which recognize known antigens specifically bind to carbohydrates. In particular, many of these antibodies recognize blood-group antigens.

The carbohydrate-specific human monoclonal antibodies have classically been isolated from patients with plasma cell dyscrasias, particularly multiple myeloma and Waldenstrom's macroglobulinemia. These patients have an expanded or neoplastic clone of plasma cells which secretes homogeneous immunoglobulins. The pathologic antibody can then be readily recovered, usually in large amounts, from the serum or plasma. The clinical features of the patient occasionally yield a clue to the antigen specificity of the antibody. Thus, the serum of a patient presenting with chronic cold hemagglutinin disease contains an antibody recognizing an oligosaccharide antigen on red blood cells.[1] In patients with plasma cell dyscrasia and peripheral neuropathy, the monoclonal antibody often binds to antigenic determinants on peripheral nerve myelin.[2] In another approach, antigen-specific monoclonal antibodies are identified by testing many sera from patients with plasma cell dyscrasias against a wide variety of known antigens.[3,4] This method generally has a low yield.[5] Several excellent reviews have appeared describing the clinical features of patients with plasma cell dyscrasias and associated syndromes, and the immunochemistry of the pathologic antibodies.[6–13]

[1] D. M. Marcus, E. A. Kabat, and R. E. Rosenfield, *J. Exp. Med.* **118,** 175 (1963).
[2] A. A. Ilyas, R. H. Quarles, T. D. MacIntosh, M. J. Dobersen, B. D. Trapp, M. C. Dalakas, and R. O. Brady, *Proc. Natl. Acad. Sci. U.S.A.* **81,** 1225 (1984).
[3] W. D. Terry, M. M. Boyd, J. S. Rea, and R. Stein, *J. Immunol.* **104,** 256 (1970).
[4] M. Freedman, R. Merrett, and W. Pruzanski, *Immunochemistry* **13,** 193 (1976).
[5] T. J. Yoo and E. C. Franklin, *J. Immunol.* **107,** 365 (1971).
[6] T. Isobe and E. F. Osserman, *Ann. N.Y. Acad. Sci.* **190,** 507 (1971).
[7] H. Metzger, *Am. J. Med.* **47,** 837 (1969).
[8] M. Seligmann and J. C. Brouet, *Semin. Hematol.* **10,** 163 (1973).
[9] D. Roelcke, *Clin. Immunol. Immunopathol.* **2,** 266 (1974).
[10] T. Feizi, *Immunol. Commun.* **10,** 127 (1981).
[11] D. Crisp and W. Pruzanski, *Am. J. Med.* **72,** 915 (1982).
[12] W. Pruzanski and K. H. Shumak, *N. Engl. J. Med.* **297,** 538 (1977).
[13] W. Pruzanski and K. H. Shumak, *N. Engl. J. Med.* **297,** 583 (1977).

TABLE I
CARBOHYDRATE-SPECIFIC HUMAN MONOCLONAL ANTIBODIES[a]

Name (isotype)	Specificity	Oligosaccharide specificity	Comments	Ref.[b]
— (μ)	Heparin	—	No binding to chondroitin sulfate	34
CA ($\gamma_1\kappa$)	Heparin	—	Binds by Fab fragment, also binds to agar	35
CAR ($\gamma_3\lambda$)	Heparin	—	—	35
CH ($\gamma_1\kappa$)	Heparin	—	—	35
LE ($\gamma_3\kappa$)	Heparin	—	Also binds to agar	35
MA ($\gamma_1\lambda$)	Heparin	—	Binds by Fab fragment	35
OA ($\gamma_1\kappa$)	Heparin	—	Also binds to agar	35
SU ($\gamma_1\kappa$)	Heparin	—	Also binds to agar	35
VA ($\gamma_1\kappa$)	Heparin	—	Also binds to agar	35
ED ($\gamma\kappa$)	Heparin	—	Binds by Fab fragment; antibody forms crystals	36
— ($\mu\kappa$)	Heparin	—	—	37
CK (μ)	Heparin	—	Desulfation of aminosulfate groups resulted in loss of antigenicity	4
BRU (γ)	Heparin	—	Desulfation of aminosulfate groups resulted in loss of antigenicity	4
Sab ($\alpha\lambda$)	Heparin	—	—	38
EH ($\mu\kappa$)	Dextran sulfate	—	Binds by Fab fragment; also recognizes multiple other antigens, including ribitol teichoic acid from *Staphylococcus aureus* and a ribose–phosphate polymer from *Haemophilus influenzae* Type B	39

(continued)

TABLE I (*continued*)

Name (isotype)	Specificity	Oligosaccharide specificity	Comments	Ref.[b]
MAC ($\mu\kappa$)	Chondroitin sulfate C	2-Acetamido-2-deoxy-3-O-(α-L-*threo*-hex-4-enopyranosyl-uronic acid)-D-galactose 6-sulfate	The nonsulfated disaccharide was also a good inhibitor	40, 41
FIS ($\mu\kappa$)	Chondroitin sulfate A, B, C	2-Acetamido-2-deoxy-3-O-(α-L-*threo*-hex-4-enopyranosyl-uronic acid)-D-galactose 4-sulfate	The nonsulfated disaccharide was also a good inhibitor	40, 41
TEM (γ)	Pneumococcal polysaccharides SI, SIII, SVIII, SX	—	Binds by Fab fragment; antibody forms crystals	4
16M3C8 (γ)	*Haemophilus influenzae* Type B capsular polysaccharide	-3)βD-Ribf(1-1)D-ribitol-5-phosphate-	—	18, 42
16M4F2 (γ)	*Haemophilus influenzae* Type B capsular polysaccharide	-3)βD-Ribf(1-1)D-ribitol-5-phosphate-	—	18, 42
Anti-PRP (γ)	*Haemophilus influenzae* Type B capsular polysaccharide	-3)βD-Ribf(1-1)D-ribitol-5-phosphate-	—	42, 43
RO (μ)	*Klebsiella* type 12 and 13 acid polysaccharides	—	—	44
Th (μ)	*Klebsiella* type 35 acid polysaccharide	—	—	44
We ($\mu\lambda$)	*Klebsiella* type 13 acid polysaccharide	—	Binds by Fab fragment	44
MAY ($\mu\kappa$)	*Klebsiella* K33 polysaccharide	3,4-Pyruvylated β-D-galactose	—	45–47

WEA (μκ)	Klebsiella K33 polysaccharide	3,4-Pyruvylated β-D-galactose	Binds by Fab fragment	45–48
NAE (μκ)	Klebsiella K33 polysaccharide	3,4-Pyruvylated β-D-galactose	Binds by Fab fragment	47, 49
ROS (μκ)	Klebsiella K33 polysaccharide	3,4-Pyruvylated β-D-galactose	—	47
CLE (μκ)	Klebsiella K14 polysaccharide	Methyl-4,6-O-(1-Carboxyethylidene)-α-D-glucose (S isomer)	—	47
TO (μκ)	Klebsiella ozaenae Type AE, strain 047 acid polysaccharide	—	Binds by Fab fragment	47, 49, 50
— (μ)	Peripheral nerve myelin	GlcUAβ1-3Galβ1-4GlcNAcβ1-3Galβ1-4Glcβ1-1 ceramide	This is a tentative structure; the glycolipid is also sulfated	2, 51, 52
OFA-1-1 (μκ)	G_{M2}	GalNAcβ1-4(NeuAcα2-3)Galβ1-4Glcβ1-1 ceramide	—	53, 54
OFA-1-2 (μκ)	G_{D2}	GalNAcβ1-4(NeuAcα2-8 NeuAcα2-3)Galβ1-4Glcβ1-1 ceramide	—	26, 53, 55, 56
McG (μκ)	—	Nonreducing terminal αGalNAc or βGalNAc	Originally described as recognizing Forssman antigen	57–59
H1-C4 (μκ)	Forssman	GalNAcα1-3GalNAcβ1-3Galα1-4Galβ1-4Glcβ1-1 ceramide	—	60
WOO (μλ)	—	Galβ1-4GlcNAcβ1-3Galβ1-4Glc	—	61
Fl (μ)	Fl blood-group antigen	NeuAcα2-3Galβ1-4GlcNAcβ1-3(Fucα1-2Galβ1-4GlcNAcβ1-6)Galβ1-4GlcNAcβ1-3Galβ1-4Glcβ1-1 ceramide	Antibody is a cold hemagglutinin	25, 62, 63
Vo (μλ)	Vo blood-group antigen	NeuAcα2-3Galβ1-4GlcNAc-R	Tentative antigen structure; antibody is a cold hemagglutinin	64
Sa (μκ)	Sa blood-group antigen	NeuAcα2-3Galβ1-4GlcNAcβ1-3Galβ1-4GlcNAcβ1-3Galβ1-4Glcβ1-1 ceramide	Antibody is a cold hemagglutinin	63, 65–67

(continued)

TABLE I (*continued*)

Name (isotype)	Specificity	Oligosaccharide specificity	Comments	Ref.[b]
MAT ($\mu\kappa$)	Gd blood-group antigen	NeuAcα2-3Galβ1-4Glc	Antibody is a cryoglobulin and a cold hemagglutinin; NeuAcα2-3Glc also inhibits antibody binding	67, 68
Kn ($\mu\kappa$)	Gd blood-group antigen	NeuAcα2-3Galβ1-4GlcNAcβ1-3Galβ1-4GlcNAcβ1-3Galβ1-4Glcβ1-1 ceramide	Antibody also recognizes NeuAcα2-3Galβ1-4Glcβ1-1 ceramide, NeuAcα2-3Galβ1-4GlcNAcβ1-3Galβ1-4Glcβ1-1 ceramide, and NeuAcα2-3Glc; antibody is a cold hemagglutinin	63, 65–67, 69, 70
Nei ($\mu\kappa$)	Gd blood-group antigen	NeuAcα2-3Galβ1-4GlcNAcβ1-3Galβ1-4GlcNAcβ1-3Galβ1-4Glcβ1-1 ceramide	Antibody is a cold hemagglutinin	65–67, 69, 70
MKV ($\mu\kappa$)	Pr_2 blood-group antigen	NeuAcα2-3Galβ1-4Glcβ1-1 ceramide (G_{M3}): NeuAcα2-3Galβ1-4GlcNAcβ1-3Galβ1-4Glcβ1-1 ceramide	Antibody is a cold hemagglutinin and a cryoglobulin	71
Mou ($\mu\kappa$)	Pr_2 blood-group antigen	NeuAcα2-3Galβ1-4Glcβ1-1 ceramide (G_{M3})	Binds by Fab fragment; antibody is a cold hemagglutinin and a cryoglobulin	72
#33 ($\mu\kappa$)	Pr_2 blood-group antigen	NeuAcα2-3Galβ1-4Glc	—	73
L.Th ($\mu\kappa$)	Pr_2 blood-group antigen	NeuAcα2-3Galβ1-4Glc-R; NeuAcα2-3Galβ1-3(NeuAcα2-6)GalNAcitol; NeuAcα2-3Galβ1-4GlcNAcβ1-3Galβ1-4GlcNAcβ1-1 ceramide	Antibody is a cold hemagglutinin; it also binds to several other related gangliosides; treatment of antigen with NaIO4 enhances antibody binding	63, 65, 69, 74, 75
#8 ($\mu\kappa$)	Pr_1 blood-group antigen	NeuAcα2-3/6Galβ1-4Glc	Antibody is a cold hemagglutinin	73
#20 (μ)	Pr_1 blood-group antigen	NeuAcα2-3/6Galβ1-4Glc	Antibody is a cold hemagglutinin	73

Antibody	Specificity	Structure	Comments	References
#21 ($\mu\kappa$)	Pr$_1$ blood-group antigen	NeuAcα2-3/6Galβ1-4Glc	Antibody is a cold hemagglutinin	73
#22 (μ)	Pr$_1$ blood-group antigen	NeuAcα2-3/6Galβ1-4Glc	Antibody is a cold hemagglutinin	73
Stef (μ)	i blood-group antigen	Galβ1-4GlcNAcβ1-3Glcβ1-4Galβ1-3Galβ1-4Glcβ1-1 ceramide	Antibody is a cold hemagglutinin	76
Den (μ)	i blood-group antigen	Galβ1-4GlcNAcβ1-3Galβ1-4GlcNAcβ1-3Galβ1-4Glcβ1-1 ceramide	Antibody is a cold hemagglutinin	77–81
Galli (μ)	i blood-group antigen	Galβ1-4GlcNAcβ1-3Galβ1-4GlcNAcβ1-3Galβ1-4GlcNAcβ1-O-methyl	Antibody is a cold hemagglutinin; it does not bind to Galβ1-4GlcNAcβ1-3Galβ1-4GlcNAcβ1-3Galβ1-4Glcβ1-1 ceramide	77–80
McDon (μ)	i blood-group antigen	Galβ1-4GlcNAcβ1-3Galβ1-4GlcNAcβ1-3Galβ1-4Glcβ1-1 ceramide	Antibody is a cold hemagglutinin	77–80
Tho (μ)	i blood-group antigen	Galβ1-4GlcNAcβ1-3Galβ1-4GlcNAcβ1-3Galβ1-4Glcβ1-1 ceramide	Antibody is a cold hemagglutinin	77–80
Nic (μ)	i blood-group antigen	Galβ1-4GlcNAcβ1-3Galβ1-4GlcNAcβ1-3Galβ1-4Glcβ1-1 ceramide	Antibody is a cold hemagglutinin	79
Hog (μ)	i blood-group antigen	Galβ1-4GlcNAcβ1-3Galβ1-4GlcNAcβ1-3Galβ1-4Glcβ1-1 ceramide	Antibody is a cold hemagglutinin	77, 78
McC ($\mu\kappa$)	i blood-group antigen	Galβ1-4GlcNAcβ1-3Galβ1-4GlcNAcβ1-3Galβ1-4Glcβ1-1 ceramide	Antibody is a cold hemagglutinin	77–80, 82
Step (μ)	I blood-group antigen	Galβ1-4GlcNAcβ1-3(GlcNAcβ1-5)Galβ1-4GlcNAcβ1-3Galβ1-4Glcβ1-1 ceramide	Antibody is a cold hemagglutinin	10, 77–80, 83–85

(continued)

TABLE I (*continued*)

Name (isotype)	Specificity	Oligosaccharide specificity	Comments	Ref.[b]
Gra (μ)	I blood-group antigen	Galβ1-4GlcNAcβ1-3(Galβ1-4GlcNAcβ1-6)Galβ1-4GlcNAcβ1-3Galβ1-4Glcβ1-1 ceramide	Antibody is a cold hemagglutinin	77, 78, 83
Sch (μ)	I blood-group antigen	Galβ1-4GlcNAcβ1-3(Galβ1-4GlcNAcβ1-6)Galβ1-4GlcNAcβ1-3Galβ1-4Glcβ1-1 ceramide	Antibody is a cold hemagglutinin	78, 83
Phi (μ)	I blood-group antigen	Galβ1-4GlcNAcβ1-3(Galβ1-4GlcNAcβ1-6)Galβ1-4GlcNAcβ1-3Galβ1-4Glcβ1-1 ceramide	Antibody is a cold hemagglutinin	10, 77, 79, 83
Ver (μ)	I blood-group antigen	Galβ1-4GlcNAcβ1-3(Galβ1-4GlcNAcβ1-6)Galβ1-4GlcNAcβ1-3Galβ1-4Glcβ1-1 ceramide	Antibody is a cold hemagglutinin	79, 80, 83
Ful (μ)	I blood-group antigen	Galβ1-4GlcNAcβ1-3(Galβ1-4GlcNAcβ1-6)Galβ1-4GlcNAcβ1-3Galβ1-4Glcβ1-1 ceramide	Antibody is a cold hemagglutinin	83
Da (μ)	I blood-group antigen	Galβ1-4GlcNAcβ1-3(Galβ1-4GlcNAcβ1-6)Galβ1-4GlcNAcβ1-3Galβ1-4Glcβ1-1 ceramide	Antibody is a cold hemagglutinin	77, 83
Low (μ)	I blood-group antigen	Galβ1-4GlcNAcβ1-3(Galβ1-4GlcNAcβ1-6)Galβ1-4GlcNAcβ1-3Galβ1-4Glcβ1-1 ceramide	Antibody is a cold hemagglutinin	80, 83

Antibody	Antigen	Structure	Notes	Ref.
Nay (μ)	I blood-group antigen	Galβ1-4GlcNAcβ1-3(Galβ1-4GlcNAcβ1-6Galβ1-4GlcNAcβ1-3Galβ1-4Glcβ1-1 ceramide	Antibody is a cold hemagglutinin	79, 80
Bannister (μ)	I blood-group antigen	Galβ1-4GlcNAcβ1-3(Galβ1-4GlcNAcβ1-6)Gal-R	Antibody is a cold hemagglutinin	84, 85
Bumeister (μ)	I blood-group antigen	Galβ1-4GlcNAcβ1-3(Galβ1-4GlcNAcβ1-6)Gal-R	Antibody is a cold hemagglutinin	84, 85
Demandox (μ)	I blood-group antigen	Galβ1-4GlcNAcβ1-3(Galβ1-4GlcNAcβ1-6)Gal-R	Antibody is a cold hemagglutinin	84, 85
Jones (μ)	I blood-group antigen	Galβ1-4GlcNAcβ1-3(Galβ1-4GlcNAcβ1-6)Gal-R	Antibody is a cold hemagglutinin	84, 85
McDonald (μ)	I blood-group antigen	Galβ1-4GlcNAcβ1-3(Galβ1-4GlcNAcβ1-6)Gal-R	Antibody is a cold hemagglutinin	84, 85
Wal (μ)	I blood-group antigen	Galβ1-4GlcNAcβ1-3(Galβ1-4GlcNAcβ1-6)Galβ1-4GlcNAcβ1-3(Galβ1-6)Galβ1-4GlcNAcβ1-3Galβ1-4Glcβ1-1 ceramide	Antibody is a cold hemagglutinin	76
Lin (μλ)	I blood-group antigen	Galα1-3Galβ1-4GlcNAcβ1-3(Galα1-3Galβ1-4GlcNAcβ1-6)Galβ1-4GlcNAcβ1-3Galβ1-4Glcβ1-1 ceramide	Antibody is a cold hemagglutinin	86
Hor (μκ)	I blood-group antigen	Galα1-3Galβ1-4GlcNAcβ1-3(Galα1-3Galβ1-4GlcNAcβ1-6)Galβ1-4GlcNAcβ1-3Galβ1-4Glcβ1-1 ceramide	Antibody is a cold hemagglutinin	86
Hir (μκ)	I blood-group antigen	Galα1-3Galβ1-4GlcNAcβ1-3(Galα1-3Galβ1-4GlcNAcβ1-6)Galβ1-4GlcNAcβ1-3Galβ1-4Glcβ1-1 ceramide	Antibody is a cold hemagglutinin	86

(continued)

TABLE I (*continued*)

Name (isotype)	Specificity	Oligosaccharide specificity	Comments	Ref.[b]
Sat ($\mu\kappa$)	I blood-group antigen	Galα1-3Galβ1-4GlcNAcβ1-3(Galα1-3Galβ1-4GlcNAcβ1-6)Galβ1-4GlcNAcβ1-3Galβ1-4Glcβ1-1 ceramide	Antibody is a cold hemagglutinin	86
Scha ($\mu\lambda$)	—	Galα1-3Galβ1-4GlcNAcβ1-3(Galα1-3Galβ1-4GlcNAcβ1-6)Galβ1-4GlcNAcβ1-3Galβ1-4Glcβ1-1 ceramide	Antibody is a cold hemagglutinin	86
Ma ($\mu\kappa$)	I blood-group antigen	Galβ1-4GlcNAcβ1-6-R	Antibody is a cold hemagglutinin	10, 19, 77–81, 83, 87–89
Woj (μ)	I blood-group antigen	Galβ1-4GlcNAcβ1-6-R	Antibody is a cold hemagglutinin	10, 19, 78, 83
Sti (μ)	I blood-group antigen	Galβ1-4GlcNAcβ1-6-R	Antibody is a cold hemagglutinin	10

[a] All sugars are in the D configuration and the pyranose form unless otherwise noted. Sugars are abbreviated as follows: Glc, glucose, GlcNAc, N-acetylglucosamine; Gal, galactose, GalNAc, N-acetylgalactosamine; NeuAc, N-acetylneuraminic acid; GlcUA, glucuronic acid; Ribf, D-ribose in furanose form.

[b] Numbers refer to text footnotes.

   Recently, other methods have become available for obtaining antigen-specific human monoclonal antibodies. These depend on immortalizing human B lymphocytes by forming human–human hybridomas,[14] human–murine heterohybridomas,[15] or by infecting cells with Epstein–Barr virus.[16] These techniques have several advantages including continuous and specific production of antibody, potential production of antibody with any desired specificity following immunization *in vitro*[17] or *in vivo*,[18] availability of the continuous cell line as a source of mRNA, and ability to select for mutants with identical antigenic specificity but variable isotype.

   Carbohydrate-specific human monoclonal antibodies are a useful model system for studying protein–carbohydrate interactions.[19] They have also been used to study the relationship between cross-reactive idiotypes and antigen specificity,[20–22] antibody amino acid sequence and antigen specificity,[23] to characterize defined carbohydrate structures on normal and malignant cells,[24] to isolate and characterize novel oligosaccharide antigens,[25] and for immunotherapy of bacterial infections[18] and human tumors.[26]

   The antibodies described in Table I are monoclonal; polyclonal carbohydrate-specific human antibodies are not included. Data were presented in the original papers describing antibody interactions with well-defined carbohydrate antigens. Ideally, studies should also have been performed with monoclonal antibody Fab fragments to ensure that a true antigen–antibody interaction was examined.[7,27] If this is not done, it may lead to erroneous results.[28] However, this was not done in all cases. Many antibodies have been described which most probably bind to carbohydrates,

[14] L. Olsson and H. S. Kaplan, *Proc. Natl. Acad. Sci. U.S.A.* **77,** 5429 (1980).
[15] J. F. Schwaber and F. S. Rosen, *J. Exp. Med.* **148,** 974 (1978).
[16] M. Steinitz, G. Klein, S. Koskimies, and O. Mäkelä, *Nature (London)* **269,** 420 (1977).
[17] H. M. Dosch and E. W. Gelfand, *J. Immunol.* **118,** 302 (1977).
[18] F. Gigliotti, L. Smith, and R. A. Insel, *J. Infect. Dis.* **149,** 43 (1984).
[19] E. A. Kabat, J. Liao, and R. U. Lemieux, *Immunochemistry* **15,** 727 (1978).
[20] R. C. Williams, Jr., H. G. Kunkel, and J. D. Capra, *Science* **161,** 379 (1968).
[21] J. Lecomte and T. Feizi, *Clin. Exp. Immunol.* **20,** 287 (1975).
[22] M. Pfreundschuh, B. Dorken, D. Roelcke, W. Romer, M. Poll, and H. Poliwoda, *Blut* **46,** 111 (1983).
[23] A. C. Wang, H. H. Fudenberg, J. V. Wells, and D. Roelcke, *Nature (London)* **243,** 127 (1973).
[24] A. Kapadia, T. Feizi, D. Jewell, J. Keeling, and G. Slavin, *J. Clin. Pathol.* **34,** 320 (1981).
[25] R. Kannagi, D. Roelcke, K. A. Peterson, Y. Okada, S. B. Levery, and S. I. Hakomori, *Carbohydr. Res.* **120,** 143 (1983).
[26] M. Katano, M. Jien, and R. F. Irie, *Eur. J. Cancer Clin. Oncol.* **20,** 1053 (1984).
[27] H. Metzger, *Biochemistry* **57,** 1490 (1967).
[28] P. Burtin and M. C. Gendron, *Immunochemistry* **8,** 423 (1971).

but detailed structural information is not yet available.[29-33] The antigen specificities were often defined by hapten-inhibition studies with defined carbohydrates using precipitin reactions, agglutination, radioimmunoassay, or ELISA as the test systems. In some cases direct binding of antibody to carbohydrate was examined using precipitin reactions, radioimmunoassay, ELISA, complement fixation, or direct binding to glycolipids separated by thin-layer chromatography.

[29] Y. S. Schoenfeld, S. C. Hsu-Lin, J. E. Gabriels, L. E. Silberstein, B. C. Furie, B. Furie, B. D. Stollar, and R. S. Schwartz, *J. Clin. Invest.* **70**, 205 (1982).

[30] A. E. G. K. von dem Borne, J. J. Mol, N. Joustra-Maas, J. G. Pegels, M. M. A. C. Langenhuijsen, and C. P. Engelfreit, *Br. J. Haematol.* **50**, 345 (1982).

[31] D. Roelcke, *Vox Sang.* **48**, 181 (1985).

[32] H. Rochant, H. Tonthat, M. F. Etievant, L. Intrator, R. Sylvestre, and B. Dreyfus, *Vox Sang.* **22**, 45 (1972).

[33] D. Roelcke, *Vox Sang.* **41**, 316 (1981).

[34] D. Miller, *Blood* **16**, 1313 (1960).

[35] D. E. Levy, A. A. Horner, and A. Solomon, *J. Exp. Med.* **153**, 883 (1981).

[36] H. I. Glueck, M. R. MacKenzie, and C. J. Glueck, *J. Lab. Clin. Med.* **79**, 731 (1972).

[37] E. Pogliani, E. Cofrancesco, and C. Praga, *Acta Haematol.* **53**, 249 (1975).

[38] J. L. Beaumont and N. Lemort, *Pathol. Biol.* **22**, 67 (1974).

[39] H. Tolleshaug and K. Hannestad, *Immunochemistry* **12**, 173 (1975).

[40] W. H. Sherman, N. Latov, A. P. Hays, M. Takatsu, R. Nemni, G. Galassi, and E. F. Osserman, *Neurology* **33**, 192 (1983).

[41] E. A. Kabat, J. Liao, W. H. Sherman, and E. F. Osserman, *Carbohydr. Res.* **130**, 289 (1984).

[42] R. M. Crisel, R. S. Baker, and D. E. Dorman, *J. Biol. Chem.* **250**, 4926 (1975).

[43] K. W. Hunter, Jr., V. G. Hemming, G. W. Fischer, S. R. Wilson, R. J. Hartzman, and J. N. Woody, *Lancet* **2**, 798 (1982).

[44] M. Harboe, J. Deverill, and J. Eriksen, *Acta Pathol. Microbiol. Scand., Sect. C* **83C**, 97 (1975).

[45] E. A. Kabat, J. Liao, H. Bretting, E. C. Franklin, D. Geltner, B. Frangione, M. E. Koshland, J. Shyong, and E. F. Osserman, *J. Exp. Med.* **152**, 979 (1980).

[46] A. S. Rao, E. A. Kabat, W. Nimmich, and E. F. Osserman, *Mol. Immunol.* **19**, 609 (1982).

[47] A. S. Rao, J. Liao, E. A. Kabat, E. F. Osserman, M. Harboe, and W. Nimmich, *J. Biol. Chem.* **259**, 1018 (1984).

[48] F. Goni and B. Frangione, *Proc. Natl. Acad. Sci. U.S.A.* **80**, 4837 (1983).

[49] K. Hannestad and K. Sletten, *J. Biol. Chem.* **246**, 6982 (1971).

[50] K. Hannestad, J. Eriksen, T. Chistensen, and M. Harboe, *Immunochemistry* **7**, 899 (1970).

[51] A. A. Ilyas, R. H. Quarles, M. C. Dalakas, and R. O. Brady, *Proc. Natl. Acad. Sci. U.S.A.* **82**, 6697 (1985).

[52] K. H. Chou, A. A. Ilyas, J. E. Evans, R. H. Quarles, and F. B. Jungalwala, *Biochem. Biophys. Res. Commun.* **128**, 383 (1985).

[53] R. F. Irie, L. L. Sze, and R. E. Saxton, *Proc. Natl. Acad. Sci. U.S.A.* **79**, 5666 (1982).

[54] T. Tai, J. C. Paulson, L. D. Cahan, and R. F. Irie, *Proc. Natl. Acad. Sci. U.S.A.* **80**, 5392 (1983).

[55] L. D. Cahan, R. F. Irie, R. Singh, A. Cassidenti, and J. C. Paulson, *Proc. Natl. Acad. Sci. U.S.A.* **79**, 7629 (1982).

[56] T. Tai, L. D. Cahan, J. C. Paulson, R. E. Saxton, and R. F. Irie, *JNCI, J. Natl. Cancer Inst.* **73**, 627 (1984).

[57] K. C. Joseph, C. R. Alving, and R. Wistar, *J. Immunol.* **112**, 1949 (1974).

[58] C. R. Alving, K. C. Joseph, and R. Wistar, *Biochemistry* **13**, 4818 (1974).

[59] M. Naiki and D. M. Marcus, *J. Immunol.* **119**, 537 (1977).

[60] R. Nowinski, C. Berglund, J. Lane, M. Loström, I. Bernstein, W. Young, S. I. Hakomori, L. Hill, and M. Cooney, *Science* **210**, 537 (1980).

[61] E. A. Kabat, J. Liao, J. Shyong, and E. F. Osserman, *J. Immunol.* **128**, 540 (1982).

[62] D. Roelcke, *Vox Sang.* **41**, 98 (1981).

[63] K. Uemura, D. Roelcke, Y. Nagai, and T. Feizi, *Biochem. J.* **219**, 865 (1984).

[64] D. Roelcke, H. Kreft, and A. M. Pfister, *Vox Sang.* **47**, 236 (1984).

[65] D. Roelcke, W. Pruzanski, W. Ebert, W. Romer, E. Fischer, V. Lenhard, and E. Rauterberg, *Blood* **55**, 677 (1980).

[66] D. Roelcke, R. Brossmer, and W. Ebert, *Protides Biol. Fluids* **29**, 619 (1981).

[67] D. Roelcke and R. Brossmer, *Protides Biol. Fluids* **31**, 1075 (1984).

[68] R. J. Weber and L. W. Clem, *J. Immunol.* **127**, 300 (1981).

[69] D. Roelcke, R. Brossmer, and W. Riesen, *Scand. J. Immunol.* **8**, 179 (1978).

[70] S. K. Kundu, D. M. Marcus, and D. Roelcke, *Immunol. Lett.* **4**, 263 (1982).

[71] C. M. Tsai, D. A. Zopf, R. K. Yu, R. Wistar, Jr., and V. Ginsburg, *Proc. Natl. Acad. Sci. U.S.A.* **74**, 4591 (1977).

[72] P. Yeni, M. L. Harpin, B. Habibi, A. Billecocq, M. J. Morelec, J. P. Clauvel, F. Danon, N. Baumann, and J. C. Brouet, *J. Clin. Invest.* **74**, 1165 (1984).

[73] D. Roelcke and H. Kreft, *Transfusion (Philadelphia)* **24**, 210 (1984).

[74] E. Lisowska and D. Roelcke, *Blut* **26**, 339 (1973).

[75] W. Ebert, J. Fey, C. Gartner, H. P. Geisen, U. Rautenberg, D. Roelcke, and H. Weicker, *Mol. Immunol.* **16**, 413 (1979).

[76] H. Egge, M. Kordowicz, J. Peter-Katalinic, and P. Hanfland, *J. Biol. Chem.* **260**, 4927 (1985).

[77] H. Niemann, K. Watanabe, and S.-I. Hakomori, *Biochem. Biophys. Res. Commun.* **81**, 1286 (1978).

[78] K. Watanabe, S.-I. Hakomori, R. A. Childs, and T. Feizi, *J. Biol. Chem.* **254**, 3221 (1979).

[79] K. Uemura, R. A. Childs, P. Hanfland, and T. Feizi, *Biosci. Rep.* **3**, 577 (1983).

[80] H. C. Gooi, A. Veyrieres, J. Alais, P. Scudder, E. F. Hounsell, and T. Feizi, *Mol. Immunol.* **21**, 1099 (1984).

[81] Y. Okada, R. Kannagi, S. B. Levery, and S.-I. Hakomori, *J. Immunol.* **133**, 835 (1984).

[82] C. M. Tsai, D. A. Zopf, R. Wistar, Jr., and V. Ginsburg, *J. Immunol.* **117**, 717 (1976).

[83] T. Feizi, R. A. Childs, K. Watanabe, and S.-I. Hakomori, *J. Exp. Med.* **149**, 975 (1979).

[84] A. Gardas, *Eur. J. Biochem.* **68**, 177 (1976).

[85] A. Gardas, *Eur. J. Biochem.* **68**, 185 (1976).

[86] P. Hanfland, H. Egge, U. Dabrowski, S. Kuhn, D. Roelcke, and J. Dabrowski, *Biochemistry* **20**, 5310 (1981).

[87] T. Feizi, E. A. Kabat, G. Vicari, B. Anderson, and W. L. Marsh, *J. Immunol.* **106**, 1578 (1971).

[88] E. A. Kabat, J. Liao, M. H. Burzynska, T. C. Wong, N. Thogersen, and R. U. Lemieux, *Mol. Immunol.* **18**, 873 (1981).

[89] R. U. Lemieux, T. C. Wong, J. Liao, and E. A. Kabat, *Mol. Immunol.* **21**, 751 (1984).

[41] Low Molecular Weight Phosphomannosyl Receptor from Bovine Testes

By JACK J. DISTLER and GEORGE W. JOURDIAN

Introduction[1]

Intracellular translocation of lysosomal enzymes is believed mediated by binding of these enzymes to membrane-associated receptors called phosphomannosyl receptors. The receptors bind ligands containing mannose 6-phosphate residues, a feature characteristic of the carbohydrate portion of lysosomal enzymes.[2] Receptor preparations from mammalian tissues exhibit a principal form (PMR-1) with a molecular weight of approximately 215,000 on SDS–PAGE. Smaller molecular weight proteins related to the phosphomannosyl receptor by their biological or immunochemical properties have been demonstrated in homogenates of monkey brain[3] and hamster liver[4] and have been obtained from PMR-1 after long periods of incubation at 4° in the absence of mannose 6-phosphate.[5] We have also observed small amounts of a protein with a molecular weight of 42,000 (PMR-2) in affinity-purified PMR-1 from bovine liver. In contrast, preparations of receptor from bovine testes consistently contain large amounts of PMR-2.[6] The isolation and properties of testes PMR-2 are described below. While the ligand-binding characteristics of testes PMR-2 are similar to bovine liver PMR-1, PMR-2 exhibits immunological properties that are distinct from those of PMR-1.

Assay Method

Principle. A radioimmunoassay based on the ability of unlabeled PMR-2 to inhibit binding of ^{125}I-labeled PMR-2 to antiPMR-2 is described.

[1] This work was supported in part by Grant AM-10531 from the National Institute of Arthritis, Diabetes, Digestive and Kidney Diseases, and by the Arthritis Foundation, Michigan Chapter.

[2] G. W. Jourdian, G. G. Sahagian, and J. J. Distler, *Biochem. Soc. Trans.* **9**, 510 (1982).

[3] K. Alvares and A. S. Balasubramanian, *Biochem. Biophys. Res. Commun.* **112**, 398 (1983).

[4] G. W. Jourdian, D. Mitchell, T. Maler, and J. J. Distler, *in* "Molecular Basis of Lysosomal Storage Disorders" (J. A. Barringer and R. O. Brady, eds.), p. 195. Academic Press, New York, 1984.

[5] D. C. Mitchell, T. Maler, and G. W. Jourdian, *J. Cell. Biochem.* **24**, 319 (1984).

[6] J. Distler and G. W. Jourdian, *Fed. Proc., Fed. Am. Soc. Exp. Biol.* **44**, 1435 (1985).

Antisera prepared against PMR-2 are 50 times more reactive toward PMR-2 than to PMR-1, allowing estimation of PMR-2 in the presence of a 10- to 20-fold excess of PMR-1. Similarly, a previously described radioimmunoassay for PMR-1[7] specifically measures PMR-1 in a 10- to 20-fold excess of PMR-2. It has not been established whether the observed interference in the radioimmunoassays is attributable to cross-reactivity of the antigens or whether it is due to trace contamination of one receptor with the other.

Reagents

Suspending medium: One volume of 0.1 M Tris–HCl, pH 7.0, is mixed with 1 volume of physiological saline containing 0.1% Triton X-100, 0.04% sodium azide, and 0.1% bovine serum albumin.

IgGsorb suspension: 10% lyophilized, formalin-fixed *Staphylococcus aureus* cells, Cowan's strain A (The Enzyme Center Inc., Boston, Massachusetts).

AntiPMR-2: AntiPMR-2 antibodies are raised in New Zealand white rabbits (3–4 kg). Each animal is injected intramuscularly with 1 ml of an emulsion composed of 1 volume of complete Freund's adjuvant and 1 volume physiological saline containing 125 μg of purified PMR-2. Initially, injections are administered every week for 4 weeks. Thereafter, injections are made at 6-week intervals using incomplete adjuvant and 125 μg of purified PMR-2 in saline. The antiserum is diluted with saline to a titer sufficient to bind 25–50% of the ^{125}I-labeled PMR-2 in the assay described below (in the absence of unlabeled PMR-2).

[125]I-labeled PMR-2 (1–5 μCi/μg): PMR-2 (100 μg, purified as described below, in 0.5 ml binding solution containing 5 mM mannose 6-phosphate) is iodinated using carrier-free sodium [^{125}I]iodide and Iodo-beads (Pierce Chemical Co., Rockford, Illinois). After iodination the solution is dialyzed at 4° for 12 hr against 1.5 liters binding solution and subjected to affinity chromatography in the manner described below (Step 5); the column size and elutant volumes are scaled down 10-fold. The ^{125}I-labeled material eluting with 1 mM mannose 6-phosphate is collected. Only PMR-2 (42,000 molecular weight constituent) is detectable on radioautograms on SDS-PAGE gels. Greater than 95% of the radioactivity binds to an IgGsorb–antiPMR-2 complex when undiluted antiserum is added.

Unlabeled PMR-2: 2–20 μg/ml, purified as described below.

[7] D. C. Mitchell, G. G. Sahagian, J. J. Distler, R. Wagner, and G. W. Jourdian, this series, Vol. 98 [25].

Procedure. Each assay mixture is prepared in duplicate. The mixtures contain (in a final volume of 225 μl) 50 μl of ^{125}I-labeled PMR-2 (approximately 5 × 10^4 cpm), 100 μl of unlabeled PMR-2 (0.1–2.0 μg), 50 μl of diluted antiPMR-2, and 25 μl of IgGsorb. Each assay mixture is incubated at 25° for 2 hr, and 3 ml of suspending medium are added. The mixture is centrifuged for 2 min at 2200 g, the supernatant removed by aspiration, and the ^{125}I content of the pellet measured in a gamma spectrometer. The PMR-2 content of each sample is determined from a standard curve plotted on semilog paper. A similar assay for estimation of PMR-1 is described in a previous volume.[7]

Purification Procedure

All operations are conducted at 0–4° unless otherwise stated. Protein content is estimated by the procedure of Peterson.[8]

Preparation of an Affinity Matrix for PMR-2. An oligosaccharide fraction enriched in phosphopentamannose is prepared by alcohol fractionation of an autohydrolysate of phosphomannan isolated from *Hansenula holstii.*[9] The oligosaccharide fraction is coupled to aminoethyl-substituted agarose by minor modification of the procedure of Gray.[10] ω-Aminoethyl-agarose (50 ml, 25–43 μmol diaminoethane/ml, Sigma Chemical Co., St. Louis, Missouri) is washed in a sintered glass funnel with 500 ml 0.2 M sodium phosphate, pH 7.0. Phosphopentamannose (10 g lyophilized material) and 800 mg sodium cyanoborohydride (freshly prepared from a dioxane complex[11]) are added to the agarose beads along with approximately 20 ml of phosphate buffer to form a very thick slurry. The reaction vessel is sealed with parafilm and the reaction allowed to proceed at 37° for 2–3 weeks with occasional mixing.

To terminate the reaction, the beads are washed with 50 volumes of isotonic saline and unsubstituted amino groups are N-acetylated. The beads are suspended in 500 ml of ice-cold 0.5 M sodium bicarbonate; 115 ml methanol and 25 ml acetic anhydride are added with stirring. The pH is maintained at 7.0 by the addition of 1 M sodium bicarbonate. After 30 min the substituted matrix is collected over a sintered glass filter and washed copiously with isotonic saline. Approximately 0.5–1.0 μmol inorganic phosphate is released per milliliter of packed substituted matrix on treatment with alkaline phosphatase. The matrix may be stored indefinitely at 4° in phosphate buffer containing 0.02% sodium azide.

[8] G. L. Peterson, *Anal. Biochem.* **83,** 346 (1977).
[9] R. K. Bretthauer, G. J. Kaczorowski, and M. J. Weise, *Biochemistry* **12,** 1251 (1973).
[10] G. R. Gray, *Arch. Biochem. Biophys.* **163,** 426 (1974).
[11] R. F. Borch, M. D. Bernstein, and H. D. Durst, *J. Am. Chem. Soc.* **93,** 2897 (1971).

Step 1. Preparation of Testes Extract. Freshly excised or frozen testes (500 g) are decapsulated, diced, and homogenized for 1 min in a Waring blender in 600 ml of a solution containing 0.5 mM calcium chloride and 1 mM sodium bicarbonate. The homogenate is filtered through a double layer of cheesecloth, adjusted to pH 5.0 by the dropwise addition of 4 N acetic acid, and centrifuged for 15 min at 10,000 g. The pellet is suspended in 1.2 liters of the same solution, the pH adjusted to 5.0, and the suspension recentrifuged.

Step 2. Acetone Powder Preparation. The washed pellet from Step 1 is homogenized for 1 min in a Waring blender in 3 liters of acetone held at −20°. The suspension is rapidly filtered through Whatman 3MM filter paper on a Büchner funnel under vacuum, and the residue is washed twice with acetone and once with 2 liters of diethyl ether (−20°). Residual traces of ether are removed under vacuum. The resulting powder (45 g) may be stored at −70° for at least 6 months.

Step 3. Extraction with Triton X-100. Acetone powder (5 g) is suspended in 140 ml of a solution containing 0.2 M sodium chloride, 10 mM EDTA, and 0.1 M sodium acetate, pH 6.0, with the aid of a glass–Teflon tissue grinder fitted with a loose pestle. After centrifugation for 15 min at 10,000 g, the supernatant is discarded and the pellet is suspended in 140 ml of distilled water and recentrifuged. The washed precipitate is resuspended in 150 ml of a solution consisting of 0.4 M potassium chloride, 50 mM imidazole–HCl, pH 7.0, and 1% Triton X-100. The suspension is stirred for 30 min, then centrifuged for 30 min at 10,000 g, and the supernatant immediately subjected to affinity chromatography in the manner described below.

Step 4. Affinity Chromatography I. This step removes glycoproteins that compete with the affinity matrix for the binding sites present on PMR-1 and PMR-2.[12] A 2 × 15 cm column is equilibrated with 2 liters of a binding solution consisting of 0.15 M sodium chloride, 0.1% Triton X-100, 0.05 M sodium phosphate, pH 7.0, and 0.02% sodium azide. The supernatant from Step 3 is applied to the column at a flow rate of 250 ml/hr. The column is washed with 500 ml of binding solution, and PMR-1 and PMR-2 are eluted with 2 column volumes of McIlvaine buffer (10 volumes of 0.1 M citric acid adjusted to pH 4.3 with approximately 7 volumes of 0.2 M Na$_2$HPO$_4$) containing 0.1% Triton X-100. The latter eluant is adjusted immediately to pH 7.0 by dropwise addition of 1 N sodium hydroxide.

Step 5. Affinity Chromatography II. The affinity column is reequilibrated with 10 column volumes of binding buffer. Up to 5 batches of the affinity-purified receptor mixture from Step 4 may be combined and ap-

[12] J. Distler, V. Hieber, G. Sahagian, R. Schmickel, and G. W. Jourdian, *Proc. Natl. Acad. Sci. U.S.A.* **76**, 4235 (1979).

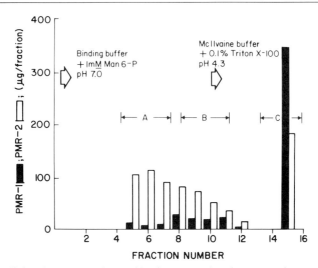

FIG. 1. Affinity chromatography II of bovine testes phosphomannosyl receptors. Receptors from affinity chromatography I (derived from 5 g acetone powder) were reapplied to a 2 × 15 cm column of the phosphopentamannose-substituted affinity matrix. The column was washed and eluted as described in the text. Aliquots of the fractions were assayed for PMR-1 and PMR-2 by their respective radioimmunoassays.

plied to the column. The column is washed as described above and eluted with binding buffer containing 1 mM mannose 6-phosphate. Ten 10-ml fractions are collected and analyzed for PMR-2 by the radioimmunoassay procedure. Suitable aliquots of each fraction may be subjected to SDS–PAGE to establish the presence and ratio of PMR-1 and PMR-2 in each fraction, or PMR-1 may be estimated by a previously described radioimmunoassay.[7] A typical elution profile is shown in Fig. 1. Small amounts of PMR-1 and PMR-2 remaining bound to the column may be eluted with McIlvaine buffer, pH 4.3, containing 0.1% Triton X-100 as described in Step 4. Fraction A (Fig. 1) contains only PMR-2. Fractions B and C may be combined, dialyzed against binding buffer to remove mannose 6-phosphate, and rechromatographed to recover additional PMR-2. Fractions containing PMR-2 are combined and concentrated over an Amicon YM 10 filter or, alternatively, concentrated with an Amicon Centricon 10 as recommended by the manufacturer. The purification and yield of PMR-2 in a typical preparation is shown in Table I.

Properties

Stability. Purified PMR-2 is stable for several months when stored at −20° at a concentration of 1 mg/ml in binding solution containing 1 mM mannose 6-phosphate.

TABLE I
PURIFICATION OF BOVINE TESTES PHOSPHOMANNOSYL RECEPTORS[a]

Fraction	Volume (ml)	PMR-1[b]		PMR-2[c]	
		mg	Yield (%)	mg	Yield (%)
Crude Triton X-100 extract (Step 3)	300	0.95	100	1.10	100
Affinity chromatography I (Step 4)	30	0.70	74	0.82	75
Affinity chromotography II (Step 5)					
Fraction A	30	<0.01	<1	0.25	23
Fraction B	30	0.18	19	0.13	12
Fraction C	30	0.35	37	0.18	16

[a] Based on extraction of 5 g acetone powder.
[b] Radioimmunoassay with ^{125}I-labeled PMR-1 and antiPMR-1 antiserum.[7]
[c] Radioimmunoassay with ^{125}I-labeled PMR-2 and antiPMR-2 antiserum.

Specificity. PMR-2 is not eluted from the affinity matrix by 50 mM mannose, N-acetylglucosamine, D-glucose 1-phosphate, or EDTA. Bovine testicular β-galactosidase, containing covalently bound mannose 6-phosphate residues, binds to PMR-2 before but not after treatment with alkaline phosphatase. Binding of β-galactosidase to PMR-2 is inhibited by 5 mM mannose 6-phosphate. The enzyme–receptor complex dissociates rapidly on dilution of the ligand. The complex exhibits a half-life of approximately 2 min in ligand-free medium.

pH Optimum. PMR-2 binds to the affinity matrix at pH values between 6 and 7 but does not bind at pH values below pH 5.5.

Activators. No absolute requirements for cofactors have been demonstrated. Binding of PMR-2 to β-galactosidase does not require the presence of divalent cations in the assay mixtures. However, 20 mM MnCl$_2$ improves binding of β-galactosidase at low β-galactosidase concentrations. (Tris buffer is substituted for phosphate buffer in these experiments.)

Physical Properties. PMR-2 exhibits a diffuse Coomassie Blue-positive band on SDS–PAGE gels and has an apparent average molecular weight of 42,000.

[42] Soluble Lactose-Binding Lectins from Chicken and Rat

By SAMUEL H. BARONDES and HAKON LEFFLER

Soluble dimeric lactose-binding lectins with subunit molecular weights in the range of 15,000 have been found in many vertebrate tissues. They have been purified from fish,[1] amphibians,[2] reptiles,[3] birds,[4–6] and mammals.[7–10] Those from chicken and rat will be referred to here as chicken lactose–lectin I (CLL-I)[4–6] and rat lectin 14.5 (RL-14.5).[8,10] Chicken and rat also contain other lactose-binding lectins—chicken lactose–lectin II (CLL-II)[6] and rat lectins with apparent molecular weights of 18,000 and 29,000 (RL-18, RL-29).[10] We here describe the isolation and some properties of these lectins. All can be purified by affinity chromatography on asialofetuin–Sepharose, using modifications of the method of DeWaard *et al.*[7] The individual lectins can then be resolved either by preparative isoelectric focusing or by ion-exchange chromatography.

Preparation of Asialofetuin–Sepharose 4B

To remove sialic acid from fetuin, 2.5 g of this glycoprotein (available from a number of commercial sources) is dissolved in 125 ml 0.15 M NaCl and H_2SO_4 is added to 0.025 N. The solution is incubated at 80° for 60 min. After neutralization with NaOH, the solution is dialyzed extensively against 0.1 M NaHCO₃, 0.5 M NaCl, pH 8.6 (BS).

Sepharose 4B (Pharmacia, Piscataway, New Jersey) is washed on a coarse sintered glass funnel with 4 volumes cold 0.15 M NaCl, 4 volumes cold H_2O, and 2 volumes cold 2 M Na₂CO₃. About 200 ml is suspended in 300 ml 2 M Na₂CO₃ and kept on ice. Cyanogen bromide (12.5 g) is dissolved, in a hood, in 30 ml acetonitrile immediately before use. Solution of the cyanogen bromide is facilitated by heating with a lamp, and it is

[1] G. Levi and V. I. Teichberg, *J. Biol. Chem.* **256**, 5735 (1981).
[2] N. C. Bols, M. M. Roberson, P. L. Haywood-Reid, R. F. Cerra, and S. H. Barondes, *J. Cell Biol.* **102**, 492 (1986).
[3] T. K. Gartner and M. L. Ogilvie, *Biochem. J.* **224**, 301 (1984).
[4] H. Den and D. A. Malinzak, *J. Biol. Chem.* **252**, 5444 (1977).
[5] T. P. Nowak, D. Kobiler, L. Roel, and S. H. Barondes, *J. Biol. Chem.* **252**, 6026 (1977).
[6] E. C. Beyer, S. E. Zweig, and S. H. Barondes, *J. Biol. Chem.* **255**, 4236 (1980).
[7] A. DeWaard, S. Hickman, and S. Kornfeld, *J. Biol. Chem.* **251**, 7581 (1976).
[8] J. T. Powell, *Biochem. J.* **187**, 123 (1980).
[9] R. A. Childs and T. Feizi, *Biochem. J.* **183**, 755 (1979).
[10] R. F. Cerra, M. A. Gitt, and S. H. Barondes, *J. Biol. Chem.* **260**, 10474 (1985).

generally done in the bottles in which it is packed, to minimize handling. The Sepharose and cyanogen bromide solutions are mixed in a hood, stirred with a spatula for 2 min on ice, then poured onto a sintered glass funnel and washed with 0.5 liter 0.1 M NaHCO$_3$, pH 9.5, then 1 liter H$_2$O, then 1 liter BS. All the washing is done as rapidly as possible. The activated Sepharose is then mixed with 2.5 g of asialofetuin in BS (prepared as above) and agitated gently overnight on a gyratory shaker at 4°.

After coupling, the derivatized Sepharose is washed with 1.5 liters of 0.1 M sodium acetate, 0.5 M NaCl, pH 4 (AS), followed by 1.5 liters of BS. Then 500 ml 1 M ethanolamine, pH 9, is added to bind up any remaining activated residues. After incubation for 4 hr at room temperature the gel is washed with 1.5 liters AS, 1.5 liters BS, and then extensively with 75 mM NaCl, 75 mM Na$_2$HPO$_4$/KH$_2$PO$_4$, pH 7.2 (PBS). The column is stored at 4° in PBS containing 0.02% sodium azide. It can be reused indefinitely if washed with 2 column volumes of PBS containing sodium azide after lectin has been eluted. We generally repack the column after 5 uses.

Purification of Rat Lung Lectins by Affinity Chromatography

About 100 g rat lung (purchased frozen from Pel Freeze, Rogers, Arkansas) are homogenized in 200 ml PBS containing 4 mM 2-mercaptoethanol, 2 mM EDTA (this mixture is referred to as MEPBS) containing 0.15 M lactose in a Sorvall Omni-Mixer. The homogenate is centrifuged at 4000 g for 30 min and the supernatant is then centrifuged at 100,000 g for 60 min. The lactose, which is critical for efficient solubilization of the lectins, is then removed from the supernatant by extensive dialysis against MEPBS (3 changes of 9 liters over a 24-hr period) such that the concentration of lactose is brought below 0.1 mM.

The dialyzed supernatant is again centrifuged at 100,000 g for 1 hr to remove any precipitate that formed, and the supernatant is pumped through the asialofetuin–Sepharose column (approximately 2.5 × 40 cm column, containing approximately 200 ml asialofetuin–Sepharose) which had been equilibrated with MEPBS. The column is then washed with 4 column volumes MEPBS, 1 column volume MEPBS containing 0.1 M sucrose, and then eluted with MEPBS containing 0.1 M lactose. The eluate is collected in 7-ml fractions, and the eluted lectin is located by monitoring the optical density at 280 nm of the eluate. The composition of each fraction is monitored by polyacrylamide slab gel electrophoresis and silver staining. A typical preparation yields about 3 mg of a mixture of RL-14.5, RL-18, and RL-29 in the ratio of approximately 4 : 2 : 4. RL-14.5 is eluted more readily than the other lectins and is relatively concentrated in the earliest eluted fractions. Elution may also be done with as little as

10 mM lactose. In this case RL-14.5 spreads out over about twice the number of fractions. Although such dilution is disadvantageous, the lower concentrations of lactose are easier to remove.

Dialysis and concentration of the lectins by ultrafiltration lead to considerable losses of lectin, especially RL-18 and RL-29. We found that recoveries are best if the lectin is concentrated by centrifugation in a Centricon 10 microconcentrator (Amicon, Danvers, Massachusetts). Lactose can be removed by several cycles of dilution with MEPBS and reconcentration.

The three lectins can be resolved by DEAE–ion-exchange chromatography.[10] For this purpose the mixture of lectins is transferred into a buffer containing 10 mM NaCl, 10 mM Tris–HCl, pH 7.5, 10 mM lactose, and 4 mM 2-mercaptoethanol, by using the Centricon 10 microconcentrator as described above. It is then applied to a DEAE column equilibrated with this buffer, then eluted with a linear gradient of 10–210 mM NaCl in 10 mM Tris–HCl, pH 7.5, containing 10 mM lactose and 4 mM 2-mercaptoethanol. Very good resolution is obtained on a Bio-Rad TSK-5 PW column using a Perkin–Elmer Series 4 liquid chromatograph. RL-29 is not retained by this column, RL-18 is eluted at 10–40 mM NaCl, and RL-14.5 at about 180 mM NaCl. Each of the resolved lectins retains its carbohydrate binding activity since it can bind specifically to a small asialofetuin–Sepharose column and be specifically eluted with lactose. However, the activity of RL-14.5 that had been purified by ion-exchange chromatography was impaired when measured in a solid phase binding assay.[10] Activity is also retained on freezing at $-20°$. We generally freeze the lectin aliquots in 1 mg/ml bovine serum albumin (radioimmunoassay grade).

The presence of 2-mercaptoethanol is critical during the purification procedure. If the reducing agent is eliminated from the homogenizing buffer, little or no lectin is recovered by affinity chromatography. Powell and Whitney[11] found that incubation of RL-14.5 with 50 mM Tris, pH 8.6, containing 0.1 M iodoacetamide and 0.1 M lactose yielded carboxyamidomethylated RL-14.5 which retained carbohydrate binding activity in the absence of reducing agents. We found that the mixture of three lectins could be treated with iodoacetamide and that all retained carbohydrate binding activity as judged by specific binding to asialofetuin and elution with lactose. For this purpose the mixture of lectins eluted from the asialofetuin column (in MEPBS containing lactose) was made 40 mM in iodoacetamide, kept at 4° for 4 hr, and then dialyzed against PBS to remove the reagent as well as lactose.

[11] J. T. Powell and P. L. Whitney, *Biochem. J.* **223,** 769 (1984).

Alkylated RL-14.5 may be iodinated with ^{125}I-labeled Bolton–Hunter reagent (Amersham, Arlington Heights, Illinois) with retention of carbohydrate-binding activity.[12] Alkylated lectin, that had been dialyzed extensively against PBS, is concentrated to 0.8 mg/ml in a Centricon 10 microconcentrator, and the buffer is changed to 0.1 M sodium borate, pH 8.5, by 10-fold dilution with this buffer and reconcentration. ^{125}I-Labeled Bolton–Hunter reagent (0.5 mCi, approximately 5×10^8 counts/min) is brought to dryness in the vial in which it is supplied, and about 12 μg of lectin in 15 μl is added. After incubation for 15 min on ice, with occasional vortexing, the reaction is terminated with 50 μl of 1 M glycine in borate buffer. The mixture is then diluted with 100 μl PBS containing 2 mg/ml bovine serum albumin (PBS–BSA) and applied to a 0.2 ml asialofetuin–Sepharose column. The column is washed with 5 ml PBS–BSA, and the bound lectin is eluted in 50-μl fractions by passing 2 ml 10 mM lactose in PBS–BSA through the column. About 80% of the bound lectin can be recovered in the 5 peak fractions comprising a total volume of 0.25 ml. In a typical preparation the iodinated lectin had incorporated 5×10^7 counts/min (about 10% of the starting radioactivity), and the specific activity of the eluted lectin was in the range of 5×10^6 counts/min/μg. The iodinated lectin may be stored at $-20°$ in aliquots in the solution in which it had been eluted from the asialofetuin–Sepharose column. It can be used for binding studies after appropriate dilution to reduce the concentration of lactose below 0.1 mM. Alkylated RL-18 and RL-29 can also be iodinated and stored frozen with retention of carbohydrate binding activity.[12]

Purification of Chicken Lactose-Binding Lectins

Although a number of chicken tissues contain both CLL-I and CLL-II, especially in development,[13] adult chicken liver contains almost exclusively CLL-I and adult chicken intestine contains almost exclusively CLL-II.[6] Each lectin can be purified by affinity chromatography on asialofetuin–Sepharose using the procedures described above. Intestinal contents are carefully removed by repeated washing in PBS before homogenization. We generally recover about 1 mg pure CLL-I per gram soluble liver protein applied to the affinity column (derived from about 50 g frozen chicken liver). We recover about 1 mg CLL-II from about 0.2 g soluble intestinal protein. Since CLL-II is localized in the secretory vesicles of the mucosal goblet cells, the starting material can be further enriched by scraping the mucosa from carefully washed intestine with a glass slide.

---

[12] H. Leffler and S. H. Barondes, *J. Biol. Chem.* **261,** 10119 (1986).
[13] E. C. Beyer and S. H. Barondes, *J. Cell Biol.* **92,** 23 (1982).

Although the liver gives virtually pure CLL-I and the intestine virtually pure CLL-II, traces of the other lectin as well as contaminants may be found. They can be removed by preparative isoelectric focusing[6] in an LKB isoelectric focusing apparatus (LKB Instruments, Pleasant Hill, California) using pH 3–10 ampholytes as described in the manual for this apparatus, with the exception that the ampholyte solutions contain 0.3 M lactose, 4 mM 2-mercaptoethanol. CLL-I has an isoelectric point of 4.1 and CLL-II an isoelectric point of 6.3. An alternative method for final purification is gel filtration on Sephadex, since CLL-I behaves as a dimer with apparent molecular weight of 31,000 and CLL-II is virtually completely monomeric with molecular weight 14,000. The gel filtration is done in MEPBS containing 0.3 M lactose.[6]

Biological Studies

The cellular localization of CLL-I,[14,15] CLL-II,[16] and RL-14.5[17] in certain tissues has been determined by immunohistochemical studies using highly specific antisera. Each is found both intracellularly and at specific extracellular sites, and localization may change with tissue differentiation. For example, CLL-I is very prominent within embryonic skeletal muscle cells and becomes concentrated extracellularly with further differentiation.[15] RL-14.5, which is synthesized in developing rat lung during elastic fiber maturation,[18] accumulates extracellularly in these fibers.[17] These findings suggest that these lectins may function in specific types of extracellular matrix, possibly by interacting with glycoconjugates found there,[19] and that this may be especially significant during development.

CLL-II is localized in the secretory vesicles of the goblet cells of the intestinal mucosa,[16] along with intestinal mucins. The lectin is secreted onto the mucosal surface where it presumably interacts with these mucins, and secretion is stimulated by cholinergic agents.[16] CLL-II may also have specific developmental functions since it is relatively abundant in embryonic kidney but is found at much lower levels in this organ in the adult.[13]

Aside from intestinal mucins, which colocalize with and bind CLL-II,[16] little is known about endogenous glycoconjugate ligands for these lectins. Several lung glycoproteins bind RL-14.5 that had been coupled to

[14] E. C. Beyer, K. Tokuyasu, and S. H. Barondes, *J. Cell Biol.* **82,** 565 (1979).
[15] S. H. Barondes and P. L. Haywood-Reid, *J. Cell Biol.* **91,** 568 (1981).
[16] E. C. Beyer and S. H. Barondes, *J. Cell Biol.* **92,** 28 (1982).
[17] R. F. Cerra, P. L. Haywood-Reid, and S. H. Barondes, *J. Cell Biol.* **98,** 1580 (1984).
[18] J. T. Powell and P. L. Whitney, *Biochem. J.* **188,** 1 (1980).
[19] S. H. Barondes, *Science* **223,** 1259 (1984).

Sepharose.[11] A β-galactoside-binding lectin from chick skin, which is probably CLL-II, interacts with a polylactosamine proteoglycan from that tissue.[20] RL-14.5 and RL-29 are localized in subsets of dorsal root ganglion neurons that also tend to bind monoclonal antibodies specific for glycolipids with a lacto series rather than a globo series backbone.[21] Although RL-14.5, RL-18, and RL-29 show similar affinities for lactose, they interact differently with certain substituted β-galactosides,[12] suggesting selective interactions with naturally occurring mammalian glycoconjugates.

Like certain plant lectins, vertebrate lectins have mitogenic effects on lymphocytes. Both CLL-I and CLL-II can stimulate DNA synthesis in spleen cells of athymic mice, suggesting that they are B-cell mitogens.[22] Electrolectin, a β-galactoside-binding lectin from *Electrophorus electricus* which appears to be related to CLL-I and RL-14.5, stimulates DNA synthesis of lymphocytes from rabbit inguinal nodes.[23]

There have been recent reports of cloning of genes that encode vertebrate β-galactoside binding lectins. Ohyama[24] isolated and sequenced the cDNA that encodes a 14 kDa lectin found in chick skin which may be CLL-II. Two cDNA clones have been isolated from a human hepatoma cDNA library which apparently encode two variants of a human lectin that is closely related to RL-14.5.[25] Based on amino acid sequences deduced from these clones and partial sequences determined for peptides derived from a human lung lectin, there are apparently at least three variants of this protein expressed in human tissues.[25]

Acknowledgments

Supported by grants from the National Institutes of Health and The Swedish Medical Research Council (MFR 6644) and by the Veterans Administration Medical Center.

[20] Y. Oda and K. I. Kasai, *Biochem. Biophys. Res. Commun.* **123**, 1215 (1984).
[21] L. J. Regan, J. Dodd, S. H. Barondes, and T. M. Jessell, *Proc. Natl. Acad. Sci. U.S.A.* **83**, 2248 (1986).
[22] J. S. Lipsick, E. C. Beyer, S. H. Barondes, and N. O. Kaplan, *Biochem. Biophys. Res. Commun.* **97**, 56 (1980).
[23] G. Levi, R. Tarrab-Hazdai, and V. I. Teichberg, *Eur. J. Immunol.* **13**, 500 (1983).
[24] Y. Ohyama, J. Hirabayashi, Y. Oda, S. Ohno, H. Kawasaki, K. Suzuki, and K. Kasai, *Biochim. Biophys. Res. Comm.* **134**, 51 (1986).
[25] M. A. Gitt and S. H. Barondes, *Proc. Natl. Acad. Sci. U.S.A.* **83**, 7603 (1986).

[43] *Xenopus laevis* Lectins

By SAMUEL H. BARONDES and MARIE M. ROBERSON

Xenopus laevis is an amphibian which is used extensively for a variety of biological studies, especially of early embryonic development. Lectin activities have been detected in oocytes,[1] unfertilized eggs,[1,2] and early embryos[1] as well as in adult tissues, including serum[3] and skin.[4,5] We here describe the purification and some properties of three lectins from *Xenopus laevis*. The one from skin shares many properties with soluble β-galactoside-binding-lectins from other vertebrates[6] (see this volume [42]). In contrast the lectins from oocytes and serum differ markedly from other soluble vertebrate lectins. They both have considerably larger subunit molecular weights, tend to form fairly large multimers, and bind α-galactosides as well as β-galactosides.

Hemagglutination Assay

Each of these lectins was originally detected as a hemagglutinin using appropriate erythrocytes and assay conditions.[1] However, we do not recommend using this assay to monitor purification since it is so easy to purify each lectin by affinity chromatography and to verify purification by polyacrylamide gel electrophoresis.

Purification of Oocyte Lectin

The oocyte lectin is readily purified from homogenates of whole ovaries obtained by laparotomy.[1] Homogenates are prepared in 10 volumes of ice-cold acetone for 1 min in a Sorvall Omni-Mixer and filtered on Whatman #1 paper. The material trapped by the filter is washed 3 times with

[1] M. M. Roberson and S. H. Barondes, *J. Biol. Chem.* **257**, 7520 (1982).

[2] L. C. Greve and J. L. Hedrick, *Gamete Res.* **1**, 13 (1978).

[3] M. M. Roberson, A. P. Wolffe, J. R. Tata, and S. H. Barondes, *J. Biol. Chem.* **260**, 11027 (1985).

[4] K. Nitta, G. Takayanagi, Y. Terasaki, and H. Kawauchi, *Experientia* **40**, 712 (1984).

[5] N. C. Bols, M. M. Roberson, P. L. Haywood-Reid, R. F. Cerra, and S. H. Barondes, *J. Cell Biol.* **102**, 492 (1986).

[6] S. H. Barondes, *Science* **223**, 1259 (1984).

cold acetone and, after the acetone cake is dry, is rehomogenized in 9 volumes (relative to the initial volume of tissue) of TCS (10 mM Tris–hydrochloride, pH 7.6, 10 mM CaCl$_2$, 150 mM NaCl) containing 0.3 M galactose for 1 min in a Sorvall Omni-Mixer at 4° followed by centrifugation at 100,000 g for 1 hr. The galactose, which facilitates solubilization of lectin, is removed by extensive dialysis against TCS (at least 2 changes of 100 volumes each) before affinity chromatography. Ca^{2+} is required for efficient binding of the lectin to the affinity column.

About 10 ml of extract (about 10–20% of the material from one *Xenopus*) is poured over a 2 ml column of aminoethylated polyacrylamide gel beads conjugated with melibiose (Selectin 4, Pierce Chemical Co., Rockford, Illinois) equilibrated with TCS. The gel beads are poured in a 5-in. polystyrene disposable column (Evergreen Scientific, Los Angeles, California). If much larger volumes of crude extract are used, the column may become plugged and will not run. Therefore, for larger scale preparations we find it most convenient to run several small columns in parallel. Each column is washed with 50 ml TCS, 50 ml TCS containing 0.3 M sucrose, and again with 50 ml TCS. Then the lectin is eluted with 2–3 ml of TCS containing 0.3 M melibiose. The gel beads can be reused after extensive washing with 1 M NaCl and TCS.

Eluted lectin consists of two fuzzy bands with apparent molecular weights of 43,000 and 45,000, upon polyacrylamide gel electrophoresis in sodium dodecyl sulfate.[1] The fuzziness of the bands reflects the glycosylation of the lectin. Recovery is very efficient and, based on hemagglutination activity, exceeds that found in the crude extract.[1] The lectin recovered is about 1% of the protein in the crude extract, and the ovaries from a single *Xenopus* give about 3–5 mg pure lectin. If necessary, trace contaminants can be removed by preparative isoelectric focusing[1] in an LKB isoelectric-focusing apparatus (LKB Instruments, Pleasant Hill, California) using pH 3.5–5 ampholytes. This is recommended if the lectin preparation will be used for immunization to raise a specific antiserum.

The lectin has also been purified from unfertilized eggs and embryos.[1] On gel filtration of the native lectin on Sepharose 6B, the apparent molecular weight is 480,000. There is heterogeneity on isoelectric focusing with pI values ranging from 4.4 to 4.9. Ca^{2+} is required for efficient binding to the affinity column. Melibiose, an α-galactoside, and lactose, a β-galactoside, are approximately equally effective in inhibiting hemagglutination activity. Each inhibits this activity by 50% at 1.3 mM. Galactose also reacts with the lectin, but a variety of other sugars, including N-acetylgalactosamine, do not. The lectin can be stored frozen without loss of activity.

Purification of Serum Lectin

Since estrogen stimulates levels of circulating serum lectin,[3] it is best purified either from females or from males that were injected (about 10 days prior to bleeding) in the dorsal lymph sac with 1 mg of 17β-estradiol suspended in 0.2 ml of sterile olive oil. Blood is obtained by heart puncture and 2 ml of serum is usually collected per animal. The serum is applied directly to the same affinity column used for purification of the oocyte lectin as described above, and all subsequent procedures are identical. Two milliliters of serum are usually applied to a 2 ml affinity column, and we generally recover about 50 μg pure lectin per ml serum.

The lectin eluted from the affinity column migrates as a single band with apparent molecular weight 69,000 on polyacrylamide gel electrophoresis in sodium dodecyl sulfate.[3] On gel filtration on Sepharose 6B it behaves as a discrete multimer with apparent molecular weight in the range of 500,000. Like the oocyte lectin, carbohydrate-binding activity is dependent on Ca^{2+}. The relative potency of saccharides as competitive inhibitors of hemagglutination activity of the serum lectin is virtually identical with that of the oocyte lectin.[3] Both react about equally well with melibiose and lactose. The purified serum lectin can be stored at $-20°$ without loss of activity. It appears to be glycosylated, based on concanavalin A binding to the lectin on a polyacrylamide gel.

Purification of Skin Lectin

Skin lectin was initially purified from skin homogenates.[5] However, we found that the animals could be induced to secrete virtually all their skin lectin into the surrounding medium upon injection with epinephrine. We find it much easier to collect starting material in this way and the yield is actually better than from skin homogenates. However, should the reader prefer the homogenization procedure, it is described elsewhere.[5]

Collection of washings from female *Xenopus laevis* skin is done by immersing the animal for 15 min in TEM (75 ml 50 mM Tris–HCl, pH 7.6, 10 mM EDTA, 4 mM 2-mercaptoethanol) in a plastic bag (we use 22 cm × 18 cm Ziploc bags) immediately after injection into the dorsal lymph sac of 100 μg of L-epinephrine in 300 μl of 150 mM NaCl. The washings are poured over a 2-ml column of agarose gel beads conjugated with lactose (Lactose II, Pierce Chemical Co., Rockford Illinois) that had been equilibrated with TEM. We use a 5-in. polystyrene disposable column (Evergreen Scientific, Los Angeles, California) for this purpose. The entire 75 ml flows through the column within 2 hr. The solution that passed through the column is collected and passed over the column again, to assure fairly complete binding. Afterward the column is washed with 80 ml TEM, 25 ml 0.3 *M* sucrose in TEM, and again with 80 ml TEM. The lectin is eluted with 4 ml of 0.3 *M* lactose in TEM. About 2 mg pure lectin is recovered from a single animal.

Lectin purified in this way usually contains no contaminants as judged by polyacrylamide gel electrophoresis in sodium dodecyl sulfate.[5] However, to remove trace contaminants, should they appear, further purification can be achieved by DEAE–ion-exchange chromatography.[5] For this purpose we use a Bio-Rad TSK 5 PW column and a Perkin Elmer series 4 liquid chromatograph. The material eluted from the affinity column is first concentrated by ultrafiltration and dialyzed against 10 m*M* NaCl, 10 m*M* Tris–HCl, pH 7.0, containing 10 m*M* lactose and 4 m*M* 2-mercaptoethanol and applied to the column. Elution is performed with a linear gradient in 10–150 m*M* NaCl in 10 m*M* Tris–HCl, pH 7.0, containing 10 m*M* lactose and 4 m*M* 2-mercaptoethanol at a 1 ml/min flow rate. The lectin is detected as a sharp peak on the recording integrator used to monitor elution.

The purified lectin has a subunit molecular weight of about 16,000 and resembles soluble vertebrate lectins purified from other tissues. Like those lectins, it reacts well with β-galactosides but not α-galactosides.[5] If 2-mercaptoethanol is removed, the purified lectin will not rebind to the affinity column. These properties, as well as the lack of Ca^{2+} dependence of this lectin, are features that distinguish it from the other two *Xenopus laevis* lectins described above.

Biological Studies

Rabbit antisera have been raised against each of these lectins and have been useful for biological studies. The oocyte and skin lectins have been localized in the tissues that make them by immunohistochemistry using light and electron microscopy. The oocyte lectin is largely intracellular in oocytes and unfertilized eggs but becomes extracellular with embryonic development.[7] Early in development it is first secreted into the cleavage furrows[7] and later it is concentrated extracellularly at sites of extensive embryonic cellular migration.[8] These findings suggest that the lectin may interact with glycoconjugates on and/or between embryonic cells and influence morphogenesis. The skin lectin is concentrated in the cytoplasm of the granular gland cells and is secreted as a cytoplasmic protein along with multiple toxins contained within these cells.[5] The serum lectin is synthesized by hepatocytes under the influence of estrogen and is secreted from these cells.[3]

---

[7] M. M. Roberson and S. H. Barondes, *J. Cell Biol.* **97,** 1875 (1983).
[8] M. M. Roberson and S. H. Barondes, unpublished.

Recommended Nomenclature

Although each *Xenopus* lectin is highly enriched in a specific tissue, it is also found in others. It is, therefore, potentially confusing to refer to the lectins on the basis of the tissues from which they were purified. It seems more reasonable to refer to them on the basis of their subunit molecular weights. Thus, skin lectin would be *Xenopus* lectin 16,000 (XL-16); oocyte lectin would be *Xenopus* lectin 43,000 and 45,000 (XL-43, 45); and serum lectin would be *Xenopus* lectin 69,000 (XL-69).

Acknowledgments

Supported by grants from the U.S. Public Health Service and by the Veterans Administration Medical Center.

[44] *O*-Acetylsialic Acid-Specific Lectin from the Crab *Cancer anntenarius*

By M. H. RAVINDRANATH and JAMES C. PAULSON[1]

The blood or hemolymph of the Californian coastal crab *Cancer antennarius* Stimpson contains a sialic acid-specific lectin with high affinity for 9-*O*-acetyl- and 4-*O*-acetyl-*N*-acetylneuraminic acids.[2,3] A method of purification, using bovine submaxillary mucin–Sepharose as an affinity adsorbent, and the properties of the lectin are described here.

Assay Method

Principle. The activity of the crab lectin is assayed by measuring its ability to agglutinate horse (rat or mouse) erythrocytes. The assay is done by the conventional procedure of serial 2-fold dilution of the lectin on microtiter plates and visual estimation of erythrocyte agglutination 1 hr after adding the cells. The method is described below.

[1] MHR, Recipient of Fulbright Fellowship; JCP, Recipient of an American Cancer Society Faculty Research Award. Supported by Grant GM-27904 from the National Institutes of Health.
[2] M. H. Ravindranath and E. L. Cooper, *Prog. Clin. Biol. Res.* **157**, 83 (1984).
[3] M. H. Ravindranath, H. H. Higa, E. L. Cooper, and J. C. Paulson, *J. Biol. Chem.* **260**, 8850 (1985).

Reagents

TBS–BSA (Tris-buffered saline with BSA): 50 mM Tris; 100 mM NaCl; 10 mM CaCl$_2$, pH 7.2, containing 0.05% bovine serum albumin.

AMAB (modified Alsevier's medium containing antibiotics): 30 mM sodium citrate, pH 6.1; 70 mM NaCl; 114 mM glucose; 100 μg/ml of neomycin sulfate; 330 μg/ml of chloramphenicol.

Erythocyte suspension: Blood of horse, rat, or mouse is collected directly in an equal volume of AMAB solution. The volume of the cells are measured by hematocrit, and the suspension may be stored for up to 1 week at 4°C. For hemagglutination assays the cell suspension is diluted to 1.5% in TBS–BSA.

Procedure. Twofold serial dilutions of 25 μl of lectin solution are made in microtiter test plates (Sarstedt: No. 82.600), using TBS–BSA as dilutent and microdiluters (Dynatech Lab, Va) which are dipped in TBS–BSA and wiped dry before use. To each well is added 25 μl of a 1.5% suspension of erythrocytes, washed thrice in TBS–BSA and suspended in TBS–BSA. After adding erythrocytes, the microtiter plate is covered with parafilm and gently agitated on low-speed vortex for 30 sec. The degree of agglutination is scored after 1 hr at room temperature. Hemagglutination (HA) units are defined as the hemagglutination titer (per 25 μl), or the reciprocal of the highest dilution which gives complete agglutination. Specific activity is defined as HA units/mg protein; i.e., the reciprocal of the lowest amount of protein giving complete agglutination. Due to inherent variability in hemagglutination assays, all samples to be compared should be done on the same day.

Purification of Crab Lectin

Principle. Because the lectin exhibits high binding affinity for the O-acetylated sialic acids in bovine submaxillary mucin (BSM), the crab lectin is purified using an affinity adsorbent of BSM coupled to Sepharose. For adsorption and elution, advantage is taken of the divalent cation (Ca^{2+}, Mg^{2+}) requirement for lectin binding and reduced activity of the lectin at pH 8.0–8.5.

Materials

Crab hemolymph

Selection of specimens. Not all species of crabs belonging to the genus *Cancer* contain the lectin with high titers and specific agglutination properties. Therefore, identification of the crab *C. anten-*

narius Stimpson is an important prerequisite. Schmitt[4] outlines distinguishing features and distribution of the species. Important distinguishing features include smooth granulated cheliped (hand), carapace widest at eighth anterolateral tooth, underparts spotted or blotched with reddish and hairy lower surface and legs. The crabs are collected in the coastal waters of Redondo, California, and brought to the market (Quality Sea Foods) at Redondo pier. Only male, uninjured, intermolt crabs are used for collection of blood to avoid possible physiological and pathological variations in the lectin concentration and activity.

Collection of sera. For the purpose of collecting a large volume of hemolymph, the crabs are placed on ice for 3–5 minutes to prevent degranulation or clumping of blood cells, which contain lytic enzymes. Prior to collection of blood, the first two legs are cleaned with a cotton swab. The tip of the leg, the dactylus, is cleaned with 70% ethanol and water. After wiping dry, this region is cut with a a pair of scissors, and the blood is allowed to drip. After the initial two drops, the blood is allowed to drip directly into 50-ml sterile polypropylene centrifuge tubes on ice. Crabs weighing 200–300 g yield about 50 ml of blood. The clot and cellular elements were immediately removed by centrifugation at 2000 g at 4°. The sera can be stored at $-20°$ for about 6 months without affecting its hemagglutination activity. However, the HA activity may gradually reduce if the sera is frozen and thawed repeatedly.

BSM–Sepharose. Bovine submaxillary mucin (BSM) is coupled to cyanogen bromide-activated Sepharose 4B following the procedure of Kohn and Wilchek.[5] For coupling, BSM is disolved in 0.1 M sodium pyrophosphate buffer, pH 8.0. The concentration of BSM coupled to Sepharose is 3 mg/ml (550 nmol sialic acid/ml). BSM–Sepharose is stored in cold TBS at pH 7.2 containing 0.02% sodium azide. It is important that the pH of the storing buffer be neutral, to avoid de-O-acetylation of BSM.

Buffers

TBS (Tris-buffered saline): 50 mM Tris–HCl, pH 7.2; 100 mM NaCl; 10 mM, CaCl$_2$.

TBS–BSA: TBS, pH 7.2, containing 0.05% BSA.

---

[4] W. L. Schmitt, *in* "The Marine Decapod Crustacea of California," Vol. 23, p. 217. Univ. of California Press, Berkeley, 1921.

[5] J. Kohn and M. Wilchek, *Biochem. Biophys. Res. Commun.* **107,** 878 (1982).

HSB (high salt buffer): 50 mM Tris–HCl, pH 7.2; 1 M NaCl; 10 mM CaCl$_2$.

LSB (low salt buffer): 50 mM Tris–HCl, pH 7.2; 300 mM NaCl; 10 mM CaCl$_2$

EB (elution buffer): 50 mM Tris–HCl, pH 8.2; 300 mM NaCl; 2 mM EDTA.

Purification Procedure

Step 1. Ultracentrifugation to remove hemocyanin: The major protein component of the blood (90–98%) of crabs is hemocyanin, the subunit (2–6) molecular weights ranging from 67 to 84 kDa.[6] Removal of a major fraction of hemocyanin is important for optimal purification of the lectin. The serum is centrifuged at 1.5×10^5 g for 5 hr at 4°. The supernatant (clarified sera) constitutes 90% of the total sera with low hemocyanin content.

Step 2. Clarified serum (120 ml) is applied to a column (Econo-polypropylene column containing 2 ml of BSM–Sepharose equilibrated with TBS at 4°). The eluant is collected at a rate of 0.6 ml/min. The column is washed with HSB until the A_{280} of the effluent was <0.002. An example of the elution profile and summary of purification are shown in Fig. 1 and Table I, respectively.

Step 3. The column is further washed with LSB at 4° until the A_{280} of the effluent was <0.002 and is then transferred from 4° to a 32° water bath. After 10 min, the column is washed again with warm (32°) LSB until the A_{280} of the effluent was <0.002. This step elutes additional inert protein and is necessary for obtaining homogenous lectin. Through this step all buffers contain calcium required for lectin binding to BSM–Sepharose.

Step 4. Elution of lectin is done with EB which contains EDTA. One-milliliter fractions are collected in polypropylene tubes containing 10 μl of 100 mM CaCl$_2$ at a rate of 0.3 ml/min. The CaCl$_2$ is necessary since the lectin is extremely unstable in the presence of EDTA. Fractions are vortexed immediately after collection and kept on ice.

Step 5. Fractions containing lectin are pooled on the same day and dialyzed against 1 mM CaCl$_2$ at 4° for 18 hr. Dialysis is continued for 3 hr with fresh 1 mM CaCl$_2$, and the dialysate is then aliquoted, lyophilized (Speed-vac, Savant), and stored at -20°.

---

[6] H. D. Ellerton, N. F. Ellerton, and H. A. Robinson, *Prog. Biophys. Mol. Biol.* **41,** 143 (1983).

TABLE I
PURIFICATION OF *Cancer antennarius* LECTIN[a]

Preparation	Volume (ml)	Protein (mg)	Total activity (HA units)	Specific activity (HA units/mg)	Purifi-cation	Yield[b] (%)
Serum	154	7650	3.0×10^7	0.4×10^3	1	100
Clarified serum	120	1020	2.5×10^7	2.4×10^3	6	83
BSM–Sepharose	90	2.46	0.9×10^7	37.0×10^3	925	30

[a] The purification shown is from native serum. Serum removed of hemocyanin is called clarified serum.
[b] Higher yields (60–75%) with low purity have been obtained by loading the column with crude serum.

TABLE II
CORRELATION BETWEEN THE PRESENCE OF *O*-ACETYL GROUPS ON
ERYTHROCYTES AND HEMAGGLUTINATION BY *C. antennarius* LECTIN[a]

Erythrocyte preparation	*O*-acetyl sialic acid[b]	Lectin source[c]		
		Cancer	Limax	Limulus
Rabbit	+	16	512	128
Rat	+	32	128	8
Mouse (DBA/J)	+	2048	128	32
Horse	+	512	128	64
Chicken	−	0	64	4
Dog	−	0	64	32
Sheep	−	0	64	16
Goat	−	0	64	2
Monkey	−	0	64	32
Human (A, B, O)	−	0	64	32
Human asialo	−	0	8	0
Human resialylated				
NeuAcα2,3Gal	−	2	128	128
NeuAcα2,6Gal	−	2	128	32
NeuGcα2,6Gal	−	2	8	1024
9-O-Ac-NeuAcα2,6Gal	+	512	32	2048

[a] Adapted from Ravindranath *et al.*[3]
[b] +, present; −, absent.
[c] Lectins were reconstituted in TBS containing 0.05% BSA to obtain the given HA titer for horse erythrocytes.
[d] Human asialo (sialidase-treated) and resialylated erythrocytes were prepared as described.[3,7]

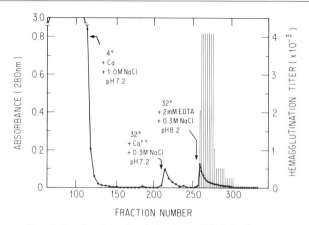

FIG. 1. Purification of crab lectin on BSM–Sepharose.

Properties of the Lectin

Specificity. The lectin exhibits unique specificity for O-acetylated sialic acid, 9-O-Ac-NeuAc, and 4-O-Ac-NeuAc. The lectin agglutinates the erythrocytes containing 9-O-Ac-NeuAc (rabbit, rat, and mouse) or 4-O-AcNeuAc (horse) but not those containing NeuAc or NeuGc only (Table II).[3,7] Moreover, human erythrocytes, which contain only NeuAc and are not normally agglutinated by the crab lectin, are agglutinated effectively after sialidase treatment and resialylation to introduce 9-O-Ac-NeuAc onto cell surface glucoconjugates (Table II). In addition, lectin-mediated hemagglutination is inhibited by O-acetylsialic acid containing glycoproteins (bovine and equine submaxillary mucins) and not by other sialoglycoproteins.[3] Removal of the O-acetyl groups from these mucins abolished their inhibitory property. Similarly, of the 24 free sugars tested,[3] only 9-O-Ac-NeuAc and 4-O-Ac-NeuAc and NeuAc inhibited the lectin activity with relative potential of 33 : 11 : 1, respectively.

Molecular Weight. Sodium dodecyl sulfate–polyacrylamide gel electrophoresis of the purified enzyme (see Fig. 2)[8] shows a doublet of M_r 37,000. Under nonreducing conditions, the protein shows a single prominent band of M_r 70,000, suggesting that the native protein is at least a dimer.

Metal Requirement. The lectin has an absolute requirement for divalent cations (Ca^{2+}, Mg^{2+}). Activity is abolished by EDTA but can be restored by the addition of calcium.

[7] J. C. Paulson and G. N. Rogers, this volume [11].
[8] U. K. Laemmli, *Nature (London)* **227**, 680 (1970).

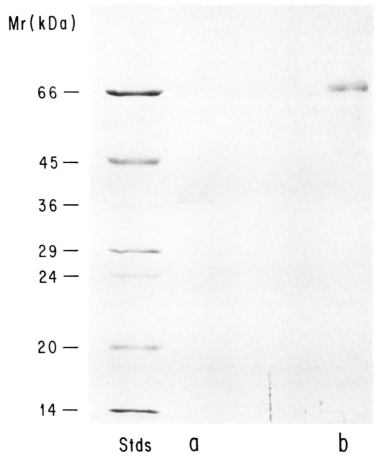

Mr (kDa)

66 —

45 —

36 —

29 —
24 —

20 —

14 —

Stds   a   b

FIG. 2. SDS–Gel electrophoresis of purified crab lectin. Electrophoresis was carried out on a SDS–polyacrylamide 15% slab gel and stained with Coomassie Blue.[8] Samples were heated for 3 min at 100° in sample buffer containing 2% 2-mercaptoethanol and 4% SDS except for the sample in lane b which was treated in the same way without 2-mercaptoethanol. Standards (Stds) of known molecular weight (M_r): bovine serum albumin (66K); egg albumin (45K); glyceraldehyde-3-phosphate dehydrogenase (36K); carbonate dehydratase (29K); trypsinogen (24K); trypsin ihibitor (20K); and α-lactalbumin (14K). Lane a represents lectin under reduced conditions; lane b represents the same in unreduced state. Extra lanes between a and b contained no samples to avoid contamination of sample with 2-mercaptoethanol.

pH Optimum. The lectin binds effectively in the pH range of 6.5–7.5 but not at or above pH 8.0.

Stability. The lectin is rapidly inactivated in the absence of $CaCl_2$. It is also inactivated in dilute solution by glass or metal surfaces. Thus, the lectin is routinely handled in plastic containers, and buffers with BSA are used for hemagglutination assays. The lyophilized lectin is stable for at least 9 months stored at $-20°$. The lectin can also be stored in solution in Tris-buffered saline (TBS) at $-20°$ in the presence of 20% glycerol, 10 mM $CaCl_2$, pH 7.2. Under these conditions, lectin activity is least affected by freezing and thawing. If stored at 4°, the lectin loses 50% of its activity in about 2 weeks. At room temperature the lectin in TBS is stable for 5 hr.

Inhibitors. Hemagglutination by the crab lectin is inhibited by sulfhydryl blockers (*p*-chloromercuric benzoate, mercuric chloride, and *N*-ethylmaleimide, at 0.1 *M*) but not by reducing agents (mercaptoethanol and dithiothreitol).

Comparison with Other Sialic Acid-Specific Lectins. The specificity of the crab lectin distinguishes it from other sialic acid-specific lectins purified from horseshoe crabs (*Limulus polyphemus, Tachypleus tridentatus, Carcinoscorpius rotundicauda*) and slug (*Limax flavus*) which show a much broader specificity. Lectins from *Limax* and *Limulus* agglutinate erythrocytes of all species tested (Table II) and is inhibited by ovine and bovine submaxillary mucins, with or without base treatment,[3] and by free *N*-acetyl- and *N*-glycolylneuraminic acids.[9]

---

[9] R. L. Miller, J. F. Collawn, Jr., and W. W. Fish, *J. Biol. Chem.* **257,** 7574 (1982).

[45] Properties of a Sialic Acid-Specific Lectin from the Slug *Limax flavus*

By Ronald L. Miller

Lectins are a group of non-immunoglobulin-like proteins which exhibit a high binding specificity toward carbohydrate residues of glycoproteins and glycolipids. Because they interact with carbohydrates in a highly specific manner, they have found wide application in studies on glycoproteins. They are found widely distributed in nature; however, those which have been purified have been obtained primarily from leguminous seeds and the hemolymph of invertebrate animals.[1]

In spite of the ubiquitous presence of lectins in nature, few have been characterized which are specific for sialic acid. Lectins which bind sialic

---

[1] I. J. Goldstein and C. E. Hayes, *Adv. Carbohydr. Chem. Biochem.* **35,** 453 (1978).

acid residues of glycoproteins have been purified from the American[2] and Indian[3] horseshoe crab, the lobster,[4] and wheat germ.[5] However, due to their macromolecular properties and/or because their specificity is not limited to sialic acid, these lectins have not proved to be fully satisfactory for the study of sialoproteins.

In 1970, Pemberton[6] reported that extracts of the slug *Limax flavus* contain an agglutinin for red blood cells. No carbohydrate specificity was defined for the agglutinin. We have recently demonstrated that this slug contains a lectin which is highly specific for sialic acid residues of glycoproteins[7] and have purified this lectin, LFA, to homogeneity.[8] In contrast to earlier studies on the purification of invertebrate lectins which used hemolymph as the source material, we have purified LFA from the body tissues of the slug.

Hemagglutination Assay

Fifty milliliters of a 0.9% NaCl solution was added to 10 ml of freshly drawn blood, and the cells were sedimented by centrifugation at 2000 *g* for 5 min. The cells were washed 4 more times with 50 ml of the saline solution in the same manner. The washed erythrocytes were diluted with the saline solution to an absorbancy of 2.0 at 620 nm. Identical hemagglutination curves were obtained with erythrocytes from A, B, and O blood types.

Aliquots of the slug lectin preparation were diluted to 0.50 ml with a Tris–saline solution (0.05 *M* Tris–Cl, 0.10 *M* NaCl, pH 7.5) containing 2 mg of gelatin per ml. One-half milliliter of washed and diluted erythrocytes was added to the diluted lectin solution, and after 30 min at room temperature the cells were pelleted by centrifugation. The cells were resuspended in the same solution and aggregated cells allowed to settle for 5 min before absorbancy at 620 nm of the upper 0.50 ml of the assay solution was determined. The data were plotted as percentage hemagglutination versus micrograms of lectin preparation added to the assay (Fig. 1). One hemagglutination unit (HU) is defined as the amount of lectin required to give 50% reduction in the absorbancy at 620 nm of the erythrocyte suspension as described above. Inhibition of hemagglutination was

[2] J. J. Marchalonis and G. M. Edelman, *J. Mol. Biol.* **32,** 453 (1968).
[3] S. Bishayee and D. T. Dorai, *Biochim. Biophys. Acta* **623,** 89 (1980).
[4] J. L. Hall and D. T. Rowlands, Jr., *Biochemistry* **13,** 828 (1974).
[5] V. P. Bhavanadan and A. W. Katlic, *J. Biol. Chem.* **254,** 4000 (1979).
[6] R. T. Pemberton, *Vox Sang.* **18,** 74 (1970).
[7] R. L. Miller, *J. Invertebr. Pathol.* **39,** 210 (1982).
[8] R. L. Miller, J. F. Collawn, Jr., and W. W. Fish, *J. Biol. Chem.* **257,** 7574 (1982).

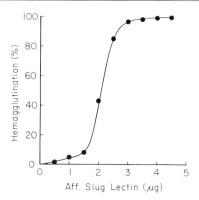

Fɪɢ. 1. Hemagglutination assay. Type A human erythrocytes were used in the assay. The assay was performed with the affinity-purified LFA as described in the text.

determined by adding potential carbohydrate inhibitors to assays containing a sufficient amount of lectin to give 90–100% agglutination and determining loss of hemagglutinin activity due to the presence of inhibitors.

Purification Procedures

It is recommended that eviscerated slugs be used as the starting material. Initial homogenates were prepared from whole slugs which had been frozen and thawed. However, the resultant solution was highly viscous and difficult to work with. Subsequent homogenates were prepared from eviscerated slugs. Using eviscerated slug tissues resulted in a much less viscous homogenate with little loss of agglutinin activity. Purification of the sialic acid-specific lectin (LFA) from the homogenates involved ammonium sulfate fractionation and affinity chromotography on a column of bovine submaxillary mucin coupled to Sepharose 4B.

Preparation of Bovine Submaxillary Mucin–Sepharose 4B

The BSM–Sepharose was prepared as previously reported by Miller *et al.*[8] Sepharose 4B was washed with deionized water by filtration, followed by a wash with an equal volume of 2.0 M potassium phosphate, pH 12. Twenty milliliters of the washed Sepharose 4B was suspended in 20 ml of the phosphate buffer, and 1.0 g of cyanogen bromide dissolved in 2.5 ml of dioxane was added over a 6-min period while stirring the Sepharose 4B in an ice bath. After stirring for an additional 10 min, 10 ml of Sigma, Type I, bovine submaxillary mucin (BSM)—10 mg/ml in 1.0 M NaCl—and 10 ml of 0.2 M NaHCO$_3$, pH 9.0, was added to the activated Sepharose 4B, and the mixture was stirred for 2 hr at 22°. The BSM–Sepharose 4B was washed with 250 ml of 1.0 M NaCl and 250 ml of Tris–saline. Approximately 80% of the BSM was covalently coupled to the Sepharose 4B indicating that the BSM–Sepharose 4B contained approximately 4.0 mg of BSM per milliliter.

Purification of LFA

All purification procedures were carried out at approximately 5°. Slugs that had been frozen were thawed, and a ventral slit was made with scissors. The viscera and rudimentary shell were removed and the remaining body tissues were washed with deionized water and Tris–saline (10 ml/g tissue). One hundred grams of tissue were added to 400 ml of Tris–saline which also contained a total of 20 mg of phenylmethanesulfonyl fluoride. The tissue was first minced with scissors and then homogenized with a Polytron at full speed for 3 min. The homogenate was centrifuged at 16,000 g for 15 min. The resultant supernatant fraction, SF-16, was adjusted to 40% saturation by the addition of ammonium sulfate, with stirring, and the precipitated proteins (AS-0–40) were collected by centrifugation at 16,000 g for 15 min, assayed to be sure they contained no LFA activity, and then discarded. The resultant supernatant fraction was adjusted to 80% saturation with ammonium sulfate and the precipitated proteins (AS-40–80) collected as described above. The proteins in the AS-40–80 fraction were suspended in 200 ml of Tris–saline and dialyzed against 3 1-liter volumes of the same buffer.

A column (1.6 × 10 cm) of the BSM–Sepharose 4B was prepared and equilibrated with the Tris–saline buffer. One-half (100 ml) of the AS-40–80 fraction was pumped onto the column at a rate of 50 ml/hr. Tris–saline buffer was pumped through the column until nonbound proteins were eluted as evidenced by the A_{280} of the effluent approaching baseline levels. This usually required about 50 ml of buffer. The bound proteins were eluted with a solution containing 0.05 M Tris–Cl, 0.30 M NaCl, and 10 mM N-acetylneuraminic acid (AcNeu) and adjusted to a pH of 7.5 at 22°. This is a higher salt concentration than originally used by Miller et al.[8] to elute bound proteins. We have observed that elution of the BSM–Sepharose 4B with 0.05 M Tris–Cl, 0.30 M NaCl, pH 7.5, does not elute any of the bound proteins and that elution with this solution containing 10 mM AcNeu results in a 2-fold increase in recovery of agglutinin activity when compared to the original procedure.[8] The LFA obtained in this manner has the same specific activity, molecular properties, and carbohydrate-binding specificity as previously reported by Miller et al.[8] Fractions containing proteins displaced by AcNeu were combined. The concentra-

TABLE I
PURIFICATION OF LFA

Fraction[a]	Volume (ml)	Protein (mg)	Specific activity (units/mg)	Total activity[b] (units × 10⁻⁶)	Activity recovered (%)
SF-16	538	3569[c]	134	0.49	100
AS-40–80	74	1686[c]	261	0.44	90
LFA	25	60[d]	7167	0.43	88

[a] One hundred grams of eviscerated slug tissues were used in the preparation. The following notations were used: SF-16, the supernatant fraction obtained from a 16,000 g centrifugation of the tissue homogenate; AS-40–80, proteins which precipitated between 40 and 80% saturation of SF-16 with ammonium sulfate; LFA, the sialic acid-specific lectin which was eluted from the BSM–Sepharose 4B column with 0.05 M Tris–Cl, 0.30 M NaCl, and 10 mM AcNeu, pH 7.5.

[b] One unit of activity is defined as that amount of LFA preparation (~0.24 μg of purified LFA) which gave 50% agglutination as determined by the hemagglutination assay described above.

[c] Protein concentration was estimated by use of the formula mg protein/ml = $A_{280} \times 1.45 - A_{260} \times 0.74$.

[d] LFA concentration was determined by use of $E_{280}^{mg/ml} = 2.10$.

tion of LFA in the combined fractions was usually in the range of 1.0–2.0 mg/ml; the $E_{280}^{mg/ml}$ of LFA is 2.1. Concentration determinations are most easily performed by measuring the solution absorbance at 280 nm. The LFA can be concentrated by reprecipitation with ammonium sulfate at 80% saturation followed by resolubilization in the desired volume of 0.05 M Tris–Cl, 0.3 M NaCl, pH 7.5. The higher salt concentration is recommended since LFA at concentrations greater than 1.0 mg/ml has a tendency to aggregate in 0.1 M NaCl solutions.[8] The LFA should then be dialyzed against a buffer of pH 7.5 which contains either 0.1 M NaCl (for LFA concentrations greater than 1.0 mg/ml) or 0.30 M NaCl (for higher concentrations of LFA).

As indicated in Table I, approximately 60 mg of LFA can be obtained from 100 g of eviscerated slug tissue. The overall recovery of agglutinin activity was 88% with a 55-fold purification. Although no significant amount of agglutinin activity could be detected in the proteins which did not bind to BSM–Sepharose 4B (Fig. 2), carbohydrate inhibition studies indicate that the AS-40–80 fraction contains more than one lectin. Of the many carbohydrates tested, agglutination of erythrocytes with the affinity-purified LFA was inhibited by only N-acetylneuraminic acid and N-glycolylneuraminic acid (Table II). These results indicate that LFA has a very high specificity for sialic acid and, in this respect, is quite different

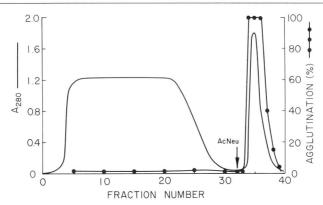

FIG. 2. Affinity purification of LFA. One hundred milliliters of the AS-40–80 fraction was pumped at 50 ml/hr onto a 1.6 × 10 cm column of BSM–Sepharose 4B that had been equilibrated with 50 mM Tris–Cl, 100 mM NaCl, pH 7.5. The column was washed with the above buffered solution until A_{280} of the effluent approached baseline; this required approximately 50 ml. The bound LFA was eluted with approximately 30 ml of 50 mM Tris–Cl, 300 mM NaCl, 10 mM AcNeu, pH 7.5, and 4.0-μl aliquots of those fractions indicated were assayed for agglutinin activity.

from the other sialic acid-binding lectins reported in the literature to date. Agglutination of erythrocytes with the AS-40–80 fraction was inhibited by high concentrations of N-acetylgalactosamine, N-acetylglucosamine, galactosamine, and glucosamine, in addition to the inhibition observed at relatively low concentrations of the sialic acids (Table II). Thus it appears that the AS-40–80 fraction from the slug extract contains a mixture of lectins with LFA being, by far, the predominant lectin present.

Stability and Storage

The LFA is stable in the Tris–saline solutions either frozen or refrigerated. However, if it is to be refrigerated for more than a few days, sodium azide (0.05% w/v) should be added to the solution in order to prevent bacterial growth. The LFA can be lyophilized in the 0.30 M NaCl solution, stored in a freezer, and at a subsequent time (we have monitored for up to 1 year) resolubilized in an equal volume of deionized water without any apparent loss in agglutinin activity.

Macromolecular Properties of LFA

The macromolecular properties of LFA are summarized in Table III. It has a native molecular weight of 44,000 and consists of two equal-sized

TABLE II

INHIBITION OF HEMAGGLUTININ ACTIVITY OF SLUG LECTIN
BY CARBOHYDRATES[a]

Carbohydrate	Carbohydrate concentration (mM)	Inhibition (%)	
		AS-40–80 fraction	Purified LFA
N-Acetylneuraminic acid	0.05	6	1
	0.10	17	28
	0.15	27	54
	0.20	94	100
N-Glycolylneuraminic acid	0.25	3	0
	0.50	20	2
	0.75	56	14
	1.0	88	85
N-Acetyl-D-galactosamine	10.0	75	0
N-Acetyl-D-glucosamine	10.0	53	0
Galactosamine	25.0	45	0
Glucosamine	25.0	19	0
D-(+)-Galactose	25.0	0	0
D-(−)-Glucose	25.0	0	0
D-(+)-Mannose	25.0	0	0
α-(+)-Arabinose	25.0	0	0
D-(+)-Xylose	25.0	0	0
α-L-(−)-Fucose	25.0	0	0
D-(−)-Glucuronic acid	25.0	0	0
α-Methyl-D-glucoside	25.0	0	0
α-Methyl-D-mannose	25.0	0	0
Lactose	25.0	0	0
Sucrose	25.0	0	0

[a] The standard hemagglutination assay was used in these comparative studies. Twenty micrograms of protein from the 40–80% ammonium sulfate fraction (AS-40–80) or 0.34 μg of the purified LFA was utilized in each assay. Taken from Miller et al.[8]

subunits with molecular weights of 22,000. It has a $s^{\circ}_{w,20}$ of 3.4. The affinity-purified LFA gave a single band upon sodium dodecyl sulfate–polyacrylamide gel electrophoresis[9] (Fig. 3). Aspartic acid and glutaminic acid make up approximately 20% of its total amino acids,[8] thus it has a relatively high content of acidic amino acids. However, many of its acidic amino acids must be amidated since LFA is a basic protein with a pI of 9–9.5. It also contains 10 residues of tyrosine and 6 residues of tryptophan

[9] U. K. Laemmli, Nature (London) **227,** 680 (1970).

TABLE III
MACROMOLECULAR PROPERTIES OF LFA

Physical measurement	Magnitude
Native molecular weight by sedimentation equilibrium centrifugation	44,000 ± 2,000
Subunit molecular weight by sedimentation equilibrium centrifugation in 6 M guanidine–Cl after reduction	22,000–26,000
Subunit molecular weight by sodium dodecyl sulfate gel electrophoresis	22,500
$E_{280}^{mg/ml}$	2.1 ± 0.2
$s°$ by sedimentation velocity centrifugation	3.45 ± 0.25
$\bar{V}$ from amino acid composition	0.724 ml/g
pI from isoelectric focusing	9–9.5

[a] Data taken from Miller et al.[8]

per subunit, and this is reflected in its high $E_{280}^{mg/ml}$ value of 2.1.[8] LFA, at concentrations greater than 1.0 mg/ml, has a propensity to aggregate if stored in a buffered solution containing 0.1 M NaCl. Aggregation can be prevented by storing LFA in a 0.3 M NaCl solution buffered at a pH of 7.5 with 50 mM Tris–Cl or 50 mM sodium phosphate.

Preparation of LFA–Sepharose 4B

The Sepharose 4B was washed and activated as described above. Twenty-five milliliters of LFA (2 mg/ml) was dialyzed against 2 1-liter volumes of 50 mM NaPO$_4$, 0.10 M NaCl, pH 6.8. Powered sodium bicarbonate and N-acetylneuraminic acid was added to a final concentration of 0.20 M and 1.0 mM, respectively, before final adjustment of the pH of the LFA solution to 9.0. The LFA solution was gently mixed with 20 ml of activated Sepharose 4B for 2 hr at 22°. The LFA–Sepharose 4B was washed with 25 ml of 1.0 M NaCl and 25 ml of Tris–saline. Approximately 85% of the LFA was coupled to the Sepharose 4B, as determined from the absorbance of the washes at 280 and 260 nm.

Use of LFA–Sepharose 4B in Affinity Chromotography

LFA coupled to Sepharose 4B maintains its sialic acid-specific binding activity and can be used for the affinity purification of sialoglycoproteins.[10] Affinity chromotography of sialoproteins on LFA Sepharose 4B

[10] R. L. Miller and J. D. Cannon, Jr., in "Recognition Proteins Receptors, and Probes: Invertebrates" (E. Cohen, ed.), p. 31. Alan R. Liss, Inc., New York, 1984.

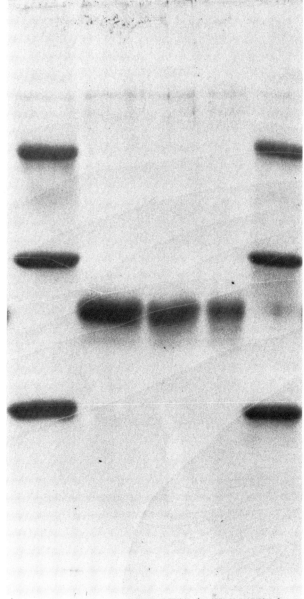

FIG. 3. Sodium dodecyl sulfate–polyacrylamide gel electrophoresis of purified LFA. Electrophoresis was carried out on 1.5-mm slab gels (9%) by the reducing and denaturation method of Laemmli.[9] Lanes 1 and 5, 5 μg each of myoglobin (M_r 17,200), chymotrypsinogen A (25,000), and ovalbumin (43,000); lanes 2, 3, and 4 contained 10, 5, and 2.5 μg of LFA, respectively.

has been performed by the following procedure. The LFA–Sepharose 4B column (1 × 16 cm) was equilibrated with 0.05 M Tris–Cl, 0.10 M NaCl, pH 7.5. Serum samples which were to be fractionated were diluted to 15 mg protein/ml with 0.05 M Tris–Cl, 0.10 M NaCl, pH 7.5. Commercially available proteins which were to be fractionated were dissolved in the above buffered solution. The protein solutions were run onto the column at a rate of 0.5 ml/min. Elution of nonbound proteins required approximately 2 column volumes or 25 ml of the above buffered solution. Bound proteins were eluted with 2 column volumes of 0.5 M Tris–Cl, 0.30 M NaCl, 10 mM AcNeu, pH 7.5. From preliminary studies,[10] we have observed that nonsialylated proteins, such as bovine serum albumin, do not bind to the LFA–Sepharose 4B, whereas sialylated proteins such as bovine fetuin bind and require elution with the AcNeu solution. LFA–Sepharose 4B which contains 2 mg LFA/ml binds approximately 0.25 mg fetuin/ml of packed gel. The amount of a sialoprotein bound will probably vary with its molecular weight and sialic acid content.

Acknowledgments

These studies were supported by Grant BC-510 from the American Cancer Society and by an appropriation for biomedical research from the state of South Carolina.

[46] *Vicia villosa* Lectins

By Sherida Tollefsen and Rosalind Kornfeld

The seeds of *Vicia villosa,* commonly known as hairy winter vetch, contain a family of tetrameric lectins composed of two different subunits (designated A and B) with distinct carbohydrate-binding activities.[1,2] The A$_4$ lectin agglutinates human blood group A erythrocytes which contain terminal α-linked N-acetylgalactosamine, the blood group A determinant, but does not agglutinate human type B or O erythrocytes which lack this determinant. The B$_4$ lectin, the predominant lectin in *V. villosa* seeds, does not agglutinate human blood group A, B, or O erythrocytes but does bind to and agglutinate human Tn-exposed erythrocytes. The Tn antigen is the N-acetylgalactosamine residue α-linked to serine and/or threonine residues in the polypeptide chain of cell-surface glycoproteins, and this

---

[1] S. E. Tollefsen and R. Kornfeld, *J. Biol. Chem.* **258**, 5165 (1983).

[2] S. E. Tollefsen and R. Kornfeld, *J. Biol. Chem.* **258**, 5172 (1983).

antigen is exposed on erythrocytes, leukocytes, and platelets of rare individuals whose cells are deficient in UDP-Gal : GalNAc-β-3-D-galactosyltransferase activity.[3-5] The B_4 lectin also binds to and agglutinates human erythrocytes with blood-group Cad specificity.[6] The Cad determinant is expressed on erythrocyte glycophorins A and B[7] and on an erythrocyte ganglioside[8] in Cad individuals. On glycophorin, the O-linked oligosaccharide bearing the Cad determinant has the structure[9] GalNAcβ1,4(NeuNAcα2,3)Galβ1,3(NeuNAcα2,6)GalNAcα → Ser/Thr, which differs from the oligosaccharide structure on normal erythrocytes by the presence of the terminal GalNAcβ1-4 residue. The hybrid A_2B_2 lectin shares the properties of A_4 and B_4 lectins.

Lectin prepared from *V. villosa* seeds was originally described by Kimura *et al.*[10] as human blood-group A specific. This preparation bound specifically to a 145-kDa glycoprotein selectively expressed by mouse Ly1$^-$2$^+$ cytotoxic T lymphocytes. In subsequent studies by other groups[11,12] the specificity of the interaction between the *V. villosa* lectin and cytotoxic T lymphocytes was questioned. The presence of lectins with different carbohydrate-binding specificities in *V. villosa* lectin preparations may account in part for the conflicting results. In addition, Grubhoffer *et al.*[13] have reported that the A_1 erythrocyte-agglutinating activity of seed extracts prepared from different cultivars within the species *V. villosa* varies widely.

Assay Methods

Erythrocyte Agglutination. Hemagglutination assays are carried out on a glass plate as previously described[14] using 40-μl droplets containing

[3] D. J. Anstee, *Semin. Hematol.* **18,** 13 (1981).
[4] J.-P. Cartron, G. Andreu, J. Cartron, G. W. G. Bird, C. Salmon, and A. Gerbal, *Eur. J. Biochem.* **92,** 111 (1978).
[5] J.-P. Cartron and A. T. Nurden, *Nature (London)* **282,** 621 (1979).
[6] S. E. Tollefsen and R. Kornfeld, *Biochem. Biophys. Res. Commun.* **123,** 1099 (1984).
[7] J.-P. Cartron and D. Blanchard, *Biochem. J.* **207,** 497 (1982).
[8] D. Blanchard, F. Piller, B. Gillard, D. Marcus, and J.-P. Cartron, *J. Biol. Chem.* **260,** 7813 (1985).
[9] D. Blanchard, J.-P. Cartron, B. Fournet, J. Montreuil, H. van Halbeek, and J. F. G. Vliegenthart, *J. Biol. Chem.* **258,** 7691 (1983).
[10] A. Kimura, H. Wigzell, G. Holmquist, B. Ersson, and P. Carlsson, *J. Exp. Med.* **149,** 473 (1979).
[11] H. R. MacDonald, J.-P. Mach, M. Schreyer, P. Zaech, and J.-C. Cerottini, *J. Immunol.* **126,** 883 (1981).
[12] V. T. Braciale, H. P. Friedman, and T. J. Braciale, *J. Immunol. Methods* **43,** 241 (1981).
[13] L. Grubhoffer, M. Ticha, and J. Kocourek, *Biochem. J.* **195,** 623 (1981).
[14] R. Kornfeld, W. T. Gregory, and S. A. Kornfeld, this series, Vol. 28 [40].

25 μl of diluted lectin and 15μl of a 4% suspension of washed erythrocytes in phosphate-buffered saline (PBS: 0.137 M NaCl, 2.7 mM KCl, 8.1 mM Na$_2$HPO$_4$, 1.5 mM KH$_2$PO$_4$, pH 7.4). One hemagglutinating unit is defined as the amount of lectin which causes 4+ agglutination in 3 min. Human A$_1$, B, and O erythrocytes are obtained by venipuncture, and Tn-exposed erythrocytes and erythrocytes with Cad specificity can be obtained from John Moulds, gamma Biologicals, Inc., Houston, Texas.

[125]I-Labeled Lectin Binding and Hapten Inhibition. Purified lectins (25–95 μg) are radioiodinated by the chloramine-T method[15] in the presence of 10 mM N-acetylgalactosamine (to protect the carbohydrate-binding site), using 250 μCi Na[125]I and a 10-sec exposure to chloramine-T. Free ^{125}I is removed by dialysis or gel filtration, and ^{125}I-labeled lectins are stored in PBS containing 1–5 mg/ml bovine serum albumin at −20°. The binding of ^{125}I-labeled lectins to erythrocytes is measured as previously described.[1] The inhibition of binding of ^{125}I-labeled B$_4$ lectin to Tn-exposed erythrocytes in the presence of varying amounts of a competing saccharide or glycopeptide is performed as previously described[2] to determine the concentration of inhibitor which causes 50% inhibition of lectin binding. In the inhibition assay, the concentration of ^{125}I-labeled B$_4$ lectin added is near the apparent K_d value (2.4 × 10^{-8} M).

Purification

Principle

The N-acetylgalactosamine-binding lectins from *V. villosa* are isolated from a seed extract by adsorption to a porcine blood-group substance affinity column and sequential elution with different concentrations of N-acetylgalactosamine. The oligosaccharides found in porcine blood-group substance are heterogeneous and include structures with A and O blood-group reactivity. The B$_4$, A$_4$, and A$_2$B$_2$ lectins are not purified to homogeneity by affinity chromatography and must be purified further by anion-exchange chromatography.

Procedure

Preparation of the Affinity Column. Affi-Gel 15 (25–35 ml) is washed as described by the manufacturer (Bio-Rad Laboratories) and added to an equal volume of 0.1 M NaHCO$_3$ containing 40 mg/ml porcine blood-group substance purified by the method of Kabat[16] from crude porcine stomach

[15] F. C. Greenwood, W. M. Hunter, and J. S. Glover, *Biochem. J.* **89,** 114 (1963).
[16] E. A. Kabat, "Blood Group Substances: Their Chemistry and Immunochemistry," 2nd ed., p. 135. Academic Press, New York, 1956.

mucin (Sigma Chemical Co.). The suspension is gently agitated for 2–4 hr at room temperature and for an additional 1 hr in 0.1 M ethanolamine–HCl, pH 8.0, to block unreacted groups. The adsorbent is washed thoroughly with 0.1 M NaHCO$_3$ and with PBS, transferred to a column, and washed with 10 ml of 10 mM N-acetylgalactosamine in PBS and then with PBS.

Preparation of an Extract of Vicia villosa Seeds. One hundred grams of *V. villosa* seeds (Schweizer Samen AG, Thun, Switzerland[17]) are finely ground and extracted overnight in 600 ml PBS. The pH is then lowered to 4.5 with 4 M acetic acid, and the solution is centrifuged at 20,000 g for 20 min. The supernatant fluid is filtered through glass wool, and the pH is readjusted to 7.4 with NaOH. Solid ammonium sulfate to 80% saturation (51.6 g/100 ml) is added, and the mixture is stirred for 2 hr. The precipitate is collected by centrifugation at 2000 g for 20 min and redissolved in 200 ml PBS. All steps are performed at 4°.

Affinity Chromatography. The partially purified seed extract is centrifuged at 2000 g for 20 min to remove insoluble material and applied in 75–100 ml portions to the porcine blood-group substance affinity column equilibrated in PBS at room temperature. The column is washed with PBS at a flow rate of 85 ml/hr until the A_{280} of the effluent is <0.06. The column is eluted with 1 mM and then with 10 mM N-acetylgalactosamine in PBS, each time until the A_{280} of the effluent falls to <0.02. The column is washed extensively with PBS before the next portion of seed extract is applied. After bound protein from the last portion applied has been displaced with 1 and 10 mM N-acetylgalactosamine, the column is eluted with 100 mM N-acetylgalactosamine. Column fractions are dialyzed extensively against PBS to remove N-acetylgalactosamine and assayed for A_1 erythrocyte agglutinating activity, as described above. The run-through fractions can be reapplied to the column to recover additional lectin. Essentially all of the bound protein from the run-through fractions can be displaced with 1 mM N-acetylgalactosamine. The amount of lectin in these fractions depends on the size and capacity of the affinity column used.

Elution of the column with 1, 10, and 100 mM N-acetylgalactosamine results in a sequential increase in the A_1 erythrocyte-agglutinating specific activity of the lectins in the eluates. The total yield of lectins from this procedure, determined after reapplication of the run-through material to the affinity column, is 75–80 mg/100 g seeds.

---

[17] *Vicia villosa* seeds from this supplier have reproducibly contained the amounts of the isolectins described herein. Different batches of seeds from a local supplier (Manglesdorf Seed Co., St. Louis, Missouri) have contained varying amounts of the lectins. This variability is presumably explained by variation in lectin content among the different cultivars within the species *V. villosa* and has also been noted by Grubhoffer *et al.* [13]

Anion-Exchange Chromatography. The B_4, A_4, and A_2B_2 lectins are purified from the N-acetylgalactosamine eluates by anion-exchange chromatography.

PURIFICATION OF B_4 LECTIN. The B_4 lectin can be readily purified from the 1 mM N-acetylgalactosamine eluate. The 1 mM N-acetylgalactosamine eluate is dialyzed against 10 mM Tris–HCl buffer, pH 8.0, and applied to a DEAE–cellulose column (Whatman preswollen microgranular anion exchanger DE-52) equilibrated in the same buffer. About 1.5 mg of protein is applied per milliliter of bed volume. The column is eluted with 85 mM NaCl buffered with 10 mM Tris–HCl, pH 8.0, at a flow rate of 40 ml/hr. All of the protein in the 1 mM N-acetylgalactosamine eluate binds to the column, and 80–85% of the protein applied is displaced with 85 mM NaCl. The protein displaced with 85 mM NaCl is purified B_4 lectin. The B_4 lectin constitutes at least 75% of the total lectin present in the seed extract.

PURIFICATION OF A_4 LECTIN. Purification of the A_4 lectin is less straightforward, since complete separation of the A_4 lectin from the residual B_4 lectin in the 10 mM N-acetylgalactosamine eluate by anion-exchange chromatography has not been possible. The A_4 lectin can be purified from the 100 mM N-acetylgalactosamine eluate. This eluate is dialyzed extensively against 10 mM sodium phosphate buffer, pH 7.2, and applied to a DEAE–cellulose column equilibrated in the same buffer. The column is eluted with a linear gradient from 0 to 0.5 M NaCl buffered with 10 mM sodium phosphate, pH 7.2, at a flow rate of 25 ml/hr. The protein in the 100 mM N-acetylgalactosamine eluate is displaced in two peaks at about 105 mM and 170 mM NaCl. The first peak displaced is purified A_4 lectin.

PURIFICATION OF A_2B_2 LECTIN. The 10 mM N-acetylgalactosamine eluate is dialyzed extensively against 10 mM sodium phosphate buffer, pH 7.2, containing 0.05 M NaCl, and applied to a DEAE–cellulose column equilibrated in 10 mM sodium phosphate buffer, pH 7.2. The column is eluted with a linear gradient from 0.05 to 0.5 M NaCl buffered with 10 mM sodium phosphate, pH 7.2, at a flow rate of 32 ml/hr. The protein in the 10 mM N-acetylgalactosamine eluate is displaced in two peaks at about 90 and 160 mM NaCl. The second peak displaced is purified A_2B_2 lectin.

The other two possible isolectins, A_1B_3 and A_3B_1, have not been purified but may also be present in the seed extract.

Comments. Other affinity adsorbents have been used to purify the lectins from *V. villosa* seeds.[10,11,18] In our experience, chromatography on

[18] A. Conzelmann, R. Pink, O. Acuto, J.-P. Mach, S. Dolivo, and M. Nabholz, *Eur. J. Immunol.* **10**, 860 (1980).

a column of *N*-acetylgalactosamine coupled to epoxy-activated cross-linked 4% beaded agarose (Pierce Chemical Co., Selectin 5) or on an asialo bovine submaxillary mucin column has not been useful in the initial isolation of the isolectins. Because the lectins do not display differential affinity for these adsorbents, all are eluted with the same concentration of *N*-acetylgalactosamine.

Properties

Physicochemical Properties. The B_4 lectin migrates as a single band on polyacrylamide gel electrophoresis in the presence of sodium dodecyl sulfate (SDS–PAGE) with an apparent molecular weight of 35,900. The molecular weight of the intact B_4 lectin, calculated from sedimentation equilibrium analysis, is 108,300, and the molecular weight of the dissociated subunits, determined by sedimentation equilibrium analysis in the presence of 6 *M* guanidine–HCl, is 25,600. The A_4 lectin migrates as a single band on SDS–PAGE with an apparent molecular weight of 33,600. The molecular weight of the intact A_4 lectin, calculated from sedimentation equilibrium analysis, is 109,500. SDS–PAGE of the A_2B_2 lectin resolves two subunits of approximately equal intensity that comigrate with the A and B subunits from the purified A_4 and B_4 lectins. The molecular weight of the A_2B_2 lectin, calculated from sedimentation equilibrium analysis, is 94,000. These analyses suggest that the purified lectins are tetramers, and they have been designated B_4, A_4, and A_2B_2 to indicate their apparent subunit compositions. Both the B_4 and A_4 lectins are rich in the acidic amino acids, aspartic acid and glutamic acid, and the hydroxylic amino acids, serine and threonine; neither lectin contains cysteine or methionine. The B_4 lectin contains 9.5 mannose, 3.2 *N*-acetylglucosamine, and 3.2 fucose residues/subunit, and the A_4 lectin contains 4.7 mannose, 2.4 galactose, 1.7 *N*-acetylglucosamine, and 2.3 fucose residues/subunit; both may also contain some xylose. The carbohydrate content of A_4 and B_4 lectins may account for the anomalously high apparent molecular weight of the A and B subunits on SDS–PAGE.

Specificity and Binding Properties. All three *V. villosa* lectins interact with terminal N-acetylgalactosamine residues, but the A_4 and B_4 lectins have distinctly different carbohydrate-binding activities. ^{125}I-labeled A_4 lectin binds to human blood-group A_1 erythrocytes (273,000 sites/cell) with an association constant of $1.8 \times 10^7 \ M^{-1}$ but does not bind to human B or O erythrocytes. ^{125}I-Labeled A_2B_2 lectin also binds to A_1 erythrocytes with a similar number of binding sites per cell and an association constant of approximately half of that of the A_4 lectin. The B_4 lectin does not bind to A_1 erythrocytes but binds to Tn-exposed erythrocytes (1.4–2.0 $\times 10^6$ sites/cell) with an association constant of $2.5–4.2 \times 10^7 \ M^{-1}$. The B_4 lectin also binds to Cad erythrocytes which, depending on the donor individual, have $0.4–2.8 \times 10^6$ B_4 lectin binding sites/cell. The association constant of binding (K_a $0.84–0.61 \times 10^7$ M^{-1}) indicates that the B_4 lectin has a better affinity for the Tn antigen than for the Cad determinant. Papain treatment of normal group O erythrocytes exposes B_4 lectin binding sites (1.2×10^6 sites/cell) with an association constant of 2.2×10^6 M^{-1}.[19]

A variety of saccharides and glycopeptides have been tested for their ability to inhibit the binding of ^{125}I-labeled B_4 lectin to Tn-exposed erythrocytes,[2] and the results are summarized in Table I. Of the monosaccharides tested, N-acetylgalactosamine is the most potent inhibitor and its α- and β-p-nitrophenyl derivatives are as potent, suggesting that the B_4 lectin does not prefer either anomeric configuration and that a hydrophobic substituent does not contribute importantly to the lectin–carbohydrate interaction. Galactose and its β-methyl derivative are 200 times less potent than N-acetylgalactosamine. Although methyl-α-galactoside is 10 times more potent than methyl-β-galactoside, suggesting that B_4 lectin may prefer the α anomeric configuration, the α- and β-p-nitrophenylgalactosides are equally potent inhibitors. Since galactosamine is also a poor inhibitor, it appears that the acetamido group on C-2 is an important determinant in B_4 lectin–carbohydrate interaction. Similarly, the inability of N-acetylglucosamine to significantly inhibit B_4 lectin binding suggests that the axial 4-hydroxyl group is also essential for strong interaction. The disaccharide Galβ1,3GalNAc has the same inhibitory potency as the α- and β-p-nitrophenyl galactosides. Glycopeptides containing this disaccharide (asialofetuin C_1 and asialofetuin B_4) are much more potent inhibitors than the disaccharide alone. Removal of the galactose residues from these glycopeptides to expose GalNAcα $\rightarrow$ Ser-R increases their inhibitory potency significantly. The best glycopeptide inhibitor is the asialo,agalacto-B_4-fetuin glycopeptide which contains both a GalNAcα $\rightarrow$ Ser and GalNAcα $\rightarrow$ Thr. The spacing of the GalNAc residues on this glycopeptide may be particularly favorable for interaction with B_4 lectin since this glycopeptide is 8 times more potent than the Smith periodate-degraded IgA glycopeptide I which contains 4 GalNAcα $\rightarrow$ Ser with a different spacing. Alternatively, the B_4 lectin may preferentially interact with saccharide units α-linked to threonine residues. The B_4 lectin-binding site on the Tn-exposed erythrocyte probably closely resembles the asialo,agalacto-B_4-fetuin glycopeptide since the concentration of glycopeptide required for 50% inhibition of binding (3.4×10^{-7} M) is within an order of magnitude of the apparent K_d of lectin binding to Tn-exposed erythrocytes (2.4×10^{-8} M).

---

[19] P. Bailly, S. E. Tollefsen, and J. P. Cartron, *Glycoconjugate J.* **2**, 401 (1985).

TABLE I
SACCHARIDE INHIBITION OF B_4 LECTIN BINDING TO Tn-EXPOSED ERYTHROCYTES

Inhibitor	Concentration causing 50% inhibition (μM)
Saccharides	
N-Acetylgalactosamine	40
p-Nitrophenyl-N-acetyl-α-galactosaminide	47
p-Nitrophenyl-N-acetyl-β-galactosaminide	38
Galactose	8,300
Methyl-α-galactopyranoside	790
Methyl-β-galactopyranoside	7,500
p-Nitrophenyl-α-galactopyranoside	2,800
p-Nitrophenyl-β-galactopyranoside	2,700
Galactosamine-HCl	51,000
N-Acetylglucosamine	>200,000
Mannose	>200,000
Galβ1,3GalNAc	2,800
Glycopeptides[a]	
Fetuin C1: NeuNAcα2,3Galβ1,3GalNAcα → Ser-R	200
Asialo-C1: Galβ1,3GalNAcα → Ser-R	47
Asialo,agalacto-C1: GalNAcα → Ser-R	4.5
Asialofetuin B_4: Galβ1,3GalNAcα Galβ1,3GalNAcα ↓ ↓ (SerPro₂Gly₁Ala₂Thr)	75
Asialo,agalacto-B_4: GalNAcα GalNAcα ↓ ↓ (SerPro₂Gly₁Ala₂Thr)	0.34

IgA glycopeptide I:

Galβ1,3 Galβ1,3   Galβ1,3 Galβ1,3
  ↓   ↓   ↓   ↓
GalNAcα GalNAcα   GalNAcα GalNAcα   GalNAcα
  ↓   ↓   ↓   ↓   ↓
  SerProSerProThrProProThrSerProSerProThrProProThrSer   13

Smith periodate-degraded IgA I:

GalNAcα GalNAcα   GalNAcα GalNAcα
  ↓   ↓   ↓   ↓
  SerProSerProThrProProThrSerProSerProThrProProThrSer   2.7

[a] Sources of these glycopeptides are detailed in Tollefsen and Kornfeld.[2]

In experiments not shown in the table, the B_4 lectin was found to interact with the oligosaccharide structure found on glycoproteins and ganglioside expressing the Cad determinant and with the globoside GalNAcβ1,3Galα1,4Galβ1,4Glc → Cer exposed by papain treatment of normal group O erythrocytes.[19] In hemagglutination inhibition assays, the Sd[a] active Tamm–Horsfall glycoprotein containing the serologically active structure, GalNAcβ1,4(NeuNAcα2,3)Galβ1,4GlcNAcβ1,3Gal,[20,21] inhibited B$_4$ lectin agglutination of both Cad and Tn-exposed erythrocytes. In addition, a tetrasaccharide, GalNAcβ1,3Galα1,4Galβ1,4Glc, significantly inhibited B$_4$ lectin binding to both untreated Tn-exposed and papain-treated group O erythrocytes. Taken together, these results indicate that the B$_4$ lectin can also interact with terminal β-linked N-acetylgalactosamine residues on glycoproteins and glycolipids.

Saccharide inhibition studies have not been performed on the A$_4$ lectin prepared by this procedure. The V. villosa lectin purified by Grubhoffer et al.[13] using an affinity column of N-(ε-aminohexanoyl)-D-galactosamine coupled to Sepharose 4B agglutinates human blood-group A$_1$ erythrocytes at 15 μg/ml and probably corresponds to the A$_4$ lectin. In hemagglutination inhibition assays this lectin was inhibited by GalNAc $\gg$ GalN · HCl = p-nitrophenyl-βGal > methyl-α-Gal = Galα1,6Glc > Galβ1,4Glc > methyl-β-Gal = Gal = Man.

[20] A. S. R. Donald, C. P. C. Soh, W. M. Watkins, and W. T. J. Morgan, Biochem. Biophys. Res. Commun. 104, 58 (1982).
[21] A. S. R. Donald, A. D. Yates, C. P. C. Soh, W. T. J. Morgan, and W. M. Watkins, Biochem. Biophys. Res. Commun. 115, 625 (1983).

[47] Erythrina Lectins

By HALINA LIS and NATHAN SHARON

Trees and shrubs of the genus Erythrina (family Leguminosae, tribe Phaseoleae) are widely distributed throughout the tropics and subtropics. The genus consists of approximately 110 species and is distinct from all other genera of the legume family. The presence of hemagglutinating activity in extracts of seeds from different species of Erythrina has been known for a long time.[1] It was also noted quite early that this activity is inhibited by galactose. However, prior to 1980, none of the lectins had been purified. Since then, more than a dozen Erythrina lectins have been isolated in different laboratories and characterized to varying degrees (Table I).

Isolation and Purification

To obtain Erythrina seeds, they should be collected at the appropriate season, since they are difficult to obtain commercially. The starting mate-

[1] O. Mäkelä, Ann. Med. Exp. Biol. Fenn. 35, Suppl. 11.1 (1957).

TABLE I
Erythrina Species from Which Lectins Have Been Purified

Species	Origin	Ref.
E. aborescens	Continental Asia	a
E. caffra	South Africa (Cape Province)	b
E. corallodendron	West Indies	b, c
E. cristagalli	South America	d
E. edulis	South America	e
E. flabelliformis	Northwest Mexico and Arizona	b
E. humeana	South Africa	b
E. indica	Continental Asia	a, f
E. latissima	Tropical Africa	b
E. lithosperma (syn. E. subumbrans)	Continental Asia	a
E. lysistemon	Tropical Africa	b
E. perrieri	Madagascar and Mauritius	b
E. stricta	Continental Asia	b
E. suberosa	Continental Asia	a
E. variegata	Continental Asia	g
E. zeyheri	South Africa	b

[a] L. Bhattacharyya, P. K. Das, and A. Sen, *Arch. Biochem. Biophys.* **211,** 459 (1981).

[b] H. Lis, F. J. Joubert, and N. Sharon, *Phytochemistry* **24,** 2803 (1985).

[c] N. Gilboa-Garber and L. Mizrahi, *Can. J. Biochem.* **59,** 313 (1981).

[d] J. L. Iglesias, H. Lis, and N. Sharon, *Eur. J. Biochem.* **123,** 247 (1982).

[e] G. Perez, *Phytochemistry* **23,** 1229 (1984).

[f] V. Horèjsï M. Tichà, J. Novotny, and J. Kocourek, *Biochim. Biophys. Acta* **623,** 439 (1980).

[g] T. K. Datta and P. S. Basu, *Biochem. J.* **197,** 751 (1981).

rial for the isolation of the lectins is the seed flour, preferably after extraction with petroleum ether. The purification procedure consists essentially of three steps—extraction with buffer, fractionation with ammonium sulfate, and affinity chromatography. The affinity adsorbents used include acid-treated Sepharose,[2,3] β-galactosylpolyacrylamide matrix,[4,5] and galactose[6] or lactose[7] coupled to divinyl sulfone-activated Sepharose 4B.

[2] L. Bhattacharyya, P. K. Das, and A. Sen, *Arch. Biochem. Biophys.* **211,** 459 (1981).

[3] T. K. Datta and P. S. Basu, *Biochem. J.* **197,** 751 (1981).

[4] V. Horèjsï, M. Tichà, J. Novotny, and J. Kocourek, *Biochim. Biophys. Acta* **623,** 439 (1980).

[5] G. Perez, *Phytochemistry* **23,** 1229 (1984).

[6] J. L. Iglesias, H. Lis, and N. Sharon, *Eur. J. Biochem.* **123,** 247 (1982).

[7] H. Lis, F. J. Joubert, and N. Sharon, *Phytochemistry* **24,** 2803 (1985).

SEED MEAL

| Petroleum ether extraction in Soxhlet

DEFATTED MEAL (60 g)

| Phosphate-buffered saline (10 volumes) for 3–4 hr at room temperature

EXTRACT

| Precipitation with $(NH_4)_2SO_4$ (60% saturation)

CRUDE LECTIN

| Affinity chromatography on column (3 × 20 cm) of lactose coupled to DVS-activated
| Sepharose; elution with 0.2 M galactose

PURIFIED LECTIN (yield 60–120 mg)

SCHEME 1. Purification of *Erythrina* Lectins. The defatted meal of *E. latissima* seeds was extracted with cold acetone and air dried before extraction with buffer.

Elution is done with galactose or lactose. The procedure routinely employed in our laboratory is described below, as illustrated by the purification of the lectin from *Erythrina corallodendron,* and summarized in Scheme 1.

Preparation of Affinity Column

Lactose-derivatized Sepharose 4B is prepared by the divinyl sulfone method of Porath and Ersson.[8] Packed Sepharose (100 g wet weight, Pharmacia) is suspended in 100 ml of 0.5 M sodium carbonate buffer, pH 11, and 10 ml of divinyl sulfone (Polysciences, Warrington, Pennsylvania) is added. The suspension is kept at room temperature for 70 min with slow stirring, and the activated gel is thoroughly washed on a glass filter with distilled water. It is then suspended in 100 ml of a 10% (w/v) solution of lactose in the carbonate buffer and left overnight in the cold room (4–6°). The resulting product is washed on a glass filter with 1 liter carbonate buffer, followed by 2 liters water, and suspended in phosphate-buffered saline, pH 7.2 (PBS).

Purification Procedure

Unless otherwise stated, all operations are carried out in the cold. Locally collected seeds of *E. corallodendron* are finely ground and extracted in a Soxhlet with petroleum ether (bp 40–60°). The air-dried meal

[8] J. Porath and B. Ersson, *in* "Proceedings of the Symposium on New Approaches for Inducing Natural Immunity to Pyrogenic Organisms" (J. B. Robbins, R. E. Horton, and R. M. Drause, eds.), p. 101. Winter Park, Florida, 1973.

(60 g) is extracted with 600 ml PBS for 4 hr with stirring, the extract is filtered through cheesecloth and clarified in a Sorvall centrifuge at 13,000 rpm for 10 min. Ammonium sulfate (35.2 g/100 ml) is added, and the mixture is kept overnight. The precipitate is collected by centrifugation as above, suspended in distilled water (100–120 ml) and dialyzed extensively against distilled water (3 × 5 liters) and finally against PBS. Any precipitate that forms during dialysis is removed by centrifugation, and the clear supernatant is applied to a column (3 × 20 cm) of lactose-derivatized Sepharose equilibrated with PBS. The column is washed with the same buffer until the absorbance of the effluent at 280 nm is less than 0.05, and the bound lectin is eluted with 0.2 M galactose in PBS. Elution is followed by monitoring the absorbance at 280 nm. The fractions containing protein are collected, dialyzed against distilled water and lyophilized. About 100 mg of lectin is obtained, representing a yield of 70–75% of the hemagglutinating activity of the crude extract. The product migrates on sodium dodecyl sulfate–polyacrylamide gel electrophoresis (10% gel, buffer system of Laemmli[9]) as a single band of M_r about 30,000 and gives a single symmetrical peak of M_r about 60,000 upon gel filtration on a column of Sephadex G-150. The molecular weight determined in the ultracentrifuge is 60,246. The purified lectin agglutinates O-type human erythrocytes at a minimal concentration of 10–20 μg/ml.

Determination of Hemagglutinating Activity

The hemagglutinating activity of the lectins is assayed by the serial dilution method on conical microtiter plates, using 50 μl of lectin and 50 μl of a 4% suspension of erythrocytes.[10] One unit of activity is defined as the lowest amount of lectin giving visible agglutination under the experimental conditions used. The inhibitory activity of sugars is measured by mixing serial dilutions of the inhibitor with 4 units of the lectin before the addition of erythrocytes and determining the lowest concentration that gives complete inhibition of agglutination.

Physicochemical Properties

The *Erythrina* lectins studied have molecular weights in the range 56,000–68,000 and are composed of two noncovalently linked subunits of identical or nearly identical molecular weight. All are glycoproteins containing between 3 and 10% carbohydrate. Whenever determined,[6,7] the carbohydrate was found to consist of glucosamine, mannose, fucose, and

[9] U. K. Laemmli, *Nature (London)* **227**, 680 (1970).
[10] E. R. Gold, *Vox Sang.* **15**, 222 (1968).

xylose in molar ratios $1–2:3–4:1:1$. A similar carbohydrate composition has been shown to be present in several other plant lectins,[11] as well as in bromelain from pineapple.[12] Xylose has not been found together with mannose and N-acetylglucosamine in carbohydrate units of animal glycoproteins. Such a carbohydrate composition thus appears to be characteristic of a class of plant glycoproteins. The structure of the carbohydrate unit of bromelain, as well as that of the lectin from *Tora* bean[13] has been elucidated and shown to contain the Manβ4GlcNAcβ4GlcNAcAsn linking region, characteristic of N-linked carbohydrate chains, to which mannose, fucose, and xylose are attached. It is likely that the carbohydrate units of the *Erythrina* lectins are similar to those above.

The *Erythrina* lectins are metalloproteins containing Mn^{2+} and Ca^{2+}.[2,4–6] Dialysis of *E. indica* and *E. edulis* lectins against acid and/or chelating agents[2,5] did not affect their hemagglutinating activity, indicating that the metals are very strongly bound or that they are not necessary for biological activity.

Isoelectric focusing,[2] as well as discontinuous electrophoresis and affinity electrophoresis,[4] of *E. indica* lectin indicated the presence of 3 molecular species; it was suggested that this lectin consists of various combinations of the two dissimilar subunits, i.e., α_2, β_2, and $\alpha\beta$.[4] Two components with different pI values were also found in the lectin from *E. edulis*.[5]

The *Erythrina* lectins all contain high proportions of acidic and hydroxylic amino acids, no cysteine, and little methionine. This amino acid composition is characteristic of many legume lectins. The N-terminal amino acid sequence (determined in 10 *Erythrina* lectins for up to 14–15 amino acids)[7] is nearly identical (Table II); there are extensive homologies also with other legume lectins,[14] two of which—soybean agglutinin and peanut agglutinin—are included in Table II. It is noteworthy that several positions in the sequence are highly conserved. Thus phenylalanine in positions 6 and 11 has been found in all legume lectins tested to date (close to 30, not counting the *Erythrina* lectins), and serine in position 5 has been found in 23 of these. This is in line with the suggestion[15] that all legume lectins have evolved from the same genetic ancestor.

[11] H. Lis and N. Sharon, *in* "Biology of Carbohydrates" (V. Ginsburg and P. W. Robbins, eds.) Vol. 2, p. 1. Wiley, New York, 1984.

[12] H. Ishihara, N. Takahashi, S. Oguri, and S. Tejima, *J. Biol. Chem.* **254,** 10715 (1979).

[13] K. Ohtani and A. Misaki, *Carbohydr. Res.* **87,** 275 (1980).

[14] A. D. Strosberg, M. Lauwereys, and A. Foriers, *in* "Chemical Taxonomy, Molecular Biology and Function of Plant Lectins" (I. J. Goldstein and M. E. Etzler, eds.), p. 7. Alan R. Liss, Inc., New York, 1983.

[15] A. Foriers, C. Wuilmart, N. Sharon, and A. D. Strosberg, *Biochem. Biophys. Res. Commun.* **75,** 980 (1977).

TABLE II
NH$_2$-TERMINAL SEQUENCES OF LECTINS FROM *Erythrina*, SOYBEAN,
AND PEANUT

Lectin[a]	NH$_2$-Terminal sequence
ECL[b,c]	Val Glu Thr Ile Ser Phe Ser Phe Ser Glu Phe Glu Pro Gly Asn
EFL[b]	Ala Glu Thr Ile Ser Phe Ser Phe Ser Glu Phe Glu Pro Gly
EPL[b]	Val Glu Thr Ile Ser Phe Ser Phe Ser Lys Phe Glu Ala Gly
SBA[d]	Ala Glu Thr Val Ser Phe Ser Trp Asn Lys Phe Val Pro Lys Glu
PNA[d]	Ala Glu Thr Val Ser Phe Asn Phe Asn Ser Phe Ser Glu Gly Asn

[a] Abbreviations used: ECL, *E. cristagalli* lectin; EFL, *E. flabelliformis* lectin; EPL, *E. perrieri* lectin; SBA, soybean agglutinin; PNA, peanut agglutinin.

[b] Data for the *Erythrina* lectins are from Lis *et al.*[7]

[c] The same sequence was found in the lectins from *E. caffra, E. corallodendron, E. humeana, E. latissima, E. lysistemon, E. stricta,* and *E. zeyheri.*

[d] Data for SBA and PNA are from Strosberg *et al.*[14]

Interactions with Carbohydrates

Erythrina lectins are specific for *N*-acetylgalactosamine and galactose; the relative inhibitory activity of these sugars is nearly the same (*N*-acetylgalactosamine being somewhat better than galactose) with slight variations observed for the different lectins. The lectins appear to lack anomeric specificity, since there are no large differences in the inhibitory power of α- and β-galactosides. The stronger inhibitory activity of the *p*-nitrophenyl-β-galactoside, as compared with that of the corresponding methyl derivative, indicates the presence of a hydrophobic region in or near the binding sites of the lectins. The best disaccharide inhibitor of *E. cristagalli* lectin,[6,16] as well as of several other *Erythrina* lectins,[7] is Galβ4GlcNAc (*N*-acetyllactosamine). Representative data are given in Table III.

Equilibrium dialysis studies with *E. indica* lectin and lactose[2] and with *E. cristagalli* lectin and methylumbelliferyl β-galactoside[17] have shown that each lectin contains two homogenous, noninteracting combining sites per dimeric molecule with $K_a = 2.2 \times 10^3 \ M^{-1}$ and $1.7 \pm 0.2 \times 10^4 \ M^{-1}$, respectively.

[16] P. M. Kaladas, E. A. Kabat, J. L. Iglesias, H. Lis, and N. Sharon, *Arch. Biochem. Biophys.* **217**, 624 (1982).

[17] H. De Boeck, F. G. Loontiens, H. Lis, and N. Sharon, *Arch. Biochem. Biophys.* **234**, 297 (1984).

TABLE III
INHIBITION BY VARIOUS SUGARS OF THE HEMAGGLUTINATING ACTIVITY OF *Erythrina*
LECTINS TESTED WITH HUMAN ERYTHROCYTES[a]

Sugar	Relative inhibitory activity				
	ECL	ECorL	ELatL	ELysL	EPL
Galactose	1.0	1.0	1.0	1.0	1.0
N-Acetylgalactosamine	2.4	1.9	5.0	1.2	5.0
p-Nitrophenyl α-galactoside	3.0	1.6	3.3	2.5	1.7
p-Nitrophenyl β-galactoside	7.1	3.2	10.0	6.0	7.0
Methyl α-galactoside	3.0	1.5	1.8	1.5	1.0
Methyl β-galactoside	1.4	1.0	1.4	1.3	1.0
Lactose	7.1	2.7	7.5	1.8	4.0
N-Acetyllactosamine	33	19	35	20	17
Galβ4GlcNAcβ2Manα6 ⟍ Manβ4GlcNAc ⟋ Galβ4GlcNAcβ2Manα3	80	40	80	40	35

[a] Four agglutinating doses of the lectin were routinely used. The inhibitory activity of galactose was arbitrarily set as 1. The minimal galactose concentrations required to give complete inhibition in a range of 3–4 experiments were as follows: ECL (*E. cristagalli* lectin), 7–14 mM; ECorL (*E. corallodendron* lectin), 4–14 mM; ELatL (*E. latissima* lectin), 2–7 mM; ELysL (*E. lysistemon* lectin), 3.3–5 mM; EPL (*E. perrieri* lectin), 7–14 mM. (Data from Lis *et al.*[7])

The binding parameters of several mono- and oligosaccharides to *E. cristagalli* lectin have been determined in substitution titrations with N-dansylgalactosamine as fluorescent indicator ligand.[17] In the series N-acetylgalactosamine, methyl β-galactoside and lactose, $-\Delta H°$ increases from 24 to 41 kJ mol^{-1}; it increases further for N-acetyllactosamine and then remains unchanged (55 ± 1 kJ mol^{-1}) for the N-acetyllactosamine-containing oligosaccharides Galβ4GlcNAcβ6Man, Galβ4GlcNAcβ2Man and Galβ4GlcNAcβ6(Galβ4GlcNAcβ2)Man. This indicates that the site specifically accomodates Galβ4GlcNAc with an important contribution from the 2-acetamido group of the penultimate sugar. No additional contacts seem to be formed when the lectin binds higher oligosaccharides.

Using fast reaction techniques, the rate constants for the binding of N-dansylgalactosamine to *E. cristagalli* lectin were found to be $k_+ = 4.8 \times 10^4 M^{-1} sec^{-1}$ and $k_- = 0.45 sec^{-1}$.[18] Similar slow rates have been obtained

[18] H. De Boeck, R. B. MacGregor, R. M. Clegg, N. Sharon, and F. G. Loontiens, *Eur. J. Biochem.* **149**, 141 (1985).

with other lectins as well.[19] Unlike enzymes, which for efficient catalysis require rapid dissociation, the slow kinetics of carbohydrate binding may be of advantage for the interaction of lectins with cells. It allows time for polyvalent binding of the lectin to receptors on the cell surface and for topological reorganization of the cross-linked receptor complex within the viscous cell membrane.

Biological Activities

The *Erythrina* lectins agglutinate human erythrocytes of all blood types at minimal concentrations of 2–20 μg/ml, with some of them, e.g., the lectins of *E. indica*[2] and *E. variegata*,[3] exhibiting a slightly higher activity toward cells of blood type O. Significant differences were, however, found in the hemagglutinating activity of different *Erythrina* lectins with regard to rabbit erythrocytes.[7] Whereas most of them are almost as active with the latter cells as they are with human erythrocytes, those of *E. humeana, E. perrieri,* and *E. zeyheri* are much weaker agglutinins when tested with rabbit erythrocytes. Of the *Erythrina* lectins tested for mitogenic activity,[6,7] most were found to stimulate human peripheral blood lymphocytes. Notable exceptions are *E. humeana* and *E. zeyheri* lectins, which are poor mitogens if at all.[7] Interestingly, these are two of the three lectins that interact only weakly with rabbit erythrocytes. These results show that lectins with the same specificity for mono- and oligosaccharides may differ in their cell specificity.

Erythrina cristagalli lectin binds specifically to immature thymocytes, to the inner mass of preimplantation embryos, and to undifferentiated teratocarcinoma cells, but not to the differentiated cells.[20] The lectin may thus be a useful marker to follow changes in surface carbohydrates during differentiation of lymphocytes and embryonic cells.

[19] F. G. Loontiens, H. De Boeck, and R. M. Clegg, *J. Biosc.* **8,** (Suppl. 1,2), 425 (1985).
[20] C. G. Webb, M. Popliker, H. Lis, and N. Sharon, *in* "Cell Membranes and Cancer" (T. Galeotti *et al.,* eds.), p. 13. Elsevier, Amsterdam, 1985.

[48] *Evonymus europaea* Lectin

By JERZY PETRYNIAK and IRWIN J. GOLDSTEIN

Introduction

The arils from seeds of the spindle tree *Evonymus europaea* contain a lectin which agglutinates A_2, B, and O erythrocytes,[1-4] stimulated murine and guinea pig peritoneal cells,[5] and Ehrlich ascites tumor cells.[6] The B and H specificity is an intrinsic property of a single lectin binding site, since absorption and elution from a blood-group type H immunoadsorbent gives an agglutinin with B as well as H specificity.[4] The *E. europaea* lectin is not inhibited by any monosaccharide. The lowest molecular weight inhibitors are disaccharides; the most convenient, lactose.

Assay Methods

Ouchterlony Gel Diffusion

The Ouchterlony double diffusion method is simple and convenient for qualitative testing of the reaction of the lectin with macromolecular substances. The most convenient glycoprotein used for assaying the lectin is a 0.1% solution of blood-group A + H substance obtained from porcine stomach mucin. A plate containing 1.5% agar in 0.05 M sodium barbital buffer, pH 8.2, or in 1% agar in 0.1 M phosphate buffer, pH 7.2, containing 0.02% NaN_3 is prepared.

Precipitin Reactions

A modification of the quantitative microprecipitin technique described by Kabat[7] can be used. Reactions are carried out in 1.5-ml polypropylene microcentrifuge tubes with caps (obtained from Curtin Matheson Scien-

[1] G. Schmidt, *Z. Immunitätsforsch.* **111**, 432 (1954).
[2] M. Krüpe, "Blutgruppenspezifische Pflanzliche Eiweisskörper (Phytagglutinine)," p. 45. Enke, Stuttgart, 1956.
[3] F. Pacák and J. Kocourek, *Biochim. Biophys. Acta* **400**, 374 (1975).
[4] J. Petryniak, M. E. A. Pereira, and E. A. Kabat, *Arch. Biochem. Biophys.* **178**, 118 (1977).
[5] J. Petryniak, D. Duś, and J. Podwińska, *Eur. J. Immunol.* **13**, 459 (1983).
[6] Unpublished data.
[7] E. A. Kabat, *in* "Kabat and Mayer's Experimental Immunochemistry," 2nd ed., p. 542. Thomas, Springfield, Illinois, 1961.

tific, Inc., Houston, Texas). Approximately 40 μg of lectin protein is mixed with varying amounts of glycoconjugates in a final volume 250 μl of 0.01 M sodium phosphate buffer, pH 7.1, in 0.15 M NaCl with 0.02% azide (PBS). The tubes are incubated at 37° for 1 hr and kept at 4° for 5–7 days. Precipitates are centrifuged at 15,600 g for 10 min at 4°, and supernatant solutions are decanted. The protein in the washed precipitates is determined by the procedure of Lowry *et al.*[8]

Hemagglutination Procedure

Hemagglutination assays are conducted using a hemagglutination plate and a 25-μl microdiluter (Cooke Engineering Co.). After serial dilution of the lectin solution with phosphate-buffered saline (0.15 M NaCl, 0.01 M sodium phosphate, pH 7.0), 25 μl of a 3% suspension of type B or O erythrocytes is added to each well. The degree of agglutination is determined after 1 hr at room temperature. The titer is defined as the reciprocal of the highest dilution showing detectable agglutination.

Purification Procedures

The purification of *Evonymus* lectin by a conventional technique was described by Pacák and Kocourek.[3] There are also two procedures based upon the principle of affinity chromatography using (1) poly (L-leucine) hog A + H blood group substance as affinity adsorbent[4] and (2) asialoglycophorin coupled to Sepharose 2B as affinity adsorbent.[9] The first affinity method in which lectin is bound to blood group H determinants allows purification of relatively large quantities of lectin. Thus 2 g of polyleucyl hog A + H substance was employed to purify 230 mg of lectin. However, this technique is laborious because of the batchwise procedure and the necessity of repeated high speed centrifugations.

Although the structure of the lectin-reactive oligosaccharide on glycophorin which binds *Evonymous* lectin is unknown,[10] this method is more convenient and faster than the first one, due to the use of column chromatography and the low lactose concentration required for elution of the lectin. The disadvantage of this method is the lower yield of lectin when compared with the first technique. Thus, 22–26 mg of lectin can be obtained from 1 run on the 150-ml column.

[8] O. H. Lowry, N. J. Rosebrough, A. L. Farr, and R. J. Randall, *J. Biol. Chem.* **193,** 265 (1951).

[9] J. Petryniak, M. Janusz, E. Markowska, and E. Lisowska, *Acta Biochim. Pol.* **28,** 267 (1981).

[10] J. Petryniak, B. Petryniak, K. Waśniowska, and H. Krotkiewski, *Eur. J. Biochem.* **105,** 335 (1980).

Preparation of Extract

Extract can be prepared from whole seeds or from arils of the seeds.
Whole Seeds Extract. The seeds of *Evonymus europaea* are ground in
a mortar to a fine slurry and extracted with 9 volumes of phosphate-
buffered saline, pH 7.1, containing 0.2% sodium azide. The suspension is
kept overnight at 4° with occasional stirring. The precipitate is centrifuged
at 2000 rpm for 1 hr at 4°, and the supernatant solution is decanted and
cleared of fat by filtration on a Büchner funnel, using Whatman No. 1
paper, and then through a 0.45-μm Millipore filter. The precipitate is
reextracted twice with PBS; a fourth extract is free of hemagglutinating
activity. The three extracts show a decrease in agglutination of human B
and O erythrocytes from an initial titer of 16.

Arils Extract. Arils are peeled off from seeds by pressing seeds in
hands, followed by shaking on a Petri dish. Fifty grams of arils is defatted
by treatment with 150 ml of cold methanol at 4° for 1 hr with occasional
stirring. The defatted arils are extracted with 9 ml of PBS per gram of
arils. The initial four extracts, containing over 90% of the hemagglutinating activity, were combined and used for lectin purification. The hemagglutinating titer of the combined extracts varied from 1 : 64 to 1 : 512.

Preparation of Affinity Absorbents

Poly(L-leucine) Hog A + H Blood-Group Substance. Hog A + H
blood-group substance is purified from hog gastric mucin (Sigma) by an
ethanol precipitation procedure.[11] Insoluble poly (L-leucine) hog A + H
blood-group substance is prepared by the copolymerization of the blood-
group substance with L-leucine N-carboxyanhydride according to Tsuyuki *et al.,*[12] as described by Kaplan and Kabat.[13] The polyleucyl hog A +
H is mixed with Celite in the ratio 1 : 1 w/w, and washed extensively with
PBS until $A_{280\ nm} \leq 0.03$. Washings are separated from the immunoadsorbent by centrifugation at 12,000 rpm for 1 hr at 15° in a Beckman L-2
centrifuge using an LW 19 rotor; this avoids loss of fine particles.

Preparation of Asialoglycophorin A–Sepharose 2B Conjugate. Glycophorin was obtained from human blood-group O,MN erythrocyte membranes by phenol–water extraction.[14] Sialic acid is released by hydrolysis

[11] E. A. Kabat, "Blood Group Substances: Their Chemistry and Immunochemistry." Academic Press, New York, 1956.
[12] H. Tsuyuki, H. Van Kley, and M. A. Stahmann, *J. Am. Chem. Soc.* **78,** 764 (1956).
[13] M. E. Kaplan and E. A. Kabat, *J. Exp. Med.* **123,** 1061 (1966).
[14] T. Baranowski, E. Lisowska, A. Morawiecki, E. Romanowska, and K. Stróżecka, *Arch.
Immunol. Ther. Exp.* **7,** 15 (1959).

in 0.025 M H_2SO_4 at 80° for 40 min and removed by dialysis.[15] Sepharose 2B (Pharmacia Fine Chemicals, Uppsala, Sweden) was activated by the cyanogen bromide technique.[16] To the activated Sepharose (150 ml) is added 150 mg of the asialoglycophorin. The coupling procedure is carried out according to the instructions of Pharmacia Fine Chemicals; 92% of the desialized glycophorin is coupled to the gel. The conjugate is stored at 4°.

Affinitiy Chromatography

Affinity Chromatography on Poly(L-leucine) Hog A + H Blood-Group Substance.[4] A batchwise technique is used. The washed affinity adsorbent (2 g of polyleucyl hog A + H mixed with 2 g of Celite) is mixed with 1.2 liters of crude extract and left for 1 hr in the cold with occasional stirring. The suspension is then subjected to ultracentrifugation at 12,000 rpm for 1 hr, and the supernatant is discarded, if free of hemagglutinating activity. The lectin is completely adsorbed. Additional *Evonymus* extract is added to the 2 g of immunoadsorbent, but saturation is not reached even with 10 liters of crude extract. The polyleucyl hog A + H substance containing the lectin is then washed with PBS, pH 7.1, and centrifuged at 12,000 rpm until the $A_{280 \text{ nm}}$ washings decreased to 0.03.

Specific elution is carried out by suspending in 90 ml of 0.5 M lactose in PBS containing 0.02% sodium azide, pH 7.1, and incubation for 15 min at 37°, followed by centrifuging at 20,000 rpm for 1 hr at 15° in a Beckman L-2 centrifuge using an SW 25.1 rotor. The eluates are decanted and combined. Elution is repeated until the $A_{280 \text{ nm}}$ of the eluate is decreased to 0.05 above that of the lactose.

The yield of lectin varied for different preparations from 120 to 230 mg per 370 g of *Evonymus* seeds. Eluates are concentrated and dialyzed against 4 changes each of 50 volumes of PBS until periodate-positive material was not found in a 24-hr dialyzate; three additional 24-hr dialyzates were carried out before the material was used. During dialysis 24–56% of the protein precipitated.

Affinity Chromatography on Asialoglycophorin A–Sepharose 2B Conjugate.[9] All procedures were performed at 4°. Before each run, the column containing the conjugate is washed consecutively with 0.15 M NaCl/0.01 M $NaHCO_3$, 0.15 M NaCl/0.01 M acetate buffer, pH 4, and finally with PBS. The *E. europaea* extract is centrifuged for 1 hr at 2000 rpm directly before application to the affinity column to remove traces of precipitate which might impair or stop the flow of extract on the column.

[15] E. Lisowska and M. Duk, *Arch. Immunol. Ther. Exp.* **24**, 39 (1976).
[16] P. Cuatrecasas, *J. Biol. Chem.* **245**, 3059 (1970).

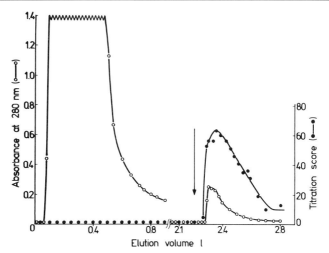

Fig. 1. Affinity chromatography of the *E. europaea* lectin on an asialoglycophorin A–Sepharose 2B column.[9] The crude lectin extract (400 ml, hemagglutinating titer 1 : 64) was applied to the 1.6 × 55 cm column, followed by washing with PBS until the absorbance at 280 nm was less than 0.005; the flow rate was 56 ml/hr, and 14-ml fractions were collected. The lectin was eluted with 0.1 *M* lactose in PBS at a flow rate of 30 ml/hr; 7.5-ml fractions were collected. The arrow shows the point of application of the lactose solution. For determination of hemagglutinating activity the aliquots of fractions eluted with lactose were dialysed against PBS. Agglutination score was counted as follows: 4+ = 10; 3+ = 8; 2+ = 5; 1+ = 3.

After exhaustive washing of the column with PBS, the bound lectin was eluted with 0.1 *M* lactose (Sigma). Active fractions showing hemagglutinating titer not lower than 1 : 2 were pooled, concentrated and dialyzed in a Diaflo chamber, using a UM 10 membrane Amicon (Holland), until no periodate-positive material[7] was found in the ultrafiltrate. The maximal lectin binding capacity of the adsorbent was about 4 ml of the extract with hemagglutinating titer 1 : 256 per 1 ml of the conjugate.

The elution pattern of *E. europaea* extract from the affinity column is shown in Fig. 1, and results of purification are presented in Table I.

Properties of *Evonymus europaea* Lectin

Stability

During 8 years of storage at 4° (the longest period observed) the purified lectin solution retained its activity. Lyophilized preparations are also stable for at least 2 years.

TABLE I

PURIFICATION OF THE *E. europaea* LECTIN ON AN ASIALOGLYCOPHORIN A–SEPHAROSE 2B COLUMN[9]

Purification step	Volume (ml)	Total protein (mg)	Total hemagglutinating activity (units)	Specific activity (units/mg protein)	Recovery of activity (%)	Purification factor
Crude extract	400	820	25 600	31	100	
Lactose eluate[a]	18	22	18 400	836	72	27

[a] After dialysis and concentration.

Homogeneity

Immunochemical. Evidence was obtained supporting immunochemical homogeneity of *Evonymus* lectin. The lectin displaced by elution with lactose from poly(L-leucine) hog A + H blood-group substance was 78% specifically precipitated by blood-group B substances,[4] the remainder probably being in the form of soluble complexes since such precipitates have appreciable solubility.

Physicochemical. By physicochemical criteria the lectin shows several distinct populations of protein molecules. In the analytical ultracentrifuge using schlieren optics, the purified lectin in PBS, pH 7.1, sedimented as 2 peaks: a large symmetrical peak with $S^0_{20,w} = 7.8$ (calculated by extrapolation to zero concentration) and a small, diffuse, slow-moving peak. The slow-moving peak is a subunit of the fast-moving peak as it was observed that an increase of pH of the lectin solution up to pH 9 caused the disappearance of fast-moving peak and an increase in the amount of the slow-moving peak.[6] Thus the lectin shows pH-dependent dissociation into subunits.

Upon immunoelectrophoresis against rabbit antisera to the crude extract, the isolated hemagglutinin gave three bands as compared with nine in the crude extract. These bands all migrated toward the anode.

The intrinsic viscosity η was 0.057 dl/g.

Disk gel electrophoresis in polyacrylamide, pH 9.4, showed three distinctive bands; two intense fast-moving bands, the faster being more diffuse, and one slow-moving band. Three additional very faint bands of intermediate mobility were seen.

Analytical isoelectric focusing revealed heterogeneity of the purified lectin. Six very close bands with isoelectric points in the range 4.3–4.7 were obtained on ampholine, pH 3–6 and 3–10.

Molecular Weight. The molecular weight of the *E. europaea* lectin calculated from the intrinsic viscosity 0.057 dl/g, sedimentation coefficient $S_{20,w}^0$ 7.8, and $\bar{v}$ of 0.71 was approximately 166,000.[4] However, Pacák and Kocourek,[3] using the Yphantis method, reported 126,700 and 119, 200 for two preparations of *Evonymus:* phytohemagglutinin I and a mixture of isophytohemagglutinins, respectively.

Subunit Structure

On sodium dodecyl sulfate–polyacrylamide gels the purified lectin revealed two intense bands of M_r 17,000 and 35,000 and two faint bands of 53,000 and 67,000. This pattern changed on reduction with 0.1% 2-mercaptoethanol. The 53,000 and 67,000 bands disappeared, the intensity of the 35,000 decreased, and the 17,000 band increased, indicating some subunit structure partly covalently and partly noncovalently linked.

Composition. The *Evonymus* lectin is an acid glycoprotein containing 4.8% D-galactose, 2.9% D-glucose, and 2.8% *N*-acetyl-D-glucosamine. The amino acid composition is noteworthy for its high content of aspartic acid (13.5%) and glutamic acid (7.1%) as well as threonine, proline, glycine, alanine, valine, leucine, and half-cystine.

Specificity

Evonymus lectin exhibits a complex type of carbohydrate-binding specificity. The purified lectin precipitates well with B, H, and A_2 blood-group substances but not with A_1 substance. Inhibition of precipitation with milk and blood-group oligosaccharides showed the lectin to be most specific for blood-group B oligosaccharides having the basic structure αDGal(1 $\rightarrow$ 3)[αLFuc(1 $\rightarrow$ 2)]βDGal(1 $\rightarrow$ 3 or 4)βDGlcNAc. The lectin does not discriminate between type 1 and type 2 oligosaccharide chains, having the core βDGal(1 $\rightarrow$ 3)βDGlcNAc or βDGal(1 $\rightarrow$ 4)βDGlcNAc, respectively, and both types are equally active, both in inhibition and precipitation tests.[4,17] *Evonymus* lectin is also inhibited by blood-group H oligosaccharides, but to a lesser degree. For 50% inhibition of precipitation, 3.5, 850, and 290,000 nmol of B and H oligosaccharides and lactose, respectively, are required.[4]

Studies with synthetic glycoconjugate shows that *Evonymus europaea* lectin precipitated with αDGal(1 $\rightarrow$ 3)βDGal(1 $\rightarrow$ 4)βDGlcNAc–BSA, αLFuc(1 $\rightarrow$ 2) βDGal(1 $\rightarrow$ 3)βDGlcNAc–BSA, αLFuc(1 $\rightarrow$ 2)βDGal(1 $\rightarrow$

[17] J. Petryniak and I. J. Goldstein, *Biochemistry,* **25,** 2829 (1986)

TABLE II

STRUCTURES OF *Evonymus* LECTIN DETERMINANTS[4,17,a]

Glycoconjugates that do precipitate with *Evonymus* lectin		Glycoconjugates that do not precipitate with *Evonymus* lectin
Best lectin inhibitors	**Good lectin inhibitors**	**Poor lectin inhibitors**
αLFuc (1 → 2) αDGal(1 → 3)βDGal(1 → 3 or 4)βDGlcNAc	αLFuc (1 → 2) βDGal(1 → 3 or 4)βDGlcNAc	αLFuc(1 → 2)βDGal
	αDGal(1 → 3)βDGal(1 → 4)βDGlcNAc αLFuc (1 → 2) αDGal(1 → 3)βDGal	βDGal(1 → 4)βDGlcNAc
		αDGal(1 → 3)βDGal
Tetrasaccharides with blood-group B activity	Trisaccharides derived from blood-group B tetrasaccharide	Disaccharides derived from blood-group B tetrasaccharide

[a] Reprinted with permission from *Biochemistry* **25**, 2829 (1986). Copyright 1986 American Chemical Society.

4)DGlcNAc, and αDGal(1 → 3)[αLFuc(1 → 2)]βDGal–BSA. However, the lectin neither precipitated with αLFuc(1 → 2)βDGal–BSA, αDGal(1 → 3)βDGal–BSA, or βDGal(1 → 4)βDGlcNAc–BSA nor agglutinated erythrocytes of O$_h$ phenotype having multiple terminal βDGal(1 → 4) βDGlcNAc residues.[17] These results indicate that the minimal structural requirement for glycoprotein precipitation or for cell agglutination by the lectin includes any of the 3 trisaccharides (fucosylated or nonfucosylated) derived from the blood-group B tetrasaccharide (see Table II). The monosaccharides linked to the β-D-galactosyl residue in the blood-group B tetrasaccharide (α-D-galactose, α-L-fucose, and β-N-acetyl-D-glucosamine) participate almost equally in binding to the lectin inasmuch as removal of any one of these sugars reduces the inhibiting potency of the resulting trisaccharide to the same degree. αLFuc(1 → 2)βDGal(1 → 3)βDGlcNAc–BSA (H type 1) and αLFuc(1 → 2)βDGal(1 → 4)βDGlcNAc (H type 2) were precipitated to the same extent.

The *E. europaea* lectin neither precipitated αDGal(1 → 4)βDGal(1 → 4)βDGlcNAc–BSA, Le^a–BSA, Le^b–BSA, or βDGlcNAc(1 → 4)[αLFuc-(1 → 6)]βDGlcNAc–BSA nor agglutinated O$_h$,Le^a and O$_h$,Le^b erythrocytes, demonstrating that terminal D-galactose linked α(1 → 4) to subterminal β-D-galactose or α-L-fucose linked to N-acetylglucosamine prevents lectin binding. CPK molecular models, built on the basis of data from ^{1}H-NMR and HSEA calculations provided by Lemieux *et al.*,[18] show that these α-D-galactosyl and α-L-fucosyl groups act to sterically hinder lectin binding to these oligosaccharides; these observations also suggest that the lectin binds to the β-side of these oligosaccharides. The β-sides, both on the blood-group H type 1 and type 2 oligosaccharides, provide a similar contour which can fully account for their equal reactivity with *E. europaea* lectin.

Among the L-fucose binding lectins (*Evonymus europaea, Lotus tetragonolobus,* and *Ulex europaeus* I) which react with blood-group H oligosaccharides, *Evonymus* lectin is the only one which shows the unique ability to react with H type 1 oligosaccharides. *Evonymus* lectin also precipitates a number of other glycoproteins containing some oligosaccharides of unknown structure. These include asialoglycophorin from human erythrocytes of both O and O$_h$ ("Bombay") phenotypes,[10,19] laminin from EHS sarcoma tumor cells,[19] as well as many blood-group glycoproteins obtained from human ovarian cysts of HLe^b,Le^a activity, I active precursor blood-group substance.[4]

[18] R. U. Lemieux, K. Bock, L. T. J. Delbaere, S. Koto, and V. S. Rao, *Can. J. Chem.* **58,** 631 (1980).
[19] In preparation.

Binding of E. europaea Lectin to Glycolipids

Evonymus lectin binds to blood-group B and H active glycolipids as detected by autoradiography, with ^{125}I-labeled lectin, on thin-layer chromatograms.[19]

[49] Lectin from Hog Peanut, *Amphicarpaea bracteata*

By M. MALIARIK, D. ROBERTS, and I. J. GOLDSTEIN

The anti-A seed lectin from *Amphicarpaea bracteata*[1] is one of several Leguminosae lectins specific for terminal nonreducing *N*-acetyl-D-galactosamine (D-GalNAc)[2] groups. The species *Amphicarpaea* exhibits two types of seeds, one above ground and one which develops at the base of the stem, near or below the soil surface. The lectin purified from both seeds appears to be identical. The lectin, originally purified by affinity chromatography on *N*-acetyl-D-galactosamine coupled to epoxy–Sepharose 6B, was reported to be homogeneous by SDS–polyacrylamide electrophoresis, and an initial characterization was reported.[3] We describe here a new purification procedure and further characterization of the physical and chemical properties of the *A. bracteata* lectin.

Assay Methods

Hemagglutination. Hemagglutination assays are conducted in a hemagglutination titer plate using 50-μl serial dilutions of lectin solution in phosphate-buffered saline (PBS: 0.1 M sodium phosphate, 0.15 M NaCl, pH 7.1). Fifty microliters of 2–3% suspension of type A_1 erythrocytes in 0.9% NaCl, 10 mM potassium phosphate, pH 7.4, are added to each well. The agglutination titer is determined after 1 hr at room temperature and defined as the reciprocal of the highest dilution causing agglutination.

Quantitative Precipitation Inhibition Assay. The carbohydrate-binding specificity of the lectin was determined by a precipitation inhibition assay. Lectin (30 μg) was mixed with type A substance (20 μg) to which

---

[1] O. Mäkelä and P. Mäkelä, *Ann. Med. Exp. Biol. Fenn.* **34**, 403 (1956).
[2] Abbreviations: D-GalNAc, 2-acetamido-2-deoxy-D-galactopyranose; Meα- and Meβ-GalNAc, methyl 2-acetamido-2-deoxy-α- and -β-D-galactopyranosides, respectively; PBS, phosphate-buffered saline.
[3] L. J. Blacik, M. Breen, H. G. Weinstein, B. A. Sittig, and M. Cole, *Biochim. Biophys. Acta* **538**, 225 (1978).

varying concentrations of haptenic sugars were added in a final volume of 0.2 ml. After incubation at 25° for 48 hr, the tubes were centrifuged and the precipitates washed with PBS (0.4 ml). The washing procedure was repeated twice. The precipitates were dissolved by addition of 0.1 ml 0.5 M NaOH, and after standing for 1 hr protein was determined by a micro Lowry assay.[4]

Purification of Lectin

Extraction. Dried seeds (collected along roadsides in New Hampshire and Michigan) (50 g) are ground to a fine meal and delipidated by extraction with 200 ml methanol for 1 hr. The methanol extraction procedure is repeated twice. The mixture is filtered on a Büchner funnel and washed with 200 ml dichloromethane, then air dried. The delipidated material is extracted with 200 ml PBS for 1 hr at 4°. The suspension is centrifuged at 16,000 g for 30 min, the supernatant removed, and the extraction repeated using 150 ml PBS overnight at 4°.

Affinity Chromatography. The combined extracts are applied to an affinity column of High Capacity Synsorb A (Chembiomed, Alberta, Canada) which contains the synthetic blood-group A trisaccharide, α-D-GalNAc-(1,3)-[α-L-Fuc-(1,2)]-β-D-Gal-Synsorb. Alternatively, GalNAc–Bio-Glas 1000 (Chembiomed) may be used as an affinity support. The column is washed extensively with PBS to remove unbound protein. The lectin is eluted with 0.1 M N-acetyl-D-galactosamine in PBS and dialyzed against PBS. A final wash with 0.1 M acetic acid, pH 3.0, removes a small amount of material which absorbs at 280 nm but does not stain on gel electrophoresis. Yields of the lectin range from 100 to 800 μg lectin per gram of seeds, depending upon the source. The subterranean seeds appear to give somewhat higher yields than the aboveground seeds.

Properties of the Lectin from *Amphicarpaea bracteata*

Homogeneity. Affinity-purified lectin gives a single symmetrical peak upon gel filtration on Sephadex G-200 with $M_r = 135,000$. A single diffuse band is obtained on polyacrylamide gel electrophoresis at pH 8.9 and 4.3. Sodium dodecyl sulfate–PAGE shows four bands of M_r 32,000, 30,000, 28,000, and 27,500. Based on the molecular weight determined by gel filtration and on SDS electrophoresis data, subunits in the native lectin are probably noncovalently assembled as a tetramer. The lectin is an acidic protein, giving several bands with apparent pI values of 5.5–6.5 in

[4] R. Mage and S. Dray, *J. Immunol.* **95,** 525 (1965).

isoelectric focusing in 8 M urea, suggesting charge heterogeneity among subunits. However, the lectin subunits are not completely dissociated in urea.[5]

Composition. Amino acid analysis of the protein shows the absence of cysteine and less than 1 methionine per subunit. The lectin contains a high content of aspartate, glycine, and serine. The lectin contains 4% carbohydrate consisting of glucosamine, mannose, galactose, fucose, arabinose, and xylose residues in the ratio $4:3:2:1:2:2$.

Carbohydrate-Binding Specificity. The lectin from *A. bracteata* is specific for terminal, nonreducing D-GalNAc units. Equilibrium dialysis indicates four equivalent, noninteracting sugar-binding sites per molecule (M_r 135,000), with $K_a = 4.0 \times 10^4 \, M^{-1}$. The lectin tolerates a number of alterations at the C-2 position: 2-N-benzamidomethyl α-D-galactosaminide is as good an inhibitor of precipitation as methyl N-acetyl-α-D-galactosaminide, and methyl 2-O-acetyl-α-D-galactoside is only slightly less inhibitory.

Precipitation inhibition studies also show the lectin to have a marked preference for the α-anomeric form of GalNAc. However, there is evidence for additional hydrophobic interactions near the carbohydrate-binding site. The α- and β-anomers of p-nitrophenyl N-acetylgalactosaminide have similar inhibition constants in precipitin inhibition; 1-[4-(N-acetyl-β-D-galactosaminyl)phenyl]-3-(2,2,6,6-tetramethylpiperidin-1-oxyl-4-yl)-2-thiourea binds much more tightly than the α-anomer in electron spin resonance studies.[6] These data suggest there is a hydrophobic region adjacent to the carbohydrate-binding site that can interact with the phenyl group in the β-anomeric position to enhance binding. It is also apparent that the nitroxyl group sterically hinders binding of the α-anomer.

Examination of disaccharide-binding specificity indicates the Forssman hapten (GalNAcα1,3GalNAc) to be a 2-fold better inhibitor of precipitation than D-GalNAc. The lectin appears to accommodate α1,2 and α1,3 linkages better than 1,6 linkages.

Comments

The lectin from *A. bracteata* has also been observed to bind adenine.[7] Equilibrium dialysis shows the protein to have a single adenine-binding site per molecule with $K_a = 1.3 \times 10^6 \, M^{-1}$. This places *A. bracteata* lectin

[5] D. D. Roberts, M. J. Maliarik, and I. J. Goldstein, *Biochemistry* **25**, 4457 (1986).
[6] L. J. Berliner, G. Musci, M. J. Maliarik, N. R. Plessas, and I. J. Goldstein, to be published.
[7] M. J. Maliarik and I. J. Goldstein, unpublished results.

in a class of several plant lectins that contain an adenine-binding site in addition to the carbohydrate site.[8,9] Limited N-terminal sequencing of *A. bracteata* lectin resulted in heterogeneity at 2 of 10 amino acid positions. Along with electrophoresis and isoelectric focusing data this also suggests that the lectin may be present in multiple subunit forms which differ in size and charge and which are bound by noncovalent interactions. The sequence data also indicate homology with the related lectins from soybean and *Dolichos biflorus* lectins, which also have been shown to exhibit subunit heterogeneity.

[8] D. D. Roberts and I. J. Goldstein, *J. Biol. Chem.* **258**, 13820 (1983).

[9] D. D. Roberts and I. J. Goldstein, *in* "Chemical Taxonomy, Molecular Biology, and Function of Plant Lectins" (M. Etzler and I. Goldstein, eds.), p. 131. Alan R. Liss, Inc., New York, 1983.

Section IV

Biosynthesis

[50] Glycosyltransferase Assay[1]

By Bo E. Samuelsson

Introduction[2]

Glycosphingolipids constitute a class of glycoconjugates where the oligosaccharide chain is attached with its reducing end to a lipid part, ceramide. Experiments concerned with the biosynthesis of these lipid-linked oligosaccharide chains has mostly been carried out by studying the transfer of a sugar unit, unlabeled or labeled, from a sugar nucleotide donor to a labeled or unlabeled glycosphingolipid precursor in the presence of an enzyme preparation, detergents, and buffer.[3,4] After lipid extraction the labeled product is quantified by liquid scintillation and identified by autoradiography of a thin-layer chromatogram with appropriate references. Identification is sometimes also brought about by studying the degradation with specific enzymes or the appearance of, e.g., blood-group activity.[3,4] As the molecular variability of complex glycosphingolipids seems like a "never ending story" the availability of highly purified and structurally characterized glycosphingolipid acceptors is increasing, though limited. To make use of only partly purified fractions for the same purpose will invalidate the conclusions drawn. To exploit a thin-layer plate after chromatographic development as a solid matrix for the glycosphingolipid precursors in biosynthetic experiments would be one possibility to use mixtures, purified or nonpurified, and still be able to interpret the results. It would also solve several methodological difficulties in the handling of amphipatic molecules.

The procedure described in this chapter is partly based on the methodology developed in conjunction with different thin-layer chromatographic binding assays[5–8] (antibodies, lectins, toxins). It has so far been used to

---

[1] This is number III in a series of papers entitled: "Solid-Phase Biosynthesis on High Performance Thin-Layer Plates of Blood-Group Glycosphingolipids."

[2] Supported by a grant (No. 6521) from the Swedish Medical Research Council. Anita Jacobsson is gratefully acknowledged for important help and Drs. M. Breimer, K.-A. Karlsson, and G. Larson for gift of samples.

[3] Y.-T. Li and S.-C. Li, *Adv. Carbohydr. Chem. Biochem.* **40**, 235 (1982).

[4] S. Basu and M. Basu, *in* "The Glycoconjugates" (M. I. Horowitz, ed.), Vol. 3, p. 265. Academic Press, New York, 1982.

[5] J. L. Magnani, D. F. Smith, and V. Ginsburg, *Anal. Biochem.* **109**, 399 (1980).

[6] M. Brockhaus, J. L. Magnani, M. Blaszczyk, Z. Steplewski, H. Koprowski, K.-A. Karlsson, G. Larson, and V. Ginsburg, *J. Biol. Chem.* **256**, 13223 (1981).

study the biosynthesis of blood-group-active glycosphingolipids with enzyme preparations from human saliva,[9] pig intestinal mucosa,[10] and cultured human cell lines.[11]

Assay Method

Principle

Silica-gel thin-layer plates are used as solid matrix for precursor glycosphingolipids after chromatographic development. A sufficient amount of glycosphingolipid is thought to be adsorbed onto silicic acid particles through its ceramide part with its carbohydrate chain exposed to the environment and available for biosynthetic reactions.

Thin-Layer Chromatography

HPTLC plates, precoated with silica gel 60 (Merck) on glass plates or aluminium sheets were used. Glass plates give a better resolution, but they are more difficult to cut and the silica gel layer is often released from the glass plates in the subsequent washing procedures.[6]

Samples are applied in duplicate, one series for later chemical detection and one series for biosynthesis. The amount spotted for biosynthetic purposes ranges between 2 ng and 2 μg for each glycolipid component. After chromatographic development with an appropriate solvent the plate is allowed to dry in the air. The plate is then divided into 2 sections, one of which is visualized with the anisaldehyde reagent[12] and the other is thereafter reduced in size according to the anisaldehyde-visualized section in order to minimize the thin-layer surface area.

Incubation Procedure

The "trimmed" thin-layer plate is rapidly dipped into a diethyl ether : hexane (1 : 1, by volume) solution of poly(isopropyl methacrylate) (Plexigum P28, Röhm GmbH, Chemische Fabrik, Darmstadt, West Ger-

[7] T. Momoi, T. Tokunaya, and Y. Nagai, *FEBS Lett.* **141,** 6 (1982).
[8] G. C. Hansson, K.-A. Karlsson, G. Larson, J. M. McKibbin, M. Blaszczyk, M. Herlyn, Z. Steplewski, and H. Koprowski, *J. Biol. Chem.* **258,** 4091 (1983).
[9] B. E. Samuelsson, *Proc. Int. Symp. Glycoconjugates, 7th, 1983,* p. 425 (1983).
[10] B. E. Samuelsson, *FEBS Lett.* **167,** 47 (1984).
[11] G. C. Hansson, personal communication (1984).
[12] E. Stahl, "Dunnschichtchromatographie," p. 817. Springer-Verlag, Berlin and New York, 1967.

many), 0.3 g in 100 ml, and left to soak for 1 min. The plate is then allowed to dry and the poly(isopropyl methacrylate) allowed to polymerize overnight in the air. The next day, the plate is sprayed with 2% bovine serum albumin in phosphate-buffered saline, pH 7.3, containing NaN_3 to obtain a homogeneous wetting, followed by incubation for 2 hr with the albumin solution in a Petri dish. After 2 hr, the albumin solution is discarded, and the plate is covered with incubation medium, approximately 1 ml/10 cm^2 (see below), and left at 37° for 3–4 hr in a saturated humid atmosphere. A practical way is to put the thin-layer plate on top of a small Petri dish and to put this into a large Petri dish equipped with pieces of filter paper soaked with buffer to create the saturated humid atmosphere.[13] Little or no additional biosynthesis takes place after 6 hr.

After incubation, the plate is washed carefully 6 times with phosphate-buffered saline and dried in the air. Washing is especially important when very crude enzyme preparations are used. The dry plate is then subjected to autoradiography using Kodak XAR-5 X-ray film.

Enzyme Preparation and Enzyme Sources

The method has so far successfully been applied using enzyme preparations from human saliva,[9] intestinal mucosa,[10] and human cultured cells.[11] Human saliva has been used as a source for the Lewis gene-coded enzyme without any purification or treatment.[9] Microsomal preparations from pig intestinal mucosa have been used as a source for $Fuc\alpha1 \rightarrow 2$ and $GalNAc\alpha1 \rightarrow 3$ transferases,[10] and a similar preparation from a human colocarcinoma cell line SW1116 as a source for the Lewis gene-coded transferase.[11] The microsomal preparations follow standard procedures.[10]

Serum has been used as source for $Fuc\alpha1 \rightarrow 3$ and $GalNAc\alpha1 \rightarrow 3$ transferases without success. The reason for this is still unknown to us, although the presence of substrate antibodies have been ruled out. Purification of the enzyme may overcome this problem.

Incubation Media

The incubation media were composed essentially as described.[9,10,14,15] For fucosyltransferases the incubation medium contained in a final vol-

[13] G. C. Hansson, K.-A. Karlsson, G. Larson, B. E. Samuelsson, J. Thurin, and L. M. Bjursten, *J. Immunol. Methods* **83,** 37 (1985).
[14] T. Pacuszka and J. Kościelak, *Eur. J. Biochem.* **64,** 499 (1976).
[15] C. A. Tilley, M. C. Crookston, J. H. Crookston, J. Shindman, and H. Schachter, *Vox Sang.* **34,** 8 (1978).

ume of 1 ml:

    650 μl of enzyme preparation

    1.3 × 10⁶ cpm of GDP-L-[¹⁴C]fucose

    Tris–HCl buffer, 25 μmol, pH 7.5

    NaN₃, 10 μmol

    ATP, 10 μmol

    MgCl₂, 7.5 μmol

    Triton X-100, 0.1%

For galactosaminyltransferases the incubation medium contained in a final volume of 1 ml:

    650 μl of enzyme preparation

    1.3 × 10⁶ cpm of UDP-N-acetyl-D-[¹⁴C]galactosamine

    Tris–HCl, 25 μmol, pH 7.5

    NaN₃, 10 μmol

    ATP, 10 μmol

    MnCl₂, 20 μmol

    Triton X-100, 0.1%

Preparation of Glycosphingolipids

Total nonacid glycosphingolipid fractions from erythrocytes and plasma of individual donors were prepared as described in this volume.[16]

Reference Glycosphingolipids

Lactosylceramide was obtained from human erythrocytes. Lactotetraosylceramide and a blood-group H, type 1 chain pentaglycosylceramide were obtained from human meconium.[17] A blood-group H, type 1 chain pentaglycosylceramide was also prepared from pig intestinal mucosa. Neolactotetraosylceramide was obtained by sialidase treatment of N-acetylneuraminosylneolactotetraosylceramide from human erythrocytes. The product was purified by DEAE–column chromatography and silicic acid chromatography.

Gangliotetraosylceramide was obtained by acid treatment of G_{M1} ganglioside from bovine brain. The major product was purified to homogeneity by DEAE–column chromatography and silicic acid chromatography. Galβ1 → 3-globotetraosylceramide was prepared from human kidneys.[18]

---

[16] K.-A. Karlsson, this volume [16].

[17] K.-A. Karlsson and G. Larson, *J. Biol. Chem.* **254**, 9311 (1979).

[18] M. E. Breimer and P.-Å. Jovall, *FEBS Lett.* **179**, 165 (1985).

ANISALDEHYDE   AUTORADIOGRAM

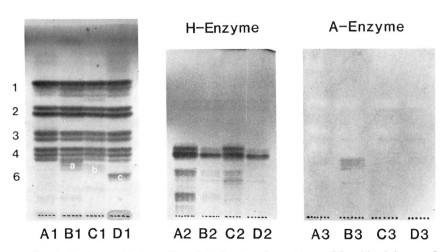

FIG. 1. Total nonacid glycosphingolipid fractions from plasma of four blood donors of different blood groups [A, O Le(a−b−)nonsecretor; B, O Le(a−b−)secretor; C, O Le(a+b−)nonsecretor; D, O Le(a−b+)secretor] were applied 3 times on a thin-layer plate. The amount applied was 30 μg (A1, B1, C1, D1) or 10 μg of each (A2, B2, C2, D2 and A3, B3, C3, D3). After development in chloroform : methanol : water (60 : 35 : 8, by volume), one section of the plate (A1, B1, C1, D1) was visualized with the anisaldehyde reagent. The figures on the left-hand side indicate the approximate number of sugars in the glycolipid carbohydrate chain. The letters a, b, and c indicate tentatively the thin-layer position of the blood-group H, type 1 chain pentaglycosylceramide, the blood-group Le^a pentaglycosylceramide, and the blood-group Le^b hexaglycosylceramide, respectively (see Ref. 10 and references cited therein). The remaining sections were incubated with enzyme preparations from intestinal mucosa of a blood-group H type pig in the presence of GDP-L-[^{14}C]fucose (lanes A2, B2, C2, D2) or with a similar enzyme preparation from a blood-group A type pig in the presence of UDP-N-acetyl-D-[^{14}C]galactosamine (lanes A3, B3, C3, D3). After 3 hr of incubation, the plates were washed and subsequently subjected to autoradiography for 8 days. (Part of this figure is reprinted from Samuelsson,[10] with permission of the publisher.)

Applications

Glycosphingolipid Mixtures

Figure 1 shows the results of two biosynthetic experiments on the thin-layer plates after chromatographic development of glycosphingolipid precursor fractions. Total nonacid glycosphingolipid fractions from plasma of four different blood donors (for ABO, Lewis, and secretor typing, see legend to Fig. 1) were used as precursor fractions. Microsomal

ANISALDEHYDE

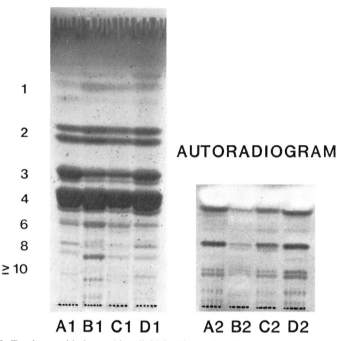

AUTORADIOGRAM

A1 B1 C1 D1   A2 B2 C2 D2

FIG. 2. Total nonacid glycosphingolipid fractions of erythrocyte membranes of 4 different blood donors (A, blood group O; B, blood group A_1; C, blood group A_2; D, blood group A_{Finn}) were applied twice on a thin-layer plate. The amount applied was 40 μg (A1, B1, C1, D1) or 20 μg of each (A2, B2, C2, D2). After chromatographic development 1 section of the thin-layer plate (A1, B1, C1, D1) was visualized with the anisaldehyde reagent. The figures on the left-hand side indicate the approximate number of sugars in the glycolipid carbohydrate chain. The remaining section (A2, B2, C2, D2) was incubated with an enzyme preparation from a blood-group A type pig in the presence of UDP-N-acetyl-D-[^{14}C]galactosamine for 3 hr and subjected to autoradiography for 4 days.

preparations from intestinal mucosa of a blood-group H type pig[10] and a blood-group A type pig were used as enzyme sources together with GDP-L-[^{14}C]fucose and UDP-N-acetyl-D-[^{14}C]galactosamine as activated sugars, respectively. The clear and distinct bands on the autoradiograms show the biosynthesized, fucosylated or galactosaminylated, products with R_f values coinciding with their corresponding precursors. The autoradiograms also show the specificity of the reactions with qualitatively different patterns between the different fractions and with the two different enzyme preparations and nucleotide sugars.

Figure 2 shows a similar experiment using total nonacid glycosphingo-

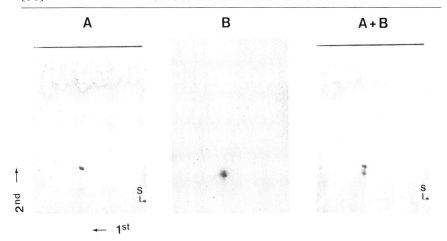

Fig. 3. Thin-layer chromatogram developed in two dimensions with biosynthesis solid phase on the thin-layer plate intervening between the two chromatographic developments. Two micrograms of a blood-group H, type 1 chain pentaglycosylceramide from meconium was applied at S. The thin-layer plate was developed in the first dimension using chloroform : methanol : water (60 : 35 : 8, by volume). After incubation with saliva from a Le(a−b+) secretor in the presence of GDP-L-[^{14}C]fucose, the plate was developed in the second dimension using the same conditions as above. The thin-layer plate was then subjected to autoradiography for 4 days (B), subsequently extracted several times with diethyl ether to remove the plastic coating, and finally visualized with the anisaldehyde reagent (A). (A + B) A composite picture with the autoradiogram centered on top of the anisaldehyde-visualized thin-layer plate.

lipids from erythrocyte membranes from four different blood donors of blood groups, O, A_1, A_2, and A_{Finn}.[19] A microsomal preparation from intestinal mucosa of a blood-group A type pig was used as enzyme source and UDP-N-acetyl-D-[^{14}C]galactosamine as activated sugar. The autoradiogram shows that all individuals carry a number of precursor glycolipids that could be galactosaminylated. The different precursor glycolipids seem to be present in all persons but the amount differs (O $\simeq$ A_{Finn} > A_2 > A_1).

Two-Dimensional Thin-Layer Chromatography
with Intervening Biosynthesis

Figure 3 shows how to use chromatographic development in a second dimension after biosynthesis solid phase with the purpose to separate precursor from product. In this experiment a highly purified blood-group H pentaglycosylceramide with a type 1 carbohydrate chain was used as

[19] R. R. Race and R. Sanger, "Blood Groups in Man." Blackwell, Oxford, 1975.

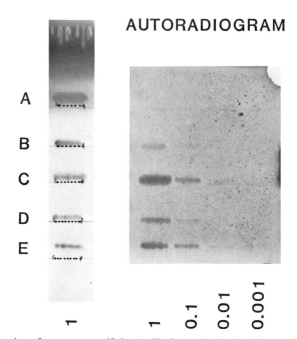

FIG. 4. Estimation of acceptor specificity. A dilution series (1, 0.1, 0.01, 0.001 μg) of five different possible precursor glycosphingolipids were applied at regular intervals on top of each other on a thin-layer plate. A, lactosylceramide; B, neolactotetraosylceramide; C, lactotetraosylceramide; D, Galβ1-3-globotetraosylceramide; E, gangliotetraosylceramide. The 1-μg samples were applied in duplicate, one of which was intended as a reference lane, visualized using the anisaldehyde reagent. The thin-layer plate was then developed in chloroform : methanol : water (65 : 25 : 4, by volume) with the purpose to move the samples just a short distance from the application lines. GDP-L-[^{14}C]Fucose and an enzyme preparation from intestinal mucosa of a blood-group H type pig were used as activated sugar and enzyme source, respectively. Autoradiography was done for 4 days.

the precursor,[17] saliva from a Le(a−b+) secretor as the enzyme source, and GDP-L-[^{14}C]fucose as the activated sugar. After incubation, the thin-layer plate was developed in the second dimension. The autoradiogram shows clearly a product with a slower thin-layer mobility than the precursor. The fucosylated product has a mobility identical (not shown) to an authentic sample of a Le^b-active hexaglycosylceramide. Hence, the above experiment illustrates the biosynthesis of the Le^b blood-group antigen from its presumed precursor.

Estimation of Acceptor Specificity

Figure 4 shows how the technique could be applied to a dilution series of several possible glycosphingolipid precursors on the same thin-layer plate with the purpose of estimating enzyme acceptor specificity. The results can be interpreted visually on the autoradiogram. The same experiment could also be repeated under varying experimental conditions. The presented results indicate on a semiquantitative basis that the enzyme preparation used is very similar to that described by Beyer and Hill.[20]

Comments

The presented method has several advantages compared to conventional techniques. It is simple, fast, and sensitive. It minimizes the risk of adding unwanted exogenous precursors, and it abolishes the need for tedious purification of products. The method gives an opportunity to use mixtures of glycosphingolipids in biosynthetic experiments. The need for relatively large volumes of incubation media and ^{14}C-labeled activated sugars may be considered a drawback. At the present stage its universal use together with different enzyme preparations cannot be proved.

[20] T. A. Beyer and R. L. Hill, *J. Biol. Chem.* **255**, 5373 (1980).

[51] Glycolipids

By MANJU BASU, TRIPTI DE, KAMAL K. DAS, JOHN W. KYLE, HUNG-CHE CHON, ROBERT J. SCHAEPER, and SUBHASH BASU

Introduction

The glycosphingolipids appear to be ubiquitous on eukaryotic cell surfaces.[1] More than 50 different complete structures of glycosphingolipids (GSLs) of the ganglio, globo, and lacto families[2] have been determined[3-6] in the last three decades. Stepwise biosynthesis *in vitro* of about

[1] L. Svennerholm, *in* "Cholera and Related Diarrheas" (O. Ouchterlony and J. Holmgren, eds.), p. 80. Karger, Basel, 1980.

[2] IUPAC–IUB Commission of Biochemical Nomenclature, *Lipids* **12**, 455 (1977).

[3] H. Wiegandt, *Adv. Neurochem.* **4**, 149 (1982).

[4] S. Hakomori, *Annu. Rev. Biochem.* **50**, 733 (1981).

[5] S. Roseman, *Chem. Phys. Lipids* **5**, 270 (1970).

[6] C. C. Sweeley, Y. K. Fung, B. A. Macher, J. R. Moskal, and H. Nunez, *ACS Symp. Ser.* **80**, 47 (1978).

20 GSLs, using membrane-bound or solubilized enzymes, has been reported.[7-58a] Based on the kinetic parameters, substrate-specificity studies, and sources from which these GSLs have been isolated, it is suggested that each linkage (anomeric and positional) in the oligosaccharide portion of the glycolipids in animal systems may be catalyzed by a specific glycosyltransferase (GLT). A better understanding of these reactions (see Table I) will depend on substrate-specificity studies with functionally purified GLTs. However, the concentration of GLTs in tissue has been found to be low compared with other metabolic enzymes.

The available source of enzyme in some cases (e.g., cultured cells) is very limited. The short-chain glycosphingolipid substrates are amphiphilic molecules that aggregate in buffers at pH 6.0–8.0, whereas GSLs with long-chain carbohydrates tend to form micelles. Either in aggregates

[7] S. Basu, M. Basu, J. W. Kyle, and H. C. Chon, in "Ganglioside Structure and Function" (R. Ledeen, ed.), p. 249. Plenum, New York, 1984.

[8] S. Basu and M. Basu, in "The Glycoconjugates" (M. I. Horowitz, ed.), Vol. 3, p. 265. Academic Press, New York, 1982.

[9] Y. T. Li and S. C. Li, Adv. Carbohydr. Chem. Biochem. 40, 235 (1982).

[10] Y. Kishimoto, in "The Enzymes" (P. D. Boyer, ed.), 3rd ed., Vol. 16, p. 358. Academic Press, New York, 1983.

[11] S. Basu, M. Basu, H. Higashi, and C. H. Evans, in "New Vistas in Glycolipid Research" (A. Makita, S. Handa, T. Taketomi, and Y. Nagai, eds.), p. 131. Plenum, New York, 1982.

[12] S. Basu, M. Basu, J. L. Chien, and K. A. Presper, in "Structure and Function of Gangliosides" (L. Svennerholm, H. Dreyfus, and P. F. Urban, eds.), p. 213. Plenum, New York, 1980.

[13] P. H. Fishman and R. O. Brady, Science 194, 906 (1976).

[14] B. Kaufman, S. Basu, and S. Roseman, in "Inborn Disorders of Sphingolipid Metabolism" (S. M. Aronson and B. W. Volk, eds.), p. 193. Pergamon, Oxford, 1967.

[15] S. Basu, B. Kaufman, and S. Roseman, J. Biol. Chem. 248, 1388 (1973).

[16] R. R. Vunam and N. S. Radin, Biochim. Biophys. Acta 573, 73 (1979).

[17] P. Morell and N. S. Radin, Biochemistry 8, 506 (1969).

[18] S. Basu, A. Schultz, M. Basu, and S. Roseman, J. Biol. Chem. 246, 4272 (1971).

[19] S. Basu, B. Kaufman, and S. Roseman, J. Biol. Chem. 243, 5802 (1968).

[20] J. Hildebrand and G. Hauser, J. Biol. Chem. 244, 5170 (1969).

[21] S. Basu, B. Kaufman, and S. Roseman, J. Biol. Chem. 240, 4115 (1965).

[22] S. Basu, Ph.D. Thesis, University of Michigan, Ann Arbor (1966).

[23] M. Basu and S. Basu, J. Biol. Chem. 247, 1489 (1972).

[24] M. Basu, D.Sc. Thesis, University of Calcutta, Calcutta, India (1974).

[25] J. R. Moskal, Ph.D. Thesis, University of Notre Dame, Notre Dame, Indiana (1977).

[26] J. W. Kyle, Ph.D. Thesis, University of Notre Dame, Notre Dame, Indiana (1985).

[27] M. Basu and S. Basu, J. Biol. Chem. 248, 1700 (1973).

[28] N. Taniguchi, K. Yanagisawa, A. Makita, and M. Naiki, J. Biol. Chem. 260, 4908 (1985).

[29] J. C. Steigerwald, S. Basu, B. Kaufman, and S. Roseman, J. Biol. Chem. 258, 6727 (1975).

[30] M. Basu, J. L. Chien, and S. Basu, Biochem. Biophys. Res. Commun. 60, 1097 (1974).

[31] J. L. Chien, T. J. Williams, and S. Basu, J. Biol. Chem. 248, 1778 (1973).

or in micelles, glycosphingolipids are thought to be packed tightly and thus inaccessible to either membrane-bound or soluble GLTs. However, suitable mixed micelles can be prepared in the presence of a detergent and a glycosphingolipid that are specific for the *in vitro* assay system.

Until the natural glycolipid-binding proteins are discovered the following assay systems have been developed for identification of the activities of different glycosyltransferases using optimum detergent concentrations. Detergents were matched to specific glycosphingolipid : glycosyltransferases (GSL : GLTs) by screening a number of detergents available commercially. The exact correlation between enzyme protein and detergent has not exactly been elucidated.

When an enzyme is membrane bound, the ratio of protein to detergent is critical to the optimum reaction rate.[33,37,42] The detergent producing

[32] N. Taniguchi and A. Makita, *J. Biol. Chem.* **259**, 5637 (1984).

[33] K. Das, M. Basu, S. Basu, and C. Evans, *Carbohydr. Res.* **149**, 119 (1986).

[34] K. Yeung, J. R. Moskal, J. L. Chien, D. A. Gardner, and S. Basu, *Biochem. Biophys. Res. Commun.* **59**, 252 (1974).

[35] N. Taniguchi, N. Tokosawa, S. Gasa, and A. Makita, *J. Biol. Chem.* **257**, 10631 (1982).

[36] S. Basu, M. Basu, H. Den, and S. Roseman, *Fed. Proc., Fed. Am. Soc. Exp. Biol.* **29**, 410 (1970).

[37] M. Basu and S. Basu, *J. Biol. Chem.* **259**, 12557 (1984).

[38] S. Basu, M. Basu, and J. L. Chien, *J. Biol. Chem.* **250**, 2956 (1975).

[39] T. Pacuszka and J. Koscielak, *Eur. J. Biochem.* **64**, 499 (1976).

[40] K. A. Presper, M. Basu, and S. Basu, *J. Biol. Chem.* **257**, 169 (1982).

[41] K. A. Presper, M. Basu, and S. Basu, *Proc. Natl. Acad. Sci. U.S.A.* **75**, 289 (1978).

[42] S. Basu, T. De, J. W. Kyle, and M. Basu, *Proc. Int. Symp. Biomol. Struct. Interact., Suppl. J. Biosci.* **8**, 413 (1985).

[43] E. H. Holmes, G. K. Ostrander, and S. Hakomori, *J. Biol. Chem.* **260**, 7619 (1985).

[44] T. Taki, R. Kamada, and M. Matsumoto, *J. Biochem.* (*Tokyo*) **91**, 665 (1983)

[45] E. H. Holmes and S. Hakomori, *J. Biol. Chem.* **258**, 3706 (1983).

[46] R. K. Yu and S. H. Lee, *J. Biol. Chem.* **251**, 198 (1976).

[47] S. Basu and B. Kaufman, *Fed. Proc., Fed. Am. Soc. Exp. Biol.* **24**, 479 (1965).

[48] B. Kaufman and S. Basu, this series, Vol. 8, p. 365.

[49] A. Arce, H. F. Maccioni, and R. Camputto, *Arch. Biochem. Biophys.* **116**, 52 (1966).

[50] A. Pretti, A. Fiorilli, A. Lombardo, L. Caimi, and G. Tettamanti, *J. Neurochem.* **35**, 261 (1980).

[51] M. Basu, K. Das, H. C. Chon, and S. Basu, *in* "Enzymes of Lipid Metabolism" (L. Freysz and S. Gatt, eds.), p. 228. Plenum, New York, 1986.

[52] B. Kaufman, S. Basu, and S. Roseman, *J. Biol. Chem.* **243**, 5804 (1968).

[53] P. Stoffyn and A. Stoffyn, *Carbohydr. Res.* **78**, 327 (1980).

[54] H. Higashi, M. Basu, and S. Basu, *J. Biol. Chem.* **260**, 824 (1985).

[55] M. Basu, S. Basu, A. Stoffyn, and P. Stoffyn, *J. Biol. Chem.* **257**, 12765 (1982).

[56] S. S. Ng and J. A. Dain, *J. Neurochem.* **29**, 1075 (1977).

[57] J. L. Chien, M. Basu, and S. Basu, *Fed. Proc., Fed. Am. Soc. Exp. Biol.* **33**, 1225 (1974).

[58] J. L. Chien, Ph.D. Thesis, University of Notre Dame, Notre Dame, Indiana (1975).

[58a] G. C. Hansson and D. Zopf, *J. Biol. Chem.* **260**, 9388 (1985).

maximal enzyme activity is also useful for solubilization of a GLT from a specific membrane,[37,42] and after solubilization, the concentration of detergent required for the maximum rate is also lowered. It has also been observed that a specific soluble glycosyltransferase, GalT-4 from embryonic chicken brain, can catalyze the transfer of galactose to the terminal N-acetylglucosaminyl moiety of α_1-acid glycoprotein in the absence of any detergent, and to LcOse3Cer (GlcNAcβ1-3Galβ1-4GlcCer) only in the presence of low concentrations of detergents.[26]

Almost all the GalTs, GalNAcTs, and GlcNAcTs require Mn^{2+} for optimum activity. Four different galactosyltransferase activities (GalT-1, GalT-2, GalT-3, and GalT-4) have been distinguished, based on divalent cation requirements and their pH optima in embryonic chicken brain.[8]

The structure,[3–5] function,[7] and biosynthesis[8–13] of various glycosphingolipids have been reviewed extensively. Utilization of both oligosaccharide and glycoprotein substrates by the purified glycosyltransferases has recently been reported.[59] We restrict our discussion to reports on glycosyltransferase reactions using specific glycosphingolipid acceptors (Table I). This article is devoted mainly to a description of the assay systems (Table II) for six families of glycosphingolipid:glycosyltransferases and their merits.

Brain NFA-Ceramide(β1-1)glucosyltransferase (GlcT)

A glucosyltransferase that catalyzes the synthesis of glucosylceramide (glucocerebroside) from NFA-ceramide (normal fatty acylsphingosine) and UDP-glucose has been isolated from 13- to 14-day-old embryonic chicken brain[15] and mouse brain.[16] A survey of various animal brains showed high GlcT activity (1–2 nmol/mg protein in homogenate/hr) in embryonic chicken brain and in young rat, guinea pig, and fetal pig brains. The activities observed in adult brains were very low (0.2–0.4 nmol/mg protein in homogenate/hr), but are considered significant in view of the specificity of the assay.[15,19] The reaction is as follows:

$$\text{UDP-[}^{14}\text{C or }^3\text{H]Glc} + \text{NFA-Cer} \xrightarrow{\text{GlcT}} \text{[}^{14}\text{C or }^3\text{H]Glc}\beta\text{1-1Cer} + \text{UDP}$$

Ceramide (containing nonhydroxy fatty acids) is the most effective lipid acceptor, and a variety of potential donors could not replace UDP-Glc. Metal ions do not stimulate the reaction, and it is not inhibited by EDTA. The most potent inhibitor is 2-decanoylamino-3-morpholino-1-phenylpropanol. The inactivation is produced perhaps by covalent reaction with the enzyme's active site.[16] It is believed that the enzyme has four recognition

[59] J. E. Sadler, T. A. Beyer, C. L. Oppenheimer, J. C. Paulson, J. P. Prieels, J. I. Rearick, and R. L. Hill, this series, Vol. 83, p. 458.

TABLE I
GLYCOSPHINGOLIPID GLYCOSYLTRANSFERASES (GSL : GLTs)[a]

Enzyme name	Abbreviation	Refs.
Glucosyltransferase		
UDP-Glc : NFA-Cer (β1-1)	GlcT	15, 16
Galactosyltransferases		
UDP-Gal : HFA-Cer (β1-1)	GalT-1	17,18
UDP-Gal : GlcCer (β1-4)	GalT-2	19, 20
UDP-Gal : G_{M2} (β1-3)	GalT-3	14, 21, 22
UDP-Gal : Lc3 (β1-4)	GalT-4	23–26
UDP-Gal : nLc4(α1-3)	GalT-5	24, 27
UDP-Gal : Lc2(α1-4)	GalT-6	28
N-Acetylgalactosaminyltransferases		
UDP-GalNAc : G_{M3} (β1-4)	GalNAcT-1	29, 30
UDP-GalNAc : Gb3 (β1-3)	GalNAcT-2	11, 31–33
UDP-GalNAc : Gb4 (α1-3)	GalNAcT-3	11, 33–35
N-Acetylglucosaminyltransferases		
UDP-GlcNAc : Lc2 (β1-3)	GlcNAcT-1	36
UDP-GlcNAc : nLc4 (β1-3)	GlcNAcT-2	37
UDP-GlcNAc : nLc4 (β1-6)	GlcNAcT-3	37
Fucosyltransferases		
GDP-Fuc : nLc4 or nLc5 (α1-2)	FucT-2	38–40
GDP-Fuc : Lc3 (α1-3)	FucT-3	41, 42
GDP-Fuc : 8-GSL (β1-3)	FucT-3	43
GDP-Fuc : G_{M1}	FucT-4	44, 45
GDP-Fuc : IV^{3}NeuAc-LcOse4Cer (α1-4)	FucT-3'	58a
Sialyltransferases		
CMP-NeuAc : GalCer (α2-3)	SAT-1 (?)	46
CMP-NeuAc : Lc2 (α2-3)	SAT-1	14, 22, 47, 51
CMP-NeuAc : G_{M3} or L_{M1} (α2-8)	SAT-2	52, 54
CMP-NeuAc : nLc4	SAT-3	14, 22, 48, 55
CMP-NeuAc : G_{M1} (α2-3) or Gg4 (α2-3)	SAT-4	8, 14, 22, 56

[a] Revised and updated since published in Ref. 8.

sites: an anionic UDP-Glc binding site, an oxygen-binding region oriented toward the third carbon atom of ceramide, a region that binds the alkyl chain of the fatty acid moiety, and a region that binds the hydrocarbon chain of the sphingosine base moiety. N-Octanoylsphingosine proved to be the best.[16] A high rate of turnover of this glucosyltransferase in mouse and rat livers has been suggested.[60]

[60] N. S. Radin, in "Gaucher Disease: A Century of Delineation and Research," p. 357. Alan R. Liss, Inc., New York, 1982.

TABLE II
RECOMMENDED ASSAY METHODS FOR GSL : GLTs

Procedure	Abbre-viation	GLT tested
Whatman SG-81 paper descending chromatography (when ^{3}H- or ^{14}C-labeled sugar nucleotide is a donor)	SGDC	GalTs, GlcNAcTs, GalNAcTs, FucTs, SATs
Whatman 3MM paper descending chromatography (when ^{14}C-labeled sugar nucleotide is a donor)	WPDC	GalTs, GlcNAcTs, GalNAcTs, FucTs, SATs
High-voltage paper electrophoresis	HVPE	GalTs, FucTs, SATs
Whatman SG-81 paper ascending chromatography	SGAC	GlcT, GalT-1, GalT-2, GlcNAcT-1
Reverse direction ascending chromatography	RDAC	GalT-4, FucT-3
Sephadex G-25 column chromatography	G25CC	GalT-3, GalNAcT-1, SAT-1, SAT-4
Sep-Pak column chromatography	SPKCC	GalNAcT-1
Monoclonal antibody binding assay	MABA	GalT-6, GalNAcTs
Iodoglycolipid ascending chromatography	IGLAC	SAT-1
Short column followed by thin-layer chromatography	SCTLC	SAT-1, SAT-2
Radioactive glycolipid and monoclonal antibody binding assay	RAGAB	FucT-3

Reagents

Ceramide (natural or synthetic), 10 mM in chloroform–methanol (2 : 1)

Cutscum–Triton X-100 (2 : 1), 20 mg/ml in chloroform–methanol (2 : 1)

Bicine buffer, pH 8.1, 1.0 M

UDP-[^{14}C]Glc, 10 mM (3.0–5.0 × 10^6 dpm/μmol)

0.25 M EDTA, pH 7.0

0.5 M KCl

Whatman SG-81 paper

$Na_2B_4O_7$, 1.0%, pH 9.1

Chloroform–methanol (2 : 1)

Incubation Mixture

Complete incubation mixtures contain the following components in final volumes of 0.1 ml: ceramide, 0.2 μmol; Cutscum–Triton X-100 (2 : 1), 0.5 mg [20 μl of ceramide solution and 25 μl of detergent solution in chloroform–methanol (2 : 1) are first mixed in an incubation tube (6 × 50 mm) and dried completely in a Savant rotary vacuum desiccator at room temperature]; Bicine buffer, pH 8.1, 15 μmol; UDP-[^{14}C]Glc, 0.1 μmol (3.0–5.0 × 10^6 dpm/μmol); and enzyme fraction, 0.4–0.8 mg of protein (if the enzyme source is membrane bound). The mixtures are incubated for 1 hr at 30° instead of 37°.[15]

Assay Method

Whatman SG-81 Paper Ascending Chromatography (SGAC). The reactions are stopped by adding 10 μl of 0.25 M EDTA (pH 7.0), 10 μl of 0.5 M KCl, and 0.5 ml of chloroform–methanol (2 : 1). After mixing in a Vortex mixer, the lower layer (0.35 ml) is washed with 0.2 ml of the theoretical upper phase (chloroform–methanol, 0.1 M KCl, 3 : 47 : 48), dried under N_2, and applied to Whatman SG-81 silica gel paper impregnated with 1% sodium tetraborate ($Na_2B_4O_7$). Ascending chromatograms are developed with chloroform–methanol–water (60 : 17 : 2). The standard glucosylceramide is detected by exposure to iodine vapor or by spraying with 0.04% bromothymol blue reagent (in 0.01 N NaOH). The dried paper strips are cut into 1-in segments and counted in a toluene system in liquid scintillation spectrometer.

The apparent K_m values for ceramide and UDP-[^{14}C]Glc are 80 and 120 μM, respectively, with embryonic chicken brain glucosyltransferase. However, using porcine submaxillary gland Golgi membranes, the apparent K_m values were 24 and 54 μM with UDP-[^{14}C]Glc and ceramide, respectively.[61] Instead of making a mixed micelle in buffer, a thin layer of ceramide coat on Celite can also be used as the source of substrate in the incubation mixture, and UDP-[^{3}H]Glc can also be used as donor.[16]

Brain HFA-Ceramide (β1-1)galactosyltransferase (GalT-1)

Galactosylceramide (galactocerebroside, GalCer)[2] is widely distributed in animal tissues, particularly in brain.[62] Following the discovery of this GalCer by Thudicum[63] from human white matter, many laboratories

[61] H. Coste, M. B. Martel, G. Azzar, and R. Got, *Biochim. Biophys. Acta* **814**, 1 (1985).

[62] H. E. Carter, P. Johnson, and R. J. Weber, *Annu. Rev. Biochem.* **34**, 109 (1965).

[63] J. L. W. Thudicum, *in* "Treatise on the Chemical Constitution of Brain." Baillière, London, 1884 (new reprint: Archon Books).

have studied its chemical structure, distribution in animal tissues, biosynthesis, and biological functions in myelin membranes.

A galactosyltransferase (GalT-1) that catalyzes the synthesis of GalCer from ceramide containing 2-hydroxy fatty acids (cerebronic acid) and UDP-galactose has been isolated from the brains of 19- to 20-day-old embryonic chickens,[18,64] young mice,[17] and rats.[65] The GalT-1 activity from rat brain has been solubilized.[66,67] It is highest (2–3 nmol/mg protein in homogenate/hr) in 19-day-old embryonic chicken and 25-day-old rat brains, but very low in 7- to 16-day-old embryonic chicken, guinea pig (3 days), rat (3 days), and fetal pig (20 cm) brains (0.05–0.15 nmol/mg protein in homogenate/hr). Because of the specificity of the assay method,[18] even low activity (100–200 pmol/mg of protein/hr) is detectable, and the values are reproducible.

The reaction is as follows:

$$\text{UDP-[}^{14}\text{C or }^{3}\text{H]Gal} + \text{HFA-Cer} \xrightarrow{\text{GalT-1}} \text{[}^{14}\text{C or }^{3}\text{H]Gal}\beta\text{1-1Cer} + \text{UDP}$$

Ceramide (containing 2-hydroxy long-chain fatty acids, e.g., cerebronic acid) is the most effective lipid acceptor, and a variety of potential donors cannot replace UDP-Gal. The rate of reaction in the absence of a divalent cation (Mg^{2+}, Mn^{2+}, or Ca^{2+}) is about 20% of that in the presence (5–10 mM).[18] pH optima of 7.7 and 7.4 have been observed in Bicine[18] and Tris–HCl[68] buffers, respectively.

Reagents

HFA-Ceramide (containing cerebronic acid), 10 mM in chloroform–methanol (2 : 1)

Triton X-100–Cutscum (1 : 2), 20 mg/ml in chloroform–methanol (2 : 1)

Bicine buffer, 1.0 M, pH 7.7

$MgCl_2$ (or $MnCl_2$), 0.1 M

UDP-[^{14}C or ^{3}H]Gal, 10 mM (2.0–3.0 × 10^6 dpm/μmol)

EDTA, 0.25 M

KCl, 0.5 M

Whatman SG-81 paper

$Na_2B_4O_7$, 1.0%, pH 9.1

Chloroform–methanol (2 : 1)

[64] S. Basu, A. Schultz, and M. Basu, *Fed. Proc., Fed. Am. Soc. Exp. Biol.* **28**, 540 (1969).

[65] N. M. Neskovic, L. L. Sarlieve, and P. Mandel, *Biochim. Biophys. Acta* **334**, 309 (1974).

[66] N. M. Neskovic, L. L. Sarlieve, and P. Mandel, *Biochim. Biophys. Acta* **429**, 324 (1976).

[67] N. M. Neskovic, P. Mandel, and S. Gatt, *in* "Enzymes of Lipid Metabolism" (S. Gatt, L. Freysz, and P. Mandel, eds.), p. 613. Plenum, New York, 1977.

[68] E. Costantino-Ceccarini and K. Suzuki, *Arch. Biochem. Biophys.* **167**, 646 (1975).

Incubation Mixture

Complete incubation mixtures contain the following components in final volumes of 0.1 ml in microtubes (6 × 50 mm) (the ceramide and detergent are mixed and dried first before addition of other reagents, as described for the GlcT incubation mixture): HFA-ceramide, 0.1 μmol (10 μl); Triton X-100–Cutscum (1 : 2), 0.4 mg (20 μl); Bicine buffer, pH 7.7, 10 μmol; MgCl$_2$ (or MnCl$_2$), 0.5 μmol (5 μl); UDP-[^{14}C or ^{3}H]Gal, 0.05 μmol (5 μl) (2.0–3.0 × 10^6 dpm/μmol); and enzyme fraction, 0.2 to 0.6 mg of protein (if the enzyme source is membrane bound). The mixtures are incubated for 1 hr at 37°.

Assay Method

The reactions are stopped by adding 10 μl of 0.25 M EDTA (pH 7.0), 10 μl of 0.5 M KCl, and 0.5 ml of chloroform–methanol (2 : 1). After mixing in a Vortex mixer, the [^{14}C or ^{3}H]GalCer enzymatic product is partitioned in the lower layer (0.35 ml), washed with 0.2 ml of the theoretical upper phase (chloroform–methanol–0.1 M KCl 3 : 47 : 48), and dried under N$_2$.

Whatman SG-81 Paper Ascending Chromatography (SGAC). The dried samples are dissolved in 50 μl of chloroform–methanol (2 : 1) and transferred quantitatively to dry Whatman SG-81 silica gel paper treated with 1% sodium tetraborate (Na$_2$B$_4$O$_7$). If UDP-[^{14}C]Gal is used as a donor, Whatman 3MM paper can replace SG-81 without any significant quenching. Maximum quenching (90%) is observed with [^{3}H]GalCer on Whatman 3MM paper, compared with 25–30% efficiency using Whatman SG-81 paper. Ascending chromatography is performed using chloroform–methanol–water (60 : 17 : 2). The standard GalCer is always spotted on a side lane and migration is visualized by exposure to iodine vapor or by spraying with 0.04% bromothymol blue reagent (in 0.01 N NaOH).

High-Voltage Paper Electrophoretic Assay (HVPE). The dried lower layer, containing [^{14}C]GalCer product (not ^{3}H-labeled), is dissolved in 50 μl of chloroform–methanol (2 : 1) and spotted 20 cm from the top of a 22-in-long strip of Whatman 3MM paper. A maximum of 10–12 samples can be spotted on a 46 × 57 cm sheet of Whatman 3MM paper. The paper is carefully wetted by dipping or spraying it with 1.0% Na$_2$B$_4$O$_7$ from both sides of the origin. Care should be taken to leave no dry spot at the origin. Any dry regions in the sample-spotted area should be soaked with borate solution using a capillary tube. Dry spots may cause charring of the paper. The wet Whatman 3MM paper is then carefully placed on the bridge of a Savant high-voltage electrophoresis apparatus and lowered into a tank containing 1% Na$_2$B$_4$O$_7$ at the electrode area and nonflammable organic coolant on the top. To reduce drying of the paper, the coolant is kept at

15° by constant circulation of cold tap water through a coil immersed in it. The paper is subjected to high-voltage electrophoresis at 2000–3000 V for 50–60 min (40 V/cm). The paper is dried under a hood, the origin, along with 2-in. areas toward the anode, is cut out, and radioactivity is quantitated in a toluene system in a liquid scintillation spectrometer.

The apparent K_m values for HFA-ceramide and UDP-[14C]Gal are 110 and 40 μM, respectively, with embryonic chicken brain GalT-1. Exogenous phosphatidylcholine and ethanolamine stimulate both Triton X-100-solubilized[67] and membrane-bound[68] GalT-1 from rat brain.

HFA-Ceramide has been adsorbed on a lyophilized enzyme preparation by evaporation from a solution in benzene.[69] This procedure and the coating of ceramide on Celite[17] makes it possible to measure the reaction rate. However, the kinetic studies are hampered by the inaccessibility of an insoluble substrate to soluble enzyme. A dispersion of ceramide with suitable detergent of liposomes with proper phospholipids is always recommended for the GalT-1 assay. Triton CF-54 or Cemulsol NP12 can also be used in place of Triton X-100 for the GalT-1 assay.[68]

Brain UDP-Gal: GlcCer (β1-4)galactosyltransferase (GalT-2)

The galactosyltransferase (GalT-2) catalyzing the synthesis of lactosylceramide (Galβ1-4GlcCer) from glucosylceramide has been studied in a number of tissues, including 13-day-old embryonic chicken brain,[8,19] rat spleen,[20] human renal cells,[70] and in bovine milk.[70a] Beside bovine milk, the enzyme activity from embryonic chicken brain has also been solubilized recently[7,26] using Triton X-100 and Nonidet P-40 detergents. However, unlike milk lactose synthase,[71–73] GalT-2 is not activated in the presence of α-lactalbumin for transfer of galactose from UDP-galactose to its substrate GlcCer. The Mg^{2+} requirement distinguishes this brain GalT-2 activity from other known galactosyltransferases including milk lactose synthase[71–73] and human lactose synthase A,[74] which require Mn^{2+} for optimal activity and are not active with Mg^{2+}. Unlike other glycolipid: glycosyltransferases involved in ganglioside biosynthesis in brains, this GalT-2 activity is high (200–300 pmol/mg protein in homogenate/hr) in early embryonic chicken brain (age 7–17 days). The activity observed in adult chicken brains is low (one-tenth of the activity in embryonic brains)

[69] A. Brenkert and N. S. Radin, *Brain Res.* **36**, 183 (1972).
[70] E. Castiglione and S. Chatterjee, *Fed. Proc., Fed. Am. Soc. Exp. Biol.* **45**, 1823 (1986).
[70a] A. A. Bushway and T. W. Keenan, *Biochim. Biophys. Acta* **572**, 146 (1979).
[71] R. L. Hill and K. Brew, *Adv. Enzymol. Relat. Areas Mol. Biol.* **43**, 411 (1975).
[72] K. Takase and K. E. Ebner, *J. Biol. Chem.* **256**, 7269 (1981).
[73] R. Richardson and K. Brew, *J. Biol. Chem.* **255**, 3377 (1980).
[74] K. Yamato and A. Yoshida, *J. Biochem.* (*Tokyo*) **92**, 1123 (1982).

but is measurable if a suitable assay method is used. The reaction is as follows:

$$\text{UDP-[}^{14}\text{C or }^3\text{H]Gal} + \text{GlcCer} \xrightarrow[\text{Mg}^{2+}\text{ and/or Mn}^{2+}]{\text{GalT-2}} \text{Gal}\beta\text{1-4GlcCer} + \text{UDP}$$

Glucosylceramide is the most effective glycolipid acceptor. A variety of potential donors could not replace UDP-Gal.

Kinetic studies indicate the optimum pH to be 6.8 in cacodylate–HCl, Bicine, or HEPES, and the apparent K_m values for glucosylceramide and UDP-[^{14}C or ^{3}H]galactose are 166 and 50 μM, respectively.

Reagents

Glucosylceramide (natural or synthetic), 10 mM in chloroform–methanol (2 : 1)

Cutscum–Triton X-100 (2 : 1), 20 mg/ml in chloroform–methanol (2 : 1)

HEPES buffer, pH 6.8, 1.0 M

UDP-[^{14}C or ^{3}H]Gal, 10 mM (2.0–3.0 × 10^6 dpm/μmol)

MnCl$_2$ and MgCl$_2$ mixture, 0.1 M each

0.25 M EDTA, pH 7.0

0.5 M KCl

Whatman SG-81 paper

Na$_2$B$_4$O$_7$, 1.0%, pH 9.1

Chloroform–methanol (2 : 1)

Incubation Mixture

Complete incubation mixtures contain the following components in final volumes of 0.1 ml in glass microculture tubes (6 × 50 mm) : glucosylceramide, 0.1 μmol (10 μl); Cutscum–Triton X-100 (2 : 1), 0.4 mg (20 μl) (glucosylceramide and detergent are first mixed and dried in a vacuum rotary desiccator); HEPES buffer, pH 6.8, 10 μmol (10 μl); mixture of MnCl$_2$ and MgCl$_2$, 0.5 μmol each; UDP-[^{14}C or ^{3}H] Gal, 0.05 μmol (2.0–3.0 × 10^6 dpm/μmol). The mixtures are incubated for 2 hr at 37° after addition of the enzyme fraction, 0.2 to 0.5 mg of protein (if the enzyme source is membrane bound). The rate of reaction remains constant for 3 hr under the suggested reaction conditions.

Assay Method

The reactions are stopped by adding 10 μl of 0.25 M EDTA (pH 7.0), 10 μl of 0.5 M KCl, and 0.5 ml of chloroform–methanol (2 : 1). After mixing on a Vortex mixer, the lower layer (0.32 ml) is washed with 0.2 ml of the theoretical upper phase (chloroform–methanol–0.1 M KCl,

3 : 47 : 48) and dried under N_2. The ^{14}C- or ^{3}H-labeled product is applied to Whatman SG-81 paper for ascending chromatography (SGAC) as described for GalT-1, except that the ascending solvent is chloroform–methanol–water (60 : 25 : 4). The incorporation of [^{14}C]Gal (not [^{3}H]Gal) into GlcCer can also be quantitatively determined by high-voltage paper electrophoresis (HVPE).

It has been suggested that a β-galactosyltransferase from human blood[74,75] which synthesizes nLcOse4Cer is also capable of synthesizing LacCer from GlcCer and is identical with human lactose synthase A protein. Further studies with solubilized GalT-2 are needed before this relationship can be established.

Brain UDP-Gal : $G_{M2}(\beta 1$-3)galactosyltransferase (GalT-3)

Ganglioside G_{M2} [GalNAc$\beta 1$-4(NeuAc$\alpha 2$-3)Gal$\beta 1$-4GlcCer] is an acceptor for the GalT-3 [UDP-Gal : $G_{M2}(\beta 1$-3)galactosyltransferase] that yields ganglioside G_{M1} [Gal$\beta 1$-3GalNAc$\beta 1$-4(NeuAc$\alpha 2$-3)Gal$\beta 1$-4Glc-Cer]. GalT-3 was first studied in a membrane fraction from 13-day-old embryonic chicken brain[8,14,21,22] and has been found in rat brain,[76] frog brain,[77] BHK cells,[78] embryonic chicken neuroretinal tissue,[79,80] and rat liver.[81–83] Recently GalT-3 has been solubilized and functionally purified from GalT-4 activity.[7,26,85] Studies with a cell line (TSD) derived from the cerebrum of an infant suffering from Tay–Sachs disease demonstrated very low GalT-3 activity[84] but a high activity of GalT-4 (see next section). No inhibitor for GalT-3 activity has been detected in extracts of TSD cells. The reaction is as follows:

$$\text{UDP-}[^{14}\text{C or }^3\text{H}]\text{Gal} + G_{M2} \xrightarrow[\text{Mn}^{2+}]{\text{GalT-3}}$$

$$\text{Gal}\beta 1\text{-3GalNAc}\beta 1\text{-4(NeuAc}\alpha 2\text{-3)Gal}\beta 1\text{-4GlcCer} + \text{UDP}$$
$$\text{(ganglioside } G_{M1})$$

[75] J. Zielenski and J. Koscielak, *Eur. J. Biochem.* **125,** 323 (1982).
[76] G. B. Yip and J. A. Dain, *Biochim. Biophys. Acta* **206,** 252 (1970).
[77] M. C. M. Yip and J. A. Dain, *Biochem. J.* **118,** 247 (1970).
[78] H. Den, A. M. Schultz, M. Basu, and S. Roseman, *J. Biol. Chem.* **246,** 2721 (1971).
[79] M. Pierce, *J. Cell Biol.* **93,** 76 (1982).
[80] P. Panzetta, H. J. F. Maccioni, and R. Caputto, *J. Neurochem.* **35,** 100 (1980).
[81] T. W. Keenan, D. J. Morré, and S. Basu, *J. Biol. Chem.* **249,** 310 (1974).
[82] F. Kaplan and P. Hechtman, *J. Biol. Chem.* **258,** 770 (1983).
[83] H. K. M. Yusuf, G. Pohlentz, and J. Schwarzmann, *Eur. J. Biochem.* **134,** 47 (1983).
[84] M. Basu, K. A. Presper, S. Basu, L. M. Hoffman, and S. E. Brooks, *Proc. Natl. Acad. Sci. U.S.A.* **76,** 2218 (1979).
[85] S. Basu, M. Basu, J. W. Kyle, T. De, K. Das, and R. J. Schaeper, *in* "Enzymes of Lipid Metabolism" (L. Freysz and S. Gatt, eds.), p. 233. Plenum, New York, 1986.

An increase in GalT-3 specific activity up to 19 days of embryonic chicken brain development has been observed.[26] The distinction between GalT-3 and GalT-4 is based on kinetic studies[8,12] and, finally, on their separation by DEAE–Sepharose CL-6B and α-lactalbumin affinity column chromatography.[7,26]

Reagents

Ganglioside G_{M2} (natural), 5mM in chloroform–methanol (2 : 1) Triton X-100 or Nonidet P-40, 10 mg/ml in chloroform–methanol (2 : 1)
Cacodylate–HCl or HEPES buffer, pH 7.2, 1 M
$MnCl_2$, 0.1 M
UDP-[^{14}C]Gal, 10 mM (2–3 $\times$ 10^6 dpm/μmol)
0.25 M EDTA, pH 7.0
Whatman 3MM paper
Whatman SG-81 paper
$Na_2B_4O_7$, 1.0%, pH 9.1
Chloroform–methanol (2 : 1)

Incubation Mixture

Complete incubation mixtures contain the following components in final volumes of 0.05 ml: G_{M2}, 0.025 μmol (5 μl) and 0.1 mg detergent (10 μl), dissolved in chloroform–methanol (2 : 1), mixed in an incubation tube (6 $\times$ 50 mm) and dried completely in a rotary vacuum desiccator at room temperature; then the following ingredients are added: cacodylate–HCl buffer, pH 7.2, 10 μmol; $MnCl_2$, 0.5 μmol (5 μl); UDP-[^{14}C]Gal, 0.02 μmol (2.0–3.0 $\times$ 10^6 dpm/μmol); and enzyme fraction, 0.1–0.2 mg of protein. If the enzyme source is detergent-solubilized supernatant, a ratio of protein to detergent of 1 : 2 is maintained, and the initial detergent added to form mixed micelles is reduced to 25 μg. The mixtures are incubated for 1 hr at 37°. The rate of reaction is constant up to 2–3 hr under the present assay conditions.

Assay Method

The reactions are stopped by adding 10 μl of 0.25 M EDTA (pH 7.0) and spotted on chromatography paper.

Whatman SG-81 Paper Descending Chromatography (SGDC). When UDP-[^{3}H]Gal is used as donor, then the incubation mixtures are spotted on Whatman SG-81 paper and subjected to descending chromatography with 1% $Na_2B_4O_7$. The radioactive product stays at the origin, and UDP-[^{3}H]Gal, [^{3}H]Gal-1-PO_3^{2-}, and [^{3}H]Gal move almost with the solvent front.

The papers are dried, and the ^{3}H-labeled origin areas are quantitated in a toluene system (4.0 g PPO and 0.15 g dimethyl-POPOP/liter of toluene) in the liquid scintillation spectrometer.

Whatman, 3MM Paper Assays. When UDP-[^{14}C]Gal is used as donor, the incubation mixtures are spotted on Whatman 3MM paper and subjected to either descending chromatography (WPDC) as above or HVPE as described for the GalT-1 assay.

The apparent K_m values for G_{M2} and UDP-[^{14}C]Gal are 200 and 130 μM, respectively, with detergent-solubilized embryonic chicken brain GalT-3.[7,26] The V_{max} is lower (10%) with GgOse3Cer (GalNAcβ1-4Galβ1-4GlcCer) than with G_{M2}. The rat liver GalT-3[81,82] can be resolved in two different forms based on their pH optima (enzyme I, 7.0; enzyme II, 6.0). GalT-3 is active only in the presence of Mn^{2+}.

Bone Marrow and Brain UDP-Gal : LcOse3Cer(β1-4)galactosyltransferase (GalT-4)

Lactotriaosylceramide (LcOse3Cer; GlcNAcβ1-3Galβ1-4GlcCer) is the substrate for GalT-4 [UDP-Gal : LcOse3Cer(β1-4)galactosyltransferase], which yields nLcOse4Cer (Galβ1-4GlcNAcβ1-3Galβ1-4GlcCer), the core structure of many blood-group-active glycosphingolipids. GalT-4 was first studied in rabbit bone marrow[23,24] and recently has been solubilized and purified from 13- to 19-day-old embryonic chicken brain.[7,26,85] In addition to the above sources, GalT-4 activity using LcOse3Cer has been detected in human, rabbit, and bovine sera,[25] rat prostate, and mammary adenocarcinoma.[86]

In 19-day-old embryonic chicken brain, the activities of GalT-3 and GalT-4 are comparable. This has not been found in any other tissues. Rabbit bone marrow has very little GalT-3 activity in comparison with GalT-4 activity. The ratio of GalT-4 to GalT-3 activity is approximately 20 : 1 in chicken liver.

Both GalT-3 and GalT-4 have been solubilized in the presence of Nonidet P-40 or Triton X-100, with detergent-to-protein ratios of 1 : 1 and 2 : 1, respectively. GalT-3 and GalT-4 have been separated[7,26,85] using a variety of techniques based on different physicochemical properties, including BioGel A1.5m gel filtration, DEAE–Sepharose CL-6B ion-exchange chromatography, and α-lactalbumin affinity chromatography. A simple microisoelectric focusing is also used for further purification of embryonic chicken brain GalT-4.[7,26,85]

Unlike lactose synthase, the purified GalT-4 from embryonic chicken brain is insensitive to α-lactalbumin (up to 3 mg/ml) when tested with

---

[86] D. M. Jenis, S. Basu, and M. Pollard, *Cancer Biochem. Biophys.* **6,** 37 (1982).

LcOse3Cer as substrate. It is inhibited only 50% when GlcNAc is used as substrate. The Mn^{2+} requirement (K_m, 0.5 mM) could not be replaced by any other divalent cation. The K_m values for LcOse3Cer and UDP-Gal are 0.75 mM and 35 μM, respectively. Two simple synthetic acceptors can be used as substrates for GalT-4: β-O-Me-GlcNAc (K_m = 0.4 mM) and p-nitrophenyl-β-D-GlcNAc (K_m = 0.3 mM). From substrate competition studies it appears that both glycolipid and glycoprotein acceptors containing N-acetylglucosamine at the terminal chain can act as substrate for this GalT-4.[26,85] The reaction catalyzed by GalT-4 is as follows:

$$UDP\text{-}[^{14}C]Gal + GlcNAc\beta1\text{-}3Gal\beta1\text{-}4GlcCer \xrightarrow[Mn^{2+}]{GalT\text{-}4} Gal\beta1\text{-}4GlcNAc\beta1\text{-}3Gal\beta1\text{-}4GlcCer$$
$$\text{(nLcOse4Cer)}$$

[^{14}C]GlcNAc and nonradioactive UDP-Gal have also been used as substrates for the synthesis of Galβ1-4[^{14}C]GlcNAc using purified embryonic chicken brain GalT-4. The anomeric linkage is proved by cleavage with purified β-galactosidase, and the positional linkage is proved by the combination of NaBH$_4$ reduction and periodate oxidation followed by characterization of D-xylosaminitol.[26,85]

Reagents

LcOse3Cer, 10 mM in chloroform–methanol (2:1)   or
   p-NO$_2$-phenyl-β-D-GlcNAc (PNP-GlcNAc), 10 mM in H$_2$O
Nonidet P-40, 5 mg/ml in chloroform–methanol (2:1)
Cacodylate–HCl or HEPES buffer, pH 6.8, 1 M
MnCl$_2$, 0.1 M
5'-AMP, 0.05 M
UDP-[^{14}C]Gal, 5 mM (2.0 × 10^6 dpm/μmol)
EDTA, 0.25 M, pH 7.0
Whatman 3MM paper
Na$_2$B$_4$O$_7$, 1.0%, pH 9.1
Chloroform–methanol (2:1)

Incubation Mixture

When glycolipid is used as substrate, glycolipid and the detergent are mixed in chloroform–methanol (2:1) and dried in a rotary vacuum desiccator. When PNP-GlcNAc or GlcNAc-α_1-acid glycoprotein (SĀ, GaĪ, α_1AGP) is used as substrate for the membrane-bound GalT-4, only the detergent is dried under N$_2$ in the tube. For solubilized GalT-4 and soluble substrates such as PNP-GlcNAc or GlcNAc-α_1-acid glycoprotein, no detergent is required. Complete incubation mixtures contain the following components in final volumes of 0.05 ml in microculture tubes (6 × 50 mm): LcOse3Cer, 0.025 μmol (or PNP-GlcNAc, 0.1 μmol); Nonidet P-40, 0.1 mg; cacodylate–HCl or HEPES buffer, pH 6.8, 10 μmol; MnCl$_2$, 0.5 μmol; 5'-AMP, 0.25 μmol; UDP-[^{14}C]Gal, 0.02 μmol (2.0 × 10^6 dpm/ μmol); and detergent-free solubilized enzyme, 0.05–0.1 mg of protein. The mixtures are incubated for 2 hr at 37°. Under the reaction conditions the rate remains constant up to 3 hr.

Assay Method

The reactions are stopped by adding 10 μl of 0.25 M EDTA (pH 7.0), and the mixture is spotted on Whatman 3MM paper and subjected to either HVPE (as described for GalT-1) or ascending chromatography after electrophoresis as described below.

Reverse Direction Ascending Chromatography (RDAC). A combination of electrophoresis and chromatography assay is used[26,85] to separate the glycoprotein and glycolipid enzymatic products and to determine whether the presence of one substrate inhibits the utilization of the second substrate by the purified GalT-4. The incubation mixtures are first spotted on a full-length Whatman 3MM paper (46 × 57 mm) 25 cm from one side and subjected to high-voltage electrophoresis in Na$_2$B$_4$O$_7$ buffer (HVPE), as described for the GalT-1 assay. The Whatman 3MM paper is then air dried and cut 3 in. from the origin on the anode side, away from the unreacted UDP-[^{14}C]Gal and degraded products such as [^{14}C]Gal-1-PO$_3^{2-}$ and [^{14}C]Gal. The radioactive products (glycolipid and glycoprotein) remain at the origin. Ascending chromatography is then run in the opposite direction using chloroform–methanol–water (60 : 35 : 8, v/v/v). This ascending chromatography is repeated once more after the paper is air dried to obtain better resolution between ^{14}C-labeled glycolipid (which moves up) and ^{14}C-labeled glycoprotein (which stays at the origin). After air drying, the appropriate areas of the chromatogram are cut out, and the radioactivity is quantitated in a toluene system in a liquid scintillation spectrometer.

Bone Marrow and Spleen
UDP-Gal : nLcOse4Cer(α1-3)galactosyltransferase (GalT-5)

Neolactotetraocylceramide (nLcOse4Cer; Galβ1-4GlcNAcβ1-3Galβ1-4GlcCer) is the substrate for GalT-5 [UDP-Gal : nLcOse4Cer(α1-3)galactosyltransferase], which yields nLcOse5Cer (Galα1-3Galβ1-4GlcNAcβ1-3Galβ1-4GlcCer), the core structure of human blood-group B-active glycosphingolipid. GalT-5 was first studied in rabbit bone marrow[24,27] and has also been solubilized from bovine spleen.[25,85]

Using nLcOse4Cer as acceptor, GalT-5 activity has been detected in mouse neuroblastoma,[12,87] monkey kidney,[88] guinea pig tumor,[41,89] and human neuroblastoma cells.[41] A continuous culture from Tay–Sachs-diseased brain also contains considerable activity of GalT-5.[84] Among many murine B and T lymphomas tested, a B-lymphocytic tumor, TEPC-15, had the highest GalT-5 activity.[90] Using lactosamine as substrate, an (α1-3)galactosyltransferase activity has been found in ascites tumor cells.[90a] It is expected that the α-galactosyltransferase from ascites tumor cells is similar to that of the above- mentioned GalT-5 utilizing nLcOse4Cer because GalT activities in all of these sources have an absolute requirement for lactosamine (Galβ1-3 or 4 GlcNAc) terminals.

The reaction catalyzed by GalT-5 is as follows:

$$\text{UDP-[}^{14}\text{C]Gal} + \text{Gal}\beta\text{1-4GlcNAc}\beta\text{1-3Gal}\beta\text{1-4GlcCer} \xrightarrow[\text{Mn}^{2+}]{\text{GalT-5}}$$

$$\text{Gal}\alpha\text{1-3Gal}\beta\text{1-4GlcNAc}\beta\text{1-3Gal}\beta\text{1-4GlcCer} + \text{UDP}$$
$$\text{(nLcOse5Cer)}$$

The K_m values for nLcOse4Cer and UDP-Gal are 1.67 mM and 143 μM, respectively,[27] with rabbit bone marrow GalT-5.

Reagents

nLcOse4Cer, 10 mM in chloroform–methanol (2 : 1)   or
    lactosamine, 10 mM in H_2O
Triton CF-54, 12 mg/ml in chloroform–methanol (2 : 1)
Cacodylate–HCl buffer, pH 7.2, 1.0 M
MnCl$_2$, 0.05 M
AMP, 0.05 M
UDP-[^{14}C]Gal, 10 mM (2.0–3.0 × 10^6 dpm/μmol)
EDTA, 0.25 M, pH 7.0
Whatman 3MM paper
Na$_2$B$_4$O$_7$, 1.0%, pH 9.1
Chloroform–methanol (2 : 1)

Incubation Mixture

When nLcOse4Cer is used as substrate, the glycolipid and the detergent are mixed in chloroform–methanol (2 : 1) and dried in a rotary vac-

[87] J. R. Moskal, D. A. Gardner, and S. Basu, *Biochem. Biophys. Res. Commun.* **61,** 751 (1974).
[88] M. Basu, J. R. Moskal, D. A. Gardner, and S. Basu, *Biochem. Biophys. Res. Commun.* **66,** 1380 (1975).
[89] M. Basu, S. Basu, W. G. Shanabruch, J. R. Moskal, and C. H. Evans, *Biochem. Biophys. Res. Commun.* **71,** 385 (1976).
[90] M. Basu, S. Basu, and M. Potter, *ACS Symp. Ser.* **128,** 187 (1980).
[90a] M. J. Elices, D. A. Blake, and I. J. Goldstein, *J. Biol. Chem.* **261,** 6064 (1986).

uum desiccator. When lactosamine (Galβ1-3 or 4 GlcNAc), lacto-N-tetraose (Galβ1-3GlcNAcβ1-3Galβ1-4Glc), or desialized α_1-acid glycoprotein (Galβ1-4GlcNAc-α_1-acid GP) is used as an acceptor for the membrane-bound GalT-5, only the detergent is dried under N_2 in the tube. For solubilized GalT-5 and soluble substrates, no detergent is required. Complete incubation mixtures contain the following components in final volumes of 0.025 ml in microculture tubes (6 × 50 mm): nLcOse4Cer, 0.1 μmol (10 μl); Triton CF-54, 240 μg (20 μl); cacodylate–HCl buffer, pH 7.2, 5.0 μmol; $MnCl_2$, 0.25 μmol; AMP, 0.25 μmol; UDP-[^{14}C]Gal, 0.05 (2.0–3.0 × 10^6 dpm/μmol); and enzyme fraction, 0.1–0.2 mg of protein. The mixtures are incubated for 1 hr at 37°. Under these reaction conditions, the rate remains constant up to at least 1 hr.

Assay Method

The reactions are stopped by adding 10 μl of 0.25 M EDTA (pH 7.0), and then the mixtures are spotted on Whatman 3MM paper and subjected to HVPE (when oligosaccharide, glycolipid, or glycoprotein is used as substrate). When only glycolipid is used as substrate, the incubation mixtures may be assayed either by HVPE or by SGDC, as described for GalT-3.

Liver UDP-Gal : LacCer(α1-4)galactosyltransferase (GalT-6)

Lactosylceramide (Galβ1-4GlcCer, LacCer) is the substrate for GalT-6 [UDP-Gal : LacCer(α1-4)galactosyltransferase], which yields GbOse-3Cer (Galα1-4Galβ1-4GlcCer), the core structure of human and porcine globoside. GalT-6 was first studied in rat spleen[20] and has recently been purified from rat liver.[28] Both the rat spleen and rat liver enzymes appear to utilize lactosylceramide as substrate. However, the iGbOse3Cer obtained from rat spleen appears to have terminal α1-3-linked galactose instead of the α1-4 linkage present in GbOse3Cer synthesized by rat liver or kidney GalT-6. The enzyme is also distinct from GalT-5, which catalyzes the addition of galactose to nLcOse4Cer to form a blood-group B-type glycolipid.[24,27] The reaction catalyzed by GalT-6 is as follows:

$$\text{UDP-[}^{14}\text{C]Gal} + \text{Gal}\beta\text{1-4GlcCer} \xrightarrow[\text{Mn}^{2+}]{\text{GalT-6}} \underset{\text{(GbOse3Cer)}}{\text{Gal}\alpha\text{1-4Gal}\beta\text{1-4GlcCer}} + \text{UDP}$$

The enzyme is inhibited by EDTA (10 mM) and can be activated in the presence of 10 mM divalent cations (Mn^{2+}, Cu^{2+}, Ni^{2+}). Of the other cations tested, the presence of Mn^{2+}, Co^{2+}, or Ca^{2+} activate the enzyme (1.5-fold), whereas Ni^{2+}, Zn^{2+}, Mg^{2+}, or Cu^{2+} inhibits the reaction.

Reagents

LacCer, 10 m*M* in chloroform–methanol (2 : 1)   or
  lactose, 10 m*M* in H$_2$O
Triton X-100, 15 mg/ml in chloroform–methanol (2 : 1)
Cacodylate–HCl buffer, pH 7.2, 1.0 *M*
MnCl$_2$, 0.1 *M*
UDP-[^{14}C]Gal, 10 m*M* (2.0–3.0 × 10^6 dpm/μmol)
EDTA, 0.25 *M*, pH 7.0
Whatman 3MM paper
Na$_2$B$_4$O$_7$, 1.0%, pH 9.1

Incubation Mixture

When LacCer is used as substrate, the glycolipid and the detergent are mixed in chloroform–methanol (2 : 1) and dried in a rotary vacuum desiccator. When lactose (Galβ1-4Glc) is used as an acceptor, only the detergent is dried first. Complete incubation mixtures contain the following components in final volumes of 0.1 ml in microculture tubes (6 × 50 mm): LacCer, 0.1 μmol; Triton X-100, 300 μg; MnCl$_2$, 1.0 μmol; UDP-[^{14}C]Gal, 0.05 (2.0–3.0 × 10^6 dpm/μmol); and enzyme fraction, 50–500 μg of protein. After incubation at 37° for 1 hr, the reaction is stopped by the addition of 5 μmol of EDTA.

Assay Method

The reaction mixtures can be assayed by any of the following assay methods.

High-Voltage Paper Electrophoresis (HVPE). This is the same method as described for GalT-1.

Monoclonal Antibody Binding Assay (MABA). After incubation, the reaction mixtures are centrifuged at 10,000 *g* for 5 min and heated in a boiling water bath for 1 min. The supernatant (500 μl) is incubated with monoclonal anti-GbOse3Cer rat ascites fluid (diluted according to need) for 2 hr at 37°. Then 5 μl of rabbit anti-rat IgM is added and the mixtures are incubated for another 2 hr, followed by the addition of 20 μl of Ig-Gsorb (0.5 mg) and a 10-min wait at 37°. The precipitate is quantitatively transferred to nitrocellulose filters. The filters are washed, and the amount of radioactivity is determined in a toluene liquid scintillation system. With this method either GbOse3Cer (Galα1-4Galβ1-4GlcCer) or iG-bOse3Cer (Galα1-3Galβ1-4GlcCer) formation can be quantitatively determined if a specific antibody is available. This assay is limited by availability and specificity of the antibody.[28] A tight-binding antibody is preferred.

UDP-GalNAc : GSL *N*-Acetylgalactosaminyltransferases

The following three *N*-acetylgalactosaminyltransferases catalyze the biosynthesis of glycosphingolipids of the ganglio (Gg) and globo (Gb) families. The reaction conditions are similar, except for their three different pH optima, and all three require detergent and Mn^{2+} for optimum activity. The reagent, incubation mixture, and assay method are described together for all three activities.

Brain UDP-GalNAc : $G_{M3}(\beta 1$-$4)$N-acetylgalactosaminyltransferase (GalNAcT-1)

Ganglioside G_{M2} [GalNAcβ1-4(NeuAcα2-3)Galβ1-4GlcCer] is widely distributed in animal brains[3,4] and other animal tissues.[5,6] Biosynthesis *in vitro* of ganglioside G_{M2} from UDP-GalNAc and ganglioside G_{M3} was first achieved using a membrane preparation isolated from embryonic chicken brain (ECB).[22,29,31] Recently, GalNAcT-1 activity has been solubilized and separated from the GalNAcT-2 activity (see next section) of 19-day-old ECB.[85,91,91a] In addition to ECB, using G_{M3} as acceptor, the GalNAcT-1 activity has been detected in rat liver,[81] rat brain,[92] mouse liver,[93] fetal pig brain,[22,29] guinea pig bone marrow,[30] NIL hamster cells,[94] mouse neuroblastoma cells,[95] and 3T3 cells.[13] The reaction is as follows:

$$UDP\text{-}[^3H \text{ or } {}^{14}C]GalNAc + G_{M3} \xrightarrow[Mn^{2+}]{GalNAcT\text{-}1}$$

$$GalNAc\beta1\text{-}4(NeuAc\alpha2\text{-}3)Gal\beta1\text{-}4GlcCer + UDP$$
$$(G_{M2})$$

The apparent K_m values for G_{M3} and UDP-GalNAc are 0.4 and 0.16 mM, respectively, with solubilized ECB GalNAcT-1 free from GalNAcT-2 activity.

Brain, Spleen, and Tumor Cell UDP-GalNAc : GbOse3Cer(β1-3)N-acetylgalactosaminyltransferase (GalNAcT-2)

Biosynthesis *in vitro* of globoside from globotriaosylceramide (GbOse-

[91] K. Das, R. J. Schaeper, M. Basu, and S. Basu, *Fed. Proc., Fed. Am. Soc. Exp. Biol.* **43,** 1566 (1984).

[91a] R. J. Schaefer, K. K. Das, and S. Basu, *Fed. Proc., Fed. Am. Soc. Exp. Biol.* **45,** 1822 (1986).

[92] J. L. Dicesare and J. A. Dain, *Biochim. Biophys. Acta* **231,** 385 (1971).

[93] Y. Hashimoto, M. Abe, Y. Kiuchi, and T. Yamakawa, *J. Biochem. (Tokyo)* **95,** 1543 (1984).

[94] M. W. Lockney and C. Sweeley, *Biochim. Biophys. Acta* **712,** 234 (1982).

[95] M. A. Scheideler, N. W. Lockney, and G. Dawson, *J. Neurochem.* **41,** 1261 (1983).

3Cer, Galα1-4Galβ1-4GlcCer) was first achieved by the use of ECB membrane-bound N-acetylgalactosaminyltransferases (GalNAcTs). Recently, two distinct GalNAcTs (GalNAcT-1 and GalNAcT-2) have been separated from embryonic chicken brain[85,91,91a] and guinea pig 104Cl cells.[33] GalNAcT-2   [UDP-GalNAc : Gb3(β1-3)N-acetylgalactosaminyltransferase] catalyzes the synthesis of globoside (GalNAcβ1-3Galα1-4Galβ1-4GlcCer) in embryonic chicken brain[31,91,91a] guinea pig tumor 104Cl cells,[11,33] and mouse adrenal Y-1-K cells.[34] It has recently been highly purified (18,000-fold) from canine spleen.[35] The purified GalNAcT-2 migrates as two major bands (MW 64,000 and 57,000) on SDS–polyacrylamide gel electrophoresis. Using guinea pig tumor cell 104Cl GalNAcT-2, the terminal GalNAc linkage has been established as β1-3.[33] It is expected that the spleen enzyme catalyzes the same linkage. The reaction is as follows:

$$\text{UDP-[}^3\text{H or }^{14}\text{C]GalNAc} + \text{GbOse3Cer} \xrightarrow[\text{Mn}^{2+}]{\text{GalNAcT-2}}$$

GalNAcβ1-3Galα1-4Galβ1-4GlcCer + UDP
(GbOse4Cer)

Irrespective of tissue source, the pH optimum is 7.0, and the enzyme has an absolute requirement for Mn^{2+} (K_m = 1.5 mM). The K_m values for GbOse3Cer and UDP-GalNAc for solubilized ECB GalNAcT-2 are 160 and 200 μM. The corresponding values for the spleen enzyme are one-tenth of those for the ECB enzyme.

Spleen and Tumor Cell UDP-GalNAc : GbOse4Cer (α1-3)N-acetylgalactosaminyltransferase (GalNAcT-3)

An α-N-acetylgalactosaminyltransferase that catalyzes the biosynthesis *in vitro* of Forssman-active GSL from globoside (GbOse4Cer, GalNAcβ1-3Galα1-4Galβ1-4GlcCer) has been characterized in guinea pig kidney,[96] mouse adrenal Y-1-K cells,[34] and guinea pig tumor 104Cl cells.[33] This GalNAcT-3 activity [UDP-GalNAc : GbOse4Cer(α1-3)N-acetylgalactosaminyltransferase] has recently been purified from canine spleen.[35] Of all tissues examined, human lung cancer cells contain the highest activity of GalNAcT-3.[97] The reaction is as follows:

$$\text{UDP-[}^3\text{H or }^{14}\text{C]GalNAc} + \text{GbOse4Cer} \xrightarrow[\text{Mn}^{2+}]{\text{GalNAcT-3}}$$

[^{3}H or ^{14}C]GalNAcα1-3GalNAcβ1-3Galα1-4Galβ1-4GlcCer + UDP
(Forssman GSL)

The purified (3500-fold) enzyme from canine spleen shows two major bands (MW 66,000 and 56,000) on SDS–polyacrylamide gel electrophore-

[96] T. Ishibashi, S. Kijimoto, and A. Makita, *Biochim. Biophys. Acta* **337**, 92 (1974).
[97] N. Taniguchi, N. Yokosawa, M. Narita, T. Misuyama, and A. Makita, *JNCI, J. Natl. Cancer Inst.* **67**, 577 (1981).

sis. Both canine spleen and guinea pig 104C1 cell GalNAcT-3 activities show pH optima at 7.0, and the enzyme has an absolute requirement for Mn^{2+} (K_m = 0.45 mM). Relative to Mn^{2+} (100%), three other cations, Zn^{2+} (83%), Fe^{2+} (76%), and Cu^{2+} (31%), also stimulate activity. The K_m values for GbOse4Cer and UDP-GalNAc with canine spleen GalNAcT-3 are 2.0 and 10 μM, respectively.[35] Under saturating concentrations of UDP-GalNAc, the K_m value for GbOse4Cer with solubilized GalNAcT-3 from guinea pig tumor cells is 160 μM.[33] UDP is a competitive inhibitor with respect to UDP-GalNAc, and a noncompetitive inhibitor with respect to globoside.

Reagents

G_{M3}, GbOse3Cer, or GbOse4Cer, 10 mM in chloroform–methanol (2 : 1)

Detergent (GalNAcT-1, Zwittergent, and Triton CF-54; GalNAcT-2, Triton CF-54, and Triton DF-12; GalNAcT-3, Triton X-100), 10 mg/ml in chloroform–methanol (2 : 1)

HEPES buffer (GalNAcT-1, pH 8.0; GalNAcT-2, pH 7.0) 1.0 M; or cacodylate–HCl buffer (GalNAcT-3, pH 7.0), 1.0 M

$MnCl_2$, 0.1 M

UDP-[^{3}H or ^{14}C]GalNAc, 0.005 mM (4.0–6.0 × 10^6 dpm/μmol)

EDTA, 0.25 M, pH 7.0

Whatman SG-81 paper

$Na_2B_4O_7$, 1.0%, pH 9.1

Incubation Mixture

Complete incubation mixtures contain the following components in final volumes of 0.05 ml in microculture tubes (6 × 50 mm) (the glycosphingolipid and detergent are mixed and dried first before other reagents are added as described before): glycosphingolipid, 0.05 μmol; detergent, 100–200 μg; buffer, 10 μmol; $MnCl_2$, 0.5 μmol; UDP-[^{3}H or ^{14}C]GalNAc, 0.02 μmol (4.0–6.0 × 10^6 dpm/μmol); and enzyme, 10–200 μg of protein. After 1 hr at 37° (under these conditions the rates of all three reactions remain constant), the reactions are stopped by the addition of 10 μl of 0.25 M EDTA and assayed by any of the following methods.

Assay Methods

SGDC. When UDP-[^{3}H]GalNAc is used as a donor, Whatman SG-81 paper descending chromatography (see the GalT-3 assay) is recommended. This method has a counting efficiency of 28–35%.

WPDC. When UDP-[^{14}C]GalNAc is used as donor, then the Whatman

3MM paper (instead of the costly SG-81 paper) is used to spot the reaction mixtures. Descending chromatography[31] with 1.0% $Na_2B_4O_7$ will remove [^{14}C]GalNAc-1-PO$_3^{2-}$ and [^{14}C]GalNAc from the origin. The radioactivity in appropriate areas is quantitatively determined in a toluene liquid scintillation system (80% counting efficiency).

UDP-GlcNAc : GSL N-Acetylglucosaminyltransferases

At least three different N-acetylglucosaminyltransferases have been characterized that catalyze the biosynthesis of glycosphingolipids of the lacto- (Lc) and I/i-related families. The reaction conditions are described together.

Bone Marrow UDP-GlcNAc : LacCer(β1-3)N-Acetylglucosaminyltransferase (GlcNAcT-1)

The biosynthesis *in vitro* of lactotriaosylceramide (GlcNAcβ1-3Galβ1-4GlcCer) has been shown with a membrane-bound GlcNAcT-1 [UDP-GlcNAc : LacCer(β1-3)N-acetylglucosaminyltransferase] from rabbit bone marrow.[8,36] The rabbit bone marrow GlcNAcT-1 activity is quite specific for LacCer as substrate, and different from that of P-1798 GlcNAcT-2 and GlcNAcT-3,[37] as described below. The reaction is as follows:

$$\text{UDP-[}^3\text{H or }^{14}\text{C]GlcNAc + LacCer} \xrightarrow[\text{Mn}^{2+}]{\text{GlcNAcT-1}} \text{GlcNAc}\beta1\text{-3Gal}\beta1\text{-4GlcCer + UDP}$$
$$\text{(LcOse3Cer)}$$

The enzyme has an absolute requirement for Mn^{2+}, which cannot be replaced by Co^{2+}, Ni^{2+}, Cu^{2+}, Fe^{3+}, or Zn^{2+}. Cd^{2+}, Mg^{2+}, or Ca^{2+} give only 15–20% of optimum LcOse3Cer formation. EDTA completely inhibits the reaction. A pH optimum of 7.0–7.2 is exhibited in cacodylate–HCl buffer. The K_m values of lactosylceramide and UDP-GlcNAc are 80 and 500 μM, respectively. Lactose is also an acceptor. However, lacto-*N*-triose I (Galβ1-3GlcNAcβ1-3Gal) is not a good acceptor.

Mouse Lymphoma UDP-GlcNAc : nLcOse4Cer(β1-3 and 6)N-acetylglucosaminyltransferase (GlcNAcT-2 and GlcNAcT-3)

A recent report[37] establishes the presence of at least two glycolipid : N-acetylglucosaminyltransferases (GlcNAcT-2 and GlcNAcT-3) in mouse P-1798 lymphoma, which are involved in the biosynthesis *in vitro* of I/i-core glycosphingolipids. The reactions are as follows:

UDP-[^{3}H or ^{14}C]GlcNAc + nLcOse4Cer $\xrightarrow[\text{Mn}^{2+}]{\text{GlcNAcT-2}}$

[^{3}H or ^{14}C]GlcNAcβ1-3Galβ1-4GlcNAcβ1-3Galβ1-4GlcCer + UDP
(i-core GSL)

UDP-[^{3}H or ^{14}C]GlcNAc + nLcOse4Cer $\xrightarrow[\text{Mn}^{2+}]{\text{GlcNAcT-3}}$

[^{3}H or ^{14}C]GlcNAcβ1-6Galβ1-4GlcNAcβ1-3Galβ1-4GlcCer
(I-core GSL)

The detergent-extracted membrane supernatant isolated from P-1798 mouse lymphoma contains both activities.[37] The terminal GlcNAc linkages have been established by the identification of 2,4,6-tri-O-methyl-[6-^{3}H]galactose and 2,3,4-tri-O-methyl[6-^{3}H]galactose obtained from the i- and I-core enzymatic products. They have broad pH optima, pH 7.0–8.0, in HEPES. Mn^{2+} (K_m = 1.25 mM for GlcNAcT-2) is essential for both activities, and EDTA completely inhibits them. With GlcNAcT-2 activity, Mn^{2+} could not be replaced by Mg^{2+}, Zn^{2+}, Cu^{2+}, or Cd^{2+}, but Ca^{2+} and Co^{2+} are 48 and 52% as effective, respectively, as Mn^{2+}. The K_m values for nLcOse4Cer and UDP-GlcNAc are 90 and 330 μM, respectively, with the solubilized GlcNAcT-2.

Reagents

nLcOse4Cer, 5 mM in chloroform–methanol (2 : 1)
Triton CF-54, 5 mg/ml in chloroform–methanol (2 : 1)
HEPES or cacodylate–HCl buffer, pH 7.2, 1.0 M
MnCl$_2$, 0.05 M
UDP-[^{3}H or ^{14}C]GlcNAc, 5 mM (3.0–4.0 × 10^6 dpm/μmol)
KCl, 1.0 M
EDTA, 0.25 M, pH 7.0
Chloroform–methanol (2 : 1)

Incubation Mixture

Complete incubation mixture contains the following components in final volumes of 0.05 ml in glass microculture tubes (6 × 50 mm) [the glycosphingolipid (0.02 μmol) and the detergent (25–50 μg) in chloroform–methanol are first mixed and dried in a rotary vacuum desiccator]: cacodylate–HCl buffer, pH 7.2, 1.0 M, or HEPES buffer, pH 7.5, 1.0 M; MnCl$_2$, 0.25 μmol; UDP-[^{3}H or ^{14}C]GlcNAc, 0.025 (3.0–4.0 × 10^6 dpm/μmol); and enzyme, 100–200 μg of protein. After 2 hr of incubation at 37°, 5 μl of EDTA (0.25 M), 10 μl of KCl (1.0 M), and 200 μl of chloroform–methanol are added to the incubation mixtures. The lower layer, containing the radioactive product, is washed with Folch's upper phase (chloroform–methanol–0.1 M KCl, 3 : 47 : 48, v/v/v).

Assay Methods

The lower layer is dried under N_2, resuspended in 50 μl of chloroform–methanol (2 : 1, v/v), and either spotted directly on the glass fiber disks (GF/A) and counted in a toluene-based system or subjected to descending chromatography on Whatman SG-81 paper (SGDC) as described for the GalT-3 assay.

GDP-Fuc : GSL Fucosyltransferases (FucTs)

Three different fucosyltransferases have been reported to catalyze the synthesis of glycosphingolipids of the lacto (Lc) family. The reaction conditions are described together.

Spleen and Serum GDP-Fuc : nLcOse4Cer(α1-2)fucosyltransferase (FucT-2)

The biosynthesis *in vitro* of blood-group H-active GSLs has been reported with a solubilized FucT-2 [GDP-Fuc : nLcOse4Cer(α1-2)fucosyltransferase] from bovine spleen,[38] IMR-32 neuroblastoma cells,[41] and human serum.[39] The reaction is as follows:

$$\text{GDP-[}^{14}\text{C]Fuc} + \text{nLcOse4Cer} \xrightarrow{\text{FucT-2}} \text{Fuc}\alpha\text{1-2Gal}\beta\text{1-4GlcNAc}\beta\text{1-3Gal}\beta\text{1-4GlcCer} + \text{GDP}$$
$$\text{(H-active GSL)}$$

The solubilized FucT-2 also catalyzes the synthesis of B-active GSLs from rabbit pentaglycosylceramide (nLcOse5Cer, Galα1-3Galβ1-4GlcNAcβ1-3Galβ1-4GlcCer). A cationic detergent, G-3634-A, is required for optimal activity, and the enzyme does not require an exogenous metal ion for activation. FucT-2 activity has been separated completely from the FucT-3 activity described in the next section. A terminal Fucα1-2 linkage has also been established in the enzymatic product, using a permethylation technique.[42] The enzyme appears to have a specific requirement for a lactosamine (Galβ1-4GlcNAc-) terminal oligosaccharide moiety and does not catalyze the transfer of fucose to a Galβ1-3GalNAc- or lactose moiety.[40] The competition studies suggest that the same enzyme catalyzes the transfer of fucose to a Galβ1-4GlcNAc-containing glycoprotein (e.g., desialized α_1-acid glycoprotein).[97a] The apparent K_m values for purified FucT-2 with nLcOse4Cer and GDP-Fuc are 35 and 40 μM, respectively. A synthetic substrate (p-NO$_2$-phenyl-β-D-Gal, K_m = 550 μM) can also be used as substrate during purification steps.

[97a] T. De, M. Basu, S. Basu, and T. Brown, *Fed. Proc., Fed. Am. Soc. Exp. Biol.* **44,** 1088 (1985).

Spleen and Human Serum
  GDP-Fuc : LcOse3Cer(α1-3)fucosyltransferase (FucT-3)

The Le^x core (Fucα1-3GlcNAcβ1-3Galβ1-4GlcCer) has been bio-synthesized from LcOse3Cer and GDP-Fuc in the presence of IMR-32 membrane bound[41] bovine spleen purified FucT-3.[42,97a] The bovine spleen and human serum[39] FucT-3 enzymes also appear to catalyze the synthesis of Le^x from nLcOse4Cer. A FucT-3 activity that catalyzes the transfer of fucose to polylactosamine-containing GSLs has been solubilized from human lung cancer cells (NCI-H69).[43] However, the acceptor activity of FucT-3 from human lung cancer cells with LcOse3Cer has not been tested. The reactions are as follows:

GDP-[^{14}C]Fuc + LcOse3Cer $\xrightarrow{\text{FucT-3}}$ [^{14}C]Fucα1-3GlcNAcβ1-3Galβ1-4GlcCer + GDP
                                                          (Le^x core)

GDP-[^{14}C]Fuc + nLcOse4Cer $\xrightarrow{\text{FucT-3}}$
                          Galβ1-4([^{14}C]Fucα1-3)GlcNAcβ1-3Galβ1-4GlcCer + GDP
                                                          (Le^x)

The FucT-3 activity (pH optimum 7.8) has been completely separated from FucT-2 (pH optimum 7.0) activity with bovine spleen as an enzyme source. When pretreated at 55° for 0–120 sec, FucT-2 lost 50% of its activity at 2 sec and FucT-3 exhibited the same result at 30 sec. The K_m values of the GSL substrate LcOse3Cer and the synthetic substrate PNP-β-GlcNAc were 100 and 400 μM, respectively.

GDP-Fuc : G$_{MI}$(α1-2)fucosyltransferase (FucT-4)

Rat bone marrow[44] and chemically transformed rat cells in culture contain a GSL : fucosyltransferase (pH optimum 7.5) that preferentially transfers fucose from GDP-Fuc to GSLs containing a terminal Galβ1-3GalNAc moiety rather than the Galβ1-4GlcNAc portion of nLcOse4Cer. The enzyme catalyzes the transfer of fucose to GgOse4Cer and ganglioside G$_{MI}$ at almost equal rates.[45] It is activated in the presence of Ca^{2+} or Mg^{2+} and maximally stimulated by Mn^{2+} (20 mM). The K_m values of G$_{MI}$ and GDP-[^{14}C]Fuc are 63 and 45 μM, respectively. The exact linkage of the terminal fucose has not been established, but the expected reaction is as follows:

GDP-[^{14}C]Fuc + G$_{MI}$ $\xrightarrow[\text{Mn}^{2+}]{\text{FucT-4}}$ Fucα-G$_{MI}$

Reagents

Glycosphingolipids (FucT-2, nLcOse4Cer; FucT-3, LcOse3Cer), 2 mM in chloroform–methanol (2 : 1)

G-3634A, 5 mg/ml in chloroform–methanol (2 : 1)
Cacodylate–HCl buffer (for FucT-2), pH 7.0, 1.0 M or HEPES,
    1.0 M (FucT-3, pH 7.8; FucT-4, pH 7.2)
MnCl$_2$, 0.05 M
GDP-[^{14}C]Fuc, 1 mM (6–8 × 10^6 dpm/μmol)
Chloroform–methanol (2 : 1)
Whatman 3MM paper
Na$_2$B$_4$O$_7$, 1.0%, pH 9.1

Incubation Mixture

Complete incubation mixtures contain the following components in final volumes of 0.05 ml in glass microculture tubes (6 × 50 mm) [after the glycosphingolipid (0.01 μmol) and detergent (10–20 μg) in chloroform–methanol have been mixed and dried in a rotary vacuum desiccator]: buffer, 10 μmol; MnCl$_2$, 0.125 μmol; GDP-[^{14}C]Fuc, 0.005 μmol (6–8 × 10^6 dpm/μmol); and enzyme, 50–100 μg of protein. The mixtures are incubated at 37° for 1 hr. Under these conditions the reaction rates for both FucT-2 and FucT-3 are constant up to 2 hr.

Assay Methods

The reactions are stopped by the addition of 20 μl of chloroform–methanol (2 : 1), and the mixtures are quantitatively transferred onto Whatman 3MM paper. Unreacted GDP-[^{14}C]Fuc and degraded [^{14}C]Fuc and [^{14}C]Fuc-1-PO$_3^{2-}$ are separated either by high-voltage electrophoresis in 1.0% Na$_2$B$_4$O$_7$ (HVPE) or by descending chromatography with 1.0% Na$_2$B$_4$O$_7$ (WPCD) as described for the GalT-3 assay. The enzymatic products, radiocarbon-labeled H, B, Le^x core, or Le^x GSLs, all stay at the origin. The appropriate areas are cut, and the radioactivity is quantitated in a toluene system. A specific monoclonal antibody against H or Le^x also can be used for the immunoprecipitation assay (MABA) as described for GalT-6. A tightly binding monoclonal antibody is essential for this assay.

CMP-NeuAc : GSL Sialyltransferases (SATs)

Biosynthetic methods for 10 different gangliosides (sialo-GSLs) have been reported using membrane-bound sialyltransferases from various sources.[8] Kinetic studies of substrate specificity and substrate inhibition indicate that more than four GSL : sialyltransferases may catalyze the synthesis of all these gangliosides.[8,14,22,46–58] Recently, 4 GSL : sialyltransferases (SAT-1, SAT-2, SAT-3, and SAT-4) have been solubilized and characterized in embryonic chicken brain.[7,54,85] Most have pH optima between 6.0 and 6.5. A divalent cation is not essential for these reactions. Detergent requirements may vary with the stage of purification and the solubility of the substrate. Each reaction is described separately first; then collective assay conditions follow.

CMP-NeuAc : LacCer(α2-3)sialyltransferase (SAT-1)

Membrane-bound 9- to 11-day-old embryonic chicken brain SAT-1 [CMP-NeuAc : LacCer(α2-3)sialyltransferase][14,22,47,48] has been shown to catalyze the transfer of sialic acid from CMP-N-acetylneuraminic acid to lactosylceramide to form ganglioside G_{M3}. The pH optimum is 6.0 in cacodylate buffer, but the enzyme usually is assayed at pH 6.2–6.5 to conserve the stability of the expensive donor CMP-[14C]NeuAc. SAT-1 activity has also been characterized in 15-day-old rat brain[49] and calf brain cortex.[50] It is increased in butyrate-induced HeLa cells[13] and dibutyryl cyclic AMP-treated mouse neuroblastoma cells.[25,87] The reaction is as follows:

$$\text{CMP-[}^{14}\text{C]NeuAc + LacCer} \xrightarrow{\text{SAT-1}} \text{NeuAc}\alpha\text{2-3Gal}\beta\text{1-4GlcCer + CMP}$$
$$(G_{M3})$$

The terminal NeuAcα2-3 linkage has been established by permethylation studies of the enzymatic product of SAT-1 from embryonic chicken brain (ECB).[53] SAT-1 has been solubilized from ECB[54,85] and liver,[85,98] using nonionic detergents (Triton CF-54 or Nonidet P-40), and the K_m value for lactosylceramide with the solubilized SAT-1 from 9-day-old ECB is 315 μM. The same SAT-1 may also catalyze the synthesis of sialosylgalactosylceramide (G_{M4}) in mouse brain microsomes.[46] Pretreatment of the mouse brain SAT-1 at 55° for 50 sec reduced the reaction rate by 50% with either GalCer or LacCer as a substrate. SAT-1 may recognize the terminal Galβ unit of any glycosphingolipid or glycoconjugate.

CMP-NeuAc : L_{MI}(α2-8)sialyltransferase (SAT-2)

A sialyltransferase involved in the synthesis of L_{D1c} (NeuAcα2-8NeuAcα2-3nLcOse4Cer) has been characterized from 9- to 14-day-old embryonic chicken brains.[54] The SAT-2 [CMP-NeuAc : L_{MI}(α2-8)sialyltransferase] activity sediments at the junction between 0.75 and 1.2 M on a discontinuous sucrose density gradient when still membrane bound. Nonidet P-40 (0.4%) solubilizes the SAT-2 activity. Substrate inhibition studies suggest that the same SAT-2 catalyzes the conversion of G_{M3} to G_{D3}.[52] The K_m values are 70 and 63 μM with CMP-[14C]NeuAc and L_{MI}

---

[98] H. C. Chon, M. S. Thesis, University of Notre Dame, Notre Dame, Indiana (1984).

(NeuGcα2-3Galβ1-4GlcNAcβ1-3Galβ1-4GlcCer), respectively, for the solubilized SAT-2. The reaction is as follows:

$$\text{CMP-[}^{14}\text{C]NeuAc} + L_{MI} \xrightarrow{\text{SAT-2}} \text{NeuAcα2-3NeuGcα2-3nLcOse4Cer} + \text{CMP}$$
$$(L_{Dic})$$

N-Ethylmaleimide at 0.5 mM inhibits SAT-2 activity 50% but does not affect SAT-1 (described above) or SAT-3.[54]

CMP-NeuAc : nLcOse4Cer(α2-3)sialyltransferase (SAT-3)

A sialyltransferase[7,8,55] activity present in 7- to 12-day-old embryonic chicken brain (ECB) catalyzes the transfer of sialic acid from CMP-NeuAc to the terminal galactose residue of [³H]nLcOse4Cer([6-³H]Galβ1-4GlcNAcβ1-3Galβ1-4GlcCer) to form [³H]L_{MI} (NeuAcα2-3[6-³H]Galβ1-4GlcNAcβ1-3Galβ1-4GlcCer). The SAT-3 [CMP-NeuAc : nLcOse4Cer-(α2-3)sialyltransferase] activity sediments (90%) at the junction of 1.2 and 1.5 M on a discontinuous sucrose density gradient when still membrane bound. SAT-3 has been solubilized[85,98] from 13-day-old embryonic chicken liver (20%) and brain (60%), using 0.4–0.8% nonionic detergents, Triton CF-54 and Nonidet P-40, respectively. The reaction is as follows:

$$\text{CMP-[}^{14}\text{C]NeuAc} + \text{nLcOse4Cer} \xrightarrow{\text{SAT-3}}$$
$$\text{[}^{14}\text{C]NeuAcα2-3Galβ1-4GlcNAcβ1-3Galβ1-4GlcCer} + \text{CMP}$$
$$(L_{MI})$$

Substrate competition studies using membrane-bound[55] and solubilized ECB SAT-3[89,98] suggest that the same enzyme also catalyzes the synthesis of G_{MIb} (NeuAcα2-3Galβ1-3GalNAcβ1-4Galβ1-4GlcCer)[48,53] from GgOse4Cer. Whether ECB SAT-3 also catalyzes the synthesis of NeuAcα2-3LcOse4Cer (NeuAcα2-3Galβ1-3GlcNAcβ1-3Galβ1-4GlcCer) has not been established. A crude microsomal system from SW1116 cancer cells has yielded biosynthetic sialyl-Le[a] antigen.[58a] The product, NeuAcα2-3Galβ1-3GlcNAc-R, has not been characterized by chemical methods. However, specific monoclonal antibodies have been used to establish the linkage.

It is predicted that a sialyltransferase from SW1116 cancer cells to catalyze the synthesis of NeuAcα2-3LcOse4Cer, which is a substrate for sialyl-Le[a] antigen. Substrate specificity studies will be essential for further identification of the sialyltransferase (SAT-3′) predicted to catalyze the synthesis of NeuAcα2-3Galβ1-3GlcNAc-R[58a,99] in SW1116 cancer cells.

---

[99] J. Weinstein, U. de Souza-e-Silva, and J. C. Paulson, *J. Biol. Chem.* **257**, 13835 (1982).

SAT-3' activity has been purified from rat liver.[99] Although it has been demonstrated with specific oligosaccharides and disaccharides, it appears not to use LcOse4Cer or nLcOse4Cer as substrate.

The K_m values for nLcOse4Cer and GgOse4Cer with solubilized ECB SAT-3 are 500 and 27 μM, respectively. At pH 6.5 the K_m value for CMP-NeuAc is 100 μM for this sialyltransferase.[85] Data presented in the next section suggest that ECB SAT-3 and SAT-4 are closely associated and that GgOse4Cer acceptor (Galβ1-3GalNAcβ1-4Galβ1-4GlcCer) activity is due to a separate protein.

CMP-NeuAc : $G_{MI}(\alpha 2$-3)sialyltransferase (SAT-4)

The biosynthesis of G_{D1a} ganglioside from G_{M1} was first reported with membrane-bound sialyltransferases from embryonic chicken brain[8,14,23,48] and rat brain.[56] Heat inactivation[8,14,48] at 55° and substrate competition studies indicate that the SAT-4 activity [CMP-NeuAc : $G_{M1}(\alpha 2$-3)sialyl-transferase] is different from the other three SATs described previously and that perhaps it is similar to the sialyltransferase that catalyzes the synthesis of sialyllactose from lactose.[100] The SAT-4 activity appears to be ubiquitous in animal tissues. SAT-4 activity has been purified (20,000-fold) from human placenta.[101] SDS–PAGE under reducing conditions reveals that the enzyme consists of a major polypeptide of M_r 41,000. Acceptor specificity studies indicate that the enzyme catalyzes the incorporation of sialic acid into Galβ1-3GalNAc-containing glycolipid, glycoprotein, or oligosaccharide. The reaction is as follows:

$$\text{CMP-[}^{14}\text{C]NeuAc} + G_{M1} \xrightarrow{\text{SAT-4}}$$
$$[^{14}\text{C]NeuAc}\alpha 2\text{-3Gal}\beta 1\text{-3GalNAc}\beta 1\text{-4(NeuAc}\alpha 2\text{-3)Gal}\beta 1\text{-4GlcCer} + \text{CMP}$$
$$(G_{D1a})$$

Using purified SAT-4 from human placenta, the lowest apparent K_m (0.3–0.7 mM) has been observed with Galβ1-3GalNAc-containing oligosaccharides. However, the V_{max} value of the reaction using lactose is 25% of that using Galβ1-3GalNAc-threonine.[101]

Reagents

Glycophingolipids (SAT-1, LacCer; SAT-2, G_{M3} or L_{M1}; SAT-3, nLcOse4Cer; SAT-4, G_{M1}), 5 mM in chloroform–methanol (2 : 1)
Triton CF-54 : Tween-80 (2 : 1), 10 mg/ml in chloroform–methanol
Cacodylate–HCl buffer, pH 6.5, 1.0 M

---

[100] D. M. Carlson, G. W. Jourdian, and S. Roseman, *J. Biol. Chem.* **248,** 5742 (1973).
[101] D. H. Joziasse, M. L. E. Bergh, H. G. J. ter Hart, P. L. Koppen, G. J. M. Hoghwinkel, and D. H. Van den Eijnden, *J. Biol. Chem.* **260,** 4941 (1985).

MgCl$_2$, 0.1 M
CMP-[^{14}C]NeuAc, 5 mM (3.0–4.0 × 10^6 dpm/μmol)

Incubation Mixture

Complete incubation mixtures contain the following components in final volumes of 0.05 ml in glass microculture tubes (6 × 50 mm) [after the glycosphingolipid (LacCer, 0.02 μmol; G$_{M3}$, L$_{M1}$, nLcOse4Cer, or G$_{M1}$, 0.01 μmol) and detergent (100–200 μg) in chloroform–methanol have been mixed and dried in a rotary vacuum desiccator]: cacodylate–HCl buffer, pH 6.5, 10 μmol; MgCl$_2$, 0.5 μmol; CMP-[^{14}C]NeuAc, 0.02 (3.0–4.0 × 10^6 dpm/μmol); and enzyme, 50–300 μg of protein. The mixtures are incubated at 37° for 2 hr. When the detergent-to-protein ratio is 0.5, the rate remains constant until at least 2 hr for all 3 sialyltransferases (SAT-1, SAT-2, and SAT-3). The mixtures are incubated at 37° for 2 hr. The reactions are stopped with 20 μl of methanol, and the mixtures are stored at −18° or assayed immediately using any of the following methods.

Assay Methods

SGDC, WPDC, or HVPE Assay. When the donor is CMP-[^{14}C]NeuAc, the reaction mixtures are spotted either on Whatman SG-81 or on Whatman 3MM paper and then subjected to descending 1.0% Na$_2$B$_4$O$_7$ chromatography (SGDC or WPDC) as described for GalT-3. The mixtures may also be assayed by quicker high-voltage electrophoresis (HVPE) as described for GalT-1.

Sephadex G-25 Column Chromatography (G25CC).[13] The reactions are stopped by adding 10 μl of 0.1 M EDTA and 0.2 ml chloroform–methanol (2:1, v/v). After standing for 1 hr at room temperature, the samples are applied quantitatively to Sephadex G-25 superfine small columns (bed volume, 1.5 ml) which are freshly developed in chloroform–methanol–water (24:12:1.6, v/v/v). The ^{14}C-labeled glycolipids are eluted with 3 × 1 ml of chloroform–methanol (2:1, v/v). Each fraction is dried under N$_2$, dissolved in 100 μl of chloroform–methanol (2:1) and spotted on Whatman 3MM paper. The radioactivity is quantitatively determined in a toluene scintillation system.

Certain ion-exchange Sephadex or Sepharose columns can be used instead of G-25, but G-25 is more versatile because it is based on the separation of ^{14}C-labeled GSL micelles from radiolabeled sugar nucleotides. However, any ion-exchange Sepharose or Sephadex column[101a]

[101a] K. M. Walton and L. Schnaan, *Anal. Biochem.* **152**, 154 (1986).

method is limited to binding of charged glycosphingolipids (gangliosides or sulfatides); it cannot be used for large neutral glycolipids (the lacto and globo families).

Sep-Pak Column Chromatography (SPKCC).[93,102] The reactions are terminated with 400 μl of 0.1 M KCl containing 0.05 M EDTA and 24 μg of egg lecithin. The mixtures are applied to a C_{18} Sep-Pak cartridge. The cartridge is washed with 5 ml of 0.1 M KCl and 25 ml of distilled water. The ^{14}C-labeled GSL fraction is eluted with 5 ml of methanol followed by 10 ml of chloroform–methanol (1 : 1). The solvents are evaporated, and the radioactive GSLs are either directly quantitated by spotting on Whatman 3MM paper and counting in a toluene system or subjected to any one of the SGDC or HVPE methods described earlier. The recovery of ^{14}C-labeled ganglioside is over 90% in the presence of lecithin from Sep-Pak columns.

Thin-Layer Chromatography after Short-Column Chromatography (SCTLC). The samples (^{14}C-labeled GSLs) from Sephadex G-25 or Sep-Pak columns can also be spotted on precoated TLC (silica gel G) plates and developed with chloroform–methanol–0.1% $CaCl_2$ (55 : 40 : 10) against standard gangliosides. After drying, the appropriate areas are scraped and counted directly in a toluene or dioxane-containing scintillation system for maximum efficiency. Direct spotting of the incubation mixture on TLC is not recommended because of the streaking effect of both radioactive donors and the radioactive products.

Use of ^{125}I-Labeled Glycolipid and Whatman SG-81 Paper Ascending Chromatography for SAT-1 Assay (IGLAC). The ceramide moiety of lactosylceramide (Galβ1-4GlcCer) is iodinated with Na^{125}I in the presence of chloramine-T[51,98] and purified by Sephadex G-25, BioSil-A, and Sep-Pak C_{18} column chromatography. Using [^{125}I]LacCer, nonradioactive CMP-NeuAc, and the detergent-solubilized SAT-1, the mixtures are incubated under the conditions described previously for SAT-1. After 2 hr at 37° the mixtures are spotted on 9-in.-long strips of Whatman SG-81 paper (9 in. high) and subjected to ascending chromatography with chloroform–methanol–water (60 : 25 : 4). Under these conditions, [^{125}I]LacCer moves almost with the solvent front, leaving the enzymatic product, iodinated G_{M3}, near the origin. After drying, the appropriate areas are counted directly on a gamma counter. This assay method is simple and cheaper, but the separation of ^{125}I-labeled substrate from the radiolabeled product is its major limitation.

Lactosylceramide containing 100% saturated fatty acids (C_{16}, palmitic) has also been chemically prepared and iodinated according to the proce-

---

[102] M. A. Williams and R. H. McCluer, *J. Neurochem.* **35,** 266 (1980).

dure described above.[51,98] It appears that many glycosphingolipids with saturated fatty acids can be iodinated according to this procedure. The final assay method must be selected according to the nature of the glycolipid substrates and enzymatic products.

Radioactive Glycolipid and Monoclonal Antibody Binding Assay (RAGAB)[85]

The ^{14}C-labeled GSL enzymatic product obtained from FucT-3 [GDP-Fuc : nLcOse4Cer(α1-3)fucosyltransferase] has been identified as Le[x] antigen based on its reactivity against SSEA-1 monoclonal antibody. Radiolabeled GSL binding to the specific monoclonal antibody (RAGAB) has been assayed by the use of a microassay method developed recently.[85] Approximately 20–200 pmol (200–2000 cpm) of purified ^{14}C-labeled GSL products were dissolved in methanol and transferred to polystyrene microtiter wells. After drying under N_2 in each well, 0.1 ml of 1.0% BSA (in PBS and 0.15 M NaCl, pH 7.2) was added and kept at 4° for 30 min. The BSA was washed 6 times using 0.1 ml PBS (pH 7.2) each time. Then 50 μl of SSEA-1 monoclonal antibody (diluted 1 : 500 with PBS–NaCl and containing 5 mM benzamidine) was added in each well. After 12 hr the SSEA-1 was removed, the wells were washed 10 times with PBS, and 0.025 ml of ^{125}I-labeled protein A was added in each well and allowed to stand for 8 hr. The excess ^{125}I-labeled protein A was removed, the wells were punched out directly into plastic minivials, and the radioactivity was quantitatively determined using a Beckman-4000 gamma counter. The results showed that SSEA-1 monoclonal antibody bound specifically to the ^{14}C-labeled Le[x] product and not to the radiolabeled H or Le[x] core. This RAGAB (radioactive glycolipid–antibody binding) microassay appears to be applicable to all GSL enzymatic products irrespective of ^{3}H or ^{14}C labeling, provided a specific monoclonal antibody is available.

[52] Acetyl-CoA : α-glucosaminide *N*-Acetyltransferase from Rat Liver

By Karen J. Bame and Leonard H. Rome

Acetyl-CoA : α-glucosaminide *N*-acetyltransferase (*N*-acetyltransferase) is a lysosomal enzyme deficient in the hereditary mucopolysaccharide storage disease known as Sanfilippo C syndrome (mucopolysaccharidosis III C).[1] Patients with this disorder accumulate heparan sulfate in lysosomes of numerous tissues and excrete heparan sulfate fragments in the urine.[2] The enzyme catalyzes the acetylation of terminal glucosamine residues of heparan sulfate, using acetyl-CoA as the acetyl donor. Once acetylated, the glucosamine can be hydrolyzed off the polysaccharide by the next enzyme in the degradation pathway, α-N-acetylglucosaminidase.

Assay Method

Enzyme activity is measured, using [^{14}C]glucosamine as the acetyl acceptor, by a modification of the procedure of Hopwood.[3] The charged substrate, [^{14}C]glucosamine, is separated from the neutral product N-[^{14}C]acetylglucosamine by passing the reaction mixture over a 1-ml Dowex column.

Reagents

[^{14}C]Glucosamine, purchased from ICN, specific activity 40–50 mCi/ mmol
Glucosamine, 50 mM
Acetyl-CoA, 50 mM, made up in 1 mM HCl, and stored frozen
Sodium citrate, 25 mM, sodium phosphate buffer, 50 mM, pH 6.0
HCl, 1 mM
Dowex, 50W H$^+$ form, 8% cross-linked, washed 3 times with 1 mM HCl, suspended in a 1 : 1 slurry with 1 mM HCl.

Procedure

Purified lysosomal membranes (15 μg) are mixed with 1 mM acetyl-CoA, 1 mM [^{14}C]glucosamine (10 mCi/mmol), 25 mM sodium citrate, and 50 mM sodium phosphate, pH 6.0, in a final volume of 20 μl. The mixture is incubated at 37° for 60 min, and the reaction is stopped by the addition of 200 μl 1 mM HCl.

Twenty microliters of the reaction mixture is used to measure total radioactivity by liquid scintillation counting. The remaining 200 μl is applied to a 1-ml Dowex column and eluted with 1.8 ml of 1 mM HCl. The total radioactivity eluted, divided by the total radioactivity applied to the

[1] U. Klein, H. Kresse, and K. von Figura, *Proc. Natl. Acad. Sci. U.S.A.* **75**, 5185 (1978).
[2] V. A. McKusick and E. F. Neufeld, *in* "The Metabolic Basis of Inherited Disease" (J. B. Stanbury, J. B. Wyngaarden, D. S. Fredrickson, J. L. Goldstein, and M. S. Brown, eds.), 5th ed., p. 751. McGraw-Hill, New York, 1983.
[3] J. J. Hopwood and H. Elliot, *Clin. Chim. Acta* **112**, 67 (1981).

column, represents the percentage of product formed. The amount in nmol of product formed is calculated by multiplying the percentage of product formed by the nmol of substrate in the reaction mixture.

The Dowex columns are made by pipetting 2.0 ml of the 1 : 1 Dowex : HCl slurry into 2-cm² chromatography columns (5 in. polypropylene; filter porosity 35 μm, Evergreen Scientific, Los Angeles, CA). These columns can be regenerated after each use by washing as follows: 1 volume 1 M NaOH, 1 volume H_2O, 1 volume 1 M HCl, 2 volumes H_2O, 2 volumes 1 mM HCl. The columns, plugged and stoppered, are stored at room temperature in 1 mM HCl.

Lysosomal Membrane Purification

The enzyme can be partially purified by isolating lysosomal membranes from rat liver, following the procedures of Ohsumi et al.[4] For best results the following protease inhibitors should be added to all the purification buffers: 5 μM leupeptin, 5 μM pepstatin, 5 μg/ml chymostatin, 5 μg/ml aprotinin, and 1 mM phenylmethylsulfonyl fluoride. Livers from Sprague–Dawley rats, weighing 200–300 g, are homogenized in 0.25 M sucrose, 0.2 M KCl (1 ml/g), using a loose-fitting Dounce homogenizer. The homogenate is diluted to 4 ml/g, filtered through cheesecloth, and centrifuged at 650 g for 10 min. The resulting pellet is rehomogenized (1 ml/g) with a tight-fitting Dounce homogenizer, and recentrifuged at 650 g for 15 min. To pellet the mitochondria and the lysosomes, the two supernatants are combined and centrifuged at 11,000 g for 20 min. This pellet (ML pellet) is resuspended in one-half the volume of the cytoplasmic extract and centrifuged again, at 11,000 g for 20 min. The extracted ML pellet is resuspended in 0.025 M sucrose (2 ml/g) and allowed to stand at 0° for 30 min; the hypotonic sucrose solution causes the lysosomes to rupture. The suspension is centrifuged at 11,000 g to pellet the mitochondria, while the lysosomal membranes remain in the supernatant. The mitochondrial pellet is reextracted with 0.025 M sucrose, 0.02 M KCl (1 ml/g liver), and recentrifuged at 11,000 g for 30 min. The supernatants from both spins are combined, and 1 M $CaCl_2$ is added to make a final concentration of 10 mM $CaCl_2$. This mixture is incubated at 0° for 30 min and then spun at 5000 g for 10 min. The 5000-g supernatant is centrifuged for 30 min at 50,000 g to pellet the lysosomal membranes, which are washed with 5 mM HEPES, pH 7.6, 0.1 mM EDTA, and resuspended in the HEPES buffer to give a protein concentration between 5 and 10 mg/ml. Usually, 30 mg of lysosomal membrane protein can be purified from 250 g of rat liver. Purifi-

[4] Y. Ohsumi, T. Ishikawa, and K. Kato, J. Biochem. (Tokyo) 93, 547 (1983).

cation of the membranes results in a 50 to 125-fold enrichment of the *N*-acetyltransferase specific activity.

The *N*-acetyltransferase is an integral lysosomal membrane protein as evidenced by two criteria: 88% of the enzyme activity remains associated with the membranes after extraction with 2 *M* KCl and 7 *M* urea, and enzyme activity is extracted into the detergent phase by Triton X-114.[5] The transferase can be solubilized from the membrane by detergents. The rat liver enzyme can be solubilized with 1% Lubrol PX or 1% Triton X-100. Enzyme from human placenta can be solubilized with 3% β-octylglucoside.[6]

Properties

Stability. The membrane-bound enzyme is extremely stable; the enzyme is still active after being stored at −70° for over 12 months. When assayed, the membrane-bound activity remains linear for over 8 hr, and at low enzyme concentrations remains active for over 20 hr. The solubilized enzyme, however, is not stable. Activity remains if the enzyme is stored frozen, but at 4° enzyme activity declines by 25% in 7 days. In the absence of substrates, the solubilized enzyme loses one-half of its activity after 5 min at 37°. In the presence of substrates, the solubilized activity is linear for only 4 hr.

Inhibitors and Activators. *N*-Acetyltransferase activity is inhibited by *p*-chloromecuribenzoate, but iodoacetamide and *N*-ethylmaleimide, two other thiol-modification reagents, have no effect. Another thiol-modification reagent, methyl methanethiolsulfonate, activates the membrane-bound enzyme 1.5-fold. The membrane-bound enzyme is insensitive to DTT or 2-mercaptoethanol; however, both reagents inactivate the solubilized enzyme.

Various sugars are slight inhibitors of the enzyme reaction. These include the monosaccharides α-methylglucose and glucosamine 6-phosphate and the disaccharides sucrose, cellobiose, and maltose. The polysaccharides chitosan (repeating β1 → 4-linked glucosamine units) and chitin (repeating β1 → 4-linked *N*-acetylglucosamine units) inhibit the enzyme reaction to a much greater extent.

The activity of *N*-acetyltransferase is affected by detergents. Incubation with low concentrations (0.1%) of taurodeoxycholate or Triton X-100 results in a 2- to 3-fold increase in enzyme activity. Incubation with 1% Zwittergent or *N*-laurylsarcosine completely destroys activity.

[5] K. J. Bame and L. H. Rome, *J. Biol. Chem.* **260**, 11293 (1985).
[6] R. Pohlmann, U. Klein, H. G. Fromme, and K. von Figura, *Hoppe-Seyler's Z. Physiol. Chem.* **362**, 1199 (1981).

Catalytic Properties. N-Acetyltransferase has a broad pH optimum, with maximal activity seen above pH 5.5. The enzyme is fairly stable at high temperatures; the membrane-bound form is most active at 45° and remains active at 55°. Studies with isolated intact lysosomes indicate that the acetyl-CoA used in the *in vivo* reaction is supplied from the cytosol.[7] Michaelis–Menton analysis gives a K_m for acetyl-CoA of 0.55 mM and a K_m for glucosamine of 0.29 mM. Double reciprocal plots and product inhibition patterns indicate that the transferase works by a Di-Iso Ping-Pong mechanism.[5] Both CoA and N-acetylglucosamine are noncompetitive inhibitors against either substrate. The K_i for CoA is 3–7 mM, and the K_i for N-acetylglucosamine is 15–17 mM.

Acknowledgments

The research was supported by U.S. Public Health Service Grant GM-31565. KJB is supported by a U.S. Public Health Service National Research Service Award (GM-07104). LHR is a recipient of an American Cancer Society Faculty Research Award.

[7] L. H. Rome, D. F. Hill, K. J. Bame, and L. R. Crain, *J. Biol. Chem.* **258**, 3006 (1983).

[53] Sialic Acids: Metabolism of O-Acetyl Groups

By ROLAND SCHAUER

Introduction

O-Acetylated sialic acids are often components of oligo- or polysaccharides, glycoproteins, and gangliosides from microorganisms, animals, and humans.[1,2] Acetic acid ester groups are most frequently found at O-9 of sialic acids (N-acetyl- or N-glycolylneuraminic acid), followed by O-7. They occur rarely at O-4. Some sialic acid species contain several O-acetyl groups, and they may be combined with O-methyl or O-lactyl groups. Unsaturated O-acetylsialic acids (2-eno derivatives) are also known to occur in biological secretions.[3] For further details about the nature and identification of different sialic acids, see elsewhere in this volume.[4] The 18 natural O-acetylated sialic acids known are listed, together with their distribution in nature, in Table I.

[1] A. P. Corfield and R. Schauer, *Cell Biol. Monogr.* **10**, 5 (1982).
[2] R. Schauer, *Adv. Carbohydr. Chem. Biochem.* **40**, 131 (1982).
[3] R. Schauer, C. Schröder, and A. K. Shukla, *Adv. Exp. Med. Biol.* **174**, 75 (1983).
[4] R. Schauer, this volume [10].

TABLE I
NATURAL O-ACETYLATED SIALIC ACIDS AND THEIR OCCURRENCE[a]

Name	Recommended abbreviation	Occurrence
N-Acetyl-4-O-acetylneuraminic acid	Neu4,5Ac$_2$	Horse, donkey, echidna
N-Acetyl-7-O-acetylneuraminic acid	Neu5,7Ac$_2$	Cow, human, etc.
N-Acetyl-8-O-acetylneuraminic acid	Neu5,8Ac$_2$	Cow
N-Acetyl-9-O-acetylneuraminic acid	Neu5,9Ac$_2$	Vertebrates, bacteria
N-Acetyl-4,9-di-O-acetylneuraminic acid	Neu4,5,9Ac$_3$	Horse
N-Acetyl-7,9-di-O-acetylneuraminic acid	Neu5,7,9Ac$_3$	Cow, human, etc.
N-Acetyl-8,9-di-O-acetylneuraminic acid	Neu5,8,9Ac$_3$	Cow
N-Acetyl-7,8,9-tri-O-acetylneuraminic acid	Neu5,7,8,9Ac$_4$	Cow
N-Acetyl-4-O-acetyl-9-O-lactylneuraminic acid	Neu4,5Ac$_2$9Lt	Horse
N-Acetyl-9-O-acetyl-2-deoxy-2,3-didehydroneuraminic acid	Neu2en5,9Ac$_2$	Cow, rat
N-Glycolyl-4-O-acetylneuraminic acid	Neu4Ac5Gc	Horse
N-Glycolyl-7-O-acetylneuraminic acid	Neu7Ac5Gc	Cow
N-Glycolyl-8-O-acetylneuraminic acid	Neu8Ac5Gc	Cow
N-Glycolyl-9-O-acetylneuraminic acid	Neu9Ac5Gc	Cow, horse, pig, rabbit
N-Glycolyl-7,9-di-O-acetylneuraminic acid	Neu7,9Ac$_2$5Gc	Cow
N-Glycolyl-8,9-di-O-acetylneuraminic acid	Neu8,9Ac$_2$5Gc	Cow
N-Glycolyl-7,8,9-tri-O-acetylneuraminic acid	Neu7,8,9Ac$_3$5Gc	Cow
N-Glycolyl-9-O-acetyl-8-O-methylneuraminic acid	Neu9Ac5Gc8Me	Starfish

[a] Data are taken from Refs. 1–3. For the basic sialic acid structure, see Fig. 1 of Chapter 10, this volume.[4]

Interest in O-acetylated sialic acids has increased in the last few years, as evidence is accumulating that these neuraminic acid derivatives contribute to the biological role of complex carbohydrates. Thus, sialic acid *O*-acetyl groups may influence the conformation and arrangement of glycoconjugates in cell membranes, as these ester groups have been shown to modify hydrogen bonding in sialic acids,[5,6] and they may reduce the hydrophilic properties of sialic acids. They also influence the antigenicity of bacterial polysaccharides (summarized in Ref. 2) and the capacity of human erythrocytes to activate the complement pathway[7]; they represent differentiation antigens[8] and seem to play a role in environmental adaptation.[9] *N*-Acetyl-9-*O*-acetylneuraminic acid has been identified as an essential component of a tumor-specific antigen (disialoganglioside G_{D3}) from human melanoma cells.[10]

Probably the most important general role of *O*-acetyl groups in sialic acids is to retard the degradation of glycoconjugates. The decomposition of glycan chains is initiated by the action of sialidases (EC 3.2.1.18, neuraminidase), and the activity of these enzymes is hindered by the presence of *O*-acetyl groups. Thus, *O*-acetyl groups in the sialic acid side chain retard sialidase action by more than 50% when compared with the hydrolysis rates of unsubstituted *N*-acetyl- or *N*-glycolylneuraminic acids.[2,11] An *O*-acetyl residue in the pyranose ring, at *O*-4, completely prevents the action of this enzyme on the sialic acid glycosidic linkage.[2,11,12] This behavior was observed with all viral, bacterial, and mammalian sialidases tested so far and seems to mirror similar conformations of the polypeptide chains at the sialic acid binding sites of sialidases. However, one exception has been described: sialidase purified from *Streptococcus sanguis* isolated from the human oral cavity is not significantly affected by *O*-acetyl groups present in the sialic acid glycerol side chain.[13] Further degradation of free sialic acids by bacterial and mammalian acylneuraminate pyruvate-lyase (EC 4.1.3.3, *N*-acetylneuraminate

[5] J. Haverkamp, H. van Halbeek, L. Dorland, J. F. G. Vliegenthart, R. Pfeil, and R. Schauer, *Eur. J. Biochem.* **122**, 305 (1982).

[6] J. F. G. Vliegenthart, L. Dorland, H. van Halbeek, and J. Haverkamp, *Cell Biol. Monogr.* **10**, 127 (1982).

[7] A. Varki and S. Kornfeld, *J. Exp. Med.* **152**, 532 (1980).

[8] R. Schauer, *Biochem. Soc. Trans.* **11**, 270 (1983).

[9] H. Rahmann, R. Hilbig, W. Probst, and M. Mühleisen, *Adv. Exp. Med. Biol.* **174**, 395 (1983).

[10] D. A. Cheresh, A. P. Varki, N. M. Varki, W. B. Stallcup, J. Levine, and R. A. Reisfeld, *J. Biol. Chem.* **259**, 7453 (1984).

[11] A. P. Corfield and R. Schauer, *Cell Biol. Monogr.* **10**, 195 (1982).

[12] R. Schauer and H. Faillard, *Hoppe-Seyler's Z. Physiol. Chem.* **349**, 961 (1968).

[13] A. Varki and S. Diaz, *J. Biol. Chem.* **258**, 12465 (1983).

lyase) is also hampered by *O*-acetyl groups present in both the side chain or the pyranose ring.[2,14]

Thus, in most cases, sialic acid ester groups extend the lifetime and concomitantly the masking function of sialic acids.[15] Correspondingly, the biological role of soluble or cell membrane-bound, sialylated glycoconjugates is strengthened. This mechanism may provide an explanation why serum glycoproteins and cell surface glycoconjugates, e.g. in erythrocytes,[4,16] are often highly O-acetylated. After partial loss of sialic acids, serum glycoproteins and blood cells, for example, are rapidly removed from the circulation and destroyed by hepatocytes or macrophages with the aid of galactose-specific lectins.[2,15] Sialidase resistance of the O-acetylated sialic acids of rabbit colonic epithelial glycoproteins is considered to be part of the mechanism of protection of colon mucosa against the action of fecal stream.[17]

Research on O-acetylated sialic acids has received a further impulse from virology. Hemagglutination of, e.g., chicken erythrocytes by influenza C virus can only be inhibited by glycoconjugates containing O-acetylated sialic acids. The best inhibitors are those O-acetylated at O-9.[18] During incubation with this virus, the inhibitory potency is lost due to the presence of a sialate-*O*-acetylesterase (see below) on the virus. It represents a new receptor-destroying enzyme, which is distinct from the receptor-destroying enzymes of other influenza viruses with sialidase activity.[2,11] From these experiments it was assumed that the receptor for influenza C virus is *N*-acetyl-9-*O*-acetylneuraminic acid. This was confirmed by specific binding of this virus to human erythrocytes which were partially resialylated with *N*-acetyl-9-*O*-acetylneuraminic acid after sialidase treatment.[18a] On the other hand there exist influenza virus strains which cannot bind to O-acetylated sialic acids but only to unsubstituted *N*-acetylneuraminic acid.[19]

Due to these observations, knowledge of the metabolism of O-acetylated sialic acids is important. The following three systems (*O*-acetyltransferases, esterase, and nonenzymatic *O*-acetyl migration) involved

[14] R. Schauer, M. Wember, F. Wirtz-Peitz, and C. Ferreira do Amaral, *Hoppe-Seyler's Z. Physiol. Chem.* **352,** 1073 (1971).

[15] R. Schauer, *Trends Biochem. Sci.* **10,** 357 (1985).

[16] A. K. Shukla and R. Schauer, *Hoppe-Seyler's Z. Physiol. Chem.* **363,** 255 (1982).

[17] A. A. Al-Suhail, P. E. Reid, C. F. A. Culling, W. L. Dunn, and M. G. Clay, *Histochem. J.* **16,** 543 (1984).

[18] G. Herrler, R. Rott, H.-D. Klenk, H.-P. Müller, A. K. Shukla, and R. Schauer, *Embo. J.* **4,** 1503 (1985).

[18a] G. N. Rogers, G. Herrler, J. C. Paulson, and H.-D. Klenk, *J. Biol. Chem.* **261,** 5947 (1986).

[19] J. C. Paulson, G. N. Rogers, S. M. Carroll, H. H. Higa, T. Pritchett, G. Milks, and S. Sabesan, *Pure Appl. Chem.* **56,** 797 (1984).

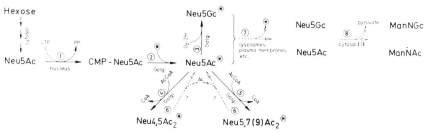

FIG. 1. Metabolism of *N,O*-acylneuraminic acids in mammals (⟶, anabolic ---→, catabolic reactions). Scheme modified and simplified from Corfield and Schauer.[11,15] Abbreviations: Neu5Ac, *N*-acetylneuraminic acid; Neu5Gc, *N*-glycolylneuraminic acid; Neu4,5Ac$_2$, *N*-acetyl-4-*O*-acetylneuraminic acid; Neu5,7(9)Ac$_2$, *N*-acetyl-7(or 9)-*O*-acetylneuraminic acid; ManNAc, *N*-acetylmannosamine; ManNGc, *N*-glycolylmannosamine. Enzymes: 1, CMP-acylneuraminate synthase (EC 2.7.7.43, acylneuraminate cytidylyltransferase); 2, sialyltransferase (EC 2.4.99.1, CMP-*N*-acetylneuraminate–β-galactoside α-2,6-sialyltransferase, and other sialyltransferases); 3, *N*-acetylneuraminate monooxygenase (EC 1.14.99.18); 4, *N*-acetylneuraminate 4-*O*-acetyltransferase (EC 2.3.1.44); 5, *N*-acetylneuraminate 7(9)-*O*-acetyltransferase (EC 2.3.1.45); 6, *N*-acyl-*O*-acetylneuraminate *O*-acetylhydrolase (EC 3.1.1.53, sialate *O*-acetylesterase); 7, sialidase (EC 3.2.1.18, neuraminidase); 8, *N*-acylneuraminate pyruvate-lyase (EC 4.1.3.3, *N*-acetylneuraminate lyase). Glycoprotein acceptor of sialic acids. Neu5Ac modification mainly occurs at the macromolecular level. Enzymes 4–6 also react with sialic acids containing *N*-glycolyl groups (not shown in the scheme). Enzyme 5 probably links *O*-acetyl groups to C-7, followed by migration of this residue to C-9. Sialic acids released by sialidases or resynthesized from *N*-acylmannosamines[2,11] may recycle after activation with CTP.

in the metabolism of these sialic acids will be described. A simplified scheme of the metabolism of sialic acids is given in Fig. 1. The metabolism of sialic acids has been discussed in detail.[2,11]

Migration of *O*-Acetyl Groups in Sialic Acids

In the enzyme reactions described below nonenzymatic isomerization reactions of O-acetylated sialic acids are involved. These are spontaneous migrations of *O*-acetyl groups within the sialic acid side chain, which have the consequence that specific enzyme reactions are not required for the attachment of *O*-acetyl groups to all hydroxyl groups of the side chain where such residues are found, or for the removal of *O*-acetyl groups from all these positions.

It was observed in many isolation procedures of sialic acids from bovine submandibular glands carried out in our laboratory[4,20] that the relative yields of 7- and 9-O-acetylated sialic acids, respectively, were rather variable. We originally thought that this was due to genetic differ-

[20] R. Schauer and A. P. Corfield, *Cell Biol. Monog.* **10**, 51 (1982).

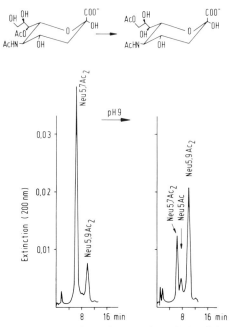

FIG. 2. Intramolecular migration of acetyl groups from O-7 to O-9 under slightly alkaline conditions, accompanied by the formation of a small amount of N-acetylneuraminic acid (Neu5Ac). Neu5,7Ac2, N-acetyl-7-O-acetylneuraminic acid; Neu5,9Ac2, N-acetyl-9-O-acetylneuraminic acid. Sialic acids were analyzed by HPLC under conditions described elsewhere in this volume.[4]

ences between the animals from which the glands were taken. However, it turned out that some purified fractions of N-acetyl-7-O-acetyl-neuraminic acid and N-acetyl-7,9-di-O-acetylneuraminic acid were converted to N-acetyl-9-O-acetylneuraminic acid and N-acetyl-8,9-di-O-ace-tylneuraminic acid during storage. Detailed investigation of this phenomenon revealed that the 7-O-acetyl group migrates to O-9 leading to quantitative conversion of N-acetyl-7-O-acetylneuraminic acid to N-ace-tyl-9-O-acetylneuraminic acid.[21] Isomerization in the reverse direction was not observed. This reaction was followed by different analytical techniques including high-performance liquid chromatography (Fig. 2) and 360-MHz proton-NMR spectroscopy. These two methods allow kinetic measurements of the O-acetyl migration. The proton-chemical shifts and the relative intensities of the different N- and O-acetyl signals are excel-

[21] J. P. Kamerling, H. van Halbeek, J. F. G. Vliegenthart, R. Pfeil, A. K. Shukla, and R. Schauer, Proc. Int. Symp. Glycoconjugates, 7th, 1983 p. 160 (1983).

lent structural parameters to monitor the migration from O-7 to O-9[21,21a] It could not clearly be established whether the O-acetyl group migrates from O-7 directly to O-9 or via O-8. Migration to O-8 is, however, possible as was shown by the isomerization of N-acetyl-7,9-di-O-acetylneuraminic acid to N-acetyl-7,8-di-O-acetylneuraminic acid. These reactions occur both in N-acetylated and N-glycolylated sialic acids. The 4-O-acetyl group seems to be stable, as no conversion of this sialic acid to other O-acetylated species was observed.

The rate of migration is independent of the concentration of sialic acid, which demonstrates a reaction of zero order, i.e., an intramolecular reaction. It increases with temperature and especially with pH value. The slowest rate was observed between pH 5 and 6; however, under physiological conditions the migration rate is significant and may be of metabolic importance (see below). For example, at pH 7.0 (50 mM Tris–HCl buffer) and 37° 50% of N-acetyl-7-O-acetylneuraminic acid was transformed to its 9-O isomer after 600 min incubation ($t_{1/2}$). At pH 7.5 the corresponding value was 270 min, 40 min at pH 8.0, 6 min at pH 8.5, and about 1 min at pH 9.0. At higher pH values, a significant loss of O-acetyl groups occurred in addition. At pH values below 3 both migration and de-O-acetylation are also fast. Similar observations with sialic acid mixtures from bovine submandibular gland glycoprotein using gas–liquid chromatography were reported by Varki and Diaz.[22]

The apparent irreversibility of the migration from O-7 to O-9 shows that N-acetyl-9-O-acetylneuraminic acid, in which the ester group is located at the primary alcohol group of neuraminic acid, is more stable than the 7-O-acetyl isomer. Correspondingly, 9-O-acetylated sialic acids can be stored even in aqueous solutions for several months at −18° without significant decomposition, while 7-O-acetylated sialic acids can be kept intact for the same time only in a carefully purified and freeze-dried state at low temperatures. The maximum time for which we could store N-acetyl-7-O-acetylneuraminic acid under these conditions so far was 12 years.

Enzymatic Incorporation of O-Acetyl Groups into Sialic Acids

Two O-acetyltransferases are known to be involved in the biosynthesis of O-acetylated sialic acids: acetyl-CoA:N-acylneuraminate 4-O-acetyltransferase (EC 2.3.1.44, N-acetylneuraminate O^4-acetyltransferase) and acetyl-CoA:N-acylneuraminate-9(7)-O-acetyltransferase (EC

[21a] J. P. Kamerling, R. Schauer, A. K. Shukla, S. Stoll, H. van Halbeek, and J. F. G. Vliegenthart, Eur. J. Biochem., in press.

[22] A. Varki and S. Diaz, Anal. Biochem. 137, 236 (1984).

2.3.1.45, *N*-acetylneuraminate O^7- (or O^9-) acetyltransferase). These enzymes were discovered in equine and bovine submandibular glands, respectively.[2,11,23-25] The assays of these enzymes, their subcellular localization, substrate specificities, and kinetic properties have been described in this series.[25] While the 4-*O*-acetyltransferase seems to be a rare enzyme and restricted to the few animals containing 4-O-acetylated sialic acids [so far horse, donkey, and the Australian monotreme echidna (*Tachyglossus aculeatus*) are known],[1,2] the 9(7)-*O*-acetyltransferase seems to be widespread, as 9-O-acetylated sialic acids have been found in bacteria and many animal species including humans.[1,2] So far, 9-*O*-acetyltransferase activity has been studied only in bovine submandibular gland[23-25] and in rat liver.[26] In both tissues, most of the enzyme activity is bound to the Golgi apparatus. Rat liver Golgi vesicles were shown to concentrate acetyl-CoA over 100-fold before using it for sialic acid O-acetylation.[26]

While 4-*O*-acetyltransferase is specific for the hydroxyl group at C-4 of the pyranose ring of neuraminic acid, the primary attachment site of the acetyl residue transferred by 9(7)-*O*-acetyltransferase could not yet unequivocally be determined. It is known that it is the glycerol side chain of sialic acid, but it is unknown whether it is the hydroxyl group at C-7 or C-9. Due to the wide occurrence of 9-O-acetylated sialic acid, which is also the main O-acetylated sialic acid usually obtained from bovine submandibular gland glycoprotein,[1,2,4] the enzyme was assumed to be specific for the hydroxyl at C-9. In the early days of investigation of sialate *O*-acetyltransferases[23,24] it was not possible to discriminate clearly between 7- or 9-O-acetylated sialic acids enzymatically formed on a micro scale, in order to denominate the enzyme correctly. There also exists, however, the possibility of the existence of more than one *O*-acetyltransferase incorporating sialic acids into different positions of the sialic acid side chain.[4]

In contrast to this possibility, a hypothesis is presented for the requirement of only one transferase specific for the 7-OH group but also enabling O-acetylation of the other hydroxyl groups of the sialic acid side chain. After insertion of the *O*-acetyl group at O-7 the ester residue may migrate to O-9. This is possible under physiological, intracellular conditions, as discussed above. Afterward, a second *O*-acetyl group can be transferred by the same enzyme to O-7, which may migrate to O-8 and open the way for linking a third acetyl group to O-7. By this mechanism, further dis-

[23] R. Schauer, *Hoppe-Seyler's Z. Physiol. Chem.* **351**, 595 (1970).
[24] R. Schauer, *Hoppe-Seyler's Z. Physiol. Chem.* **351**, 749 (1970).
[25] R. Schauer, this series, Vol. 50, p. 374.
[26] A. Varki and S. Diaz, *J. Biol. Chem.* **260**, 6600 (1985).

cussed below, complete or partial O-acetylation of sialic acid side chain with only one enzyme would be possible.

The following observations may support this hypothesis. In bovine submandibular gland, which contains a relatively high O-acetyltransferase activity, six sialic acid species have been found with one O-acetyl group either at position 7 or 8 or 9, with two O-acetyl groups either at positions 7 and 9 or 8 and 9, and with three O-acetyl groups at 7, 8, and 9 (Table I).[1,2] The 9-O-acetylsialic acid preparations isolated from many tissues, cells, and fluids often additionally contain sialic acids with one O-acetyl residue at O-7 or two O-acetyl groups at the side chain in smaller relative amounts. The strongest support for the hypothesis may come from the observation that the yield of N-acetyl- or N-glycolyl-7-O-acetylneuraminic acids from bovine submandibular gland increases and that of the corresponding 9-O-acetylated sialic acids decreases, if a fresh gland is taken or the conditions of sialic acid isolation applied are suitable to keep the migration rate of O-acetyl groups low. The isolation of relatively large proportions of N-acetyl-7-O-acetylneuraminic acid from this tissue was described.[12,27] Considering the lability of this sialic acid, this observation may indicate that the 7-O-acetylated sialic acid prevails in native bovine submandibular gland glycoproteins and not the 9-O-acetylated species. Final proof of the assumption that the primary reaction product of bovine sialate-O-acetyltransferase are 7-O-acetylated sialic acids as precursors of the other sialic acids containing O-acetyl groups in their side chains must await purification of this enzyme and analysis of its reaction products by modern methods of sialic acid analysis described elsewhere in this volume.[4]

Assay Method

A prerequisite for the isolation of O-acetyltransferases is the availability of a suitable assay. Tests for the estimation of soluble and microsome-bound sialate-O-acetyltransferase activities have been described earlier in this series.[25] Free radioactive N-acetylneuraminic acid and unlabeled acetyl-CoA or microsome-bound, nascent N-acetylneuraminic acid and labeled acetyl-CoA were used as substrates. As these tests are rather time-consuming and therefore not suited for monitoring enzyme activity, e.g., in column eluates during enzyme purification, the following assay was developed, allowing a quick and sensitive test of the enzymes.[27a]

---

[27] G. Reuter, R. Pfeil, S. Stoll, R. Schauer, J. P. Kamerling, C. Versluis, and J. F. G. Vliegenthart, *Eur. J. Biochem.* **134**, 139 (1983).
[27a] R. Schauer, A. P. Corfield, and M. Wember, unpublished results.

Reagents

Sodium phosphate, 0.1 M–EDTA, 1 mM–mercaptoethanol, 1 mM, buffer, pH 6.9

Edible bird nest glycopeptides bound to Sepharose 4B via the bromocyan method to give 2.3 mg N-acetylneuraminic acid/ml packed gel.

Acetyl-CoA, 0.54 mM, dissolved in the phosphate buffer

[1-^{14}C]Acetyl-CoA, specific radioactivity 5.7 μCi/μmol

Procedure. The following assay has been developed to determine soluble O-acetyltransferase activity in a variety of biological sources and to follow isolation of this enzyme from bovine submandibular gland: 400 μl total incubation mixture in the phosphate buffer contain 75 μl packed glycopeptides–Sepharose 4B (0.55 μmol N-acetylneuraminic acid), 23 nmol acetyl-CoA, 44,000 dpm [1-^{14}C]acetyl-CoA, and 100 μl enzyme solution. The blank contains the same substances with the exception of acetyl-CoA which is added at the end of incubation. After incubation for 30 min at 37°, Sepharose beads are successively washed on filters 3 cm in diameter with 5 ml each of ice-cold phosphate buffer, 2 M sodium chloride, and water. Radioactivity on the filters was found in the sialic acid fraction only. It is therefore, after subtraction of the blank value, taken as measure of enzyme activity.

This assay is suited only for enzyme preparations in solution, as microsomes having endogenous substrate cannot be completely separated from Sepharose beads and therefore may indicate falsely high enzyme values. Soluble enzyme activity is found in supernatants of tissue homogenates.[25] Enzyme activity bound to subcellular membranes can be solubilized by detergents. CHAPS was found to be best suited for this purpose, especially for solubilization of the O-acetyltransferase from bovine submandibular gland. Using the new assay system, it was possible to enrich the enzyme from this tissue about 100-fold so far, by applying anion–exchange chromatography and affinity chromatography on 2-mercaptoethylamine bound to Sepharose.[27a]

Some properties of the crude N-acylneuraminate-9(7)-O-acetyltransferase as well as of the corresponding 4-O-acetyltransferase have been described.[25,28]

Enzyme Hydrolysis of O-Acetyl Groups of Sialic Acids

The existence of enzymes hydrolyzing sialic acid ester groups in the course of sialic acid catabolism has been assumed for a long time,[2] as O-

[28] A. P. Corfield, C. Ferreira do Amaral, M. Wember, and R. Schauer, *Eur. J. Biochem.* **68,** 597 (1976).

acetyl groups hinder or even prevent the removal of sialic acids from glycoconjugates by sialidases (see above). Strikingly, horse liver sialidase cannot hydrolyze the glycosidic bonds of 4-O-acetylated sialic acids representing a large proportion of the sialic acids of this animal.[29] As this fact does not lead to storage of sialoglycoconjugates and to sialidosis disease, an esterase was expected to remove *O*-acetyl groups in order to initiate the degradation of sialoglycoconjugates. An esterase as the "missing link" of sialic acid metabolism has now been discovered in a variety of vertebrates and microorganisms. It releases *O*-acetyl groups with higher activity from position 9 of sialic acid and with lower rate from O-4. It has systematically been denominated as *N*-acyl-*O*-acetylneuraminate *O*-acetylhydrolase (EC 3.1.1.53); the recommended name is sialate *O*-acetylesterase.[18,30–33]

Assay Method

Principle. The most specific esterase assay is the analysis of de-O-acetylation of O-acetylated sialic acids. This reaction can be followed either with glycosidically bound sialic acids, e.g., in bovine submandibular gland glycoprotein, or with free O-acetylated sialic acids. In the first case liberation of acetate is followed enzymatically using a commercial test kit.[18] In the other method, free *N*-acetyl-9-*O*-acetylneuraminic acid, or other O-acetylated sialic acids, isolated by procedures described elsewhere in this volume,[4] are used as substrates. The reaction can be followed conveniently by high-pressure liquid chromatography.[4,18,31,34] A very rapid and sensitive estimation of sialate *O*-acetylesterase activity can be carried out by using 4-methylumbelliferyl acetate, as introduced by Varki *et al.*[34] The latter two tests will be described in the following section.

Reagents

Tris–HCl, 0.1 *M*, pH 7.5
N-Acetyl-*O*-acetylneuraminic acids, 5.12 m*M*, dissolved in the Tris buffer

[29] R. W. Veh, M. Sander, J. Haverkamp, and R. Schauer, *in* "Glycoconjugate Research" (J. D. Gregory and R. W. Jeanloz, eds.), p. 557. Academic Press, New York, 1979.
[30] A. K. Shukla, S. Stoll, and R. Schauer, *Hoppe-Seyler's Z. Physiol. Chem.* **365,** 1065 (1984).
[31] A. K. Shukla and R. Schauer, *Falk Symp.* **34,** 565 (1983).
[32] A. K. Shukla and R. Schauer, *Proc. Int. Symp. Glycoconjugates, 7th, 1983* p. 436 (1983).
[33] R. Schauer and A. K. Shukla, *Proc. Int. Symp. Glycoconjugates, 8th, 1985* Vol. 1, p. 264 (1985).
[34] A. Varki, E. Muchmore, and S. Diaz, *Proc. Int. Symp. Glycoconjugates, 8th, 1985* Vol. 2, p. 603 (1985).

4-Methylumbelliferyl acetate, 1 mM, dissolved in acetone–water (1 : 1, v/v)

Sodium sulfate, 0.75 M, dissolved in bidistilled water Anion-exchange resin Aminex A-29.

Procedure. Incubation mixtures (100 μl or less, Tris buffer) for routine assays of esterase activity by HPLC contain O-acetylated sialic acids (1.28 mM) and an appropriate amount of enzyme solution. After 10 min or longer an aliquot of 10 μl is withdrawn and analyzed on HPLC as described elsewhere in this volume.[4] Spontaneous O-acetyl hydrolysis is checked by running a blank with inactivated or without enzyme. The amounts of residual substrate and product and thus enzyme activity are calculated from the HPLC peaks (see Fig. 3, and Table II for retention times of different sialic acids, of Chapter 10, this volume[4]).

When using the fluorimetric assay, the enzyme reaction is started by adding 4-methylumbelliferyl acetate solution to the incubation mixture (100 μl, Tris buffer) to yield a final concentration of 0.2 mM. After 10 min at 37° the reaction is stopped by addition of 100 μl ethanol, and the samples are put into ice. They are then diluted with 800 μl of water and measured against a blank without enzyme activity (excitation at 365 nm and fluorescence at 450 nm). The amount of hydrolyzed substrate is calculated from a calibration curve obtained from the fluorescence of 4-methylumbelliferone. With our equipment, 1000 scale units correspond to 9.1 μmol of 4-methylumbelliferone liberated.

Definition of Unit. One enzyme unit is defined as the quantity that converts 1 μmol of N-acetyl-9-O-acetylneuraminic acid or 4-methylumbelliferyl acetate to N-acetylneuraminic acid and 4-methylumbelliferone, respectively, per minute, under the conditions described.

Isolation of Sialate O-Acetylesterase

Sialate O-acetylesterase has so far been isolated from equine liver,[32,35] bovine brain,[36] and influenza C virus.[37] In contrast to the viral enzyme, the esterase from both mammalian sources is a soluble enzyme. It was partially purified from the 100,000 g supernatants of liver and brain, respectively, using ammonium sulfate precipitation, gel filtration, anion-exchange chromatography, and isoelectric focusing. The purification factors were about 900 for the liver and 2,600 for the brain enzyme. The esterase from influenza C virus was purified to electrophoretic homogeneity.

[35] R. Schauer, A. K. Shukla, and S. Stoll, unpublished results.
[36] R. Schauer, G. Reuter, and S. Stoll, unpublished results.
[37] R. Schauer, G. Reuter, S. Stoll, G. Herrler, and H.-D. Klenk, unpublished results.

Properties

Stability. The crude enzyme activities are stable for several months in the frozen state, but the isolated enzyme preparations lose activity rapidly also at low temperatures. Freezing and thawing of the crude enzyme does not significantly reduce its activity.

pH Optimum. Maximum activity of the esterase from different sources was observed at pH 7.8–8.0.

Cofactors. No cofactors or metal ions were found to be required for full activity. The presence of thiol compounds, however, e.g., mercaptoethanol, is essential for retaining enzyme activity, especially during the isolation procedure.

Inhibitors. Diisopropyl fluorophosphate and diethyl 4-nitrophenyl phosphate at concentrations of 1 mM completely inhibit esterase activities.[34–37] This shows that sialate O-acetylesterase is a serine-type esterase. Mercury, copper, and zinc ions also inhibit enzyme activity.

Kinetic Properties. The de-O-acetylation reaction using the natural or fluorescent substrates is linear up to about 20 min. The K_m value of the enriched esterase from equine liver is 1 mM for N-acetyl-9-O-acetylneuraminic acid.

Substrate Specificity. The esterase from the different sources seems to be specific for acetic acid esters. It hydrolyzes with highest rate both free N-acetyl-9-O-acetylneuraminic acid and 4-methylumbelliferyl acetate. The action on N-acetyl-4-O-acetylneuraminic acid is much slower, and there seem to be differences between the esterases from the various sources. While the hydrolysis rate of this sialic acid is very low with influenza C virus esterase (Fig. 3), it is about one-third of the rate observed with N-acetyl-9-O-acetylneuraminic acid in the case of the esterase from bovine brain.

Sialic acid O-acetyl groups are hydrolyzed both in free and glycosidically bound sugars. The esterase acts on oligosaccharides, as was tested with N-acetyl-9-O-acetylneuraminic acid-α(2-3)-lactose, on glycoproteins,[18] and gangliosides. Sialate O-acetylesterase acts only very slowly on 4-nitrophenyl acetate and 1-naphthyl acetate. Strikingly, it is inactive with N-acetyl-7-O-acetylneuraminic acid (Fig. 3),[30–33] the β-methyl glycoside of N-acetyl-9-O-acetylneuraminic acid,[34] and with acetic acid ethyl ester. However, 9-O-acetyl groups in N-acetyl-7,8,9-tri-O-acetylneuraminic acid or N-acetyl-7,9-di-O-acetylneuraminic acid are susceptible to enzyme action.

Esters other than acetyl esters tested so far are not hydrolyzable, as was checked with the methyl ester of N-acetylneuraminic acid, phenylbutyrate, and 4-methylumbelliferyl butyrate. The latter substrate proved to be useful to be applied in parallel with 4-methylumbelliferyl acetate during

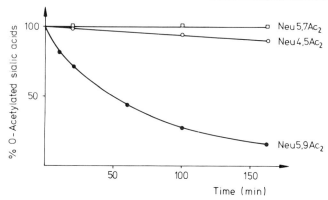

Fig. 3. Rate of de-O-acetylation of N-acetyl-9-O-acetylneuraminic acid (Neu5,9Ac$_2$) and N-acetyl-4-O-acetylneuraminic acid (Neu4,5Ac$_2$) by sialate O-acetylesterase from influenza C virus.[18] N-Acetyl-7-O-acetylneuraminic acid (Neu5,7Ac$_2$) is not a substrate for this esterase. The reactions were followed by HPLC.[4]

the isolation of the esterase from tissues. In impure enzyme preparations both fluorescent substrates are well hydrolyzed, but the hydrolysis rate of the butyryl ester rapidly decreases when compared to the acetyl ester in the course of purification of sialate O-acetylesterase.

Other Properties. The molecular weight of the esterase from equine liver was estimated to be 54,000. Multiple forms of esterase were obtained from both equine liver and bovine brain, exhibiting isoelectric points at pH values between 4.8 and 6.3.

Occurrence of the Enzyme. Sialate O-acetylesterase activity has so far been found in influenza C virus,[18] in liver of human, cow, horse,[32] and rat, in brain of calf,[38] cow,[36] and horse, in erythrocytes of human[34] and rat, and in feces of rat.[39,40] An especially rich source of this enzyme is bovine brain, which exhibits a much higher activity than, e.g., the liver of this animal. In vertebrates, the enzyme appears to be cytosolic, as it is found in the supernatants of particle-free homogenates. It is not yet known whether the esterase of feces is of mammalian or bacterial origin.

Concluding Remarks

The discovery of the esterase exhibiting preferred hydrolysis of sialic acid O-acetyl groups contributes much to our understanding of the metab-

[38] H. H. Higa and J. C. Paulson, *J. Biol. Chem.* **260,** 8838 (1985).
[39] H. Poon, P. E. Reid, C. W. Ramey, and W. L. Dunn, *Can. J. Biochem.* **61,** 868 (1983).
[40] A. P. Corfield and R. Schauer, unpublished results.

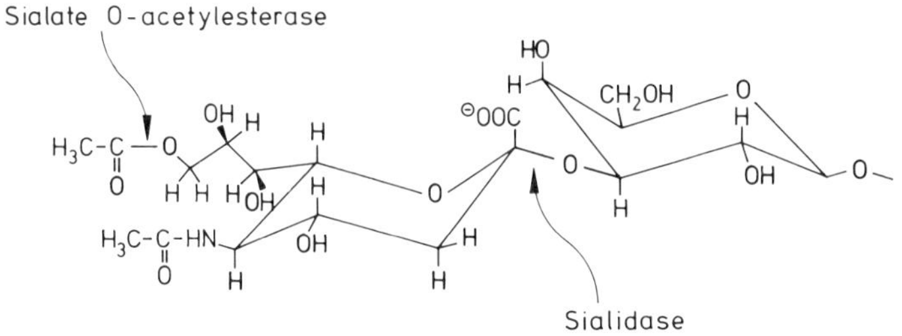

FIG. 4. The sites of action of the receptor-destroying enzymes sialidase and sialate O-acetylesterase, respectively, of influenza viruses on N-acetyl-9-O-acetylneuraminic acid-α(2–3)-galactosyl residues.[18]

olism of sialic acids. The enzyme is believed to be involved in the regulation of the catabolic site of this metabolism, by assisting the action of sialidases on O-acetylated sialic acids. Correspondingly, it participates in the regulation of the biological functions of O-acetyl groups of sialic acids. An excellent example for this is influenza C virus esterase, which is involved in binding of the virus to cell membranes,[18] but it simultaneously, similar to the sialidase of other influenza virus types, destroys the receptor (N-acetyl-9-O-acetylneuraminic acid) for the virus. Thus, this esterase represents a second receptor-destroying enzyme (RDE), as shown in Fig. 4.

The esterase acting only on the 4-O-acetyl group of the pyranose ring of sialic acid and, with much higher efficiency, on the primary hydroxyl residue of the side chain of neuraminic acid, but not on 7-O-acetyl groups, is able to lead to complete de-O-acetylation of sialic acids. It was discussed above that the acetyl group from O-7 easily migrates to O-9 under physiological conditions, where it can be hydrolyzed. These reactions are summarized in Fig. 5. The scheme also shows O-acetylation of N-acetylneuraminic acid by two distinct O-acetyltransferases (T4 and T9), which, in combination with spontaneous migration of O-acetyl groups, can lead to complete O-acetylation of the sialic acid molecule by successive enzymatic transfer of ester groups. Similarly, including the same nonenzymatic reaction, hydrolysis of all O-acetyl groups can be explained with a single esterase exhibiting specificity for acetyl groups at O-4 (E4) and O-9 (E9). This scheme is also valid for N-glycolylneuraminic acid.

Evidence is accumulating that the esterase described here is responsible for de-O-acetylation of sialic acids in the organism and indeed is a new enzyme. Acetylcholinesterase (EC 3.1.1.7) and nonspecific esterases, e.g., carboxylesterases (EC 3.1.1.1) from pig liver or esterases from rat liver microsomes, were shown not to remove O-acetyl groups from sialic acids. Varki et al. present evidence that sialate O-acetylesterase is identi-

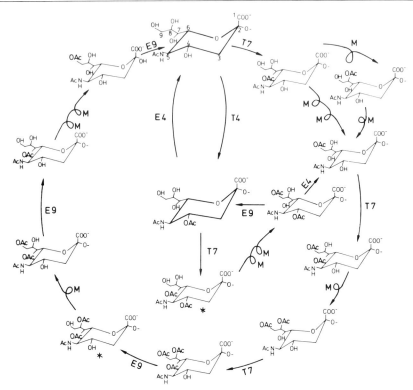

FIG. 5. Metabolism of O-acetylated sialic acids. Cooperative effects of two sialate O-acetyltransferases, sialate O-acetylesterase, and nonenzymatic migration of O-acetyl groups.[33] T4, reaction catalyzed by N-acetylneuraminate 4-O-acetyltransferase (EC 2.3.1.44); T7, N-acetylneuraminate 7(9)-O-acetyltransferase (EC 2.3.1.45); E4, sialate O-acetylesterase acting on 4-O-acetyl residues; E9, the same esterase acting on 9-O-acetyl groups. M indicates one or two migration steps of O-acetyl groups. Sialic acids marked with an asterisk have not yet been found, probably because they are very labile intermediates. For further details see the text.

cal with esterase D isolated from human erythrocytes, the function of which so far was unknown.[34] Thus, the detection of sialate O-acetylesterase is one of the rare cases where a specific biological substrate has been found for an esterase.

Acknowledgments

The expert technical assistance in these studies of Margret Wember and Sabine Stoll is gratefully acknowledged. This research was supported by the Deutsche Forschungsgemeinschaft and the Fonds der Chemischen Industrie.

[54] CMP-NeuNAc: Poly-α-2,8-Sialosyl Sialyltransferase in Neural Cell Membranes[1]

By RONALD D. MCCOY and FREDERIC A. TROY

Golgi-enriched membranes from fetal rat brain contain a membrane-associated CMP-NeuNAc: poly-α-2,8-sialosyl sialyltransferase (poly-α-2,8-sialosyl sialyltransferase)[2] that catalyzes the incorporation of [14C]NeuNAc from CMP-[14C]NeuNAc into polymeric products:

$$n(\text{CMP-NeuNAc}) + [\rightarrow 8)\text{-}\alpha\text{-NeuNAc-}(2\rightarrow]\rightarrow X \longrightarrow$$
$$[\rightarrow 8)\text{-}\alpha\text{-NeuNAc-}(2\rightarrow]_n\rightarrow X + n(\text{CMP})$$

In embryonic neural membranes, X is presumed to be N-asparaginyl-linked glycan chains of the complex type that are believed to be attached to neural cell adhesion molecules (N-CAM).[1,3-6] In N-CAM, n is oligo- or polysialosyl chains containing at least 12–20 sialyl units (DP 12–20).[3-5]

In sodium dodecyl sulfate–polyacrylamide (SDS–PA) gels, the major radiolabeled species migrates with a mobility expected for N-CAM. A bacteriophage-derived endo-N-acetylneuraminidase (Endo-N) specific for catalyzing the depolymerization of polysialic acid is used to demonstrate that at least 20–30% of the [14C]NeuNAc incorporated is in α-2,8-linked polysialosyl chains. This is confirmed by structural studies which show that the Endo-N-sensitive brain material consists of multimers of sialic acid. Addition of chick N-CAM to the membranous sialyltransferase stimulates sialic acid incorporation 3.3-fold. The product of this reaction is also sensitive to Endo-N, showing that N-CAM can serve as an exogenous acceptor for sialylation *in vitro*. Sialic acid incorporated into adult rat brain membranes is resistant to Endo-N, indicating that expression of

[1] R. D. McCoy, E. R. Vimr, and F. A. Troy, *J. Biol. Chem.* **260**, 12695 (1985).

[2] Abbreviations used: poly-α-2,8-sialosyl sialyltransferase, CMP-NeuNAc: poly-α-2,8-sialosyl sialyltransferase; NeuNAc, N-acetylneuraminic acid (sialic acid); CMP-NeuNAc, cytidine 5′-monophospho-N-acetylneuraminic acid; N-CAM, neural cell adhesion molecules; DP, degree of polymerization (refers to the number of sialic acid units in a sialyl oligomer or polymer); SDS–PA, sodium dodecyl sulfate–polyacrylamide; Endo-N, endo-N-acetylneuraminidase (endoneuraminidase); PNGase F, peptide: N-glycosidase F.

[3] E. R. Vimr, R. D. McCoy, H. F. Vollger, N. C. Wilkison, and F. A. Troy, *Proc. Natl. Acad. Sci. U.S.A.* **81**, 1971 (1984).

[4] J. Finne, *J. Biol. Chem.* **257**, 11966 (1982).

[5] J. Finne, U. Finne, H. Deagostini-Bazin, and C. Gordis, *Biochem. Biophys. Res. Commun.* **112**, 482 (1983).

[6] J. Finne and P. H. Mäkelä, *J. Biol. Chem.* **260**, 1265 (1985).

the poly-α-2,8-sialosyl sialyltransferase is restricted to an early stage in development.[1]

This chapter describes the *in vitro* conditions that permit detection of the sialyltransferase and procedures to quantitate the amount of NeuNAc incorporated specifically into α-2,8-linked polysialosyl units in N-CAM.

Assay Methods

Principle. Determination of the reaction catalyzed by the membranous poly-α-2,8-sialosyl sialyltransferase is based on the transfer of [14C]NeuNAc from CMP-[14C]NeuNAc to high molecular weight endogenous acceptors believed to be N-CAM molecules.[1] Formation of polymeric products is followed by measuring the incorporation of [14C]NeuNAc into chromatographically immobile polymers. Identification of poly-α-2,8-linked sialosyl carbohydrate units is based on immunoreactivity to polyclonal IgM antibodies specific for α-2,8-linked polysialic acid and by susceptibility of the radiolabeled polymers to depolymerization by the Endo-N specific for hydrolyzing oligo- or polysialic acid in α-2,8-ketosidic linkage. Proof that the products of Endo-N digestion are oligomers of sialic acid is determined by chromatographic and structural analyses.

Procedure for Preparing Membranes. Fetal rats are killed *in utero* at 20 days of gestation by CO_2 asphyxiation. Ten brains are quickly removed and homogenized in 5 ml ice-cold MET buffer (50 mM MES, 1 mM dithiothreitol, pH 6.0) using a Dounce homogenizer (B pestle) at 7–10 strokes. Newborn rats are killed by decapitation within 24 hr of birth and their brains processed identically to the fetal brains. Adult rats, greater than 3 months old, are killed by cervical dislocation and the brains processed as described above. Golgi-enriched membrane fractions are prepared essentially as described by Morré.[7] MET buffer is used in all steps to isolate the Golgi-enriched fraction.

Procedure for Measuring CMP-NeuNAc : Poly-α-2,8-Sialosyl Sialyltransferase Activity. Incorporation of [14C]NeuNAc from CMP-[14C]NeuNAc is determined by incubating membrane fractions in MET buffer at 33° as previously described.[8] To determine the amount of [14C]NeuNAc incorporated into polymeric products, samples are removed at various times and spotted onto Whatman 3MM paper. The papers are subjected to descending chromatography in ethanol : 1 M ammonium acetate, pH 7.5 (7 : 3). Radioactivity remaining at the origin is

---

[7] D. J. Morré, *in* "Molecular Techniques and Approaches in Developmental Biology" (M. J. Chrispeels, ed.), p. 1. Wiley (Interscience), New York, 1973.

[8] F. A. Troy, I. K. Vijay, M. A. McCloskey, and T. E. Rohr, this series, Vol. 83, p. 540.

quantitated by liquid scintillation counting. Specific activity is calculated after protein is determined by a modified Lowry protein assay.[9]

Endoneuraminidase. The soluble form of the Endo-N specific for cleaving α-2,8-linked polysialic acid is obtained from a bacteriophage, designated K1F, as described in Chapter 12.

Procedure to Determine the Amount of [^{14}C]NeuNAc Incorporated into α-2,8-Linked Polysialic Acid. The percentage of [^{14}C]NeuNAc incorporated into α-2,8-linked polysialic acid is determined by measuring susceptibility of the polymeric products to depolymerization by Endo-N. After completion of the sialyltransferase reaction, membranes containing radiolabeled sialyl polymers are divided into two aliquots. One aliquot is incubated with Endo-N and the other with buffer. After incubation at 33°, samples are removed at different times, spotted on Whatman 3MM paper, and chromatographed in ethanol : ammonium acetate as described above. The difference in [^{14}C]NeuNAc remaining at the origin between control and Endo-N-treated samples represents the minimum amount of sialic acid in α-2,8-linked polysialic acid. The limit digestion products of Endo-N are chromatographically mobile under these conditions.

Procedure to Characterize [^{14}C]NeuNAc-Labeled Sialyl Oligomers Synthesized and Released from Brain Membranes by Endo-N. Radiolabeled sialyl oligomers synthesized *in vitro* and released by treatment with Endo-N are characterized by DEAE–Sephadex A-25 chromatography. A modification of the procedure of Nomoto *et al.*,[10] as previously described,[3] is carried out. Authentic oligomers of sialic acid are prepared from colominic acid after hydrolysis at pH 2.0 and 80° for 1 hr by ion-exchange chromatography.[3]

Preparation of Embryonic Chick Brain N-CAM. N-CAM is purified from 100 14-day-old embryonic chick brains as described by Hoffman *et al.*[11] Purification is carried out up to the octyl-agarose chromatography. The presence of N-CAM is confirmed at each step by monitoring for polysialic acid by Western blot analysis using the antibody against α-2,8-linked polysialic acid, as previously described.[3]

Distribution of Poly-α-2,8-Sialosyl Sialyltransferase in Membrane Fractions from Fetal Rat Brains

As shown in Table I, the poly-α-2,8-sialosyl sialyltransferase in fetal rat brain is enriched in a membrane fraction. Using a standard method to

[9] O. H. Lowry, N. J. Rosebrough, A. L. Farr, and R. J. Randall, *J. Biol. Chem.* **193**, 265 (1951).

[10] H. Nomoto, M. Iwasaki, T. Endo, S. Inoue, Y. Inoue, and G. Matsumara, *Arch. Biochem. Biophys.* **218**, 335 (1982).

[11] S. Hoffman, B. C. Sorkin, P. C. White, R. Brackenbury, R. Mailhammer, U. Rutishauser, B. A. Cunningham, and G. M. Edelman, *J. Biol. Chem.* **257**, 7720 (1982).

TABLE I
LOCALIZATION OF CMP-NeuNAc : POLY-α-2,8-SIALOSYL SIALYLTRANSFERASE IN
RAT BRAIN MEMBRANES

Fraction[a]	Sialyltransferase activity[b]			NeuNAc in poly-α-2,8-sialosyl units[c]
	Total units (pmol NeuNAc incorporated)	Recovery (%)	Specific activity (pmol NeuNAc/ mg protein/hr)	
Homogenate	6300	100	30	33
Total membranes (P-1)	4400	70	31	11
Supernatant I	<50	<1	<5	0
Membrane pellet (P-2)	100	2	26	30
Golgi-enriched (P-3)	4200	67	84	27
Golgi-enriched from adult	960	ND[d]	16	0

[a] Refers to fraction designation as described by Morré.[7]

[b] Brains from 13 newborn rats are homogenized and fractionated as described in the text. The adult rat used in this experiment was >3 months old and was the mother of the newborn rats. Sialyltransferase activity is determined in standard incubation mixtures containing 112 nmol of CMP-[14C]NeuNAc (7.9×10^3 dpm nmol^{-1}) in 0.32 ml MET buffer. Protein content is about 3.9 mg per incubation mixture.

[c] Designates the percentage of [14C]NeuNAc in polymeric products that is sensitive to digestion by the endoneuraminidase specific for α-2,8-linked polysialosyl units, as described in the text.

[d] N.D., not determined.

enrich in Golgi apparatus,[7] the specific activity is nearly 3-fold higher in this fraction (84 pmol NeuNAc/mg protein/hr), than in the homogenate (30 pmol NeuNAc/mg protein/hr). Approximately two-thirds of the sialyltransferase activity is recovered in the Golgi-enriched fraction. The data in Table I also show that about one-third of the [14C]NeuNAc incorporated is into α-2,8-linked polysialic acid, based on its susceptibility to depolymerization by Endo-N (see below). A membrane fraction isolated from an adult rat brain by the identical procedure contains only about 20% of the sialyltransferase activity of fetal brain (Table I). None of the [14C]NeuNAc incorporated by the adult brain is in polysialic acid, since these polymeric products are resistant to Endo-N treatment. The poly-α-2,8-sialosyl sialyltransferase responsible for catalyzing polysialylation thus appears restricted to an early stage of development.

Effect of Endo-N on the SDS–Polyacrylamide Gel Electrophoretic
Profile of ^{14}C-Labeled Polysialic Acid Synthesized Endogenously by
Fetal Rat Brain

To determine the molecular weight profile of [^{14}C]NeuNAc incorporated into polysialic acid, Golgi-enriched membranes are allowed to incorporate maximal levels of [^{14}C]NeuNAc into endogenous acceptor molecules (see below). The radiolabeled membranes are subjected to SDS–PA
gel electrophoresis before and after treatment with Endo-N. As shown in
Fig. 1, the major peak of radioactivity exists as a broad band suggestive of
molecular weight heterogeneity ranging from about 120,000 to 240,000. N-
CAM shows a similar molecular weight heterogeneity.[3,12–14] Control experiments (Fig. 1, insert) show that immunoreactivity to anti-polysialosyl
antibodies is in the same high molecular weight region as the ^{14}C-radiola-
beled peak (lane 1) and that this immunoreactivity is abolished by pretreatment with Endo-N (lane 2).

Stimulation of Polysialic Acid Synthesis by the Exogenous Addition of
N-CAM

Golgi-enriched sialyltransferase from fetal rat brain catalyzes the incorporation of [^{14}C]NeuNAc from CMP-[^{14}C]NeuNAc into endogenous
acceptor molecules, as shown in Fig. 2 (bottom). In confirmation of the
results described above, about 30% of the sialic acid incorporated is in
polysialic acid, based on its sensitivity to treatment with Endo-N (Fig. 2).
At pH 6.0, 84 pmol of NeuNAc per mg protein per hr are incorporated.
Polysialic acid synthesis is stimulated about 3-fold by the addition of
embryonic chick brain N-CAM as an exogenous acceptor (Fig. 2, top).
These results also show that at least 60–70% of the product of this reaction is sensitive to Endo-N, suggesting that N-CAM can serve as an
exogenous acceptor for polysialylation *in vitro*. The relationship between
the degree of oligo- or polysialylation of N-CAM and its efficacy as an
acceptor is not known. Colominic acid, a homooligomer of sialic acid
containing about 10–20 α-2,8-linked units does not serve as an exogenous

[12] J. M. Lyles, D. Linnemann, and E. Bock, *J. Cell. Biol.* **99**, 2082 (1984).
[13] J. B. Rothbard, R. Brackenbury, B. A. Cunningham, and G. M. Edelman, *J. Biol. Chem.* **257**, 11064 (1982).
[14] U. Rutishauser, *Nature (London)* **310**, 549 (1984).
[15] U. K. Laemmli, *Nature (London)* **227**, 680 (1970).
[16] H. T. Towbin, T. Staehelin, and J. Gordon, *Proc. Natl. Acad. Sci. U.S.A.* **76**, 4350 (1979).
[17] F. A. Troy and M. A. McCloskey, *J. Biol. Chem.* **254**, 7377 (1979).
[17a] P. C. Hallenbeck, E. R. Vime, F. Yu, B. L. Bassler, and F. A. Troy, *J. Biol. Chem.*, in press (1987).

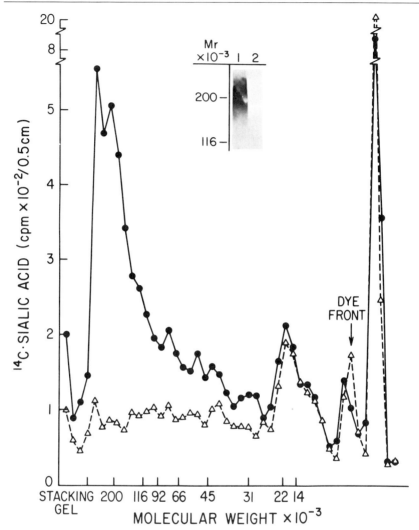

FIG. 1. Effect of Endo-N on the SDS–PA gel profile of ^{14}C-labeled polysialic acid synthe-sized endogenously by fetal rat brain. The Golgi-enriched fraction (10.5 mg protein), isolated as described in the text, is incubated with 280 nmol CMP-[^{14}C]NeuNAc (7.9 × 10[3] dpm nmol[-1]) in 1.0 ml of MET buffer at 33°. After 3 hr, the membranes are sedimented (17,000 g for 30 min) and residual substrate removed by washing 3 times in MET buffer. Washed membranes in 0.9 ml MET buffer are divided in equal aliquots. To one aliquot is added Endo-N (△) and to the second, an equal volume of MET buffer (●). Both samples are incubated at 28° for 1.5 hr then sedimented at 17,000 g for 30 min and washed once. Membranes are then resuspended in 90 μl Laemmli sample buffer.[15] Ten microliters of each sample containing approximately 300 μg protein are electrophoresed in 5–15% polyacryl-amide gradient gels (1.5 mm) and electrophoretically transferred to nitrocellulose, as de-

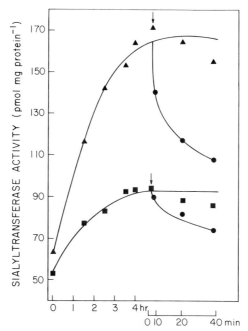

FIG. 2. *In vitro* synthesis of polysialic acid in fetal rat brains is stimulated by the exogenous addition of N-CAM. Standard incubation mixtures containing 1.38 mg of Golgi-enriched membrane protein and 113 nmol of CMP-[^{14}C]NeuNAc (15,000 cpm nmol^{-1}) are incubated at 33° in the absence (■) and presence (▲) of 66 μg of partially purified N-CAM prepared from 14-day-old chick embryo brains, as exogenous acceptor. After a 4-hr incubation, the amount of [^{14}C]NeuNAc incorporated into α-2,8-linked polysialic acid is determined by quantitating the amount of radiolabeled polymer sensitive to hydrolysis by Endo-N (●) as described in the text. The arrows denote the time at which Endo-N or buffer was added to each incubation mixture.

acceptor. This oligosialic acid is effective as an exogenous acceptor for the membranous sialyltransferase from *E. coli* K1 strains that synthesize α-2,8-linked sialyl polymers containing about 200 sialyl residues.[17]

Endo-N does not remove all of the poly-α-2,8-linked [^{14}C]NeuNAc from the exogenous N-CAM (Fig. 2) because the enzyme does not cleave sialyl oligomers containing less than five residues.[3,6,17a] After treating brain

<hr />

scribed by Towbin *et al.*[16] Electroblotting is carried out at 200 mA for 16 hr. Immune blotting using the anti-polysialic acid antibody H.46 is carried out as previously described.[3] Insert, lane 1 is the control sample that was not treated with Endo-N before immune blotting, while lane 2 was pretreated with Endo-N. The amount of [^{14}C]NeuNAc in polysialic acid is quantitated after SDS–PA gel electrophoresis of the remaining 80-μl portions of each sample by slicing the gel into 0.5 × 2.5 cm sections and determining the radioactivity by scintillation counting, as previously described.[17]

polysialosyl glycopeptides with a similar endoneuraminidase, Finne and Mäkelä showed that 5 sialyl units still remained attached to the glycopeptide.[6] For this reason, our results with both the endogenous and exogenous acceptors give only a minimum estimate of the percentage of sialic acid in poly-α-2,8 linkage.

Endoneuraminidase Treatment of Fetal Membranes Incubated with CMP-[^{14}C]NeuNAc and Exogenous N-CAM Releases ^{14}C-Labeled Sialyl Oligomers

Structural confirmation that Endo-N treatment of Golgi-enriched membranes incubated with CMP-[^{14}C]NeuNAc and N-CAM release oligomers of sialic acid is shown in Fig. 3.[18] In this experiment, [^{14}C]NeuNAc from CMP-[^{14}C]NeuNAc is incorporated into fetal rat brain membranes in the presence of exogenous N-CAM, as described in the legend to Fig. 2. After washing to remove excess substrates, the membranes are treated exhaustively with Endo-N. ^{14}C-Labeled sialyl oligomers are recovered after centrifugation and fractionated by chromatography on DEAE–Sephadex A-25 as previously described.[3] As shown in Fig. 3, the Endo-N-solubilized ^{14}C-labeled sialyl oligomers elute predominately with a DP of 3. This is an expected result because the primary products of a limited Endo-N digestion of rat brain polysialic acid are oligomers containing three to four sialyl units.[3] Thus, these data support the conclusion that N-CAM can serve as an exogenous acceptor for sialylation *in vitro*. Treatment of the sialyl trimer with exoneuraminidase quantitatively converts the label to [^{14}C]NeuNAc, thus proving that the oligomers are composed of sialic acid.

The possible involvement of a sialyl dolichol-linked intermediate in the poly-α-2,8-sialosyl sialyltransferase reaction has not been investigated. The role of sialylmonophosphorylundecaprenol as an intermediate in the synthesis of the polysialic acid capsule in *E. coli* K235 and *Neisseria meningitidis* serogroup B has been documented.[19,20]

Comments. The relationship between the biosynthetic pathway and the mechanism regulating the developmental reduction in polysialosyl units during the embryonic to adult conversion is unknown. Our preliminary studies to investigate the effect of chick N-CAM as an exogenous acceptor in a homologous system (embryonic chick N-CAM plus embryonic chick brain membranes) may be relevant to this point. These studies

[18] L. Warren, *J. Biol. Chem.* **234**, 1971 (1959).
[19] F. A. Troy, I. K. Vijay, and N. Tesche, *J. Biol. Chem.* **250**, 156 (1975).
[20] L. Masson and B. E. Holbein, *J. Bacteriol.* **161**, 861 (1985).

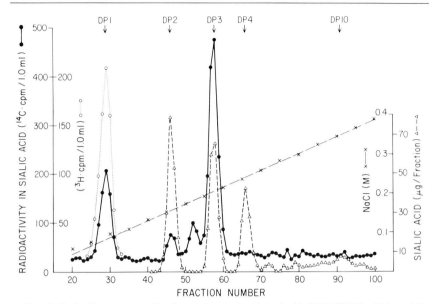

FIG. 3. Endo-N treatment of brain membranes incubated with CMP-[14C]NeuNAc and N-CAM releases sialyl oligomers. Golgi-enriched membranes (15.6 mg of protein) from 20-day-old fetal rat brains are incubated with 0.96 mg of purified chick brain N-CAM in 1.0 ml of MET buffer containing 280 nmol of CMP-[14C]NeuNAc (7.9 × 10³ dpm nmol⁻¹). After 6 hr at 33°, membranes are sedimented (17,000 g for 30 min), washed in 100 mM MES, pH 6.0, and treated with Endo-N at 33°. After 5.5 hr, additional Endo-N is added and incubated at 33° for 1 hr. Endo-N-treated membranes are sedimented at 17,000 g for 30 min. Oligomers of sialic acid released by the enzyme are lyophilized, resuspended in 0.35 ml of water, and fractionated on DEAE–Sephadex A-25 with a 0–0.4 M gradient of NaCl in 10 mM Tris, pH 7.6 (X). Authentic oligomers of sialic acid with DP of 2, 3, and 4 (3 mg each), and 10 (0.5 mg) are added as markers (△), and their elution volume determined by the thiobarbituric acid method.[18] A tracer amount of [3H]NeuNAc (○) is used to determine the elution position of sialic acid.

indicate that under some conditions, the exogenous addition of N-CAM can inhibit the incorporation of [14C]NeuNAc into endogenous acceptors and that the inhibition appears to be specific for the poly-α-2,8-sialosyl sialyltransferase.[21] This suggests that the extent of polysialylation of N-CAM, or its state of aggregation,[11] may determine its efficacy as an acceptor in the homologous system. This possibility is suggested because no inhibition is observed in the heterologous system (embryonic chick N-CAM plus embryonic rat brain membranes), as shown in Fig. 2. Accessibility of the membranous sialyltransferase to N-CAM added exogenously may also be different between rat and chick membrane preparations.

[21] R. D. McCoy and F. A. Troy, unpublished results.

Deglycosylation of Polysialylated N-CAM by Peptide : N-Glycosidase F

The polysialylated, embryonic form of N-CAM is deglycosylated by peptide : N-glycosidase F (PNGase F). Since PNGase F catalyzes the hydrolysis of glycosylamine linkages in N-asparaginyl-linked glycans,[22] these data provide additional evidence that the polysialosyl carbohydrate units on N-CAM are associated with asparaginyl-linked oligosaccharide chains. PNGase F has been used to deglycosylate the polysialylated oligosaccharide chains from purified, embryonic N-CAM and N-CAM associated with neural membranes. Deglycosylation is determined by showing the loss of immunoreactivity to antibodies against α-2,8-linked polysialic acid in Western blot experiments. These results are confirmed by demonstrating the sensitivity of PNGase F-solubilized sialyl polymers to Endo-N.[21]

Procedure to Radiolabel and Deglycosylate Polysialylated N-CAM from Neural Membranes. Six 16-day-old chick brains are homogenized in MET buffer and the membranous poly-α-2,8-sialosyl sialyltransferase prepared as described above. [^{14}C]NeuNAc is incorporated into endogenous N-CAM molecules in standard incubation mixtures containing 630 nmol CMP-[^{14}C]NeuNAc (7×10^3 dpm nmol^{-1}) and 35 mg membrane protein. After incubation at 33° for 9 hr, the membranes are sedimented at 20,000 g for 30 min and washed with MET buffer until the amount of radioactivity in the supernatant fraction reaches a baseline level. The ^{14}C-sialyl-labeled membranes are then resuspended in 2.0 ml of MET buffer containing 0.02% sodium azide and divided in two aliquots. The experimental sample receives 17 units of PNGase F[22] and the control an equivalent volume of buffer. After incubation at 33° for 15 hr, both samples are sedimented at 20,000 g for 30 min. The membranes are resuspended in MET buffer, treated for an additional 48 hr with 6.8 units of PNGase F, and recovered by centrifugation.

Radioactivity released from the membranes by PNGase F is quantitated by scintillation counting of the supernatant fractions. Approximately 52% of the [^{14}C]NeuNAc incorporated into the membranes is solubilized after treatment with PNGase F. In contrast, about 30% of the radioactivity in the control is soluble. Chromatography in ethanol : ammonium acetate and radiochromatographic scanning shows that about half of this radioactivity is either CMP-[^{14}C]NeuNAc or [^{14}C]NeuNAc. Essentially all of the [^{14}C]NeuNAc in polymeric products is sensitive to treatment with Endo-N, showing that PNGase F releases polysialylated N-CAM molecules from neural membranes. This result is confirmed by immunoblotting experiments of the membrane fractions before and after

[22] A. L. Tarentino, C. M. Gomez, and T. H. Plummer, Jr., *Biochemistry* 24, 4665 (1985).

treatment with PNGase F. Before treatment, the major immunoreactivity characteristic of N-CAM at molecular weight 180,000–240,000 is present. After PNGase F treatment, there is nearly a complete loss of immunoreactivity in this high molecular weight region. Similar results from immunoblotting experiments using purified chick N-CAM have recently been obtained.[21]

Comments. In some experiments, we have observed an increase in immunoreactivity in lower molecular weight species after treatment with PNGase F. This may reflect limited proteolysis of polysialylated glycoproteins not susceptible to PNGase F or detection of oligosialylated chains rendered accessible to the antibody by PNGase F-mediated removal of polysialosyl carbohydrate units. Presumably such oligopolysialylated chains would be resistant to deglycosylation by PNGase F.

Tarentino *et al.* have reported that PNGase F hydrolyzes a broad spectrum of asparagine-linked glycans.[22] Our results extend further the classes of *N*-asparaginyl-linked glycoproteins that appear sensitive to PNGase F.

Acknowledgments

We thank C. C. Sweeley for consultation regarding the enzyme nomenclature used in this work. We would like to thank A. L. Tarentino for a sample of PNGase F, E. R. Vimr for preparing Endo-N, and J. B. Robbins for the anti-polysialosyl antibody. We are indebted to L. A. Troy for excellent editorial and secretarial assistance in preparing this manuscript. This study was supported by U.S. Public Health Service Research Grant AI-09352 (to FAT) from the National Institutes of Health.

[55] Fungal 1,3-β-Glucan Synthase

By Enrico Cabib and Mohinder S. Kang

$$2n \text{ UDP-Glc} \longrightarrow 2n \text{ UDP} + [\text{Glc } \beta(1 \rightarrow 3)\text{Glc}]_n$$

Assay Method

Principle. The incorporation of radioactivity from UDP-[^{14}C]Glc into acid-insoluble product is measured.

Reagents

UDP-[^{14}C]Glc, commercially available, specific activity 250,000 cpm/ μmol, concentration 0.1 *M*

Tris–chloride, 1 M, pH 7.5

Bovine serum albumin (Pentex fraction V from Miles Laboratories, Inc., Naperville, Illinois), 10% (w/v)

EDTA, 10 mM

KF, 0.5 M

Guanosine 5'-(γ-thio)triphosphate (GTPγS, Boehringer Mannheim Biochemicals, Indianapolis, Indiana), 0.4 mM

Trichloroacetic acid, 10% (w/v)

Glacial acetic acid

95% Ethanol

66% Ethanol

Procedure. The reaction mixture contains variable amounts of enzyme, usually between 5 and 20 μl, 2 μl of UDP-[^{14}C]Glc, 3 μl of Tris–chloride, 3 μl of bovine serum albumin, 2 μl of KF, 3 μl of EDTA, 2 μl of GTPγS, and water, in a total volume of 40 μl. The mixture is usually incubated at 30° for 30 min or 1 hr, depending on the activity of the preparation. The incorporation of radioactivity into $\beta(1 \rightarrow 3)$-glucan is measured either by filtration through glass filters after addition of trichloroacetic acid (Method a) or by ascending chromatography on silica-impregnated glass filter strips, after addition of glacial acetic acid (Method b), as described elsewhere in this volume [56].

Note. The enzyme preparation contains 33% glycerol, which stimulates the activity. If amounts of enzyme below 10 μl are used, the difference to 10 μl should be made up with 33% glycerol.

Enzyme Preparation

Principle. Glucan synthase is a membrane-bound enzyme. Active preparations can be obtained from *Saccharomyces cerevisiae* either by osmotic lysis of protoplasts or by disruption of intact cells with glass beads, followed by centrifugation. With other fungi, only the breakage with glass beads has been successful.

From Saccharomyces cerevisiae[1]

Yeast Strains and Culture Conditions. The yeast strain used is *S. cerevisiae* GS-1-36, a glycogen synthase-defective mutant,[2] available from this laboratory. The organism is grown in YEPD medium [1% yeast

[1] E. M. Shematek, J. A. Braatz, and E. Cabib, *J. Biol. Chem.* **255**, 888 (1980).

[2] L. B. Rothman-Denes and E. Cabib, *Proc. Natl. Acad. Sci. U.S.A.* **66**, 967 (1970).

extract (Difco), 2% peptone (Difco), and 2% glucose] at 30° and harvested in midlogarithmic phase, at a cell density of about 1 mg (dry weight)/ml.

Protoplast Preparation. Cells are converted into protoplasts with snail intestinal juice (Cytohelicase, LKB, Gaithersburg, Maryland) as described in Vol. 22 [14], except that 0.8 M sorbitol is used as osmotic stabilizer. Protoplasts are stored overnight at 5° in 0.8 M sorbitol, containing citrate–phosphate buffer at one-twentieth the original concentration (Vol. 1 [16]) and 50 μg/ml of aureomycin, in a final volume of 1 ml per gram (wet weight) of yeast cells used.

Protoplast Lysis and Isolation of Particulate Glucan Synthetase. The procedure is described for 10 ml of protoplast suspension. To 0.5-ml portions of suspension, 9.5 ml of 1 mM EDTA, pH 7, are added at room temperature, and the mixture is immediately stirred for 1 min in a Lourdes Multimix homogenizer at half-maximal speed (another blender, such as a Sorvall Omnimixer may be used). During homogenization and in subsequent steps, the preparation is kept at 0–2°. The pooled homogenate is centrifuged for 45 min at 100,000 g. The pellet is washed once with 25 ml of 0.05 M Tris–chloride, pH 7.5, containing 1 mM EDTA and 1 mM 2-mercaptoethanol (Buffer A), and finally suspended in a mixture of 67 parts, by volume, of Buffer A and 33 parts of glycerol, to a final volume of 10 ml. The preparation is stored at −80°.

Note. Use of zymolyase (Miles Laboratories, Inc., Naperville, Illinois) to prepare protoplasts resulted in preparations almost devoid of glucan synthase activity. On the other hand, the chitin synthase of these preparations was at a normal level.

From Saccharomyces cerevisiae and Other Fungi

Fungal Strains and Culture Conditions. The same strain of *S. cerevisiae* mentioned above is grown as described. For the other fungi, the strains used are as follows: *Hansenula anomala* NIH 2990 and *Cryptococcus laurentii* NIH 3763, both available from K. J. Kwon-Chung, National Institute of Allergy and Infectious Diseases, Bethesda, Maryland; *Neurospora crassa* csp1 (allele USLA37) mating type A, from the Fungal Genetics Stock Center (No. 2554); *Schizophyllum commune* isogenic 699, mating type A43B43 (ATCC 26069); *Wangiella dermatitidis* Me13 (ATCC 44504); and *Achlya ambisexualis* GR1, male, available from T. J. Mullins, University of Florida, Gainesville, Florida.

Cells of *H. anomala, C. laurentii,* and *W. dermatitidis* are grown as described above for *S. cerevisiae.* Hyphae of the filamentous fungi are also grown at 30° in 800 ml of broth media with shaking.

Neurospora crassa is maintained on slants of Vogel minimal medium[3] with 2% agar and is transferred with a loop to liquid Vogel medium. For Vogel's minimal medium the following solutions are prepared: solution A, containing, per liter, 25 g sodium citrate, 50 g KH_2PO_4, 20 g NH_4NO_3, 2 g $MgSO_4 \cdot 7H_2O$, and 1 g $CaCl_2 \cdot 2H_2O$; solution B, containing, in 100 ml, 11 mg of boric acid, 48 mg of ammonium molybdate, 11 mg of $CuSO_4 \cdot 5H_2O$, and 1.67 g of $ZnSO_4 \cdot 7H_2O$; solution C, 190 mg $FeCl_3 \cdot 6H_2O$ in 100 ml of water; and solution D, 1 mg/liter biotin. For 1 liter of medium, 100 ml of solution A, 3 ml of solution B, and 1 ml of solution C are mixed, and the volume is made up with water. This mixture is sterilized by autoclaving, whereas solution D is sterilized by filtration. Finally, 1 ml of solution D is added per liter of mixture. For solid medium, 2% agar is added to the mixture before autoclaving.

In order to obtain maximal activity of glucan synthase and maximal stimulation by guanosine nucleotide, it is necessary to subculture *N. crassa* 3–5 times. Each time, about 1 ml of mycelial suspension is transferred to a flask containing 800 ml of medium and incubated on a shaker at 30° for 24 hr. The yield is 12–15 g (wet weight).

For *S. commune*, Raper complete medium[4] is used (per liter, 20 g glucose, 2 g peptone, 0.5 g $MgSO_4$, 0.46 g KH_2PO_4, and 120 μg thiamin–HCl). Inoculum is prepared by cutting a Raper agar culture (Raper's medium with 2% agar) into small cubes that are inoculated in 80 ml of Raper complete medium. Mycelium is allowed to grow out of the cubes for 2–5 days. These subcultures are then blended in a chilled and previously sterilized Waring blender microcup with 30 ml of Raper's medium, and used as inoculum for 800 ml of Raper complete medium. Overnight growth results in a yield of 5–7 g of mycelium (wet weight).

Achlya ambisexualis is maintained on PYG agar [0.125% peptone (Difco), 0.125% yeast extract (Difco), 0.3% glucose, and 2% agar]. An encysted zoospore inoculum is obtained by first washing small cubes of the PYG agar with 0.5 mM $CaCl_2$ for 2 hr at room temperature and then incubating overnight with a second wash of 0.5 mM $CaCl_2$. The dense spore suspension obtained is inoculated into 800 ml of PYG liquid medium. After overnight culture at 30°, about 20 g of wet mycelium is obtained.

In all cases, hyphal cultures are harvested by filtration through nylon mesh.

Cell Disruption and Isolation of Particulate Glucan Synthetase.[5] All operations are carried out at 0–5°. Yeast cells or hyphae of the different

---

[3] H. J. Vogel, *Microb. Gen. Bull.* **13,** 42 (1956).

[4] J. R. Raper and G. S. Krongelb, *Mycologia* **50,** 707 (1958).

[5] P. J. Szaniszlo, M. S. Kang, and E. Cabib, *J. Bacteriol.* **161,** 1188 (1985).

fungi (7 g, wet weight), suspended in 16 ml of 1 mM EDTA, are added to 25 g of glass beads 0.5 mm in diameter (B. Braun, Melsungen, West Germany) in the 30-ml vessel of a Bead-Beater (Biospec Products, Bartlesville, Oklahoma). The Bead-Beater is operated with ice cooling for two periods of 2 min each, with a 1-min cooling period in between. This treatment results in about 90% breakage of the cells, as observed in the phase-contrast microscope. The extract is aspirated from the glass beads with a long-tipped Pasteur pipette connected to an evacuated filter flask, and the beads are washed several times with small portions of 1 mM EDTA. The pooled extracts are centrifuged at 1000 g in a swinging bucket rotor to sediment cell walls and the supernatant fluid is recentrifuged at 100,000 g for 40 min. The pellet is washed with 17.5 ml of Buffer A, suspended in Buffer A–glycerol (67 : 33, v/v) in a total volume of 7 ml and homogenized in a Dounce homogenizer. The enzyme is stored at −80°.

Properties

Stability. The preparations are stable for several months when stored at −80°. At 30°, activity is lost to a variable extent, depending on the organism. At least with the enzymes from *S. cerevisiae, N. crassa,* and *H. anomala,* the inactivation is greatly accelerated by the presence of 1–2 mM EDTA, although no stimulation of the activity by metals has been observed. Guanosine nucleotides prevent the inactivation. The most effective compound is GTPγS, which affords almost 100% protection at 20 μM.

Kinetics. The kinetic properties of fungal 1,3-β-glucan synthases are summarized in Table I. The data for V_{max} have only indicative value. Detailed data for the pH optimum are available only for the enzyme from *S. cerevisiae,* which shows a maximum between pH 7.5 and 8.

Stimulators and Inhibitors. The enzymatic activity is enhanced in most fungi by bovine serum albumin. The synthase from *S. cerevisiae* is additionally stimulated by glycerol, but there is no information on the effect of glycerol on the other preparations. The most important stimulators from the physiological point of view appear to be nucleoside triphosphates.[5–7] The most active compounds are GTP and its analogs, which are effective in the micromolar range. Stimulations between 4- and 40-fold have been observed, depending on the organism and the preparation. In all organisms tested, GTPγS behaves as the most potent stimulator; a concentration of 10 μM is usually sufficient for maximal activity.

[6] E. M. Shematek and E. Cabib, *J. Biol. Chem.* **255,** 895 (1980).

[7] V. Notario, H. Kawai, and E. Cabib, *J. Biol. Chem.* **257,** 1902 (1982).

TABLE I
KINETIC PROPERTIES OF 1,3-β-GLUCAN
SYNTHASES FROM DIFFERENT FUNGI

Organism	K_m [a] (mM)	V_{max} [a,b]
Saccharomyces cerevisiae	3.8	2,500
Achlya ambisexualis	7.1	500
Hansenula anomala	0.67	120
Neurospora crassa	2.9	104
Cryptococcus laurentii	0.86[c]	129[d]
Schizophyllum commune	0.8[c]	100[d]
Wangiella dermatitidis	1.8	172

[a] Determined in the presence of 20 μM GTP and 0.75 mM EDTA.
[b] Nanomoles of glucose incorporated per hour per milligram of protein.
[c] Determined in the absence of GTP.
[d] Stimulation by GTP was measured after the enzyme had been stored for some time and lost some activity. V_{max} was calculated by multiplying the original activity of the fresh preparation, as determined in the absence of nucleotide, by the stimulation factor.

Most glucan synthases are partially inhibited by the sulfhydryl reagents N-ethylmaleimide (0.16 mM) and showdomycin (0.1 mM).

The glucan synthase from A. ambisexualis differs sharply in its properties from all the others so far tested.[5] It shows a high K_m value (Table I), no stimulation by bovine serum albumin or nucleotides, and no inhibition by sulfhydryl reagents.

[56] Chitin Synthase from *Saccharomyces cerevisiae*[1]

By Enrico Cabib, Mohinder S. Kang, and Janice Au-Young

Chitin synthase zymogen
↓ protease
Active chitin synthase

$2n$ UDP-GlcNAc ⟶ $2n$ UDP + [GlcNAc $\beta(1 \rightarrow 4)$GlcNAc]$_n$

Fungal chitin synthetase is an integral membrane protein, the major portion of which is found in the plasma membrane in *Saccharomyces cerevisiae*.[2,3] The enzyme is present in the cell in a zymogen form that can be transformed into active form(s) by partial proteolysis.

Assay Method

Principle. Zymogen is converted into active form by incubation with trypsin. The incorporation of radioactivity from UDP-[14C]GlcNAc into water-insoluble chitin is then measured.

Reagents

UDP-[14C]GlcNAc, commercially available, specific activity 400,000 cpm/μmol, concentration 10 mM

Tris–chloride, 0.5 M, pH 7.5, or imidazole chloride, 0.5 M, pH 7

Crystalline trypsin freshly prepared at concentrations ranging between 0.125 and 1 mg/ml, in 0.05 M phosphate buffer, pH 7

Soybean trypsin inhibitor, at concentrations ranging between 0.19 and 1.5 mg/ml, in 0.05 M phosphate buffer, pH 7

Magnesium acetate, 40 mM

N-Acetyl-D-glucosamine, 0.8 M

Phosphatidylserine, 3 mg/ml. A commercial preparation of the phospholipid in chloroform is evaporated to dryness in a rotary evaporator. A solution containing 25 mM Tris–chloride, pH 7.5, and 0.1% digitonin is added, and phosphatidylserine is emulsified by repeated and vigorous vortexing.

Digitonin, 1%. A stock solution of 10% digitonin is made by bringing a suspension of the saponin in water to boiling with vigorous stir-

---

[1] See also Vol. 28 [80].

[2] A. Duran, B. Bowers, and E. Cabib, *Proc. Natl. Acad. Sci. U.S.A.* **72**, 3952 (1975).

[3] E. Cabib, B. Bowers, and R. L. Roberts, *Proc. Natl. Acad. Sci. U.S.A.* **80**, 3318 (1983).

METHODS IN ENZYMOLOGY, VOL. 138

ring. Sodium azide, 0.02%, is added as a preservative. Upon storage at room temperature, some material usually precipitates out. Paradoxically, the more concentrated the digitonin solution, the less precipitation is observed.

Trichloroacetic acid, 10% (w/v)
Glacial acetic acid
Ethanol, 95%
Ethanol, 66%

Procedure

ACTIVATION STEP. The incubation mixture contains 5–10 μl of enzyme, 3 μl of Tris or imidazole buffer, 2 μl of magnesium acetate, 2 μl of trypsin solution, and water, in a total volume of 38 μl. When particulate preparations are assayed, the mixture also contains 5 μl of 1% digitonin, which increases the activity by about 100%. The amount of trypsin required for maximal activation varies from preparation to preparation and is roughly inversely proportional to the protein content of the preparation. Usually, three or four different concentrations of trypsin are used, with concentrations increasing each time by a factor of 2. An excess of trypsin destroys the activity.

The activation mixture is incubated at 30° for 15 min. The reaction is stopped by addition of 2 μl of soybean trypsin inhibitor, at a concentration 1.5 times that of the corresponding trypsin solution, and the tubes are immediately cooled in ice.

ASSAY STEP. To each incubation mixture from the previous step, 2 μl of N-acetylglucosamine, 3 μl of phosphatidylserine, and 5 μl of UDP-[^{14}C]GlcNAc are added. The mixtures are incubated 30 min at 30°. The incorporation of radioactivity into chitin may be measured by two different procedures.

METHOD A. The reaction is stopped by adding to each tube 1 ml of 10% trichloroacetic acid, and the suspension is filtered under suction through a Gelman glass fiber filter Type A/E (Gelman Sciences, Inc., Ann Arbor, Michigan). The filter is washed 4 times with 1 ml of 10% trichloroacetic acid and 2 times with 1 ml of 95% ethanol. The filter is transferred to a scintillation vial, 10 ml of scintillation fluid (Hydrofluor, National Diagnostics, Somerville, New Jersey) is added, and the radioactivity is counted.

METHOD B. The reaction is stopped by addition of 5 μl of glacial acetic acid. Aliquots of the reaction mixture (40 μl) are applied to 1.5 × 9 cm strips cut from silica gel-impregnated glass fiber sheets (Gelman Sciences, Inc., Ann Arbor, Michigan), 1.5 cm above the bottom edge. The strips are dried with warm air and developed ascendingly, with the use of 66%

ethanol as solvent. Mason jars are convenient containers for the development. The strips are lined up around the wall of the jar, with the upper end touching the wall. After 15 min of development the strips are taken out and again dried with warm air. The unreacted substrate has moved near the solvent front whereas the chitin formed in the reaction remains at the origin. A 2.3-cm segment of each strip, from 1.3 cm above to 1 cm below the starting line, is cut out and inserted into a scintillation vial. After adding 10 ml of Hydrofluor, the samples are counted.

Method b may be preferred with dilute solutions of highly purified enzyme, because it yields somewhat lower blanks and more uniform results. Furthermore, with this method losses of even minute amounts of chitin are virtually impossible. The procedure is essentially the same as previously reported for glycogen synthase.[4]

Enzyme Purification[5]

All operations are carried out at 1–5°, unless otherwise stated.

Principle. Chitin synthase, in the zymogen form, is solubilized from yeast membranes with digitonin. The solubilized zymogen is activated with Sepharose-immobilized trypsin. When the enzyme is incubated with the components of the assay mixture, chitin is formed and precipitates out, trapping part of the activity. The synthase is eluted from the chitin precipitate with buffer.

Yeast Strains and Culture Conditions. The preferred organism is a diploid strain of *S. cerevisiae,* Matα/Mata prb1-1122/prb-1122, available from Dr. E. W. Jones, Carnegie-Mellon University, Pittsburgh, Pennsylvania. This strain lacks proteinase *b,* which may cause accidental activation of the chitin synthetase zymogen during enzyme preparation. Nevertheless, other strains, such as *S. cerevisiae* X2180 (ATCC 26109), may be used. The yeast is grown at 30° in a minimal medium [2% glucose, 0.7% yeast nitrogen base (Difco)] to the late logarithmic phase, and harvested at a cell density of about 1 mg (dry weight)/ml.

Cell Disruption and Enzyme Solubilization. Yeast cells are usually processed in 0.5-kg batches, but the preparation can be scaled down to any desired size without difficulty. In a typical preparation, 557 g of cells, wet weight, are suspended in Buffer A (20 m*M* Tris–chloride, pH 7.5, containing 2 m*M* magnesium acetate) to a final volume of 1200 ml. A 200-ml portion of the suspension is added to 306 g of glass beads (0.5-mm diameter, B. Braun, Melsungen, West Germany) in the vessel of a Bead-

[4] K.-P. Huang and E. Cabib, *Biochim. Biophys. Acta* **302,** 240 (1973).
[5] M. S. Kang, N. Elango, E. Mattia, J. Au-Young, P. W. Robbins, and E. Cabib, *J. Biol. Chem.* **259,** 14966 (1984).

Beater (Biospec Products, Bartlesville, Oklahoma) with ice cooling. The Bead-Beater is operated for two 4-min periods, with a 1-min cooling period in between. The temperature of the suspension rises to ~25°, and cell breakage is usually about 90%, as monitored with a phase-contrast microscope.

The extract is aspirated from the glass beads with a long-tipped Pasteur pipette connected to an evacuated filter flask, and the beads are washed several times with small portions of Buffer A. Each portion of beads may be used to disrupt three 200-ml batches of cell suspension. The final volume of the extract is 1570 ml.

The crude extract is centrifuged at 100,000 g for 30 min. The supernatant liquid is discarded, and the pellet is suspended with a Dounce homogenizer in Buffer A to a final volume of 1 liter. After centrifugation as above, the supernatant fluid is discarded and the pellet is suspended in a solution containing 1% digitonin (w/v), 20 mM Tris–chloride, pH 7.5, 5 mM magnesium acetate, and 0.2 M NaCl, to a final volume of 1600 ml. After homogenization with a Dounce homogenizer, the suspension is incubated at 30° for 45 min with shaking, then centrifuged as above. The supernatant fluid, 1480 ml, containing the solubilized chitin synthase zymogen, is stored at −80°.

Sephadex G-75 Filtration and Zymogen Activation. The gel filtration is carried out with 150 ml of solubilized zymogen at a time. The column of Sephadex G-75 (Pharmacia, Piscataway, New Jersey), 4 × 85 cm, is equilibrated with 20 mM Tris–chloride, pH 7.5, containing 0.1% digitonin; the same buffer is used for elution. The absorbance of the effluent at 280 nm is recorded with an LKB UV monitor or measured with a spectrophotometer, and the large peak emerging at the void volume is collected in a total volume of 220–250 ml.

Immobilized trypsin is prepared by coupling crystalline trypsin (400 mg) to 10 g of activated CH–Sepharose 4B (Pharmacia, Piscataway, New Jersey) according to the manufacturer's directions. After coupling, no trypsin should be detected in the supernatant fluid with the Hide Powder Azure assay.[6] The gel is stored at 5° in 50 mM Tris–chloride, pH 7.5, containing 0.02% azide. Before and after use, the gel is washed twice with 50 mM Tris–chloride, pH 7.5. The same preparation of trypsin–Sepharose can be used at least 10 times for activation of chitin synthase zymogen without apparent loss of activity. In fact, aged preparations often perform better than fresh ones.

For zymogen activation, the Sephadex G-75 eluate is mixed with 20–30 ml (settled gel volume) of previously centrifuged trypsin–Sepharose

[6] R. E. Ulane and E. Cabib, *J. Biol. Chem.* **249**, 3418 (1974).

4B, and the suspension is incubated with shaking for 30 min at 30°. The trypsin–Sepharose is pelleted in a clinical centrifuge and washed with 15 ml of 0.05 M Tris–chloride, pH 7.5. The wash fluid is added to the first supernatant liquid, and the pooled solutions are concentrated to 80 ml in an ultrafiltration cell fitted with an Amicon YM100 membrane (Amicon, Danvers, Massachusetts).

The optimal amount of immobilized trypsin for activation is ascertained in small-scale trials for each batch of zymogen. With some batches, it is necessary to add more digitonin to the zymogen, to a final concentration of 0.35%, to obtain maximal activation with immobilized trypsin. No such requirement has been observed for activation with soluble trypsin.

Chitin-Entrapment Step. To the bottom of each of four centrifuge tubes for the Beckman SW27 rotor are added 6 ml of a mixture containing 12% glycerol, 6 mM UDP-GlcNAc, 40 mM N-acetyl-D-glucosamine, and 4 mM magnesium acetate. To each of four 20-ml portions of concentrated enzyme from the previous step are added 3 ml of 50 mM UDP-GlcNAc, 1.2 ml of 0.8 M N-acetyl-D-glucosamine, and 0.1 ml of 1 M magnesium acetate. After mixing each portion, the solution is immediately layered on the glycerol-containing cushion. The tubes are incubated for 15 min in a 30° bath before being placed on ice for 2.5 hr. A precipitate of chitin

TABLE I
PURIFICATION OF CHITIN SYNTHASE

Step	Total volume (ml)	Total protein (mg)	Total activity[a] (units)	Specific activity (units/mg protein)	Purification (-fold)	Recovery (%)
Crude extract	1570	69,640	292	0.004	1	100
Particulate fraction[b]		18,500	274	0.015	3.5	94
Solubilized zymogen	1480	10,800	164	0.015	3.5	56
Activated and concentrated Sephadex G-75 eluate	(800)[c]	(2,500)	(171)	0.068	17	58
Chitin entrapment						
Step 1	(400)	(19.5)	(102)	5.2	1,300	35
Step 2	(200)	(1.08)	(46)	42.7	10,700	16

[a] In the first three steps, chitin synthase activity was measured after activation with optimal amounts of trypsin. One unit corresponds to the incorporation of 1 μmol of N-acetylglucosamine per minute at 30°.

[b] An aliquot of the crude extract was separately centrifuged, and the pellet was suspended in Tris buffer for assay.

[c] Values in parentheses have been recalculated as if the whole preparation had been used in the chitin-entrapment steps. In fact, only 10% of the total was used each time.

appears almost immediately upon mixing of the reaction components and becomes heavy and flocculent subsequently. The tubes are centrifuged in an SW27 Beckman rotor for 25 min at 20,000 rpm (53,000 g). The supernatant fluid is carefully aspirated and the chitin pellets are stored overnight at $-80°$. For extraction, each of the four pellets is suspended in 10 ml of 20 mM Tris–chloride, pH 7.5, containing 0.1% digitonin, and homogenized with a Dounce homogenizer. After incubation for 5 min at 30°, the suspension is centrifuged for 10 min at 16,000 g, and the supernatant is saved.

Second Chitin-Entrapment Step. If further purification is desired a second chitin-entrapment step may be carried out, immediately after the first. If preparations are stored at $-80°$ before carrying out the second chitin step, the yield of enzymatic activity decreases dramatically.

The second chitin-entrapment step is performed in the same manner as the first one, but halving all quantities and using tubes for the Beckman SW40 rotor. The purified enzyme is stored in polypropylene tubes at $-80°$.

A summary of a preparation is given in Table I. The overall purification has varied, with different batches, between 3,000- and 10,000-fold.

Notes. The Bead-Beater can be used with vessels of different volumes to accommodate smaller batches. Nevertheless, the optimal ratio of glass beads to yeast cells may vary depending on the size of the vessel. In all cases, it is important for maximum breakage to fill the container completely with glass beads plus yeast cell slurry.

If desired, cell walls can be sedimented from the crude extract by centrifugation at 1000 g for 10 min in a swinging bucket rotor. Membranes are then sedimented from the supernatant fluid by high-speed centrifugation as described above.

Properties

Purity. Upon sodium dodecyl sulfate–polyacrylamide gel electrophoresis and staining with Coomassie Blue or silver salts, purified preparations give rise to a major and a minor band, corresponding to molecular weights of 63,000 and 74,000, respectively. After longer development with silver salts, some additional minor bands appear. It is not possible at this time to attribute with certainty the enzymatic activity to a particular band.

Stability. The digitonin-solubilized enzyme is stable for several months at $-80°$. The purified enzyme is stable for at least a few weeks when stored at the same temperature.

Kinetics. The trypsin-activated synthase shows a broad optimum between pH 6.5 and 7.5. With purified preparations, the apparent K_m for UDP-GlcNAc is 0.7 mM.

Activators and Inhibitors. The enzyme has an absolute requirement for divalent cations, Mg^{2+}, Mn^{2+}, or Co^{2+}. Mg^{2+} shows a broad optimum, between 1 and 10 mM. With particulate preparations, N-acetyl-D-glucosamine further stimulates the activity between 3- and 5-fold, with a K_m of 4.7 mM. Purified preparations, however, are stimulated only 10–20% by the amino sugar.

There is no requirement for an added primer to the reaction mixture. Chitin oligosaccharides are slightly inhibitory and are not intermediates in the reaction.

Crude preparations are stimulated by digitonin (see Assay Method); purified fractions also require phosphatidylserine. With highly purified enzyme, the optimal concentration of phospholipid depends somewhat on that of the saponin. A good combination appears to be 0.5 mg/ml of digitonin and 0.05 mg/ml of phosphatidylserine.

Antibiotics of the polyoxin and nikkomycin family are potent competitive and specific inhibitors of chitin synthase, with K_i values in the micromolar range. UDP, a reaction product, is inhibitory at millimolar concentrations.

[57] Dextransucrase: A Glucosyltransferase from *Streptococcus sanguis*

By ROBERT M. MAYER

Introduction

Dextransucrase (EC 2.4.1.5) is a glucosyltransferase (GTF) produced by a variety of oral streptococci and other microorganisms. It utilizes sucrose as a donor substrate, and yields fructose and dextran, a D-glucan with predominantly $\alpha(1\rightarrow6)$ bonds, as the products:

Several organisms produce multiple types of glucosyltransferases which are distinguished from one another by the structural characteristics of the polymers they form. Some produce soluble polymers whereas others yield insoluble ones. The terms GTF(S) and GTF(I) have been used to designate the enzymes responsible for the formation of soluble and insoluble products.[1] Since solubility is related to the proportion of $\alpha(1{\to}6)$ bonds, those enzymes that form soluble polysacharides are more specifically designated as dextransucrases.

The reaction catalyzed by GTFs may play an important role in the formation of dental caries, since both products are necessary participants in the process. Glucans are major constituents of dental plaque, and appear to be important for microbial colonization of the plaque. Fructose is the major energy source for these organisms, and it is metabolized to form a series of organic acids that serve as erosive agents in the cariogenic process. Many of the oral streptococci produce more than one GTF, and may in fact have both GTF(S) and GTF(I) present. The dextransucrase from *Streptococcus sanguis* is a good model for these enzymes since the organism produces only one GTF. The purification of this enzyme and its properties will be described.

Assay Procedure

Principle

Direct or indirect analyses may be employed for the measurement of dextransucrase activity. Direct procedures utilize [^{14}C]sucrose and measure the incorporation of isotope into products.[2,3] Paper chromatographic analysis of the reaction mixtures permits one to distinguish between polymeric (nonmobile) and oligomeric products. The latter are formed in reactions that include added low molecular weight acceptor substrates. Indirect procedures take advantage of the fact that fructose is formed stoichiometrically in relation to the amount of glucose incorporated into products. Fructose can be quantitated by a variety of procedures including colorimetric and enzymatic analysis.[4,5] The convenience of the latter permits rapid screening and quantitation during purification and will be

[1] J. E. Ciardi, *in* "Glucosyltransferases, Glucans, Sucrose and Dental Caries" (R. J. Doyle and J. E. Ciardi, eds.), Spec. Suppl., p. 51. IRL Press, Arlington, Virginia, 1983.

[2] G. R. Germaine, C. F. Schachtele, and A. M. Chludzinski, *J. Dent. Res.* **53**, 355 (1974).

[3] R. M. Mayer, M. M. Mathews, C. L. Futerman, V. K. Parnaik, and S. M. Jung, *Arch. Biochem. Biophys.* **208**, 178 (1981).

[4] E. Newbrun and J. Carlsson, *Arch. Oral Biol.* **14**, 751 (1982).

[5] G. A. Luzio, D. A. Grahame, and R. M. Mayer, *Arch. Biochem. Biophys.* **216**, 751 (1982).

described in detail. The procedure involves the conversion of the generated fructose to glucose 6-phosphate using hexokinase and phosphoglucose isomerase and the oxidation of glucose 6-phosphate with glucose-6-phosphate dehydrogenase with the formation of stoichiometric amounts of NADPH, which can be measured by its absorbance at 340 nm. The analysis is carried out in two steps; the first is the dextransucrase-catalyzed reaction, while the second is the coupled enzyme measurement of fructose.

Reagents

Substrate solution. A 100 ml solution of 1×10^{-2} M Dextran T-10 (Pharmacia, dialyzed and lyophilized), 0.2 M sucrose, and 0.1 M phosphate buffer, pH 6.0, is prepared and stored in a freezer after dividing into 10 vessels.

Assay mixture. Solutions (10 ml) containing 0.1 M Tris–HCl, pH 7.0, 5 mM MgCl$_2$, 5 mM ATP, and 5 mM NADP$^+$ can be prepared and stored at $-20°$ indefinitely. Prior to use, 25 units of hexokinase, phosphoglucose isomerase, and glucose-6-phosphate dehydrogenase are added to the stock solution to make a working solution. After the addition of the enzymes the solution may be stored frozen for several days with retention of sufficient activity to be utilized in routine assays.

Procedure

Step 1. To a sample (0.1 ml) of active or reactivated enzyme is added 0.1 ml of the substrate solution. For each analysis a control is run, which is a duplicate reaction mixture, but for which the enzyme has been heated at 100° for 2 min prior to the addition of the substrate. Reactions are maintained at 37° for 15 min, heated at 100° for 2 min, and then allowed to cool.

Step 2. Aliquots (up to 0.1 ml) from the reactions described in Step 1 are transferred to 0.1 ml of the working solution of the assay mixture. The reactions are maintained at 37° for 20 min, diluted to 1 ml, and the absorbance at 340 nm is measured. The ΔA_{340} is calculated from the absorbances for the experimental and the respective control reaction.

A unit is defined as that amount of enzyme that will form 1 μmol of fructose per minute at 37°. The units per milliliter of an enzyme solution can be calculated as follows:

$$\text{units/ml} = (\Delta A_{340}) \: [1 \: \text{mol}/(6.2 \: \text{absorbance units/ml})] \: (0.2 \: \text{ml}/X)$$
$$\times \: (1.0 \: \text{ml}/Y) \: (1/Z)$$

or $\text{units/ml} = [(\Delta A_{340}) \: (0.0322)]/XYZ$

where X is the volume (ml) of the enzyme used in Step 1, Y the volume (ml) of the enzyme taken from Step 2, and Z the time of incubation (min) used in Step 1.

Enzyme preparations that have been inactivated with sodium dodecyl sulfate (SDS) (see below), can be reactivated by the addition of Triton X-100 (Rohm and Haas) prior to the assay.[6,7] This is done in a 10-min incubation at room temperature of 90 μl of 1% Triton and 10 μl of enzyme.[8] The reactivated enzyme is used directly in the assay procedure described above. At the levels employed, the detergents have no effect on the enzymes employed in Step 2.

Method of Purification

General Remarks

Dextransucrase produced by *S. sanguis* is a constituitive exocellular enzyme and can be purified starting with the culture fluids of late log or early stationary phase cells grown on glucose or other simple carbon sources. The enzyme forms large aggregates, which make purification in high yield difficult. To circumvent this, the strategy for purification exploits the observation by Russell[6] that SDS-inhibited enzyme can be reactivated with nonionic detergents such as Triton X-100. SDS causes the dissociation of the aggregates,[8] and it has the additional advantage of inhibiting proteases, thereby preventing degradation during purification. The procedure involves the use of SDS at all stages of purification which include gel permeation and hydroxylapatite chromatography.[7] The addition of Triton X-100 following purification restores enzyme activity.

Purification

Crude Enzyme. *Streptococcus sanguis* ATCC 10558 is grown at 37° in 12 liters of sterilized ultrafiltrate (Amicon H1P10-20) of brain heart infusion (Difco). A 9% innoculum of a freshly grown culture along with 1 liter of sterile 10% glucose are added to initiate growth, and the pH of the culture is maintained at 6.5 with the addition of KOH. Additional solid glucose is added at appropriate intervals to maintain the concentration between 0.05 and 0.2%. When cell growth (A_{660}) begins to diminish, the cells are harvested by centrifugation, and, with the use of two ultrafiltration devices (Amicon H1P100-20), the culture fluid is concentrated ap-

[6] R. R. B. Russell, *Anal. Biochem.* **97**, 173 (1978).
[7] D. A. Grahame and R. M. Mayer, *Carbohydr. Res.* **142**, 285 (1985).
[8] D. A. Grahame and R. M. Mayer, *Biochim. Biophys. Acta* **786**, 42 (1984).

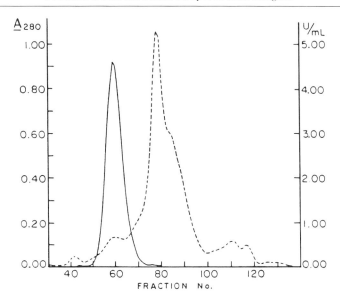

FIG. 1. Separation of dextransucrase on Sepharose CL-6B-200.

proximately 30-fold and dialyzed by the continuous addition of distilled water until the conductivity of the ultrafiltrate is <100 μmho/cm. Following centrifugation to remove residual turbidity, the concentrated crude enzyme can be stored (see Storage of the Enzyme) or can be carried through the purification procedure.

Dissociation of Aggregate and Gel Filtration. At 10°, 3.33 ml of 10% SDS is added to 80 ml of concentrated crude enzyme to make a final concentration of 0.4% SDS which is necessary to dissociate the aggregate forms to the monomeric unit. The mixture is gently stirred for 1 hr, during which the temperature is allowed to rise to 25°. This solution is applied to a Sepharose CL-6B 200 column (5.0 × 115 cm) that has been equilibrated with a solution of 0.1 M lithium phosphate, 0.1% SDS, pH 6.3, at 4°. Elution is carried out with the same buffer at a flow rate of 1.5 ml/min. The collected fractions are monitored for enzyme activity and absorbance at 280 nm, and those fractions with the highest activity to A_{280} ratio are pooled. Typical results are shown in Fig. 1.

Hydroxylapatite Chromatography. Chromatography on hydroxylapatite is carried out at room temperature using the pooled enzyme from the Sepharose column. It is applied to 30 g of BioGel HTP (Bio-Rad Laboratories) in a column (2.5 × 14.4 cm) that has been equilibrated with a 0.1 M lithium phosphate buffer, pH 6.3, containing 0.1% SDS. The column is

TABLE I
SUMMARY OF PURIFICATION OF DEXTRANSUCRASE

Purification step	u/ml	Units	Specific activity	Relative purification	Recovery[a] (%)	Yield[b] (%)
Culture fluid	0.33	973	0.035	1	100	100
Concentrated culture fluid	11.0	880	3.21	91	90.4	90.4
Sepharose/SDS	3.61	758	30.9	882	86.2	77.9
Hydroxylapatite/SDS I[c]	5.71	560	93.0	2657	95.1	57.5
Hydroxylapatite/SDS II[d]	4.43	284	106.7	3048	56.5	29.2

[a] Recovery, recovery for individual step.
[b] Yield, overall recovery.
[c] 588 units were applied.
[d] The pool of the central portion of the peak.

washed with 1 bed-volume of the same buffer, and elution is achieved with a linear gradient (900 ml) of sodium phosphate, pH 6.4 (0.1–0.5 M) containing 0.1% SDS. The conductivity of the collected fractions is determined on a 25-μl aliquot that is diluted to 2.7 ml. Enzyme activity and protein[9] are measured, and the fractions with the highest specific activity are pooled. The pooled sample is then diluted with 3 volumes of 0.1% SDS and rechromatographed on a second but identical hydroxylapatite column, as described above. The final enzyme peak may be divided into three pools; the central portion of the peak which has the highest specific activity, and the leading and trailing edges, both of which have slightly lower specific activities. Typical separations on the hydroxylapatite columns are shown in Fig. 2,A and B.

Table I is a summary of the purification procedure. It can be seen that the overall yield in the central fractions of the peak from the last column is 29% and that the enzyme undergoes a 3000-fold purification. SDS–gel electrophoretic analysis of this fraction shows the presence of one major band (94% homogeneity) with a second minor component. The second component is a proteolyzed form of the native enzyme that is completely active (see below). The combination of the native and proteolyzed forms constitute better than 99% of the proteins present.

Storage of the Enzyme. The enzyme is best stored either after concentration of the culture fluids or following the second hydroxylapatite column. In the presence of SDS the enzyme is relatively stable and can be conveniently maintained at 0° without significant loss over a period of several days. Long-term storage is best achieved at −70° after the enzyme

[9] O. H. Lowry, N. J. Rosebrough, A. L. Farr, and R. J. Randall, *J. Biol. Chem.* **193,** 262 (1951).

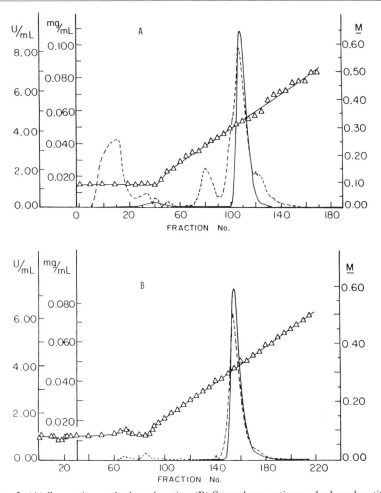

FIG. 2. (A) Separation on hydroxylapatite. (B) Second separation on hydroxylapatite.

solution has been dialyzed and quick frozen by allowing it to slowly drip into liquid nitrogen. The frozen droplets retain 80 to 85% of their activity after storage for more than 1 year at $-70°$.

Properties

Physical Properties

Dextransucrase from *S. sanguis* is a protein that forms aggregates whose weight can be as large as 5×10^6.[5] Dissociation of the aggregates can be achieved by the use of moderate to high concentrations of SDS to yield a monomeric polypeptide with a molecular weight of 174,000.[5] As described previously, the SDS-treated enzyme is inactive and may be completely reactivated by the addition of nonionic detergents such as Triton X-100.[6,8] In the presence of low concentrations of SDS the enzyme is partially dissociated to form a series of oligomeric forms.[5] The organisms also produce an exocellular protease that cleaves the native enzyme to yield a fully active proteolyzed form with a molecular weight of 156,000. If proteolysis has occurred the aggregate and oligomeric forms are hybrids of the native and proteolyzed monomers.[8]

The amino acid compositions of the native and proteolyzed enzyme have been determined.[7] Both forms are rich in acidic and basic amino acids, and are devoid of cysteine. Proteolysis results in the loss of 177 amino acids, of which 114 are polar or ionic. No amino sugars are observed in the amino acid analysis, and a determination of neutral sugars indicates that the hexose content is below the limit of detectability. Under the conditions of analysis this means that there is less than one residue of sugar per molecule of enzyme and that it is not a glycoprotein.

The enzyme has a broad pH optimum between 6.0 and 6.5, and an isoelectric point of 5.15. Since the enzyme can catalyze the formation of products in the presence and absence of acceptor substrates (see below), the catalytic constants have been determined under both sets of conditions. The K_m for sucrose in the presence and absence of acceptors (dextran T-10) is 5.09 and 0.93 mM, respectively. The maximal velocities under the same conditions were determined to be 109 and 23 μmol/min/mg of protein, respectively. The proteolyzed enzyme has very similar constants.

Substrate Specificity

Since the enzyme has the capacity to transfer glucosyl residues to acceptors, there are two types of substrates: glucosyl donors and glucosyl acceptors. The donor substrate specificity is rather narrow, since sucrose is the only naturally occurring substrate known. Lactulosucrose and glucosyl sorboside, both of which have been prepared enzymatically, have also been shown to serve as donors. α-1-Fluoroglucose has been shown to donate glucosyl residues to acceptors, and to have a K_m similar to that of glucose, which suggests that the fructosyl moiety is not critical for activity.[10] α-1-Fluoro derivatives of other hexoses were unable to serve as donor substrates, which supports the idea that the glucopyranosyl configuration is necessary for donor activity.[11] Substitution with a halogen for

[10] S. M. Jung and R. M. Mayer, *Arch. Biochem. Biophys.* **208**, 288 (1981).
[11] T. J. Grier and R. M. Mayer, *Arch. Biochem. Biophys.* **212**, 651 (1981).

one or more of the primary hydroxyl groups of sucrose converts sucrose from a donor substrate to a weak competitive inhibitor.[12] *p*-Nitrophenyl glucoside has also been shown to be an active donor substrate[13] for enzyme derived from other organisms. In addition, there is some indication that glucose oligosaccharides can serve as donors to a limited degree, since slow disproportionation reactions have been observed.[14]

Acceptor substrate specificity is extremely broad and ranges from simple alcohols, such as ethanol, to glucose oligosaccharides and dextran. While dextran is probably the best acceptor for the enzyme, the oligosaccharides are the most widely investigated. It is generally recognized that the most effective acceptors contain α-D-glucopyranosyl residues at nonreducing positions. For example, methyl α-glucoside is a better acceptor than glucose, and maltose is better than cellobiose.[15] Studies on analogs of methyl α-glucoside indicate that compounds that have axial hydroxyl groups at positions 2 and 4 are poor acceptors, while the enzyme can tolerate significant changes at other positions.[16] This includes halogen substitution at C-6, the elimination of oxygen at that position, or the elimination of C-6 altogether.

Fructose also serves as an acceptor; however, it generates two types of products. In short-term reactions it can serve to form sucrose in an isotope-exchange reaction,[3] while in long-term reactions[15] leucrose (5-*O*-α-D-glucopyranosyl-D-fructopyranose) and maltulose (4-*O*-α-D-glucosyl-D-fructose) are the major products.

Mechanism

Dextransucrase catalyzes a group transfer reaction, in which a glucosyl residue is transferred from sucrose to a developing product molecule. A series of related reactions that the enzyme catalyzes is as follows:

1. Glucosyl transfer
$$GF + Acceptor \rightarrow G\text{-}Acceptor + F$$
2. Polymerization
$$n\,GF \rightarrow G_n + nF$$
3. Hydrolysis
$$GF + H_2O \rightarrow G + F$$
4. Isotope exchange
$$GF + F^* \rightarrow GF^* + F$$

where GF is sucrose, G glucose, F fructose, and F* [^{14}C]fructose.

[12] M. K. Bhattacharjee and R. M. Mayer, *Carbohydr. Res.* **142**, 277 (1985).

[13] T. P. Binder and J. F. Robyt, *Carbohydr. Res.* **124**, 287 (1983).

[14] T. P. Binder, and G. L. Cote, and J. F. Robyt, *Carbohydr. Res.* **124**, 275 (1983).

[15] J. F. Robyt and S. H. Eklund, *Carbohydr. Res.* **121**, 279 (1983).

[16] M. K. Bhattacharjee and R. M. Mayer, *Carbohydr. Res.*, in press.

Each reaction has specific characteristics that will be discussed. The first two represent reaction pathways that can lead to the formation of polymers. Reaction (1) is dependent on the availability of acceptors. The characteristics of acceptor substrates have been described. Low molecular weight acceptors yield homologous series of oligosaccharides, which suggests that the initial products can serve as acceptor substrates for subsequent steps.[3] Thus chain growth occurs by a series of single glucosyl transfers. This suggests that, under these conditions, the enzyme operates by a multichain mechanism. The addition of new sugars by reaction (1) is at nonreducing positions, which makes this pathway analogous to the formation of glycogen and starch. Chemical analysis and *nmr* spectroscopic data indicate that the new bonds are $\alpha(1 \rightarrow 6)$.

The polymerization pathway [reaction (2)] is distinguished by the fact that it can proceed in the absence of added acceptors. A critical issue is whether endogenous acceptors are present in the enzyme preparation. The manner by which the *S. sanguis* enzyme is isolated tends to preclude the copurification of bound saccharides. The utilization of high concentrations of SDS would have the effect of denaturing the enzyme and destroying any noncovalent sugar binding sites. It has been observed that all detectable sugar is separated from the enzyme during gel permeation in SDS. As already discussed in the Properties section, analyses of the purified enzyme indicate that there is less than one residue of sugar per molecule of enzyme. Thus it appears unlikely that there is any significant sugar present with the enzyme. In the absence of added acceptors the initial products appear to remain tightly associated with the enzyme and may in fact be covalently attached.[17] The products formed in reaction (1), on the other hand, are readily dissociated.

Short reaction times (10 sec) in the absence of acceptors yield moderate length oligosaccharides (average chain length, 17–18). The direction of chain growth also is different from that observed in reaction (1). Isotope at the reducing end of the oligomers formed in a pulse with [^{14}C]sucrose and during a chase with cold sucrose was measured subsequent to quantitative release of the sugars from the enzyme.[18] The results showed a rapid decrease in the radioactivity at the reducing terminus as a function of time of the chase. This indicates that addition of new sugars occurs at the reducing end and probably proceeds by a mechanism whereby the new glucosyl residues are inserted between the growing chain and the enzyme. A mechanism of this type was proposed by Ebert and Schenk[19]

[17] V. K. Parnaik, G. A. Luzio, D. A. Grahame, S. L. Ditson, and R. M. Mayer, *Carbohydr. Res.* **121,** 257 (1983).

[18] S. L. Ditson and R. M. Mayer, *Carbohydr. Res.* **126,** 170 (1984).

[19] K. H. Ebert and G. Schenk, *Adv. Enzymol.* **30,** 179 (1968).

and supported by observations of Robyt *et al.*,[20] who utilized the enzyme from *Leuconostoc mesenteroides*.

Reaction (2) must represent a route for the *de novo* formation of dextran, and hence must involve initiation, propagation, and termination steps. The tight association between the growing polymer and the enzyme suggests that the protein may serve as the primer for chain growth. Initiation would occur when the first glucose residue becomes attached to the enzyme. The propagation of the polymer appears to proceed by a single-chain, or multirepetitive, mechanism, whereby once the chain is initiated, multiple additions are made prior to release. The termination step is not understood, but evidence suggests that the enzyme-bound saccharides are slowly released. The free oligosaccharides, under typical reaction conditions, could then serve as acceptors in glucosyl transfer steps [reaction (1)]. This would mean that polymerization [reaction (2)] represents a route by which the enzyme generates its own acceptors for glucosyl transfer.

Reaction (3) may be considered to have strong analogies to reaction (1) if water is viewed as the acceptor. The rate of this reaction is slow relative to the others, but occurs under all sets of conditions.[21] Reaction (4) might be considered analogous to reactions (1) and (3); however, the fact that sucrose is reformed distinguishes it to some degree. Bond formation at the hemiketal carbon of fructose is quite different than bond formation at any of the other hydroxyl groups since the product is isoenergetic with the substrate and has a high energy bond. This reaction occurs predominantly at short reaction times and in the presence of high fructose concentrations.

The four reactions compete with one another. For example, it has been observed that reaction (2) diminishes as acceptors are added to reaction mixtures and that glucose is incorporated into low molecular weight products [reaction (1)]. The fact that they are competitive suggests that there is an intermediate common to all the reactions. Furthermore, the nature of the reactions suggests that such an intermediate would be a glucosylated form of the enzyme.

A glucosylated enzyme derivative has been isolated by using enzyme that has been immobilized on hydroxylapatite.[21] Enzyme immobilized in this way is completely active and offers the operational advantage of easy and rapid removal of substrates from reaction mixtures by centrifugation or filtration. Incubation of the immobilized enzyme with [[14]C-*glucose*]sucrose results in extensive labeling of the protein within 10–15 sec. Studies

[20] J. F. Robyt, B. K. Kimble, and T. F. Walseth, *Arch. Biochem. Biophys.* **165,** 634 (1974).
[21] G. A. Luzio and R. M. Mayer, *Carbohydr. Res.* **111,** 311 (1983).

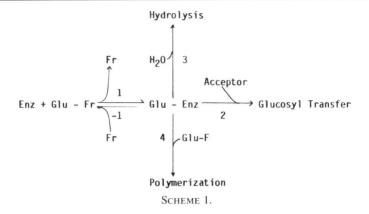

Scheme 1.

on the labeled (charged) enzyme have shown that two types of saccharides are covalently bound: monomeric glucose and glucose oligosaccharides with average chain lengths of 17–18. The latter saccharides have already been described. The charged enzyme was utilized in a series of studies designed to determine if the monomeric glucosyl residues could participate in the four reactions described above. When acceptors, such as maltose, are added to the labeled enzyme, the amount of monomeric glucose declined and a corresponding amount of the labeled trisaccharide was formed.[22] Similarly, the addition of cold sucrose resulted in the rapid transfer of the radioactive glucose residues to the oligosaccharide fraction, and an increase in the average chain length. Increasing the reaction times to 15 min resulted in the formation of chains with degrees of polymerization of greater than 100 units.

It has been suggested that the enzyme contains two equivalent active sites to which glucosyl residues may become bonded.[20] Polymerization would then occur by the sequential alternating transfer of glucosyl, isomaltosyl, isomaltotriosyl, etc., residues between the two sites as new monomeric glucose moieties become bound. When the charged enzyme was incubated in buffer a slow release of free glucose was observed. On the other hand, the addition of fructose to the charged enzyme yielded radioactive sucrose. The results of these studies support the idea that glucosylation of dextransucrase is the first step in the catalytic pathway. This forms an intermediate that is central to all the described reactions. In Scheme 1, which summarizes these results, it can be seen that reaction (1) occurs by the combination of steps 1 and 2, while reaction (2) is the sum of

[22] G. A. Luzio, V. K. Parnaik, and R. M. Mayer, *Carbohydr. Res.* **121,** 269 (1983).

steps 1 and 4. The isotope exchange reaction (3) is represented by steps 1 and -1 and hydrolysis (4) by steps 1 and 3.

Glucosylated enzyme intermediates have been described for other glycosyltransferases such as sucrose phosphorylase[23] and maltase (α-glucosidase).[24] A fructosyl enzyme also has been observed for levansucrase,[25] a fructosyltransferase analogous to dextransucrase.

The dextran formed by *S. sanguis* contains between 70 and 75% $\alpha(1 \rightarrow 6)$ bonds and 25–30% $\alpha(1 \rightarrow 3)$ bonds. The mechanism by which the two bond types are generated has not been established; however, studies with other systems suggest two possible pathways. Robyt and Taniguichi[26] have suggested that branch formation occurs when a nascent dextranyl chain is transferred from the enzyme to an acceptor. Subsequently, the same laboratory[27] reported on the presence of a single enzyme in *Leuconostoc mesenteroides* that catalyzes the formation of both $\alpha(1 \rightarrow 6)$ and $\alpha(1 \rightarrow 3)$ bonds. In long-term reactions the enzyme isolated from *S. sanguis* produces a polymer that contains both $\alpha(1 \rightarrow 6)$ and $\alpha(1 \rightarrow 3)$ bonds. The structure of the polymer formed on the enzyme must yet be examined. However, in experiments using charged immobilized enzyme, transfer of dextranyl chains to acceptors was not observed. It would, therefore, appear likely that the formation of $\alpha(1 \rightarrow 3)$ bonds does not proceed by dextranyl transfer.

[23] J. G. Voet and R. H. Abeles, *J. Biol. Chem.* **245,** 1020 (1970).
[24] H. L. Lai, L. G. Butler, and B. Axelrod, *Biochem. Biophys. Res. Commun.* **60,** 635 (1974).
[25] R. Chambert, G. Treboul, and R. Dedonder, *Eur. J. Biochem.* **41,** 285 (1971).
[26] J. F. Robyt and H. Taniguchi, *Arch. Biochem. Biophys.* **165,** 634 (1976).
[27] G. L. Cote and J. F. Robyt, *Carbohydr. Res.* **101,** 57 (1982).

[58] Glycosylation Inhibitors for N-Linked Glycoproteins

By ALAN D. ELBEIN

Introduction

Inhibitors of specific metabolic reactions can be valuable tools for studying complex biochemical pathways. Thus, a given inhibitor may stop the metabolic pathway at a specific step and allow the accumulation of an intermediate in the pathway, or such an inhibitor may cause the formation of a product with altered structure. In addition, if these inhibitors can be used in cell culture, valuable information can be obtained

METHODS IN ENZYMOLOGY, VOL. 138

regarding the importance of a given pathway to the cell or the necessity for a given structure to be produced intact and without variation. Some of the studies described in this review on the N-linked glycoproteins demonstrate the usefulness of this type of approach to examine physiological functions. In this chapter, the various inhibitors that affect glycosylation of the N-linked glycoproteins will be considered. Since N-linked glycosylation involves two rather distinct series of reactions, i.e., the formation of lipid-linked saccharides, which donate carbohydrate to protein, and then modification of the oligosaccharide chains on the protein, the various inhibitors will be considered in separate sections depending on which of these pathways they affect. In addition, certain other inhibitors, such as those affecting protein synthesis or those affecting the formation of the polyisoprenyl-lipid carrier will be considered because they also affect the formation of N-linked glycoproteins. For purposes of clarity, the appropriate pathway of biosynthesis will be briefly outlined in each section so that specific sites of inhibition can be discussed.

Inhibitors of the Lipid-Linked Saccharide Pathway

The first series of reactions in the biosynthesis of the N-linked oligosaccharides, frequently referred to as the dolichol cycle, involves the formation of a series of lipid-linked saccharide intermediates leading to the production of $Glc_3Man_9GlcNAc_2$-pyrophosphoryldolichol.[1,2] Then, in the final stage of this dolichol cycle, the oligosaccharide chain is transferred to asparagine residues on the protein.[3] The sequence Asn-X-Ser (Thr) is necessary, but not sufficient, for the protein to be glycosylated, since not all sequences of this type carry carbohydrate chains.[4,5] Probably protein conformation, especially around the potential glycosylation sites, is also critical. Figure 1 outlines the reactions of the dolichol cycle.

The formation of lipid-linked saccharides begins with the transfer of GlcNAc-1-P, from UDP-GlcNAc, to dolichyl-P, catalyzed by the membrane-bound enzyme, UDP-GlcNAc-dolichyl-P : GlcNAc-1-P transferase.[6,7] This enzyme and the other enzymes of the dolichol cycle are located in the endoplasmic reticulum membranes, and perhaps in other membranes as well. The GlcNAc-1-P transferase is the site of action of the antibiotic tunicamycin (see below). The second GlcNAc is

---

[1] E. Li, I. Tabas, and S. Kornfeld, *J. Biol. Chem.* **253**, 7762 (1978).

[2] T. Liu, B. Stetson, S. J. Turco, S. C. Hubbard, and P. W. Robbins, *J. Biol. Chem.* **254**, 4554 (1979).

[3] M. Kiely, G. S. McKnight, and R. T. Schimke, *J. Biol. Chem.* **251**, 5490 (1976).

[4] R. Kornfeld and S. Kornfeld, *Annu. Rev. Biochem.* **45**, 217 (1976).

[5] R. D. Marshall, *Annu. Rev. Biochem.* **41**, 673 (1974).

[6] A. Heifetz, R. W. Keenan, and A. D. Elbein, *Biochemistry* **18**, 2186 (1979).

[7] R. K. Keller, D. Y. Boon, and F. C. Crum, *Biochemistry* **18**, 3946 (1979).

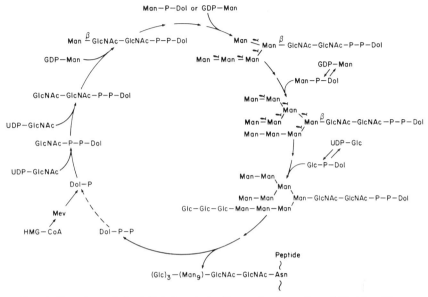

FIG. 1. Reactions of the dolichol cycle leading to the formation of $Glc_3Man_9GlcNAc_2$-pyrophosphoryldolichol.

then transferred to the GlcNAc-pyrophosphoryldolichol to form GlcNAcβ1,4GlcNAc-pyrophosphoryldolichol.[8,9] A β-linked mannose residue is next added from GDP-mannose to the terminal GlcNAc residue to give a Manβ1,4GlcNAcβ1,4GlcNAc-pyrophosphoryldolichol.[10–12] Four α-linked mannose residues are then added (one α1,3; one α1,6; two α1,2) to give the Man$_5$GlcNAc$_2$-pyrophosphoryldolichol [i.e., Manα1,2Manα-1,2Manα1,3(Manα1,6)ManβGlcNAcGlcNAc-pyrophosphoryldolichol]. All four of these α-linked mannose residues have been shown to come directly from GDP-mannose.[13–17] Four more α-linked mannose

[8] A. Hercovics, C. D. Warren, B. Bugge, and R. W. Jeanloz, *J. Biol. Chem.* **253**, 160 (1978).

[9] W. W. Chen and W. J. Lennarz, *J. Biol. Chem.* **252**, 3473 (1977).

[10] J. A. Levy, H. Carminatti, A. I. Cantarella, N. H. Behrens, L. F. Leloir, and E. Tabora, *Biochem. Biophys. Res. Commun.* **60**, 118 (1974).

[11] J. F. Wedgewood, C. D. Warren, R. W. Jeanloz, and J. L. Strominger, *Proc. Natl. Acad. Sci. U.S.A.* **71**, 5022 (1974).

[12] A. Heifetz and A. D. Elbein, *Biochem. Biophys. Res. Commun.* **75**, 20 (1977).

[13] W. W. Chen and W. J. Lennarz, *J. Biol. Chem.* **252**, 3473 (1977).

[14] A. Chapman, E. Li, and S. Kornfeld, *J. Biol. Chem.* **254**, 10243 (1979).

[15] A. Chapman, K. Fujimoto, and S. Kornfeld, *J. Biol. Chem.* **255**, 4441 (1980).

[16] J. P. Chambers, W. T. Forsee, and A. D. Elbein, *J. Biol. Chem.* **252**, 2498 (1977).

[17] W. T. Forsee, J. A. Griffin, and J. S. Schutzbach, *Biochem. Biophys. Res. Commun.* **75**, 799 (1977).

residues are then added from dolichylphosphorylmannose to produce the Man$_9$GlcNAc$_2$-pyrophosphoryldolichol [Manα1,2Manα1,2Manα-1,3(Manα1,2Manα1,3(Manα1,2Manα1,6)Manα1,6)ManβGlcNAcGlcNAc-pyrophosphoryldolichol].[16,17] Three glucose residues are finally added from dolichylphosphorylglucose (two in α1,3 linkages and the terminal one in 1,2 linkage) to give the Glc$_3$Man$_9$GlcNAc$_2$-pyrophosphoryldolichol.[18,19] It is not absolutely clear whether all of the mannose residues have to be added to the lipid carrier before any of the glucose residues can be attached. Since the 1,3-branch is biosynthesized first, it seems possible that it could act as a glucose acceptor even in the absence of complete mannosylation. In the final step of this dolichol cycle, the oligosaccharide is transferred to protein, while the protein is being synthesized on membrane-bound polysomes.[20–22]

Tunicamycins, Streptovirudins, and Related Antibiotics

In 1971, Takatsuki, Arima, and Tamura described the isolation from *Streptomyces lysosuperificus* of a glucosamine-containing antibiotic that showed antiviral activity.[23] This antibiotic, which they named tunicamycin because it appeared to affect the cell coat or tunic, has been extensively used as a biochemical tool to examine the effect of protein glycosylation (or lack of glycosylation) on the role of carbohydrate in glycoprotein function.

Tunicamycin is a nucleoside antibiotic whose structure is shown in Fig. 2. The chemistry of the antibiotic was elucidated by the elegant studies of Tamura and associates, largely through characterization of the products resulting from acid hydrolysis.[23,24] As shown in Fig. 2, the antibiotic is composed of uracil, fatty acid, and two glycosidically linked sugars. The sugars are *N*-acetylglucosamine and an unusual 11-carbon aminodeoxydialdose called tunicamine. The tunicamine is linked to the uracil in an N-glycosidic bond, and is itself substituted at two positions. At the anomeric carbon, the *N*-acetylglucosamine is bound in an O-glycosidic bond, while at the amino group a long-chain fatty acid is attached in an amide bond.

[18] T. L. Huffaker and P. W. Robbins, *Proc. Natl. Acad. Sci. U.S.A.* **80,** 7466 (1983).

[19] R. J. Staneloni, R. A. Ugalde, and L. F. Leloir, *Eur. J. Biochem.* **105,** 275 (1980).

[20] E. Rodriguez-Boulan, G. Kreibich, and D. Sabatini, *J. Cell Biol.* **78,** 874 (1978).

[21] V. R. Lingappa, J. R. Lingappa, R. Prasad, K. Ebner, and G. Blobel, *Proc. Natl. Acad. Sci. U.S.A.* **75,** 2338 (1978).

[22] R. C. Das and E. C. Heath, *Proc. Natl. Acad. Sci. U.S.A.* **77,** 3811 (1980).

[23] A. Takatsuki, K. Arima, and G. Tamura, *J. Antibiot.* **24,** 215 (1971).

[24] A. Takatsuki, K. Kawamura, M. Okima, Y. Kodama, T. Ito, and G. Tamura, *Agric. Biol. Chem.* **41,** 2307 (1977).

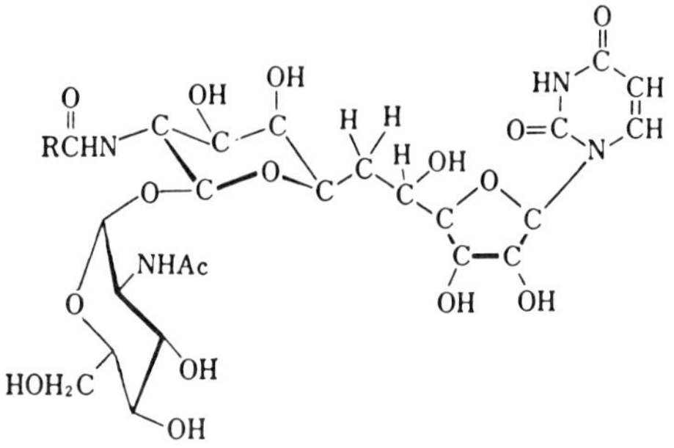

FIG. 2. Structures of the various tunicamycin components. R refers to the various fatty acids shown below.

I : R= (CH₃)₂CH(CH₂)₇CH=CH—   VI : R= (CH₃)₂CH(CH₂)₁₁—
II :   (CH₃)₂CH(CH₂)₈CH=CH—   VII :   (CH₃)₂CH(CH₂)₁₀CH=CH—
III :   CH₃(CH₂)₁₀CH=CH—   VIII :   CH₃(CH₂)₁₂CH=CH—
IV :   CH₃(CH₂)₁₁CH=CH—   IX :   CH₃(CH₂)₁₃CH=CH—
V :   (CH₃)₂CH(CH₂)₉CH=CH—   X   :   (CH₃)₂CH(CH₂)₁₁CH=CH—

Tunicamycin is actually produced by a number of different *Strepto-myces* species as a mixture of a number of closely related structures. Using high-performance liquid chromatography with a reverse-phase column, tunicamycin has been separated into as many as 10 different components that differ in molecular weight from 802 to 872.[25-27] These differences in molecular weight have been shown to be attributable to differences in the fatty acid components, which may vary in chain length from C_{13} to C_{17}, in structure from normal to branched, or from saturated to unsaturated. The streptovirudins are chemically related to the tunicamy-cins but generally have shorter chain fatty acids.[28] Thus, the molecular weights of the streptovirudins vary from 790 to 816, but other biological and chemical properties of the streptovirudins are similar to those of tunicamycin.[29] However, the streptovirudins are not quite as effective as inhibitors as the tunicamycins (see below). Several other related antibiot-ics have been described. These are called mycospocidin,[30] antibiotic

[25] T. Ito, A. Takatsuki, K. Kawamura, K. Soto, and G. Tamura, *Agric. Biol. Chem.* **44**, 695 (1980).

[26] W. C. Mahoney and D. Duksin, *J. Chromatogr.* **198**, 506 (1980).

[27] R. W. Keenan, R. L. Hamill, J. L. Occolowitz, and A. D. Elbein, *Biochemistry* **70**, 2968 (1980).

[28] K. Eckardt, H. Thrum, G. Bradler, E. Tonew, and M. Tonew, *J. Antibiot.* **28**, 274 (1975).

[29] A. D. Elbein, J. Gafford, and M. S. Kang, *Arch. Biochem. Biophys.* **196**, 311 (1979).

[30] J. Tkacz and A. Wong, *Fed. Proc., Fed. Am. Soc. Exp. Biol.* **37**, 1766 (1978).

24010,[31] and antibiotic MM 19290.[32] The site of action of these compounds appears to be the same as that of tunicamycin,[29] but the detailed chemistry of these compounds has not been reported.

In some studies, the tunicamycin complex was shown to inhibit the incorporation of [³H]leucine into protein, while in other cases, no inhibition was observed, or protein synthesis was not examined. Thus, at tunicamycin levels of 2–10 µg/ml, [³H]glucosamine incorporation is usually inhibited 65–90%, while [³H]leucine incorporation may be blocked by 10–30%. However, it is still not clear whether this inhibition of amino acid incorporation is due to one, or all, of the tunicamycin components, or whether the complex is contaminated with another compound that is an inhibitor of protein synthesis. In some studies, the various tunicamycin analogs were separated by HPLC, and none of the individual components inhibited leucine incorporation.[33] However, in some other cases, one of the HPLC components was reported to inhibit protein synthesis.[34] These disparities indicate that additional studies are needed to clarify this important point. In addition, it should be pointed out that the amount of tunicamycin necessary to block protein glycosylation should be determined for each biological system, and the lowest amount giving about 90% inhibition should be used. It is also advisable to examine protein synthesis at these tunicamycin levels to be certain that this is not a factor in considerations on carbohydrate function.

Using a particulate enzyme fraction from animal cells, Tkacz and Lampen showed that tunicamycin inhibited the first step in the lipid-linked saccharide pathway, i.e., the formation of dolichylpyrophosphoryl-GlcNAc.[35] The formation of dolichylpyrophosphoryl-GlcNAc was also demonstrated in particulate enzyme fractions of plants, and the plant UDP-GlcNAc-dolichyl-P : GlcNAc-1-P transferase was also very sensitive to tunicamycin.[36] However, the enzyme that adds the next GlcNAc from UDP-GlcNAc to dolichyl-PP-GlcNAc to form dolichyl-PP-GlcNAcβ1,4GlcNAc was not inhibited by tunicamycin,[37] nor were the GlcNAc transferases that add terminal GlcNAc residues to the mannose chains.[38]

[31] M. Mizano, Y. Shimojima, T. Sugawara, and L. Takeda, *J. Antibiot.* **24**, 896 (1971).
[32] M. Kenig and C. Reading, *J. Antibiot.* **32**, 549 (1979).
[33] A. D. Elbein, J. L. Occolowitz, R. L. Hamill, and K. Eckardt, *Biochemistry* **20**, 4210 (1981).
[34] D. Duksin and W. C. Mahoney, *J. Biol. Chem.* **257**, 3105–3109 (1982).
[35] J. S. Tkacz and J. O. Lampén, *Biochem. Biophys. Res. Commun.* **65**, 248 (1975).
[36] M. Ericson, J. Gafford, and A. D. Elbein, *J. Biol. Chem.* **252**, 7431 (1977).
[37] L. Lehle and W. Tanner, *FEBS Lett.* **71**, 167 (1976).
[38] D. K. Struck and W. J. Lennarz, *J. Biol. Chem.* **252**, 1007 (1977).

The mechanism of action of tunicamycin was examined with the solubilized GlcNAc-1-P transferase from hen oviduct,[7] pig aorta,[6] and chick embryo microsomes.[39] Structurally, tunicamycin bears a striking resemblance to the sugar nucleotide substrate, UDP-GlcNAc, and therefore it might be expected to be a competitive inhibitor with respect to this substrate. However, the inhibition could not be due to this structural relationship alone, since tunicamycin does not affect various other GlcNAc transferases. With the aorta GlcNAc-1-P transferase, competitive inhibition could not be demonstrated probably because of the strong affinity of the inhibitor for this enzyme.[6] Interestingly enough, when the substrate concentration studies were done with UDP-GlcNAc in the presence of various amounts of streptovirudin, it was possible to show that this inhibition was of a competitive type.[29] Probably the reason for this is that streptovirudin has a much lower affinity for the transferase than does tunicamycin and therefore the streptovirudin inhibition is much easier to reverse with high concentrations of UDP-GlcNAc. On the other hand, with the oviduct enzyme[7] or with chick embryo microsomes,[39] the data were more indicative of the tunicamycin being a reversible, competitive inhibitor with respect to UDP-GlcNAc. It was suggested that tunicamycin might act as a substrate–product transition state analog, binding irreversibly to the enzyme.

In oviduct slices or in whole cells, tunicamycin inhibited the incorporation of [^{3}H]glucosamine into lipid-linked monosaccharides and lipid-linked oligosaccharides, and greatly reduced incorporation into protein.[40] The incorporation of [^{3}H]mannose into lipid-linked oligosaccharides and into protein in the presence of tunicamycin was also greatly reduced, but incorporation into dolichylphosphorylmannose was not altered.[41] Cells incubated for sufficient periods of time with tunicamycin generally produce the nonglycosylated form of the N-linked glycoprotein. This can readily be shown by comparing the gel electrophoresis patterns of the tunicamycin-induced proteins to the normal proteins, before and after treatment with endo-β-N-acetylglucosaminidases, such as endoglucosaminidase H.[41,42] Tunicamycin has been used by some investigators to prove that certain proteins have N-linked oligosaccharides. The problem

[39] A. Takatsuki and G. Tamura, in "Tunicamycin" (G. Tamura, ed.), p. 35. Jpn Sci. Press, Tokyo, 1982.
[40] D. K. Struck and W. J. Lennarz, in "The Biochemistry of Glycoproteins and Proteoglycans." (W. J. Lennarz, ed.), p. 35, Plenum Press, New York, 1980.
[41] M. M. Sveda, L. J. Markoff, and C.-J. Lai, Cell (Cambridge, Mass.) 30, 649 (1982).
[42] J. P. Merlie, R. Sebbane, S. Tzartos, and J. Linström, J. Biol. Chem. 257, 2694 (1982).

in some of these cases is that the tunicamycin complex (i.e., the mixture of tunicamycin components isolated and purified from *Streptomyces* and available commercially) inhibits protein synthesis to variable degrees depending on the cell system in question. Therefore, one should check the incorporation of leucine, or some other amino acid, into protein in the presence of tunicamycin to be certain that this is not inhibited, before deciding whether the oligosaccharide is involved in glycoprotein function.

In most studies with tunicamycin, the protein portion of the molecule is still synthesized in the presence of antibiotic, but it may be subjected to more rapid proteolysis (see below). Although these results would indicate that the synthesis of the protein is not linked to glycosylation, a regulatory link between protein synthesis and glycosylation has been proposed in some studies. For example, in studies on yeast carboxypeptidase Y[43] and barley α-amylase,[44] the inhibition of glycosylation by tunicamycin led to decreased amounts of the protein. The unglycosylated form of the yeast carboxypeptidase was found to be as stable in the cell extracts as was the normal glycoprotein, strongly suggesting that the decreased levels of protein observed in the presence of inhibitor were not due to proteolysis. A significant difference between *in vivo* and *in vitro* inhibition of protein synthesis was observed with regard to thyroglobulin synthesis.[45] In these studies, tunicamycin did not significantly inhibit protein synthesis *in vitro* (>4%), but *in vivo* it was inhibited more than 30%. One explanation for some of these results is that there could be a decreased stability of certain proteins in the absence of glycosylation, but another possibility is that a regulatory link does exist between the synthesis of carbohydrate and the synthesis of the protein.

The consequences of the protein not being glycosylated may vary widely, depending on the glycoprotein in question. Thus, in many cases, the absence of carbohydrate caused marked alterations in the fate and properties of the molecule. For example, the immunoglobulins IgM and IgA are inefficiently secreted when these proteins are synthesized in the presence of tunicamycin.[46] In these experiments, electron microscopic examination and direct immunofluoresence showed the presence of distended rough endoplasmic reticulum (RER) that contained a dense granular precipitate that was identified as immunoglobulin. The authors suggested that the absence of sugar resulted in an alteration in the physical properties of the immunoglobulins and led to decreased solubility and

[43] A. Hasilik and W. Tanner, *Eur. J. Biochem.* **91,** 567 (1978).
[44] H. Schwaiger and W. Tanner, *Eur. J. Biochem.* **102,** 375 (1979).
[45] M. J. Seagar, R. D. Miquelis, and C. Simon, *Eur. J. Biochem.* **113,** 91 (1980).
[46] S. Hickman, A. Kulczycki, Jr., R. G. Lynch, and S. Kornfeld, *J. Biol. Chem.* **252,** 4402 (1977).

increased aggregation. Another study using tunicamycin suggested that the specific carbohydrate residues and polypeptide portions of a large and multiglycosylated glycoprotein can serve different roles in the processing and secretion of that glycoprotein.[47] In that study, tunicamycin largely blocked the secretion of IgM in the wild-type cells and in three mutant cell lines (producing abnormal μ chains), but it caused less inhibition in the secretion in two other mutant cell lines. The immunoglobulin IgG_{2A} secreted by the hybridoma line M31 was found to contain covalently linked sulfate. The sulfate appeared to be in the heavy chain bound to N-linked oligosaccharides that were located in the Fab portion. Strangely enough, the unglycosylated IgG secreted in the presence of tunicamycin was not unsulfated but contained 4 times as much sulfate on the heavy chain as control IgG. In this case, the sulfate was localized in the Fc portion and was bound to tyrosine residues.[48]

At 5 μg/ml, tunicamycin blocked glycosylation and markedly affected thyroxine-binding globulin secretion, almost doubling the time required for the secretion of 50% of the protein.[49] The migration of the envelope proteins of vesicular stomatitis virus (G protein) and Sindbis virus from the RER to the plasma membrane is also impaired when the proteins are synthesized in the presence of the antibiotic.[50] In chick cells, tunicamycin abolished the glycosylation of the acetylcholine receptor and also inhibited the accumulation of unglycosylated receptors in intact cells.[51] In addition, the acetylcholine receptors that were incorporated into cell membranes of tunicamycin-treated cells had a greatly increased rate of degradation. In 3T3 adipocytes, tunicamycin rapidly decreased the number of insulin receptors on the cell surface and this resulted in decreased hexose transport and decreased sensitivity toward antibody against the insulin receptor.[52]

Depending on culture conditions, exposure to tunicamycin caused either a marked inhibition of induction of hydroxymethylglutaryl-CoA reductase (HMG-CoA reductase) activity, or, under steady-state conditions, a marked reduction in enzymatic activity.[53] In one study, tunicamycin treatment of fibroblasts impaired the high affinity binding of low density lipoprotein and also the intracellular degradation,[54] but an-

[47] C. Sidman, M. J. Potash, and G. Kohler, *J. Biol. Chem.* **256,** 13180 (1981).

[48] P. A. Bauerle and W. B. Huttner, *EMBO J.* **3,** 2209–2215 (1984).

[49] L. Bartalena and J. Robbins, *J. Biol. Chem.* **259,** 13610 (1984).

[50] R. Leavitt, S. Schlesinger, and S. Kornfeld, *J. Biol. Chem.* **252,** 9018 (1977).

[51] J. Prives and D. Bar-Sagi, *J. Biol. Chem.* **258,** 1775 (1983).

[52] O. M. Rosen, G. H. Chia, C. Fung, and C. S. Rubin, *J. Cell. Physiol.* **99,** 37 (1979).

[53] J. J. Volpe and R. I. Goldberg, *J. Biol. Chem.* **258,** 9220 (1983).

[54] O. Filipovic and K. von Figura, *Biochem. J.* **186,** 373 (1980).

other study concluded that the drug decreased the number of LDL receptors, but did not affect their function nor the internalization or degradation of LDL.[55] In terms of the lysosomal enzyme systems, tunicamycin interfered with the expression of phosphomannose receptors on the cell surface of fibroblasts and the synthesis of phosphomannosyl residues on lysosomal enzymes after an initial lag phase.[56] However, tunicamycin did not change the amount of the major cell surface glycoprotein appearing at the surface of chick embryo fibroblasts, but it did decrease the amount present in fibroblast cultures.[57] The decrease in total cell surface glycoproteins was apparently due to an increased degradation in cells treated with tunicamycin.

The hypothesis that the oligosaccharide chains may play a key role in protecting the protein from proteolytic attack is indicated from a number of studies. Thus, the carbohydrate-free forms of chick fibronectin,[57] of the ACTH-endorphin precursor,[58] and of the hemagglutinin of fowl plague virus[59] are unusually susceptible to proteolytic digestion in vivo. In the case of the fowl plague virus, the unglycosylated hemagglutinin could be isolated when a protease inhibitor such as N-α-p-tosyl-L-lysine chloromethyl ketone (TLCK) was included in the incubation medium. In other studies along similar lines, it was shown that IgM produced in a mouse β-lymphoma in the presence of tunicamycin was much more sensitive to trypsin digestion than was the normal glycoprotein,[60] while removal of the carbohydrate chains of invertase by digestion with endoglucosaminidase H also reduced the stability of this protein to trypsin.[61] In the absence of carbohydrate, some proteins are less soluble[62] or have altered conformations,[61,63] or may be altered in their intracellular transport properties.[64,65] On the other hand, the envelope protein of murine leukemia virus is not proteolytically processed unless it has been glycosylated.[66] Also, in mouse pituitary cells inhibited with tunicamycin the α- and β-subunits of

[55] S. Chatterjee, C. S. Sekerke, and P. O. Kwiterovich, Jr., Eur. J. Biochem. 101, 103 (1981).
[56] K. Olden, J. B. Parent, and S. L. White, Biochim. Biophys. Acta 650, 209 (1982).
[57] K. Olden, R. M. Pratt, and K. M. Yamada, Cell (Cambridge, Mass.) 13, 461 (1978).
[58] Y. P. Lou and H. Gainer, Endocrinology (Baltimore) 105, 474 (1979).
[59] R. J. Schwarz, J. M. Rohrschneider, and M. F. G. Schmidt, J. Virol. 19, 782 (1976).
[60] C. H. Sibley and R. A. Wagner, J. Immunol. 126, 1868 (1981).
[61] F. K. Chu, R. B. Trimble, and F. Maley, J. Biol. Chem. 253, 8691 (1978).
[62] A. L. Tarentino, T. H. Plummer, Jr., and F. Maley, J. Biol. Chem. 249, 818 (1974).
[63] J. H. Pazur, H. R. Knull, and D. L. Simpson, Biochem. Biophys. Res. Commun. 40, 110 (1970).
[64] F. Melchers, Biochemistry 12, 1471 (1973).
[65] I. S. Trowbridge, R. Hyman, and C. Mazauskas, Cell (Cambridge, Mass.) 14, 21 (1978).
[66] M. Bielinska, G. A. Grant, and I. Boime, J. Biol. Chem. 253, 7117 (1978).

thyroid-stimulating hormone are synthesized, but these subunits apparently require the carbohydrate to form active hormone.[67]

However, in a number of cases, lack of carbohydrate had little or no effect on the location of function of the protein in question. For example, the secretion of the serum glycoproteins, transferrin and very low density lipoproteins, by cultured hepatocytes was not prevented or inhibited by tunicamycin, even though these proteins were shown to be devoid of carbohydrate.[68,69] Likewise, the secretion of interferon,[70,71] IgG,[72] ovalbumin[73] and the α-subunit of human chorionic gonadotropin[74] are not impaired in tunicamycin-treated cells. In addition, the signal peptides of placental prelactogen and the α-subunit of human chorionic gonadotropin are proteolytically processed in the presence of tunicamycin, even though they lack carbohydrate.[66] Various glycoprotein enzymes, such as alkaline phosphatase of yeast, are still active even in the unglycosylated form.[75]

Surveying the above results, one is led to the conclusion that the oligosaccharide chains must have different functions in different proteins. In some cases, the carbohydrate may participate in, and be essential for, the proper folding of the protein. In other cases, the oligosaccharide apparently increases the solubility of the protein, or protects the protein from proteolytic degradation. In still other situations, the carbohydrate may be involved in intracellular trafficking to target a specific protein or group of proteins to selected sites in the cell. An excellent example of this latter situation is the lysosomal enzyme system where the recognition event involves mannose 6-phosphate residues on the high-mannose oligosaccharides. Other types of recognition signals may be involved in the targeting of other N-linked glycoproteins such as membrane proteins to their ultimate destination.

Perhaps the important point in all of this is that the role played by carbohydrate may depend to a great extent on the amino acid sequence of

[67] B. D. Weintraub, B. S. Stannard, D. Linnekin, and M. Marshall, *J. Biol. Chem.* **255,** 5715 (1980).

[68] K. Edwards, M. Nagashina, H. Dryburgh, A. Wykes, and G. Schreiber, *FEBS Lett.* **100,** 269 (1979).

[69] D. K. Struck, P. R. Suita, M. D. Lane, and W. J. Lennarz, *J. Biol. Chem.* **253,** 5332 (1978).

[70] A. Mizrahi, J. A. O'Malley, W. A. Carter, A. Takatsuki, G. Tamura, and E. Sulkowski, *J. Biol. Chem.* **253,** 7612 (1978).

[71] J. I. Fujisawa, Y. Iwakura, and Y. Kawade, *J. Biol. Chem.* **253,** 8677 (1978).

[72] S. Hickman and S. Kornfeld, *J. Immunol.* **121,** 990 (1978).

[73] R. K. Keller and G. D. Swank, *Biochem. Biophys. Res. Commun.* **85,** 762 (1978).

[74] G. S. Cox, *Biochemistry* **20,** 4893 (1981).

[75] H. R. Onishi, J. S. Thacz, and J. O. Lampén, *J. Biol. Chem.* **254,** 11943 (1979).

the protein in question. Experimental evidence to support this idea was obtained using various temperature-sensitive mutants of vesicular stomatitis virus.[76] With one of these viral strains, called the San Juan strain, tunicamycin prevented viral replication by greater than 90% when the virus was grown at 38° and almost the same when grown at 30°. However, with the Orsay strain of virus, replication was again inhibited by 85–90% at 38°, but in this case inhibition was only 30–50% at 30°. When the Orsay strain was grown at 30° in the presence of tunicamycin, the unglycosylated G protein could be detected at the host cell surface indicating that it had been synthesized and transported in the normal manner. However, when the Orsay strain was grown at 37°, no G protein was found at the cell surface. In addition, differences were observed in the physical properties of the viral glycoproteins formed at various temperatures in the presence of tunicamycin. Thus, the Orsay G proteins formed at 30° in the presence of antibiotic could be solubilized by Triton X-100 whereas the protein formed at 37° remained insoluble.[76] These data suggest that different proteins have different requirements for carbohydrate in order to assume the proper conformation.[77]

Peptide Antibiotics

The structure of amphomycin, a lipopeptide antibiotic, is shown in Fig. 3. This antibiotic is produced by *Streptomyces canus*[78] and was shown to be an undecapeptide containing either 3-isododecanoic or 3-anteisododecanoic acid attached to an N-terminal aspartic acid by an amide bond.[79] In gram-positive bacteria, amphomycin inhibited the synthesis of the cell wall peptidoglycan by preventing the transfer of phospho-N-acetylmuramoylpentapeptide from its UMP derivative to the lipid carrier, undecaprenyl phosphate.[80] Tsushimycin is another peptide antibiotic that also belongs to the amphomycin–glumamycin group.[81] While the complete structure of this inhibitor is not known, it has been shown to differ from amphomycin in amino acid and fatty acid composition. It also appears to act as an inhibitor of peptidoglycan synthesis.

Because amphomycin was shown to inhibit the formation of lipid-linked saccharides in bacteria, it was tested as an inhibitor of the forma-

[76] R. Gibson, S. Schlesinger, and S. Kornfeld, *J. Biol. Chem.* **254,** 3600 (1979).

[77] R. Gibson, S. Kornfeld, and S. Schlesinger, *Trends Biochem. Sci.* **5,** 290 (1980).

[78] B. Heinemann, M. A. Kaplan, R. D. Muir, and I. R. Hooper, *Antibiot. Chemother.* **3,** 1239 (1953).

[79] M. Bodansky, G. F. Sigler, and A. Bodansky, *J. Am. Chem. Soc.* **95,** 2352 (1973).

[80] H. Tanaka, R. Oiwa, S. Matsukura, J. Inokoshi, and S. Omura, *J. Antibiot.* **35,** 1216 (1982).

[81] T. Shoji, *Adv. Appl. Microbiol.* **24,** 187 (1978).

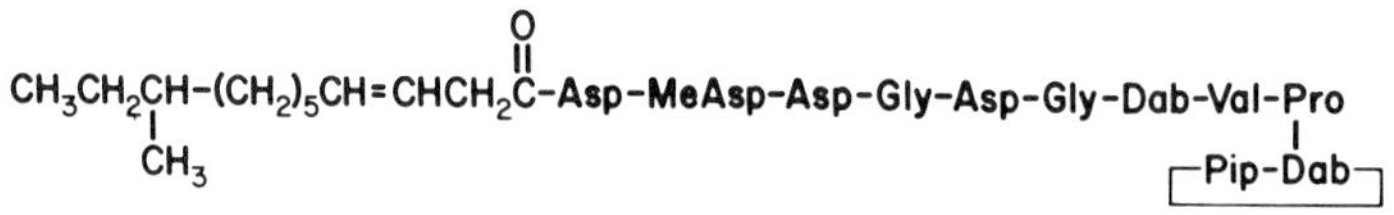

$$CH_3CH_2CH-(CH_2)_5CH=CHCH_2\overset{\overset{O}{\|}}{C}-\text{Asp-MeAsp-Asp-Gly-Asp-Gly-Dab-Val-Pro}$$

CH₃ group on the third carbon; Pip-Dab bracket below Pro.

FIG. 3. Structure of amphomycin.

tion of lipid-linked saccharides in animal cells.[82] Cell-free extracts of pig aorta have the capacity to transfer mannose from GDP-mannose, GlcNAc from UDP-GlcNAc, and glucose from UDP-glucose into lipid-linked monosaccharides (dolichylphosphorylmannose, dolichylpyrophosphoryl-GlcNAc, etc.) and lipid-linked oligosaccharides.[83] With this enzyme preparation, amphomycin was found to be a much better inhibitor of mannose incorporation into dolichylphosphorylmannose than of mannose incorporation into lipid-linked oligosaccharides. Thus, at antibiotic concentrations of 50–200 μg per incubation mixture, the transfer of mannose from GDP-mannose to dolichyl-P was inhibited more than 90%, while mannose incorporation into lipid-linked oligosaccharides was only inhibited 60–70%. This experiment suggested that all of the mannose residues in the lipid-linked oligosaccharides could not come from dolichylphosphorylmannose. Since all of the mannose incorporated into lipid-linked oligosaccharides in the presence of amphomycin was in a single oligosaccharide characterized as a Man₅GlcNAc₂, it seemed most likely that the first five mannose residues were derived from GDP-mannose rather than from dolichylphosphorylmannose.[82]

Studies from other laboratories have shown that the mannose units in the Man₅GlcNAc₂–lipid are derived from GDP-mannose. Thus, a mutant lymphoma cell line was isolated that had lost the capacity to produce dolichylphosphorylmannose.[84] This mutant organism synthesized the Man₅GlcNAc₂–lipid intermediate but was not able to add the next four mannose residues, presumably because it lacked the mannosyl donor (i.e., dolichylphosphorylmannose) for these mannose units. However, cell-free extracts of the mutant could elongate the Man₅GlcNAc₂–lipid when supplemented with dolichylphosphorylmannose. In addition, several studies have now shown the direct transfer of mannose from GDP-mannose to lipid-linked oligosaccharides by solubilized and partially purified enzyme preparations from various animal tissues.[85,86]

Amphomycin has also been used as an inhibitor of lipid-linked sac-

[82] M. S. Kang, J. P. Spencer, and A. D. Elbein, *J. Biol. Chem.* **253**, 8860 (1978).

[83] J. P. Chambers and A. D. Elbein, *J. Biol. Chem.* **250**, 6904 (1975).

[84] A. Chapman, K. Fujimoto, and S. Kornfeld, *J. Biol. Chem.* **255**, 4441 (1980).

[85] J. P. Spencer and A. D. Elbein, *Proc. Natl. Acad. Sci. U.S.A.* **77**, 2542 (1980).

[86] J. W. Jensen and J. S. Schutzbach, *J. Biol. Chem.* **256**, 12899 (1981).

charide formation in membrane preparations from brain tissue.[87] These studies were generally similar to those described above, and also showed that the antibiotic inhibited glucose transfer from UDP-glucose to dolichyl phosphate. However, the transfer of glucose from UDP-glucose to particle-bound glucan or to ceramide was not blocked by amphomycin. Since amphomycin also inhibited transfer of GlcNAc-1-P from UDP-GlcNAc to dolichyl phosphate, these workers suggested that amphomycin may form a complex with dolichyl phosphate and in this way obstruct the glycosylation reactions. Regardless of the mechanism, amphomycin also inhibits the same types of glycosylation reactions in plants.[88]

Tsushimycin is another antibiotic that is closely related to amphomycin chemically, and it appears to have the same general effect on the synthesis of lipid-linked saccharides. Thus, at concentrations of 50–200 μg/ml of this antibiotic, the formation of dolichylphosphorylmannose was almost completely inhibited but mannose was still transferred from GDP-mannose to the $Man_5GlcNAc_2$-pyrophosphoryldolichol.[89]

Bacitracin is another peptide antibiotic that is produced by certain strains of *Bacillus licheniformis*. The structure of this compound is shown in Fig. 4. Bacitracin was shown to inhibit the biosynthesis of cell wall peptidoglycan in bacteria by blocking the dephosphorylation of undecaprenyl pyrophosphate.[90] This undecaprenyl pyrophosphate must be converted to undecaprenyl phosphate in order to recycle as a carrier of sugars in peptidoglycan synthesis. Bacitracin has also been reported as an inhibitor of the biosynthesis of squalene and sterols from mevalonate,[91] formation of ubiquinones,[92,93] and degradation of thyrotropin-releasing factor and leutinizing hormone-releasing factor.[94]

Bacitracin has also been tested on a number of mammalian systems as an inhibitor of N-linked oligosaccharide synthesis. The results in these studies have been quite variable and may depend on the system in question. For example, when hen oviduct membranes were incubated with UDP-[^{14}C]GlcNAc in the presence of 1 mM bacitracin, a trisaccharide-lipid, characterized as ManβGlcNAc$_2$–lipid, accumulated.[95] This suggested that the antibiotic inhibited the addition of the first α-linked mannose residue. On the other hand, 0.2–1 mM bacitracin was reported to

[87] D. K. Bannerjee, M. G. Scher, and C. J. Waechter, *Biochemistry* **20**, 1561 (1981).

[88] M. G. Ericson, J. Gafford, and A. D. Elbein, *Arch. Biochem. Biophys.* **191**, 598 (1980).

[89] A. D. Elbein, *Biochem. J.* **193**, 477 (1981).

[90] K. J. Stone and J. L. Strominger, *Proc. Natl. Acad. Sci. U.S.A.* **68**, 3223 (1971).

[91] K. L. Stone and J. L. Strominger, *Proc. Natl. Acad. Sci. U.S.A.* **69**, 1287 (1972).

[92] N. Schechter, K. Momose, and H. Rudney, *Biochem. Biophys. Res. Commun.* **48**, 83 (1972).

[93] D. R. Storm and J. L. Strominger, *J. Biol. Chem.* **249**, 1823 (1974).

[94] J. F. McKelvy, P. LeBlanc, C. Laudes, L. Perrie, Y. Grumm-Jorgensen, and C. Kordon, *Biochem. Biophys. Res. Commun.* **73**, 507 (1976).

[95] W. W. Chen and W. J. Lennarz, *J. Biol. Chem.* **251**, 7802 (1976).

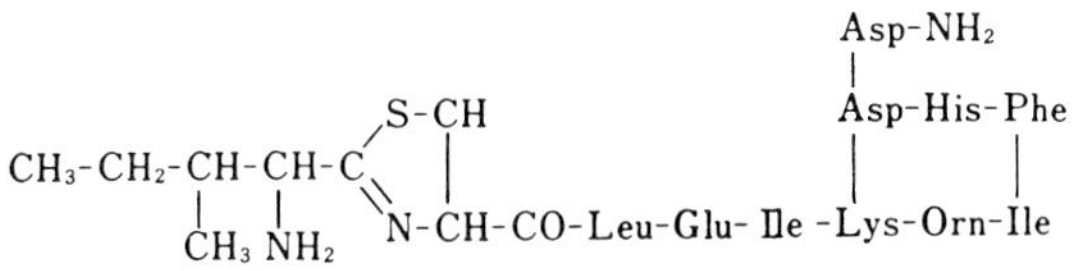

$$
\begin{array}{c}
\text{Asp-NH}_2 \\
|\\
\text{Asp-His-Phe}
\end{array}
$$

$$
\begin{array}{c}
\qquad\qquad\qquad\quad\diagup\text{S-CH} \qquad\qquad |\qquad\quad | \\
\text{CH}_3\text{-CH}_2\text{-CH-CH-C}\diagdown\quad |\\
\qquad |\quad\ |\qquad\qquad \text{N-CH-CO-Leu-Glu- Ile -Lys-Orn-Ile}\\
\qquad \text{CH}_3\ \text{NH}_2
\end{array}
$$

Fig. 4. Structure of bacitracin.

inhibit the formation of dolichylpyrophosphoryl-GlcNAc by calf pancreas microsomes, but had no effect on the synthesis of dolichylpyrophosphoryl-GlcNAc$_2$, dolichylphosphorylmannose, or dolichylphosphorylglucose.[96] In yeast membrane preparations, 0.33 mM bacitracin caused an inhibition in the formation of dolichylpyrophosphoryl-GlcNAc$_2$, but in this case no inhibition of dolichylpyrophosphoryl-GlcNAc was reported.[97] Thus, the above studies suggest at least three different sites of action on the dolichol pathway.

Most likely, the action of bacitracin in the eukaryotic systems is due to the formation of a complex with the lipid carrier, dolichyl phosphate, as has been shown for the microbial systems (i.e., with undecaprenyl phosphate). Thus, using either the particulate enzyme preparation or a solubilized enzyme system from pig aorta, 0.1–0.2 mM bacitracin blocked the incorporation of both mannose and GlcNAc from their sugar nucleotides into the lipid-linked monosaccharides and lipid-linked oligosaccharides.[98] The inhibition of dolichylphosphorylmannose formation could be overcome by the addition of high concentrations of dolichyl phosphate, but this addition did not reverse the inhibition of dolichylpyrophosphoryl-GlcNAc synthesis. Bacitracin also inhibited the transfer of mannose from GDP-mannose to lipid-linked oligosaccharides by the particulate enzyme as well as the transfer of mannose from dolichylphosphorylmannose to lipid-linked oligosaccharides.[98] Similar kinds of inhibitory reactions were observed when bacitracin was tested with a particulate enzyme preparation from mung bean seedlings.[99] These data suggest that bacetracin may block many of the steps in the lipid-linked saccharide pathway.

Showdomycin

Showdomycin is a broad-spectrum, nucleoside antibiotic that is elaborated by *Streptomyces showdoensis*.[100] Its structure is presented in Fig. 5.

[96] A. Hercovics, B. Bugge, and R. W. Jeanloz, *FEBS Lett.* **82**, 800 (1977).

[97] F. Reuvers, P. Boer, and E. P. Steyn-Parve, *Biochem. Biophys. Res. Commun.* **82**, 800 (1978).

[98] J. P. Spencer, M. S. Kang, and A. D. Elbein, *Arch. Biochem. Biophys.* **190**, 829 (1978).

[99] M. C. Ericson, J. Gafford, and A. D. Elbein, *Plant Physiol.* **62**, 373 (1978).

[100] H. Nishimura, M. Mayama, Y. Komatsu, H. Kato, N. Shimaoka, and Y. Tanaka, *J. Antibiot.* **17**, 148 (1964).

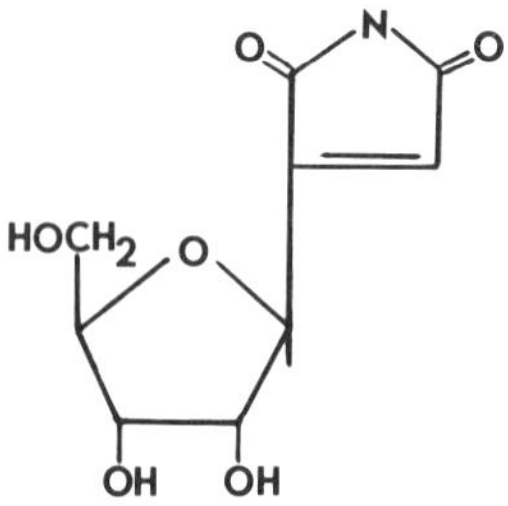

FIG. 5. Structure of showdomycin.

Showdomycin is moderately active against gram-positive and gram-negative bacteria and also shows considerable activity against Ehrlich acites tumors in mice and against HeLa cells.[101] This antibiotic was a potent inhibitor of bovine liver UDP-glucose dehydrogenase, and this activity was attributed to its alkylating action on the enzyme. Preincubation of showdomycin with cysteine completely blocked the inhibitory action.

Using a particulate enzyme preparation from porcine aorta, it was found that showdomycin effectively inhibited the formation of dolichylphosphorylglucose, but this inhibition was much more pronounced in the presence of detergents such as NP-40. At 0.25% NP-40, 50% inhibition of this activity required about 10 μg/ml of antibiotic. Showdomycin also inhibited the transfer of mannose from GDP-mannose to dolichylphosphorylmannose and to lipid-linked oligosaccharides, but in both of these cases, inhibition was only evident in the presence of detergent and much larger amounts of antibiotic were needed here as compared to glucose inhibition. On the other hand, relatively little inhibition of GlcNAc transfer was observed either in the presence or absence of detergent.[102] While showdomycin inhibited glucolipid formation in aorta extracts, it greatly stimulated glucose incorporation into lipid in yeast membrane preparations. However, the glucolipid formed in yeast had chemical and chromatographic properties like those of glucosylceramide rather than of glucosylphosphoryldolichol. The stimulation of glucosylceramide formation appeared to be due to protection of the substrate, UDP-glucose, from degradation.[102] Thus, showdomycin may inhibit one or more of the enzymes that are involved in the catabolism of UDP-glucose.

When membrane preparations of *Volvox cartere × nagaraensis* were tested with showdomycin, the formation of both dolichylphosphorylglucose and dolichylpyrophosphoryl-GlcNAc were sensitive to antibiotic to

[101] S. Roy Burman, P. Roy Burman, and D. W. Visser, *Cancer Res.* **28**, 1605 (1968).
[102] M. S. Kang, J. P. Spencer, and A. D. Elbein, *J. Biol. Chem.* **254**, 10037 (1979).

about the same extent in the presence of Triton X-100.[103] This inhibitory effect was lost when an excess of dithiothreitol was added. These enzymes were also inactivated by N-ethylmaleimide, and this inhibition was comparable to that of showdomycin. Furthermore, the inhibition of showdomycin or N-ethylmaleimide could be prevented by adding UMP to the incubation mixtures. The conclusion of these experiments was that the inhibition by showdomycin is probably due to irreversible reaction of the maleimide structure of the antibiotic with thiol groups on the protein.

Showdomycin was also used as a tool to examine the regulation of dolichylpyrophosphoryl-GlcNAc formation from dolichylphosphoryl-mannose.[104] Previous studies by Kean had shown a great stimulation in GlcNAc–lipid synthesis when excess dolichylphosphorylmannose was added to the incubation mixtures. In these studies, showdomycin (and duimycin, see below) was added to incubation mixtures at concentrations that partially inhibited the formation of dolichylphosphorylmannose. Under those conditions where formation of dolichylphosphorylmannose was depressed, the stimulation of GlcNAc–lipid synthesis was also blocked. However, when dolichylphosphorylmannose was added along with the showdomycin, stimulation of GlcNAc–lipid formation still occurred. On the other hand, in the presence of showdomycin or duimycin, formation of the chitobiosyl- and mannose-containing trisaccharide–lipids was completely suppressed while dolichylpyrophosphoryl-GlcNAc synthesis was not. The author suggests that the target of activation by dolichylphosphorylmannose is the GlcNAc-1-P transferase that synthesizes dolichyl-pyrophosphoryl-GlcNAc.[104]

Antibiotics of the Moenomycin Group

Duimycin belongs to the same group of antibiotics as moenomycin, prasinomycin, macarbomycin, 8036RP, 11837RP, and 19402RP, all of which affect the lipid-linked saccharide pathway that participates in assembly of bacterial cell wall peptidoglycan.[105,106] Duimycin was originally isolated from fermentation broths of *Streptomyces umbrinus*[107] and was shown to inhibit cell wall biosynthesis in *Staphylococcus aureus*.[108] Al-

[103] T. Muller, E. Bause, and L. Jaenicke, *FEBS Lett.* **128,** 208 (1981).

[104] E. L. Kean, *J. Biol. Chem.* **260,** 12561 (1985).

[105] H. J. Rogers, H. R. Perkins, and J. B. Ward, "Microbiol Cell Walls and Membranes," p. 298. Chapman & Hall, London, 1980.

[106] W. E. Brown, V. Seinerova, W. M. Chan, A. L. Laskin, P. Linnet, and J. L. Strominger, *Ann. N.Y. Acad. Sci.* **235,** 399 (1974).

[107] E. Meyers, D. S. Slusarchyk, T. L. Bouchard, and F. L. Weisenborn, *J. Antibiot.* **22,** 490 (1969).

[108] E. J. J. Lugtenberg, J. A. Hellings, and G. J. van de Berg, *Antimicrob. Agents Chemother.* **2,** 485 (1972).

though its structure is not known, it has been shown to liberate glucosamine, glucose, ammonia, acetic acid, phosphate, and a C_{25} fatty acid by acid hydrolysis.[109]

Diumycin inhibited the formation of dolichylphosphorylmannose when either a membrane fraction or a solubilized enzyme from *Saccharomyces cerevisiae* was used.[110] Inhibition was somewhat more pronounced with the solubilized enzyme, and 90% inhibition of the above reaction required about 150 μg/ml of antibiotic. The transfer of mannose to preformed dolichylpyrophosphoryl-GlcNAc$_2$ still occurred in the presence of duimycin, but the transfer of mannose from dolichylphosphorylmannose to serine or threonine residues on the protein was inhibited. However, the transfer of the mannose portion of GDP-mannose directly to serine or threonine residues was not blocked by antibiotic. With a soluble enzyme preparation from *Acanthamoeba,* duimycin inhibited the transfer of mannose from GDP-mannose and GlcNAc from UDP-GlcNAc into the lipid-linked monosaccharides. In this study, 250 μg/ml of antibiotic inhibited dolichylphosphorylmannose formation by 90% and dolichylpyrophosphoryl-GlcNAc formation by 70%. The synthesis of dolichylphosphorylglucose was only slightly affected by this antibiotic. In addition, duimycin also inhibited the transfer of the second GlcNAc from UDP-GlcNAc to dolichylpyrophosphoryl-GlcNAc to form dolichylpyrophosphoryl-GlcNAc$_2$, and this reaction was even more sensitive than was the formation of dolichylpyrophosphoryl-GlcNAc. The kinetics of inhibition of dolichylphosphorylmannose formation were of the mixed type, suggesting that this antibiotic may bind at a site other than the active site and may alter the affinity of the enzyme for the substrates.[110]

Flavomycin is also a phosphoglycolipid antibiotic of the moenomycin group that is produced by various species of *Streptomyces.* The antibiotic is mainly active against gram-positive bacteria, presumably by interfering with the synthesis of peptidoglycans of the bacterial cell wall.[111] Moenomycin A, the main component of flavomycin, was recently shown to contain the C_{25} lipid alcohol, moenocinol, which is bound via an ether bridge to the C-2 hydroxy group of 3-phosphoglyceric acid, while the phosphate group is linked by a phosphoacetal ester bond to a branched carbohydrate moiety.[112] Flavomycin was found to interfere with the formation of lipid-linked saccharides in membrane fractions of pig brain.[113]

[109] P. Babczinski, *Eur. J. Biochem.* **112**, 53 (1972).

[110] C. L. Villemez and P. L. Carlo, *J. Biol. Chem.* **255**, 8174 (1980).

[111] G. Huber, "Antibiotics VII, Mechanism of Action of Antimicrobial Agents" (E. Hahn, ed.), pp. 135, Springer-Verlag, Berlin, 1979.

[112] P. Welzel, F. J. Witteler, D. Muller, and W. Riemer, *Angew. Chem., Int. Ed. Engl.* **93**, 130 (1981).

[113] E. Bause and G. Legler, *Biochem. J.* **201**, 481 (1982).

The formation of dolichylphosphorylglucose was most sensitive, being inhibited by 50% at about 0.2 mM antibiotic, whereas synthesis of dolichylpyrophosphoryl-GlcNAc, dolichylpyrophosphoryl-N',N'-diacetylchitobiose, and dolichylpyrophosphoryl-GlcNAc$_2$Man required higher concentrations of flavomycin for comparable amounts of inhibition. At 1 mM concentrations of antibiotic, dolichylphosphorylmannose formation from GDP-[^{14}C]mannose was not inhibited, but the lipid-linked oligosaccharides that accumulated only had five to seven mannose residues. Thus, the mode of action of flavomycin remains obscure but it may be an interesting inhibitor for some studies on the lipid-mediated pathway.

Amino Sugars

Although N-acetylglucosamine is normally found in all N-linked glycoproteins, glucosamine is an inhibitor of the formation of N-linked oligosaccharides. Thus, millimolar concentrations of glucosamine were found to inhibit viral multiplication of a variety of enveloped viruses.[114] When mammalian cells were grown (or incubated) in the presence of glucosamine, the entire oligosaccharide chain of the glycoprotein was missing, indicating that this inhibitor might be affecting the formation of lipid-linked saccharides.[114] However, in one study, no aberrant forms of the lipid-linked oligosaccharides were observed.[115] Furthermore, at inhibitory concentrations of glucosamine, no unusual metabolites were found and there was no evidence for intermediates of glucosamine such as UDP-glucosamine.[116] These studies indicated that glucosamine itself was necessary for the inhibition and that this inhibition could only be demonstrated in whole cells or tissues.

In another study, glucosamine (at millimolar concentrations) did cause dramatic alterations in the composition of the lipid-linked oligosaccharides.[117] Normally, in influenza virus-infected MDCK cells, the major oligosaccharide associated with the lipid-linked oligosaccharides is a Glc$_3$Man$_9$GlcNAc$_2$ structure. However, at lower concentrations of glucosamine (0.5–1 mM), the Glc$_3$Man$_9$GlcNAc$_2$ structure was replaced by a lipid-linked oligosaccharide having a smaller-sized oligosaccharide, characterized as a Man$_7$GlcNAc$_2$-pyrophosphoryldolichol. As the glucosamine level was raised to 2–10 mM, the Man$_7$GlcNAc$_2$–lipid declined and was replaced by a Man$_3$GlcNAc$_2$–lipid. These results indicated that these two oligosaccharides, or the reaction leading to their formation, may

[114] R. T. Schwarz, M. F. G. Schmidt, U. Answer, and H. D. Klenk, *J. Virol.* **23**, 217 (1977).
[115] R. Datema and R. J. Schwarz, *Biochem. J.* **184**, 113 (1979).
[116] V. Kooh, R. T. Schwarz, and C. Scholtissek, *Eur. J. Biochem.* **94**, 515 (1979).
[117] Y. T. Pan and A. D. Elbein, *J. Biol. Chem.* **257**, 2795 (1982).

represent control points in the lipid-linked saccharide pathway. However, the nature of these regulatory sites, if they exist, has not been elucidated. In these studies, as in others described above, the inhibition by glucosamine was reversible. Thus, removal of glucosamine from the culture medium and washing the cell monolayers with glucosamine-free medium led to a restoration of normal lipid-linked oligosaccharide synthesis.

The inhibition by glucosamine was specific and could not be mimicked by other amino sugars such as galactosamine, mannosamine, or N-acetylglucosamine. However, mannosamine also proved to be an inhibitor of the lipid-linked saccharide pathway, but the inhibition in this case appeared to be at a different site in the pathway.[118] Thus, when MDCK cells were incubated in the presence of 1–10 mM mannosamine and labeled with [2-^{3}H]mannose, the major oligosaccharides associated with the dolichol were Man$_5$GlcNAc$_2$ and Man$_6$GlcNAc$_2$. Strangely enough, both of these structures were susceptible to digestion by endoglucosaminidase H, indicating that the Man$_5$GlcNAc$_2$ structure induced by mannosamine must be different than the usual biosynthetic intermediate (since this is resistant to Endo H). However, pulse–chase studies done in the presence of mannosamine showed that the inhibition was in the biosynthetic pathway, suggesting that mannosamine might inhibit an α1,2-mannosyl transferase. No evidence for this inhibition could be obtained in studies with cell-free extracts. Even though the lipid-linked oligosaccharides produced in the presence of mannosamine did contain smaller sized oligosaccharides, these oligosaccharides were still transferred to protein, and, furthermore, the oligosaccharides were processed to hybrid and complex structures after this transfer. However, in MDCK cells in the absence of mannosamine, no hybrid chains were found but only complex and high-mannose structures. Mannosamine was found to inhibit the processing α1,2-mannosidase of plants, in vitro.[119]

Sugar Analogs

Various sugar analogs such as 2-deoxy-D-glucose and 2-fluoro-2-deoxy-D-glucose have been found to interfere with protein glycosylation. Thus, these compounds have been widely used and have provided valuable insight into the mechanisms of biosynthesis and function of N-linked oligosaccharides.[120] However, some of these compounds have shown a

---

[118] Y. T. Pan and A. D. Elbein, *Arch. Biochem. Biophys.* **242**, 447 (1985).

[119] T. Szumilo, G. P. Kaushal, H. Hori, and A. D. Elbein, *Plant Physiol.* **81**, 383 (1986).

[120] R. Schwarz and R. Datema, *Adv. Carbohydr. Chem. Biochem.* **40**, 287 (1982).

lack of specificity and they have been found to inhibit other metabolic reactions. It is interesting that 2-deoxyglucose was initially discovered as an inhibitor of protein glycosylation because of its antiviral activity against enveloped viruses, and the observation that this inhibition could be reversed by mannose.[121] Later it was shown that this antiviral activity was due to inhibition of glycosylation of the viral envelope glycoproteins, preventing the formation of viral envelope.[122] Usually, 2-deoxyglucose is used at concentrations of 0.1–5 mM.

When cultured animal cells are incubated with 2-deoxyglucose, the sugar is converted to both UDP-2-deoxyglucose and GDP-2-deoxyglucose, as well as the dolichyl derivative, dolichylphosphoryl-2-deoxyglucose. Apparently the inhibition of protein glycosylation is not the result of depletion of the sugar nucleotide pool since the level of GDP-mannose and UDP-GlcNAc were actually found to increase in the presence of 2-deoxyglucose. In fact, GDP-2-deoxyglucose appears to be the major nucleotide involved in the inhibition since the inhibition could be reversed by mannose and addition of this sugar led to a reduction in the levels of GDP-2-deoxyglucose in the cells.[122,123]

The actual site of inhibition is probably at the level of lipid-linked saccharides.[124] Thus, GDP-2-deoxyglucose was found to inhibit the formation of lipid-linked oligosaccharides in crude membrane preparations of chick embryo cells. Under these conditions, dolichylpyrophosphoryl-N,N'-diacetylchitobiosyl-2-deoxyglucose accumulated. It was not possible to add additional mannose residues to this trisaccharide–lipid, and protein glycosylation was inhibited. On the other hand, the effects of GDP-2-deoxyglucose were also examined under partially inhibitory conditions which were obtained using a mixture of GDP-2-deoxyglucose and GDP-mannose. In this case, the major glycolipid formed by these membranes was dolichylpyrophosphoryl-GlcNAc$_2$-mannose-2-deoxyglucose. This oligosaccharide apparently could not be transferred to protein, nor could it be further elongated. Since the levels of dolichylphosphate in the cell appear to be limiting, high levels of 2-deoxyglucose may tie up all the available dolichyl phosphate and result in inhibition of protein glycosylation. This is one plausible explanation for the inhibition by 2-deoxyglucose, since it could be partially reversed by the addition of dolichyl phosphate.[124] However, since reversal was not complete, there may be other

[121] G. Kaluza, M. F. G. Schmidt, and C. Scholtissek, *Virology* **54**, 179 (1973).
[122] L. Lehle and R. Schwarz, *Eur. J. Biochem.* **67**, 239 (1976).
[123] R. Datema and R. Schwarz, *Eur. J. Biochem.* **90**, 505 (1978).
[124] R. Datema, R. Pont Lezica, P. W. Robbins, and R. Schwarz, *Arch. Biochem. Biophys.* **206**, 65 (1981).

facets to the inhibition beside the levels of dolichyl phosphate. *In vivo*, it was shown that the normally glycosylated K-46 chain produced by myeloma cells was not glycosylated in the presence of 2-deoxyglucose.[125]

Fluoroglucose and fluoromannose (i.e., 2-deoxy-2-fluoro-D-glucose or -mannose) are other analogs of glucose and mannose that have been found to inhibit protein glycosylation, as do other fluoro analogs such as 4-fluoroglucose or 3-fluoroglucose.[126] In chick embryo cells treated with 2-fluoroglucose, the formation of lipid-linked oligosaccharides did not go to completion and oligosaccharides with decreased amounts of glucose and mannose were formed.[127] This inhibition could be reversed by adding glucose or mannose to the culture medium. In contrast to 2-deoxyglucose, the fluoroglucose was not incorporated into the lipid-linked oligosaccharides. However, this sugar analog did inhibit the formation *in vivo* of dolichylphosphorylglucose and dolichylphosphorylmannose, but it did not prevent the transfer of these sugars from their dolichyl derivatives to the lipid-linked oligosaccharides. In the presence of fluoroglucose, the pool size of UDP-glucose, but not that of GDP-mannose or UDP-GlcNAc, was decreased. It seems likely that the smaller sized lipid-linked oligosaccharides produced in the presence of inhibitor are not transferred efficiently to protein.[127]

Effects of Glucose Starvation or Energy Deprivation

When Chinese hamster ovary (CHO) cells are incubated in the absence of glucose (i.e., starved for glucose), they do not synthesize the $Glc_3Man_9GlcNAc_2$-pyrophosphoryldolichol, but instead they accumulate a $Man_5GlcNAc_2$-pyrophosphoryldolichol which then becomes glucosylated to a $Glc_3Man_5GlcNAc_2$-pyrophosphoryldolichol.[128-130] In addition, some animal cells may accumulate smaller amounts of $Man_2GlcNAc_2$-pyrophosphoryldolichol.[131] In several animal cell lines, the effect of glucose starvation was evident within 20 min of glucose removal as long as the cells were kept at low to moderate densities, but at high cell densities the effect was not evident. The addition of glucose, but not pyruvate, glutamine, galactose, inositol, or glycerol, could prevent the effects of glucose starvation.[131] The glycosylation of the G protein of vesicular stomatitis virus was also altered when virus-infected BHK cells were placed

[125] P. K. Eagon and E. C. Heath, *J. Biol. Chem.* **252**, 2372 (1977).
[126] R. Datema, R. Schwarz, and A. W. Jankowski, *Eur. J. Biochem.* **109**, 331 (1980).
[127] M. F. G. Schmidt, R. Schwarz, and H. Ludwig, *J. Virol.* **18**, 819 (1976).
[128] S. Turco, *Arch. Biochem. Biophys.* **205**, 330 (1980).
[129] B. M. Sefton, *Cell (Cambridge, Mass.)* **10**, 659 (1977).
[130] J. J. Rearick, A. Chapman, and S. Kornfeld, *J. Biol. Chem.* **256**, 6255 (1981).
[131] H. Gershman and P. W. Robbins, *J. Biol. Chem.* **256**, 7774 (1981).

in the absence of glucose. The $Glc_3Man_5GlcNAc_2$ was transferred from its lipid carrier to the G protein.[132] Glucose-starved rat hepatoma cells produced a lower molecular weight form of α_1-acid glycoprotein. This reduction in size is due not only to the transfer of truncated oligosaccharide (i.e., $Glc_3Man_5GlcNAc_2$) to protein, but also to the attachment of a reduced number of oligosaccharide chains. The authors speculate that glucose starvation might reduce the supply of oligosaccharides available for transfer, or that glucose withdrawal results in an immediate energy depletion that influences oligosaccharide synthesis.[133]

In regard to the possible effect of energy depletion on the formation of lipid-linked oligosaccharides, a number of studies have been done using an uncoupler of oxidative phosphorylation, CCCP (carbonyl cyanide *m*-chlorophenylhydrazone). This compound causes energy depletion of cells when it is added to the culture media. When mammalian cells were placed in the presence of 10 mM CCCP, they did not produce dolichylphosphorylmannose but could still synthesize dolichylphosphorylglucose, dolichylpyrophosphoryl-GlcNAc, and dolichylpyrophosphoryl-N,N'-diacetylchitobiose. These cells incubated in the presence of CCCP produce a $Man_5GlcNAc_2$–lipid and transfer this oligosaccharide to protein. Some of this lipid may become glucosylated to $Glc_3Man_5GlcNAc_2$, and this may be the species that is actually transferred.[134] On the other hand, when thyroid slices were incubated in the presence of CCCP, there was a disappearance of $Glc_3Man_9GlcNAc_2$-pyrophosphoryldolichol, but instead of a $Man_5GlcNAc_2$–lipid appearing, these workers reported the presence of a $Man_9GlcNAc_2$-pyrophosphoryldolichol. A smaller amount of $Man_8GlcNAc_2$–lipid was also accumulated. In addition, there was a concomitant decrease in N-glycosylation of proteins. Several inhibitors of respiration, i.e., N_2 or antimycin A, gave similar effects.[135] Since uncouplers of oxidative phosphorylation have been reported to disrupt the recycling of the glucose transport carrier, these compounds may limit the entry of glucose into cells and mimic the glucose starvation effect. However, these differences observed with inhibitors of energy metabolism await further clarification.

Inhibitors of Glycoprotein Processing

After the $Glc_3Man_9GlcNAc_2$ has been synthesized on the lipid carrier and transferred to protein (see Fig. 1), the oligosaccharide portion of the

[132] S. J. Turco and J. L. Picard, *J. Biol. Chem.* **257**, 8674 (1982).
[133] H. Baumann and G. P. Jahreis, *J. Biol. Chem.* **258**, 3942 (1983).
[134] R. Datema and R. T. Schwarz, *J. Biol. Chem.* **256**, 11191 (1981).
[135] R. G. Spiro, M. J. Spiro, and V. D. Bhoyroo, *J. Biol. Chem.* **258**, 9469 (1983).

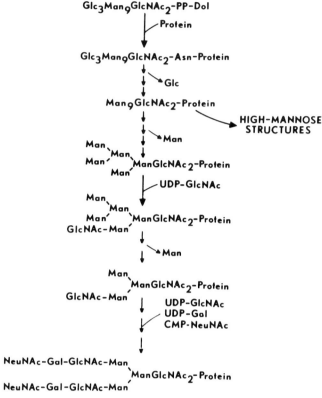

Fig. 6. Reactions involved in the processing of the oligosaccharide chains of the N-linked glycoproteins.

protein is subjected to a number of processing reactions. These reactions are outlined in Fig. 6. First of all, two different membrane-bound glucosidases remove all three glucose units. Thus, glucosidase I removes the terminal α1,2-linked glucose residue, and then glucosidase II can cleave the next two α1,3-linked glucoses. These reactions result in the formation of a $Man_9GlcNAc_2$–protein, which is the substrate for mannosidase action. An α-mannosidase in the endoplasmic reticulum and/or an α-mannosidase in the cis region of the Golgi apparatus can remove all four α1,2-linked mannose units. The $Man_5GlcNAc_2$–protein resulting from the above action is the substrate for a GlcNAc transferase, called GlcNAc transferase I, which adds a GlcNAc residue to the mannose unit that is linked α1,3 to the β-linked mannose. This GlcNAc → Man-R appears to be a signal for another mannosidase, called mannosidase II, to remove the

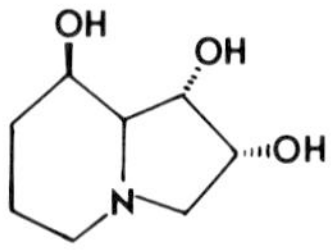

FIG. 7. Structure of swainsonine.

α1,3 and α1,6-linked mannoses that are attached to the α1,6-mannose branch. Following the removal of these two mannose residues, other sugars found in the complex types of oligosaccharides (i.e., GlcNAc, galactose, sialic acid, and fucose) can be added.[136]

Swainsonine

The first studies showing a biological role for the indolizidine alkaloids were those done with swainsonine, and this compound was the first of the glycoprotein processing inhibitors. Swainsonine [(1S,2R,8R,8aR)-1,2,8-trihydroxyoctahydroindolizine] is an indolizidine alkaloid that was first isolated from the Australian wild plant, *Swainsona canescens*.[137] The structure of swainsonine is shown in Fig. 7. Livestock that eat this plant develop symptoms similar to those of humans with α-mannosidosis,[138] and swainsonine was found to be a very potent inhibitor of lysosomal and jack bean α-mannosidase.[139] The alkaloid was also isolated from several locoweed species (*Astragalus*) that grow in the southwestern United States[140,141] and also from the fungus *Rhizoctonia leguminicola*.[142] The chemical synthesis of swainsonine has recently been reported by several research groups starting from derivatives of mannose or glucose.[143–145] This alkaloid has also been synthesized from a 4-carbon precursor.[146] In addition, several isomers of swainsonine were also synthesized chemi-

[136] S. C. Hubbard and R. C. Ivatt, *Annu. Rev. Biochem.* **50**, 555 (1981).
[137] S. M. Colegate, P. R. Dorling, and C. R. Huxtable, *Aust. J. Chem.* **32**, 2257 (1979).
[138] P. R. Dorling, C. P. Huxtable, and S. M. Colegate, *Neuropathol. Appl. Neurobiol.* **4**, 285 (1978).
[139] P. R. Dorling, C. R. Huxtable, and S. M. Colegate, *Biochem. J.* **191**, 649 (1980).
[140] R. J. Molyneux and J. P. James, *Science* **216**, 190 (1982).
[141] D. Davis, P. Schwarz, T. Hernandez, M. Mitchell, B. Warnock, and A. D. Elbein, *Plant Physiol.* **76**, 972 (1984).
[142] M. J. Schneider, F. S. Ungemach, H. Broquist, and T. M. Harris, *Tetrahedron* **39**, 29 (1983).
[143] T. Suami, K. Tadano, and Y. Iimura, *Carbohydr. Res.* **136**, 67 (1985).
[144] N. Yasuda, H. Tsutsumi, and T. Tanaka, *Chem. Lett.* p. 1201 (1984).
[145] M. H. Ali, L. Hough, and A. C. Richardson, *Carbohydr. Res.* **136**, 225 (1985).
[146] C. E. Adams, F. J. Walker, and K. B. Sharpless, *J. Org. Chem.* **50**, 420 (1985).

cally, but so far nothing is known concerning their biological activity.[147,148]

The first studies to show that swainsonine inhibited glycoprotein processing were done with various mammalian cell lines using [2-^{3}H]mannose to label the N-linked glycoproteins. Cells grown in the presence of swainsonine showed a great increase in the amount of radiolabeled glycopeptide that was susceptible to digestion by endoglucosaminidase H as compared to control cells. These studies indicated that swainsonine blocked the processing pathway and prevented the formation of complex types (i.e., Endo H-resistant) of oligosaccharides.[149] Swainsonine also prevented the formation of complex chains in the influenza viral hemagglutinin and led to viral glycoproteins with endoglucosaminidase H-sensitive oligosaccharides.[150]

In these studies, it was mistakenly thought that swainsonine inhibited all of the mannosidase activity since the hybrid structure migrated on BioGel P-4 columns in the same position as the Man$_9$GlcNAc$_2$. However, when the effect of swainsonine was examined directly on mannosidase I and mannosidase II, swainsonine was shown to be specific for the latter enzyme.[151] This *in vitro* finding was verified in several *in vivo* systems. Thus, when vesicular stomatitis virus was grown in BHK cells in the presence of swainsone, the G protein of this virus contained hybrid types of oligosaccharides, rather than the typical complex structures. These hybrid chains were susceptible to the action of endoglucosaminidase H.[152] In this system, the effect of swainsonine was reversible as long as the inhibitor was removed before the G protein was synthesized in the cells. Swainsonine did not appear to effect the production of virus, i.e., the same number of viral particles were observed in the media of swainsonine-grown and control cells. Also the virus formed in the presence of swainsonine appeared to retain its infectivity. Swainsonine also caused fibroblasts to produce hybrid types of oligosaccharides. Thus, at 10 µg/ml of alkaloid, most of the complex chains of the fibroblast N-linked glycoproteins were replaced by hybrid chains.[153]

Swainsonine was also utilized to study the processing of the asparagine-linked carbohydrate chains of α_1-antitrypsin,[154] as well as transferrin

[147] N. Yasuda, H. Tsutsumi, and T. Takaya, *Chem. Lett.* p. 31 (1985).

[148] C. Adams and K. Sharpless, unpublished observations.

[149] A. D. Elbein, R. Solf, P. D. Dorling, and K. Vosbeck, *Proc. Natl. Acad. Sci. U.S.A.* **78**, 7393 (1981).

[150] A. D. Elbein, P. R. Dorling, K. Vosbeck, and M. Horisberger, *J. Biol. Chem.* **257**, 1573 (1982).

[151] D. P. R. Tulsiani, T. M. Harris, and O. Touster, *J. Biol. Chem.* **257**, 7936 (1982).

[152] M. S. Kang and A. D. Elbein, *J. Virol.* **46**, 60 (1984).

[153] D. P. R. Tulsiani and O. Touster, *J. Biol. Chem.* **258**, 7578 (1983).

[154] V. Gross, T.-A. Tran-Thi, K. Vosbeck, and P. Heinrich, *J. Biol. Chem.* **258**, 4032 (1983).

and α_1-antichymotrypsin,[155] in cultured rat hepatocytes. Although swainsonine caused the oligosaccharide chains of these proteins to be altered from complex structures to hybrid chains, these alterations did not affect the rate of secretion of these serum proteins from the hepatocytes. These investigators examined the secreted proteins by SDS gel electrophoresis before and after digestion by endoglucosaminidase H. Thus, the secreted proteins from normal hepatocytes have complex oligosaccharides, while those produced in the presence of swainsonine had carbohydrates susceptible to endoglucosaminidase H. Similar results were obtained on the effects of swainsonine on the secretion of IgM and IgD.[156]

Fibronectin is an N-linked glycoprotein that is secreted by fibroblasts and is necessary for the adhesion of some cells to the substratum. The protein contains 4–10% carbohydrate mostly as complex chains of the biantennary type.[157] Fibronectin was secreted into the medium by swainsonine-treated fibroblasts and found to contain endoglucosaminidase H-sensitive structures, rather than the usual endoglucosaminidase H-resistant chains. The structure of the swainsonine-modified, Endo H-sensitive oligosaccharide was deduced to be GalβGlcNAcβManα[Manα(Manα)-Manα]ManβGlcNAcβ($\pm$Fuc)GlcNAc.[158] This study also showed that L-fucose was still incorporated into the hybrid types of oligosaccharides that were produced in the presence of swainsonine, and that the fucosylated hybrid fibronectin was still susceptible to endoglucosaminidase H.[158] Swainsonine also did not prevent the fucosylation or the sulfation of the hemagglutinin of influenza virus.[159] In the case of the fucose-labeled hemagglutinin, endoglucosaminidase H released a fucose-free oligosaccharide, while the GlcNAc remaining with the peptide did contain the label. This experiment showed that the innermost GlcNAc was the site of fucose attachment.[159a] In the case of sulfate-labeled hemagglutinin, endoglucosaminidase H released a ^{35}S-labeled oligosaccharide and left an unlabeled protein. The sulfate was found to be attached to the second GlcNAc.[159] The above studies indicate that significant changes in oligosaccharide structure have little effect upon the routing and cellular release of the N-linked glycoproteins.

Swainsonine, at concentrations as high as 1 μg/ml, did not affect the growth of MDCK, Chinese hamster ovary, SV101, B-16 melanoma, or Intestine 407 cells, as measured by changes in cell numbers over a 5-day period. There was also no apparent change in cell size or shape in cells

[155] H. F. Lodish and N. Kong, *J. Cell. Biol.* **98,** 1720 (1984).

[156] N. Peyrieras, E. Bause, G. Legler, R. Vasilov, L. Claeson, P. Peterson, and H. Ploegh, *EMBO J.* **2,** 823 (1983).

[157] M. Fukuda and S. Hakomori, *J. Biol. Chem.* **254,** 5442 (1979).

[158] R. G. Arumugham and M. L. Tanzer, *J. Biol. Chem.* **258,** 11883 (1983).

[159] R. Merkle, A. D. Elbein, and A. Heifetz, *J. Biol. Chem.* **260** (1985).

[159a] P. M. Schwarz and A. D. Elbein, *J. Biol. Chem.* **260,** 14452 (1985).

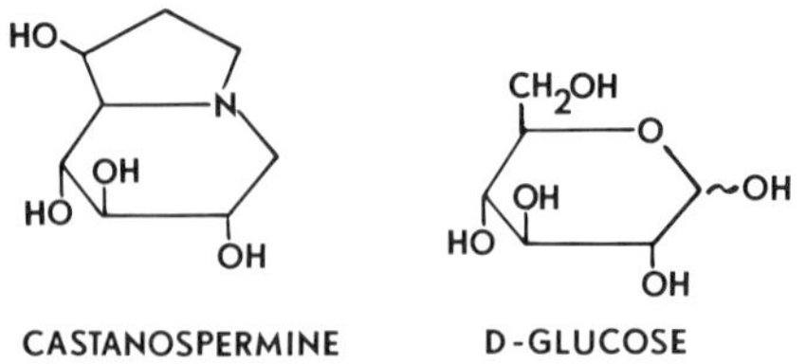

CASTANOSPERMINE   D-GLUCOSE

FIG. 8. Structure of castanospermine and its structural relationship to D-glucose.

grown in the presence of alkaloid. Swainsonine did not appear to be cytotoxic, nor to cause alterations in cell morphology as evidenced by comparisons of electron micrographs of normal cells or cells grown in alkaloid for up to 5 days.[160] These cell lines, after growth for 5–7 days in swainsonine, were able to bind almost twice as much ³H-labeled concanavalin A as untreated cells, while the binding of wheat germ agglutinin was diminished in treated cells. These studies support the site of action of swainsonine and indicate that these cells have an increase in mannose-containing oligosaccharides at their cell surface and a decrease in complex structures.

Castanospermine

Castanospermine [(1S, 6S, 7R, 8R, 8aR)-1,6,7,8-tetrahydroxyindolizine] is another plant alkaloid that was isolated from the seeds of the Australian tree *Castanospermum australe*.[161] The structure of this alkaloid is shown in Fig. 8. Castanospermine has been synthesized chemically from D-glucose,[162] and it was found to be a potent inhibitor of almond emulsin β-glucosidase while having no effect on yeast α-glucosidase, α- or β-galactosidase, α-mannosidase, β-N-acetylhexosaminidase, β-glucuronidase, or α-L-fucosidase.[163] Interestingly enough, when castanospermine was tested against lysosomal glycosidases from fibroblast extracts, it was found to be a good inhibitor of β-glucosidase, β-glucocerebrosidase, and α-glucosidase.[163] It is still not clear why some α-glucosidases, such as yeast α-glucosidase, are not sensitive to castanospermine.

The mechanism of inhibition by castanospermine on fungal amyloglucosidase (an exo-1,4-α-glucosidase) and emulsin β-glucosidase was exam-

[160] A. D. Elbein, Y. T. Pan, R. Solf, and K. Vosbeck, *J. Cell. Physiol.* **115**, 265 (1983).

[161] L. D. Hohenschutz, E. A. Bell, P. J. Jewess, D. P. Leworthey, R. J. Pyrce, E. Arnold, and J. Clardy, *Phytochemistry* **20**, 811 (1981).

[162] R. C. Bernotas and B. Ganem, *Tetrahedron Lett.* **25**, 165 (1984).

[163] A. Saul, J. P. Chambers, R. J. Molyneux, and A. D. Elbein, *Arch. Biochem. Biophys.* **221**, 265 (1983).

ined.[164] Castanospermine proved to be a competitive inhibitor of amyloglucosidase at both pH 4.5 and 6.0 when assayed with p-nitrophenyl-α-D-glucopyranoside as substrate. It was also a competitive inhibitor of almond emulsin β-glucosidase at pH 6.5, but previous studies had indicated that at pH 4.5 inhibition was of the mixed type. During these studies, it was observed that the pH of the incubation mixture had a marked effect on the inhibition. Thus, in all cases, castanospermine was a more potent inhibitor at pH 6.0–6.5 than it was at lower pH values. The pK for castanospermine was found to be 6.09, indicating that this alkaloid is probably more active in the unprotonated form. This observation was supported by the finding that the N-oxide of castanospermine, while still a competitive inhibitor of amyloglucosidase, was 50–100 times less active, and its activity was unaffected by the pH of the incubation mixture.

When influenza virus was raised in MDCK cells in the presence of castanospermine, at 50 μg/ml or higher, 80–90% of the viral glycopeptides became susceptible to the action of endoglucosaminidase H, whereas in the normal virus 70% of the glycopeptides are resistant to this enzyme.[165] The major oligosaccharide released by this enzyme sized like a hexose$_{10}$GlcNAc on calibrated BioGel P-4 columns. This oligosaccharide was characterized as a Glc1,2Glc1,3Glc1,3Manα1,2Manα1,2Manα1,3[Manα-1,3(Manα1,6)Manα1,6]ManβGlcNAc by a variety of enzymatic and chemical treatments, including methylation analysis. These data indicate that one or two mannose residues can still be removed as the Glc$_3$-Man$_9$GlcNAc$_2$–protein is transported to the locations of the processing α-mannosidases. Castanospermine, even at 100 μg/ml, did not inhibit the incorporation of [^{3}H]leucine or [^{14}C]alanine into protein, nor did it alter the infectivity of the virus raised in its presence. The castanospermine did, however, alter the surfaces of cells, since MDCK cells grown for 3 or 4 days in its presence (10 μg/ml) were able to bind almost twice as much ^{3}H-labeled concanavalin A as were control cells. In the presence of castanospermine, [^{3}H]fucose and [^{35}S]sulfate were not incorporated into the influenza viral hemagglutinin, indicating that the oligosaccharide chain had not been modified to the proper extent, or that some other necessary signals for these additions had not occurred.[159a]

Castanospermine also inhibited glycoprotein processing in suspension-cultured soybean cells.[166] Soybean cells were pulse-labeled with [2-^{3}H]mannose and chased for varying periods of time in unlabeled medium.

[164] R. Saul, R. J. Molyneux, and A. D. Elbein, *Arch. Biochem. Biophys.* **230,** 668 (1984).
[165] Y. T. Pan, H. Hori, R. Saul, B. A. Sanford, R. J. Molyneux, and A. D. Elbein, *Biochemistry* **22,** 3975 (1983).
[166] H. Hori, Y. T. Pan, R. J. Molyneux, and A. D. Elbein, *Arch. Biochem. Biophys.* **228,** 525 (1984).

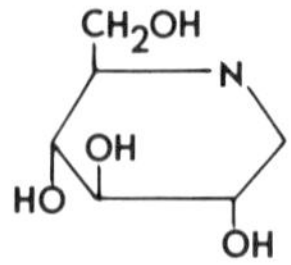

FIG. 9. Structure of deoxynojirimycin.

In normal cells, the initial glycopeptides (i.e., at 0 min of chase) contained oligosaccharides having $Glc_3Man_9GlcNAc_2$ structures, and these were trimmed with time to $Man_9GlcNAc_2$ and down to $Man_7GlcNAc_2$. In the presence of castanospermine, no trimming of glucose residues occurred, although some mannose units were apparently still removed. Thus, the major oligosaccharide in the glycopeptides of castanospermine-incubated cells after a 90-min chase was the $Glc_3Man_7GlcNAc_2$ structure. Thus, in plants, castanospermine also appears to inhibit glucosidase I.[166] The glucosidase I was purified from mung bean seedlings and shown to be specific for the removal of the terminal $\alpha 1,2$ linked glucose unit from the $Glc_3Man_9GlcNAc_2$. This enzyme was strongly inhibited by castanospermine, with 50% inhibition requiring about 5 μM concentrations of drug.[167] Castanospermine also prevented the processing of the oligosaccharide chains of various N-linked glycoprotein enzymes secreted by the fungus *Aspergillus fumigatus*.[168] Here also the oligosaccharide chains were mostly $Glc_3Man_{7-9}GlcNAc_2$ structures. In these studies, it was found that castanospermine caused some inhibition in the rate of secretion of β-N-acetylhexosaminidase but did not appear to effect the secretion of some of the other glycoprotein enzymes, such as β-glucosidase or β-galactosidase. On the other hand, castanospermine had no effect on the secretion of glycoproteins by MDCK cells, nor did it inhibit the formation of lipid-linked saccharides in these cells.[169]

Nojirimycin and Deoxynojirimycin

The antibiotic nojirimycin is produced by several strains of *Streptomyces,* while its reduced form, called 1-deoxynojirimycin, is synthesized by *Bacillus* species. These compounds are glucose analogs with an NH-group substituting for the oxygen atom in the pyranose ring (Fig. 9).[170] These compounds have been shown to inhibit intestinal α-glucosidases

[167] T. Szumilo, G. P. Kaushal, and A. D. Elbein, *Arch. Biochem. Biophys.* **247,** 261 (1986).
[168] A. D. Elbein, M. Mitchell, and R. J. Molyneux, *J. Bacteriol.* **160,** 67 (1984).
[169] Y. T. Pan and A. D. Elbein, *Arch. Biochem. Biophys.* **242,** 447 (1985).
[170] S. Inouye, T. Tsuruoka, T. Ito, and T. Nuda, *Tetrahedron* **24,** 2125 (1968).

and pancreatic α-amylase, both *in vitro* and *in vivo*.[171,172] These antibiotics also inhibited sucrase from small intestine[173] and lysosomal α-glucosidase from human liver.[174,175] The complete chemical synthesis of deoxynojirimycin has been reported.[162,175]

Deoxynojirimycin inhibited partially purified glucosidase I (I_{50}, 20 μM) and glucosidase II (I_{50}, 2 μM) from *Saccharomyces cerevisiae*.[176] However, when the glucosidases from calf liver were examined, deoxynojirimycin showed a preferential inhibition for glucosidase I (50% inhibition at 3 μM) over glucosidase II (50% inhibition at 20 μM).[177] Perhaps these differences reflect differences in the trimming enzymes from different sources. In IEC-6 intestinal epithelial cells labeled with [^{3}H]mannose, 5 mM deoxynojirimycin caused a decrease in complex types of oligosaccharides and an increase in high-mannose oligosaccharides. These high-mannose oligosaccharides were less susceptible to the action of α-mannosidase, suggesting that they might contain blocking glucose residues.[176]

The effects of deoxynojirimycin on the biosynthesis of several well-defined secretory and membrane glycoproteins were studied by several groups. The secretion of IgD, but not IgM, was blocked in deoxynojirimycin-treated cells.[156] The secretion of α_1-antitrypsin in fibroblasts was inhibited by about 50%,[178] and also the rate of secretion of α_1-antitrypsin and α_1-antichymotrypsin by human hepatoma HepG2 cells was greatly reduced.[155] The α_1-antitrypsin and α_1-antichymotrypsin apparently accumulated in the rough endoplasmic reticulum. The authors determined these results by the localization of proteins in subcellular fractions and their sensitivity to endoglucosaminidase H. On the other hand, deoxynojirimycin showed marginal effects on the secretion of the glycoproteins C-3 or transferrin and on the protein albumin. When the deoxynojirimycin was removed from the HepG2 cells, its inhibitory effect was reversible, and α_1-antitrypsin was secreted normally.[155]

[171] W. Frommer, B. Junge, L. Mueller, and E. Truscheit, *Planta Med.* **35**, 195 (1979).

[172] D. D. Schmidt, W. Frommer, L. Mueller, and E. Truscheit, *Naturwissenschaften* **66**, 584 (1979).

[173] G. Hanozet, H. Pircher, P. Vanni, B. Oesch, and G. Semenza, *J. Biol. Chem.* **256**, 3703 (1981).

[174] J. P. Chambers, A. D. Elbein, and J. P. Williams, *Biochem. Biophys. Res. Commun.* **107**, 1490 (1982).

[175] G. Kinast and M. Schadel, *Angew. Chem.* **9**, 799 (1981).

[176] B. Saunier, R. D. Kilker, J. S. Tkacz, Jr., A. Quaroni, and A. Hercovics, *J. Biol. Chem.* **257**, 14155 (1982).

[177] H. Hettkamp, E. Bause, and G. Legler, *Biosci. Rep.* **2**, 899 (1982).

[178] V. Gross, T. Andus, T.-A. Tran-Thi, R. Schwarz, K. Decker, and P. Heinrich, *J. Biol. Chem.* **258**, 12203 (1983).

These data suggest that for certain secretory glycoproteins, removal of glucose residues is necessary for the protein to move from the endoplasmic reticulum to the Golgi apparatus. In addition, the IgD from deoxynojirimycin-treated cells appeared to be degraded intracellularly during long periods of chase, rather than secreted.[156] However, surface expression of human class I antigens,[156] the G protein of vesicular stomatitis virus, and the hemagglutinin of influenza virus were not affected by deoxynojirimycin.[179] In addition, this drug did not interfere with the rate of production of influenza or VSV virus, or with the infectivity of these viruses.[180] In many of these studies, some of the glycoprotein oligosaccharides became resistant to endoglucosaminidase H even in the presence of deoxynojirimycin.[155,156,179] These data suggest that either the concentration of drug was not sufficient to completely inhibit all of the glucosidase I or perhaps there is an alternate route of processing.

The lipid-linked oligosaccharides synthesized in the presence of deoxynojirimycin and N-methyldeoxynojirimycin were compared in IEC-6 intestinal epithelial cells. HPLC analysis of the oligosaccharides obtained before and after exhaustive jack bean α-mannosidase digestion indicated that the major oligosaccharide in control and methyldeoxynojirimijcin-treated cells was $Glc_3Man_9GlcNAc_2$. However, in the presence of 5 mM deoxynojirimycin, the formation of this oligosaccharide was greatly reduced, and instead the major lipid-linked oligosaccharide was a $Man_9GlcNAc_2$-pyrophosphoryldolichol. The authors suggest that this decreased availability of the preferred donor for protein glycosylation may account for the impaired glycosylation and secretion of certain glycoproteins in the presence of deoxynojirimycin.[180]

N-Methyldeoxynojirimycin and N,N'-dimethyldeoxynojirimycin are obtained through monomethylation or through dimethylation with methyliodide.[181] N-Methyldeoxynojirimycin inhibits the glucose release in a cell-free system of rat liver microsomes incubated with the Glc_3-$Man_9GlcNAc_2$.[182] The N-methylderivative was almost twice as effective as deoxynojirimycin. Since the derivative caused the formation of Glc_3-$Man_xGlcNAc_2$ oligosaccharides *in vivo*, it seems likely that it inhibits glucosidase I. The inhibition by the permanently cationic derivative N,N-dimethyldeoxynojirimycin was similar to that by the monomethyl compound.

[179] B. Burke, K. Matlin, E. Bause, G. Legler, N. Peyrieras, and H. Ploegh, *EMBO J.* **3,** 551 (1984).
[180] P. A. Romero, P. Freidlander, and A. Hercovics, *FEBS Lett.* **183,** 29 (1985).
[181] U. Fuhrmann, E. Bause, and H. Ploegh, *Biochim. Biophys. Acta* **825,** 95 (1985).
[182] P. A. Romero, R. Datema, and R. T. Schwarz, *Virology* **130,** 238 (1983).

Bromoconduritol

Bromoconduritol (6-bromo-3,4,5-dihydroxycyclohex-1-ene) is an active-site-directed covalent inhibitor of glucosidases.[183,184] In *in vitro* and *in vivo* studies, this compound was shown to inhibit the trimming of the innermost glucose residue from the $Glc_3Man_9GlcNAc_2$ precursor.[185] Therefore, it was suggested that bromoconduritol inhibits glucosidase II.[185] However, since the enzyme is believed to be responsible for removing the two $\alpha 1,3$-linked glucose residues, it is not clear why bromoconduritol does not give rise to a $Glc_2Man_9GlcNAc_2$ oligosaccharide or to oligosaccharides with two glucose and nine or fewer mannoses, (i.e., $Glc_2Man_9GlcNAc$, etc.). In chick embryo cells infected with influenza virus, the glycopeptides produced in the presence of bromoconduritol contained oligosaccharides of the type $GlcMan_9GlcNAc$, $GlcMan_8$-$GlcNAc$, and $GlcMan_7GlcNAc$ (after endoglucosaminidase H treatment).[185] These results may indicate that there is a third trimming glucosidase. In these studies, bromoconduritol inhibited the release of infectious virus particles, in contrast to studies with castanospermine. However, bromoconduritol has a half-life of only 15 min in water. Perhaps, it or some degradation product inhibits other reactions beside those of trimming. At any rate, the short half-life of this inhibitor will certainly limit its usefulness in biological systems.

Deoxymannojirimycin

1-Deoxymannojirimycin (1,5-dideoxy-1,5-imino-D-mannitol) is the mannose analog of deoxynojirimycin. This compound was synthesized chemically and shown to be a mannosidase inhibitor.[186] This inhibitor was also tested *in vivo* with a hybridoma cell line to determine its effects on the biosynthesis and secretion of the N-linked glycoproteins IgM and IgD. After preincubation with inhibitor, the cells were pulse-labeled for 10 min with [^{35}S]methionine and then chased for various times in the presence of unlabeled methionine, plus inhibitor. The IgM and IgD present within the cells and that secreted into the medium were examined by SDS gel electrophoresis, and control cells were compared to cells incubated with inhibitor. The deoxymannojirimycin appeared to inhibit mannosidase I, thereby blocking the conversion of high-mannose chains to complex oligosaccharides. Thus, in the presence of inhibitor, the major oligosac-

[183] G. Legler, this series, Vol. 46, p. 368.

[184] G. Legler and W. Lotz, *Hoppe-Seyler's Z. Physiol. Chem.* **354**, 243 (1973).

[185] R. Datema, P. A. Romero, G. Legler, and R. J. Schwarz, *Proc. Natl. Acad. Sci. U.S.A.* **79**, 6787 (1982).

[186] G. Legler and E. Julich, *Carbohydr. Res.* **128**, 61 (1984).

charide released by endoglucosaminidase H corresponded to a Man₉GlcNAc standard. Deoxymannojirimycin did not inhibit the secretion of IgM and IgD although these proteins had an altered oligosaccharide chain.[187] These results, coupled with studies using other inhibitors, indicate that glycoproteins with high-mannose or hybrid chains are still secreted at normal rates, whereas some proteins carrying glucose residues may be retarded in their secretion.

Deoxymannojirimycin was also utilized as an inhibitor of the G protein of vesicular stomatitis virus, the hemagglutinin of influenza virus, and the histocompatability antigens.[179,188] Deoxymannojirimycin greatly altered the oligosaccharide structures of membrane and secretory proteins in these various systems. However, it did not interfere with surface expression, nor did it inhibit the production of mature, infectious virus. The secretion of α_1-proteinase inhibitor and α_1-acid glycoprotein by rat hepatocytes was not inhibited in the presence of the drug.[189] On the other hand, secretion of glycoproteins by MDCK cells was severely retarded in the presence of deoxymannojirimycin even though the other inhibitors such as castanospermine and swainsonine did not display this effect.[190] This inhibition was not due to the use of an abnormally high concentration of deoxymannojirimycin since in these experiments inhibition of secretion was seen at 10–25 μg/ml, whereas other studies reported no effects even at 4 mM. Further studies in this system indicated that deoxymannojirimycin also severely inhibited the formation of lipid-linked oligosaccharides.[190] Based on this finding, a possible explanation for the inhibition of secretion is that the glycoproteins are underglycosylated in the presence of deoxymannojirimycin as a result of insufficient amounts of oligosaccharide donor. Thus, this case may be analogous to that previously reported for deoxynojirimycin which also inhibited lipid-linked oligosaccharide synthesis.[180] However, in that study, 5 mM deoxynojirimycin was used, whereas in the above study only about 0.05–0.1 mM deoxymannojirimycin was necessary.

2,5-Dihydroxymethyl-3,4-dihydroxypyrrolidine

2,5-Dihydroxymethyl-3,4-dihydroxypyrrolidine (DMDP) was isolated from the plants *Lonchocarpus sericeus* and *Derris elliptica*[191] and shown

[187] U. Fuhrmann, E. Bause, G. Legler, and H. Ploegh, *Nature (London)* **307,** 755 (1984).

[188] A. D. Elbein, G. Legler, A. Tlusty, W. McDowell, and R. Schwarz, *Arch. Biochem. Biophys.* **235,** 579 (1984).

[189] V. Gross, K. Steube, T.-A. Tran-Thi, W. McDowell, R. Schwarz, K. Decker, W. Gerok, and P. Heinrich, *Eur. J. Biochem.* **150,** 41 (1985).

[190] Y. T. Pan and A. D. Elbein, in preparation.

[191] A. Welter, J. Jadot, G. Dardenne, M. Marlier, and T. Casimir, *Phytochemistry* **15,** 747 (1976).

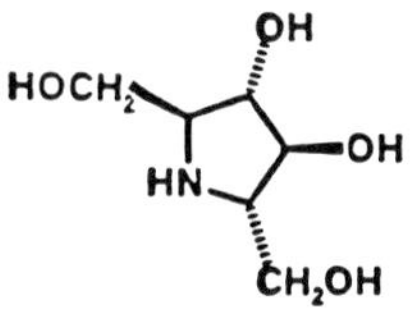

FIG. 10. Structure of 2,5-dihydroxymethyl-3,4-dihydroxypyrrolidine (DMDP).

to have the structure presented in Fig. 10. DMDP was tested as an inhibitor of various glycosidases using the appropriate p-nitrophenylglycoside as substrate. DMDP was a potent inhibitor of almond emulsin β-glucosidase ($K_i = 7 \times 10^{-6}$ M), yeast α-glucosidase ($K_i = 6.5 \times 10^{-6}$ M), and insect trehalase ($K_i = 5.5 \times 10^{-5}$ M). The kinetics of inhibition of these enzymes by DMDP were not presented, so it is not known whether inhibition was competitive.[192]

DMDP was tested as an inhibitor of glycoprotein processing to determine whether it prevented the formation of complex types of oligosaccharides in the influenza viral hemagglutinin. At 250 μg/ml of DMDP, more than 80% of the [^{3}H]mannose-labeled glycopeptides became susceptible to the action of endoglucosaminidase H. The major oligosaccharide found in these glycopeptides sized like a hexose$_{11}$GlcNAc. Both the [^{3}H]mannose-labeled oligosaccharide and the [^{3}H]glucose-labeled oligosaccharide were characterized by methylation analysis. The results indicated that the oligosaccharide was the typical Glc$_3$Man$_8$GlcNAc structure, consistent with the idea that DMDP inhibits glucosidase I.[193] However, in another study, different results were reported with DMDP. In IEC-6 intestinal epithelial cells in culture, 5 mM DMDP inhibited complex chain formation about 80%. HPLC showed similar endoglucosaminidase H-sensitive oligosaccharides for both control and treated cells. Thus, an increase in Man$_{7-9}$GlcNAc types of oligosaccharides was observed as well as the appearance of glucosylated oligosaccharides. These workers suggest that since the major oligosaccharides found in the DMDP-treated cells were nonglucosylated, its primary effect is not due to inhibition of glucosidases.[194] However, DMDP has been tested directly on the purified processing glucosidase I and the processing α-mannosidase from plants. This compound clearly inhibited the glucosidase activity although it was a much poorer inhibitor than was castanospermine or deoxynojirimycin. Thus, 50% inhibition required about 6–8 μg/ml of

[192] I. Cenci, D. Bello, P. Dorling, S. Evans, L. Fellows, and B. Winchester, *Biochem. Trans.* **13,** 1127 (1985).

[193] A. D. Elbein, M. Mitchell, B. A. Sanford, L. E. Fellows, and S. V. Evans, *J. Biol. Chem.* **259,** 12409 (1984).

[194] P. A. Romero, P. Friedlander, L. Fellows, S. V. Evans, and A. Hercovics, *FEBS Lett.* **184,** 197 (1985).

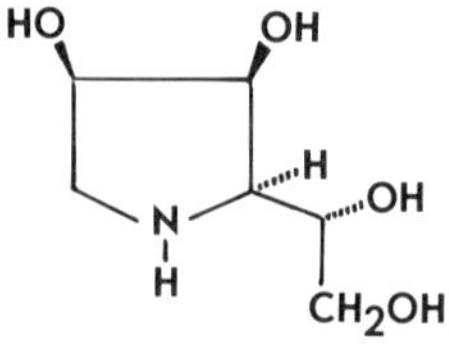

FIG. 11. Structure of 1,4-dideoxy-1,4-imino-D-mannitol.

DMDP as compared to 0.5–1.0 μg/ml for castanospermine. However, DMDP had no effect on the mannosidase activity.[119]

1,4-Dideoxy-1,4-imino-D-mannitol

1,4-Dideoxy-1,4-imino-D-mannitol (Fig. 11) was synthesized chemically from benzyl-α-D-mannopyranoside and shown to be an inhibitor of jack bean α-mannosidase, causing 50% inhibition at 5×10^{-7} M. This inhibition was of the competitive type.[195] These results, along with those from DMDP studies, demonstrate that simple nitrogen analogs of furanose sugars are a class of compounds that allow the design of specific glycosidase inhibitors. Furthermore, the synthesis of polyhydroxylated piperidines and pyrrolidines may provide a general and predictive method for controlling glycosidase, glycosyltransferase, and other enzyme-catalyzed reactions that involve carbohydrate substrates. These types of compounds could be valuable tools for probing the mechanism of action of such enzymes.

1,4-Dideoxy-1,4-iminomannitol also proved to be an inhibitor of glycoprotein processing.[196] Thus, in the presence of this drug, at 100–200 μg/ml, the formation of complex types of oligosaccharides in the influenza viral hemagglutinin was inhibited more than 75%. The major oligosaccharide in the viral glycoprotein in the presence of the iminomannitol was susceptible to endoglucosaminidase H, and it sized like a Man$_9$GlcNAc on columns of BioGel P-4. This oligosaccharide was completely susceptible to digestion by α-mannosidase and gave rise to free mannose and ManβGlcNAc. Thus this compound appears to be an inhibitor of mannosidase I, but it has not been tested directly on that enzyme. Although the iminomannitol is much less effective as an inhibitor than is swainsonine, it represents a new class of glycosidase inhibitors that are relatively easy to synthesize chemically. Thus it should be possible to make various iso-

[195] G. W. Fleet, P. W. Smith, S. V. Evans, and L. E. Fellows, *J. Chem. Soc., Chem. Commun.* p. 1240 (1984).
[196] G. Palamartczyk, M. Mitchell, P. W. Smith, G. W. J. Fleet, and A. D. Elbein, *Arch. Biochem. Biophys.* **243,** 35 (1985).

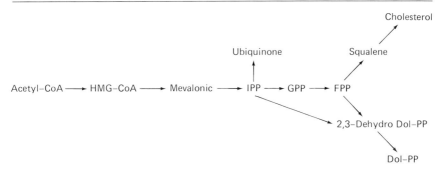

FIG. 12. Reactions involved in the biosynthesis of cholesterol, dolichyl phosphate, and ubiquinone.

mers and derivatives of this structure that may inhibit other enzymes, or may be better inhibitors than the existing ones.

Inhibitors of Dolichol Synthesis

The amount of dolichyl phosphate (and probably of dolichol) in most membranes is quite low, and thus these lipids are considered to be limiting. In fact, dolichyl phosphate levels have been postulated to be one of the control points in glycoprotein biosynthesis.[197,198] Although the turnover rate for dolichyl phosphate has not been measured in most membranes, one would expect inhibitors that prevent the biosynthesis of dolichyl phosphate to have a profound effect on the formation of the oligosaccharide portion of the asparagine-linked glycoproteins.

The biosynthesis of dolichyl phosphate involves the same initial series of reactions as those that participate in the formation of cholesterol. These reactions are outlined in Fig. 12 and involve the conversion of acetate to hydroxymethylglutaryl-CoA and of mevalonate to farnesyl pyrophosphate. At this point, the pathway branches either to form squalene and then cholesterol or to form 2,3-dehydrodolichyl pyrophosphate and then dolichyl pyrophosphate.[199] The biosynthesis of cholesterol, and probably also of dolichyl phophate, is regulated at the level of the hydroxymethylglutaryl-CoA reductase (HMG-CoA reductase). There are probably also other regulatory spots at the branches that are specific for each of the pathways. Several inhibitors of the HMG-CoA reductase that inhibit cholesterol biosynthesis have also been reported to block the for-

[197] J. J. Lucas and E. Levin, *J. Biol. Chem.* **252**, 4330 (1977).
[198] J. R. Harford, C. J. Waechter, and F. L. Earl, *Biochem. Biophys. Res. Commun.* **76**, 1036 (1977).
[199] D. K. Grange and W. L. Adair, Jr., *Biochem. Biophys. Res. Commun.* **79**, 734 (1977).

mation of the oligosaccharide portion of the asparagine-linked glycoproteins.

25-Hydroxycholesterol

When aortic smooth muscle cells were grown in culture in the presence of 25-hydroxycholesterol, at 10 μg/ml, there was an 80–90% inhibition in the incorporation of radioactive acetate into both cholesterol and dolichol. In addition, the incorporation of glucose into lipid-linked oligosaccharides and into N-linked glycoproteins was also greatly decreased.[200] On the other hand, the incorporation of mevalonic acid into cholesterol and dolichol was not altered by growth in 25-hydroxycholesterol. In fact, the addition of mevalonic acid to these inhibited cells restored the normal synthesis of the dolichyl-linked saccharides and the N-linked glycoproteins. At 1–5 μM, 25-hydroxycholesterol also inhibited the incorporation of acetate into both cholesterol and dolichol in L-cell cultures, but in this study the relationship between the concentration of hydroxycholesterol and the extent of inhibition was different for the two lipids. Thus, large fluctuations in cholesterol synthesis were observed under conditions that only slightly affected the synthesis of dolichol. However, under conditions where sterol synthesis was repressed to levels below 25% of control cultures, further inhibition of cholesterol biosynthesis was accompanied by a proportional decline in dolichol formation.[201] The authors postulated that a rate-limiting enzyme unique to the dolichol pathway may be saturated at a lower level of intermediate than is necessary to saturate the cholesterol pathway at the next rate-limiting step beyond the branch point. Therefore, HMG-CoA reductase activity and the rate of cholesterol synthesis may fluctuate with little change in dolichol synthesis, as long as the levels of these intermediates are sufficient to saturate the dolichol branch of the biosynthetic pathway.

Nevertheless, such studies do indicate that the rate of synthesis of dolichyl phosphate, and thus the levels of dolichyl phosphate in the cell, may play a role in the regulation of the biosynthesis of the oligosaccharide portion of glycoproteins. It seems likely that in some tissues, the synthesis of dolichyl phosphate is regulated at steps in its biosynthetic pathway that are unique to the dolichol branch in addition to the regulation at the HMG-CoA reductase step. Thus, feeding animals a diet high in cholestyramine, a compound that causes increased activity of the HMG-CoA reductase, resulted in a great stimulation in acetate incorporation into cholesterol, but not into dolichyl phosphate. This finding led to the con-

---

[200] J. T. Mills and A. Adamany, *J. Biol. Chem.* **253**, 5270 (1978).
[201] M. J. James and A. A. Kandutsch, *J. Biol. Chem.* **254**, 8442 (1979).

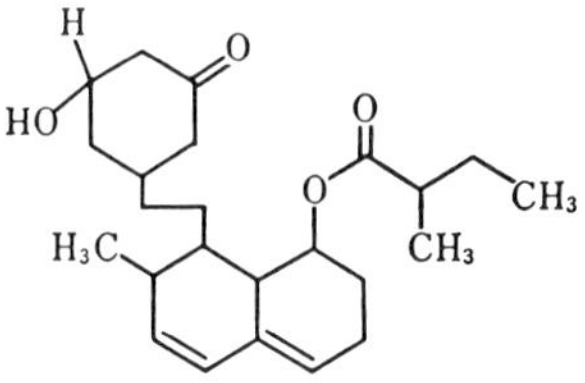

FIG. 13. Structure of compactin.

clusion that in rat liver, the rate of dolichyl phosphate synthesis is not regulated at the HMG-CoA reductase step.[202] It was suggested that regulation of this lipid may be at the level of the dolichyl phosphate synthetase. However, animals fed a high-cholesterol diet, which should suppress the biosynthesis of cholesterol, show increased incorporation of mevalonate into dolichol and dolichyl pyrophosphoryl oligosaccharides as well as increased activity of some of the glycosyltransferses associated with glycoprotein synthesis.[203,204] In this case, inhibiting the cholesterol branch of the pathway may lead to increased concentrations of various intermediates (farnesyl pyrophosphate, isopentenyl pyrophosphate) and therefore to a stimulation of dolichyl phosphate biosynthesis.

Compactin

Compactin is a fungal metabolite that blocks the cholesterol and polyisoprenol biosynthetic pathways at the hydroxymethylglutaryl-CoA-to-mevalonate step. Compactin, or ML-236B, is produced by *Penicillium brevicompactin*[205] and other strains of fungi. The structure of this compound as shown in Fig. 13 includes a lactonized ring that resembles the lactone form of mevalonic acid.[206] In cultured Chinese hamster ovary cells, 1–10 μM compactin blocked the synthesis of cholesterol by inhibiting the HMG-CoA reductase. Such cells are therefore dependent on exogenous cholesterol.[207] Because of the importance of cholesterol synthesis in atherosclerosis and the potential significance of an inhibitor of cholesterol biosynthesis, a great deal of effort has been directed at the chemical

[202] R. K. Keller, W. L. Adair, Jr., and G. C. Ness, *J. Biol. Chem.* **254**, 9966 (1979).

[203] I. A. Tavares, T. Coolbear, and F. W. Hemming, *Arch. Biochem. Biophys.* **207**, 427 (1981).

[204] D. A. White, B. Middleton, S. Pawson, J. P. Bradshaw, R. J. Clegg, F. W. Hemming, and G. D. Bell, *Arch. Biochem. Biophys.* **208**, 30 (1981).

[205] A. Endo, M. Kuroda, and Tanza *FEBS Lett.* **72**, 323 (1976).

[206] A. G. Brown, T. C. Smale, T. T. King, R. Hosenkamp, and R. H. Thompson, *J. Chem. Soc., Perkins Trans.* **1**, p. 1165 (1976).

[207] J. L. Goldstein, J. A. S. Helgeson, and M. S. Brown *J. Biol. Chem.* **254**, 5403 (1979).

synthesis of compactin. The synthesis of compactin and related compounds has been accomplished, and references and a summary of these studies are given.[208]

Since the dolichol biosynthetic pathway is also believed to involve the HMG-CoA reductase and therefore may be regulated in a manner similar to cholesterol, compactin was tested as an inhibitor of glycoprotein synthesis. Thus, when $1–5$ μM compactin was given to sea urchin embryos, it induced abnormal gastrulation. This effect was apparently due to an inhibition of dolichyl phosphate synthesis since the embryos cultured in compactin showed a decreased capacity to synthesize mannose-labeled glycolipids and asparagine-linked glycoproteins. In support of this idea, the inhibitory effect of compactin on development and on glycoprotein biosynthesis could be overcome by supplementing the embryos with exogenous dolichol or dolichyl phosphatae, but not with cholesterol or coenzyme Q.[209] Thus, compactin and 25-hydroxycholesterol are examples of inhibitors that can be used to study some of the steps in dolichyl phosphate synthesis and also its role in glycoprotein synthesis. It would, of course, be valuable to have other inhibitors that block the biosynthetic pathway of dolichol formation after the branch point, since such compounds would be much more specific in their action.

Inhibitors of Glycoprotein Transport

The synthesis of the oligosaccharide chain (i.e., $Glc_3Man_9GlcNAc_2$) via the lipid-linked saccharide pathway occurs in the endoplasmic reticulum, which is also the location where transfer of oligosaccharide to protein occurs. Following the attachment of carbohydrate, the newly formed glycoprotein is transported, probably in vesicles, to the Golgi apparatus. During this transport through the Golgi cisternae, the glycoprotein undergoes a number of processing or trimming reactions (outlined in Fig. 6), and the sugars of the complex types of oligosaccharides are added. The protein is then transported, probably also in vesicles, to the cell surface where it may be inserted into the cell membrane or secreted from the cell. Although the signals involved in targeting these proteins to specific locations are not known, a number of compounds, called ionophores, have been found to perturb the movement of glycoproteins from the endoplasmic reticulum to the plasma membrane.

Ionophores were first recognized through their effect of stimulating energy-linked transport in mitochondria. These studies thus provided

[208] T. Rosen and C. H. Heathcock, *J. Am. Chem. Soc.* **107**, 3731 (1985).
[209] D. D. Carson and W. J. Lennarz, *Proc. Natl. Acad. Sci. U.S.A.* **76**, 5709 (1979).

valuable tools for experiments on the link between metabolism and transport, and also stimulated extensive *in vivo* studies to determine the molecular basis of ionophore action. Ionophores are compounds of low to moderate molecular weight (200–2000) that form lipid-soluble complexes with polar cations. The polar cations of most significance from a biological standpoint are K^+, Na^+, Ca^{2+}, and Mg^{2+} as well as the biogenic amines. Although the ionophores were first isolated from fermentation broths of microorganisms, several synthetic ionophores were discovered later and found to have equivalent molecular properties.[210] Some of these ionophores have been utilized to study the intracellular movement of glycoproteins.

The effect of the ionophores A23187, monesin, and nigeracin on immunoglobulin secretion by plasma cells was examined.[211] A23187 has been shown to lower the intracellular Ca^{2+} concentration, whereas monensin and nigeracin induce partial Na^+/K^+ equilibration. In the plasma cell study, the Ig molecules were labeled by pulsing the plasma cells with [^{3}H]leucine for 1 hr and then following the secretion of ^{3}H-labeled immunoglobulin (Ig) during a 20-hr chase. The secretion of Ig was markedly inhibited by these ionophores, and striking alterations in the ultrastructural appearance of the Golgi complex were observed. The authors postulated that when the Ca^{2+} level is depleted (i.e., in the presence of A23187), the radioactive immunoglobulin is transported to Golgi vesicles, but these vesicles are incapable of fusion or migration. Therefore, such vesicles accumulate in exaggerated numbers in the Golgi apparatus. On the other hand, when Na^+/K^+ equilibration occurs in the presence of monensin or nigeracin, the labeled Ig is also transported to Golgi elements, but in this case large vacuoles are formed that contain increasing amounts of immunoglobulin. These data are consistent with the hypothesis that Ig secretion is mediated by the continual movement of smooth vesicles containing Ig to the cell surface where Ig is discharged by exocytosis. The inhibition of this process by ionophores is accompanied by dramatic alterations in the Golgi complex that can be ascribed to disturbances of vesicular traffic between the RER, the Golgi, and the plasma membrane.[211]

Procollagen and fibronectin are major glycoprotein products of cultured fibroblasts, and both are secreted from cells. In the presence of monensin (at $0.1–1$ mM), an ionophore having an affinity for monovalent cations, the secretion of these two proteins was reduced to 20% of normal fibroblasts. Electron microscopy showed that the inhibition of secretion

[210] B. C. Pressman, *Annu. Rev. Biochem.* **45**, 501 (1979).

[211] A. M. Tartakoff and P. Vassalli, *J. Exp. Med.* **146**, 1332 (1977).

was accompanied by an accumulation of membrane vacuoles. The authors suggested that the monensin affected the secretory structures rather than the proteins themselves.[212] In support of this idea, the ionophore had no effect on the hydroxylation or glycosylation of procollagen, nor on the glycosylation of fibronectin. There were also no significant changes in the incorporation of amino acids into cellular proteins. Pulse–chase studies indicated that the rates of secretion were impaired by the ionophore without enhancing intracellular degradation. The removal of monensin from the culture medium led to a restoration of the secretion of these glycoproteins. The distribution of procollagen and fibronectin were compared in control cells and in cells treated with monensin using immunofluoresence microscopy. In the control cells, both proteins were present throughout the cytoplasm and in the Golgi region. On the other hand, in monensin-treated cells both procollagen and fibronectin accumulated within Golgi vacuoles, while normal Golgi complexes were not found. These proteins were also found in more peripheral deposits corresponding to the RER area of the cells.[213]

Monensin and the other ionophores have also been found to inhibit the appearance of viral glycoproteins at the surface of the host cell, or to affect their normal maturation.[214,215] In studies with Friend murine leukemia virus[214] or with Mason-Pfizer monkey virus,[216] animal cells treated with monensin, at 10^{-6} to 10^{-7} M, continued to synthesize viral particles and these particles continued to bud normally from the cell surface. However, the particles released from the cells had an altered bouyant density on sucrose gradients and were noninfectious. While these noninfectious particles had a normal compliment of proteins, they showed a significant reduction in the amount of glycosylated proteins. It appeared that the precursor proteins for the viral glycoproteins were not cleaved to the mature glycoproteins. The uncleaved molecules were found at the cell surface.

However, the glycoproteins of Sindbis and vesicular stomatitis virus grown in BHK and chick embryo fibroblasts were synthesized in normal amounts in the presence of monensin, but these glycoproteins did not appear at the cell surface of the infected cells. Proteolytic cleavage of Sindbis viral glycoprotein PE2 to E2 was inhibited in the ionophore-

[212] N. Uchida, H. Smilowitz, and M. L. Tanzer, *Proc. Natl. Acad. Sci. U.S.A.* **76,** 1868 (1979).
[213] P. W. Ledger, N. Uchida, and M. L. Tanzer, *J. Cell. Biol.* **87,** 663 (1980).
[214] R. V. Srinivas, L. R. Melsen, and R. W. Compans, *J. Virol.* **42,** 1067 (1982).
[215] S. Chaterjee, J. A. Bradac, and E. Hunter, *J. Virol.* **44,** 1003 (1982).
[216] L. Kaariainen, K. Hashimoto, J. Saraste, I. Virtanen, and K. Penttinen, *J. Cell Biol.* **87,** 783 (1980).

treated cells, but the attachment of fatty acid to PE2 proceeded normally. Also fatty acid attachment to the G protein of vesicular stomatitis virus and processing of the oligosaccharide portion occurred in drug-treated cells. By electron microscopy, it was determined that the ionophores prevented the movement of viral glycoproteins from the Golgi apparatus to the cell surface membrane where budding and release of virus particles occur.[217]

In MDCK cells, influenza virus is assembled at the apical surface, whereas VSV virus buds from the basolateral surface. Monensin, at 10^{-6} M, reduced the yield of VSV in MDCK cells more than 90%, whereas influenza virus yields were unaffected. On the other hand, in BHK cells where no polarity of maturation occurs, monensin inhibited production of both viruses.[218] These results suggest that at least two distinct pathways of transport of glycoproteins to the plasma membrane exist in MDCK cells, and only one of these pathways is blocked by monensin. Evidence for a different pathway for secretory and membrane glycoproteins was also obtained in cultured avian myotubes using monovalent ionophores.[219] This tissue produces two glycoproteins, the integral membrane glycoprotein, acetylcholine receptor, and the secretory glycoprotein, acetylcholinesterase. Nigericin and monensin had no effect on the appearance of acetylcholine receptor, but acetylcholinesterase secretion was severely inhibited.

Monensin has also been valuable to study the transport of proteins through intracellular compartments. Studies on the assembly of Semliki Forest virus glycoproteins in BHK cells in the presence of monensin indicated that the Golgi stacks could be divided into functionally distinct cis, medial, and trans compartments, each comprising one or two adjacent cisternae.[220] Monensin blocked intracellular transport of the glycoproteins but not their synthesis. As a result, the glycoproteins accumulated in the medial cisternae. The cisternae containing these intracellular caspids, called intracellular caspid-binding membranes, sedimented at a higher density in sucrose gradients and thus could be separated from other Golgi cisternae. Thus, monensin was of considerable value for the isolation of specific Golgi cisternae.[221] Labeling the cells with [^{3}H]palmitate in the presence of monensin showed that most of the label was in the intracellular caspid-binding membranes, indicating that fatty acid acylation occurs in the cis or medial Golgi. In contrast, the distribution of $\alpha 1,2$-

[217] D. C. Johnson and M. J. Schlesinger, *Virology* **103,** 407 (1980).
[218] F. V. Alonso and R. W. Compans, *J. Cell Biol.* **89,** 700 (1981).
[219] H. Smilowitz, *Cell (Cambridge, Mass.)* **19,** 237 (1980).
[220] G. Griffiths, P. Quinn, and G. Warren, *J. Cell Biol.* **96,** 835 (1983).
[221] P. Quinn, G. Griffiths, and G. Warren, *J. Cell. Biol.* **96,** 851 (1983).

mannosidase and galactosyl transferase was not affected by monensin. The authors suggest that these data indicate that these two activities are in the trans cisternae.[221] Other studies have indicated that the galactosyl transferase is a late acting processing enzyme and is usually thought to be in the trans Golgi. However, the α1,2-mannosidase is an early acting enzyme and evidence from other studies has localized this enzyme to the ER or cis Golgi.

Effects on Protein Synthesis

β-Hydroxynorvaline

The sequence of amino acids at the asparagine residue that becomes glycosylated has been found to be . . . Asn-X-Ser (Thr) . . . . The importance of threonine in this recognition sequence was assessed by examining the effects of the threonine analog, β-hydroxynorvaline, on cotranslational glycosylation. β-Hydroxynorvaline inhibited the glycosylation of the α-subunit of human chorionic gonadotropin and the β-subunit of bovine luteinizing hormone. The effect was prevented by threonine, indicating that β-hydroxynorvaline acted via its incorporation into protein.[222] The results suggest that the threonine site may be sensitive to steric hindrance. In the presence of this inhibitor, fibroblasts synthesized cathepsin D with two, one, or zero oligosaccharides. The nonglycosylated cathepsin was degraded within 45 min of its synthesis, probably in the ER. The data indicate that in the absence of carbohydrate, the cathepsin D is rapidly degraded and that replacement of threonine by β-hydroxynorvaline results in an enhanced degradation of mature catepsin D in the lysosomes.[223] The β-hydroxynorvaline also caused the formation of α_1-acid glycoprotein with zero to six oligosaccharide chains in cultured hepatocytes. Partially glycosylated glycoproteins that contain from one to five oligosaccharides, or unglycosylated protein produced in the presence of tunicamycin, exited the cells more slowly than normal α_1-acid glycoprotein. The results suggest that hydroxynorvaline-induced changes, either in the extent of glycosylation or in the peptide sequence of α_1-acid glycoprotein, can interfere with its transport through the cell.[224]

Other Inhibitors of Protein Synthesis

Several inhibitors of protein synthesis have been found to also affect the synthesis of lipid-linked oligosaccharides. In one study MDCK cells were grown in [2-^{3}H]mannose in the presence of cycloheximide or puro-

[222] G. Hortin and I. Boime, *J. Biol. Chem.* **255**, 8007 (1981).
[223] M. Hentze, A. Hasilik, and K. von Figura, *Arch. Biochem. Biophys.* **230**, 375 (1984).
[224] P. A. Docherty and N. N. Aronson, Jr., *J. Biol. Chem.* **260**, 10847 (1985).

mycin, and the effects of these inhibitors on the synthesis of lipid-linked saccharides and proteins were examined.[225] In this study, the inhibition of protein synthesis resulted in a substantial inhibition in the incorporation of [³H]mannose into lipid-linked oligosaccharides. However, under these conditions, the formation of dolichylphosphorylmannose was only slightly affected. Cycloheximide had no effect on the *in vitro* incorporation of mannose from GDP-[¹⁴C]mannose into lipid-linked saccharides by membrane preparations of pig aorta. The inhibition of lipid-linked oligosaccharide formation did not appear to be caused by a decrease in the amount of the various glycosyltransferases as a result of the inhibition of protein synthesis, nor was it the result of more rapid degradation of lipid-linked oligosaccharides. The most plausible explanations for these results were either that the amount of dolichyl phosphate was limiting and therefore this lipid was not available as a carrier for the oligosaccharide, or that the synthesis of lipid-linked oligosaccharide is subject to feedback control. However, since the formation of dolichylphosphorylmannose was not inhibited in the presence of cycloheximide, the levels of dolichyl phosphate did not appear to be limiting, unless there are different pools of this lipid.[225] Thus, the formation of the lipid-linked oligosaccharides may be regulated, at least in part, by end-product inhibition.

In another study, oligosaccharide–lipid synthesis was examined in cells incubated in the presence of 1 μg/ml of actinomycin D to depress levels of mRNA, or in the presence of 100 μg/ml cycloheximide to abolish protein synthesis. The results indicated that the synthesis of lipid-linked oligosaccharides was proportional to the rate of protein synthesis.[226] The regulated step appeared to be prior to the formation of $Man_5GlcNAc_2$-pyrophosphoryldolichol, leading these workers to suggest that the most likely control point was the availability of dolichyl phosphate.

Inhibitors of Glycosyltransferases

There are relatively few inhibitors known that act on glycosyltransferases, although such compounds could be of extreme benefit to studies in many areas of complex carbohydrates. The following two inhibitors, polyoxin and papalucandin, have been studied in some detail and will serve as examples of these compounds.

The polyoxins are a series of peptidyl antibiotics that have the general structure shown in Fig. 14. These antibiotics strongly inhibit chitin synthase from a spectrum of fungi[227] and anthropods.[228] The polyoxins

[225] J. W. Schmitt and A. D. Elbein, *J. Biol. Chem.* **254**, 12291 (1979).
[226] S. C. Hubbard and P. W. Robbins, *J. Biol. Chem.* **255**, 11782 (1980).
[227] M. Hori, K. Kakaki, S. Suzuki, and T. Misato, *Agric. Biol. Chem.* **38**, (1974).
[228] M. J. Gijswift, D. H. Deul, and B. Dejong, *J. Pestic. Biochem. Physiol.* **12**, 87 (1979).

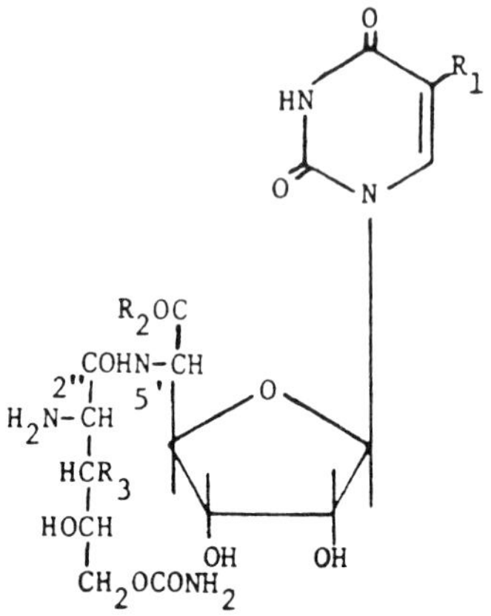

FIG. 14. General structure of the polyoxins.

were shown to represent structural anologs of UDP-N-acetylgluco-samine[229] and these compounds were found to be competitive inhibitors of cell wall chitin synthetase from *Neurospora*[230] and *Saccharomyces*. A variety of polyoxin derivatives have been synthesized, and a number of these have been shown to be competitive inhibitors. The inhibitory activities were dependent on pH, and it was concluded that the ionized amino group at the C-2″ positions had a very important role for binding the polyoxins to the chitin synthase.[230] Apparently the free carboxyl group is also necessary for inhibitory activity.

Papalucandin B is an antibiotic produced by the deuteromycete, *Papulana sphaerosperma*.[231] This compound contains a glucose and galactose residue as well as two fatty acids and an aromatic ring.[232] The antibiotic appears to be relatively specific for inhibiting the formation of alkali-insoluble $\beta(1 \rightarrow 3)$-glucans.[233] It has also been shown to inhibit the glucan synthase in some organisms. In one study, it was found that the inhibition of glucan synthase became less pronounced as the substrate concentration of UDP-glucose was decreased. In fact, at low levels of UDP-glucose, the enzymes from *Saccharomyces cerevisiae* or *Wangiella dermatitidis* were stimulated by papulacandin.[234] It was suggested that the glucan synthase might exist in more than one intraconvertible form and

[229] K. Isono, T. Azuma, and S. Suzuki, *Chem. Pharm. Bull.* **19,** 505 (1971).

[230] M. Hori, K. Kakiki, and T. Misato, *Agric. Biol. Chem.* **38,** 699 (1974).

[231] B. C. Baguley, G. Rommele, J. Gruner, and W. Wehrle, *Eur. J. Biochem.* **97,** 345 (1974).

[232] P. Traxler, H. Fritz, and W. Richter, *Helv. Chim. Acta.* **60,** 578 (1977).

[233] P. Perez, I. Garcia-Acha, and A. Duran, *J. Gen. Microbiol.* **129,** 245 (1983).

[234] M. S. Kang, P. J. Szanislo, V. Notario, and E. Cabib, *Carbohydr. Res.* (in press).

that stimulation by papulacandin could be due to preferential binding to the active form of the enzyme.

Conclusions

A great deal of information is available concerning the pathway of biosynthesis of the oligosaccharide portion of the N-linked glycoproteins. In terms of the lipid-linked saccharide pathway, some of the information has come through the use of the inhibitors that block specific steps in the pathway. For example, tunicamycin inhibits the first step in the pathway by preventing the formation of dolichylpyrophosphoryl-GlcNAc. As a result it inhibits the synthesis of $Glc_3Man_9GlcNAc_2$-pyrophosphoryldolichol and prevents protein glycosylation since this lipid is the donor of oligosaccharide to protein. The antibiotics amphomycin and tsushimycin, and some of the sugar analogs, inhibit the formation of dolichylphosphorylmannose which is the mannosyl donor for the last four mannose residues of the $Glc_3Man_9GlcNAc_2$-pyrophosphoryldolichol. Thus, in the presence of these inhibitors, a $Man_5GlcNAc_2$-pyrophosphoryldolichol accumulates and cannot be further elongated because of the absence of dolichylphosphorylmannose. Other inhibitors are also known that block these reactions, but some of these have proven less specific.

One problem with some of these compounds such as the peptide antibiotics is that they are not able to enter the cells and therefore cannot be used for *in vivo* studies. Thus, the ideal inhibitor is one that is specific for a single step in the lipid-linked saccharide pathway and is also able to penetrate into the cells and express its effect *in vivo*. Such an inhibitor would be useful for causing alterations in the oligosaccharide composition of glycoproteins and for examining functional aspects of the indicated glycoprotein. Tunicamycin fits this definition of an ideal inhibitor but in the presence of this compound the protein is not glycosylated at all. The problem here is that many proteins have such an altered conformation in the presence of tunicamycin that they may become insoluble or may be rapidly degraded. While this antibiotic is still quite valuable to show that various proteins are N-glycosylated, it would be very useful to have other inhibitors that may cause alterations in the oligosaccharide chains during or after they are transferred to protein.

A number of compounds have recently been described that inhibit specific steps in the glycoprotein-processing pathway, and these inhibitors may provide a means to prevent the formation of complex chains and allow glycoproteins with altered oligosaccharides to be formed. Swainsonine, a plant indolizidine alkaloid, inhibits the processing mannosidase II that removes the α1,3- and α1,6-mannoses from the GlcNAc-$Man_5GlcNAc_2$-protein. Thus cells grown in the presence of this alkaloid produce glycoproteins that have hybrid types of oligosaccharides, with partial high-mannose and partial complex structure. Castanospermine is another plant indolizidine alkaloid that inhibits glucosidase I, the processing enzyme that removes the first $\alpha 1,2$-glucose from the $Glc_3Man_9GlcNAc_2$–protein, so the major oligosaccharide structure found in most cells is a $Glc_3Man_7GlcNAc_2$. The antibiotic, deoxynojirimycin, is another processing inhibitor that inhibits glucosidase I (and also glucosidase II) *in vitro*. Cells grown in the presence of this compound do not form complex oligosaccharides, and render the high-mannose structures less susceptible to the action of α-mannosidase (because of the presence of blocking glucose units).

A chemically synthesized glucose analog, bromoconduritol, has also been shown to inhibit the processing glucosidases and to give rise to increasing amounts of glucose-containing structures which the authors characterized as $GlcMan_9GlcNAc_2$, $GlcMan_9GlcNAc_2$, and $Glc-Man_7GlcNAc_2$. These results would suggest that bromoconduritol inhibits glucosidase II, but that it should leave two glucose residues on the oligosaccharide. The interpretation of these results are not clear. The mannose analog of deoxynojirimycin (deoxymannojirimycin) is an inhibitor of mannosidase I and causes the accumulation of $Man_9GlcNAc_2$ structures. Several pyrrolidine types of structures are inhibitors of glucosidase I or mannosidase I. Because of the ease of chemically synthesizing these five-membered ring structures, it seems likely that many other analogs will be produced that should inhibit other glycosidases.

As indicated in the Introduction, the formation of N-linked glycoproteins is a complex process and involves the synthesis of dolichyl phosphate, attachment of sugars to this lipid, transfer of oligosaccharide from lipid to protein, processing of oligosaccharide, and movement of glycoprotein through the various cellular compartments. Several inhibitors of cholesterol biosynthesis also block the biosynthesis of dolichyl phosphate (hydroxycholesterol and compactin). Since cells grown in the presence of these compounds are unable to produce dolichyl phosphate, the results of these treatments are like those with tunicamycin, i.e., unglycosylated proteins. The difficulty with these inhibitors is that they also block cholesterol biosynthesis, and this steroid is an essential membrane component. The inhibition of cholesterol can be overcome by adding exogenous cholesterol. Another group of interesting compounds are the ionophores, which have recently been shown to block the movement of glycoproteins from the Golgi to the cell surface. These compounds have already proven of value for tracing pathways of glycoprotein movement within the cell.

In closing, it should be mentioned that another approach to studies on glycoprotein biosynthesis, beside the use of inhibitors, is to use mutant cell lines that are missing specific enzymes in the biosynthetic pathways. For example, a mutant cell line was isolated that was missing the enzyme that forms dolichylphosphorylmannose. This mutant accumulated the $Man_5GlcNAc_2$-pyrophosphoryldolichol but was unable to elongate it.[235] Thus, this mutant gives similar results to those found with the antibiotic amphomycin. Another example is the recent isolation of a glucosidase II mutant which gives rise to oligosaccharides having $Glc_2Man_9GlcNAc_2$ and $Glc_2Man_8GlcNAc_2$. Many animal cell mutants have now been isolated, and in many cases isolation was simplified by screening for lectin resistance. These mutants will also be of considerable value for the role of carbohydrate in cell function. This area has recently been reviewed.[236]

[235] M. L. Reitman, I. S. Trowbridge, and S. Kornfeld, *J. Biol. Chem.* **257**, 10357 (1982).
[236] P. Stanley, *in* "The Biochemistry of Glycoproteins and Proteoglycans" (W. T. Lennarz, ed.), p. 161. Plenum, New York, 1980.

[59] Transport of Sugar Nucleotides into the Lumen of Vesicles Derived from Rat Liver Rough Endoplasmic Reticulum and Golgi Apparatus

By MARY PEREZ and CARLOS B. HIRSCHBERG[1]

Introduction

Previous methods that were used to detect the permeability of microsomal membranes to solutes, such as glucose 6-phosphate,[2] provided a rapid assessment of the penetration of a radiolabeled solute into vesicles. A complete and rapid isolation of the vesicles from the incubation mixture was achieved by centrifuging the vesicles through a layer of silicone oil. However, this method had the disadvantage that only "heavy" or Ca^{2+}-bound microsomes could be rapidly sedimented using a microcentrifuge. In this chapter we describe methods to determine whether an intact sugar nucleotide is transported into rat liver derived vesicles of the Golgi apparatus or the rough endoplasmic reticulum (RER). This assay also has the advantage that the total transport activity of a radiolabeled sugar nucleotide can be separated into two components: those radiolabeled solutes

[1] C. B. Hirschberg supported by Grant GM-30365 from the National Institutes of Health; M. Perez supported by National Institutes of Health Grant 5T 32 HL-07050.
[2] L. M. Ballas and W. J. Arion, *J. Biol. Chem.* **252**, 8512 (1977).

which are within the lumen of the vesicles, and those which have become transferred (covalently linked) to macromolecules.

Principle

To be of physiological significance, the transport assays described are applicable only to vesicles which have been shown to be sealed and of the same membrane topography as *in vivo*. Sealed vesicles are incubated with sugar nucleotides radiolabeled in the sugar along with separate control incubations of a standard penetrator, such as [2-^{3}H]deoxyglucose, and a standard nonpenetrator, such as [^{3}H-*methoxy*]inulin.[3] Following the incubation, the reaction is stopped by dilution, and the vesicles are reisolated by ultracentrifugation. The resulting vesicle pellet consists of the following radiolabeled components: the radiolabeled solutes which have been transported across the membrane, radiolabeled sugar which has been transferred to macromolecules, and radiolabel which has become adsorbed or is between the vesicles. The vesicle pellet is then resuspended and disrupted by sonication. Covalently linked macromolecules are acid precipitated and reisolated by centrifugation. The resulting supernatant solution contains those solutes which are *within* the lumen of the vesicles, plus those which are *adsorbed* and *between* the vesicles in the pellet. The concentration of the radioactive solutes *within* the vesicles [S$_i$] can be determined since the total solutes, those *outside* and *between* the vesicles, can be calculated. Total transport activity of the sugar nucleotide is the sum of the radioactive solutes within vesicles, plus the radioactive sugars in the acid-insoluble pellet.

Golgi Apparatus

Isolation of Golgi-Derived Vesicles from Rat Liver. Golgi-derived vesicles are isolated according to the method of Leelavathi *et al.*[4] and resuspended in 10 mM Tris–HCl, pH 7.5, plus 0.25 M sucrose using the "B" or loose pestle of a 7-ml Wheaton glass dounce to a final protein concentration of 5 mg/ml. Vesicles are preferably used fresh for transport assays. If vesicles are to be frozen, they are resuspended in the above buffer, plus 10 mg/ml of bovine serum albumin, and frozen at −70°.[5] Vesicles are thawed quickly at 37° and used immediately.

[3] Other standard penetrators are glycerol and sucrose; other standard nonpenetrators are acetate, pyruvate, and dextran.
[4] D. E. Leelavathi, L. W. Estes, D. S. Feingold, and B. Lombardi, *Biochim. Biophys. Acta* **211,** 124 (1970).
[5] Rat liver Golgi vesicles can be frozen and thawed and still retain their membrane integrity

Analytical Methods. Protein is estimated by the method of Peterson.[6] Vesicle integrity and topography is assessed by the latency of the vesicles to neuraminidase treatment of vesicles prelabeled with CMP-N-[^{14}C]acetylneuraminic acid.[7] The amount of plasma membrane contamination is determined by 5'-nucleotidase.[8] Vesicle enrichment is determined by the activity of sialyltransferase.[9] RER contamination was assessed by the activity of glucose-6-phosphatase.[8]

Transport Assay. Transport assays are carried out in 10-ml polycarbonate ultracentrifuge tubes with screw caps. The assay mixture consists of 10 mM Tris–HCl, pH 7.5, plus the appropriate additives, 3.0×10^5 dpm of sugar nucleotide or standards adjusted with unlabeled solute to the desired concentration in a total volume of 0.8 ml (see Table I). Following preincubation of the incubation mixture at the appropriate temperature, incubations are initiated with the addition of 0.2-ml aliquots of a Golgi vesicle suspension containing 0.4–1 mg of protein. Following incubation, 2 ml of ice-cold incubation buffer is added to stop the reaction, and the reaction mixture is immediately centrifuged in a Ti 50 rotor for 25 min (100,000 g at 4°). The supernatant solution is removed, and a 0.1 ml aliquot is counted by liquid scintillation spectrometry for determination of [S_m], the concentration of solute in the incubation medium.

$$[S_m] = \frac{\begin{array}{c}\text{total solute in the supernatant} \\ \text{(expressed as cpm/ml)}\end{array}}{\begin{array}{c}\text{specific activity of the sugar nucleotide} \\ \text{at the beginning of the incubation} \\ \text{(expressed as cpm/nmol)}\end{array}}$$

The surface of the pellet is washed 3 times, each with 1.5 ml of ice-cold buffer. Ice-cold water (0.5 ml) is added to each pellet, mixed in a vortex mixer, and sonicated for 25 min at 5°.[10] Following vesicle disruption, the

---

and topography if frozen in bovine serum albumin. However, their total transport activity is decreased as compared to freshly prepared vesicles. This decrease in transport activity varies with the sugar nucleotide tested.

[6] G. L. Peterson, *Anal. Biochem.* **83**, 346 (1977).

[7] D. J. Carey and C. B. Hirschberg, *J. Biol. Chem.* **256**, 989 (1981). Vesicles which prove to be at least 95% intact are used for transport assays.

[8] N. N. Aronson, Jr. and O. Touster, this series, Vol. 31, p. 90. Golgi vesicles prepared by this method are usually 2-fold enriched over homogenate in 5'-nucleotidase, and 1.4-fold enriched in glucose-6-phosphatase, an RER marker enzyme.

[9] E. B. Briles, E. Li, and S. Kornfeld, *J. Biol. Chem.* **252**, 1107 (1977). Vesicles which are at least 40-fold enriched over homogenate in sialyltransferase activity are used.

[10] Pellet disruption is performed using a bath sonicator such as a Branson 220-W bath sonicator placed at 5°. Incubation tubes are placed in 100 ml beakers filled with approximately 2 cm of ice-cold H$_2$O. During the sonication period, the pellets are mixed in a Vortex mixer vigorously several times.

TABLE I

Sugar Nucleotide Translocation into Golgi Vesicles Derived from Rat Liver

Sugar nucleotide	Incubation buffer[a]	Final sugar nucleotide concentration (μM)	Incubation period (min)	Solutes within vesicles (S_i) (pmol/mg protein)	Acid-insoluble (T) (pmol/mg protein)	Total transport ($S_i + T$) (pmol/mg protein)
GDP-fucose	TKM	7.5	10	72	125	197
CMP-sialic acid	TKM	2.5	10	155	168	323
UDP-N-acetylglucosamine	STKM	8	2.5	625	264	889
UDP-galactose	STKM	2	10	85	287	372

[a] TKM = 10 mM Tris–HCl, pH 7.5 + 0.15 M KCl + 1 mM MgCl$_2$ containing 0.5 mM 2,3-dimercaptopropanol (inhibitor of nucleotide pyrophosphatases); STKM = TKM + 0.25 M sucrose + 0.5 mM 2,3-dimercaptopropanol.

pellets are each suspended in a Vortex mixer using the highest speed of the mixer for 15 sec. Perchloric acid (8%, 0.5 ml) is added to the pellets, suspended in a Vortex mixer, and the mixture is allowed to stand on ice for 15 min. Following centrifugation (30,000 g, 15 min), a 0.5-ml aliquot of the supernatant is removed for determination of S_t, the total solutes in the pellet.

$$S_t = \frac{\text{total soluble radioactivity in the pellet}}{\begin{array}{c}\text{(expressed as cpm/mg of protein)}\\ \hline \text{specific activity of the sugar nucleotide}\\ \text{at the beginning of the incubation}\\ \text{(expressed as cpm/pmol)}\end{array}}$$

To determine S_o, the concentration of solute which is outside the vesicles in the pellet, the pellet volume which is outside the vesicles, V_o, must first be calculated. This parameter is calculated using a standard nonpenetrator such as radiolabeled inulin.

$$V_o = \frac{\begin{array}{c}\text{total soluble radioactivity in the pellet}\\ \text{for the standard nonpenetrator}\\ \text{(expressed as cpm/mg of protein)}\end{array}}{\begin{array}{c}\text{cpm}/\mu\text{l of the initial supernatant}\\ \text{for the nonpenetrator}\end{array}}$$

$$S_o = V_o \; (\mu\text{l/mg}) \times [S_m] \; (\text{pmol}/\mu\text{l})$$

The amount of solute inside the vesicles in the pellet, S_i, is the difference between the total solutes in the pellet and those solutes which are outside the vesicles.

$$S_i = S_t \; (\text{pmol/mg}) - S_o \; (\text{pmol/mg})$$

The concentration of solute inside the vesicles in the pellet, $[S_i]$, is calculated by dividing the amount of solute inside the vesicles by the pellet volume inside the vesicles, V_i. This parameter is the difference between the total pellet volume (outside plus inside), V_t, and the volume outside the vesicles in the pellet, V_o. V_t is assumed to be the volume occupied by a freely penetrating solute such as deoxyglucose.

$$V_t = \frac{\begin{array}{c}\text{total soluble radioactivity in the pellet}\\ \text{for a penetrator}\\ \text{(expressed as cpm/mg of protein)}\end{array}}{\begin{array}{c}\text{cpm}/\mu\text{l of the initial supernatant}\\ \text{for the penetrator}\end{array}}$$

$$V_i = V_t \; (\mu\text{l/mg}) - V_o \; (\mu\text{l/mg})$$

$$[S_i] = \frac{S_i \; (\text{pmol/mg of protein})}{V_i \; (\mu\text{l/mg})}$$

The total transport of the sugar nucleotide is defined as the total solutes within the lumen of the vesicles, plus the amount of solute which has been subsequently transferred to macromolecules, T. To determine the amount of sugar which has been transferred to macromolecules, the vesicle pellets are surface washed 2 times, each with 2 ml of 8% perchloric acid. The pellets are resuspended in 2 ml of 8% perchloric acid by sonication, then reisolated by centrifugation (100,000 g, 25 min). The resulting supernatant solution is discarded, and the pellet is dissolved in 1 ml of 1 N NaOH. The pellet is neutralized by the addition of 0.35 ml of 4 N HCl and 0.5 ml of water and counted in 18 ml of Aquasol 2 (New England Nuclear).

$$T \text{ (transfer)} = \frac{\text{total insoluble radioactivity in the pellet (expressed as cpm/mg of protein)}}{\begin{array}{c}\text{specific activity of the sugar nucleotide} \\ \text{at the beginning of the incubation} \\ \text{(expressed as cpm/pmol)}\end{array}}$$

$$\text{Total transport} = S_i \text{ (transport)} + T \text{ (transfer)}$$

Assay conditions and typical transport values for selected sugar nucleotides into Golgi-derived vesicles are shown in Table I.

Rough Endoplasmic Reticulum

Isolation of RER Vesicles from Rat Liver. Rat liver RER are isolated according to the method of Carey and Hirschberg,[11] a modification of the Fleischer and Kervina procedure.[12,13] Vesicles are very gently resuspended in 10 mM Tris–HCl, pH 7.5, plus 0.25 M sucrose using the loose pestle of a Wheaton glass dounce homogenizer, to a final protein concentration of 7 mg/ml. RER vesicles are very fragile and should be used immediately for transport assays.

Analytical Methods. Protein is estimated by the Peterson method.[6] Vesicle integrity and topography is assessed by the latency of mannose-6-phosphatase.[14] Enrichment of the vesicles is determined by glucose-6-phosphatase and is usually 5- to 7-fold enriched over the homogenate.

[11] D. J. Carey and C. B. Hirschberg, *J. Biol. Chem.* **255**, 4348 (1980).

[12] S. Fleischer and M. Kervina, this series, Vol. 31, p. 6.

[13] When RER, SER, and Golgi vesicles are needed from the same preparation, the procedure of Morré provides a good yield and enrichment of all three fractions. D. J. Morré *in* "Molecular Techniques and Approaches in Developmental Biology" (M. J. Chrispeels, ed.), p. 1. Wiley (Interscience), New York, 1973.

[14] W. J. Arion, L. M. Ballas, A. J. Lange, and B. K. Wallin, *J. Biol. Chem.* **251**, 4901 (1976).

Transport Assay. Transport assays are carried out as described for Golgi vesicles with slight modifications. The incubation mixture contains STKM buffer (10 mM Tris–HCl, pH 7.5, 0.25 M sucrose, 0.15 M KCl, 1 mM MgCl$_2$), 0.5 mM 2,3-dimercaptopropanol, and 3.0×10^5 dpm of sugar nucleotide in a total volume of 0.8 ml. The incubation mixture is brought to 30°, and 0.2 ml of RER vesicles (0.6–1.3 mg of protein) are added to start the reaction. After 10 min at 30°, 2.0 ml of STKM buffer is added, and the reaction mixture is centrifuged for 25 min at 60,000 g. The assay is continued as described for Golgi vesicles.

Final Comment

The following criteria are used to determine whether the sugar nucleotide is translocated across the membranes of vesicles. The penetration of the sugar nucleotide should be dependent on temperature, saturable at high concentrations of the sugar nucleotide (micromolar range), and inhibited when the integrity of the membrane is disrupted. In order to show that the intact sugar nucleotide, and not breakdown products, is being translocated across the membranes, vesicles are incubated with a mixture of the sugar nucleotide labeled with different radioisotopes in the sugar and the nucleotide. Following the incubation period, the double-radiolabeled ratio of the vesicle pellet is compared to the ratio of the incubation mixture at the beginning of the incubation period. If these two ratios are similar, it suggests that the intact sugar nucleotide was transported across the membrane.

[60] Plant Cell Wall Polysaccharides That Activate Natural Plant Defenses

By PAUL D. BISHOP and CLARENCE A. RYAN

Current hypotheses concerning the chemical bases for plant defenses against predators include the concept that plants can respond to attacks of insects, microorganisms, and viruses by activating genes that produce defensive chemicals.[1] In recent years studies of the insect-induced proteinase inhibitors in plant leaves and of fungal-elicited phytoalexin synthesis in cotyledons and seedlings have demonstrated that plant cell walls

---

[1] J. Kuč, *BioScience* **32,** 854 (1982).

contain within their polysaccharide structures informational components that can activate synthesis of these types of defensive chemicals.[2,3] The plant cell wall fragments, produced by endopolygalacturonases released from compartments in the plant cells or secreted by microorganisms, have been hypothesized to be activators of a universal recognition system in plants that can locally and/or systemically activate genes which control the synthesis of plant defense chemicals, such as the antibiotic phytoalexins and antinutritive proteinase inhibitors.[4] Preparations of pectin fragments, derived from tomato leaves and from soybean cell walls, that are inducers of defense responses in plants are described as follows:

Isolation and Purification of the Proteinase Inhibitor Inducing Factor (PIIF), a Pectic Polysaccharide from Tomato Leaves

Pectic substances in general refer to a part of the polysaccharides which form the matrix between plant cell walls. The pectic polysaccharides are highly complex and are primarily composed of a backbone consisting of $\alpha(1 \rightarrow 4)$-D-galacturonic acid with occasional $\alpha(1 \rightarrow 2)$-L-rhamnose units[5] and short branching side chain glycans of varying composition, attached to some of the 2 and 3 hydroxyl groups of the galacturonic acids.[5]

A pectic polysaccharide of M_r of 4,000–5,000 has been isolated from tomato leaves[6] that is a potent inducer of proteinase inhibitor proteins when supplied in solution to young tomato plants through their cut petioles. It is also an elicitor of isoflavinoid and terpenoid phytoalexins in bean cotyledons.[7] This polysaccharide, called the proteinase inhibitor inducing factor (PIIF) is prepared as follows.[8]

Three kilograms of mature tomato leaves are lyophilized, crushed, and sieved through 10 mesh screen to remove stem and veins, yielding approximately 175 g of dry leaf powder. The material (approximately 2.5%) is extracted with $CH_3Cl : MeOH$ (2 : 1), and air dried. The dry leaf powder

[2] R. Bishop, D. J. Makus, G. Pearce, and C. A. Ryan, *Proc. Natl. Acad. Sci. U.S.A.* **78,** 3536 (1981).

[3] M. G. Hahn, A. G. Darvill, and P. Albershem, *Plant Physiol.* **68,** 1161 (1981).

[4] C. A. Ryan, P. Bishop, G. Pearce, and M. Walker-Simmons, *in* "Cellular and Molecular Biology of Plant Stress" (J. L. Key and T. Kosuge, eds.), Vol. 22, p. 319. Alan R. Liss, Inc., New York, 1985.

[5] M. McNil, A. G. Darvill, S. C. Fry, and P. Albersheim, *Ann. Rev. Biochem.* **53,** 625 (1984).

[6] C. A. Ryan, *Trends Biochem. Sci.* **5,** 148 (1978).

[7] M. Walker-Simmons, D. Jin, C. West, L. Hadwiger, and C. A. Ryan, *Plant Physiol.* **76,** 833 (1984).

[8] P. Bishop, G. Pearce, J. B. Bryant, and C. A. Ryan, *J. Biol. Chem.* **259,** 13172 (1984).

is thoroughly dispersed in 4 liters of distilled water. The mixture is vacuum filtered through Celite 545, and the filtrate is dialyzed against 1 mM HCl overnight, then against cold running water for 1 day, lyophilized, and stored at 3°. This material is termed dialyzed PIIF. Three hundred milligrams of dialyzed PIIF is dissolved in 30 ml of 5 mM NaAc buffer (pH 4.7) and applied to a DE-52 column (2.5 × 16 cm), previously equilibrated with the buffer, and eluted with 400 ml of a 0–0.3 M NaCl gradient. The fractions (4 ml) are monitored for carbohydrate content, using the phenol–sulfuric acid assay of Dubois *et al.*,[9] and for proteinase inhibitor inducing activity using the tomato plant assay of Ryan.[10] Two large carbohydrate peaks are obtained, but only the acidic fraction, which elutes at approximately 0.2 M NaCl, possesses biological activity. The peak fractions associated with activity are pooled, dialyzed (2,000 MW cut off tubing) against distilled water, and lyophilized. This material contains over 90% carbohydrate and is referred to as DEAE PIIF.

DEAE PIIF is further fractionated by size-exclusion chromatography. A water-jacketed column (3 × 40 cm) packed with Sephadex G-50–40 and equilibrated with distilled water at 50° is employed (it should be noted that a temperature of 50° is necessary during chromatography to prevent large losses of active material due to nonspecific adsorption). Approximately 0.5 g of DEAE PIIF is dissolved in 9 ml water and chromatographed using water as eluent. The column fractions (4 ml) are monitored for carbohydrate and biological activity as above. The majority of the activity is found in the elution volume corresponding to a molecular weight of 5,000–10,000 (Fig. 1). These peak fractions (accounting for approximately 80% of the PIIF activity), when pooled and lyophilized, result in a light, fluffy white material referred to as G-50 PIIF. The degree of purification of PIIF activity by these steps is shown in Table I.

Preparation of a Galacturonic Acid-Rich PIIF by Partial Hydrolysis of Pectic Polysaccharides

G-50 PIIF is a complex pectic polysaccharide as judged by its composition and susceptibility to polygalacturonase digestion. This material can be essentially stripped of its neutral sugar residues while still retaining biological activity by partial acid hydrolysis as follows.[8]

G-50 PIIF (0.5 g) is dissolved in 10 ml of 2 N trifluroacetic acid (TFA) and incubated 1 hr at 80° *in vacuo*. This treatment results in a voluminous white precipitate which is removed by centrifugation and discarded. The

[9] M. Dubois, K. A. Gilles, J. K. Hamilton, P. A. Rebers, and F. Smith, *Anal. Chem.* **28,** 350 (1956).
[10] C. A. Ryan, *Plant Physiol.* **54,** 328 (1974).

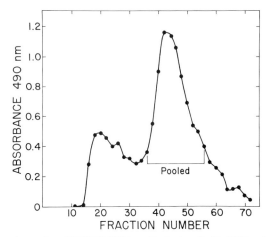

FIG. 1. Chromatography of DEAE PIIF on Sephadex G-50. PIIF polysaccharide from DE-52 chromatography (450 mg dissolved in 9 ml of H_2O) was chromatographed using a water-jacketed, Sephadex G-50–40 column (3 × 40 cm) at 50° using distilled H_2O as eluent. Aliquots (25 μl) of each 4-ml fraction were assayed for carbohydrate.[9]

supernatant is diluted with distilled water and lyophilized to removed the TFA. The lyophilized material is redissolved in a minimal quantity of water, neutralized by adding 0.1 N NaOH dropwise, and is then applied to a DE-52 column (1.8 × 16 cm) previously washed with 1 M NH$_4$HCO$_3$ and

TABLE I
PURIFICATION OF THE PROTEINASE INHIBITOR-INDUCING
FACTOR FROM TOMATO

| | | Activity | |
| | | | Total |
Purification step	g	units/mg[a]	units
Fresh tomato leaves	3000	0.018	54,000
Autoclaved, lyophilized tomato leaves	176	0.2	35,200
Fat-extracted	153	0.2	30,600
Water-soluble dialyzed, lyophilized	7.65	2.0	15,300
DEAE	3.97	3.3	13,000
G-50	3.92	3.3	10,000

[a] One unit is the concentration (mg/ml) required to give half-maximal accumulation of Inhibitor I in leaves of detached young tomato plants when supplied through the cut petioles for 30 min followed by incubation in 1000 foot-candle light for 24 hr.

equilibrated with distilled water. The column is first washed with 85 ml of water and then eluted with 400 ml of a 0–0.4 M NH$_4$HCO$_3$ gradient. The polysaccharide peak fractions (4 ml/tube) which elute at 0.3 M NH$_4$HCO$_3$ are pooled and lyophilized. This preparation is referred to as TFA PIIF. This polysaccharide is essentially a mixture of galacturonic acid polymers containing small amounts of rhamnose and fucose. The observed specific activity in inducing protease inhibitors in young tomato plants is slightly greater than G-50 PIIF.

Preparation of Oligouronides with PIIF Activity by Digestion of Pectic Polysaccharides with Tomato Polygalacturonase

Polygalacturonase digestion is performed by adjusting a solution of 50 mg TFA PIIF to pH 4.5 with 1 M acetic acid and adding 0.15 unit of tomato endopolygalactouronase.[2] The progress of this reaction is monitored by viscometry at 30°. Initially the visocity falls rapidly and the reaction must be terminated during this phase if intermediate sized galacturonide fragments are desired (i.e., near DP 6). Complete digestion requires approximately 6 hr and results in production of primarily mono-, di-, and trigalacturonic acid with much lesser amounts of galacturonide oligomers with DPs greater than 3 and some complex oligosaccharides containing neutral sugar residues.

Oligogalactouronides are purified using a DE-52 anion-exchange column (2 × 25 cm) with a 400 ml, 0–0.3 M linear NH$_4$HCO$_3$ gradient. The column fractions (4 ml) are assayed for total carbohydrate, the peak fraction pooled, and the oligomers recovered after lyophilization. With this method, the complete resolution of pure oligomers is limited to the mono- through hexameric galacturonides (Fig. 2). The proteinase inhibitor-inducing activities of these pure oligomers is shown in Fig. 3.

Higher chromatographic resolution of oligo- and polyuronides, up to 14 residues, has been achieved by Jin and West[11] by digestion of commercial polygalacturonic acid with a endopolygalacturonase isolated from the fungus *Rhizopus stolonifer*. This method of anion-exchange chromatography employs a very long (2.5 × 135 cm) column packed with DEAE-Sephadex A-25 equilibrated with 10 mM imidazole (pH 7.0) and 100 mM KCl. The added endopolygalacturonase digest is adjusted to an ionic strength lower than the starting buffer (as measured by conductivity), applied to the column, and washed with 2 liters of starting buffer. The uronides are then eluted using an 8-liter concave gradient from 0 to 300 mM KCl (Fig. 4). The peak fractions are pooled, desalted on a Sephadex G-10 column, and lyophilized. Only those oligomers above DP 9 were

[11] D. F. Jin and C. A. West, *Plant Physiol.* **74,** 989 (1984).

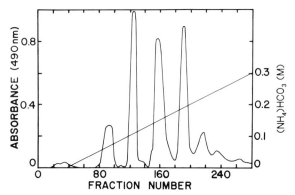

Fig. 2. Separation of galacturonic acid oligomers from tomato endopolygalacturonase digests of TFA PIIF on DE-52 cellulose. The digest (500 mg of TFA PIIF) was adjusted to pH 5.6 and diluted to 100 ml with water and applied to the column (4 × 8 cm). After washing with water, a 0–0.3 M linear gradient of ammonium bicarbonate was applied to elute the oligosaccharide fractions. Mono-, di-, tri-, tetra-, and pentagalacturonic acid eluted in that order. The fractions comprising each peak were pooled and lyophilized.

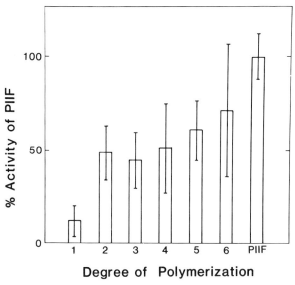

Fig. 3. Proteinase inhibitor-inducing activities of galacturonic oligomers normalized to the activity of G-50 PIIF. Oligomer activities were assayed at 2 mg/ml. Each histogram represents the normalized mean of the Inhibitor I concentrations determined from 30 assays.

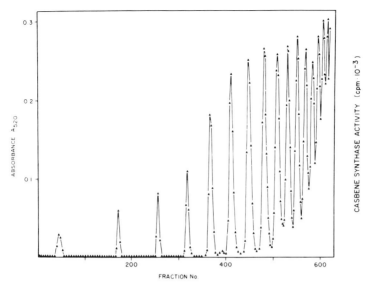

FIG. 4. Chromatographic resolution of oligo- and polyuronides from *R. stolonifer.*

active elicitors of the enzyme casbene synthase, which catalyzes the final step of synthesis of the phytoalexin casbene in castor bean cotyledons.

Preparation of Elicitor-Active Oligosaccharides from Purified Plant Cell Walls

A galacturonic acid-rich polysaccharide that possesses phytoalexin-elicitor activity in soybeans can also be prepared by partial acid hydrolysis of soybean cell walls.[3] The preparation and characterization of this polyuronide is qualitatively similar to the procedure for the preparation of TFA PIIF.

Preparation of Plant Cell Walls

Cell walls from soybean which can be used to prepare biologically active polysaccharides may be prepared as follows.[12]

Eight- to nine-day-old soybean seedlings are frozen in liquid N_2, wrapped in fine nylon mesh, crushed to a fine powder, and passed through a 2-mm wire sieve to remove residual stem pieces. Approximately 35 g of this powder is homogenized in 250 ml of 0.5 M $NaPO_4$ (pH 7.0) buffer

---

[12] K. W. Talmadge, K. Keegstra, W. D. Baver, and P. Alberstein, *Plant Physiol.* **51**, 158 (1973).

using a ground glass homogenizer. The mixture is centrifuged at 16,000 *g* for 15 min at 0° and the supernatant discarded. The pellet is washed once with the same buffer and then 3 times with distilled water by resuspending and centrifugation as above. The pellet is suspended in boiling ETOH for 1 hr to extract plant pigments, and the insoluble material is collected by filtration through a sintered glass funnel and washed with several volumes of hot EtOH. The insoluble cell walls are delipidized with several washes of chloroform : methanol (1 : 1) and finally resuspended in acetone and filtered until dry. The resulting preparation consists of fluffy white cell walls and represents about 3% by weight of the original tissue powder.

Cell walls can also be prepared from a variety of plant tissues using a modified procedure of Selvendran.[13] This method differs in its approach from the above in employing a detergent in the initial extraction. Lyophilized mature whole tomato plants (except roots) are homogenized in 0.1% sodium deoxycholate for 24 hr at 4° using a ball mill. The resulting homogenate is centrifuged, the pellet washed by resuspension and recentrifugation in ice cold distilled water, and the resulting pellet extracted twice with phenol–acetic acid–H_2O (2 : 1 : 1, w/v/v). The pellet is washed with ice cold water, then ethanol, and finally resuspended in ether followed by filtration through sintered glass. The product is then dried under reduced pressure. This method results in pure white powdered cell wall material that represents 4% of the wet starting tissue. The advantage of this method is that the majority of pectic polysaccharides that are noncovalently attached to the cell wall are still present and can be readily extracted by hot aqueous buffers.

Isolation of the Endogenous Elicitor of Phytoalexins from Soybean Cell Walls[14]

Soybean cell walls were suspended at 10 mg/ml in 2 *N* TFA. The suspensions were placed in screw-capped bottles and heated for 4 hr in an 85° water bath. After cooling, the suspensions were cleared of acid-insoluble residue by passage through Whatman GF/A glass microfiber filters. Solvent was removed from the filtrate by rotary evaporation under reduced pressure at 35°. The resulting tacky paste was suspended in methanol and evaporated to dryness. Resuspension in methanol and reevaporation were repeated 4 times. The final dried solids were suspended in water (10 ml/g of initial walls or pectin), titrated to pH 7.0 with 5.0 *M* imidazole base, and stirred overnight at 4°. The resulting mixture was cleared of particulates by progressive filtration through Whatman GF/A, Millipore Type HA 0.45 μm, and Millipore Type GS 0.22 μm filters. The final

[13] R. R. Selvendran, *Phytochemistry* **14,** 1011 (1975).
[14] E. A. Nothnagel, E. A., M. McNeil, and P. Albersheim, *Plant Physiol.* **71,** 916 (1983).

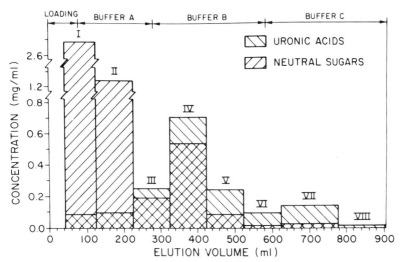

FIG. 5. First anion-exchange chromatography of crude hydrolysate. A 3.1 × 13.5 cm column of QAE–Sephadex (A-25–120) was equilibrated with buffer A (125 m*M* imidazole–HCl, pH 7.0). The crude hydrolysate was adjusted (by dilution or by addition of 5.0 *M* imidazole–HCl, pH 7.0) to a conductivity slightly less than that of buffer A and was then loaded onto the column. Elution was stepwise as shown. Buffer B was 550 m*M* imidazole–HCl (pH 7.0), and buffer C was 750 m*M* imidazole–HCl (pH 7.0). Fractions (80–150 ml) were collected as shown and assayed colorimetrically for neutral sugar (as rhamnose equivalents) and uronic acid (as GalUA equivalents) content.

filtrate was the crude cell wall hydrolysate. The acid-soluble material released from cell walls during partial acid hydrolysis of soybean cell walls obtained from 5 g of cell walls contained approximately 650 mg of total carbohydrate, 33% of which was uronic acid. The crude cell wall hydrolysate was subjected to gross, stepwise fractionation on a QAE–Sephadex anion-exchange column. Aliquots of the fractions eluting from this column were desalted and stored.

Materials eluted from the first anion-exchange column by buffers B and C (Fig. 5) were pooled and applied to a high resolution QAE–Sephadex anion-exchange column as shown in Fig. 6. Elution of this column revealed the presence of at least 15 different compounds in the hydrolysate preparation (Fig. 6). Many of the early-to-middle eluting peaks contained a considerable amount of neutral sugar.

Fractions corresponding to most of the peaks appearing in the elutions of Fig. 6 were desalted and assayed for elicitor activity at 25 μg total carbohydrate per cotyledon. Low elicitor activities were apparent in fractions 97 and 130–132. With the exception of these two peaks, none of the early eluting peaks possessed significant elicitor activity. The majority of the elicitor activity was contained in the late eluting, uronic acid-rich

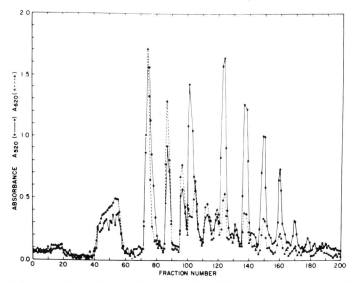

FIG. 6. Second anion-exchange chromatography of the crude hydrolysate. A 1.7 × 43.5 cm column of QAE–Sephadex (A-25–120) was equilibrated with buffer A (see Fig. 5 legend). The indicated fractions from the first anion-exchange column were pooled and diluted with deionized H_2O to give a conductivity less than that of buffer A loaded onto the column, and washed with six column volumns of buffer. The carbohydrate was eluted with a linear gradient consisting of 1,250 ml of the above buffer and 1,250 ml of buffer C (see Fig. 5 legend). Coinciding with the state of the gradient, collection of 250-drop (about 12 ml) fractions was begun.

peaks. One of these peaks was markedly more active than the others. This most active peak was present in fractions 181–185 (Fig. 6) and was several times more active than the respective crude hydrolysate. The most active peak was preceded by two or three peaks of lower (but significant) activity and was again followed by another less active peak. This procedure was also employed to isolate elicitor-active fractions from rehydrolyzed cell walls (after the first TFA hydrolysis) and from commercial pectin.[10] The elicitor-active fractions eluted from the columns at about the same positions as described above.

The most active fraction from the initial TFA hydrolysis of cell walls (fractions 181–185) contained 84.5% galacturonosyl residues, 5.8% rhamnose, 5.2% xylosyl, and 3.5% glucosyl residues. The most active fractions from the rehydrolyzed cell walls contained 98.7% galacturonosyl residues and from pectin, 98.1% galacturonosyl residues. Thus, as reported for tomato PIIF[4] and the casbene elicitor,[7] the biological activity appears to be associated with fragments containing the galactouronosyl backbone.

Section V

Degradation

[61] Enzymatic Diagnosis of Sphingolipidoses

By Kunihiko Suzuki

The sphingolipidoses are defined as disorders caused by genetic defect in catabolism of sphingosine-containing lipids. Classical sphingolipidoses fit within the conceptual framework of the inborn lysosomal disease of Hers[1] in that the underlying genetic defects are mutations in lysosomal hydrolases. However, in recent years, the concept of the sphingolipidoses has been expanded to include a group of genetic disorders caused by defective or absent activator proteins that are normally required for *in vivo* hydrolysis of the hydrophobic substrates by the hydrolases.[2] Even though the genetic defects are not in lysosomal hydrolases themselves, these activator deficiencies often closely resemble the respective hydrolase deficiencies in their clinical, pathological, and analytical biochemical manifestations. Since a separate chapter is devoted in this volume to the subject of the natural activators of sphingolipid degradation and their genetic abnormalities [66], this chapter will consider only those disorders due to genetic deficiencies of sphingolipid hydrolases. The only exception is the recently defined galactosialidosis.[3] This disorder established another new category of the sphingolipidoses, being due to a genetic defect in the "protective protein" rather than lysosomal hydrolases. Galactosialidosis is included in this chapter because the disease can be diagnosed by assays of appropriate lysosomal hydrolases, even though the genetic defect resides elsewhere. Furthermore, the possibility that the "protective protein" might in fact be a subunit of sialidase remains to be excluded.

Patients affected by the sphingolipidoses as defined above can be diagnosed definitively by appropriate enzyme assays, and heterozygous carriers can be identified in most instances with reasonable reliability. In addition, in the majority of the sphingolipidoses, affected fetuses can be diagnosed during pregnancy either with cultured amniotic fluid cells or with biopsied chorionic villi. The purpose of this section is to provide descriptions of reliable and practical enzymatic assay procedures for diagnosis of the sphingolipidoses. It is not a comprehensive collection of

---

[1] H. G. Hers, *Gastroenterology* **48**, 625 (1966).

[2] E. Conzelmann and K. Sandhoff, *in* "Recent Progress in Neurolipidoses and Allied Disorders" (M. T. Vanier, ed.), p. 77. Fondation Marcel Mérieux, Lyon, France, 1984.

[3] A. d'Azzo, A. Hoogeveen, A. J. J. Reuser, D. Robinson, and H. Galjaard, *Proc. Natl. Acad. Sci. U.S.A.* **79**, 4535 (1982).

METHODS IN ENZYMOLOGY, VOL. 138

available procedures. Alternative procedures and modifications of described procedures, with different substrates or even with different principles, almost always exist. Attempts are made to give references for alternative procedures but they are necessarily selective. For clinical, pathological, and biochemical details of the individual diseases, the standard book edited by Stanbury et al.[4] is suggested. A book that serves the similar purpose as this chapter also exists.[5]

Disease and Enzymatic Defects

Major sphingolipids involved in the genetic sphingolipidoses can be depicted systematically in simplified notations as a structurally and metabolically interrelated series of compounds (Fig. 1). Locations of the genetic degradative blocks known to occur in humans are indicated, each caused by a deficiency of the hydrolase that normally catalyzes the degradative step. The genetic diseases, described in the literature as involving deficiencies of either desialylation of ganglioside (ganglioside neuraminidase) or degradation of lactosylceramide are yet to convince this author and thus are not included. Not included in Fig. 1 is galactosialidosis, which shows simultaneous and *secondary* deficiencies of ganglioside G_{MI} β-galactosidase (step 7) and glycoprotein neuraminidase. These disorders are tabulated in Table I together with their clinicopathological features, analytical abnormalities, and the genetic defects.

Enzyme Sources and Choice of Substrates for Diagnostic Assays

A variety of tissue sources can be utilized for enzymatic diagnosis of the sphingolipidoses. Each source has its advantages and disadvantages. Postmortem solid tissues, such as brain, spleen, liver, or kidney, are generally excellent when antemortem diagnosis was not established or needs confirmation. Most enzymes involved in the sphingolipidoses are sufficiently stable to allow reliable assays even with tissues obtained 12 hr or longer after death. However, in view of other readily obtainable enzyme sources, solid tissue biopsies should no longer be permitted for the sole purpose of antemortem enzymatic diagnosis.

Serum is readily obtained and is a reliable source for some enzymes. However, the enzymatic activities in serum are generally lower than

[4] J. B. Stanbury, J. B. Wyngaarden, D. S. Fredrickson, J. L. Goldstein, and M. S. Brown, eds., "The Metabolic Basis of Inherited Disease," 5th ed. McGraw-Hill, New York, 1983.

[5] R. H. Glew and S. P. Peters, eds., "Practical Enzymology of Sphingolipidoses." Alan R. Liss, Inc., New York, 1977.

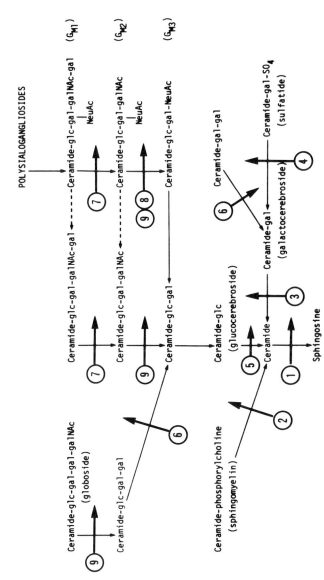

FIG. 1. Chemical and metabolic relationships among major sphingolipids and the location of the genetic blocks. The numbers correspond to those in Table I.

TABLE I
MAJOR SPHINGOLIPIDOSES

Disease[a]	Clinicopathological features	Stored material	Enzymatic defect
1. Farber disease (lipogranulomatosis)	Mostly infantile disease, tender swollen joints, multiple subcutaneous nodules, later flaccid paralysis and mental impairment	Ceramide	Acid ceramidase
2. Niemann–Pick disease	Neuropathic and nonneuropathic forms, hepatosplenomegaly, foamy cells in bone marrow, severe neurological signs in the neuropathic form (Type A)	Sphingomyelin	Sphingomyelinase
3. Globoid cell leukodystrophy (Krabbe disease)	Almost always infantile disease, white matter signs, peripheral neuropathy, high spinal fluid protein, loss of myelin, globoid cells in white matter	Galactosylceramide, galactosylsphingosine	Galactosylceramidase
4a. Metachromatic leukodystrophy	Late infantile, juvenile, and adult forms, white matter signs, peripheral neuropathy, high spinal fluid protein, metachromatic granules in the brain, nerves, kidney, and urine	Sulfatide	Arylsulfatase A ("sulfatidase")
4b. Multiple sulfatase deficiency	Similar to metachromatic leukodystrophy (4a), but additional gray matter signs, facial and skeletal abnormalities, organomegaly similar to mucopolysaccharidoses	Sulfatide, other sulfated compounds (see text)	Arylsulfatase A, B, C, (see text)
4c. Sulfatidase activator deficiency	Similar to later onset forms of metachromatic leukodystrophy (4a)	Unknown	Sulfatidase activator
5. Gaucher disease	Neuropathic and nonneuropathic forms, hepatosplenomegaly, "Gaucher cells" in bone marrow, severe neurological signs in the neuropathic form, also an intermediate form (Type III)	Glucosylceramide, glucosylsphingosine	Glucosylceramidase

Disease	Clinical signs	Stored material	Enzyme defect
6. Fabry disease (angiokeratoma corporis diffusum)	Primarily adult and nonneurological, angiokeratoma around buttocks, renal damage, X-linked	Trihexosylceramide (GalGalGlcCer), digalactosylceramide	α-Galactosidase A (trihexosylceramidase)
7a. G_{M1}-Gangliosidosis	Slow growth, motor weakness, gray matter signs, infantile form with additional facial and skeletal abnormalities and organomegaly, swollen neurons	Ganglioside G_{M1}, galactose-rich fragments of glycoproteins	Ganglioside G_{M1} β-galactosidase
7b. Galactosialidosis	Similar to later onset form of G_{M1}-gangliosidosis (7a)	Unknown	"Protective protein" (secondary defect in Ganglioside G_{M1} β-galactosidase and sialidase)
8a. Tay–Sachs disease	Severe gray matter signs, slow growth, motor weakness, hyperacusis, cherry-red spot, head enlargement, swollen neurons	Ganglioside G_{M2}	β-Hexosaminidase A
8b. Juvenile G_{M2}-gangliosidosis	Later onset, slower progression but otherwise similar to Tay–Sachs disease, milder pathology, some cases possibly B1 variant	Ganglioside G_{M2}	β-Hexosaminidase A (partial defect)
8c. B1 variant of G_{M2}-gangliosidosis	Similar to either Tay–Sachs disease (8a) or juvenile G_{M2}-gangliosidosis (8b)	Ganglioside G_{M2}	β-Hexosaminidase A (see text)
8d. AB variant of G_{M2}-gangliosidosis	Similar to Tay–Sachs disease (8a)	Ganglioside G_{M2}	β-Hexosaminidase A activator
9. Sandhoff disease	Panethnic but otherwise indistinguishable from Tay–Sachs disease (8a)	Ganglioside G_{M2}, asialo-ganglioside G_{M2}, globoside	β-Hexosaminidase A and B

[a] The numbers correspond to the steps of metabolic blocks in Fig. 1.

those in cellular sources, and some enzymes cannot be assayed reliably. Furthermore, some hydrolases in serum appear to be less stable than those in cellular sources, such as β-galactosidase.[6] The use of plasma should generally be avoided because commonly used anticoagulants are moderately inhibitory to many lysosomal enzymes. Leukocytes are generally convenient and reliable. Leukocytes isolated by any variation of the differential sedimentation procedure from 6% dextran in saline are satisfactory.[7] Lymphocytes give, by virtue of their cellular homogeneity, more consistent results with less scatters of the data, and for some purposes, such as carrier detection, they are preferable to mixed leukocytes.[8]

Cultured fibroblasts are also commonly used reliable source of lysosomal enzymes. One can obtain relatively large amounts of the material, if needed. A time lag of 3–4 weeks between skin biopsy and enzymatic assays is inevitable. The enzymatic activities can vary considerably depending on the nature of the culture medium, the phase of cell growth, and proportion of epithelial cells. These factors must be controlled appropriately. The same comments apply to cultured amniotic fluid cells for the purpose of prenatal diagnosis. The amniotic fluid itself or uncultured amniotic fluid cells have been utilized for prenatal diagnosis, but they are risky sources because of possible contamination by maternal blood and presence of varying and unknown proportions of dead cells. It is anticipated that the chorionic villi, which can be obtained much earlier in pregnancy than amniotic fluid and can be used immediately for enzymatic assays, may soon replace cultured amniotic fluid cells for prenatal diagnosis of most of the sphingolipidoses.[9,10] Other enzyme sources have been used from time to time for specific purposes in some specific diseases. They include urine, tear drops, and hair follicles. These will be noted under sections of individual diseases.

Contrary to the general rule of basic enzymology, clean enzymes do not necessarily give the cleanest results in diagnosis. In fact, crude enzyme sources, such as whole homogenate, often are the most advantageous sources for diagnostic enzyme assays, because any manipulation of the enzyme source prior to the assay involves risk of unknown and uncontrolled loss of the enzyme in question. One cannot a priori assume the same degree of loss when one deals with normal and mutant enzymes.

[6] Y. Suzuki and K. Suzuki, *Science* **171**, 73 (1971).

[7] R. A. Snyder and R. O. Brady, *Clin. Chim. Acta* **25**, 331 (1969).

[8] T. Tanaka, *Hiroshima J. Med. Sci.* **27**, 253 (1978).

[9] M. Upadhyaya, I. M. Archer, P. S. Harper, B. Jasani, A. Roberts, D. J., Shaw, N. S. T. Thomas, and H. Williams, *Clin. Chim. Acta* **140**, 39 (1984).

[10] L. Poenaru, L. Castelnau, A. Choiset, Y. Rouquet, and F. Thepot, *Hum. Genet.* **69**, 378 (1985).

Theoretically, assays with natural substrates should give more reliable diagnosis than those with chromogenic or fluorogenic artificial substrates. Even in instances in which identities of the activities determined with natural and artificial substrates have been demonstrated in normal enzyme sources, there is no a priori reason to assume that, in certain specific mutations, a discrepancy in the activities against the two categories of the substrates does not occur. We in fact do know such a mutation in β-hexosaminidase α-subunit in a variant form of G_{M2}-gangliosidosis (see G_{M2}-Gangliosidosis section). On the other hand, use of natural substrates is less convenient. Few natural glycosphingolipids are available commercially in radioactively labeled forms. Some are available unlabeled and must be radioactively labeled, purified, and characterized before they can be used for enzymatic assays. Some natural substrates are either not available commercially or are prohibitively expensive and thus must be prepared in the laboratory from appropriate natural sources. Furthermore, enzymatic assays with lipid substances are technically more complex and require greater care. Thus, for the purpose of diagnosis, which is a pragmatic application of basic enzymology, practical advantages of artificial substrates often outweigh their theoretical limitations. Use of artificial and natural lipid substrates is discussed under individual diseases whenever appropriate.

Expression of Results

The most widely applicable and most commonly used basis for expressing results of diagnostic enzyme assays is the protein content of the enzymatic source. It is relatively easy to determine accurately and provides a common basis for comparison among different enzyme sources. The procedure originated by Lowry et al.[11] and the more recent and increasingly popular dye-binding assay[12] should not be mixed in the same series because they do not give comparable results, although results are internally consistent within the respective procedures. When solid tissues are the enzyme source, wet weight can be a reliable basis for comparison within the same organ.[13] Although results obtained on leukocytes, fibroblasts, and amniotic fluid cells can be expressed on the basis of cell numbers, they are less consistent than those expressed on the basis of

[11] O. H. Lowry, N. J. Rosebrough, A. L. Farr, and R. L. Randall, *J. Biol. Chem.* **193**, 267 (1951).

[12] M. M. Bradford, *Anal. Biochem.* **72**, 248 (1976).

[13] E. Costantino-Ceccarini, T. F. Fletcher, and K. Suzuki, in "Current Trends in Sphingolipidoses and Allied Disorders" (B. W. Volk and L. Schneck, eds.), p. 127. Plenum, New York, 1976.

protein, in our experience.[14] The volume appears to be the only basis to express enzymatic activities in serum. Appropriate control enzymes should always be included for diagnostic assays. If their activities are much lower than expected for normal specimens, a generally poor condition of the enzyme source is indicated, and consequently the diagnosis will not be reliable. The ratio of the activities of the enzyme in question to that of the control enzyme is often useful in interpretation of results, although it cannot be used as the sole basis for comparison.

Individual Disorders

Farber Disease (Lipogranulomatosis)

Lipogranulomatosis was first described by Farber *et al.* in 1957.[15] Affected children develop swollen and painful joints and later numerous subcutaneous nodules. There are multiple granulomata in visceral organs with foamy cell infiltration. In the nervous system, neurons also exhibit storage of lipid material. The abnormal storage material in Farber disease was identified in 1969 as ceramide,[16] and the genetic defect underlying the disease is a deficiency of acid ceramidase (acylsphingosine deacylase, EC 3.5.1.23).[17]

Enzymatic Diagnosis. Ceramidase catalyzes the reaction

$$\text{Ceramide (= acylsphingosine)} \rightarrow \text{ceramide + fatty acid}$$

Several solid tissues have been used to demonstrate the deficiency, but for the purpose of diagnosis, both leukocytes and cultured fibroblasts can be used conveniently.[17,18] No convenient artificial chromogenic or fluorogenic substrate is available.

Principle of Assay. The enzyme source is incubated with ceramide labeled radioactively on the acyl moiety (N-[1-^{14}C]oleylsphingosine). The liberated [1-^{14}C]oleic acid is separated from unreacted ceramide by solvent partitioning and thin-layer chromatography, and its radioactivity determined by scintillation counting.

Preparation of Radioactive Ceramide. N-[1-^{14}C]Oleylsphingosine can be synthesized from [1-^{14}C]oleic acid and sphingosine.[19] The principle is

[14] Y. Suzuki, P. H. Berman, and K. Suzuki, *J. Pediatr.* **78,** 643 (1971).

[15] S. Farber, J. Cohen, and L. L. Uzman, *J. Mt. Sinai Hosp.* **24,** 816 (1957).

[16] H. W. Moser, A. L. Prensky, H. J. Wolfe, N. P. Rosman, S. Carr, and G. Ferreira, *Am. J. Med.* **47,** 869 (1969).

[17] M. Sugita, J. T. Dulaney, and H. W. Moser, *Science* **178,** 1100 (1975).

[18] J. T. Dulaney, A. Milunsky, J. B. Sidbury, N. Hobolth, and H. W. Moser, *J. Pediatr.* **89,** 59 (1976).

[19] K. C. Kopaczyk and N. S. Radin, *J. Lipid Res.* **6,** 140 (1965).

conversion of oleic acid to oleyl chloride by the reaction with thionyl chloride, which then reacts with sphingosine to form ceramide. The procedure has been described in detail.[20] Since the labeled ceramide has a single fatty acid, it is desirable to synthesize also a sufficient amount of unlabeled oleylsphingosine so that the radioactive ceramide can be adjusted to an appropriate specific activity with oleylsphingosine rather than with a natural ceramide with a heterogenous acyl moiety.

Reagents

N-[1-^{14}C]Oleylsphingosine (1000–2000 cpm/nmol), 1 mg/ml in chloroform–methanol (2 : 1, v/v)

Triton X-100, 0.5% in chloroform–methanol (2 : 1, v/v)

Tween 20, 0.5% in chloroform–methanol (2 : 1, v/v)

Sodium cholate, 2% in methanol

Sucrose, 0.25 M in 1 mM EDTA

Sodium citrate–phosphate buffer, 1 M, pH 4.0

Oleic acid, 1 mg/ml in chloroform–methanol (2 : 1, v/v)

2-Propanol–heptane–2 N NaOH (40 : 10 : 1, v/v/v)

Heptane

H_2SO_4, 1 N

Procedure. The following assay procedure has been adapted from Dulaney and Moser.[20]

To 13 × 100 mm screw-capped test tubes the following solutions are added: 40 μl of the substrate (40 μg), 50 μl of Triton X-100 (250 μg), 20 μl of Tween 20 (100 μg), and 20 μl of sodium cholate (400 μg). They are dried together under a stream of nitrogen. These materials are suspended by addition of 25 μl of the buffer and 75 μl of the sucrose–EDTA solution, followed by sonication in a water-bath-type ultrasonicator. Alternatively, a sufficient volume of the suspension can be prepared in a larger tube and 0.1 ml aliquots distributed to individual 13 × 100 mm tubes. The enzyme source, suspended and sonicated in 0.25 M sucrose–1 mM EDTA to give 1–2.5 mg protein/0.1 ml, is added, and the final volume adjusted to 0.2 ml with the sucrose–EDTA solution. The blank tube is set up with boiled enzyme source. The reaction mixture is incubaed at 37° for 2 hr with gentle shaking.

At the end of incubation, 50 μl of the oleic acid solution, 3 ml of the propanol–heptane–NaOH mixture, 1.8 ml of heptane, and 1.6 ml of water are added; the tubes are shaken and centrifuged briefly. The upper phase containing unhydrolyzed ceramide is removed, and the lower phase

---

[20] J. T. Dulaney and H. W. Moser, *in* "Practical Enzymology of the Sphingolipidoses" (R. H. Glew and S. P. Peters, eds.), p. 283. Alan R. Liss, Inc., New York, 1977.

washed twice more with 2 ml of heptane. The lower phase is acidified by adding 1 ml of 1 N sulfuric acid and extracted with 2.4 ml of heptane. The heptane phase is collected, and the lower phase is extracted twice more with 2.4 ml of heptane. All heptane phases are combined and dried under a stream of nitrogen.

The dried fraction is then chromatographed in silica gel G thin-layer plate in a solvent system of chloroform–methanol–acetic acid (94 : 1 : 5, v/v/v), and the oleic acid spot is located by exposing the plate to iodine vapor. Silica gel containing oleic acid is scraped into the counting vial. Sample radioactivity is corrected for the counts from the blank tubes that are carried through the procedure simultaneously.

Interpretation of Results. With either leukocytes or cultured fibroblasts as the enzyme source, affected patients should generally give ceramidase activity of 5% or less of the average of normal control.[20]

Niemann–Pick Disease

The term Niemann–Pick disease is currently used to cover a heterogenous group of genetic diseases with certain similarities in clinicopathological manifestations. Among the several phenotypes of Niemann–Pick disease as defined originally by Crocker,[21] only Types A and B have been clearly established as due to a primary genetic deficiency of sphingomyelinase (sphingomyelin phosphodiesterase, EC 3.1.4.12).[22,23] The underlying genetic defects of other types of Niemann–Pick disease (Types C, D, E, etc) have not been identified with certainty, but they are unlikely to be due to genetic sphingomyelinase deficiency. The eponym should preferably be used only for Types A and B in order to avoid unnecessary confusion. This section will consider only those types due to genetic sphingomyelinase deficiency.

Consistent deficiency of sphingomyelinase is observed in Type A and B patients. Patients with either form of the disease show enormous enlargement of the liver and spleen and characteristic storage cells in the bone marrow. Type A disease tends to occur in the infantile period with severe involvement of the central nervous system, whereas Type B disease is typically of later ages without neurological manifestations. Type A disease is rapidly fatal while Type B can have a course of many years.

Enzymatic Diagnosis. Sphingomyelinase catalyzes hydrolysis of sphingomyelin to ceramide and phosphorylcholine.

Sphingomyelin (= ceramide phosphorylcholine) → ceramide + phosphorylcholine

[21] A. C. Crocker, *J. Neurochem.* **7,** 69 (1961).
[22] R. O. Brady, J. N. Kanfer, M. B. Nock, and D. S. Fredrickson, *Proc. Natl. Acad. Sci. U.S.A.* **55,** 366 (1966).
[23] P. B. Schneider and E. P. Kennedy, *J. Lipid Res.* **8,** 202 (1967).

For the purpose of diagnosis of affected patients, either leukocytes or cultured fibroblasts are reliable enzyme sources.[24,25] Leukocytes, however, tend to give relatively high activities even from affected patients. Many laboratories also found leukocytes to be a poor source for heterozygote detection because of a large overlap in sphingomyelinase activities between the normal and carrier ranges. Prenatal diagnosis of an affected fetus is possible with either cultured amniotic fluid cells[26] or biopsied chorionic villi.[27]

Activity of sphingomyelinase toward conventional artificial substrates for phosphodiesterase is variable.[28] Purified human brain sphingomyelinase was relatively inactive toward bis-*p*-nitrophenyl phosphate. Activities toward bis-4-methylumbelliferyl phosphate and 2-hexadecanoylamino 4-nitrophenyl phosphorylcholine, which was specifically synthesized for the purpose of diagnosis of Niemann–Pick disease,[29] were much higher. Successful diagnosis of Niemann-Pick disease has been reported with either of the substrates.[30,31] Successful diagnosis of Niemann–Pick disease with 4-methylumbelliferyl pyrophosphate as substrate has also been reported.[32] Perhaps more promising are recently introduced sphingomyelin derivatives which are made colored or fluorescent (but not chromogenic or fluorogenic) by introduction of appropriate chemical groups (*N*-ω-trinitrophenylaminolaurylsphingosylphosphorylcholine or anthroyloxysphingomyelin).[33,34] With these substrates, the purification factor of highly pure human brain sphingomyelinase was identical with that estimated with radioactive sphingomyelin.[28] While these newer substrates avoid use of radioactivity, the assay mixture and subsequent manipulations are no less complicated than those for radioactive sphingomyelin. Here, a sphingomyelinase assay with the radioactive substrate is described.

[24] J. P. Kampine, R. O. Brady, J. N. Kanfer, M. Feld, and D. Shapiro, *Science* **155**, 86 (1967).

[25] H. R. Sloane, B. W. Uhlendorf, J. N. Kanfer, R. O. Brady, and D. S. Fredrickson, *Biochem. Biophys. Res. Commun.* **34**, 582 (1969).

[26] C. J. Epstein, R. O. Brady, E. L. Schneider, R. M. Bradley, and D. Shapiro, *Am. J. Hum. Genet.* **23**, (1971).

[27] M.-T. Vanier, personal communication.

[28] T. Yamanaka and K. Suzuki, *J. Neurochem.* **38**, 1753 (1982).

[29] A. E. Gal and F. J. Fash, *Chem. Phys. Lipids* **16**, 71 (1976).

[30] G. T. N. Besley, *Clin. Chim. Acta* **90**, 269 (1978).

[31] A. E. Gal, R. O. Brady, S. R. Hibbert, and P. G. Pentchev, *N. Engl. J. Med.* **293**, 632 (1975).

[32] A. H. Fensom, P. F. Benson, A. W. Barbarik, A. R. Grant, and L. Jacobs, *Biochem. Biophys. Res. Commun.* **74**, 877 (1977).

[33] S. Gatt, T. Dinur, and Y. Barenholz, *Biochim. Biophys. Acta* **530**, 503 (1978).

[34] S. Gatt, T. Dinur, and Y. Barenholz, *Clin. Chem. (Winston-Salem, N.C.)* **26**, 93 (1980).

Principle of Assay. [*methyl*-^{14}C]Sphingomyelin is incubated with the enzyme source. The liberated radioactive phosphorylcholine is separated from sphingomyelin on the basis of its water solubility, and its radioactivity is determined.

Radioactive Labeling of Sphingomyelin. Carbon-14 can be introduced into the methyl group of choline by demethylation by thiophenolate followed by remethylation with [^{14}C]methyl iodide.[35,36] Sphingomyelin labeled by this procedure is now commercially available (New England Nuclear Corp., Boston, MA). When appropriately diluted with unlabeled compound, it is suitable for diagnostic assays.

Reagents

[*methyl*-^{14}C]Sphingomyelin (500–1000 cpm/nmol), 1 mg/ml in chloroform–methanol (2 : 1, v/v)
Triton X-100, 0.5% in chloroform–methanol (2 : 1, v/v)
Sodium acetate buffer, 0.1 M, pH 5.0
Chloroform–methanol (2 : 1, v/v)
Pure solvent upper phase (chloroform–methanol–water, 3 : 48 : 47, v/v/v)
Pure solvent lower phase (chloroform–methanol–water, 86 : 14 : 1, v/v/v)

Procedure. The procedure described gives satisfactory results in the author's laboratory. The substrate (50 μl, 50 μg) and the Triton X-100 (20 μl, 100 μg) solutions are dried together in 13 × 100 mm screw-capped test tubes under a stream of nitrogen. They are suspended in 0.15 ml of the acetate buffer by sonication in a water-bath-type ultrasonicator. The enzyme source is suspended in distilled water, alternately frozen–thawed 4 times, and briefly sonicated for uniform dispersion. The enzyme source containing 10–100 μg of fibroblast protein or 0.1–0.2 mg of leukocyte protein is added to the tubes together with water if necessary to make the final volume to 0.2 ml.

After incubation at 37° for 1 hr with gentle shaking, the reaction is terminated by the addition of 0.2 ml of water and 2 ml of chloroform–methanol (2 : 1, v/v). The tubes are shaken and centrifuged. The clear upper layer is transferred. The lower phase is extracted twice with 1 ml each of the pure solvent upper phase, and all upper phases are combined. The combined upper phase is then washed once with 1 ml of the pure solvent lower phase. The final upper phase is transferred into the scintilla-

---

[35] W. Stoffel, D. LeKim, and T. S. Tschung, *Hoppe-Seyler's Z. Physiol. Chem.* **352**, 1058 (1971).

[36] D. A. Wenger, *in* "Practical Enzymology of Sphingolipidoses" (R. H. Glew and S. P. Peters, eds.), p. 39. Alan R. Liss, Inc., New York, 1977.

tion vial, dried in an oven at 110°, dissolved in 0.5 ml of water, and 12 ml of scintillation fluid capable of incorporating at least 0.5 ml of water without phase separation (e.g., 10% Bio-Solv, Beckman Instruments) is added. Blank tubes can be with either boiled enzyme or no enzyme source.

Comments. Human tissues, particularly brain, contain another sphingomyelinase which is magnesium dependent and has a neutral pH optimum. The biochemical and genetic relationship of this enzyme to the acid sphingomyelinase, if any, and its conceivable relevance to some forms of so-called Neimann–Pick disease are unclear. It has been well established, however, that the magnesium-dependent neutral sphingo-myelinase is normal in Type A and B of Niemann–Pick disease. Despite the dramatic phenotypic differences, enzymatic differentiation of the neuropathic (Type A) and nonneuropathic (Type B) forms is not possible at this time. The described assay procedure does not distinguish the two forms.

Globoid Cell Leukodystrophy (Krabbe Disease)

Globoid cell leukodystrophy is almost always a disease of early infancy with the typical clinical onset of 4–6 months. The nervous system, particularly the white matter in the central nervous system and peripheral nerves, is the exclusive site of clinical and pathological manifestations. At the terminal stage, the white matter is almost totally devoid of myelin, and the oligodendroglia are replaced by severe astrocytic gliosis and the unique abnormal globoid cells. Unlike in any other sphingolipidoses, abnormal accumulation of the natural substrate of the genetically defective enzyme, galactosylceramide, does not occur in the whole tissue, although the globoid cells are the site of a local accumulation of the lipid. The underlying cause of the disease is defective galactosylceramidase activity (EC 3.2.1.46).[37]

Enzymatic Diagnosis. Galactosylceramidase is a β-galactosidase that catalyzes hydrolysis of galactosylceramide to ceramide and galactose.

Galactosylceramide → ceramide + galactose

For diagnosis of affected patients, serum, leukocytes, and cultured fibroblasts have been utilized as the enzyme source with good results. Although reliable, use of serum requires special care. Galactosylceramidase activity in serum is low, and therefore the specific activity of the substrate should be increased by 5–10 times that indicated in the procedure. Furthermore, the enzyme in serum is unstable.[6] Leukocytes and fibroblasts

---

[37] K. Suzuki and Y. Suzuki, *Proc. Natl. Acad. Sci. U.S.A.* **66,** 302 (1970).

can be used with similar reliability. A small overlap is usually observed between the normal and the carrier ranges. Prenatal diagnosis is possible with either cultured amniotic fluid cells[38] or chorionic villi[39] as the enzyme source.

Although galactosylceramidase has activity toward 4-methylumbelliferyl β-galactoside,[40] this substrate cannot be used for diagnosis of globoid cell leukodystrophy, because the other lysosomal β-galactosidase in the tissue, ganglioside G_{M1} β-galactosidase, is even more active toward the artificial substrates. An artificial chromogenic substrate, 2-hexadecanoyl-4-nitrophenyl β-galactoside, appeared promising in an early report for diagnosis of this disorder.[41] Subsequent larger series of studies, however, indicated that this substrate is not as reliable as the natural substrate. More recently, colored and fluorescent derivatives of galactosylceramide, analogous to those described for sphingomyelin (see Niemann-Pick Disease section), have been reported to be more reliable substrates for diagnosis.[42,43]

Principle of the Assay. The enzyme source is incubated with galactosylceramide labeled with tritium at the galactose moiety. The liberated [³H]galactose is determined after it is separated from unreacted galactosylceramide by solvent partitioning.

Preparation of Radioactive Galactosylceramide. The common procedure used to introduce tritium into galactosylceramide is sequential oxidation by galactose oxidase and reduction by boro[³H]hydride.[44] Tritium is located on carbon 6 of galactose.

Reagents

[³H]Galactosylceramide (about 1000 cpm/nmol), 2 mg/ml in chloroform–methanol (2 : 1, v/v)
Sodium taurocholate (pure preparation, e.g., Calbiochem or Sigma synthetic), 10 mg/ml in chloroform–methanol (2 : 1, v/v)
Oleic acid, 1 mg/ml in chloroform–methanol (2 : 1, v/v)
Sodium citrate–phosphate buffer, 0.1 M pH 4.2
Galactose, 1 mg/ml in water

[38] K. Suzuki, E. Schneider, and C. J. Epstein, *Biochem. Biophys. Res. Commun.* **45**, 1363 (1971).
[39] D. A. Wenger, personal communication.
[40] H. Tanaka and K. Suzuki, *Brain Res.* **122**, 325 (1977).
[41] A. E. Gal, R. O. Brady, P. G. Pentchev, F. S. Furbish, K. Suzuki, H. Tanaka, and E. L. Schneider, *Clin. Chim. Acta* **77**, 53 (1977).
[42] G. T. N. Besley and S. Gatt, *Clin. Chim. Acta* **110**, 19 (1981).
[43] S. Okada, T. Kato, H. Yabuuchi, K. Yoshino, M. Naoi, K. Kiuchi, and K. Yagi, *Clin. Chim. Acta* **136**, 57 (1984).
[44] N. S. Radin, this series, Vol. 28 [32].

Chloroform–methanol (2 : 1, v/v)

Pure solvent upper phase (chloroform–methanol–water, 3 : 48 : 47, v/v/v)

Pure solvent lower phase (chloroform–methanol–water, 86 : 14 : 1, v/v/v)

Procedure. The assay system described here has been simplified from that described in an earlier volume[45] and has been optimized for diagnostic purposes. In 13 × 100 mm screw-capped test tubes, 30 μl (60 μg) of the substrate, 0.1 ml (1 mg) of the taurocholate, and 30 μl (30 μg) of the oleic acid solutions are dried together under a stream of nitrogen. The materials are suspended in 0.1 ml of the citrate–phosphate buffer and evenly dispersed by brief sonication. The enzyme source is suspended in water, frozen–thawed a few times, and briefly sonicated. Appropriate volumes of the enzyme source, containing up to 200 μg of leukocyte or fibroblast protein, and water are added to the final volume of 0.2 ml. For assays of serum enzyme, 0.1 ml of serum is directly added to the substrate–detergent suspension. The tubes are incubated for 2 hr at 37° with gentle shaking.

At the end of the incubation, 0.3 ml of the carrier galactose and 2.5 ml of chloroform–methanol (2 : 1, v/v) are added. The solutes are partitioned by shaking and brief centrifugation. The lower phase is removed with a disposable capillary pipette, and the upper phase is washed twice with 2 ml each of the pure solvent lower phase. The final upper phase is quantitatively transferred to the scintillation vial. The content is dried in an oven at 110°, dissolved in 0.5 ml of water, and radioactivity is determined as described above for the sphingomyelinase assay. Blank tubes contain no enzyme source.

Interpretation of the Results. Diagnosis of affected patients is generally unequivocal. False negatives—affected patients with high galactosylceramidase activity—have not been reported, but false positives (pseudo-deficiency) are known to occur.[46] When a sufficiently large number of specimens are examined, the average of known heterozygous carriers gives a statistically highly significant intermediate value between normal controls and affected patients. However, varying degrees of overlapping activities between the normal and carrier ranges are almost always observed, as in enzymatic identification of heterozygotes for all other sphingolipidoses.

Comments on Natural Substrates. There are three natural lipid substrates that can be used for assays of galactosylceramidase. Of these,

[45] K. Suzuki, this series, Vol. 28 [112].

[46] D. A. Wenger and V. M. Riccardi, *J. Pediatr.* **88**, 76 (1976).

galactosylsphingosine[47,48] and monogalactosyldiglyceride[49] are not recommended for practical reasons. Both substrates are much more difficult to prepare. The activities of the enzyme toward these substrates are lower than that toward galactosylceramide, and only limited data are available in the literature for comparison. On the other hand, lactosylceramide is a practically useful substrate.[50] It is cleaved by the enzyme at a rate several-fold higher than that for galactosylceramide. A disadvantage is that lactosylceramide is a natural substrate for both of the two lysosomal β-galactosidases, galactosylceramidase and ganglioside G_{M1} β-galactosidase.[51] Consequently the assay system must be standardized with great care in order to determine exclusively the lactosylceramide-cleaving activity due to galactosylceramidase in the presence of the other β-galactosidase. The assay system devised by Wenger et al. is excellent for this purpose.[52] The single most important factor is the amount and purity of sodium taurocholate. One must use a relatively small amount of highly pure taurocholate in order to avoid interference by ganglioside G_{M1} β-galactosidase. In view of this potential difficulty, the galactosylceramidase assay is described here as the standard procedure for diagnosis of globoid cell leukodystrophy. Procedures with all of the natural substrates are described in the literature.[53] While assays with the natural substrates are almost always done with radioactively labeled galactolipids, an enzymatically coupled and amplified reaction that permits determination of liberated unlabeled galactose with high sensitivity has been described.[54]

Metachromatic Leukodystrophy

Metachromatic leukodystrophy belongs, together with globoid cell leukodystrophy, to the category of the classical genetic leukodystrophies. There are different phenotypes with different ages of onset and rapidity of the clinical course—late infantile, juvenile, and adult forms. While predominant neurological manifestations are those of the white matter and the peripheral nerves, mental symptoms often precede neurological signs in the adult form. The different clinical phenotypes are genetically distinct

[47] T. Miyatake and K. Suzuki, *Biochem. Biophys. Res. Commun.* **48**, 538 (1972).
[48] T. Miyatake and K. Suzuki, *J. Neurochem.* **22**, 231 (1974).
[49] D. A. Wenger, M. Sattler, and S. P. Markey, *Biochem. Biophys. Res. Commun.* **53**, 680 (1973).
[50] D. A. Wenger, M. Sattler, and W. Hiatt, *Proc. Natl. Acad. Sci. U.S.A.* **71**, 854 (1974).
[51] H. Tanaka and K. Suzuki, *J. Biol. Chem.* **250**, 2324 (1975).
[52] D. A. Wenger, M. Sattler, C. Clark, and H. McKelvy, *Clin. Chim. Acta* **56**, 199 (1974).
[53] K. Suzuki, *in* "Practical Enzymology of the Sphingolipidoses" (R. H. Glew and S. P. Peters, eds.), p. 101. Alan R. Liss, Inc., New York, 1977.
[54] T. Kato and Y. Suzuki, *Proc. Jpn. Acad., Ser. B* **55**, 69 (1979).

from each other. Systemic involvements are minimal compared to the devastating pathology in the nervous system. A genetically unrelated disease, multiple sulfatase deficiency, is often included in metachromatic leukodystrophy. This disorder, however, is not allelic with the classical metachromatic leukodystrophy. Enzymatic diagnosis of multiple sulfatase deficiency is included in this section. Another metachromatic leukodystrophy-like genetic disorder due to a genetic abnormality of a natural activator protein[55,56] is discussed in another chapter in this volume [66].

The genetic defect of metachromatic leukodystrophy is a deficient activity of arylsulfatase A (cerebroside-sulfatase, EC 3.1.6.8).[57,58] As the result, an abnormal accumulation of galactosylceramide sulfate (sulfatide) occurs primarily in the white matter and the peripheral nerves in the form of the characteristic cytoplasmic inclusions.[59,60] These inclusions stain brown–yellow with acidic cresyl violet ("metachromasia"). A moderate accumulation of lactosylceramide sulfate also occurs in the kidney.[61] This compound is also a substrate for arylsulfatase A.

Enzymatic Diagnosis. Arylsulfatase A (cerebroside sulfatase) hydrolyzes its natural substrate, sulfatide, to galactosylceramide and sulfate.

Sulfatide (= galactosylceramide sulfate) → galactosylceramide + sulfate

The enzyme also hydrolyzes an artificial chromogenic substrate, *p*-nitrocatechol sulfate.

p-Nitrocatechol sulfate → *p*-nitrocatechol + sulfate

The latter reaction with the chromogenic substrate is the most commonly utilized for diagnosis of metachromatic leukodystrophy. Despite the theoretical advantage of assays with the natural substrate, sulfatide, it is rarely used for diagnosis because of technical difficulties of preparing ^{35}S-labeled sulfatide and its relatively short half-life. *p*-Nitrocatechol sulfate of satisfactory quality is readily available commercially.

The most reliable sources for diagnosis of metachromatic leukodystrophy are leukocytes and cultured fibroblasts.[62] Heterozygote detection is possible with reasonable accuracy with these enzyme sources. Cultured

[55] R. L. Stevens, A. L. Fluharty, H. Kihara, M. M. Kaback, L. J. Shapiro, B. Marsh, K. Sandhoff, and G. Fischer, *Am. J. Hum. Genet.* **33**, 900 (1981).
[56] K. Inui, M. Emmett, and D. A. Wenger, *Proc. Natl. Acad. Sci. U.S.A.* **80**, 3074 (1983).
[57] E. Mehl and H. Jatzkewitz, *Biochem. Biophys. Res. Commun.* **19**, 407 (1965).
[58] J. H. Austin, D. Armstrong, and L. Shearer, *Arch. Neurol. (Chicago)* **13**, 593 (1965).
[59] H. Jatzkewitz, *Hoppe-Seyler's Z. Physiol. Chem.* **311**, 279 (1958).
[60] J. H. Austin, *Proc. Soc. Exp. Biol. Med.* **100**, 361 (1959).
[61] E. Martensson, A. Percy, and L. Svennerholm, *Acta Paediatr. Scand.* **55**, 1 (1966).
[62] J. T. Dulaney and H. W. Moser, *in* "Practical Enzymology of the Sphingolipidoses" (R. H. Glew and S. P. Peters, eds.), p. 137. Alan R. Liss, Inc., New York, 1977.

amniotic fluid cells have been used for reliable prenatal diagnosis.[62] Urine was commonly used earlier for diagnosis of affected patients but is not suitable for carrier detection due to large variations in the normal population. Generally the activities of arylsulfatase A in the amniotic fluid and serum are too low for reliable assays, and they should be avoided for more than a preliminary screening purpose.

Principle of Assay. p-Nitrocatechol sulfate is chromogenic in that the intact compound is colorless while the hydrolysis product, p-nitrocatechol, exhibits a brick-red color. The intensity of the color thus is proportional to the enzymatic activity (see also the section Multiple Sulfatase Deficiency below regarding possible interference from other arylsulfatases).

Reagents

p-Nitrocatechol sulfate, 10 mM, in 0.5 M sodium acetate buffer, pH 5.0, containing 0.5 mM Na$_4$P$_2$O$_7$ and 10% NaCl
NaOH, 1 N
p-Nitrocatechol

Procedure. Most of the arylsulfatase A assay procedures in the literature are based on that by Baum *et al.*[63] Leukocyte or fibroblast pellets are suspended in water at a protein concentration of 100–400 μg/0.1 ml, alternately frozen and thawed a few times, and briefly sonicated. To a 13 × 100 mm screw-capped test tube are added 0.1 ml of the enzyme source and 0.1 ml of the substrate solution. Tubes are incubated at 37° for 30 min, and the reaction terminated by the addition of 1 ml of 1 N NaOH. Appropriate control tubes are essential in this assay because usual enzyme sources, particularly leukocyte preparations, often have a reddish color themselves. In the author's laboratory, the 0-time sample is used as the control (blank). The 0-time sample is prepared by adding 1 ml of 1 N NaOH immediately after the enzyme source and the substrate solution are mixed in the tube. Control tubes in duplicate are required for *each* sample. The optical density is determined at 515 nm, and the amount of free p-nitrocatechol is calculated on the basis of the standard curve obtained with the p-nitrocatechol standard.

Interpretation of Results. Arylsulfatase A exhibits anomalous kinetics with respect to the reaction time.[64] Therefore, the results obtained with different incubation periods are not necessarily comparable. Fibroblasts from individuals of different ages give significantly different activities and thus age-matching is desirable for this enzyme source.[65] Although diagno-

[63] H. Baum, K. S. Dodgson, and B. Spencer, *Clin. Chim. Acta* **4,** 453 (1973).
[64] H. Baum, K. S. Dodgson, and B. Spencer, *Biochem. J.* **69,** 567 (1958).
[65] M. M. Kaback, C. O. Leonard, and T. H. Parmley, *Pediatr. Res.* **5,** 366 (1971).

sis of affected patients is almost always unequivocal, this assay does not distinguish different phenotypes of the arylsulfatase deficiencies. With cellular sources—leukocytes, fibroblasts, cultured amniotic fluid cells—heterozygote detection can be achieved with a reasonable accuracy. However, some phenotypically normal carrier individuals have been known to show arylsulfatase A activity as low as those expected for homozygous affected patients (pseudo-deficiency).[66] This possibility must be kept in mind, particularly in prenatal diagnosis.

Comments. Arylsulfatase A hydrolyzes several natural sulfated lipids other than sulfatide, including lactosylceramide sulfate,[67] galacto-sylsphingosine sulfate,[68] ascorbic acid 2-sulfate,[69] and seminolipid.[70–72] Despite the theoretical interest, use of these lipids are impractical for diagnosis of metachromatic leukodystrophy because of the technical complexities in preparation and assay procedures.

Multiple Sulfatase Deficiency

Multiple sulfatase deficiency shares the arylsulfatase A deficiency and all of its consequent manifestations with metachromatic leukodystrophy. However, the disease is not allelic to metachromatic leukodystrophy. Enzymatically, it is characterized by additional deficiencies of other sulfatases, including arylsulfatase B and C, estrone sulfatase, dehydroepiandrosterone sulfatase, cholesterol sulfate sulfatase, and probably also the sulfatases involved in degradation of sulfated glycosaminoglycans.[73] Consequently, the clinical and pathological manifestations of this disorder are a combination of metachromatic leukodystrophy and mucopolysaccharidoses. Patients show, in addition to the features of metachromatic leukodystrophy, hepatosplenomegaly, abnormal facies, and skeletal changes. Tissues, particularly systemic organs, show additional accumulations of cholesteryl sulfate, dermatan sulfate, and heparan sulfate. The primary

[66] I. T. Lott, J. T. Dulaney, A. Milunsky, D. Hoefnagel, and H. W. Moser, *J. Pediatr.* **89,** 438 (1976).

[67] K. Harzer and H. U. Benz, *Hoppe-Seyler's Z. Physiol. Chem.* **355,** 744 (1974).

[68] Y. Eto, U. Wiesemann, and N. N. Herschkowitz, *J. Biol. Chem.* **249,** 4955 (1974).

[69] A. L. Fluharty, R. L. Stevens, R. T. Miller, S. S. Shapiro, and H. Kihara, *Biochim. Biophys. Acta* **429,** 508 (1976).

[70] K. Yamato, S. Handa, and T. Yamakawa, *J. Biochem.* (*Tokyo*) **75,** 1241 (1974).

[71] A. L. Fluharty, R. L. Stevens, R. T. Miller, and H. Kihara, *Biochem. Biophys. Res. Commun.* **61,** 348 (1974).

[72] S. Yamaguchi, K. Aoki, S. Handa, and T. Yamakawa, *J. Neurochem.* **24,** 1087 (1975).

[73] J. V. Murphy, H. J. Wolfe, E. A. Balasz, and H. W. Moser, *in* "Lipid Storage Diseases: Enzymatic Defects and Clinical Implications" (J. Bernsohn and H. J. Grossman, eds.), p. 67. Academic Press, New York, 1971.

genetic defect underlying the multiple sulfatase deficiency has not been elucidated.

Enzymatic Diagnosis. The nature of the disease requires that more than one enzyme be assayed for diagnosis of multiple sulfatase deficiency. The most practical procedure is probably to assay for arylsulfatases A and B, both with p-nitrocatechol sulfate as the substrate. Both enzymes hydrolyze it to p-nitrocatechol and sulfate. However, there are sufficient differences in the properties of the two enzymes to allow differential assays of one enzyme in the presence of the other. Assays for arylsulfatase A are done as described above for metachromatic leukodystrophy. The assay procedure for arylsulfatase B is described below. The comments on the nature of the enzyme source for arylsulfatase A apply also to arylsulfatase B.

Reagents

p-Nitrocatechol sulfate, 50 mM, in 0.5 M sodium acetate buffer, pH 6.0, containing 10 mM barium acetate
NaOH, 1 N
p-Nitrocatechol

Procedure. Three separate 13×100 mm screw-capped test tubes are set up for each sample (6 tubes in duplicate). They serve as 0-time, 30-min, and 90-min samples. The substrate solution, 0.1 ml, and 0.1 ml of the enzyme source are mixed in the assay tubes and incubated at 37° with shaking. At 0-time, 30 min, and 90 min, the reaction is stopped by the addition of 1 ml of 1 N NaOH. The optical density is determined at 515 nm. After subtracting the 0-time reading, the 30-min and 90-min readings are plotted on the graph and the line passing the two readings is extrapolated to 0-time. The intercept on the optical density axis is generally positive (X). This is the result of interference from arylsulfatase A. The optical density attributable to arylsulfatase B/hr is calculated as $OD_{90 \text{ min}} - OD_{30 \text{ min}} - 0.2X$.

Interpretation of Results. Multiple sulfatase deficiency is the only known disorder in which both arylsulfatase A and B are simultaneously deficient. Therefore, assays for additional sulfatases, such as arylsulfatase C, are not necessary for pragmatic purposes. Deficient arylsulfatase B activity alone is not sufficient for diagnosis of multiple sulfatase deficiency because one of the mucopolysaccharidoses, Maroteaux–Lamy syndrome, is due to a specific deficiency of arylsulfatase B with normal arylsulfatase A. Simultaneous determination of arylsulfatase A and B, therefore, can conveniently differentiate enzymatically the three genetic disorders, metachromatic leukodystrophy, multiple sulfatase deficiency, and Maroteaux–Lamy syndrome.

Gaucher Disease

Gaucher disease is commonly classified into three groups; Types I, II, and III. The Type I disease is prevalent among the Jewish population. Type II is panethnic. The term Type III is usually used to designate cases that fall between Types I and II in chronicity and nervous system involvement. This category thus is likely to be genetically heterogenous. Among those usually classified in Type III is a specific form of the disease probably due to a specific mutation that originated in the Norrbottnian region of Sweden. All forms share systemic manifestations of enormous enlargement of the liver, spleen, and other reticuloendothelial organs and presence of characteristic Gaucher cells in the bone marrow. Type I disease shows little or no nervous system involvement, occurs predominantly in adults, and takes a very long clinical course. Type II disease, on the other hand, occurs primarily in infants, severely affects the central nervous system, and runs its course usually in a few years. The clinical pictures of Type III patients tend to be intermediate between Types I and II. The genetic cause of Gaucher disease is a deficiency of a β-glucosidase, glucosylceramidase (EC 3.2.1.45).[74,75] The enlarged liver and spleen contain massive amounts of glucosylceramide.

Enzymatic Diagnosis. Glucosylceramidase is a β-glucosidase that catalyzes hydrolysis of glucosylceramide to glucose and ceramide.

$$\text{Glucosylceramide} \rightarrow \text{ceramide} + \text{glucose}$$

For diagnosis of Gaucher disease, cellular sources, such as leukocytes, cultured skin fibroblasts, and cultured amniotic fluid cells, are all suitable. Serum and other fluid sources generally do not contain sufficient activities of the enzyme for reliable assays.

Both the natural substrate, glucosylceramide, and artificial substrates, such as *p*-nitrophenyl or 4-methylumbelliferyl β-glucoside, can be used for reliable diagnosis of Gaucher disease. Since the assay with the natural substrate is less accessible for many laboratories due to the technical complexities of preparing radioactively labeled glucosylceramide, both procedures are described here. More recently, glucosylceramide modified with a fluorescent group in the lipid portion of the molecule was introduced as an alternate substrate, and the results are highly promising.[76,77]

[74] R. O. Brady, J. N. Kanfer, and D. Shapiro, *Biochem. Biophys. Res. Commun.* **18,** 221 (1965).
[75] A. D. Patrick, *Biochem. J.* **97,** 17C (1965).
[76] G. A. Grabowski, T. Dinur, S. Gatt, and R. J. Desnick, *Clin. Chim. Acta* **124,** 123 (1982).
[77] T. Dinur, G. A. Grabowski, R. J. Desnick, and S. Gatt, *Anal. Biochem.* **136,** 223 (1984).

Assays with Glucosylceramide

PRINCIPLE OF ASSAY. Appropriately labeled glucosylceramide is incubated with the enzyme source. When the label is in the glucose moiety, the liberated glucose is separated from the unhydrolyzed glucosylceramide by solvent partitioning, and its radioactivity determined. The principle is identical with that for the galactosylceramidase assay described earlier. When the label is in the ceramide portion of the molecule, separation of radioactive ceramide from the similarly hydrophobic intact glucosylceramide requires more tedious procedures, such as column or thin-layer chromatography.

PREPARATION OF LABELED GLUCOSYLCERAMIDE. Glucose-labeled glucosylceramide was available commercially for a brief period but has been removed from the market. Glucose-labeled glucosylceramide can be synthesized *de novo*,[78,79] but this requires considerable skills in organic chemistry. Other procedures require unlabeled glucosylceramide as the starting material. Since commercially available glucosylceramide is prohibitively expensive, glucosylceramide is usually prepared from either liver or spleen of patients with Gaucher disease by conventional lipid extraction and purification techniques. For the sake of convenience in assaying glucosylceramidase activity with glucose-labeled substrate, the procedure of choice for its radioactive labeling is that of MacMaster and Radin,[80] which allows introduction of tritium into glucose. Other procedures that introduce the label into the ceramide portion either by catalytic reduction of sphingosine or by replacing the unlabeled fatty acid with a radioactive one[81] do not offer practical advantage. Furthermore, it has been shown recently that the affinity of glucosylceramide to the enzyme can be altered drastically when the lipid portion of the molecule is modified.[82] Only the assay procedure with glucose-labeled glucosylceramide is described below. The procedure for ceramide-labeled substrate is found in a previous volume.[81]

Reagents

Labeled glucosylceramide (500–1000 cpm/nmol), 2 mg/ml in chloroform–methanol (2 : 1, v/v)
Sodium taurocholate (pure preparation, e.g., Calbiochem or Sigma synthetic), 10 mg/ml in methanol

[78] R. O. Brady, J. N. Kanfer, and D. Shapiro, *J. Biol. Chem.* **240,** 39 (1965).
[79] P. Stoffyn, A. Stoffyn, and G. Hauser, *J. Lipid Res.* **12,** 318 (1971).
[80] M. C. MacMaster, Jr., and N. S. Radin, *J. Labelled Compd. Radiopharm.* **13,** 353 (1977).
[81] K. Suzuki, this series, Vol. 50C [47].
[82] A. M. Vaccaro, M. Muscillo, and K. Suzuki, *Clin. Chim. Acta* **131,** 1 (1983).

Oleic acid, 1 mg/ml in hexane
Sodium acetate buffer, 0.1 M, pH 5.5
Glucose, 1 mg/ml in water
Pure solvent upper phase (chloroform–methanol–water, 3 : 48 : 47, v/v/v)
Pure solvent lower phase (chloroform–methanol–water, 86 : 14 : 1, v/v/v)

Procedure. In 13 × 100 mm screw-capped tubes, 30 μl (60 μg) of the substrate, 0.1 ml (1 mg) of the sodium taurocholate, and 50 μl (50 μg) of the oleic acid solutions are dried together. They are evenly suspended by addition of 0.1 ml of the acetate buffer and brief sonication. This suspension can be prepared in bulk and aliquots pipetted into individual tubes, if preferred. The enzyme source is suspended in water, frozen–thawed a few times, and briefly sonicated. Appropriate volumes of the enzyme source and water are added to the tube to make the final volume to 0.2 ml. The reaction mixture is incubated at 37° for 1 or 2 hr with shaking. The reaction is terminated by adding 0.3 ml of the glucose solution and 2.5 ml of chloroform–methanol (2 : 1, v/v). The liberated glucose can be determined in exactly the same way as described for liberated galactose determination in the galactosylceramidase assay. The results are corrected for the blank tube counts.

Assay with 4-Methylumbelliferyl β-Glucoside

Among the two procedures used to diagnose Gaucher disease with artificial substrates, the one which took advantage of the very low pH optimum of glucosylceramidase compared to other β-glucosidases[83,84] is now rarely used as routine procedure. The other procedure takes advantage of the finding that, with 4-methylumbelliferyl β-glucoside as the substrate, glucosylceramidase is stimulated by sodium taurocholate while other β-glucosidases are inhibited.[85,86] This procedure appears highly reliable for diagnosis of Gaucher disease.

PRINCIPLE OF ASSAY. While there are β-glucosidases other than glucosylceramidase that can hydrolyze 4-methylumbelliferyl β-glucoside, sodium taurocholate activates glucosylceramidase and simultaneously inhibits activity of other β-glucosidases. 4-Methylumbelliferone enzymatically liberated in the presence of taurocholate therefore represents only glucosylceramidase activity.

[83] E. Beutler and W. Kuhl, *Clin. Med.* **76**, 747 (1970).
[84] M. W. Ho, J. Seck, D. Schmidt, M. L. Veath, W. Johnson, R. O. Brady, and J. S. O'Brien, *Am. J. Hum. Genet.* **24**, 37 (1972).
[85] S. P. Peters, P. Coyle, and R. H. Glew, *Arch. Biochem. Biophys.* **175**, 569 (1976).
[86] D. A. Wenger, M. Sattler, C. Clark, and C. Wharton, *Life Sci.* **19**, 413 (1976).

Reagents

4-Methylumbelliferyl β-glucoside, 5 mM, in 0.2 M sodium citrate–phosphate buffer, pH 5.8 (for fibroblasts) or pH 5.4 (for leukocytes)

Sodium taurocholate (pure preparation, e.g., Calbiochem or Sigma synthetic), 10 mg/ml in methanol

Oleic acid, 1 mg/ml in hexane (for fibroblasts)

Triton X-100, 1% in chloroform–methanol (2 : 1, v/v) (for leukocytes)

Glycine buffer, 0.2 M, pH 10.8

4-Methylumbelliferone

Procedure. The assay system described here is according to Wenger *et al.*[86] For assays with fibroblasts, 50 μl (0.5 mg) of the sodium taurocholate and 50 μl (50 μg) of the oleic acid solutions are dried together, and 0.1 ml of the substrate solution is added in 13 × 100 mm screw-capped test tubes. The fibroblasts, previously suspended in water, frozen and thawed a few times, and briefly sonicated, and water are added to bring the final volume to 0.2 ml. For leukocyte assays, 40 μl of the Triton X-100 solution replaces the oleic acid solution. After incubation at 37° for 1 hr, 3.8 ml of the glycine buffer is added. The fluorescence is determined spectrofluorometrically with the excitation wavelength of 365 nm and the emission wavelength of 448 nm, and the amount of liberated 4-methylumbelliferone is calculated on the basis of the fluorescence of known amounts of the standard. Since taurocholate, particularly impurities in some batches, contributes significant fluorescence, its purity is critical, and the blank tubes should be set up with care.

Interpretation of Results. Diagnosis of affected patients is usually unambiguous, while diagnosis of heterozygous carriers usually involves the "gray area" where activities of the normal population and known carriers overlap. At present, enzymatic activity alone cannot differentiate clearly the three phenotypes, although some progress has been reported either with the taurocholate–4-methylumbelliferone system[87] or with an assay system with the natural activator.[88] It has also been reported that a specific monoclonal antibody could differentiate the three types of Gaucher disease.[89]

Comments. Glucosylsphingosine is also a natural substrate for gluco-

[87] R. H. Glew, L. B. Daniels, L. S. Clark, and S. W. Hoyer, *J. Neuropathol. Exp. Neurol.* **41,** 639 (1982).

[88] D. A. Wenger and S. Roth, *in* "Gaucher Disease: A Century of Delineation and Research" (R. J. Desnick and G. A. Grabowski, eds.), p. 551. Alan R. Liss, Inc., New York, 1982.

[89] E. I. Ginns, F. P. W. Tegelaers, R. Barnveld, H. Galjaard, A. J. J. Reuser, R. O. Brady, J. M. Tager, and J. A. Barranger, *Clin. Chim. Acta* **131,** 283 (1983).

sylceramidase.[90] Its accumulation in the liver and brain of patients with Gaucher disease has been demonstrated.[91,92] However, use of glucosylsphingosine does not offer any practical advantage for enzymatic diagnosis of Gaucher disease.

Fabry Disease (Angiokeratoma Corporis Diffusum)

Fabry disease is the only X-linked disorder among the sphingolipidoses. The disease is further unusual in that it is primarily nonneurological and occurs mostly in the adult population. The clinical onset is typically between 10 and 18 years, and the presenting symptom is usually keratotic skin lesions in the lower trunk region. The vascular system is heavily involved, and cardiac and renal involvements determine the prognosis. The clinical course can run for decades. Sweeley and Klionsky conclusively established in 1963 that Fabry disease is a sphingolipidosis characterized by abnormal accumulation of a trihexosylceramide (α-galactosyl-β-galactosylglucosylceramide).[93] The underlying genetic defect is a deficiency of an α-galactosidase which normally hydrolyzes the terminal α-galactosyl residue from the trihexosylceramide (galactosylgalactosylglucosylceramidase, EC 3.2.1.47).[94,95]

Enzymatic Diagnosis. Trihexosylceramidase catalyzes hydrolysis of the terminal α-galactose from the trihexosylceramide.

Trihexosylceramide → lactosylceramide + galactose

Of the two α-galactosidase isozymes, the major and heat-labile component, α-galactosidase A, constitutes approximately 90% of the total α-galactosidase activity in the tissue and is responsible for trihexosylceramide hydrolysis.[96] Leukocytes[94] and fibroblasts[97] are the common enzyme sources for diagnosis of Fabry disease but various fluid sources, such as serum, plasma, urine, and tear drops, have also been successfully

[90] S. S. Raghavan, R. A. Mumford, and J. N. Kanfer, *Biochem. Biophys. Res. Commun.* **54,** 256 (1973).

[91] S. S. Raghavan, R. A. Mumford, and J. N. Kanfer, *J. Lipid Res.* **15,** 484 (1974).

[92] O. Nilsson and L. Svennerholm, *J. Neurochem.* **39,** 709 (1982).

[93] C. C. Sweeley and B. Klionsky, *J. Biol. Chem.* **238,** 3148 (1963).

[94] R. O. Brady, A. E. Gal, R. M. Bradley, E. Martensson, A. L. Warshaw, and L. Loster, *N. Engl. J. Med.* **276,** 1163 (1967).

[95] J. A. Kint, *Science* **167,** 1268 (1970).

[96] R. J. Desnick, K. Y. Allen, S. J. Desnick, M. K. Raman, R. W. Bernlohr, and W. Krivit, *J. Lab. Clin. Med.* **81,** 157 (1973).

[97] E. Beutler and W. Kuhl, *J. Hum. Genet.* **24,** 237 (1972).

utilized.[96,98] Prenatal diagnosis has been done with cultured amniotic fluid cells.[99]

While the natural substrate would be theoretically more desirable, it is very difficult to prepare in quantity unless one has access to tissues from patients with Fabry disease. The terminal α-galactose can then be readily labelled with tritium by the galactose oxidase–sodium boro[³H]hydride procedure.[44] For routine diagnostic assays, however, the artificial fluorogenic substrate, 4-methylumbelliferyl α-galactoside, is satisfactory because α-galactosidase A is by far the major component of α-galactosidases in human tissues.

Principle of the Assay. 4-Methylumbelliferyl α-galactoside is incubated with the enzyme source, and the liberated 4-methylumbelliferone is determined spectrofluorometrically. The procedure measures the combined activity of both α-galactosidase A and B. The diagnosis is possible because the genetically unaffected B isozyme constitutes only 10% of the total activity.

Reagents

4-Methylumbelliferyl α-galactoside, 5 mM, in 0.2 M sodium citrate–phosphate buffer, pH 4.5
Glycine buffer, pH 10.8
4-Methylumbelliferone

Procedure. In 13 × 100 mm screw-capped tubes, 0.1 ml of the substrate solution and the enzyme source are mixed, and the final volume is brought to 0.2 ml with addition of an appropriate volume of water, if necessary. Either leukocytes or fibroblasts are suspended in water, frozen and thawed a few times, and briefly sonicated prior to the assay. Each tube should contain 50–200 μg of protein. The reaction mixture is incubated at 37° for 1 hr. The reaction is stopped by the addition of 3.8 ml of the glycine buffer. The liberated 4-methylumbelliferone is determined as described for the β-glucosidase assay for Gaucher disease.

Interpretation of the Results. α-Galactosidase activities in affected patients are usually in the range of 10–15% of normal because of the genetically unaffected α-galactosidase B. Detection of heterozygous females is also feasible with reasonable reliability.

G$_{M1}$-*Gangliosidosis*

G$_{M1}$-Gangliosidosis occurs in at least three relatively distinct phenotypes. Type I is the most severe form with the onset in the infantile

[98] D. L. Johnson, M. A. del Monte, E. Cotlier, and R. J. Desnick, *Clin. Chim. Acta* **63**, 81 (1975).

[99] R. O. Brady, B. W. Uhlendorf, and C. B. Jacobson, *Science* **172**, 174 (1971).

period. Severe and progressive neurological manifestations primarily involving gray matter are accompanied by facial and skeletal abnormalities and hepatosplenomegaly reminiscent of mucopolysaccharidoses. Type II disease has a later onset, prolonged course, and lacks the systemic manifestations. Some patients with the late onset form are often classified as Type III, or the adult form. In recent years it has been recognized that some patients previously classified as Morquio disease are deficient in β-galactosidase activity.[100–102] They can be considered to be at the other end of the clinical spectrum of G_{M1}-gangliosidosis in that they show severe skeletal abnormalities with no or little nervous system manifestations. Although the disease is still referred to as Morquio disease (Type B), the mutation appears to be allelic with G_{M1}-gangliosidosis. Ganglioside G_{M1} [galactosyl-N-acetylgalactosaminyl(N-acetylneuraminyl)galactosylglucosylceramide] accumulates abnormally in the brain.[103–106] The abnormal material in the systemic organs appears to be heterogeneous fragments of glycoproteins and keratan sulfate-like materials.[107–109] The underlying genetic cause of the disease is a deficiency of a β-galactosidase, which can be demonstrated by chromogenic or fluorogenic substrates,[110] ganglioside G_{M1},[111,112] asialoganglioside G_{M1},[111] lactosylceramide,[111,112] glycoprotein fractions extracted from patients' tissues,[113] or desialylated fetuin.[113]

Enzymatic Diagnosis. The β-galactosidase which is genetically deficient in G_{M1}-gangliosidosis normally hydrolyzes the terminal β-galactose residue from the varieties of substrates listed above. However, practical diagnostic assays can be carried out reliably with the use of artificial substrates, such as p-nitrophenyl or 4-methylumbelliferyl β-galactoside, because this β-galactosidase is by far the predominant of the two lyso-

[100] E. Paschke and H. Kresse, *Biochem. Biophys. Res. Commun.* **109**, 568 (1982).
[101] G. T. J. van der Horst, W. J. Kleijer, A. T. Hoogeveen, J. G. M. Huijmans, W. Blom, and O. P. van Diggelen, *Am. J. Med. Genet.* **16**, 261 (1983).
[102] J. J. van Gemund, M. A. H. Giesberts, R. F. Erdmans, W. Blom, and W. J. Kleijer, *Hum. Genet.* **64**, 50 (1983).
[103] H. Jatzkewitz and K. Sandhoff, *Biochim. Biophys. Acta* **70**, 354 (1963).
[104] J. S. O'Brien, M. B. Stern, B. H. Landing, J. K. O'Brien, and D. N. Donnell, *Am. J. Dis. Child.* **109**, 338 (1965).
[105] N. K. Gonatas and J. Gonatas, *J. Neuropathol. Exp. Neurol.* **24**, 318 (1965).
[106] K. Suzuki, K. Suzuki, and S. Kamoshita, *J. Neuropathol. Exp. Neurol.* **28**, 25 (1969).
[107] K. Suzuki, *Science* **159**, 1471 (1968).
[108] L. S. Wolfe and N. M. K. Ng Ying Kin, *in* "Current Trends in Sphingolipidoses and Allied Disorders" (B. W. Volk and L. Schenck, eds.), p. 15. Plenum, New York, 1976.
[109] G. C. Tsay and G. Dawson, *J. Neurochem.* **27**, 733 (1976).
[110] S. Okada and J. S. O'Brien, *Science* **160**, 1002 (1968).
[111] Y. Suzuki and K. Suzuki, *J. Biol. Chem.* **249**, 2113 (1974).
[112] H. Tanaka and K. Suzuki, *Clin. Chim. Acta* **75**, 267 (1977).
[113] M. C. MaBrinn, S. Okada, M. W. Ho, C. C. Hu, and J. S. O'Brien, *Science* **163**, 946 (1969).

somal β-galactosidases in all the materials commonly used for enzymatic diagnosis.

Leukocytes, cultured fibroblasts, and amniotic fluid cells are the standard enzyme sources for diagnosis of G_{M1}-gangliosidosis.[53] Serum appears to be less reliable, and urine is not recommended because of its wide range of normal values.

Reagents

4-Methylumbelliferyl β-galactoside, 2 mM, in 0.1 M sodium citrate–phosphate buffer, pH 4.0
NaCl, 0.2 M, in water
Glycine buffer, 0.2 M, pH 10.8
4-Methylumbelliferone

Procedure. In 13 × 100 mm screw-capped tubes, 0.1 ml of the substrate solution, 0.1 ml of 0.2 M NaCl, an appropriate amount of the enzyme source, and water are mixed to the final volume of 0.4 ml. The standard leukocyte or fibroblast suspension in water, which has been frozen–thawed a few times and briefly sonicated, can be used. Each tube can contain 20–50 μg of protein for the fluorogenic substrate. After incubation at 37° for 1 hr with shaking, the reaction is stopped by the addition of 3.6 ml of the glycine buffer. The released 4-methylumbelliferone is determined as described for β-glucosidase in the section for diagnosis of Gaucher disease above.

Interpretation of the Results. Since ganglioside G_{M1} β-galactosidase predominates over galactosylceramidase in leukocytes and fibroblasts, diagnosis of G_{M1}-gangliosidosis patients can be made unambiguously with the procedure described above. In the author's experience, brain white matter is the only tissue that can give equivocal results, with patients' activities ranging up to 30% of normal. This is because white matter normally contains higher activities of galactosylceramidase than ganglioside G_{M1} β-galactosidase and because galactosylceramidase is often highly elevated in G_{M1}-gangliosidosis.[51] This point should not be a concern for practical antemortem diagnosis. Similarly unambiguous results are expected for prenatal diagnosis of affected fetuses with cultured amniotic fluid cells. The above procedure does not readily distinguish the different clinical phenotypes.

Comments. Among the natural substrates mentioned above, ganglioside G_{M1}, asialoganglioside G_{M1}, and lactosylceramide are commonly used in research laboratories for diagnosis of G_{M1}-gangliosidosis. However, they are not necessary for routine diagnostic assays. Preparation, radioactive labeling, and the assay procedures for these natural glycolipid substrates have been described in detail.[53] Utility of the natural substrates for diagnosis of different forms of G_{M1}-gangliosidosis may well change in the near future, however. It has been reported recently that the mutant β-galactosidase in the adult form of the disease show much more profound deficiency toward the lipid substrate than toward the glycoprotein substrate, while that in Morquio Type B disease show the reverse.[114]

G_{M2}-Gangliosidosis

The genetic and enzymatic picture of G_{M2}-gangliosidosis has become highly complex and confusing in recent years. This state came about as the results of studies on the level of proteins and nucleic acids, rather than enzymatic activites. On this level, it is no longer feasible to discuss just a few genetic variants. For the pragmatic purpose of diagnosis on the basis of enzymatic activites, however, most of the patients with G_{M2}-gangliosidosis can be diagnosed reliably within a somewhat simplistic conceptual framework.

G_{M2}-Gangliosidosis can be classified into three major groups according to the genes in which the mutations responsible for the disease occurred: N-acetyl β-hexosaminidase ("hexosaminidase") A deficiency (mutation in the α-subunit), hexosaminidase A and B deficiency (mutation in the β-subunit), and the natural activator deficiency (mutation in the natural activator protein for ganglioside G_{M2} hydrolysis). Of these, the natural activator deficiency (G_{M2}-gangliosidosis AB variant) is discussed elsewhere in this volume [66].

The classical Tay–Sachs disease is the best known among the genetic hexosaminidase A deficiency states. The gene incidence is at least ten times higher in the Ashkenazi Jewish population of eastern European origin. It is a rapidly progressive neurological disorder primarily involving neurons. The clinical onset is usually around 6 months, and most patients die by 3 years. There is a massive accumulation of ganglioside G_{M2} [N-acetylgalactosaminyl(N-acetylneuraminyl)galactosylglucosylceramide] and, to a lesser degree, its asialo derivative in the brain.[115,116] The genetic defect of the disease is the deficiency of one of the two major N-acetyl-β-hexosaminidases—hexosaminidase A, which is a heterodimer consisting of the α- and β-subunits, as the result of a mutation in the α-subunit.[117,118] Hexosaminidase B ($\beta\beta$) is catalytically intact, or more commonly elevated. There are many phenotypic variants of the hexosaminidase α-

[114] T. Mutoh, M. Naoi, A. Takahashi, M. Hoshino, Y. Nagai, and T. Nagatsu, *Neurology* **36**, 1237 (1986).

[115] E. Klenk, *Hoppe-Seyler's Z. Physiol. Chem.* **262**, 128 (1939).

[116] L. Svennerholm, *Biochem. Biophys. Res. Commun.* **9**, 436 (1962).

[117] K. Sandhoff, *FEBS Lett.* **4**, 351 (1969).

[118] S. Okada and J. O'Brien, *Science* **165**, 698 (1969).

subunit mutations. The juvenile form can be considered to be a milder form of the disease with a later onset and prolonged clinical course. Pathological and biochemical abnormalities are milder. Hexosaminidase A activity is partially deficient.[119,120] An enzymologically exceedingly interesting mutation in the α-subunit is known that results in a complete loss of catalytic activity toward the natural lipid substrate, ganglioside G_{M2}, with preserved activity toward the standard chromogenic and fluorogenic substrates.[121,122] Some of the juvenile cases in the literature may well have had this form of the disease.

When a mutation occurs in the β-subunit of hexosaminidase, it results in inactivation of both hexosaminidase A and B, since the β-subunit is common to both forms (Sandhoff disease).[123,124] The disease is panethnic but otherwise clinically indistinguishable from Tay–Sachs disease. In addition to an accumulation of ganglioside G_{M2} similar to Tay–Sachs disease, there are additional massive accumulations of asialoganglioside G_{M2} and globoside (N-acetylgalactosaminylgalactosylgalactosylglucosylceramide) both in the brain and systemic organs.[124]

Enzymatic Diagnosis. Satisfactory diagnosis of several forms of G_{M2}-gangliosidosis is possible without use of natural lipid substrates. Both hexosaminidase A and B hydrolyze the standard artificial substrates p-nitrophenyl and 4-methylumbelliferyl N-acetyl-β-glucosaminide. Although β-hexosaminidases also hydrolyze equivalent galactosaminides, they are rarely used because the V_{max} is much lower. Hexosaminidases A and B can be assayed differentially on the same sample by taking advantage of different heat or pH stability of the two forms. More recently, however, a new substrate, p-nitrophenyl or 4-methylumbelliferyl N-acetyl-β-glucosamine 6-sulfate, has been found to be hydrolyzed only by hexosaminidase A but not by hexosaminidase B.[125] This substrate not only permits direct assay of hexosaminidase A in a mixture without pretreatment but also permits diagnosis of the unusual α-subunit mutation (B1 variant) because it closely approximates the unique differential affinity of ganglioside G_{M2} toward the two hexosaminidase subunits.[126] In-

[119] K. Suzuki, K. Suzuki, I. Rapin, Y. Suzuki, and N. Ishii, *Neurology* **20,** 190 (1970).

[120] Y. Suzuki and K. Suzuki, *Neurology* **20,** 848 (1970).

[121] S.-C. Li, Y. Hirabayashi, and Y.-T. Li, *Biochem. Biophys. Res. Commun.* **101,** 479 (1981).

[122] H. Kytzia, U. Hinrichs, I. Maire, K. Suzuki, and K. Sandhoff, *EMBO J.* **2,** 1201 (1983).

[123] K. Sandhoff, U. Andreae, and H. Jatzkewitz, *Life Sci.* **7,** 283 (1968).

[124] Y. Suzuki, J. C. Jacob, K. Suzuki, M. Kutty, and K. Suzuki, *Neurology* **21,** 313 (1971).

[125] H. Kresse, W. Fuchs, J. Glössel, D. Holtfrerich, and W. Gilberg, *J. Biol. Chem.* **256,** 12926 (1981).

[126] H. Kytzia and K. Sandhoff, *J. Biol. Chem.* **260,** 7568 (1985).

creasing use of this sulfated substrate can be predicted in the near future.[127–130]

Varieties of enzyme sources have been used for diagnosis of G_{M2}-gangliosidosis with varying degrees of reliability. They include serum,[131] cultured fibroblasts,[132] urine,[133] tear drops,[134] amniotic fluid cells either cultured or prior to culture,[135] chorionic villi,[136] and the amniotic fluid itself.[135] However, all of the fluid sources, except serum, are less reliable and are not recommended for purposes other than preliminary tests.

The most commonly utilized diagnostic procedure is by differential heat inactivation of hexosaminidase A.[131] This has been adapted to the clinical autoanalyzer for mass screening of the high-risk population for heterozygous carriers.[137] The availability of the sulfated substrate makes assays of hexosaminidase A simpler and more reliable by eliminating the need for heat inactivation. Since this substrate does not give information on hexosaminidase B activity, the standard fluorogenic substrate will have to be used for diagnosis of Sandhoff disease. Procedures for both of the substrates are given here.

Assay with 4-Methylumbelliferyl N-Acetyl-β-Glucosaminide

PRINCIPLE OF THE ASSAY. β-Hexosaminidase A is more heat labile than β-hexosaminidase B. The A component is almost totally inactivated at 50° for 3 hr at pH 4.5, while the component B is barely affected. Thus, both heated and unheated enzyme sources are incubated with 4-methylumbelliferyl N-acetyl-β-glucosaminide, and liberated 4-methylumbelliferone is determined spectrofluorometrically. The activity of the unheated sample gives the total hexosaminidase activity (A + B), and that of the heated sample, hexosaminidase B activity. The activity of hexosaminidase A is then calculated by the difference.

[127] Y.-T. Li, Y. Hirabayashi, and S.-C. Li, Am. J. Hum. Genet. 35, 520 (1983).

[128] W. Fuchs, R. Navon, M. M. Kaback, and H. Kresse, Clin. Chim. Acta 133, 253 (1983).

[129] J. Bayleran, P. Hechtman, and W. Saray, Clin. Chim. Acta 143, 73 (1984).

[130] K. Inui and D. A. Wenger, Clin. Genet. 26, 318 (1984).

[131] J. S. O'Brien, S. Okada, A. Chen, and D. L. Fillerup, N. Engl. J. Med. 283, 15 (1970).

[132] S. Okada, M. L. Veath, J. Leroy, and J. S. O'Brien, Am. J. Hum. Genet. 23, 55 (1971).

[133] R. Navon and B. Padeh, J. Pediatr. 80, 1026 (1972).

[134] P. J. Carmody, M. C. Rattazzi, and R. G. Davidson, N. Engl. J. Med. 289, 1072 (1973).

[135] J. S. O'Brien, S. Okada, D. L. Fillerup, M. L. Veath, B. Adornato, P. H. Brenner, and J. Leroy, Science 172, 61 (1971).

[136] G. A. Grabowski, J. R. Kruse, J. D. Goldberg, K. Chockkalingam, R. E. Gordon, K. J. Blakemore, M. J. Mahoney, and R. J. Desnick, Am. J. Hum. Genet. 36, 1369 (1984).

[137] E. H. Kolodny, in "Practical Enzymology of the Sphingolipidoses" (R. H. Glew and S. P. Peters, eds.), p. 1. Alan R. Liss, Inc., New York, 1977.

Reagents

4-Methylumbelliferyl N-acetyl-β-glucosaminide, 2 mM, in 0.1 M sodium citrate–phosphate buffer, pH 4.5
Sodium citrate buffer, 0.1 M, pH 4.5
Glycine buffer, 0.2 M, pH 10.8
4-Methylumbelliferone

Procedure. Serum or a water suspension of cellular enzyme sources is diluted appropriately ($\times 20$ for serum) with the 0.1 M citrate–phosphate buffer. The cell suspensions are frozen–thawed a few times and briefly sonicated prior to the assay. They should contain 5–15 μg of protein per 0.1 ml after dilution with the buffer. The diluted enzyme source is divided into two equal portions. One part is kept at 4°, while the other part is heated at 50° for 3 hr in a tightly capped test tube. The heated and unheated samples are then assayed for β-hexosaminidase activity simultaneously. The enzyme reaction is carried out in 13 × 100 mm screw-capped test tubes, in which 0.3 ml of the substrate solution and 0.1 ml of the appropriately pretreated enzyme source are mixed and incubated at 37° for 1 hr. The reaction is stopped by the addition of 3.6 ml of the glycine buffer. The liberated 4-methylumbelliferone is determined as described for the β-glucosidase assay for Gaucher disease above.

Assay with 4-Methylumbelliferyl N-Acetyl-β-Glucosamine 6-Sulfate

SYNTHESIS OF THE 6-SULFATED SUBSTRATE. As of this writing, 4-methylumbelliferyl N-acetyl-β-glucosamine 6-sulfate is not yet commercially available on a regular basis. (Inquire: HSC Res. Develop. Corp., Toronto, Canada, for availability.) The procedure for its synthesis has been described in detail.[129]

Reagents

4-Methylumbelliferyl N-acetyl-β-glucosamine 6-sulfate (potassium salt), 2 mM in 0.2 M sodium citrate–phosphate buffer, pH 4.2
Glycine buffer, 0.2 M, pH 10.8
4-Methylumbelliferone

Procedure. In 13 × 100 mm screw-capped tubes, 0.15 ml of the substrate solution and an enzyme source are mixed with the final volume adjusted with water to 0.2 ml, if necessary. Appropriate amounts of the enzyme sources are 10–25 μg protein/tube for leukocytes, 5–15 μg protein/tube for fibroblasts, and 0.02 ml/tube for serum. Tubes are incubated for 37° for 30 min and the reaction stopped by the addition of 3.8 ml of the glycine buffer. The amount of liberated 4-methylumbelliferone is determined fluorometrically as described for the β-glucosidase assay for Gaucher disease.

INTERPRETATION OF THE RESULTS. Numerous articles have been published on the subject of enzymatic diagnosis of Tay–Sachs disease. Diagnosis of affected patients and heterozygous carriers for the classical disease is generally reliable. However, phenotypically normal individuals with deficient hexosaminidase A activities (pseudo-deficiency) have been reported.[138,139] The most commonly used expression of hexosaminidase A activity is the proportion of A in the total hexosaminidase activity. It is often helpful, however, to evaluate the results also on the basis of net activity per serum volume or cellular protein. When hexosaminidase A is determined by the differential inactivation method, the margin of error for the calculated activity will be increasingly larger as its proportion/total activity becomes smaller. The assay with the sulfated substrate should eliminate this potential source of inaccuracy, since no pretreatment of samples is involved and since the procedure gives directly the activity of β-hexosaminidase A.

Diagnosis and carrier detection for Sandhoff disease is usually more straightforward. However, apparently normal individuals with deficient hexosaminidase A and B activities have also been reported,[140] a source for caution. The sulfated substrate alone is not suitable for diagnosis of G_{M2}-gangliosidosis as a whole, because the activity toward the sulfated substrate will not give any indication regarding the activity of hexosaminidase B. Furthermore, in Sandhoff disease the activity is not deficient toward the sulfated substrate. It appears that β-hexosaminidase S ($\alpha\alpha$), which has almost negligible activity toward 4-methylumbelliferyl β-glucosaminide and ganglioside G_{M2}, has relatively high catalytic activity against the sulfated substrate.[141] Due to the defect in the β-subunit, hexosaminidase S is present in Sandhoff disease at relatively high concentrations. The binding site for the sulfated substrate and ganglioside G_{M2} resides on the α-subunit.[126] The sulfated substrate is essential, short of using the natural lipid substrate, ganglioside G_{M2}, for diagnosis of the α mutation that causes defective hydrolysis of ganglioside G_{M2} and the sulfated substrate but not of the standard unsulfated fluorogenic substrate (B1 variant).

In order to differentiate the major forms of β-hexosaminidase mutations, it is necessary therefore to do both the differential heat inactivation assay with the unsulfated substrate and the assay with the sulfated substrate.

[138] R. Navon, B. Padeh, and A. Adam, *Am. J. Hum. Genet.* **25**, 287 (1973).
[139] J. Vidgoff, N. R. M. Buist, and J. S. O'Brien, *Am. J. Hum. Genet.* **25**, 372 (1973).
[140] J.-C. Dreyfus, L. Poenaru, and L. Svennerholm, *N. Engl. J. Med.* **292**, 61 (1975).
[141] H.-J. Kytzia, U. Hinrichs, and K. Sandhoff, *Hum. Genet.* **67**, 414 (1984).

COMMENTS. Use of the natural substrate, ganglioside G_{M2}, could in theory be useful for diagnosis of certain types of G_{M2}-gangliosidosis. For example, a case of juvenile G_{M2}-gangliosidosis which showed approximately half-normal hexosaminidase A activity toward 4-methylumbelliferylglucosaminide was much more severely deficient in the activity toward ganglioside G_{M2}.[142] In practice, however, ganglioside G_{M2} is difficult to prepare in quantity, difficult to label, and difficult to assay for its hydrolysis. Availability of the sulfated fluorogenic substrate further diminishes the need for ganglioside G_{M2} for diagnostic purposes. The other glycolipid substrates, asialoganglioside G_{M2} and globoside, are hydrolyzed by both hexosaminidase A and B and thus are only useful for diagnosis of Sandhoff disease. They do not offer pragmatic advantage over the fluorogenic substrates.

Galactosialidosis

In early 1970s it was recognized that some cases of G_{M1}-gangliosidosis of late onset was not allelic with other more typical G_{M1}-gangliosidosis in that genetic complementation was observed in somatic hybrids between fibroblasts from these patients and those from other types of G_{M1}-gangliosidosis.[143] Wenger then found that these patients are deficient not only in β-galactosidase activity but also in glycoprotein neuraminidase (sialidase) activity.[144] The disease is now known as galactosialidosis. It appears to be relatively common in Japan. Series of studies by Galjaard and colleagues has clarified that the genetic defect of this disease lies in the absence of a protein which normally protects ganglioside G_{M1} β-galactosidase and sialidase from degradation within the lysosome.[3,145] The clinical features are reminiscent of genuine adult G_{M1}-gangliosidosis. Thus, deficient activity of β-galactosidase in an adult patient is not sufficient for diagnosis of adult G_{M1}-gangliosidosis unless galactosialidosis is excluded by normal sialidase activity.

Enzymatic Diagnosis. Diagnosis of galactosialidosis can be made by simultaneous deficiency of β-galactosidase and sialidase. Both cultured fibroblasts and leukocytes have been used for this purpose. The assay procedure for β-galactosidase is identical with that for G_{M1}-gangliosidosis

---

[142] J. Zerfowski, and K. Sandhoff, *Acta Neuropathol.* **27**, 225 (1974).

[143] H. Galjaard, A. Hoogeveen, W. Keijzer, H. A. de Wit-Verbeek, A. J. J. Reuser, M. W. Ho, and D. Robinson, *Nature (London)* **257**, 60 (1975).

[144] D. A. Wenger, T. J. Tarby, and C. Wharton, *Biochem. Biophys. Res. Commun.* **82**, 589 (1978).

[145] H. Galjaard, A. Hoogeveen, A. d'Azzo, F. W. Verheijen, and A. J. J. Reuser, *in* "Recent Progress in Neurolipidoses and Allied Disorders" (M. T. Vanier, ed.), p. 147. Fondation Marcel Mérieux, Lyon, France, 1984.

as above. The assay procedure for glycoprotein sialidase is described below.

Several different substrates have been used. For simplicity and high sensitivity, the fluorogenic substrate, 4-methylumbelliferyl-N-acetyl-neuraminic acid (sodium salt), now available commercially, appears to be the substrate of choice for diagnostic procedure. Leukocytes and cultured fibroblasts are the most commonly used enzyme sources.

Reagents

4-Methylumbelliferyl-N-acetylneuraminic acid (sodium salt), 1 mM in 0.5 M sodium acetate buffer, pH 4.3
Glycine buffer, 0.2 M, pH 10.8
4-Methylumbelliferone

Procedure. In 13 × 100 mm screw-capped tubes, 0.1 ml of the sub-strate solution and the enzyme source are mixed, and the final volume brought with water to 0.2 ml if necessary. The leukocyte or fibroblasts are suspended in water, frozen–thawed a few times, and briefly sonicated before the assay. The reaction mixture is incubated at 37° for 1 hr, at which time 3.8 ml of the glycine buffer is added. The liberated 4-methyl-umbelliferone is determined spectrofluorometrically as described above for β-glucosidase assays for Gaucher disease.

Comments. Patients with either G_{M1}-gangliosidosis or primary sialido-sis should give deficient activity of either β-galactosidase or sialidase with normal activity of the other. When both enzymes are deficient, diagnosis of galactosialidosis is likely. Degrees of deficiency, however, vary consid-erably among different enzyme sources. Residual activities in leukocytes are generally higher than those in fibroblasts.[146–148] In one series, sialidase activity in granulocytes was nearly normal, although the activity in lym-phocytes was deficient.[146] The neuraminidase deficient in galactosialidosis was reported to be the component labile to sonication[146] or to freezing treatment.[149] The deficient activities of sialidase and β-galactosidase can be partially restored by protease inhibitors.[150,151] Sialidase activity in het-

[146] S. Tsuji, T. Yamada, A. Tsutsumi, and T. Miyatake, *Ann. Neurol.* **11**, 541 (1982).

[147] H. Sakuraba, T. Aoyagi, and Y. Suzuki, *Clin. Chim. Acta* **125**, 275 (1982).

[148] S. Tsuji, T. Yamada, T. Ariga, I. Toyoshima, H. Yamaguchi, Y. Kitahara, T. Miyatake and T. Yamakawa, *Ann. Neurol.* **15**, 181 (1984).

[149] Y. Suzuki, H. Sakuraba, M. Potier, M. Akagi, M. Sakai, and H. Beppu, *Hum. Genet.* **58**, 387 (1981).

[150] H. Galjaard, A. Hoogeveen, F. Verheijen, O. P. van Diggelen, A. Konings, A. d'Azzo, and A. J. J. Reuser, *in* "Sialidases and Sialidoses" (G. Tettamanti, P. Durand, and S. Di Donato, eds.), p. 317. Ermes, Milano, 1981.

[151] Y. Suzuki, H. Sakuraba, K. Hayashi, K. Suzuki, and K. Imahori, *J. Biochem.* (*Tokyo*) **90**, 271 (1982).

erozygous carriers is often found to be intermediate between patients and normal controls.[148] Two possibilities can be considered. If the genetically deficient "protective protein" is in fact a subunit of sialidase, the intermediate activity can be expected. On the other hand, if the "protective protein" is genetically unrelated to sialidase, the intermediate activity in heterozygotes is difficult to explain. Until this question is definitely answered, use of sialidase activity for carrier detection seems risky.

Since galactosialidosis was recognized relatively recently as an independent genetic disease, there are more uncertainties still to be clarified than in many other long-established sphingolipidoses. While the above procedure should give a reasonable means of enzymatic diagnosis of the disease, careful review of the literature is strongly suggested as an essential part of evaluation of patients.

Concluding Remarks

The enzymatic diagnostic procedures described here may not necessarily be the "best" procedures for the respective enzymes. Procedures tend to be slightly different from one laboratory to another. Nevertheless, these procedures have been utilized extensively and successfully for the pragmatic purpose of diagnosis. One must be always aware that definitions of these genetic disorders are becoming almost a matter of semantics. Extensive genetic heterogeneity is expected for each of these disorders. The procedures described in this chapter are adequate for typical cases, but are likely to be too simplistic to cope with unusual enzymatic variants. The rapid progress in recent years of the molecular biological technology is beginning to replace some of the enzymatic diagnostic procedures with those based on DNA.[152,153] The enzymatic diagnosis is in principle quantitative, while the DNA technology can be qualitative. Elucidation of exact nature of many phenotypically and enzymologically atypical forms must await approaches from the nucleic acids. cDNA and in some instances, genomic DNA, for several of the lysosomal enzymes involved in the sphingolipidoses have already been cloned, sequenced, and otherwise characterized. Definitive information of enormous complexity will soon be forthcoming concerning the nature of mutations in these disorders. This chapter limited itself to the enzymatic diagnosis of the sphingolipidoses, but the newer diagnostic technology is clearly on the horizon.

[152] H. H. Kazazian, Jr., *Clin. Chem. (Winston-Salem, N.C.)* **31,** 1509 (1985).
[153] B. R. Jordan, *BioEssays* **2,** 196 (1985).

[62] Endo-β-N-acetylglucosaminidase H from *Streptomyces plicatus*

By ROBERT B. TRIMBLE, ROBERT J. TRUMBLY, and FRANK MALEY

Earlier studies[1,2] have described the purification and properties of endo-β-N-acetylglucosaminidase H (Endo H), an important oligosaccharide-cleaving enzyme. Because of its great utility in studies on the biosynthesis of glycoproteins and the relationship of oligosaccharides to function of specific glycoproteins, demand for this endoglycosidase has escalated markedly. To satisfy this demand, we have cloned the gene for Endo H adjacent to the λ p_L promoter of the plasmid vector pKC30[3] and, on transformation of the appropriate host, have affected about a 150-fold improvement in the yield of this enzyme.[4] This increase enabled the purification procedure to be greatly simplified and the final product to be devoid of detectable protease activity.

Materials and Methods

[³H]Dansyl chloride (specific activity 21.3 Ci/mmol) and uniformly ¹⁴C-labeled L-amino acid mixture (algal profile; specific activity 50 mCi/m atom carbon) were obtained from New England Nuclear Corp., and restriction endonucleases were obtained from New England Biolabs or Bethesda Research Laboratories. DNase was obtained from Worthington, Pronase from Calbiochem-Behring, DE-52 microgranular cellulose from Whatman, and Sephadex G-75 and Protein A–Sepharose from Pharmacia. The cloning vector pKC30[3] was a generous gift of Dr. Martin Rosenberg (Smith, Kline and Beckman, Philadelphia, PA), and the Endo H gene-containing plasmid pEH7′[5] was kindly provided by Dr. P. Robbins (M.I.T. Cancer Center, Cambridge, MA). DEAE–cellulose membranes (NA-45; 0.45 μm) were purchased from Schleicher and Schuell. L-[4,5-³H]Leucine (specific activity 53 Ci/mmol), the fluorographic enhancer, Amplify, and the *in vitro* prokaryotic protein synthesiz-

[1] A. L. Tarentino and F. Maley, *J. Biol. Chem.* **249**, 811 (1974).
[2] A. L. Tarentino, R. B. Trimble, and F. Maley, this series, Vol. 50, p. 574.
[3] H. Shimatake and M. Rosenberg, *Nature (London)* **272**, 128 (1981).
[4] R. J. Trumbly, P. W. Robbins, M. Belfort, F. D. Ziegler, F. Maley, and R. B. Trimble, *J. Biol. Chem.* **260**, 5683 (1985).
[5] P. W. Robbins, R. B. Trimble, D. F. Wirth, C. Hering, F. Maley, G. F. Maley, R. Das, B. W. Gibson, and K. Biemann, *J. Biol. Chem.* **259**, 7577 (1984).

ing system were from Amersham. Growth media components were obtained from Difco. All other chemicals were reagent grade or better.

Assay Procedure for Endo H. The assay employed [^{3}H]dansyl-Asn(GlcNAc)$_4$(Man)$_6$ (ovalbumin peak C glycopeptide) and measured the cleavage of [^{3}H]dansyl-AsnGlcNAc from this compound by separating the products as described.[2] A unit is that amount of enzyme which releases 1 μmol dansyl-AsnGlcNAc/min at 37° under the assay conditions. Protein was determined by the method of Lowry *et al.*,[6] using bovine serum albumin as a standard. An $A_{280\,nm}^{1\,cm}$ of 1.0 corresponded very closely to 1 mg of protein/ml.

Protease Assay. The release of acid-soluble radioactivity from [^{14}C]glycine-labeled hemoglobin[7] by Endo H samples was used as a measure of residual protease activity.

Amino Acid Analysis and Sequence. Amino acid compositions were determined using the Beckman 119CL amino acid analyzer with norleucine as an internal standard. Automated N-terminal sequencing was performed with a Beckman 890B sequenator as described earlier for this enzyme.[5]

Plasmid Construction

Plasmid EH7′[5] was digested with *Bam*HI, and the restriction fragments were separated on a 1% agarose gel. The 1.1-kb fragment, which contained the Endo H gene fused to the *Escherichia coli lac* promoter and ribosome-binding site, was purified by electrophoresis onto strips of DEAE–cellulose and eluted by an adaptation of the method of Dretzen *et al.*[8] The 5′ protruding ends were filled in with Klenow fragment, and the blunt end was ligated with 25 ng of pKC30 that had been digested previously with *Hpa*I. The ligation was accomplished by an overnight incubation at 14° with T4 DNA ligase. To linearize undesired recircularized pKC30 molecules, the ligation mixture was digested with *Hpa*I. The remaining circular plasmids containing the Endo H gene were used to transform *E. coli* strain HB101 (λc^+).[9] Plasmid DNA was prepared from 1-ml cultures of 12 transformants and digested with *Eco*RI. Recombinant plasmids containing the Endo H gene inserted in both orientations were identified. An *E. coli* strain lysogenic for a defective λ prophage containing a

[6] O. H. Lowry, N. J. Rosebrough, A. L. Farr, and R. J. Randall, *J. Biol. Chem.* **193,** 265 (1951).

[7] R. B. Trimble and F. Maley, *Anal. Biochem.* **141,** 515 (1984).

[8] G. Dretzen, M. Bellard, P. Sassone-Corsi, and P. Chambon, *Anal. Biochem.* **112,** 295 (1981).

[9] H. W. Boyer and D. Roulland-Dussoix, *J. Mol. Biol.* **41,** 459 (1969).

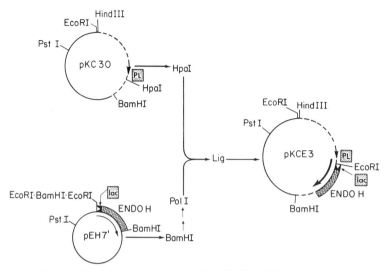

FIG. 1. Construction of expression vector pKCE3. Plasmid EH7′, containing the structural gene for Endo H fused to the β-galactosidase control region (*lac*), was digested with *Bam*HI. The 1.1-kb *Bam* fragment encompassing the Endo H gene (cross-hatched) and the *E. coli lac* promoter and its ribosome-binding site was purified from an agarose gel; the ends were filled in with DNA polymerase. This fragment was ligated to *Hpa*I-digested pKC30 to generate pKCE3 which contains the λ p_L and p_{lac} promoters in tandem. The orientation of the promoters is indicated by arrows on the plasmids. The Endo H transcripts are indicated by arrows within the plasmids.

temperature-sensitive repressor [N99*hfl*1-*recA*(λcI857*int*6*cro*27*Pam*3)][10] was transformed with both plasmid constructions. The overall cloning strategy is summarized in Fig. 1. Procedures for DNA preparation, polymerase fill-in reaction, ligation, and transformation have been described previously.[11]

Purification of Endo H

Step 1. Preparation of the Cell Extract. Five milliliters of an overnight starter culture containing transformed cells at an $A_{600\,nm}^{1\,cm}$ of about 0.15 were inoculated into each of eight 2-liter flasks containing 500 ml of LB medium (10 g Bactotryptone, 5 g Bacto-yeast extract, and 10 g NaCl/liter adjusted to pH 7.5 with NaOH) and incubated at 30° on a rotary shaker at 180 rpm. Culture density was followed until an $A_{600\,nm}^{1\,cm}$ of 0.8–0.9 was

[10] M. Belfort, G. F. Maley, and F. Maley, *Proc. Natl. Acad. Sci. U.S.A.* **80**, 1858 (1983).
[11] M. Belfort, A. Moelleken, G. F. Maley, and F. Maley, *J. Biol. Chem.* **258**, 2045 (1983).

reached (6 × 10⁸ cells/ml), then cultures were shifted to 42° for 90 min to induce synthesis of Endo H.

Step 2. Preparation of the 250,000 g Supernatant. Cells were chilled on ice and centrifuged at 16,000 g for 10 min in a Sorvall GSA rotor. The supernatant was discarded, and the viscous cell pellet resuspended in 40 ml buffer consisting of 10 mM Tris–HCl, pH 7.5, 1 mM MgCl$_2$, and 40 μg of DNase. After incubation at room temperature for 30 min to reduce the viscosity, the suspension was chilled in an ice-water bath and sonicated for three 1-min bursts with a Biosonic IV 1-in. probe at 80% intensity. All subsequent steps were performed at 4°. The broken cell suspension was diluted to 100 ml by addition of 0.1 M sodium citrate buffer, pH 5.5, and centrifuged at 250,000 g in a Beckman 60Ti rotor for 1.5 hr. The clear yellow supernatant, which contained about 90% of the starting activity, was decanted and saved while the pellet was discarded.

Step 3. Ammonium Sulfate Fractionation. The 250,000 g supernatant was diluted to 200 ml with 0.1 M sodium citrate buffer, pH 5.5, and brought to 0.4 saturation by addition of solid ammonium sulfate (22.5 g/100 ml) with stirring. After an additional 60 min of stirring, the solution was centrifuged at 13,000 g for 15 min. The supernatant fraction was adjusted to 0.8 saturation by addition of solid ammonium sulfate (25.8 g/100 ml), stirred, and centrifuged as above. The pellet was dissolved in 50 ml of 10 mM potassium phosphate buffer, pH 8.45, and dialyzed overnight against two 2-liter changes of this buffer.

Step 4. DE-52 Cellulose Chromatography. The dialyzed Step 3 material was applied at a flow rate of 40 ml/hr to a DE-52 microgranular cellulose column (2 × 20 cm) equilibrated with 10 mM potassium phosphate buffer, pH 8.45. The column was flushed with this buffer until unadsorbed protein was eluted, and the column was developed with a linear gradient formed with 500 ml of 10 mM potassium phosphate buffer, pH 8.45, in the mixing chamber and 500 ml of 180 mM potassium phosphate buffer, pH 8.45, in the reservoir. Fractions of 3.5 ml were collected at a flow rate of 40 ml/hr. The Endo H eluted at 40–50 mM potassium phosphate, and the fractions representing the central 75% of the peak were pooled and lyophilized. The residue was dissolved in a minimum volume of distilled water and dialyzed overnight against two 2-liter changes of 50 mM sodium phosphate buffer, pH 6.5.

Step 5. Sephadex G-75 Chromatography. The Step 4 material was reduced to 2.5 ml by lyophilization and chromatographed on a column (1.5 × 195 cm) of Sephadex G-75 (medium beads) which had been equilibrated with 50 mM sodium phosphate buffer, pH 6.5. The column was eluted with this buffer at a flow rate of 6 ml/hr, and fractions of 2.5 ml were collected. Those fractions containing activity were pooled (Fig. 2),

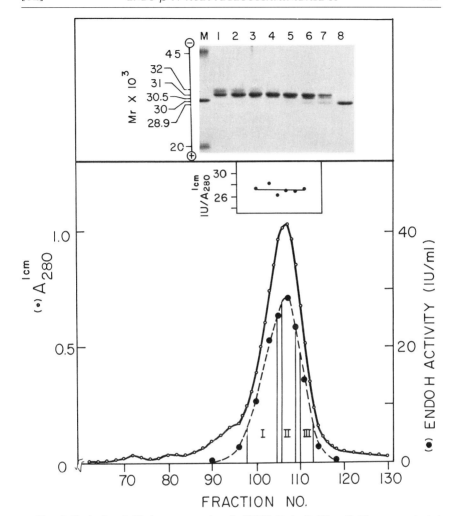

FIG. 2. Sephadex G-75 chromatography of pKCE3 Endo H (Step 5). The concentrated dialyzed Step 4 material (2.5 ml) was chromatographed on a 1.5 × 195 cm column of G-75 medium beads in 0.05 M sodium phosphate buffer, pH 6.5, containing 0.5% butanol (bottom). Despite the specific activity (bottom inset) being constant at 27 units/$A_{280\,nm}^{1\,cm}$ throughout the profile, SDS–PAGE revealed the Endo H to be a heterogeneous mixture of species ranging from 28.9 to 32 kDa (top). The gel contained 0.5 μg of protein from fraction numbers 100 (lane 1), 103 (lane 2), 105 (lane 3), 107 (lane 4), 109 (lane 5), 111 (lane 6), 113 (lane 7), and 0.5 μg of *S. plicatus* Endo H (lane 8). The markers (M) are ovalbumin (45 kDa), carbonic anhydrase (30 kDa), and soybean trypsin inhibitor (20 kDa). Amino acid analyses for pools I, II, and III are presented in Table II.

TABLE I

PURIFICATION OF ENDO H FROM A 4-LITER CULTURE OF N110 (λcIts)/pKCE3

Purification step	Volume (ml)	Protein (mg)	Activity (units)	Specific activity (units/mg)	Recovery (%)
1. Cell Extract	160	1,710	1,500	0.9	100
2. 250,000 g Supernatant	200	1,290	1,370	1.1	90
3. 0.4–0.8 Ammonium Sulfate	96	363	1,020	2.8	67
4. DE-52, pH 8.5	51	37	760	20.5	50
5. Sephadex G-75, pH 6.5	31	23	600	26.0	40

TABLE II

COMPARISON OF *S. plicatus* AMINO ACID
COMPOSITION WITH THREE POOLS OF CLONED
ENDO H FROM THE G-75 COLUMN OF FIG. 1

		Endo H preparation		
			pKCE3 (excess moles/mol)[b]	
Amino acid	*S. plicatus*[a] (moles/mol)	I	II	III
Asp	38			
Thr	16	2	1	1
Ser	17	3	2	1
Glu	21			
Pro	12	4.5	4	3.5
Gly	25	1		
Ala	33	6	3	1
$\frac{1}{2}$-Cys				
Val	25	3	1	
Met	2			
Ile	9			
Leu	20	2	1	
Tyr	16			
Phe	10			
His	3			
Lys	9			
Arg	11			
Trp	2			

[a] Deduced from the DNA sequence of the Endo H gene minus its leader sequence.[5]

[b] Excess residues represent those remaining after subtracting the amino acid composition of *S. plicatus* Endo H from the amino acid analyses for pools I–III.

```
 1                    5                         10                        15
H₂N- Met  Phe  Thr  Pro  Val  Arg  Arg  Arg  Val  Arg  Thr  Ala  Ala  Leu  Ala
                     20                        25                        30
     Leu  Ser  Ala  Ala  Ala  Ala  Leu  Val  Leu  Gly  Ser  Thr  Ala  Ala  Ser
                     35                        40                        45
     Gly  Ala  Ser  Ala  Thr  Pro  Ser  Pro  Ala  Pro  Ala  Pro  Ala  Pro  Ala
                     50                              ↑              ↑
     Pro  Val  Lys  Gln  Gly  . . .
```

SCHEME 1. *Streptomyces plicatus* Endo H Leader Sequence.

passed through a sterile Millipore GV disposable filter, and stored at
$-20°$. The purification of the pKCE3 Endo H from 4 liters of cells is summarized in Table I.

Properties of Endo H

Most of the properties, including the specificity of this enzyme, are
identical to those reported earlier,[1,2] except for the complete absence of
detectable protease as measured by the solubilization of [14C]glycine-la-
beled hemoglobin. In addition, the apparent heterogeneity of the elution
profile in Fig. 2, as evidenced by the SDS–PAGE analysis in the upper
panel, does not appear to be due to protein impurities. This is supported
by the constant specific activity throughout the elution profile (inset) and
the fact that antibody to Endo H absorbs all of the protein species in each
of the lanes. The DNA sequence of the gene for Endo H reveals that its
protein product should possess a 4.02-kDa leader sequence,[5] which if
unprocessed would yield a 33-kDa protein. This, in fact, is obtained when
an *in vitro* coupled transcription–translation system is employed with the
Endo H gene.[4] The fact that the Endo H isolated from *Streptomyces
plicatus* filtrates is about 29 kDa (Fig. 2), while none of the bands in the
elution profile of Fig. 2 approaches 33 kDa, suggests that some processing
of the cloned Endo H occurs in *E. coli*. Table II presents data to confirm
this thesis in that the Endo H in pool I possesses 21–22 additional amino
acids, while that in pool II appears to be a mixture of species retaining the
first 15–18 leader sequence amino acids. The Endo H in pool III retains an
average of 8–10 leader sequence residues (33–42) and even contains some
Endo H completely processed to the size of the *S. plicatus* enzyme (Fig.
2, lanes 6 and 7 versus lane 8). The amino acid data obtained are consistent with the leader sequence deduced from the DNA[5] (presented in
Scheme 1).

Inspection of Table II reveals that the difference between the amino
acid composition of each pool of pKCE3 Endo H and that secreted by *S.
plicatus* can be accounted for by the amino acid composition of a comparable portion of the leader sequence 5′ to the start region (Ala-Pro . . .) of

the secreted or processed enzyme. That the leader sequence terminates in the region indicated by the arrows has been verified by end-group sequence analysis of the *S. plicatus* enzyme, which revealed the repeating Ala-Pro dipeptide.[5] The heterogeneity of the amino terminus of the cloned Endo H had no effect on the enzyme's specific activity or on its specificity.

[63] Peptide-N^4-(N-acetyl-β-glucosaminyl) asparagine Amidase and Endo-β-N-acetylglucosaminidase from *Flavobacterium meningosepticum*

By ANTHONY L. TARENTINO and THOMAS H. PLUMMER, JR.

Peptide-N^4(N-acetyl-β-glucosaminyl)asparagine amidase F (PNGase F),[1] EC 3.5.1.52, catalyzes the reaction

$$\begin{array}{ccc} R^1 & & R^1 \\ | & & | \\ \text{AsnGlcNAcGlcNAc-Oligo} + H_2O \rightarrow & \text{Asp} + \text{GlcNAcGlcNAc-Oligo} + NH_3 \\ | \quad | & & | \quad\quad | \\ R^2 \ (\text{Fuc}) & & R^2 \quad (\text{Fuc}) \end{array}$$

where R = peptide chain and Oligo = high-mannose, hybrid, or bi-, tri-, or tetraantennary oligosaccharide chains. Hydrolysis is unaffected by a fucose linked $\alpha 1 \rightarrow 6$ to the proximal GlcNAc of complex oligosaccharides. PNGase F is assumed to hydrolyze the glycosylamine linkage by a mechanism similar to that of the analogous almond enzyme[2] and the aspartylglycosylamine amidohydrolase.[3]

Endo-β-N-acetylglucosaminidase F (Endo F), EC 3.2.1.96, catalyzes the reaction

$$\begin{array}{ccc} R^1 & & R^1 \\ | & & | \\ \text{AsnGlcNAcGlcNAc-Oligo} \rightarrow & \text{AsnGlcNAc} + \text{GlcNAc-Oligo} \\ | \quad | & & | \quad | \\ R^2 \ (\text{Fuc}) & & R^2 \ (\text{Fuc}) \end{array}$$

where R = H or peptide chain and Oligo = high-mannose, some hybrid, or biantennary chains with or without an internal fucose.

---

[1] T. H. Plummer, Jr., J. H. Elder, S. Alexander, A. W. Phelan, and A. L. Tarentino, *J. Biol. Chem.* **259**, 10700 (1984).

[2] N. Takahashi and H. Nishibe, *J. Biochem. (Tokyo)* **84**, 1467 (1978).

[3] M. Makino, T. Kojima, and I. Yamashina, *Biochem. Biophys. Res. Commun.* **24**, 961 (1966).

Assay Methods

Principle. PNGase F readily hydrolyzes the aforementioned N-linked oligosaccharide types in peptide linkage, releasing an intact oligosaccharide and an aspartyl peptide. The highest rate of cleavage has been observed with a triantennary pentaglycopeptide from fetuin. This glycopeptide (Leu-Ala-Asn(Oligo)-AeCys-Ser) has been didansylated with [³H]dansyl chloride and used as an assay based on the release of dansyl peptide by PNGase-type enzymes.[1] The released peptide is separated by paper chromatography, eluted from the paper, and quantitated by liquid scintillation spectrometry.[4]

Endo F hydrolyzes high-mannose oligosaccharides with the highest rate of cleavage, and its assay is based on the release of [³H]dansyl-AsnGlcNAc from [³H]dansyl-Asn(GlcNAc)$_2$(Man)$_5$. Quantitation is identical to the PNGase F assay. Each substrate is specific for the enzyme being tested.

Reagents

PNGase *F*:

[³H]Dansyl-Leu-Ala-Asn (Oligo)-([³H]dansyl-Ae) Cys-Ser, 0.36 m*M*, specific activity, 20 × 10³ dpm/nmol

Sodium phosphate, 0.25 *M*, pH 8.6

Endo *F*:

[³H]Dansyl-Asn(Oligo), 0.45 m*M*, specific activity, 16.3 × 10³ dpm/nmol

Sodium acetate, 0.25 *M*, pH 6.2

Procedure for PNGase F or Endo F. Two microliters of the appropriate dansyl glycopeptide are placed in the tip of a 0.5 ml polyethylene disposable microtube at 37°, and the reaction is initiated with 2 μl of enzyme solution appropriately diluted in 0.25 *M* buffer of the desired pH. The reaction mixture is incubated for 15–60 min and then terminated by rapidly pipetting the contents onto a sheet of Whatman 3 MM paper with a disposable micropipette. The chromatogram is folded into a circle and developed ascendingly for 45 min in 1-butanol–ethanol–water (4:2:3, v/v). Enzymatically released [³H]dansyl products migrate away from the substrates, which remain at the origin, and are located as a dansyl fluorescent spot with the use of UV light. The product is excised, placed in a scintillation vial, and extracted with 1.3 ml of 0.5% SDS. After incubation at 60° for 15 min, 10 ml of water-miscible scintillation solution is added, and the radioactivity is measured in a scintillation spectrometer.

---

[4] A. L. Tarentino, R. B. Trimble, and F. Maley, this series, Vol. 50, p. 574.

Activity. A milliunit (munit) of enzyme activity is the amount of enzyme which releases 1 nmol of [³H]dansyl product/min at 37° under the aforementioned assay conditions. The formation of dansyl product is linear with time and proportional to amount of enzyme when hydrolysis of the dansyl substrates is less than 20% of maximum. Protein is determined spectrophotometrically by the method of Warburg and Christian.[5] Specific activity is reported as milliunits per milligram of protein.

Materials

Flavobacterium meningosepticum was obtained from the American Type Culture Collection (ATCC 33958). The inoculum was grown to mid-log phase at 25° in 50 ml of 1% Tryptone–0.5% NaCl–0.5% yeast extract. Cells were harvested by low speed centrifugation, suspended in fresh media (12.5 ml) and 25% glycerol (12.5 ml), divided into 0.5 ml aliquots, and stored at −70° until needed.

Fetuin pentaglycopeptide [Leu-Ala-Asn(Oligo)-AeCys-Ser] is obtained from reduced, aminoethylated fetuin,[1] and ovalbumin Asn(GlcNAc)$_2$(Man)$_5$ is prepared as outlined before.[1] [³H]Dansyl substrates are prepared as described previously[1,6] and purified on Sephadex G-25 in 25% methanol.

Purification Procedure

The purification procedure outlined below has been modified slightly from the original description.[7] The steps described are numbered sequentially and are listed as such in Table I, which summarizes the purification of both PNGase F and Endo F.

Step 1. Culture Filtrate. A starter culture is prepared by inoculating one 0.5-ml aliquot of cells previously frozen at −70° into 100 ml of 1% tryptone–0.5% NaCl–0.5% yeast extract in a 250-ml flask. Cultures are grown aerobically at 30° for about 6 hr (early log phase). For enzyme production, each 2-liter flask containing 1 liter of M9 medium plus 0.55% casamino acids[8] is inoculated with 100 ml of starter culture and grown to stationary phase (40–48 hr) at 25° in a New Brunswick shaker. PNGase F and Endo F are secreted into the medium during growth of *Flavobacterium meningosepticum* and approach levels of 7.0 and 3.5 munits/ml, respectively, at stationary phase.

[5] O. Warburg and W. Christian, *Biochem. Z.* **310,** 384 (1941).
[6] Y. Tupuhi, D. E. Schmidt, W. Lindner, and B. L. Karger, *Anal. Biochem.* **115,** 123 (1981).
[7] A. L. Tarentino, C. M. Gomez, and T. H. Plummer, Jr., *Biochemistry* **24,** 4665 (1985).
[8] J. H. Elder and S. Alexander, *Proc. Natl. Acad. Sci. U.S.A.* **79,** 4540 (1982).

TABLE I
PURIFICATION OF *Flavobacterium meningosepticum* OLIGOSACCHARIDE
CHAIN-CLEAVING ENZYMES

Purification step	Total protein[a] (mg)	Total munits[b]	Specific activity (munits/mg protein)	Yield (%)
PNGase F				
1. Concentrated cultural filtrate[c]	N.D.[d]	45,760		100
2. Ammonium sulfate precipitation	416	37,554	90	82
3. TSK HW-55(S)	24.6	33,120	1346	72
4. Second TSK HW-55(S)	10.2	27,700	2638	61
5. SP-TRISACRYL M	4.4	22,567	5128	49
Endo-β-N-acetylglucosaminidase F				
1. Concentrated cultural filtrate[c]	N.D.[d]	22,905		100
2. Ammonium sulfate precipitation	416	20,188	49	88
3. TSK HW-55(S)	78.5	19,260	245	84
6. SP-TRISACRYL M	7.6	13,260	1740	58
7. SP-TRISACRYL M	1.4	6,241	4620	27

[a] Protein was estimated spectrophotometrically by the method of Warburg and Christian.[5]

[b] One milliunit (munit) equals 1 nmol of dansyl substrate hydrolyzed/min.

[c] A total of 7900 ml of cultural filtrate concentrated to 500 ml by ultrafiltration.

[d] N.D., not determined.

Step 2. Ammonium Sulfate Precipitation. All operations are conducted at 4°. The medium is centrifuged at 8000 g for 30 min, and the cultural filtrate (eight flasks, 7900 ml) is adjusted to 10 mM EDTA to protect the enzymes from proteolysis. The extract is concentrated to 500 ml by hollow fiber ultrafiltration (DC-2 with H1P10 cartridge, Amicon Corp.) and adjusted with constant stirring to 90% ammonium sulfate (600 g/liter of extract). After 24 hr, the concentrate is centrifuged at 12,000 g for 15 min. The ammonium sulfate pellet in each tube is resuspended in a small volume of 90% ammonium sulfate containing 10 mM EDTA, combined into a single 40-ml tube, and recentrifuged. Material can be stored at −70° at this stage. The precipitate is resuspended in about 7 ml of 20 mM Tris–Cl, pH 7.1, containing 100 mM NaCl and 5 mM EDTA (TSK column buffer), and the insoluble material is removed by centrifugation at 12,000 g for 15 min. From the cultural filtrates, an average of 85% of the Endo F and PNGase F activity are recovered in the solubilized ammonium sulfate fraction.

Step 3. TSK HW-55(S). The viscous extract obtained from Step 2 (total $A_{280 \text{ nm}}$ 416, 11.0 ml) is applied to a TSK HW-55(S) (Merck) column (2.5 × 180 cm) equilibrated in TSK column buffer. The column is devel-

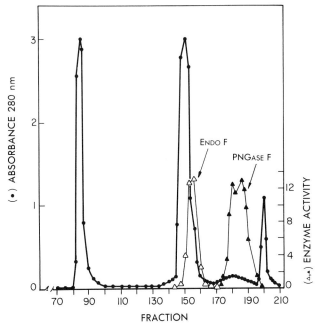

FIG. 1. Separation of Endo F and PNGase F on TSK HW-55(S). Fractions were assayed with [3H]dansyl-Asn(GlcNAc)$_2$(Man)$_5$ for Endo F ($\triangle$) and with [3H]dansyl-fetuin pentagly-copeptide for PNGase F ($\blacktriangle$). Chromatography conditions are described in the text.

oped at a flow rate of 24 ml/hr, and fractions of 3.3 ml are collected. Endo F and PNGase F are completely resolved on TSK-HW-55(S) (Fig. 1), but the basis for this separation is not size. Endo F chromatographs normally in this system and corresponds to a 32000-dalton species on SDS–PAGE[7] in fractions 151–160. PNGase F is retarded on the column, being present in a major pool in fractions 171–192. PNGase F corresponds to a 35500-dalton species in polyacrylamide gels of column fractions.[7] Sometimes a minor PNGase F pool containing approximately 15% of the activity is present between these two pools. Although PNGase F has a larger apparent molecular weight than Endo F, it elutes much later, suggesting a strong hydrophobic interaction with other proteins and the column matrix.

Step 4. Rechromatography of PNGase F on TSK HW-55(S). The broad PNGase F peak from Step 3 is concentrated by ultrafiltration (YM10 membrane, Amicon Corp.) to 3 ml. The retentate, which contains all the PNGase activity, is chromatographed on a TSK-HW55(S) column (2.0 × 165 cm). The column is developed at 17 ml/hr, and fractions of 2.2

ml are collected. On rechromatography PNGase F now elutes as a sharp, single species, but the elution position shifts to where the minor pool would normally be. Rechromatography of PNGase F increases its specific activity to 2638 munits/mg protein with a 84% recovery of applied enzyme.

Step 5. Chromatography of PNGase F on SP-TRISACRYL M (LKB). The fractions containing PNGase F activity from Step 4 are pooled, concentrated as before to 8.0 ml, and dialyzed exhaustively against 10 m*M* sodium acetate, pH 5.55. The sample is applied to a column of SP-TRIS-ACRYL M (1.0 × 20 cm) previously equilibrated in dialysis buffer. The column is washed with 10 m*M* sodium acetate until any unabsorbed protein is eluted. The column is developed with a linear sodium chloride gradient formed with 125 ml of 10 m*M* sodium acetate, pH 5.55, in the mixing chamber and 125 ml of 10 m*M* sodium acetate, pH 5.55, containing 0.25 *M* NaCl, in the reservoir. Fractions of 2.5 ml are collected at a flow rate of 25 ml/hr. PNGase F is the most retarded peak and elutes at 50 m*M* NaCl. Fractions containing enzyme activity are pooled, neutralized to pH 7.0 with 0.5 *M* sodium phosphate, pH 8.6, and concentrated as before. Specific activity is improved to over 5100 munits/mg protein.

Steps 6, 7. Chromatography of Endo F on SP-TRISACRYL M. The Endo F fractions from Step 3 are concentrated by ultrafiltration (YM10 membrane) to 8.0 ml, and dialyzed exhaustively against 10 m*M* sodium acetate, pH 5.55. The sample of Endo F is applied to a SP-TRISACRYL M column as in Step 5, but with a linear gradient of sodium chloride to 0.2 *M* (Step 6). Endo F is detected as the most retarded peak at 50 m*M* NaCl. Fractions containing Endo F activity are pooled, concentrated to 8.0 ml, and reapplied to another SP-TRISACRYL M column but with a linear gradient to 0.1 *M* sodium chloride. Fractions containing Endo F activity are pooled, neutralized to 7.0 with 0.5 *M* sodium phosphate, pH 8.6, and concentrated by ultrafiltration. By use of these two steps, the specific activity of Endo F is improved to 1700 and 4600 munits/mg protein after Steps 6 and 7, respectively. Part of the loss of enzyme activity between Steps 6 and 7 is accounted for by the very narrow selection of active fractions to be pooled after Step 7.

Properties of the Enzymes

Purity and Molecular Weight. PNGase F from Step 5 reveals a sharp band on SDS–PAGE that corresponds to a 35,500-dalton species.[7] Only one other trace band is detectable, and the enzyme is estimated to be over 95% pure and to be free of Endo F. Endo F from Step 7 appears as a sharp band on SDS–PAGE that corresponds to a 32,000-dalton species[1] with one other minor component. The enzyme is estimated to be over 90% pure and contains less than 0.1% PNGase F activity. No other glycosidase activities have been detected in either preparation.

The presence of a trace protease in either PNGase F or Endo F preparation cannot be ruled out. Whereas most glycoproteins do not show proteolytic degradation during incubation at 37° with the PNGase F preparation, SDS-denatured ovalbumin does show a disappearance of carbohydrate-free protein on SDS–PAGE with time. This degradation can be effectively suppressed by 10 mM o-phenanthroline (500 mM stock solution in methanol), but is unaffected by 60 mM EDTA. Alternatively, chromatography on hemoglobin–Sepharose completely removes this trace protease.

Stability. PNGase F and Endo F appear to be stable for months when stored at 4°. Endo F is stable to lyophilization, but PNGase F loses more than 50% of the original activity. Both enzymes when present at moderate enzyme levels (at least 0.5 munits/ml) are stable at 25° and 37° for long periods of time in the absence of detergent. Instability has been observed with PNGase F and Endo F at concentrations of less than 0.1 munits/ml at 37°. When glycoproteins denatured in 1% SDS followed by 0.6% NP-40 are used as substrates for deglycosylation, PNGase F loses most if its activity after 18 hr at 37° but is nearly fully active at 25°. When low levels of enzyme are used, deglycosylation at 25° is preferred and will go to completion.

If trace proteases are present in a glycoprotein substrate preparation, a number of protease inhibitors can be used without any effect on either PNGase F or Endo F activities. These include the aforementioned o-phenanthroline and EDTA as well as 10 mM diisopropyl fluorophosphate or phenylmethanesulfonyl fluoride, 100 μM leupeptin, and trasylol at 625 units/ml.

pH Optimum. PNGase F has a much higher pH optimum (pH 8.5) than Endo F (pH 4–6).[1] These properties allowed some selection for enzyme activities in enzyme preparations at a cruder stage or in a commercial preparation containing both activities.[1] At pH 4.0, only Endo F activity is functional, releasing peptide-GlcNAc and oligosaccharide-GlcNAc. At pH 9.3, the predominant activity is by PNGase F at the glycosylamine bond, releasing peptide and oligosaccharide-GlcNAc-GlcNAc. However, this latter oligosaccharide can in turn be hydrolyzed by Endo F activity still functional at pH 9.3, yielding oligosaccharide-GlcNAc and GlcNAc. This results in a mixture of both oligosaccharide products and free GlcNAc.

Specificity. PNGase F and Endo F have been incubated with defined glycopeptides of all major classes of N-linked oligosaccharides. The results can be summarized in Table II. Endo F hydrolyzes primarily high-

TABLE II

HYDROLYSIS OF GLYCOPEPTIDE CLASSES BY PNGASE F AND ENDO F[a]

Oligosaccharide structure	PNGase F[b]	Endo F[c]
OVB-High mannose[d]	+	+
OVB-Hybrid containing GlcNAc β1,4 Man[e]	+	−/+
OVB-Hybrid-without GlcNAc β1,4 Man[e]	+	+
Ovomucoid hybrids[f]	+	−
Biantennary complex[g]	+	+
Triantennary complex[h]	+	−
Tetraantennary complex[i]	+	−

[a] Hydrolysis of glycopeptides was followed by a difference assay on the amino acid analyser.[7]

[b] Asn-oligosaccharide required to be in peptide linkage for efficient cleavage.

[c] Asn-oligosaccharide could be free or in peptide linkage.

[d] Cyanogen bromide-released, tryptic-cleaved, Con A-retarded ovalbumin octaglycopeptide.[7]

[e] Ovalbumin Con A-nonretarded glycopeptides. Hybrid fractions containing an *N*-acetylglucosamine residue linked β1,4 to the trimannosyl core (α1,3 branch) initially appeared resistant to cleavage by Endo F. Structures were analyzed by 500-MHz ^{1}H-NMR spectroscopy (unpublished collaborative study with Drs. Robert Trimble and Paul Atkinson). Hybrids are of the type:

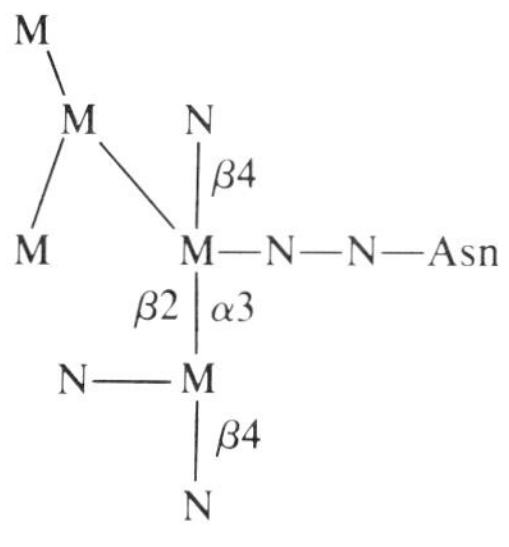

where M = mannose and N = is acetylglucosamine.

[f] Thermolytic glycopeptide mixture partially purified on Sephadex G-50 and Dowex 50-X8.

[g] Tryptic glycopeptide from IgM.[7]

[h] Thermolytic glycopeptide from fetuin.[7]

[i] Thermolytic glycopeptide from human α_1-acid glycoprotein.[7]

mannose oligosaccharides. Some structures containing peripheral *N*-acetylglucosamine substitutions (e.g., see Table II) appear resistant,[7] but will cleave if very high levels of enzyme are used. For example, a 100-fold higher Endo F level (350 munits/ml) is required to cleave an ovalbumin glycopeptide with peripheral GlcNAc (peak A^9) at a rate equal to an ovalbumin high-mannose glycopeptide (peak E^9). Endo F will slowly hydrolyze complex biantennary oligosaccharides,[7] but large amounts of enzyme are required to obtain complete deglycosylation. Complex tri- and tetraantennary asparagine-linked glycans are resistant to Endo F. It has been noted that hydrolysis by Endo F in the presence of high levels of glycerol (10% or more) can yield a nonreducible oligosaccharide addition product that can complicate quantitation of oligosaccharide chains.[10] PNGase F has deglycosylated all types of N-linked oligosaccharides tested, releasing an intact oligosaccharide chain and converting the asparaginyl–glycosylamine bond to an aspartic acid. It is very important that both the asparaginyl α-amino and carboxyl groups are in peptide linkage for significant cleavage to take place. PNGase F is, therefore, the all-purpose enzyme to hydrolyze high-mannose, hybrid, and bi-, tri-, and tetraantennary oligosaccharides. Highly purified enzyme is available commercially[7] and is to be preferred over any preparation containing both enzymes. The ability of Endo F to cleave susceptible PNGase F products creates a mixture of oligosaccharide types and free GlcNAc, and renders interpretation of oligosaccharide data difficult.[1]

Deglycosylation of native proteins by PNGase F can provide a useful approach for investigating structure–function studies of biologically active glycoproteins. With few exceptions, oligosaccharide chain-cleaving enzymes (Endo H, PNGase A) are generally inaccessible to glycosylation sites on proteins in their native state,[11] and denaturation is necessary to promote complete access to these sites.[12] PNGase F is somewhat different in that deglycosylation can be achieved in many cases (RNase B, fetuin, human transferin, lipase B, ovomucoid, and Fab fragment) in both the absence and presence of detergents. However, carbohydrate removal from native glycoproteins requires higher levels than from detergent-denatured glycoproteins, but the relative difference in the amount of PNGase F required varies considerably with the glycoprotein and must be determined empirically.[7] There is a large range in susceptibility by glycoproteins to complete deglycosylation by PNGase F. The enzyme should be used in small, concentrated reactions for greatest efficiency. Amounts have ranged from 0.6 munits/ml for RNase B to 30 munits/ml for fetuin. If insufficient enzyme is used with a glycoprotein with several oligosaccharide chains, you can get multiple bands resulting from successive removal of individual oligosaccharides.[7]

[9] C.-G. Huang, H. E. Mayer, Jr., and R. Montgomery, *Carbohydr. Res.* **13,** 127 (1970).
[10] R. B. Trimble, F. Maley, P. Atkinson, K. Tomer, T. H. Plummer, Jr., and A. L. Tarentino, *J. Biol. Chem.* **261,** 12000 (1986).
[11] F. K. Chu, F. Maley, and A. L. Tarentino, *Anal. Biochem.* **116,** 152 (1981).
[12] A. L. Tarentino and T. H. Plummer, Jr., *J. Biol. Chem.* **257,** 10776 (1982).

[64] α-Mannosidases I and II from *Aspergillus saitoi*

By Akira Kobata and Junko Amano

Ever since the α-mannosidase purified from *Aspergillus saitoi* was found to cleave specifically the Manα1 → 2Man linkage and not Manα1 → 3Man and Manα1 → 6Man linkages,[1] it has been used as an effective tool to identify high-mannose type sugar chains.[2-6] When a series of high-mannose type sugar chains, the structures of which can be written as (Manα1 → 2)$_n$ · Manα1 → 6(Manα1 → 3)Manα1 → 6(Manα1 → 3)Manβ1 → 4GlcNAcβ1 → 4GlcNAc$_{OT}$, are incubated with the *A. saitoi* α-mannosidase, they are all converted to a heptasaccharide, Manα1 → 6(Manα1 → 3)Manα1 → 6(Manα1 → 3)Manβ1 → 4GlcNAcβ1 → 4GlcNAc$_{OT}$. However, some batches of the enzyme purified according to the method of Ichishima *et al.*[7] are found to convert high-mannose type oligosaccharides to a hexasaccharide and not the heptasaccharide.

Study of the crude extract of *A. saitoi* mycelia indicated that it contains another α-mannosidase which does not act on *p*-nitrophenyl α-mannoside but cleaves Manα1 → 6(Manα1 → 3)Manα1 → 6(Manα1 → 3)Manβ1 → 4GlcNAcβ1 → 4GlcNAc$_{OT}$.[8] Therefore, one must be careful about the contamination of this α-mannosidase in order to use the Manα1 → 2Man-cleaving enzyme for the identification of high-mannose type sugar chains. Furthermore, substrate specificity of another α-mannosidase indicated that it is also useful for the structural study of complex type sugar chains. This chapter describes the isolation and characterization of the two α-mannosidases from a commercial source of *Aspergillus saitoi*, Morushin. The α-mannosidase which cleaves Manα1 → 2Man linkages only will be called α-mannosiase I in this paper and another α-mannosidase termed α-mannosidase II.

[1] K. Yamashita, E. Ichishima, M. Arai, and A. Kobata, *Biochem. Biophys. Res. Commun.* **96,** 1335 (1980).

[2] T. Mizuochi, Y. Nishimura, K. Kato, and A. Kobata, *Arch. Biochem. Biophys.* **209,** 298 (1981).

[3] H. Yoshima, M. Nakanishi, Y. Okada, and A. Kobata, *J. Biol. Chem.* **256,** 5355 (1981).

[4] H. Yoshima, N. Shiraishi, A. Matsumoto, S. Maeda, T. Sugiyama, and A. Kobata, *J. Biochem.* (*Tokyo*) **91,** 233 (1982).

[5] K. Yamashita, A. Hitoi, Y. Matsuda, A. Tsuji, N. Katunuma, and A. Kobata, *J. Biol. Chem.* **258,** 1098 (1983).

[6] T. Ohkura, K. Yamashita, Y. Mishima, and A. Kobata, *Arch. Biochem. Biophys.* **235,** 63 (1984).

[7] E. Ichishima, M. Arai, Y. Shigematsu, H. Kumagai, and R. Sumida-Tanaka, *Biochim. Biophys. Acta* **658,** 45 (1981).

[8] J. Amano and A. Kobata, *J. Biochem.* (*Tokyo*) **99,** 1645 (1986).

Assay Methods

Detection of α-Mannosidases I and II Activities

Principle. Since α-mannosidases I and II cannot cleave *p*-nitrophenyl α-mannoside, radioactive oligosaccharides are essential substrates to detect their activities. However, their strict specificities are useful for the separate detection of the two enzymes from an enzyme mixture by choosing appropriate oligosaccharides as substrates. For example, α-mannosidase I readily hydrolyzes the oligosaccharide I but cannot act on the oligosaccharide II shown below.

Oligosaccharide I:
  Manα1 → 2Manα1 → 3Manβ1 → 4GlcNAc$_{OT}$

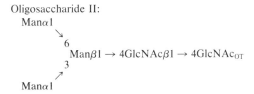

Oligosaccharide II:
  Manα1
    ↘
     6
      Manβ1 → 4GlcNAcβ1 → 4GlcNAc$_{OT}$
     3
    ↗
  Manα1

In contrast, α-mannosidase II removes one α-mannosyl residue from oligosaccharide II but acts very slowly on oligosaccharide I. Oligosaccharides I and II bind to concanavalin A (Con A)–Sepharose columns, but they do not bind to the column after one of the α-mannosyl residues is removed. Therefore, activities of α-mannosidases I and II can be assayed by the formation of radioactive oligosaccharides which pass through Con A–Sepharose column, from oligosaccharides I and II, respectively.

Reagents

Sodium acetate buffer, 0.1 M, pH 5.0, containing CaCl$_2$, 1 mM
Oligosaccharide I,[9] 0.2 mM, in the above buffer
Oligosaccharide II,[10] 0.2 mM, in the above buffer
Tris–HCl buffer, 10 mM, pH 7.5, containing NaCl, 100 mM, MnCl$_2$, 1 mM, MgCl$_2$, 1 mM, and CaCl$_2$, 1 mM
Tris–HCl buffer, 10 mM, pH 7.5, containing NaCl, 100 mM, MnCl$_2$, 1 mM, MgCl$_2$, 1 mM, CaCl$_2$, 1 mM, and methyl α-mannopyranoside, 0.1 M

Procedure. Ten microliters of one of the oligosaccharide solution is mixed with 40 μl of enzyme solution appropriately diluted with the so-

[9] K. Yamashita, Y. Tachibana, K. Mihara, S. Okada, H. Yabuuchi, and A. Kobata, *J. Biol. Chem.* **255,** 5126 (1980).
[10] K. Yamashita, J. P. Kamerling, and A. Kobata, *J. Biol. Chem.* **258,** 3099 (1983).

dium acetate buffer containing $CaCl_2$ and incubated at 37° for 10–30 min, depending on the activity of the preparation. The reaction is stopped by addition of 10 μl of 0.1 *N* NaOH. After dilution by adding 50 μl of 10 m*M* Tris–HCl buffer, pH 7.5, the reaction mixture is applied to a Con A (1 ml)–Sepharose column, and the column is washed with 10 ml of the Tris–HCl buffer. The radioactivity in the effluent is then determined by the liquid scintillation method after mixing a 1-ml aliquot of the sample with 7 ml of toluene-based liquid scintillator. The value (*A*) will afford the rough estimate of the enzyme activity. For a more accurate assay of the enzyme activities, unhydrolyzed oligosaccharides are recovered from the Con A–Sepharose column by elution with 10 ml of 10 m*M* Tris–HCl buffer containing 0.1 *M* methyl α-mannoside, and the radioactivity in the effluent (*B*) is determined in the same manner. The accurate amount of the oligosaccharide hydrolyzed can be calculated by the following equation.

$$2 \text{ nmol} \times A/(A + B)$$

Detection of Other Exoglycosidases

Principle. Preliminary study of the crude extract of Morushin indicates that it contains β-mannosidase, β-galactosidase, and β-*N*-acetylglucosaminidase which cleave *p*-nitrophenylglycosides. Therefore, the enzymes are allowed to react with the appropriate *p*-nitrophenylglycoside, and the yellow color of the *p*-nitrophenol released by enzymatic hydrolysis is estimated at pH 9.8 in a spectrophotometer at 400 nm.

Reagents

Sodium acetate buffer, 0.1 *M*, pH 5.0
p-Nitrophenyl-β-galactoside, 2 mg/ml, in the above buffer
p-Nitrophenyl-β-*N*-acetylglucosaminide, 2 mg/ml, in the above buffer
p-Nitrophenyl-β-mannoside, 2 mg/ml, in the above buffer
Na_2CO_3, 0.2 *M*

Procedure. Forty microliters of *p*-nitrophenylglycoside is incubated with the enzyme solution appropriately diluted with the sodium acetate buffer at 37° for 5–60 min, depending on the activity of the preparation. The reaction is stopped by adding 2 ml of Na_2CO_3 solution, and the intensity of the color is determined at 400 nm.

Unit

One unit (U) of enzyme is defined as the amount of the enzyme required to release 1 μmol of either mannose or *p*-nitrophenol per minute.

Purification Procedure

Unless otherwise indicated, all operations are carried out below 4°. One hundred grams of freeze-dried mixture of culture fluid and mycelia of *A. saitoi* (commercially available by the name of Morushin from Seishin Seiyaku Co. Ltd., Noda, Noda City, Japan) is suspended in 500 ml of 0.1 *M* sodium acetate buffer, pH 5.0, using a disperser (LK-21, Yamato Kagaku Ltd., Tokyo, Japan), and the mixture is kept still for 4 hr. The extract is centrifuged for 10 min at 6,000 *g*. After adjusting the pH of the supernatant to 5.0 with 1 *N* HCl, 60–80% ammonium sulfate precipitate is collected from it by centrifugation for 20 min at 10,000 *g*. The precipitate is dissolved in 600 ml of 10 m*M* sodium acetate buffer, pH 5.0, and enzymes in the solution are precipitated by adding ethanol up to 65% concentration. The precipitate is collected by centrifugation for 15 min at 10,000 *g* and dissolved in 150 ml of 10 m*M* sodium acetate buffer, pH 5.0, containing 0.2 *M* NaCl.

Gel Filtration on Sephacryl S-200. The crude enzyme solution described above (120 ml) is applied to a column of Sephacryl S-200 (5 × 90 cm) previously equilibrated with 10 m*M* sodium acetate buffer, pH 5.0, containing 0.2 *M* NaCl. The column is washed with the same buffer, and the effluent is collected in 12-ml fractions. The main purpose of this chromatography is to remove most of the indifferent proteins and facilitate the following purification procedures. However, this chromatography is also useful in removing large parts of other three exoglycosidases, β-galactosidase, β-*N*-acetylglucosaminidase, and β-mannosidase, which are eluted faster than the two α-mannosidases. α-Mannosidase I is eluted a little slower than α-mannosidase II. Therefore, α-mannosidase I contaminated only with a small amount of α-mannosidase II can be obtained by pooling the main enzyme fraction. The main fraction of α-mannosidase II is still contaminated with amounts of other glycosidases but may be free from them by the following two purification steps. The pooled fractions of α-mannosidases I and II are concentrated to 50 ml by ultrafiltration and dialyzed against 10 m*M* sodium acetate buffer, pH 5.0.

DEAE–BioGel A Column Chromatography. The enzyme solutions (50 ml each) of α-mannosidases I and II are applied to columns (2.5 × 10 cm) of DEAE-BioGel A equilibrated with 10 m*M* sodium acetate buffer, pH 5.0. The columns are eluted with linear gradient of 0 to 0.12 *M* NaCl in 10 m*M* sodium acetate buffer, pH 5.0, and the effluents are collected in 12-ml fractions. As shown in Fig. 1A, α-mannosidase I free from other exoglycosidases can be obtained by pooling the fractions indicated by a bar. α-Mannosidase II contaminated with small amounts of β-mannosidase and β-galactosidase can be obtained by pooling the fractions shown by a bar in

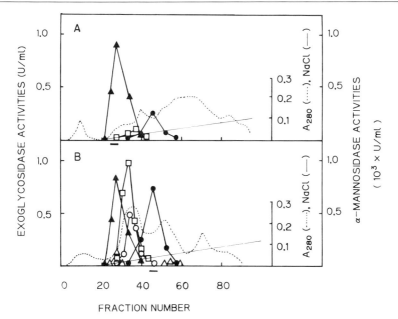

FIG. 1. DEAE–BioGel A column chromatographies of α-mannosidase I (A) and α-mannosidase II (B) preparations. Dotted lines represent the protein concentrations as determined by optical density at 280 nm. ▲——▲, α-mannosidase I; ●——●, α-mannosidase II; △——△, β-*N*-acetylglucosaminidase; ○——○, β-mannosidase; □——□, β-galactosidase. Exoglycosidase activities in the left ordinate indicate the values obtained by using *p*-nitrophenylglycosides as substrates.

Fig. 1B. The pooled sample of α-mannosidase II is concentrated to 10 ml by ultrafiltration and dialyzed against 10 m*M* sodium acetate buffer, pH 5.0.

SP-Sephadex C-50 Column Chromatography. Ten milliliters of α-mannosidase II solution is applied to a column (1.5 × 6 cm) of SP-Sephadex C-50 which is equilibrated with 10 m*M* sodium acetate buffer, pH 4.5. Elution is performed stepwise with 100 ml of 10 m*M* sodium acetate buffer, pH 4.5, and then with 50 ml of the same buffer containing 0.2 *M* NaCl. The effluent is collected in 5-ml fractions. α-Mannosidase II is eluted readily with the first buffer and completely freed from β-mannosidase and β-galactosidase, which are eluted with the second buffer. Tubes containing α-mannosidase II are pooled and dialyzed against 10 m*M* sodium acetate buffer, pH 5.0, containing 1 m*M* CaCl₂.

The procedures for preparing the purified α-mannosidases I and II are summarized in Tables I and II, respectively. Both purified enzymes are

TABLE I
PURIFICATION OF α-MANNOSIDASE I

Fraction	Total protein (mg)	Total activity (U)	Specific activity (U/μg)	Purification (-fold)
Alcohol precipitate	6279	0.30	0.048	1
Sephacryl S-200	240	0.12	0.50	10.4
DEAE–BioGel A	2.6	0.09	34.6	720.8

free from mutual contamination and β-galactosidase, β-N-acetylglucos-aminidase, β-mannosidase, α-mannosidase, and α-fucosidase when assayed with p-nitrophenylglycosides as substrates.

Properties

Stability. Purified preparations of α-mannosidases I and II are quite stable and can be stored, respectively, in 10 mM sodium acetate, pH 5.0, and 10 mM sodium acetate buffer, pH 5.0, containing 1 mM CaCl$_2$ either at 4° or at −20° for at least 4 months.

Metal Requirement and pH Optimum. α-Mannosidase I does not require metal for its activity. In contrast, more than 13 times higher activity of α-mannosidase II is obtained in the presence of Ca^{2+} (1 mM) as compared without Ca^{2+}. The activation effect of Ca^{2+} reaches a plateau at 1 mM, and the same enzyme activity is obtained in the presence of 10 mM Ca^{2+}. The pH optima of both enzymes are 5.0.

Substrate Specificities.[1,8] α-Mannosidase I cleaves the Manα1 → 2Man linkage but cannot act on Manα1 → 3Man or Manα1 → 6Man linkages. Because of this strict specificity, the enzyme can be used as a

TABLE II
PURIFICATION OF α-MANNOSIDASE II

Fraction	Total protein (mg)	Total activity (U)	Specific activity (U/μg)	Purification (-fold)
Alcohol precipitate	6279	0.24	0.038	1
Sephacryl S-200	215	0.19	0.88	23.2
DEAE–BioGel A	7.8	0.09	12.0	315.8
SP-Sephadex	2.1	0.08	38.0	1000

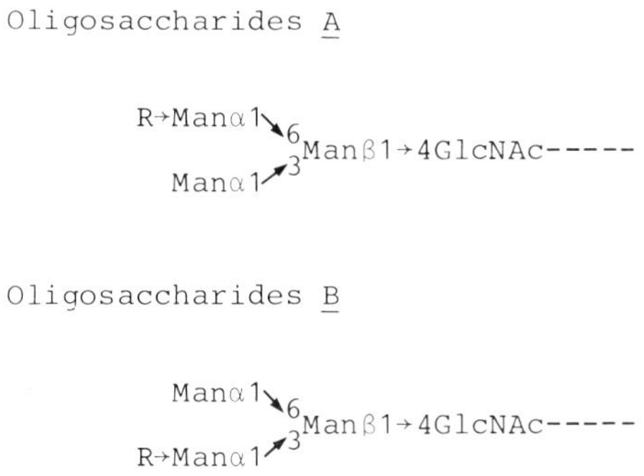

Oligosaccharides <u>A</u>

$$\begin{array}{c} R{\rightarrow}Man\alpha 1_{\diagdown 6} \\ \qquad\qquad{}_{3}Man\beta 1{\rightarrow}4GlcNAc----- \\ Man\alpha 1^{\diagup} \end{array}$$

Oligosaccharides <u>B</u>

$$\begin{array}{c} Man\alpha 1_{\diagdown 6} \\ \qquad\qquad{}_{3}Man\beta 1{\rightarrow}4GlcNAc----- \\ R{\rightarrow}Man\alpha 1^{\diagup} \end{array}$$

FIG. 2. Structures of two isomeric monoantennary oligosaccharides. R represents either monosaccharide or oligosaccharide.

very effective tool to identify the high-mannose type asparagine-linked sugar chains. The K_m and V_{max} values of the enzyme for oligosaccharide I are 0.8 mM and 10 μmol/min/mg protein, respectively.

The substrate specificity of α-mannosidase II is more complicated. Like α-mannosidase I, it cannot hydrolyze *p*-nitrophenyl-α-mannopyranoside. It cleaves the terminal α-mannosyl linkage of Manα1 → 3Manβ1 → 4GlcNAc$_{OT}$ more than 10 times faster than those of Manα1 → 6Manβ1 → 4GlcNAcβ1 → 4GlcNAc$_{OT}$ and Manα1 → 2Manα1 → 3Manβ1 → 4GlcNAC$_{OT}$. Furthermore, it cleaves the Manα1 → 6Man linkage of oligosaccharide II, only after its Manα1 → 3 residue is removed. Therefore, a mannose residue is released from oligosaccharides *A* but not from oligosaccharides *B* in Fig. 2. The K_m and V_{max} values of the enzyme for oligosaccharide II are 1.25 mM and 1.89 μmol/min/mg protein, respectively.

It is reported that the nonreducing terminal α-mannosyl residues in oligosaccharides *B* cannot be removed by jack bean α-mannosidase.[11] However, this specificity is found to be incomplete because the α-mannosyl linkage can be cleaved at high enzyme concentration (>40 units/ml).[12] In contrast, the substrate specificity of α-mannosidase II from *A. saitoi* is more strict and the Manα1 → 6Man linkages in oligosaccharides *B* cannot be cleaved even if the enzyme concentration is increased up to 6.2 munits/ml, which is more than 10 times than the concentration usually used. Therefore, α-mannosidase II can be used more reliably to discriminate oligosaccharides *A* from their isomeric oligosaccharides *B*.

[11] K. Yamashita, Y. Tachibana, T. Nakayama, M. Kitamura, Y. Endo, and A. Kobata, *J. Biol. Chem.* **255**, 5635 (1980).

[12] K. Hanaoka, S. Takasaki, A. Kobata, H. Miyamoto, T. Nakamura, and M. Mayumi, *J. Biochem. (Tokyo)* **99**, 1273 (1986).

[65] Polysialic Acid Depolymerase

By Bartłomiej Kwiatkowski and Stephan Stirm

Glycan glycanohydrolase activities catalyzing the depolymerization of bacterial surface polysaccharides are associated with the tail parts of many bacteriophages.[1,2] During the infective process, they help the viruses to penetrate the outer layers of their host cells.[3] Generally, these bacterioviral hydrolases are highly specific for one or a few structurally related heteroglycans; therefore, they contribute to the restriction of bacteriophage host range.[4]

Among the phage particles with glycanohydrolase activities, those acting on host *capsular* polysaccharides are distinguished by a special plaque morphology: The "plaque propre" is surrounded by an acapsular halo which continues to spread after the medium is exhausted.[5] This phenomenon is caused by radially diffusing free tail parts (produced in addition to complete virus) and renders it very easy to isolate phages of this type from their natural habitat, e.g., from sewage. Since *whole virus particles* function as catalysts and are readily purified by poly(ethylene glycol) precipitation and isopycnic ultracentrifugation,[6] these enzymes constitute a general means for the production of oligomers from bacterial capsular glycans.[2]

In this chapter, we describe the propagation and purification of *Escherichia coli* bacteriophage $\phi92$,[7] the depolymerization of colominic acid, $[-\alpha\text{-NeuAc-}(2 \rightarrow 8)\text{-}]_n$, by the virus particles, and the isolation of the oligosialic acid products.

Propagation and Purification of Bacteriophage

Materials

Virus: *E. coli* bacteriophage $\phi92$,[7] No. 35860-B1 of the American Type Culture Collection

[1] A. A. Lindberg and I. W. Sutherland, *in* "Surface Carbohydrates of the Procaryotic Cell" (I. W. Sutherland, ed.), pp. 209 and 289. Academic Press, New York, 1977.

[2] H. Geyer, K. Himmelspach, B. Kwiatkowski, S. Schlecht, and S. Stirm, *Pure Appl. Chem.* **55**, 637 (1983).

[3] M. E. Bayer, H. Thurow, and M. H. Bayer, *Virology* **94**, 95 (1979).

[4] D. Rieger-Hug and S. Stirm, *Virology* **113**, 363 (1981).

[5] W. Bessler, E. Freund-Mölbert, H. Knüfermann, C. Rudolph, H. Thurow, and S. Stirm, *Virology* **56**, 134 (1973).

[6] K. Yamamoto, B. Alberts, R. Benzinger, L. Lawhorne, and G. Treiber, *Virology* **40**, 734 (1970).

[7] B. Kwiatkowski, B. Boschek, H. Thiele, and S. Stirm, *J. Virol.* **45**, 367 (1983).

Host organism: *E. coli* Bos12[8] (O16 : K92 : H⁻), No. 35860 of the
American Type Culture Collection

Merck standard I medium (both nutrient agar plates and broth)

P medium[9]: (1) 25 g of D-glucose in 1 liter of distilled water, (2) 6.25 g
of casamino acids (Difco Laboratories), 0.4 g of L-tryptophan, 0.3 g
of L-cysteine, 2.5 g of KH_2PO_4, 15.6 g of $Na_2HPO_4 \cdot 12H_2O$, 1.3 g
of NH_4Cl, and 0.01 g of gelatin in 1 liter of water, (3) 5 g of $MgSO_4 \cdot 7H_2O$ in 100 ml of water, and (4) 500 mg of $CaCl_2$ in 10 ml of water;
suitable aliquots of the four solutions are separately autoclaved (30
min at 120°) and stored at 4°; before use, they are mixed in a ratio of
200 : 800 : 10 : 1, and adjusted to pH 7.2 with 1 M NaOH

Silicon antifoam

Poly(ethylene glycol) with a molecular weight of 6000–7500

Sodium chloride

0.1 M Tris–HCl buffer, pH 7.5, containing 0.5% NaCl and 0.1%
NH_4Cl

Cesium chloride (extra pure), 10 and 50 g, each dissolved in 50 ml of
Tris–HCl buffer; the solutions should have densities of 1.1–1.15
and 1.6–1.65 g/ml, respectively

Bacteriophage Stock. Drops of a 1 : 10 broth dilution series of phage
are placed on nutrient agar seeded with host bacteria. After growth at 37°
overnight (incubator), one or a few single plaques (with optimally developed halos) are touched with a platinum needle, and the phages are
washed into 2 ml of broth. Similarly, one loopfull of host bacteria (from a
fresh colony on nutrient agar) are suspended in another 2 ml of broth.
Seven test tubes, each with 5 ml of broth, are simultaneously inocculated
with 0.2 ml of suspended bacteria and are incubated at 37° under gentle
stirring (roller). At 0, 30, 60 min, etc., the cultures are consecutively
infected with 0.2 ml of virus suspension. After 3–7 hr, some of the first
cultures lyse visibly (while the rest have been infected too late). The last
two lysates are combined and centrifuged for about 10 min at 1000 g
(bench centrifuge) to remove resistant bacteria and bacterial debris. The
supernatant is sterilized by vigorous shaking with about 0.5 ml of chloroform. It generally contains 10^9–10^{10} plaque-forming units (PFU) per milliliter as determined (after removal of the chloroform by centrifugation and
aeration) by plating 25- or 50-μl aliquots of a 1 : 10 dilution series on a
bacterial lawn as above (or by the agar overlayer method[10]).

At 4° over chloroform, the virus stock doesn't lose titer appreciably

[8] W. Egan, T.-Y. Liu, D. Dorow, J. S. Cohen, J. D. Robbins, E. C. Gotschlich, and J. B. Robbins, *Biochemistry* **16**, 3687 (1977).
[9] S. Stirm and E. Freund-Mölbert, *J. Virol.* **8**, 330 (1971).
[10] M. Adams, "Bacteriophages." Wiley (Interscience), New York, 1959.

for several months. For longer periods, ϕ92 (and Bos 12) suspensions are conveniently stored in liquid nitrogen. Occasional controls by slide agglutination[11] with an *E. coli* O16 antiserum (Difco Laboratories) are recommended: Fresh colonies or lawns of *E. coli* Bos 12 are O16 inagglutinable, while the bacteria *in the halos* around the plaques of ϕ92 react with this serum.

Large-Scale Propagation of Bacteriophage. P medium—e.g., 250 ml in a *sterile* wash bottle with a sintered glass gas distributor and with stoppered (cotton wool) inlet and outlet—is warmed to 37° (water bath), strongly aerated (e.g., by suction with a water pump), and inoculated with a loopfull of *E. coli* Bos 12 from a fresh colony. When the bacterial culture has grown to an optical density of (0.9–) 1.0 at 660 nm (in a 1-cm cell with P medium as a blank), it contains 4.5×10^8 colony-forming organisms per ml and is infected with the ϕ92 stock suspension (chloroform removed) to a multiplicity of 0.1 (1 virus for 10 cells). Lysis occurs about 1 hr later and may necessitate the addition of antifoam and/or a reduction of aeration. After roughly another hour, the optical density rises again, and by this time the cultivation is stopped and the phage lysate is clarified by centrifugation (20 min at 5000 g, e.g., in a Beckman Instruments Model J2-21 centrifuge). It generally contains around 6×10^{10} PFU/ml and is stored at 4° after sterilization with chloroform as above.

Purification of Bacteriophage Particles.[6] After 12 hr or more at 4°, the lysate is centrifuged once again, and solid NaCl (to 0.5 M) and poly(ethylene glycol) (to 10%) are dissolved in the clear supernatant. Storage at 4° is continued overnight (or longer), and the precipitated virus is recovered in a small volume (e.g., 1–3 ml from 250 ml) by a third centrifugation at 5000 g, decanting most of the supernatant and resuspending the pellet(s).

Linear density gradients from about 1.1–1.5 to 1.6–1.65 g/ml are introduced into swinging bucket rotor (e.g., Beckman SW41) tubes by draining (with a Büchler polystaltic pump) 2 volumes from a beaker containing 1 volume of the dilute CsCl solution, while adding (under stirring with a magnetic stirrer) 1 volume of the concentrated one. One to three milliliters of the suspension of precipitated phage is placed on each gradient and is centrifuged for 90 min at 90,000 g (e.g., in a Beckman Model L5-65 ultracentrifuge). The opaque virus band at $g = 1.48$ g/ml is clearly visible if at least 10^{11} PFU per tube were centrifuged; the band is withdrawn by puncturing the tube with a syringe and is finally dialyzed against volatile buffer. About two-thirds of the PFU in the lysate is generally recovered, i.e., approximately 10^{13} PFU from 250 ml. As seen under the electron

---

[11] F. Ørskov and I. Ørskov, *Methods Microbiol.* **14**, 43 (1984).

microscope,[7] the ϕ92 particles thus obtained are free of contaminants. They are comparatively large bacteriophages (length ~190 nm, head diameter ~80 nm) with a contractile tail, and they have tail appendages of curly appearance with which the depolymerase is probably associated.[2] Also the enzymatic activity of the purified virus particles is stable for several months, if they are stored over chloroform at 4°.

Depolymerization of Colominic Acid

Principle. Chain internal $\alpha(2 \rightarrow 8)$-linked NeuAc units do not consume periodate. By depolymerization of colominic acid, however, oxidizable functions are produced. Therefore, the bleaching of the violet ferrous 2,4,6-tri-2-pyridyl-*s*-triazine complex [Fe(TPTZ)$_2^{2+}$] by residual periodate may serve as an assay for the virus-catalyzed reaction.[12,13]

Substrate. Colominic acid is commercially available (e.g., from Sigma Chemical Co. Ltd.), or may be isolated[14] from *E. coli* K235 (O1 : K1 : H$^-$, L+O), which can be obtained (No. 13027) from the American Type Culture Collection. Since bacterial polysialic acids may be O-acetylated[14] or otherwise esterified[15,16] to various degrees, the commercial product (e.g., 100 mg) is first dissolved and stored in ice-cold aqueous 0.1 M NaOH (e.g., 40 ml) for 45–60 min. After neutralization, under stirring, with cold 2 M HCl, the mixture is dialyzed against deionized water and lyophilized. To remove low molecular weight components, the product is then passed through a column of BioGel P-30 (conditions as below) and the fractions appearing in the void volumn are freeze-dried again.

Reagents

Volatile buffer, i.e., a solution of ammonium acetate (0.1 M) and ammonium carbonate (0.05 M), adjusted to pH 7.2 with 1 M acetic acid.

1 M sodium acetate buffer, pH 4.0

3 mM sodium metaperiodate in the sodium acetate buffer

TPTZ reagent: 1 liter of the violet reagent solution (optical density at 593 nm about 1.8 in a 1-cm cell) is prepared from 75 mg (240 μmol) of 2,4,6-tri-2-pyridyl-*s*-triazine, 46 ml of glacial acetic acid, 210 ml

[12] G. Avigad, *Carbohydr. Res.* **11**, 119 (1969).

[13] B. Kwiatkowski, B. Boschek, H. Thiele, and S. Stirm, *J. Virol.* **43**, 697 (1982).

[14] F. Ørskov, I. Ørskov, A. Sutton, R. Schneerson, W. Lin, W. Egan, G. E. Hoff, and J. B. Robbins, *J. Exp. Med.* **149**, 669 (1979).

[15] M. R. Lifely, A. S. Gilbert, and C. Moreno, *Carbohydr. Res.* **94**, 193 (1981).

[16] E. C. Gotschlich, B. A. Fraser, O. Nishimura, J. B. Robbins, and T.-Y. Liu, *J. Biol. Chem.* **256**, 8915 (1981).

of the sodium acetate buffer, and 31.4 mg (80 μmol) of ferrous ammonium sulfate (Fe[NH$_4$]$_2$[SO$_4$]$_2$ · 6H$_2$O) as described by Avigad[12]

Procedure. A suspension of purified bacteriophage ϕ92 particles and a solution of alkali-treated colominic acid, both in volatile buffer, are combined to yield a mixture (e.g., 10 ml) containing 4.5 mg of substrate (about 15 mM with respect to sialic acid units) and about 5 × 10^{11} PFU/ml. A small amount (e.g., 0.2 ml) of chloroform is added to maintain sterility, and the reaction is allowed to proceed at 37° (water bath) in a closed vessel (flask with a ground glass stopper). At intervals, e.g., after 5, 10, 15, 20, 40, 60, 90 min and after 2, 3, 5, 8, and 24 hr, aliquots of 0.4 ml are withdrawn, heated at 100° for 3 min to stop the reaction, cooled to room temperature, and analyzed for periodate consumption. For this purpose, 0.8 ml of sodium acetate buffer and 1.2 ml of periodate solution are added to each sample and the mixtures are stored for 6 hr at room temperature in the dark. Then 0.1-ml portions are finally mixed with 4.5 ml of the TPTZ reagent, filled up to 5 ml with water, and the absorbance is read at 593 nm after a few minutes (e.g., 5 min).[12] A sample withdrawn at zero time, or incubated with heat-denatured virus (3 min at 100°), serves as a blank.

Initially, i.e., up to about 40 min, approximately 5 × 10^{12} PFU produce 1 μmol of periodate-consuming functions per minute under these conditions. After around 6 hr, the increase of oxidant consumption has slowed down.

Isolation of Sialic Acid Oligomers

Reagents

Resorcinol–HCl reagent, prepared from 200 mg of recrystallized resorcinol, 80 ml of concentrated hydrochloric acid, and 0.25 ml of 0.1 M aqueous CuSO$_4$ as described by Tuppy and Gottschalk.[17]
A mixture of 1-butyl acetate (85%) and 1-butanol (15%) P$_2$O$_5$ and KOH

Procedure. For exhaustive depolymerization in a small volume, 100 mg of alkali-treated colominic acid and about 2 × 10^{12} PFU of ϕ92 in 4 ml of volatile buffer are incubated as above, but for 48 hr. After decanting and chasing of the chloroform, most of the bacteriophages are sedimented by centrifugation at 16000 g (J2-21) for 1 hr.

For removal of residual virus and virus debris, the supernatant is first passed over a column (2 × 100 cm) of BioGel P-30 (50–100 mesh; Bio-Rad

[17] H. Tuppy and A. Gottschalk, *in* "Glycoproteins: Their Composition, Structure and Function" (A. Gottschalk, ed.), Part A, p. 439. Elsevier, Amsterdam, 1972.

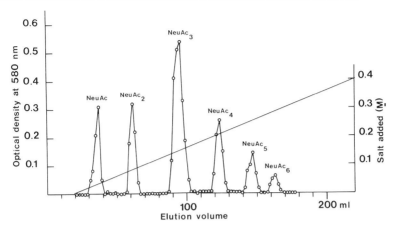

FIG. 1. Ion-exchange chromatography, on DEAE–Sephadex A-25, of sialic acid and its oligomers, as produced by depolymerization of colominic acid with particles of *Escherichia coli* bacteriophage φ92. Detection with resorcinol–HCl according to Tuppy and Gotts-chalk.[17]

Laboratories) with the volatile buffer (at 16 ml/hr). Fractions of 3.8 ml are collected, and aliquots of 0.1 ml are heated with 2 ml of the resorcinol–HCl reagent and are extracted with butyl acetate–butanol as described by Tuppy and Gottschalk.[17] The absorbance at 580 nm (in a 1-cm cell with the solvent mixture as a blank) is read, and the included sialic acid mono- and oligomers—eluting from about 50 ml after the void volume (as determined with bovine serum albumin) until a few milliliters ahead of the inclusion volume (as determined with D-glucose)—are pooled and lyophilized.

The mixture of depolymerization products is finally redissolved in 1 ml of volatile buffer and is adsorbed to a column (1.8 × 25 cm) of DEAE–Sephadex A-25 (Pharmacia Fine Chemicals) equilibrated with the volatile buffer. Sialic acid and its oligomers are consecutively eluted (at 6 ml/hr) with a linear gradient of ammonium acetate (to 0.4 M) in the same buffer, collecting fractions of 2 ml and testing 20-μl aliquots as above (see Fig. 1). The fractions containing the single oligomeric species are pooled, lyophi-lized, and dried *in vacuo* over P_2O_5 and KOH. About 4, 8, 24, 13, 12, and 6 mg, respectively, of sialic acid and its $\alpha(2 \rightarrow 8)$-linked di-, tri-, tetra-, penta-, and hexamer are obtained.

Variations of the Method

Naturally, the colominic oligosaccharides may also be separated by gel filtration,[7] and larger sialic acid oligomers are obtained after shorter periods of incubation with virus. In addition to colominic acid, the $\phi 92$ sialidase also depolymerizes the *E. coli* K92 capsular polysaccharide, [-α-NeuAc-(2 → 9)-α-NeuAc-(2 → 8)-]$_n$, but not the *Neisseria meningitidis* type C glycan, [-α-NeuAc-(2 → 9)-]$_n$. Within the K92 substrate, the viral enzyme preferentially cleaves the $\alpha(2 → 8)$ linkages, leading predominantly to the formation of α-NeuAc-(2 → 9)-NeuAc, α-NeuAc(2 → 9)-α-NeuAc-(2 → 8)-α-NeuAc-(2 → 9)-NeuAc, etc.[7]

Acknowledgments

This project was supported by Alexander von Humboldt-Stiftung, by Deutsche Forschungsgemeinschaft (SFB 47), by Fonds der Chemischen Industrie (341), and by the Polish Academy of Sciences (MR. II.17). We wish to express our gratitude to Mrs. Maria Magdalena Stein for excellent technical assistance.

[66] Activator Proteins for Lysosomal Glycolipid Hydrolysis

By Ernst Conzelmann and Konrad Sandhoff

Introduction

The final degradation of glycosphingolipids occurs in the digestive intracellular organelles known as lysosomes and is accomplished by the sequential action of specific hydrolases. Some of these enzymes appear to be membrane associated and to interact with their glycolipid substrates directly whereas others, mainly those acting also on water-soluble substrates, are water soluble and cannot attack their lipid substrates directly. Instead, this interaction has to be mediated by some small nonenzymatic proteins, so-called activator proteins.

The existence of these activator proteins was discovered in 1964, when Mehl and Jatzkewitz[1] observed that after fractionation by free-flow electrophoresis a crude preparation of porcine arylsulfatase A had lost its ability to degrade sulfatide (sulfogalactosylceramide) in the absence of detergents but remained fully active with artificial water-soluble substrates. The activity against the glycolipid could be restored by the addition of another, enzymatically inactive, fraction. A similar cofactor that promotes sulfatide catabolism was then shown to exist in normal human liver as well as in tissues from patients with metachromatic leukodystrophy[2] and was therefore designated as the "activator of cerebroside

[1] E. Mehl and H. Jatzkewitz, *Hoppe-Seyler's Z. Physiol. Chem.* **339,** 260 (1964).
[2] H. Jatzkewitz and K. Stinshoff, *FEBS Lett.* **32,** 129 (1973).

sulfatase.'' Although the name is somewhat misleading since the cofactor does not activate the enzyme (see below), such protein cofactors for the hydrolysis of lipid substrates by water-soluble enzymes are now generally referred to as activator proteins. Thus the cofactor for sulfatide hydrolysis is usually designated as sulfatide activator.

This sulfatide activator was purified by Fischer and Jatzkewitz[3] and identified as a water-soluble glycoprotein with a molecular weight of approximately 22,000. From their kinetic and binding experiments and from the fact that this cofactor stimulated the degradation of lipid substrates but not of water-soluble substrates, these authors concluded that the cofactor binds to the lipid, extracts it from the membrane (or micelle), and solubilizes it as a protein/lipid complex, which they assumed to be the true substrate of the ensuing enzymatic reaction.[4,5]

Li and co-workers found that the enzymatic hydrolysis of gangliosides G_{M1} and G_{M2} by β-galactosidase and β-hexosaminidase A, respectively, and of globotriaosylceramide by α-galactosidase A also requires the presence of low molecular weight protein cofactors.[6,7] However, the protein cofactor they isolated was later shown to mediate the degradation of globotriaosylceramide and ganglioside G_{M1} but not that of ganglioside G_{M2} and has turned out to be identical with the sulfatide activator.[8,9]

In contrast, the hydrolysis of ganglioside G_{M2} and related glycosphingolipids (such as glycolipid G_{A2}) by β-hexosaminidase A depends on the presence of another activator protein, the G_{M2} activator,[10-12] which has properties similar to the sulfatide activator but differs in glycolipid and enzyme specificity. The two proteins do not cross-react immunochemically[13] and are encoded on different chromosomes (G_{M2} activator on chromosome 5,[14] sulfatide activator on chromosome 10[15]). The G_{M2} activa-

[3] G. Fischer and H. Jatzkewitz, *Hoppe-Seyler's Z. Physiol. Chem.* **356**, 605 (1975).

[4] G. Fischer and H. Jatzkewitz, *Biochim. Biophys. Acta* **481**, 561 (1977).

[5] G. Fischer and H. Jatzkewitz, *Biochim. Biophys. Acta* **528**, 69 (1978).

[6] Y.-T. Li, M. Y. Mazzotta, C.-C. Wan, R. Orth, and S.-C. Li, *J. Biol. Chem.* **248**, 7512 (1973).

[7] S.-C. Li and Y.-T. Li, *J. Biol. Chem.* **251**, 1159 (1976).

[8] K. Inui, M. Emmett, and D. A. Wenger, *Proc. Natl. Acad. Sci. U.S.A.* **80**, 3074 (1983).

[9] S.-C. Li, H. Kihara, S. Serizawa, Y.-T. Li, A. L. Fluharty, J. S. Mayes, and L. J. Shapiro, *J. Biol. Chem.* **260**, 1867 (1985).

[10] E. Conzelmann and K. Sandhoff, *Proc. Natl. Acad. Sci. U.S.A.* **75**, 3979 (1978).

[11] E. Conzelmann and K. Sandhoff, *Hoppe-Seyler's Z. Physiol. Chem.* **360**, 1837 (1979).

[12] E. Conzelmann, J. Burg, G. Stephan, and K. Sandhoff, *Eur. J. Biochem.* **123**, 455 (1982).

[13] S.-C. Li, T. Nakamura, A. Ogamo, and Y.-T. Li, *J. Biol. Chem.* **254**, 10592 (1979).

[14] J. Burg, E. Conzelmann, K. Sandhoff, E. Solomon, and D. M. Swallow, *Ann. Hum. Genet.* **49**, 41 (1985).

[15] K. Inui, F.-T. Kao, S. Fujibayashi, C. Jones, H. G. Morse, M. L. Law, and D. A. Wenger, *Hum. Genet.* **69**, 197 (1985).

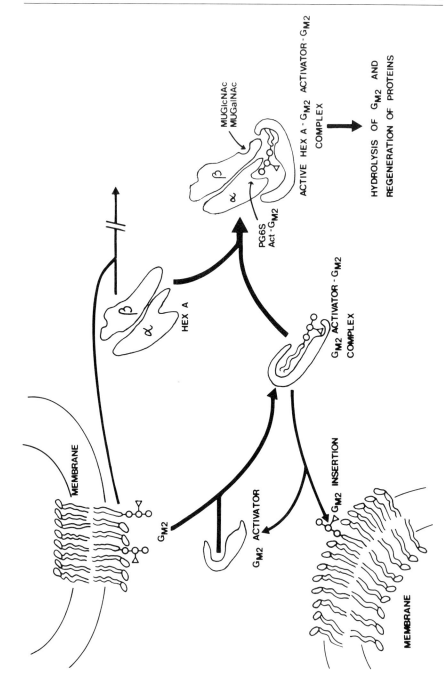

tor is also a glycoprotein, with a molecular weight of approximately 22,000.[12,16] Upon subcellular fractionation it codistributes with the lysosomal marker enzyme β-hexosaminidase[17] and, like most other lysosomal proteins studied so far, is synthesized as a form with a slightly higher molecular weight ($\sim$24,000[16]) which is probably the precursor of the mature form.

Detailed studies on the mechanism by which the G_{M2} activator promotes the enzymatic degradation of lipid substrates[11,12] showed that this activator binds to ganglioside G_{M2} and solubilizes it as a water-soluble complex with a molar ratio of one glycolipid molecule per protein. Other related glycolipids are also bound but much less strongly. As this complex formation is reversible, the G_{M2} activator acts, at least *in vitro,* also as a glycolipid transfer protein.[12] The mechanism by which ganglioside G_{M2} and related substrates are hydrolyzed in the lysosomes is schematically depicted in Fig. 1.[18,19]

The physiological significance of activator proteins is demonstrated by two fatal lipid storage diseases that are caused by deficiencies of activator proteins rather than of the degrading enzymes. Variant AB of infantile G_{M2}-gangliosidosis results from a defect of the G_{M2} activator protein[10,16,17,20,21] whereas a deficiency of the sulfatide/G_{M1} activator leads to an atypical variant of juvenile metachromatic leukodystrophy.[8,22,23]

---

[16] J. Burg, A. Banerjee, and K. Sandhoff, *Biol. Chem. Hoppe-Seyler* **366**, 887 (1985).

[17] A. Banerjee, J. Burg, E. Conzelmann, M. Carroll, and K. Sandhoff, *Hoppe-Seyler's Z. Physiol. Chem.* **365**, 347 (1984).

[18] K. Sandhoff and E. Conzelmann, *Neuropediatrics, Suppl.* **15**, 85 (1984).

[19] H.-J. Kytzia and K. Sandhoff, *J. Biol. Chem.* **260**, 7568 (1985).

[20] P. Hechtman, B. A. Gordon, and N. M. K. Ng Ying Kin, *Pediatr. Res.* **16**, 217 (1982).

[21] Y. Hirabayashi, Y.-T. Li, and S.-C. Li, *J. Neurochem.* **40**, 168 (1983).

[22] L. J. Shapiro, K. A. Aleck, M. M. Kaback, H. Itabashi, R. J. Desnick, N. Brand, R. L. Stevens, A. L. Fluharty, and H. Kihara, *Pediatr. Res.* **13**, 1179 (1979).

[23] R. L. Stevens, A. L. Fluharty, H. Kihara, M. M. Kaback, L. J. Shapiro, B. Marsh, K. Sandhoff, and G. Fischer, *Am. J. Hum. Genet.* **33**, 900 (1981).

FIG. 1. Model for the interaction between ganglioside G_{M2}, the G_{M2} activator, and β-hexosaminidase A.[18] Hexosaminidase A does not interact directly with the membrane-bound ganglioside. Instead, the activator protein binds to the ganglioside and extracts it from the membrane to form a water-soluble complex. This complex then recognizes a specific binding site on the α-subunit of β-hexosaminidase A[19] and binds in such a way that the terminal N-acetylgalactosamine moiety is correctly positioned for the enzymatic hydrolysis of the glycosidic bond by the active site of the α-subunit. This site can also cleave sulfated N-acetylglucosaminides such as the chromogenic p-nitrophenyl-6-sulfo-N-acetylglucosaminide (PG6S). In contrast, the active site of the β-subunit is specific for the neutral water-soluble substrates such as the 4-methylumbelliferyl-β-N-acetylglucosaminides and -galactosaminides (MUGlcNAc and MUGalNAc, respectively). After hydrolysis of the glycolipid, the resulting ganglioside G_{M2}–activator complex dissociates and the activator is set free for another round of catalysis.

The specific interaction between β-hexosaminidase A and the G_{M2} activator has also been taken into account in the development of assay systems for ganglioside G_{M2} hydrolysis by extracts of cultured skin fibroblasts.[24,25] With such assays the residual activities in cells from patients with different variants of G_{M2}-gangliosidosis could be determined very precisely.[25]

An entirely different type of protein cofactor for lysosomal glycolipid hydrolases had first been described by Ho and O'Brien[26] for acid β-glucosidase (glucosylceramidase). This protein accelerates the hydrolysis of glucosylceramide as well as of water-soluble artificial substrates by lysosomal β-glucosidase. It also seems to have an activating effect on acid sphingomyelinase and on galactosylceramide β-galactosidase.[27,28] In contrast to the activator proteins described above, this cofactor does not bind to the substrate[29] but activates the enzyme itself. However, the nature and the physiological relevance of this type of protein cofactor are still controversial: The preparations isolated from different sources by various groups differ widely in molecular weight,[26,30-32] specific activity,[30,33] and extent and necessity of glycosylation.[26,30,34] Recently it was also found that when the natural substrate, glucosylceramide, is incorporated into phospholipid bilayers, inclusion of a certain percentage of acidic phospholipids such as phosphatidic acid or phosphatidylserine in the phospholipid mixture leads to a high activity of glucosylceramidase and eliminates the need for any activating protein cofactor.[35]

A new type of glucosylceramidase cofactor has recently been described by Vaccaro et al.[36] This protein, which seems to be tightly associated with the enzyme, stimulates the hydrolysis of glucosylceramide but not of water-soluble artificial glucosides. Similarly, two small proteins

[24] A. Erzberger, E. Conzelmann, and K. Sandhoff, *Clin. Chim. Acta* **108,** 361 (1980).
[25] E. Conzelmann, H.-J. Kytzia, R. Navon, and K. Sandhoff, *Am. J. Hum. Genet.* **35,** 900 (1983).
[26] M. W. Ho and J. S. O'Brien, *Proc. Natl. Acad. Sci. U.S.A.* **68,** 2810 (1971).
[27] D. A. Wenger, M. Sattler, and S. Roth, *Biochim. Biophys. Acta* **712,** 639 (1982).
[28] H. Christomanou, *Hoppe-Seyler's Z. Physiol. Chem.* **361,** 1489 (1980).
[29] S. L. Berent and N. S. Radin, *Biochim. Biophys. Acta* **664,** 572 (1981).
[30] S. P. Peters, P. Coyle, C. J. Coffee, R. H. Glew, M. S. Kuhlenschmidt, L. Rosenfeld, and Y. C. Lee, *J. Biol. Chem.* **252,** 563 (1977).
[31] S. P. Peters, C. J. Coffee, R. H. Glew, R. E. Lee, D. A. Wenger, S.-C. Li, and Y.-T. Li, *Arch. Biochem. Biophys.* **183,** 290 (1977).
[32] D. A. Wenger and S. Roth, *Biochem. Int.* **5,** 705 (1982).
[33] S. S. Iyer, S. L. Berent, and N. S. Radin, *Biochim. Biophys. Acta* **748,** 1 (1983).
[34] S. L. Berent and N. S. Radin, *Arch. Biochem. Biophys.* **208,** 248 (1981).
[35] F. Sarmientos, G. Schwarzmann, and K. Sandhoff, *Eur. J. Biochem.,* in press.
[36] A. M. Vaccaro, M. Muscillo, E. Gallozzi, R. Salvioli, M. Tatti, and K. Suzuki, *Biochim. Biophys. Acta* **836,** 157 (1985).

that stimulate specifically lysosomal sphingomyelinase were recently observed by Christomanou and Kleinschmidt.[37] In both cases, however, it is yet too early to speculate on the role and significance of these proteins.

Assays for Activator Proteins

Activator proteins are generally quantified by measuring their capability to accelerate *in vitro* the hydrolysis of glycolipid substrates by the corresponding hydrolases. The relation between activator concentration and resulting reaction rate is not necessarily a linear one but may be more complex.

Those activator proteins that promote the hydrolysis of glycolipid substrates by water-soluble hydrolases, i.e., the G_{M2} activator and the sulfatide/G_{M1} activator, seem to act by similar mechanisms: they bind the glycolipid and extract it from the membrane to form a water-soluble complex which is the true substrate for the enzyme. Accordingly, the reaction rate depends on the concentration of the activator/lipid complex, with Michaelis–Menten type saturation kinetics.[11] Strictly speaking, such a saturation curve is at no point exactly linear, but for practical purposes such as quantification of activator proteins sufficient linearity can be assumed at activator concentrations below the K_m value of the enzyme for the respective activator/lipid complex as substrate.

Another factor that should be taken into account is the equilibrium of the binding reaction between activator and lipid. Under the conditions used for the quantification of activators, i.e., with pure glycolipid substrates at concentrations well above the K_D of the respective activator/lipid complex, the activator can be assumed to be saturated with the lipid, so that the activator concentration practically equals the concentration of the substrate of the reaction (the activator/lipid complex). It should, however, be borne in mind that this condition may not be fulfilled at low glycolipid concentrations or high activator concentrations. Also, the presence of other lipids such as phospholipids in the assay mixture may increase the experimental K_D by orders of magnitude since the mixed aggregates formed may be much more stable than the pure glycolipid micelles. (At a large excess of phospholipids as in the case of liposome-bound substrate, the K_D may depend linearly on the phospholipid concentration.) As a consequence the concentration of the activator/lipid complex may be far below the total activator concentration and the enzymatic reaction will accordingly be much slower than with pure lipid substrate.

[37] H. Christomanou and T. Kleinschmidt, *Biol. Chem. Hoppe-Seyler* **366**, 245 (1985).

According to the Michaelis–Menten equation the reaction rate depends linearly on the enzyme concentration. This is also true for the hydrolysis of glycolipids in the presence of activator proteins. If the absolute concentration of activator is to be estimated from the turnover rate observed, the amount of enzyme used in the assay must be known exactly. This may present problems if several isoenzymes exist of which only one is able to degrade the glycolipid substrate in the presence of the activator but of which all hydrolyze the artificial substrates employed for quantification of the enzyme. This is, for example, the case for α-galactosidase, β-hexosaminidase, and arylsulfatase. In these cases separation of the isoenzymes, e.g., by ion-exchange chromatography, is required.

It should also be noted that the Michaelis–Menten equation was derived under the assumption that the substrate concentration is always much higher than the enzyme concentration. This condition may not be met when the substrates involve macromolecules such as proteins. Thus, at low activator or high enzyme concentration some deviation from linearity may be encountered.

The mechanisms by which the cofactors for glucosylceramide and galactosylceramide catabolism stimulate the respective enzymes has not yet been elucidated. Quantification of these cofactors by their capacity to enhance enzymatic glycolipid hydrolysis *in vitro* can therefore not be based on theoretical considerations but has to rely entirely on practical experiences. These will be discussed with the respective assay systems.

As indicated above, it is in most cases not necessary to use highly purified enzymes. Usually, partially purified preparations will do, provided that they are sufficiently concentrated, essentially free of endogenous activator, and contain only the one isoenzyme that does interact with the activator/lipid complex (or the isoenzyme composition is known and the results can be corrected for it). For the water-soluble enzymes, affinity chromatography on concanavalin A–Sepharose, followed by ion-exchange chromatography and, in some cases, gel filtration, yields usually satisfactory preparations. A major exception is arylsulfatase A which has to be quite pure since the binding of sulfatide to the sulfatide activator is inhibited by many contaminating proteins.[3,38] The membrane-associated enzymes glucosylceramide β-glucosidase and galactosylceramide β-galactosidase require solubilization with detergents prior to any chromatographic separation.

Activator proteins can, of course, also be quantified with immunochemical techniques if suitable antisera are available. Enzyme-linked immunosorbent assays (ELISA) have been described for the sulfatide/G_{M1}

---

[38] M. Lee-Vaupel and E. Conzelmann, submitted for publication.

activator[39] and for the G_{M2} activator.[17] For the G_{M2} activator, ELISA was more than 10 times as sensitive as enzymatic assays and permitted the precise determination of G_{M2} activator in very dilute samples such as subcellular fractions of cultured skin fibroblasts.[17]

Sulfatide/G_{MI} Activator

Although in principle any one of the reactions promoted by the sulfatide activator can be used to assay for this cofactor, the degradation of ganglioside G_{M1} by β-galactosidase is most conveniently used since this reaction proceeds fast enough to permit sensitive measurements of activator content and is the least sensitive to disturbances by impurities. Globotriaosylceramide is much more tedious to prepare in pure form than ganglioside G_{M1}. Degradation of sulfatide by arylsulfatase A is also comparatively slow and may be strongly inhibited by contaminating proteins and other compounds so that the assay requires the use of highly purified enzyme and activator.[3,38] Reliable quantitation of the activator during purification is therefore not possible with this assay.

Assay with Ganglioside G_{MI}/β-Galactosidase

SUBSTRATE. Ganglioside G_{M1} can be purified from brain lipid extracts by any one of several published methods.[40-44] The yield of ganglioside G_{M1} can be increased severalfold by converting higher gangliosides into ganglioside G_{M1} with sialidase as follows: The crude ganglioside preparation (e.g., the Folch upper phase[45]) is taken to dryness and dissolved in 50 mM acetate buffer, pH 5.5, 1 mM CaCl$_2$, and incubated overnight with sialidase from *Clostridium perfringens* (10 U/liter).[46]

For enzymatic studies, the substrate is most conveniently labeled in the terminal galactose moiety by modifications[47,48] of the galactose oxidase/boro[³H]hydride method of Radin.[49] In this procedure, the alkalinity during the reduction step with NaB³H$_4$ appears to be very critical: In our

[39] A. Gardas, Y.-T. Li, and S.-C. Li, *Glycoconjugate J*. **1**, 37 (1984).
[40] L. Svennerholm, *Methods Carbohydr. Chem*. **6**, 464 (1972).
[41] G. Tettamanti, F. Bonali, S. Marchesini, and V. Zambotti, *Biochim. Biophys Acta* **296**, 160 (1973).
[42] T. Momoi, S. Ando, and Y. Nagai, *Biochim. Biophys. Acta* **441**, 488 (1976).
[43] S. Kundu, this series, Vol. 72, p. 174.
[44] R. W. Ledeen and R. K. Yu, this series, Vol. 83, p. 139.
[45] J. Folch, M. Lees and G. H. Sloane Stanley, *J. Biol. Chem*. **226**, 497 (1957).
[46] G. Schwarzmann, personal communication.
[47] H. R. Sloan, this series, Vol. 28, p. 868.
[48] K. Suzuki, *in* "Practical Enzymology of the Sphingolipidoses" (R. H. Glew and S. P. Peters, eds.), p. 101. Alan R. Liss, Inc., New York, 1977.
[49] N. S. Radin, this series, Vol. 28, p. 300.

experience (which has been confirmed by others[50]), at NaOH concentrations higher than 10 mM some of the ganglioside G_{M1} is converted to a highly labeled product with the chromatographic mobility of ganglioside G_{M2}. (The precise nature of this product has not yet been investigated.) Lowering the NaOH concentration to 1 mM largely suppresses this side reaction.

The labeled product is finally purified by column chromatography according to Svennerholm.[40] If only small amounts of labeled ganglioside are needed, purification by preparative thin-layer chromatography is more convenient: The ganglioside solution in organic solvent is applied to a thin-layer plate (e.g., Merck Kieselgel G 60, 0.25 mm thickness) in a 16-cm wide streak. Analytical plates are preferable over preparative ones since they give a much better resolution. Up to 5 mg of lipid can be applied to one plate (20 × 20 cm). The plate is developed in chloroform/methanol/15 mM aqueous CaCl$_2$ (55/45/10 by volume). Radioactive bands are located with a radioscanner, the plate is lightly sprayed with water, and the area containing the ^{3}H-labeled ganglioside G_{M1} is scraped off. (If water is sprayed until the plate starts to become translucent, the lipid may be visible as a white band on a darker background.) The scrapings are suspended in 5 ml chloroform/methanol/water (55/45/10 by volume), packed into a suitable column (e.g., a large Pasteur pipette with glass wool plug), and eluted with 2 × 5 ml of the same solvent. The solvent is then evaporated under a stream of N$_2$, the dry residue is taken up in distilled water (ca. 1 ml/mg lipid), dialyzed against distilled water, and lyophilized. To determine the specific radioactivity, the sialic acid content of the sample can be measured, e.g., by a modification[51] of the resorcinol method.[52] Simply weighing the product is not recommended as it may contain some silicic acid.

β-GALACTOSIDASE. Lysosomal β-galactosidase can be purified from human liver or placenta by conventional methods[47,53,54] or by affinity chromatography on immobilized p-aminophenyl- or 6-aminohexylthio-β-D-galactoside.[55,56] The most rapid and convenient procedure seems to be the two-step method of Miller et al.[55] which is reported to give a more than 20,000-fold purification with 41% yield. The affinity gel is commercially available.

[50] C. H. Wynn, personal communication.
[51] T. Miettinen and I.-T. Takki-Luukkainen, Acta Chem. Scand. 13, 856 (1959).
[52] L. Svennerholm, Biochim. Biophys. Acta 24, 604 (1957).
[53] M. Meisler, this series, Vol. 28, p. 820.
[54] T. Miyatake and K. Suzuki, J. Biol. Chem. 250, 585 (1975).
[55] A. L. Miller, R. G. Frost, and J. S. O'Brien, Biochem. J. 165, 591 (1977).
[56] J.-T. Lo, K. Mukerji, Y. C. Awasthi, E. Hanada, K. Suzuki, and S. K. Srivastava, J. Biol. Chem. 254, 6710 (1979).

The enzyme activity is usually measured with the fluorogenic substrate 4-methylumbelliferyl-β-D-galactoside, e.g., 1 mM in 50 mM citrate/ 0.1 M phosphate buffer, pH 4.0, with 0.1 M NaCl.[48] (The high ionic strength is needed to stabilize the enzyme.) One unit of β-galactosidase is generally defined as the amount of enzyme that splits 1 μmol of this substrate per minute under the above conditions, at 37°.

ASSAY. The standard assay for determination of the activator consists of 10 μl of a 100 μM aqueous dispersion of ganglioside G$_{M1}$, ^{3}H-labeled in the terminal galactose moiety (ca. 100,000 cpm), 5 mU β-galactosidase, and up to 25 μl of the suitably diluted activator sample, in a total volume of 50 μl 50 mM citrate buffer, pH 4.5. The mixtures are incubated for 1 hr at 37°, then the vials are transferred to an ice-bath, and 1 ml of an ice-cold 1 mM galactose solution is added. The assays are then loaded onto small (0.5–1 ml) columns of DEAE–cellulose (in Pasteur pipettes) that have been washed with distilled water. The columns are eluted with 2 × 1 ml of 1 mM aqueous galactose solution, the combined effluents are collected in scintillation vials, and, after addition of scintillation fluid (10–15 ml, depending on the type used), their radioactivity is measured in a liquid scintillation counter. Blanks are run with water instead of activator solution and are subtracted.

The K_m value for the interaction between β-galactosidase and the activator/ganglioside G$_{M1}$ complex is approximately 8 μM.[57] This means that the reaction depends almost linearly on the activator concentration up to about 5 μM (~5 μg of activator/assay). The K_D of the activator/lipid complex is approximately 3 μM[57] so that the activator can be assumed to be saturated with the lipid under the above conditions.

Assay with Sulfatide/Arylsulfatase A

SUBSTRATE. Sulfatides can be purchased commercially or can be prepared from crude brain lipid extracts by one of the conventional procedures.[58–62]

For the radioactive labeling of sulfatides, three different strategies have been described. (1) Catalytic reduction of the sphingoid base by a modification[63] of the method of Schwarzmann.[64] (2) Preparation of sulfo-

[57] A. Vogel, Diplom-Arbeit, Universität Bonn (1985).
[58] E. Martensson, *Biochim. Biophys. Acta* **116**, 521 (1966).
[59] G. Rouser, G. Kritchevsky, A. Yamamoto, G. Simon, C. Galli, and A. J. Baumann, this series, Vol. 14, p. 272.
[60] P. O. Kwiterovich, Jr., H. R. Sloan, and D. S. Fredrickson, *J. Lipid Res.* **11**, 322 (1970).
[61] T. Saito and S.-I. Hakomori, *J. Lipid Res.* **12**, 257 (1971).
[62] R. K. Yu and R. W. Ledeen, *J. Lipid Res.* **13**, 680 (1972).
[63] S. S. Raghavan, A. Gajewski, and E. H. Kolodny, *J. Neurochem.* **36**, 724 (1981).
[64] G. Schwarzmann, *Biochim. Biophys. Acta* **529**, 106 (1978).

galactosylsphingosine ("lysosulfatide") by alkaline hydrolysis of sulfatide and reacylation of the amino group with a radioactive fatty acid.[65] (3) Biosynthetic labeling of sulfatide by injecting [^{35}S]sulfate into the brain of a suitable laboratory animal and isolation of the sulfatide formed after a few days.[1,66]

The product of the latter procedure, [^{35}S]sulfatide, allows for a simple quantitation of the enzymatic reaction, by extracting the liberated sulfate into the aqueous phase of a two-phase system. However, the comparatively short half-life of ^{35}S (87 days) limits the usefulness of this method. Of the other two methods, which both employ long-lived isotopes (^{3}H or ^{14}C), the first one is less tedious and gives better yields.

ARYLSULFATASE A. The arylsulfatase A preparations used for assays with sulfatide and the natural activator protein have to be rather pure since contaminating proteins may interfere with this reaction in vitro. A variety of methods have been published for purification of the enzyme from human liver,[67-70] kidney,[71] placenta,[72] and urine.[73-75] In our laboratory the procedure of Stinshoff[71] is usually followed.

Although a fluorogenic substrate (4-methylumbelliferyl sulfate) exists and is used by some laboratories, the chromogenic 4-nitrocatechol sulfate is generally preferred. The most widely used assay system is that of Baum et al.[76] which was initially developed for the determination of arylsulfatase A in the presence of the B isoenzyme. The unusual time course of the hydrolysis of 4-nitrocatechol sulfate by arylsulfatase A[71,77,78] may present a problem for standardization of the enzyme preparation and should be taken into account, e.g., by using short incubation times (15–30 min). The

[65] G. Dubois, B. Zalc, F. LeSaux, and N. Baumann, Anal. Biochem. 102, 313 (1980).

[66] A. L. Fluharty, M. L. Davis, H. Kihara, and G. Kritchevsky, Lipids 9, 865 (1974).

[67] J. Collins, W. Yamada, W. Worth, and J. Austin, in "Current Trends in Sphingolipidoses and Allied Disorders" (B. W. Volk and L. Schneck, eds.), p. 225. Plenum, New York, 1975.

[68] R. K. Draper, G. M. Fiskum, and J. Edmond, Arch. Biochem. Biophys. 177, 525 (1976).

[69] G. T. James and J. H. Austin, Clin. Chim. Acta 98, 103 (1979).

[70] T. A. Sarafian, A. L. Fluharty, H. Kihara, G. Helfand, and J. Edmond, J. Appl. Biochem. 4, 126 (1982).

[71] K. Stinshoff, Biochim. Biophys. Acta 276, 475 (1972).

[72] J. Gniot-Szulzycka and M. Komoszynski, Acta Biochim. Pol. 17, 185 (1970).

[73] J. L. Breslow and H. R. Sloan, Biochem. Biophys. Res. Commun. 46, 919 (1972).

[74] R. L. Stevens, A. L. Fluharty, M. H. Skokut, and H. Kihara, J. Biol. Chem. 250, 2495 (1975).

[75] J. A. F. M. Luijten, M. C. M. Van der Heijden, G. Rijksen, and G. E. J. Staal, J. Mol. Med. 3, 213 (1978).

[76] H. Baum, K. S. Dodgson, and B. Spencer, Clin. Chim. Acta 4, 453 (1959).

[77] A. B. Roy, Biochem. J. 53, 12 (1953).

[78] H. Baum, K. S. Dodgson, and B. Spencer, Biochem. J. 69, 567 (1958).

enzyme unit is usually defined as the amount of enzyme that hydrolyzes 1 μmol of this substrate per minute.

ASSAY. The standard assay mixture consists of 5 nmol sulfatide, either tritium-labeled in the sphingoid base or ^{14}C-labeled in the fatty acid, 5 mU arylsulfatase A, and an aliquot of the activator sample to be assayed in a total volume of 100 μl of 0.1 M sodium acetate buffer, pH 5.0, with 0.1 M NaCl. After incubation for 3 hr at 37°, the reaction is stopped by adding 400 μl of chloroform/methanol (2/1 by volume). After thorough mixing, phases are separated by centrifugation. The (aqueous) upper phase is discarded, the organic phase is washed with 200 μl theoretical upper phase (chloroform/methanol/water, 3/48/47 by volume) and then loaded onto a 0.5-ml column of DEAE–cellulose (in a Pasteur pipette) equilibrated with methanol. Unbound lipids are eluted with 2 ml methanol, collected in a scintillation vial, and their radioactivity quantified by liquid scintillation counting. Blanks run with water instead of activator solution are subtracted.

As noted above, the reaction is very sensitive to the presence of contaminating proteins and other compounds. Also, arylsulfatase A is competitively inhibited by many anions such as sulfate and phosphate ($K_1 = 0.6$ mM^{79}). Precise determination of activator content with this method is therefore very difficult.

Assay with Globotriaosylceramide/α-Galactosidase A

SUBSTRATE. Globotriaosylceramide can be isolated from the neutral lipid fraction of many tissues, including kidney, spleen, and liver. Porcine intestine[80,81] or erythrocytes[82] have been recommended as convenient sources. A particularly rich source are tissues of patients with Fabry's disease,[83] especially liver and kidney. This disorder is, however, quite rare, and autopsy tissue is usually difficult to obtain.

Extraction and purification of globotriaosylceramide follows the usual methods described for neutral glycosphingolipids.[80–82,84]

Radioactive labeling of the terminal galactose moiety is usually achieved with the galactose oxidase/boro[^{3}H]hydride method of Radin,[49] e.g., as described by Suzuki and Suzuki[85] or Dean and Sweeley.[81] Experi-

[79] J. Gniot-Szulzycka, *Acta Biochim. Pol.* **21**, 247 (1974).
[80] C. Suzuki, A. Makita, and Z. Yoshizawa, *Arch. Biochem. Biophys.* **127**, 140 (1968).
[81] K. J. Dean and C. C. Sweeley, *in* "Practical Enzymology of the Sphingolipidoses" (R. H. Glew and S. P. Peters, eds.), p. 173. Alan R. Liss, Inc., New York, 1977.
[82] T. Taketomi and N. Kawamura, *J. Biochem.* (*Tokyo*) **72**, 791 (1972).
[83] C. C. Sweeley and B. Klionsky, *J. Biol. Chem.* **238**, 3148 (1963).
[84] W. J. Esselman, R. A. Laine, and C. C. Sweeley, this series, Vol. 28, p. 140.
[85] Y. Suzuki and K. Suzuki, *J. Lipid Res.* **13**, 687 (1972).

ence in our laboratory[86] indicates that, like in the case of ganglioside G_{M1}, the NaOH concentration should be carefully controlled in the reduction step since reduction with NaB^3H_4 at high pH leads to the formation of a highly labeled side product that migrates like lactosylceramide in thin-layer chromatography.

α-GALACTOSIDASE A. The enzyme can be purified from human liver or placenta with conventional techniques[87–89] or affinity chromatography.[90] The enzyme is routinely assayed with the fluorogenic substrate 4-methyl-umbelliferyl-α-D-galactoside, most assays being modifications of the method of Desnick et al.,[91] in which the substrate is used at a concentration of 2.5 mM in 0.1 M citrate/0.2 M phosphate buffer, pH 4.6. Enzyme units (μmol/min) are usually also defined with this substrate.

ASSAY. Globotriaosylceramide, ^{3}H-labeled in the terminal galactose residue, is dispersed with sonication in 230 mM sodium acetate buffer, pH 4.2, at a concentration of 0.2 mM. For the standard assay for activator protein, 10 μl of this dispersion (equal to 2 nmol of substrate, ca. 22,000 cpm) are incubated with 1 mU α-galactosidase A and a suitable dilution of the sample to be assayed in a total volume of 50 μl for 1–3 hr at 37°. (Assays with purified activator should contain some 5 μg of bovine serum albumin, to prevent nonspecific adsorption of protein on surfaces.) The reaction is terminated with 0.6 ml chloroform/methanol (2/1 by volume) and 0.1 ml of 1 mM galactose/1 M NaCl in water. After thorough mixing and centrifugation for 5 min in a benchtop centrifuge (Eppendorf), the lower phases are removed and discarded, and the remaining upper phases are washed with 0.5 ml of theoretical lower phase (chloroform/methanol/water, 86/14/1 by volume) and then transferred to scintillation vials. Radioactivity is quantified by liquid scintillation counting, after addition of 5 ml scintillation fluid. Blanks which were run with water instead of activator solution are subtracted.

The K_m value of α-galactosidase A for the activator/globotriaosyl-ceramide complex as substrate is approximately 3 μM.[92] Since under the assay conditions the activator can be assumed to be saturated with the lipid (K_D ca. 30 μM[92]), the reaction rate increases linearly with activator concentration up to 2–3 μM (ca. 2–3 μg/assay). Attention must, however,

[86] B. Röder and W. Fürst, personal communication.
[87] E. Beutler and W. Kuhl, J. Biol. Chem. 247, 7195 (1972).
[88] W. G. Johnson and R. O. Brady, this series, Vol. 28, p. 849.
[89] K. J. Dean and C. C. Sweeley, J. Biol. Chem. 254, 9994 (1979).
[90] D. F. Bishop and R. J. Desnick, J. Biol. Chem. 256, 1307 (1981).
[91] R. J. Desnick, K. Y. Allen, S. J. Desnick, M. K. Raman, R. W. Bernlohr, and W. Krivit, J. Lab. Clin. Med. 81, 157 (1973).
[92] W. Fürst, Diplom-Arbeit, Universität Bonn (1984).

be paid to the salt concentration in the samples as the reaction rate is almost inversely proportional to the ionic strength.

G_{M2} Activator

Measurement of the G_{M2} activator is based on its ability to accelerate the hydrolysis of ganglioside G_{M2} by β-hexosaminidase A.

Substrate. Ganglioside G_{M2} occurs only in traces in most mammalian tissues, so purification from normal sources is extremely tedious. The ganglioside is usually isolated by the method of Svennerholm[40] from brain tissue of patients who died from G_{M2}-gangliosidosis (Tay–Sachs disease or Sandhoff disease). A more readily available source has recently been indicated by Li and co-workers,[93] who found that ganglioside G_{M2} constitutes the main ganglioside of striped mullet (*Mugil cephalus*) roe. The ceramide portion of the fish ganglioside, however, differs somewhat from that of the mammalian ganglioside and care should be exercised when using such preparations for kinetic studies.

The procedure for radioactive labeling of the terminal N-acetylgalactosamine moiety with galactose oxidase/NaB^{3}H$_4$[85] is essentially the same as for ganglioside G_{M1} (see above). However, the specific activity attainable is considerably lower than that of ganglioside G_{M1} or other glycolipids, due to an only limited degree of oxidation of the *N*-acetylgalactosamine residue. Since nonspecific labeling of the other parts of the molecule may therefore contribute significantly to the specific activity of the labeled substrate, it is recommended to reduce the starting material with unlabeled NaBH$_4$ prior to the enzymatic oxidation.[85] The final purification of the amounts required for enzymatic assays and determination of specific activity can be done with preparative thin-layer chromatography as described for ganglioside G_{M1} in the preceding section.

β-Hexosaminidase A. The enzyme can be purified from normal human liver,[94] placenta,[95,96] kidney,[97,98] brain,[99] or urine.[100] While the enzyme

[93] Y.-T. Li, Y. Hirabayashi, R. De Gasperi, R. K. Yu, T. Ariga, T. A. W. Koerner, and S.-C. Li, *J. Biol. Chem.* **259**, 8980 (1984).

[94] K. Sandhoff, E. Conzelmann, and H. Nehrkorn, *Hoppe-Seyler's Z. Physiol. Chem.* **358**, 779 (1977).

[95] J. E. S. Lee and A. Yoshida, *Biochem. J.* **159**, 535 (1976).

[96] B. Geiger and R. Arnon, this series, Vol. 50, p. 547.

[97] J. E. Wiktorowicz, Y. C. Awasthi, A. Kurosky, and S. K. Srivastava, *Biochem. J.* **165**, 49 (1977).

[98] S. K. Srivastava, *in* "Practical Enzymology of the Sphingolipiodoses" (R. H. Glew and S. P. Peters, eds.), p. 217. Alan R. Liss, Inc., New York, 1977.

[99] R. A. Aruna and D. Basu, *J. Neurochem.* **27**, 337 (1976).

[100] D. V. Marinkovic and J. N. Marinkovic, *Biochem. Med.* **20**, 422 (1978).

preparation used does not have to be pure, it is essential to remove β-hexosaminidase B (or at least to know the proportions of the two enzymes) since only the A isoenzyme interacts with the activator/lipid complex. A suitable preparation can in our experience be obtained from liver or placenta extract (20% in water) with chromatography on (commercially available) concanavalin A–Sepharose 4B as described in Sandhoff *et al.*,[94] followed by ion-exchange chromatography on DEAE–cellulose in 10 m*M* phosphate buffer, pH 6.0 (column volume ca. 1 ml per 5 mg protein to be applied). Hexosaminidase B is not retained on the column and is washed out with 2 column volumes of buffer, then hexosaminidase A is eluted with 0.5 *M* NaCl in the same buffer. The preparation must be dialyzed against distilled water before use because the enzymatic reaction with ganglioside G_{M2} and activator is strongly inhibited by an ionic strength higher than 20 m*M*.[12]

During purification and for standardization, the enzyme is most conveniently assayed with the fluorogenic substrate 4-methylumbelliferyl-β-D-*N*-acetylglucosaminide[94,101–103] (1 m*M* in 50 m*M* citrate buffer, pH 4.5). One enzyme unit is usually defined as the amount of enzyme that splits 1 μmol of this substrate per minute under standard conditions.

Assay. For quantification of the G_{M2} activator a substrate/buffer solution is prepared by dissolving ³H-labeled ganglioside G_{M2} at a concentration of 1 m*M* in 80 m*M* citrate buffer, pH 4.0, containing 0.5 mg bovine serum albumin per ml. Five microliters of this solution are incubated with 100 mU β-hexosaminidase A and up to 25 μl of the suitably diluted activator sample in a total volume of 40 μl, for 1–4 hr at 37°. Samples are then transferred to an ice-bath and loaded onto 1-ml columns of DEAE–cellulose (in Pasteur pipettes) that have been washed with distilled water. The columns are eluted with 2 × 1 ml of a 1 m*M* aqueous *N*-acetylgalactosamine solution. The combined effluents are collected in scintillation vials, and, after addition of 10 ml scintillation fluid, their radioactivity is measured. Blanks run with water instead of activator solution are subtracted.

The reaction rate, which is very low without activator, increases practically linearly up to 1 μg of G_{M2} activator per assay corresponding to a degradation rate of about 50 nmol/(hr × U hex A). The absolute reaction rate decreases strongly with increasing ionic strength.[12] Samples should therefore be dialyzed against water when the activator content is to be determined precisely.

[101] D. H. Leaback and P. G. Walker, *Biochem. J.* **78**, 151 (1961).
[102] N. Dance, R. G. Price, D. Robinson, and J. L. Stirling, *Clin. Chim. Acta* **24**, 189 (1969).
[103] S. K. Srivastava, A. Yoshida, Y. C. Awasthi, and E. Beutler, *J. Biol. Chem.* **249**, 2043 (1974).

Cofactors for Glucosylceramide β-Glucosidase and Galactosylceramide β-Galactosidase

Since the initial report by Ho and O'Brien[26] that lysosomal β-glucosidase activity toward water-soluble substrates as well as glucosylceramide[104] is greatly stimulated by small nonenzymatic protein cofactors that are particularly abundant in Gaucher spleen, such factors have been purified by several groups from human spleen[26,30,31,33,105,106] and brain[32] and from bovine spleen.[29,34] Some of these preparations were also tested with other enzymes and were found to stimulate also the hydrolysis of galactosylceramide by the corresponding β-galactosidase.[27] The nature and the physiological significance of these cofactors and the mechanism by which they stimulate the enzymes are still unknown.

Assay with β-Glucosidase

SUBSTRATES. In normal tissues, glucosylceramide is present in only low concentrations. In tissues of patients with Gaucher's disease (particularly spleen and liver) glucosylceramide is stored in enormous amounts, due to the deficiency of the catabolizing enzyme, and Gaucher spleen is the most frequent source of the glucosylceramide used in enzymatic and other studies.

Extraction and purification can be done as described elsewhere.[107,108] Chemical synthesis, which is also feasible,[109,110] has the advantage of yielding homogeneous products with respect to fatty acid and sphingoid base but is, of course, more laborious.

Although radioactive labeling can basically be achieved by catalytic reduction of the sphingosine double bond[64] or by amidation of a radioactive fatty acid with glucosylsphingosine prepared from glucosylceramide by alkaline hydrolysis,[108,111] the most useful substrate for enzymatic assays is the one labeled in the glucose moiety. This can be obtained either by chemical synthesis with radioactive glucose[109,110] or by labeling intact glucosylceramide by the method of McMaster and Radin.[112]

[104] M. W. Ho, J. S. O'Brien, N. S. Radin, and J. S. Erickson, *Biochem. J.* **131**, 173 (1973).

[105] Y.-B. Chiao, J. P. Chambers, R. H. Glew, R. E. Lee, and D. A. Wenger, *Arch. Biochem. Biophys.* **186**, 42 (1978).

[106] N. S. Radin, in "Molecular Basis of Lysosomal Storage Disorders" (R. O. Brady and J. A. Barranger, eds.), p. 93. Academic Press, New York, 1985.

[107] N. S. Radin, *J. Lipid Res.* **17**, 290 (1976).

[108] S. P. Peters, R. H. Glew, and R. E. Lee, in "Practical Enzymology of the Sphingolipidoses" (R. H. Glew and S. P. Peters, eds.), p. 71. Alan R. Liss, Inc., New York, 1977.

[109] R. O. Brady, J. N. Kanfer, and D. Shapiro, *J. Biol. Chem.* **240**, 39 (1965).

[110] P. Stoffyn, A. Stoffyn, and G. Hauser, *J. Lipid Res.* **12**, 318 (1971).

[111] J. S. Erickson and N. S. Radin, *J. Lipid Res.* **14**, 133 (1973).

[112] M. C. McMaster, Jr. and N. S. Radin, *J. Labeled Compd. Radiopharm.* **13**, 353 (1977).

ACID β-GLUCOSIDASE. Glucosylceramide β-glucosidase (acid β-glucosidase) can, like other lysosomal enzymes, be isolated from almost any human tissue, but placenta has proved to be a good source which is readily available. An initial problem in the preparation of this enzyme, its tight association with the lysosomal membrane, has been overcome by the discovery that the enzyme can be obtained in water-soluble form after extraction of crude preparations with n-butanol or other organic solvents.[113,114] High purity with good yield can be achieved with conventional techniques[115–117] as well as with affinity chromatography.[118,119] However, synthesis of the required affinity adsorbents is laborious and offers advantages only when the isolation procedure has to be repeated frequently.

The standard substrate for quantitation of acid β-glucosidase is the fluorogenic 4-methylumbelliferyl-β-D-glucoside. Standardization of enzyme preparations is, however, still problematic since acid β-glucosidase is stimulated by acidic phospholipids[30,114,120,121] or detergents[114,122,123] so that the absolute activity depends strongly on the amount and nature of such additions.

ASSAY. Since acid β-glucosidase accepts water-soluble as well as lipid substrates, its activity and hence its activation by cofactors can be determined with either kind of substrate. Due to the rapidity and convenience of assays with fluorogenic substrates, most workers prefer to use the commercially available 4-methylumbelliferyl-β-D-glucoside. In some cases it is, however, desirable or even necessary to confirm the results with the natural substrate, glucosylceramide. Both types of assays will therefore be described here.

The definition of "optimal" conditions for the quantification of a stimulating cofactor is a central problem in both cases. As outlined above, β-glucosidase activity depends strongly on the nature and amount of addi-

[113] E. Blonder, C. Klibansky, and A. De Vries, *Biochim. Biophys. Acta* **431**, 45 (1976).
[114] G. L. Dale, D. Villacorte, and E. Beutler, *Biochem. Biophys. Res. Commun.* **71**, 1048 (1976).
[115] G. L. Dale and E. Beutler, *Proc. Natl. Acad. Sci. U.S.A.* **73**, 4672 (1976).
[116] F. S. Furbish, H. E. Blair, J. Shiloach, P. G. Pentchev, and R. O. Brady, this series, Vol. 50, p. 529.
[117] G. L. Dale, E. Beutler, P. Fournier, P. Blanc, and J. Liautaud, *in* "Enzyme Therapy in Genetic Diseases: 2" (R. J. Desnick, ed.), p. 33. Alan R. Liss, Inc., New York, 1980.
[118] P. M. Strasberg, J. A. Lowden, and D. Mahuran, *Can. J. Biochem.* **60**, 1025 (1982).
[119] G. A. Grabowski and A. Dagan, *Anal. Biochem.* **141**, 267 (1984).
[120] M. W. Ho and D. Light, *Biochem. J.* **136**, 821 (1973).
[121] F. M. Choy and R. G. Davidson, *Pediatr. Res.* **14**, 54 (1980).
[122] M. W. Ho, *Biochem. J.* **136**, 721 (1973).
[123] S. P. Peters, P. Coyle, and R. H. Glew, *Arch. Biochem. Biophys.* **175**, 569 (1976).

tions to the assay (acidic phospholipids, anionic detergents) so that it is difficult to define the activator activity in absolute terms such as increase in the amount of substrate hydrolyzed per enzyme unit. Most workers express the activation therefore as multiples of basal activity ("-fold activation"). Generally, conditions must be sought that lead to a maximal stimulation of a small basal activity by the cofactor.

It has been shown that the interaction between acid β-glucosidase and activating protein cofactors depends on the presence of acidic lipids,[120,124] so a certain amount of such lipids has to be added to the assay system. (Frequently the partially purified enzyme preparations used contain some lipids, including acidic ones.) On the other hand, higher concentrations of acidic lipids may stimulate the enzyme to almost maximal activity[29,124] and may render protein cofactors obsolete.

Assay with 4-methylumbelliferyl-β-D-glucoside. The activator sample to be assayed is incubated with 2 mM substrate in 0.2 M acetate buffer, pH 5.5,[30,31] or 50 mM acetate, pH 4.5,[29,32,125] with an appropriate amount of partially purified enzyme and, if necessary, with 5 μg phosphatidylserine, in a total volume of 100 μl. After incubation (1 hr, 37°) the reaction is terminated by the addition of 0.5 ml 0.2 M glycine/0.2 M sodium carbonate, and the liberated 4-methylumbelliferone is determined fluorimetrically (excitation 365 nm, emission 440 nm).

Assay with Glucosylceramide. Assay conditions are essentially the same as with the fluorogenic substrate except that [glucose-6-^{3}H]glucosylceramide is used as substrate (30–40 nmol/assay[32,34]). Specific radioactivity should be at least 1 mCi/mmol. After incubation, 5 volumes of chloroform/methanol (2/1 by volume) and 0.5 volume of an aqueous solution of 0.2 M glucose/0.2 M KCl are added. The mixture is thoroughly vortexed and then centrifuged, and the liberated [^{3}H]glucose is measured in an aliquot of the upper phase. [The sensitivity of the method can be improved by washing the upper phase with 2 volumes of theoretical lower phase (chloroform/methanol/water, 64/9/1 by volume).]

Assay with Galactosylceramide β-Galactosidase

SUBSTRATE. Galactosylceramide is a major lipid of myelin[126] and can be purified from the neutral fractions of brain lipid extracts.[49] The galactose moiety can easily be tritium-labeled by the galactose oxidase/sodium borohydride method, e.g., as described in a previous volume of this series.[49]

[124] A. Basu, R. H. Glew, L. B. Daniels, and L. S. Clark, *J. Biol. Chem.* **259**, 1714 (1984).
[125] N. S. Radin and S. L. Berent, this series, Vol. 83, p. 596.
[126] W. T. Norton and S. E. Poduslo, *J. Neurochem.* **21**, 759 (1973).

ENZYME PREPARATION. The stimulating effect of protein cofactors on galactosylceramide hydrolysis was studied by Wenger and co-workers, with crude tissue homogenates or extracts as enzyme sources.[27] Purification of galactosylceramide β-galactosidase, which seems to be somewhat difficult,[48] may therefore not be necessary.

If a more purified preparation is desired, the enzyme, like acid β-glucosidase, can be converted into a water-soluble form by extraction with organic solvents.[127,128] However, the purification protocols published so far[129,130] rely mainly on extraction with detergents.

ASSAY. The assay method described is that used by Wenger et al.[27] [³H]Galactosylceramide (10 nmol; specific activity ca. 1 mCi/mmol) and 25 μg pure phosphatidylserine are dispersed in 0.1 ml 0.1 M citrate/0.2 M phosphate buffer, pH 4.6, by sonication. Enzyme, activator sample, and distilled water are added to a final volume of 0.2 ml. After incubation at 37° for 1 hr, 1 ml of chloroform/methanol (2/1 by volume) is added, and the liberated [³H]galactose, which partitions into the upper phase, is determined by liquid scintillation counting.

The reaction rate was reported to increase linearly with the amount of cofactor added up to 5- to 10-fold stimulation.[27]

Purification of Activator Proteins

Sulfatide/G_{M1} Activator

The sulfatide/G_{M1} activator has been purified from a number of human tissues, including liver, kidney, and brain. Another convenient source is urine, which seems to have a comparatively high content of such small proteins. If it is not necessary to work with the human protein then porcine kidney provides a rich source of this activator.[131]

Purification of the human sulfatide activator to a preparation that yielded one band in nondenaturing electrophoresis was reported by Fischer and Jatzkewitz.[31] Li and Li purified the G_{M1} activator,[7] which later turned out to be the same protein. Both procedures exploit the unusual thermal stability of this protein which can be heated to 95° for several minutes without loss of activity. A gentler procedure that allows the simultaneous preparation of both the sulfatide/G_{M1} activator and the more

---

[127] H. Tanaka and K. Suzuki, Arch. Biochem. Biophys. **175**, 332 (1976).

[128] H. Tanaka and K. Suzuki, Brain Res. **122**, 325 (1977).

[129] N. S. Radin, this series, Vol. 28, p. 834.

[130] D. A. Wenger, M. Sattler, and C. Clark, Trans. Am. Soc. Neurochem. **6**, 151 (1975).

[131] A. L. Fluharty, personal communication.

sensitive G_{M2} activator was also suggested by Li and Li.[132] Unfortunately, neither publication contains information on yield and purity of the final preparations. Also, the steps involving chromatography on DEAE-Sephadex A-50 proved to be quite cumbersome in our hands. The method of Fischer and Jatzkewitz appears to give comparatively pure products, with reasonable yields (5.5 mg/kg tissue), but employing acetone precipitation as the first purification step leads to very large volumes that have to be centrifuged.

We use the following procedures. All steps are performed at 4° unless otherwise stated. The human autopsy material should be collected within 24 hr after death and should be stored at or below −20° until used.

Purification from Human Liver.[92] Human autopsy liver (500 g) is homogenized in 4 volumes 10 mM phosphate buffer, pH 6.5, with an Ultra-Turrax or Polytron homogenizer for 5 min. The homogenate is centrifuged at 100,000 g for 30 min. The pellet is rehomogenized in 4 volumes of the same buffer and centrifuged again. The combined supernatants are filtered through glass wool.

Solid ammonium sulfate is slowly (within 1 hr) added with stirring, up to 60% saturation, and stirring is continued overnight. Precipitated material is spun down (13,000 g, 1 hr), redissolved in approximately 100 ml of 10 mM phosphate buffer, pH 6.5, and dialyzed against two changes of 10 liters each of the same buffer. This solution is then loaded onto a DEAE–cellulose column (2.6 × 40 cm) equilibrated with the same phosphate buffer. After washing with 2 column volumes of buffer, the column is eluted with a linear gradient of 0–0.3 M NaCl in 1 liter phosphate buffer (flow rate 40 ml/hr). Fractions of 15 ml are collected and analyzed for activator content. The fractions with an activator concentration of more than 20% of maximum are pooled, dialyzed against distilled water, lyophilized, and then dissolved in 20 ml of 10 mM phosphate buffer, pH 6.5, with 100 mM NaCl. The solution is passed over a Sephadex G-75 sf column (2.6 × 86 cm) in the same buffer, at a flow rate of 10 ml/hr. Fractions of 10 ml are collected.

The activator-containing fractions are loaded onto a 15 ml octyl-Sepharose column packed in 10 mM phosphate buffer, pH 6.5. The column is washed with 2 volumes 50% ethylene glycol in water and eluted with 1% cholic acid (analytical grade) in 10 mM phosphate buffer, pH 6.5. The protein-containing fractions are pooled, dialyzed first against 2 × 2 liters of 10 mM phosphate buffer, pH 6.5, then against 2 × 2 liters of distilled water, and finally lyophilized.

[132] S.-C. Li and Y.-T. Li, this series, Vol. 83, p. 588.

Final purification is achieved by HPLC ion-exchange chromatography as follows: The sample is dissolved in a small volume of 10 mM phosphate buffer, pH 6.0, and, after removal of any insoluble material by centrifugation, (Eppendorf benchtop centrifuge), up to 10 mg per run is loaded onto a 6-ml TSK-545 DEAE column (LKB, Sweden). The column is eluted at room temperature with a linear gradient of 0–0.3 M NaCl in 60 ml of the same phosphate buffer (flow rate 0.5 ml/min). Fractions of 0.5 ml are collected and analyzed for activator protein.

The overall yield of this procedure is approximately 15% (30% before HPLC). The final product yields a single symmetrical peak in analytical HPLC gel filtration. With HPLC ion-exchange chromatography on DEAE columns, it can, however, be resolved into several peaks, presumably due to heterogeneity of its carbohydrate content.[92] The electrophoretical purity is difficult to assess since the activator, which according to gel filtration experiments seems to have a molecular weight of about 22,000, consistently yields several bands in the range between 6,000 and 10,000 on SDS electrophoresis.

Purification from Human Kidney and Urine[57]

KIDNEY. Postmortem human kidney tissue (500 g) is homogenized with an Ultra-Turrax or Polytron homogenizer in 1.5 liters of 10 mM phosphate buffer, pH 6.0, and then centrifuged at 13,000 g for 30 min. The supernatant is heated to 60° for 30 min in a water-bath. Precipitated material is centrifuged off (13,000 g, 30 min) and discarded. The supernatant is adjusted to pH 4.0 with 1 M citric acid and stirred overnight. After removal of precipitated protein by centrifugation, the extract is dialyzed against 10 mM phosphate buffer, pH 6.0. The dialyzed solution is concentrated at room temperature with a rotary evaporator to a volume of 40 ml, centrifuged (13,000 g, 30 min), applied in 20 ml portions to a Sephadex G-75 column (2.6 × 60 cm) and chromatographed in 10 mM phosphate buffer, pH 6.0. Fractions with activator content of more than 20% of maximum are pooled and loaded onto a DEAE–cellulose column (2.6 × 25 cm) equilibrated with 10 mM phosphate buffer, pH 6.0. After washing with 2 volumes of this buffer, the column is eluted with a linear gradient of 0–0.5 M NaCl in 1 liter of the same buffer. The activator-containing fractions are pooled, dialyzed overnight against 1% glycine solution, and subjected to isoelectric focusing.

The sample is mixed into a linear sucrose gradient from 35 to 0% (w/v), containing 5% (by volume) of carrier ampholyte solution, pH 4–7, in a 110-ml electrofocusing column (LKB, Sweden). The cathode space is filled with 1.5% (w/v) of ethylene diamine in a 50% (w/v) sucrose solution, and the sample space is topped off with 0.1% aqueous H_2SO_4 as anode solution. Focusing is performed for 72 hr at 500 V, then the contents are fractionated in 3-ml fractions.

Carrier ampholytes are removed from the combined activator-containing fractions by hydrophobic chromatography. The fractions are loaded onto a small (1-ml) column of octyl-Sepharose. Ampholytes are washed out with 5 ml of 10 mM phosphate buffer, pH 7, then the activator is eluted with 1% cholic acid (analytical grade) in the same buffer. After addition of 0.1 mg cytochrome c/ml (as protection against adsorption of the small amounts of protein onto dialysis tubing and vessel walls), the sample is dialyzed extensively against distilled water and then lyophilized.

Final purification by HPLC is done as described above for the liver protein. Yield and purity of the final preparation are comparable to those of the first method.

URINE. Freshly voided human urine is filtrated through ordinary filter paper in a Büchner funnel and then concentrated 10-fold by ultrafiltration (exclusion limit of the membrane 10,000 Da). The concentrate is dialyzed against distilled water, lyophilized, and taken up in 40 ml of distilled water.

Aliquots of 20 ml are then loaded onto a Sephadex G-75 column and processed as described above for the kidney extract. Yield and quality of the final product are comparable to those of the other methods described above.

During this latter procedure, the G_{M2} activator is separated from the sulfatide activator by the ion-exchange chromatography step and can be further purified as described below.

G_{M2} Activator

The richest source of human G_{M2} activator is kidney tissue[10,11] which contains some 800 ng activator/mg extract protein, as compared to 100 ng/mg for brain and 50 ng/mg for liver.[17] Another, more convenient, source is urine,[17,133] with a G_{M2} activator content between 200 and 1400 ng/mg protein.[17] The reason for the high content of this protein in urine as well as kidney is probably 2-fold. (1) Routing of this small protein to the lysosomes seems to be less efficient than for the larger lysosomal enzymes and much of it may be secreted into the extracellular space. Cultured skin fibroblasts secrete, for instance, about half of the newly synthesized activator into the culture medium.[17] (2) Small proteins filtrate through kidney

[133] Y.-T. Li, I. A. Muhiudeen, R. DeGasperi, Y. Hirabayashi, and S.-C. Li, *Am. J. Hum. Genet.* **35**, 629 (1983).

glomeruli much more easily than large ones. Thus some of the serum activator may be excreted while another portion may be recaptured by kidney cells, probably by binding of their carbohydrate residues to surface receptors of some cells.

The following purification scheme from human kidney is a simplified version of the one originally used in our laboratory.[11] Human urine, concentrated by ultrafiltration (exclusion limit ca. 10,000 Da) may be substituted for the crude kidney extract.

All steps are performed at 4° unless otherwise stated. Postmortem human kidney (1 kg) obtained within less than 24 hr after death and stored frozen ($-20°$) is homogenized in 4 liters of distilled water, with an Ultra-Turrax or a Polytron homogenizer. The homogenate is centrifuged at 13,000 g (9,000 rpm in a Sorvall GSA or GS-3 rotor) for 30 min. The supernatant is saved and the pellet is reextracted with 3 liters of water.

The combined supernatants are heated to 60° in a water bath for 2 hr and precipitated material is removed by centrifugation (as above). After cooling down to 4°, the supernatant is adjusted to pH 3.0 by slow addition of 10% trichloroacetic acid under rapid stirring and then stirred for another 2 hr (or overnight). Precipitated protein is spun down, the clear supernatant is dialyzed against 10 mM phosphate buffer, pH 6.0. (Dialysis of such large volumes can easily be done with large tubings, in 50- or 100-liter plastic tubs supported by two large magnetic stirrers.) After filtration the solution is loaded onto a column of DEAE–cellulose (2.6 × 30 cm) equilibrated with the same phosphate buffer. The column is washed with 1 liter phosphate buffer and then eluted with a linear gradient of NaCl (0–0.5 M in 1 liter phosphate buffer). Fractions of 20 ml are collected and monitored for G$_{M2}$ activator. Since this step, ion-exchange chromatography, separates the G$_{M2}$ activator from the sulfatide/G$_{M1}$ activator, the latter may also be recovered and further purified as described above, if desired.

Fractions containing the G$_{M2}$ activator (usually those with more than 20% of maximal activity) are pooled, dialyzed against distilled water, and lyophilized. At this point several preparations may be combined and processed together if larger quantities of the activator protein are to be prepared.

The dry residue is dissolved in 10–15 ml of 50 mM citrate buffer, pH 4.2, and, after removal of insoluble material by centrifugation (10,000 g, 10 min), passed over a gel-filtration column (Sephadex G-100; 2.6 × 90 cm) in the same buffer, at a flow rate of about 25 ml/hr.

Activator-containing fractions are combined and loaded onto an octyl-Sepharose column (10 ml) packed in water. The column is then washed with 3 volumes 10 mM phosphate buffer, pH 6.0, and finally eluted with a linear gradient of 0–1% sodium cholate (analytical grade) in 50 ml of the same buffer. Fractions of 1 ml are collected and analyzed for G_{M2} activator. The active fractions are pooled and dialyzed exhaustively against several changes of distilled water and then lyophilized.

This procedure yields a more than 80% pure preparation as judged by analytical HPLC gel filtration, with 30–50% recovery. If an even purer preparation is needed, final purification is achieved with HPLC gel filtration on a TSK G-2000 SW column (150 μl per run, on a 13-ml column, in 0.1 M phosphate buffer, pH 6.0, flow rate 0.5 ml/min) or HPLC anion-exchange chromatography on a TSK 545 DEAE column (both columns are from LKB, Sweden). In the latter procedure, up to 10 mg of activator (dissolved in 10 mM phosphate buffer, pH 6.0) can be applied to a 6-ml column which is then eluted with a linear gradient of 0–0.3 M NaCl in 60 ml of the same buffer.

The purity of G_{M2} activator is difficult to assess with electrophoretic methods since the material isolated from tissue or urine appears to be heterogeneous with respect to its carbohydrate content and runs on SDS electrophoresis as a broad, diffuse band, with an apparent M_r of approximately 22,000.

Co-Glucosidase

It is rather difficult to assess the various procedures that were suggested for the purification of protein or peptide cofactors that stimulate *in vitro* the activity of glucosylceramide β-glucosidase and of galactosylceramide β-galactosidase. A comparison of the products obtained from various sources with different methods (or even with the same method) indicates a tremendous heterogeneity.[26,30–34] Recent evidence[35] also suggests that acid β-glucosidase may be able to degrade its membrane-bound glycolipid substrate without the aid of any protein cofactor. (The role of the intrinsic cofactor for glucosylceramide β-glucosidase recently described by Vaccaro *et al.*[36] may be different but can not yet be assigned.)

One purification scheme for co-glucosidase, from bovine spleen, was recently described in this series.[125] Another convenient method that yields essentially pure products, from human tissue, is that of Peters *et al.*[30] However, as long as it is not known which of the stimulating factors for glucosylceramide and galactosylceramide degradation, if any, is required *in vivo*, we feel that it is impossible to decide on which one of the purification methods currently used to recommend.

[67] Keyhole Limpet Oligosaccharyl Sulfatase

By Premanand V. Wagh, Kalyan R. Anumula, and Om P. Bahl

Introduction

The occurrence of sulfated esters associated with the *N*-acetylhexosamine residues in the O-linked carbohydrate units of various mucins and proteoglycans has been well documented.[1,2] Sulfated glycolipids and glycoproteins are important constituents of animal cell membranes and viral envelopes.[3–6] It is only recently that the detailed structures of N-linked heterosaccharides containing sulfate esters have been elucidated. For instance, pituitary glycoprotein hormones, lutropin and thyrotropin, have been shown to contain sulfated GlcNAc and GalNAc at the nonreducing termini of the asparagine-linked carbohydrate units.[7–9] The significance of these sulfated hexosamines in the activity of gonadotropins and other biologically active glycoproteins, however, remains virtually unknown. Since chemical methods for desulfation of oligosaccharides are quite drastic and cause denaturation of proteins, we focused our attention here to obtain a purified preparation of an oligosaccharyl sulfatase capable of specific hydrolysis of the sulfate esters in macromolecules.

Although a wide variety of aryl sulfatases, employing low molecular weight substrates such as *p*-nitrocatechol sulfate and *N*-acetylhexosamininyl sulfates, have been purified from a number of sources,[10–12] a specific sulfatase capable of hydrolyzing the sulfate esters of N-linked carbohydrate units of glycoproteins has not been so far reported. We,

[1] L. Rodén, *in* "The Biochemistry of Glycoproteins and Proteoglycans" (W. J. Lennarz, ed.), p. 267. Plenum, New York, 1980.
[2] A. Dorfman, *in* "Cell Biology of Extracellular Matrix" (E. D. Hay, ed.), p. 115. Plenum, New York, 1981.
[3] I. Ishizuka, M. Inomata, K. Ueno, and T. Yamakawa, *J. Biol. Chem.* **253**, 898 (1978).
[4] A. Heifetz, W. H. Kinsey, and W. J. Lennarz, *J. Biol. Chem.* **255**, 4528 (1980).
[5] E. K. Merkle and A. Heifetz, *Arch. Biochem. Biophys.* **234**, 460 (1984).
[6] C.-H. Hsu and D. W. Kingsbury, *J. Biol. Chem.* **257**, 9035 (1982).
[7] G. S. Bedi, W. C. French, and O. P. Bahl, *J. Biol. Chem.* **257**, 4345 (1982).
[8] T. F. Parsons and J. G. Pierce, *Proc. Natl. Acad. Sci. U.S.A.* **77**, 7089 (1980).
[9] E. D. Green, H. van Halbeek, I. Boime, and J. U. Baenziger, *J. Biol. Chem.* **260**, 15623 (1985).
[10] J. J. Helwig, A. A. Farooqui, C. Bollack, and P. Mandel, *Biochem. J.* **165**, 127 (1977).
[11] R. Basner, H. Kresse, and K. von Figura, *J. Biol. Chem.* **254**, 1151 (1979).
[12] J. S. Bruce, M. W. McLean, F. B. Williamson, and W. F. Long, *Eur. J. Biochem.* **152**, 75 (1985).

therefore, used ³H-labeled oLH (ovine luteinizing hormone) reduced oligosaccharides and intact ³⁵SO₄-labeled oLH as substrates for the purification of an oligosaccharyl sulfatase with the anticipation that such an enzyme will be of general application and, therefore, useful in probing the role of sulfate in biologically active glycoproteins containing sulfated complex type carbohydrate structures. From among the several commercially available crude or purified sulfatase preparations, we found that the keyhole limpet acetone powder contained the desired enzyme. In this chapter, we describe the purification of such a sulfatase and some of its properties.

Reagents

Keyhole limpet acetone powder, *Megathura crenulata* (Sigma Chemical Co.)
Ovine pituitary glands
H₂³⁵SO₄
NaB³H₄
p-Nitrocatechol sulfate
Protease substrates: azocoll, azocasein, and azoalbumin
Glycosidase substrates: *p*-nitrophenylglycosides
Sephadex G-100, CM-Sephadex C-50, and SP-Sephadex C-50
Dowex AG 3-X4 and Dowex AG 1-X2
DEAE Affi-Gel Blue

Preparation of the Substrates

³⁵SO₄-Labeled oLH. Ovine pituitary gland slices were incubated in a medium containing H₂³⁵SO₄. The ³⁵SO₄-labeled oLH was purified from the 100,000 *g* supernatant fluid obtained from the tissue–medium homogenate after chromatography on Sephadex G-100 and CM-Sephadex C-50 essentially as described previously.[13]

³H-Labeled Reduced Asparagine-Linked oLH Oligosaccharides. ³H-GlcNAcOH-containing oligosaccharides were obtained by the hydrazinolysis of the intact oLH followed by reduction with NaB³H₄ according to the procedure described by Kobata.[14] The ³H-labeled oligosaccharide mixture (15.7 × 10⁶ cpm in 5 ml H₂O) containing neutral, mono-, and disulfated oligosaccharides were separated using ion-exchange resins as follows: The mixture was applied to a Dowex AG 3-X4 acetate column (0.8 × 12 cm) coupled to a Dowex 1-X2 acetate column (0.8 × 12 cm) at

¹³ K. R. Anumula and O. P. Bahl, *Arch. Biochem. Biophys.* **220,** 645 (1983).
¹⁴ S. Takasaki, T. Mizuochi, and A. Kobata, this series, Vol. 83 [17].

the bottom. After equilibration of the columns with water, the coupled columns were eluted first with 60 ml of water after which they were disconnected from each other and each column was eluted with 40 ml of 0.4 M pyridine–acetate buffer, pH 5.0. The distribution of the radioactivity in the oligosaccharides was as follows: water-eluted fraction containing neutral oligosaccharides from the coupled columns, 3.15×10^6 cpm, pyridine–acetate-eluted fraction containing monosulfated oligosaccharides from Dowex AG-1 acetate column, 6.0×10^6 cpm, and pyridine–acetate-eluted fraction containing disulfated oligosaccharides from Dowex AG-3 column, 1.6×10^6 cpm. The mono- and disulfated oligosaccharides were lyophilized and reconstituted in water and stored frozen at $-20°$.

Enzyme Assays

Arylsulfatase activity was measured by the release of p-nitrocatechol from p-nitrocatechol sulfate. The assay mixture contained 30 mM sodium acetate, pH 5.6, 5 mM p-nitrocatechol sulfate, and enzyme in a final volume of 0.5 ml. The tubes were incubated for 10 min at 37°, and the reaction was terminated by the addition of 2.5 ml of 0.4 M NaOH. The amount of liberated p-nitrocatechol ($\varepsilon_{510} = 12,600$ liter $\cdot$ mol$^{-1} \cdot$ cm^{-1}) was determined spectrophotometrically.[10] One unit of enzyme activity was defined as the amount liberating 1 μmol of p-nitrocatechol/min under the above conditions. The specific activity was expressed as units/mg protein.

The oligosaccharyl sulfatase activity employing ^{3}H-labeled monosulfated oligosaccharide (3140 cpm in 50 μl as the substrate in a final volume of 1 ml of the above buffer) was determined by estimating the ^{3}H-labeled neutral oligosaccharide formed by the enzymatic hydrolysis. The tubes were incubated at 37° for 18–20 hr. The reaction mixtures were diluted with 3 volumes of water, and the contents were passed through AG 1-X2 acetate (0.5×6 cm) columns. The columns were eluted with 2 ml of water and the effluents were dried *in vacuo*. The residues were dissolved in 0.4 ml of water, and the solutions were mixed with 3 ml of liquiscint and were counted. Under similar conditions, the nonspecific hydrolysis of the sulfate esters determined at zero time and at 18–20 hr without the addition of the enzyme and in the presence of 25 and 50 μg of bovine serum albumin was found to be approximately 10% of the total radioactivity in the original incubation mixtures of the oligosaccharide substrate. The extent of desulfation of ^{3}H-labeled oLH monosulfated oligosaccharide was expressed as percentage neutral oligosaccharide formed per hour after correcting for nonspecific hydrolysis. A unit of oligosaccharyl sulfatase activity was arbitrarily defined as the amount producing 1% of neutral oligosaccharide per hour under the above conditions.

The sulfatase activity employing $^{35}SO_4$-labeled oLH as the substrate was determined by estimating the release of $^{35}SO_4^{2-}$ from the intact hormone as described below. The substrate mixture contained 50 mM sodium acetate, pH 5.6, 12 mM magnesium acetate, $^{35}SO_4$-labeled oLH (21,872 cpm, 190 μg oLH), and 1.18 mg unlabeled oLH in a final volume of 2 ml. To this mixture was added 0.15 ml of the purified enzyme obtained after sulfopropyl–Sephadex chromatography, and the reaction mixture was incubated at 37° in the presence of toluene. Aliquots of 50 μl were removed at 1, 20, 44, and 70 hr for the estimation of the $^{35}SO_4^{2-}$ released. Fresh enzyme was added soon after the removal of aliquots. For the determination of liberated $^{35}SO_4^{2-}$, the aliquots were diluted with 3 ml of water and applied to Dowex AG-50 (H$^+$) columns prewashed with water. The effluents were collected in tubes containing 10 nmol of NaHCO$_3$. The columns were eluted twice with 1 ml of water, and the combined effluents were dried *in vacuo*. The residues were dissolved in 0.5 ml of water and counted in 3 ml liquiscint. The nature of the released radioactivity was confirmed further by high-voltage paper electrophoresis in borate buffer, pH 4.7, as described.[13]

In order to isolate the resulting desulfated hormone, the enzymatic digest following 70 hr incubation was adjusted to pH 7.0 with 0.2 M Na$_2$HPO$_4$. The solution was chromatographed on a CM-Sephadex column (1.2 × 13 cm). The column was eluted first with 0.01 M sodium phosphate, pH 6.0, then with 0.5 M sodium phosphate, pH 8.0, and finally with a continuous linear gradient between 40 ml of 0.05 M sodium phosphate, pH 8.0, buffer and 40 ml of the same buffer containing 0.4 M NaCl. The column effluent was monitored for $^{35}SO_4^{2-}$ and for desulfated hormone by radioreceptor assay employing super-ovulated rat ovarian membranes.[15]

Purification of Oligosaccharyl Sulfatase

Step 1. Preparation of Crude Extract. Keyhole limpet acetone powder (30 g) was stirred in 600 ml of 0.05 M sodium acetate buffer, pH 5.6, containing 1 mM 2-mercaptoethanol and 0.15 M NaCl for 18 hr. The supernatant fluid was collected after centrifugation at 10,400 g for 30 min. The sedimented material was reextracted with 300 ml of the same buffer for 24 hr and centrifuged as before. The supernatants from both extractions were pooled and designated as crude extract.

Step 2. Ammonium Sulfate Precipitation. Ammonium sulfate (19.4 g per 100 ml) was added to the crude extract to attain 0.35 saturation. After stirring for 2 hr, the precipitate was removed by centrifugation at 10,400 g

[15] N. K. Kalyan and O. P. Bahl, *J. Biol. Chem.* **258**, 67 (1983).

for 1 hr. To the supernatant fluid, ammonium sulfate (29.1 g per 100 ml) was added to achieve 0.80 saturation. The contents were stirred for 20 hr, and the precipitate was collected by centrifugation at 10,400 g for 1 hr. The precipitate was dissolved in 75 ml of 0.01 M Tris–HCl, pH 7.6, buffer containing 0.1 M NaCl. The precipitate solution was dialyzed against the same buffer with several buffer changes. Following dialysis, the precipitating materials in the dialyzed solutions were removed after centrifugation at 27,000 g for 30 min.

Step 3. DEAE–Cellulose Chromatography. The supernatant fluid, obtained after dialysis from the 0.35–0.80 ammonium sulfate precipitate, was chromatographed on a DEAE–cellulose column. DEAE–cellulose was thoroughly equilibrated in 0.01 M Tris–HCl, pH 7.6, buffer containing 0.1 M NaCl and was packed in a column (4.6 × 32 cm). After application of the sample, the column was first eluted with 1.34 liters of the above buffer followed by 1.74 liters of 0.01 M Tris–HCl, pH 7.6, buffer containing 0.3 M NaCl. Proteins eluted in fractions 22–68 and 95–139 (Fig. 1, peaks A and B) were precipitated with ammonium sulfate at 0.8 saturation. The precipitates were dissolved in 50 and 30 ml of 0.01 M Tris–HCl, pH 7.6, buffer containing 1 mM 2-mercaptoethanol, respectively, and dialyzed exhaustively against the same buffer. The precipitates in the dialysate were removed by centrifugation at 27,000 g for 1 hr. The supernatant fluid obtained from peak B (Fig. 1) was used for DEAE–Affi-Gel Blue chromatography as described in Step 4.

Step 4. DEAE–Affi-Gel Blue Chromatography. The protein in peak B (296 mg protein) from Step 3 was applied to a DEAE–Affi-Gel Blue column (1.5 × 95 cm) equilibrated with 0.01 mM Tris–HCl buffer, pH 7.6, containing 1 mM 2-mercaptoethanol, and the column was eluted with 420 ml of the same buffer. The column was then eluted with 800 ml of a continuous linear ionic gradient between 0 and 0.4 M NaCl in the above buffer (Fig. 2). Fractions 109–148 were pooled, and the material in the pooled fraction was collected by precipitation with ammonium sulfate at 0.8 saturation followed by centrifugation at 10,400 g for 30 min. The precipitate was dissolved in 10 ml of 0.01 M Tris–HCl buffer, pH 7.6, containing 0.05 M NaCl and 1 mM 2-mercaptoethanol. The solution was exhaustively dialyzed against the same buffer. The insoluble material was removed by centrifugation and the supernatant fluid was applied to a second DEAE–Affi-Gel Blue column (2.2 × 29 cm) equilibrated with the same buffer. The column was first eluted with 225 ml of the equilibration buffer followed by 500 ml of a continuous linear ionic gradient between 0.05 and 0.25 M NaCl in 0.01 M Tris–HCl, pH 7.6, buffer containing 1 mM 2-mercaptoethanol (Fig. 3). Fractions 125–165 and 180–195 were pooled and designated peak A and peak B, respectively.

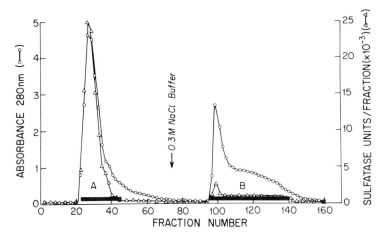

FIG. 1. DEAE–cellulose chromatography of oligosaccharyl sulfatase from keyhole limpet (Step 3). The enzyme solution from Step 2 (140 ml) was applied to a 4.6 × 32 cm column of DEAE–cellulose equilibrated with 0.01 M Tris–HCl, 0.1 M NaCl, pH 7.6. The column was eluted with 1.34 liters of the above buffer followed by 1.74 liters of the same buffer containing 0.3 M NaCl. Fraction volume and flow rate were 20 ml and 68 ml/hr, respectively. Sulfatase activity was monitored employing p-nitrocatechol sulfate as the substrate.

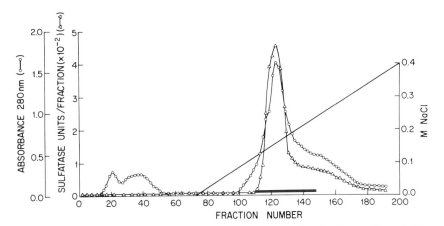

FIG. 2. DEAE–Affi-Gel Blue chromatography of oligosaccharyl sulfatase (Step 4). The enzyme solution (59 ml), peak B from DEAE–cellulose chromatography was applied to a 1.5 × 95 cm column preequilibrated with 0.01 M Tris–HCl, 1 mM 2-mercaptoethanol, pH 7.6 buffer. The column was first eluted with 420 ml of the above buffer and then with 800 ml of a continuous linear ionic gradient between 0 and 0.4 M NaCl in the same buffer. Fraction volume and flow rate were 6 ml and 50 ml/hr, respectively. Sulfatase activity was monitored using p-nitrocatechol sulfate as the substrate. The fractions containing the enzyme as shown by the solid bar were pooled and rechromatographed on a second DEAE–Affi-Gel Blue column.

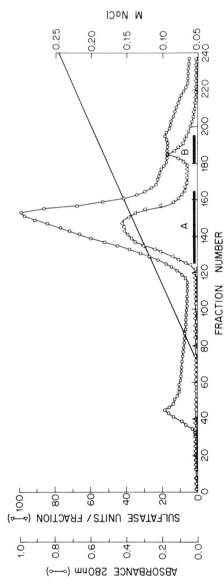

FIG. 3. Rechromatography of oligosaccharyl sulfatase on DEAE–Affi-Gel Blue (Step 4). The enzyme solution (20 ml) was applied to a 2.2 × 29 cm column equilibrated with 0.01 M Tris–HCl, 0.05 M NaCl, 1 mM 2-mercaptoethanol, pH 7.6. The column was first eluted with 225 ml of the equilibration buffer followed by 500 ml of a continuous linear ionic gradient between 0.05 and 0.25 M NaCl in 0.01 M Tris–HCl, 1 mM 2-mercaptoethanol, pH 7.6, buffer. The column was monitored for sulfatase activity using p-nitrocatechol sulfate as the substrate. Fraction volume and flow rate were 3 ml and 30 ml/hr, respectively. Fractions indicated by the bars were pooled separately.

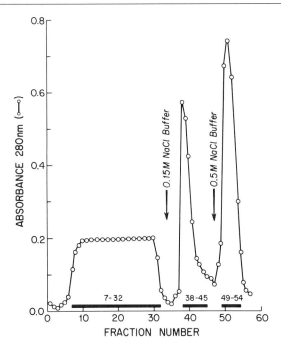

FIG. 4. SP-Sephadex chromatography of oligosaccharyl sulfatase (Step 5). The enzyme solution (35 mg protein in 100 ml), peak A from the second DEAE–Affi-Gel Blue chromatography, with 10 mM sodium acetate, 1.5 mM NaCl, pH 4.0, buffer, was applied to a 1.2 × 18 cm column equilibrated with the same buffer. The column was sequentially eluted with the equilibration buffer followed by 10 mM sodium acetate buffer, pH 4.0, containing 15 mM and 0.5 M NaCl. Fraction volume and flow rate were 4 ml and 32 ml/hr, respectively. Solid bars indicate the fractions pooled.

Step 5. SP-Sephadex Chromatography. The peak A effluent fraction from the previous step was adjusted to pH 4.0 by the addition of cold 0.2 N acetic acid under rapid stirring. The solution was exhaustively dialyzed against 10 mM sodium acetate buffer, pH 4.0, containing 1.5 mM NaCl. The precipitated material was removed by centrifugation at 27,000 g for 30 min. The supernatant fluid (100 ml, 35 mg protein) was passed through a SP-Sephadex C-50 column (1.2 × 18 cm) previously equilibrated with the above dialysis buffer. The column was sequentially eluted with the equilibration buffer followed by 10 mM sodium acetate buffer, pH 4.0, containing 15 mM and 0.5 M NaCl. The elution profile is shown in Fig. 4. Fractions indicated by the solid bars were pooled separately, adjusted to pH 5.6 with 0.5 M sodium acetate, and dialyzed against 25 mM sodium acetate, pH 5.6, buffer. They were tested for aryl and oligosaccharyl sulfatase activities. The *p*-nitrocatechol sulfatase activity recovered from the

TABLE I

PURIFICATION OF OLIGOSACCHARYL SULFATASE FROM LIMPET ACETONE POWDER (30 g)

Step	Protein (mg)	Sulfatase activities using substrates[a] (units)		Specific activities using substrates (units/mg)		Purification using substrates (-fold)	
		A	B	A	B	A	B
1. Crude extract	2682	150	2012	56	0.75	1	1
2. 35–80% (NH$_4$)$_2$SO$_4$	1322	252	2948	191	2.22	3	3
3. DEAE–cellulose	311	16	1365	51	4.39	1	6
4. DEAE–Affi-Gel Blue	37	3.9	409	105	11.05	2	15
5. SP-Sephadex	4.5	1.8	281	400	62.44	7	83

[a] Substrates: A, p-nitrocatechol sulfate; B, ^{3}H-labeled monosulfated oLH oligosaccharide. Details of the purification are in the text.

SP-Sephadex column was found to be 76% of the total units applied to the column, of which 60% of the activity was eluted with the 0.15 M NaCl buffer. On the other hand, the recovery of oligosaccharyl sulfatase activity, employing ^{3}H-labeled monosulfated oligosaccharide as the substrate, was 69% of which approximately 90% of the activity was accounted for in the 0.15 M NaCl eluted fraction. As such, we used this fraction to study some of its properties using ^{3}H-labeled monosulfated oligosaccharide and intact ^{35}SO$_4$-labeled oLH as the substrates.

Comments

Table I summarizes the data on the purification of oligosaccharyl sulfatase. Ammonium sulfate purification between 35 and 80% saturation resulted in a 1.5-fold increase in sulfatase units with both substrates, suggesting that the enhancement in total activity units may be a reflection of the removal of endogeneous inhibitors. Sequential chromatography of the ammonium sulfate fraction on DEAE–cellulose, DEAE–Affi-Gel Blue, and SP-Sephadex progressively resulted in the enrichment of the oligosaccharyl sulfatase enzyme as evidenced by increase in the specific activity. An 83-fold purification over the crude extract was achieved at the SP-Sephadex step, with a yield of 14%. However, the same fraction was only 7-fold enriched over the crude extract when p-nitrocatechol sulfate was used as the substrate, with 1% recovery. This is not unexpected since p-nitrocatechol sulfate is a poor substrate for this enzyme.

The time course of hydrolysis of sulfate ester in ^{3}H-labeled oLH monosulfated oligosaccharide was linear up to 24 hr. Due to the limited amount of the available ^{3}H-labeled oLH oligosaccharide, the K_m value of the enzyme for this substrate was not determined. The pH optimum for both substrates was found to be 5.6. Several divalent cations were tested for their effect on oligosaccharyl sulfatase activity. Whereas Ca^{2+} and Mn^{2+} did not influence the catalytic rate, significant enhancement in enzyme activity was observed in the presence of Mg^{2+}. Ca^{2+} and Cu^{2+} were found to be inhibitory. The enzyme was found to be quite stable on storage at $-20°$ over a period of at least 3 months. The purified enzyme was free from protease activity employing azocoll, azoalbumin, and azocasein as the substrates. However, the preparation was found heterogeneous by SDS–PAGE and contained traces of a few glycosidase activities.

It is clear from the above data that the limpet acetone powder contains a specific enzyme that hydrolyzes sulfate ester in the asparagine-linked oligosaccharides preferentially to p-nitrocatechol sulfate. The apparent rate of hydrolysis was slow, amounting to about 8% of the $^{35}SO_4^{2-}$ released in 24 hr. This was probably due to the loss of the enzyme activity at $37°$ since the addition of fresh enzyme to the incubation mixture after this time period resulted in further increase in the hydrolysis of sulfate ester.

When $^{35}SO_4$-labeled oLH was digested with the above purified enzyme preparation for 70 hr and the desulfated hormone was separated from the reaction mixture employing CM-Sephadex chromatography, about 73% of the total radioactivity, as free inorganic sulfate determined by paper electrophoresis, was eluted with the equilibration buffer. The desulfated α- and β-subunits were eluted from the column with the pH 8 buffer and accounted for 22% of the radioactivity applied to the column. The desulfated hormone was eluted essentially at the same place as the native hormone. The desulfated hormone showed receptor-binding activity in the ^{125}I-labeled hCG competition assay. Thus, keyhole limpet is a good source for the isolation of the sulfatase that would act on asparagine-linked carbohydrate structures containing sulfate esters. It may provide a useful tool in probing the functional role of sulfate ester in glycoproteins.

Acknowledgment

This work was supported by the National Institutes of Health, Grant Nos. HD-08766 and 12581.

Author Index

Numbers in parentheses are footnote reference numbers and indicate that an author's work is referred to although the name is not cited in the text.

Subject Index

A

monosialyl oligosaccharides, isolation, 295

sialyl oligosaccharide phosphate, isolation, 299–300

Compactin

inhibition of dolichol synthesis, 699–700

structure, 699

Concanavalin A

in lectin affinity chromatography, 239–240

source, 245

Concanavalin A-Sepharose

preparation, 245

separation of oligosaccharides with, 244–246

elution of bound material, 245–246

specificity of interaction, 244–245

Corn, glycophosphosphingolipid, 187, 190

Coronavirus, in studies of glycosylation mutant, 451

Correlated spectroscopy, 41

COSY. *See* Correlated spectroscopy

Cranostrea gigas, lectin, 160

Cryptococcus laurentii, glucan synthase, 639, 642

Cultured cells, oligosaccharide in, 96, 98

microanalysis, 110–112

Cyclic orthoesters, in oligosaccharide synthesis, 373–374

Cycloheximide, effect on protein synthesis, 704–705

Cytochrome *c*

interaction with glycosaminoglycans, 264–265

iodinated, detection of glycosaminoglycans with, 260–267

peroxidase activity, 267

D

D2-CAM. *See* D2-cell adhesion molecule

D2-cell adhesion molecule, 169

Datura stramonium

agglutinin

immobilized, preparation, 255

in lectin affinity chromatography, 239–240

source, 254–255

seeds

hemagglutinin, 254

leukoagglutinin, 254

Datura stramonium agglutinin-agarose

preparation, 254–255

separation of glycopeptides with, 254–255

elution of bound material, 255–256

specificity of interaction, 254

Deca-*O*-acetyl-α-monosialolactosyl bromide, synthesis, 328

Deglycosylation

chemical, 341–350

by endoglycosidases, 354–359

enzymatic, 350–359

of glycoproteins, 341–342

by PNGase F, 778

Dehydroepiandrosterone sulfatase, deficiency, 745

2-Deoxy-2,3-didehydro-*N*-acetylneuraminic acid, 147

2-Deoxyglucose

inhibition of glycosylation, 680–681

J-connectivity data, 48

Deoxymannojirimycin

inhibition of glycoprotein processing, 693–694

inhibition of glycosidases, 708

3-Deoxy-D-*manno*-2-octulosonic acid, 148

derivatives, 402–403

Deoxynojirimycin

inhibition of glycoprotein processing, 691–692, 708

structure, 690

2-Deoxyribose, *J*-connectivity data, 48

Dermatan sulfate, detection using iodinated cytochrome *c*, 262–267

Dextran hydrazide, 439

preparation, 440

Dextransucrase

from *L. mesenteroides*, mechanism, 659, 661

from *S. sanguis*, 649–661

assay, 650–652

mechanism, 657–661

physical properties, 655–656

properties, 655–661

purification, 652–655

reaction catalyzed, 649, 657–658

Mycoplasma, cell surface receptor determinants, 162
Mycoplasma pneumoniae, cell attachment factor, 168
Mycospocidin, 666
Myelin basic protein, interaction with sulfated lipid, 476

N

N-CAM. *See* Neural cell adhesion molecule
Neamine, synthesis, 385
Neisseria meningitidis B, polysialic acid capsule, 170
Neoglycoprotein
  from bovine serum albumin, 424
  from glycosyl-Asn derivatives, 409–413
    coupling to proteins, by reductive amination, 411–412
    materials, 410–411
    methods, 411–412
    ninhydrin reaction, protocol, 411–412
    purification, 412
  from noncovalent attachment of oligosaccharide-Asn derivative to protein, 418–424
    advantages of, 423
    stability, 422–423
    stoichiometry, 422–424
  from oligosaccharide-Asn derivatives, preparation with transglutaminase, 413–418
    methods, 414–417
    preparation of acceptor proteins, 414
    preparation of oligosaccharide donor, 415
    preparation, by transglutaminase reaction, 416–417
    advantages, 417–418
    drawbacks, 417
Neolactotetraosylceramide, source, 570
Neural cell adhesion molecule, 169, 627
  E form, 170
  exogenous addition to fetal rat brain, stimulation of polysialic acid synthesis, 631–634

polysialylated, deglycosylation by PNGase F, 636–637
Neural membrane, preparation, for immune blotting, 179–181
Neuraminic acid derivatives, 134
  on DB-5, 19
  partially methylated, methane chemical ionization mass spectrometry, nominal masses of fragments obtained, 19–20
Neuraminic acid methyl ester methylglycosides, analysis, 17–18
Neuroblastoma, plasma ganglioside with, 305–306
Neurospora crassa, glucan synthase, 639–642
Newcastle disease virus, sialidase, 136–137
NFA-ceramide(β1-1)glucosyltransferase, brain
  assay, 578–581
  reaction catalyzed, 578
Niemann–Pick disease
  clinicopathologic features, 730
  definition of, 736
  differentiation of neuropathic and nonneuropathic, 739
  enzymatic defect in, 730
  enzymatic diagnosis, 736–739
  phenotypes, 736
  stored material in, 730
Nigeracin, inhibition of glycoprotein transport, 701
Nikkomycin, inhibition of chitin synthase, 649
Nitrosochloride, in chemical synthesis of oligosaccharide, 387–388
NMR. *See* Nuclear magnetic resonance spectroscopy
NOESY. *See* Nuclear Overhauser effect spectroscopy
Nojirimycin, properties, 690
Nuclear magnetic resonance spectroscopy, 38
  chemical shift correlated spectra, 41
  J-correlated, 41
  of polysaccharides, 26
  pulsing sequences, 40
  of sialic acids, 157–158
  spectrometer-nuclei dialogues, 38
  two-dimensional, 38

applications to oligosaccharides, 41–42

correlated, 39–40

dipole correlated spectra, 40, 41

heteronuclear, 40

homonuclear, 40

J-connectivity data, identification of residue type via, 47–54

J-resolved, 39–40

to reveal through-bond connectivities, 40

to reveal through-space connectivities, 40

scalar correlated, generation of J connectivities via, 42–47

scalar correlated spectra, 40–41

Nuclear Overhauser effect spectroscopy, two-dimensional, 41

compared to 2-D SECSY, 43, 44

limitations, 57–58

oligosaccharide sequence and linkage sites via, 54–56

through-space couplings revealed by, 43

O

O antigen, 268

lipopolysaccharide isolated from, 268

synthesis, 268

O antigen-containing molecules, antibody to define, 274

Oligosaccharide

from acetolysis cleavage products, 98

affinity purification, 307–313

advantages, 312–313

column design, 309–312

flow rates, 309–312

materials, 307

preparation of affinity columns, 308–309

aldopyranosyl residues, stereochemistry, correlation of vicinal coupling constant patterns with, 47, 49

2-aminoglycosyl, synthesis, 374

anomeric residues

across the ring couplings, 54–56

interresidue couplings, 54–56

benzoylated, separation by reversed-phase HPLC, 94–116

benzoylation, time course of, 99–100

binding to amino-bonded silica gel plate, 210–212

BioGel P-4 column chromatography, 84–94

chemical synthesis, 359–404. *See also* Glycosylation

acid acceptors, 362–363

condensation reaction, 360, 362–363

electrophile, 361–362

importance of anhydrous conditions, 361

intermediates, 360

mechanistic considerations, 377–378

nature of combining species, 361–362

nucleophile, 361–362, 375–376

activation, 376

persistent and temporary blocking groups, 360–361

protected glycosylating agent, 361–375

protected intermediates, 361–362

protection–deprotection, 360

steps, 360

from cultured fibroblasts, 98

direct ammonia chemical ionization and mass spectrometry, for sequence and molecular weight determinations, 76–77

direct chemical ionization mass spectrometry, 71–79

fragmentation scheme accounting for major ions detected, 73–75

dissociation from antibodies, 313

effective sizes, expressed in glucose units, 84–94

enzymatic synthesis, by transformations to desired products, 361

exoglycosidase treatment, 194

fibroblast, microanalysis, 110–112

free, immunostaining on TLC, 208–212

glycosidic cleavage, 71

in G_{M1} fibroblasts, analysis, 114–116

high-mannose, 779

separation, 94–116

applications, 104–114

column selection, 102–104

Q

R